MODERN NONLINEAR OPTICS
Part 3

ADVANCES IN CHEMICAL PHYSICS

VOLUME LXXXV

MODERN NONLINEAR OPTICS
Part 3

Edited by

MYRON EVANS
Department of Physics
The University of North Carolina
Charlotte, North Carolina

STANISŁAW KIELICH
Nonlinear Optics Division
Institute of Physics
Adam Mickiewicz University
Poznań, Poland

ADVANCES IN CHEMICAL PHYSICS
VOLUME LXXXV

Series Editors

ILYA PRIGOGINE
University of Brussels
Brussels, Belgium
and
University of Texas
Austin, Texas

STUART A. RICE
Department of Chemistry
and
The James Franck Institute
University of Chicago
Chicago, Illinois

AN INTERSCIENCE® PUBLICATION
JOHN WILEY & SONS, INC.
NEW YORK • CHICHESTER • BRISBANE • TORONTO • SINGAPORE

This text is printed on acid-free paper.

An Interscience® Publication

Copyright © 1994 by John Wiley & Sons, Inc.

Library of Congress Catalog Number: 58-9935
ISBN 0-471-30499-9

Printed in the United States of America

10 9 8 7 6 5 4 3 2 1

CONTRIBUTORS TO VOLUME LXXXV
Part 3

LUCA BONCI, Dipartimento di Fisica dell'Università di Pisa, Pisa, Italy

DIPING CHE, Department of Chemistry and Center for Molecular Electronics, Syracuse University, Syracuse, New York

KOEN CLAYS, Laboratory of Chemical and Biological Dynamics, University of Leuven, Leuven, Belgium

LEO DE MAEYER, Max-Planck Institut fur Biophysikalische Chemie, Gottingen, Germany

V. V. DODONOV, Lebedev Physics Institute, Moscow, Russia

P. D. DRUMMOND, Physics Department, The University of Queensland, St. Lucia, Queensland, Australia

M. I. DYKMAN, Department of Physics, Stanford University, Stanford, California

TERESA B. FREEDMAN, Department of Chemistry and Center for Molecular Electronics, Syracuse University, Syracuse, New York

WOJCIECH GAWLIK, Instytut Fizyki, Uniwersytet Jagiellonski, Krakow, Poland

PAOLO GRIGOLINI, Dipartimento di Fisica dell'Università di Pisa, Pisa, Italy

STANISŁAW KIELICH, Nonlinear Optics Division, Institute of Physics, Adam Mickiewicz University, Poznań, Poland

TAKAYOSHI KOBAYASHI, Department of Physics, University of Tokyo, Bunkyo, Tokyo, Japan

GEORGE C. LIE, IBM Corporation, Midland, Michigan

D. G. LUCHINSKY, All Russian Research Institute for Metrological Service, Moscow, Russia

V. I. MAN'KO, Lebedev Physics Institute, Moscow, Russia

R. MANNELLA, Departimento di Fisica, Università di Pisa, Pisa, Italy

P. V. E. MCCLINTOCK, School of Physics and Materials, Lancaster University, Lancaster, United Kingdom

ADAM MIRANOWICZ, Nonlinear Optics Division, Institute of Physics, Adam Mickiewicz University, Poznań, Poland

LAURENCE A. NAFIE, Department of Chemistry and Center for Molecular Electronics, Syracuse University, Syracuse, New York

ANDRE PERSOONS, Laboratory of Chemical and Biological Dynamics, University of Leuven, Leuven, Belgium

M. D. REID, Department of Physics, The University of Queensland, Queensland, Australia

G. WILSE ROBINSON, SubPicosecond and Quantum Radiation Laboratory, Texas Tech University, Lubbock, Texas

N. D. STEIN, School of Physics and Materials, Lancaster University, Lancaster, United Kingdom

N. G. STOCKS, School of Physics and Materials, Lancaster University, Lancaster, United Kingdom

SURJIT SINGH, SubPicosecond and Quantum Radiation Laboratory, Texas Tech University, Lubbock, Texas

D. F. WALLS, Department of Physics, University of Auckland, Auckland, New Zealand

JIAN-MIN YUAN, Department of Physics, Drexel University, Philadelphia, Pennsylvania

SHENG-BAI ZHU, IBM Research Division, Almaden Research Center, San Jose, California

INTRODUCTION

Few of us can any longer keep up with the flood of scientific literature, even in specialized subfields. Any attempt to do more and be broadly educated with respect to a large domain of science has the appearance of tilting at windmills. Yet the synthesis of ideas drawn from different subjects into new, powerful, general concepts is as valuable as ever, and the desire to remain educated persists in all scientists. This series, *Advances in Chemical Physics*, is devoted to helping the reader obtain general information about a wide variety of topics in chemical physics, a field which we interpret very broadly. Our intent is to have experts present comprehensive analyses of subjects of interest and to encourage the expression of individual points of view. We hope that this approach to the presentation of an overview of a subject will both stimulate new research and serve as a personalized learning text for beginners in a field.

ILYA PRIGOGINE
STUART A. RICE

PREFACE

Statistical molecular theories of electric, magnetic, and optical saturation phenomena developed by S. Kielich and A. Piekara in several papers in the late 1950s and 1960s clearly foreshadowed the developments of the next thirty years. In these volumes, we as guest editors have been honored by a positive response to our invitations from many of the most eminent contemporaries in the field of nonlinear optics. We have tried to give a comprehensive cross section of the state of the art of this subject. Volume 85 (Part 1) contains fourteen review articles by the Poznań School and associated laboratories, and volume 85 (Part 2 and Part 3) contain a selection of reviews contributed from many of the leading laboratories around the world. We thank the editors, Ilya Prigogine and Stuart A. Rice, for the opportunity to produce this topical issue.

The frequency with which the work of the Poznań School has been cited in these volumes is significant, especially considering the overwhelming societal difficulties that have faced Prof. Dr. Kielich and his School over the last forty years. Their work is notable for its unfailing rigor and accuracy of development and presentation, its accessibility to experimental testing, the systemic thoroughness of the subject matter, and the fact that it never seems to lag behind developments in the field. This achievement is all the more remarkable in the face of journal shortages and the lack of facilities that would be taken for granted in more fortunate centers of learning.

We hope that readers will agree that contributors to these volumes have responded with readable and useful review material with which the state of nonlinear optics can be measured in the early 1990s. We believe that many of these articles have been prepared to an excellent standard. Nonlinear optics today is unrecognizably different from the same subject in the 1950s, when lasers were unheard of and linear physics ruled. In these two volumes we have been able to cover only a fraction of the enormous contemporary output in this field, and many of the best laboratories are not represented.

We hope that this topical issue will be seen as a sign of the ability of scientists all over the world to work together, despite the frailties of human society as a whole. In this respect special mention is due to Professor Mansel Davies of Criccieth in Wales, who was among the first in the West to recognize the significance of the output of the Poznań School.

MYRON W. EVANS

Charlotte, North Carolina
November 1993

CONTENTS

CONTENTS

MODERN NONLINEAR OPTICS
Part 3

ADVANCES IN CHEMICAL PHYSICS

VOLUME LXXXV

MOLECULAR BISTABILITY AND CHAOS

JIAN-MIN YUAN

Department of Physics, Drexel University, Philadelphia, Pennsylvania

GEORGE C. LIE*

Dow Corning Corporation, Midland, Michigan

CONTENTS

I. INTRODUCTION

One important goal in electronic devices is to miniaturize electronic components, for this will give rise to compact but powerful computer systems that have higher packing densities and lower energy consumptions. Researchers in miniaturization believe that the ultimate devices will be limited only by the size of atoms and molecules. While electronic devices at the atomic and molecular electronic level are still a dream many

*Permanent address: IBM Corporation, 2125 Ridgewood Drive, Midland, MI 48642.

Modern Nonlinear Optics, Part 3, Edited by Myron Evans and Stanisław Kielich. Advances in Chemical Physics Series, Vol. LXXXV.
ISBN 0-471-30499-9 © 1994 John Wiley & Sons, Inc.

1

years away, workshops on this subject have already been held to chart the path of reaching such a goal.[1, 2]

Three basic functions of a computer—arithmetic operations, logical operations, and the storage of information (i.e., memory)—are all done by devices that have two stable states.[3, 4] In the arithmetic operation the two states are 0 and 1, in the logical operations, they are true and false, and memory is achieved by occupying one of the two states. Thus, computers are made of devices that have two stable states, and their speeds are often limited by the rate at which these devices can switch between the two states.

As an example of electronic device at molecular level a class of hemiquinones has been investigated theoretically and experimentally as a possible candidate for a molecular information storage device, based on the tunneling phenomena of electrons and protons through a double-well potential, exhibiting a bistable property.[5] On the other hand, optical bistability, bistable behavior of a laser cavity with nonlinear media, is an important and practical mechanism for building fast computers. Significant progress has been made in this field over the last two decades.[3, 6]

Motivated by the goal of both miniaturization and fast switching it is important for us to examine the bistable behavior and memory effects exhibited by atomic and molecular systems and understand their underlying principles. One well-known type of memory phenomena in atomic and molecular systems is the one revealed in spin-echo or photon-echo experiments. Here memory is the result of time-reversal symmetry of the underlying dynamics and is limited by the transverse (or phase) relaxation time of the magnetization or polarization vector. The molecular bistability we discuss in this review belongs to the category of intrinsic optical bistability, that is, optical bistability without a resonator.[7, 8] This may be advantageous over the resonator-based optical bistability, when it comes to the fabrication of real devices. The technological aspects of the devices are not dealt with in this chapter. We focus instead only on the fundamental principles on which such devices may someday be conceived.

The purpose of this chapter is to review some of the theoretical investigations of the bistable and memory effects in driven molecular vibrational systems, resulting from their anharmonicity. These systems exhibit bistable, and even chaotic, behavior, but the nature of their bistability is intrinsically different from that of hemiquinones. In the latter case the bistability arises from tunneling through a double-well potential. In the former, the bistability has a purely dynamical origin. This phenomenon is best understood in terms of nonlinear dynamics.[9-11] When the parameters of a nonlinear dynamical system are varied, the solutions often undergo qualitative changes, called bifurcations. For example, a single stable solution (such as equilibrium, steady-state, or steady oscillations) of the systems may lose its stability and two stable solutions appear in its place. This corresponds to a pitchfork bifurcation, as it is sometimes called. More relevant to our systems is a saddle-node bifurcation, in which a stable node and an unstable saddle are generated simultaneously and seemingly out of nowhere. In general, these stable solutions do not correspond to minima of a potential system and often are time-dependent. In this sense the bistability we discuss in this chapter is dynamical in nature, not static.

Not all papers concerning the bistability of molecular systems are discussed here. For example, we do not review papers dealing with Duffing oscillators, which are favorite systems in nonlinear dynamics[9-18] and can be used to model molecular bending vibrations. Also excluded are papers on bistability of systems involving a gas of molecules.[19]

The organization of this review is as follows: In the next section we describe the bistable and chaotic behavior of molecular vibrational systems, as modeled by a Morse potential, when driven by an external field. Classical, semiclassical, and quantum mechanical approaches to the phenomena as well as experimental comparison are all reviewed. Material of this section is more up-to-date than that of a previous review.[20] In Section III, we discuss the dynamical behavior of weakly bound systems as modeled by a Lennard-Jones potential. This shows the generality of the bistable behavior among anharmonic potentials. Intuitive physical pictures of molecular bistability are provided in Section IV. In Section V we discuss the possibility of multiple stability in molecules in terms of a model system, the coupled logistic maps. We then report on some preliminary results on basin boundaries of attraction and escape rates associated with molecular vibrational systems in Section VI. Discussion and concluding remarks are included in the last section.

II. BISTABILITY AND CHAOS ON A MORSE POTENTIAL

A Morse oscillator[21, 22] is often used to describe the vibrational motion of a diatomic molecule, because its quantum problem is exactly soluble and its eigenvalues reproduce the molecular vibrational frequencies reasonably well, except for highly excited states close to the dissociation threshold.[23] Even in polyatomic molecules Morse oscillators are commonly used in the local-mode representation[24, 25] (i.e., vibrations along chemical bonds) of vibrational motions. For instance, it is reasonable to simulate the energy transfer in a large polyatomic molecule by using a Morse oscillator to describe the bright state and a bath of harmonic oscillators to describe the dark states into which energy flows.[26, 27] A driven Morse oscillator, on the other hand, is the favorite model for the study of multiphoton excitation and dissociation of diatomic molecules.[28-45] This model has been extended by including a damping mechanism to simulate the same process for polyatomic molecules or surface-absorbed molecules. In the extended model the pumped mode, which is in resonance with the external field, is described by a Morse oscillator and the rest of the rovibrational and other degrees of freedom as a reservoir—the damping mechanism. The latter model is extensively discussed in this chapter.

In the following subsections, we review the progress made in understanding the hysteretic behavior of a driven damped Morse oscillator, from classical-mechanical, semiclassical, and quantum-mechanical points of view. A subsection on experimental relevance ends the section.

A. Classical Approach

The familiar singularity in the response curve of a harmonic oscillator, driven by an external field sweeping through its resonant frequency can be smoothed out by introducing damping to the system. In an anharmonic oscillator the natural frequency changes with the amplitude.[46-48] In general, the slope $d\omega/dA$ is negative for a potential that is softer than the corresponding harmonic oscillator and positive for a potential that is harder.[49] The response curve of an anharmonic oscillator without damping tilts toward the side that the natural frequency curve does and becomes discontinuous at the tip. This curve becomes continuous when damping is introduced into the system. Furthermore, the stronger the damping, the shorter the peak becomes.[46-48] An example of such a response curve is shown schematically in Fig. 1. The λ-shaped curve of Fig. 1 is a triply valued function for some frequency range. Usually the middle branch is unstable. A bistable region exists if both the upper and lower branches are stable. Response of such a system shows hysteresis if the frequency is varied through a maximal value greater than the maximal value of the lower branch. This too is illustrated in the figure.

This type of bistability, of course, can be found in textbooks,[46-48] but usually is associated only with macroscopic systems. In this chapter we discuss the bistable behavior of molecular systems, represented by two most commonly encountered potentials, the Morse and the Lennard-Jones potentials. In both cases the potential is softer than the harmonic oscillator; thus, the natural frequency decreases as the vibrational amplitude increases. The details of the Lennard-Jones potential are discussed in the next section.

Classical nonlinear dynamical behavior of a driven damped Morse oscillator, as well as that of a Duffing oscillator, is discussed by Kapral et al.[50] They have found, occurring in a driven Morse oscillator, period-

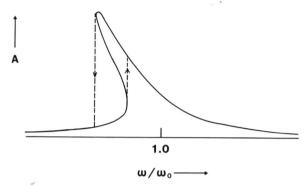

A

1.0

$\omega / \omega_0 \longrightarrow$

Figure 1. Response curve of a damped anharmonic oscillator to a sinusoidal field with frequency ω. A denotes the amplitude of oscillation of the anharmonic oscillator, which is softer than a harmonic oscillator; ω_0 is its fundamental frequency. The solid curve denotes the response and dashed lines the hysteresis loop.

doubling bifurcations leading to chaos and saddle-node bifurcations associated with the period-3 windows in the chaotic region, but they have not mentioned the existence of bistability.

We have studied the bistable and chaotic behavior of a driven damped Morse oscillator,[49] whose equation in dimensionless form can be written as

$$\ddot{x} + \gamma \dot{x} + (1 - e^{-x})e^{-x} = A\mu'(x)\cos \omega t \qquad (1)$$

where t and ω have been scaled by the Morse angular frequency $\omega_0 = \beta\sqrt{2D/m}$, and damping constant γ by the product of m and ω_0, and the amplitude A by $m\omega_0^2/\beta$, where β, D, and m are the Morse range parameter, dissociation energy, and the reduced mass.[21] The symbol $\mu'(x)$ denotes the spatial derivative of a dipole moment function $\mu(x)$, which for most cases will be assumed to be a linear function of x. This approximation will be tested later by using a dipole function calculated for HF by one of us.[51] As discussed at the beginning of Section II, Eq. (1) can be thought of as a model for infrared multiphoton dissociation of molecules. We solve it by converting it to a set of two first-order differential equations,

$$\dot{x} = p$$
$$\dot{p} = -\gamma p - (1 - e^{-x})e^{-x} + A\mu'(x)\cos \omega t \qquad (2)$$

and integrating it using the fourth-order Runge-Kutta method. The accu-

racy of the method is greater for a dissipative system than for a Hamiltonian one. This is because a dissipative system self-repairs near or on
attractors. In a chaotic regime any numerical integrator will have accuracy
problems.

We discuss below some of the numerical results obtained. The immediate objective is to show how bistability appears and then how chaos arises
in the system. For the first set of calculations γ is fixed at 0.001 and A at
0.01, while the value of the scaled frequency ω increases adiabatically
from 0 to 1.2 and decreases back down to zero. Adiabaticity here means
that equations are integrated for a sufficiently long time so that the system
reaches steady behavior and the final point of the trajectory in the phase
space is used as the initial point of a new slightly shifted value of ω. In this
way we follow the evolution of a certain attractor as a parameter (here ω)
changes. For the above chosen parameters, trajectory almost always approaches asymptotically to a periodic solution with the period equal to
that of the driving field. This type of solutions will be called a period-one
(P1) orbit, following the convention that period-n (Pn) orbit has a period
of n times that of the driving period.

In Fig. 2a we plot the maximal x value, x_{max}, of a periodic orbit as a
function of ω. As shown in the figure, a big hysteresis loop appears as the
driving frequency sweeps through a maximal value. On the decrease of ω
(from 1.0) the response of the Morse oscillator does not fall back to the
original value of the lower branch, but rises smoothly to a higher new
branch. Thus in the frequency range of $\omega = 0.36$ to 0.94, there exist two
stable period-one solutions, with the time-averaged molecular energy of
one solution greater than that of the other. They will be separately called
the upper and the lower branch. But in fact a middle branch also exists,
with the average energy lying in between. Since it is unstable, trajectories
will be repelled from it when propagated forward in time. Also shown in
Fig. 2a is a natural response curve of a field-free Morse oscillator. The
upper branch lies above this curve and follows it quite closely, presumably
until it collides with the unstable middle branch and is annihilated via a
saddle-node bifurcation.[52] The lower branch may disappear due to the
same mechanism.

In Fig. 2a (and every other calculation discussed later except Fig. 2b)
the dipole moment function $\mu(x)$ is assumed to be simply proportional to
x, the interatomic distance deviated from the equilibrium position. This
may be a good approximation for processes occurring around the potential
well region, for instance, molecular photoabsorption. However, to more
realistically simulate the laser–molecular interaction, a more accurate
dipole moment function should be used. Unfortunately, molecular dipole
moment functions are not generally available either theoretically or experimentally, except for simple molecules. In the latter category hydrogen

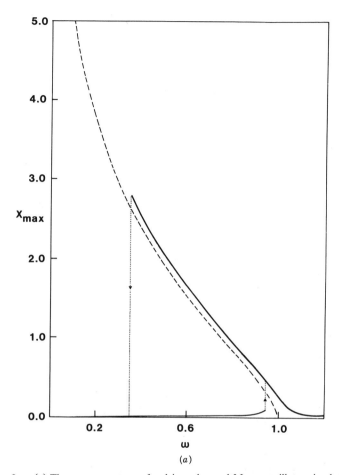

Figure 2a. (a) The response curve of a driven damped Morse oscillator simulating HF. The solid curve represents the response and dashed curve the natural frequency of the Morse oscillator with the classical frequency normalized to unity. x_{max} is the maximal amplitude of a periodic orbit (P1). The dotted lines with arrows denote the hysteresis loop. The dipole moment function $\mu(x) = x$, $\gamma = 0.001$, and $A = 0.01$ are used in obtaining this figure. (b) The same as (a) except a more realistic $\mu(x)$ function given by Eq. (3) is used.

fluoride (HF) is one of the most intensively studied molecules. Dipole moment functions for HF, obtained both from experiments and ab initio calculations, have been quite accurate for some time.[51, 53] Here we shall use the theoretical one by Lie,[51] given by

$$\mu(x) = 1.814 + 0.6919x - 0.340x^2 - 0.3685x^3 + 0.0645x^4 + 0.0261x^5$$

$$(3)$$

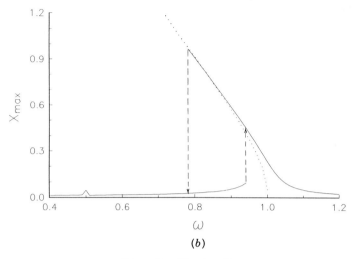

Figure 2b.   *(Continued)*

to shed light on the possible changes in the response curve when a more realistic $\mu(x)$ is used. In Eq. 3, $\mu(x)$ is in debyes, $x = r - r_e$ in bohrs and r_e is 1.730 bohrs.

We first extend the $\mu(x)$ function of Lie, Eq. (3), to infinity by fitting an exponentially decaying function to the last two calculated points. To make meaningful comparison we scale our A value below in such a way that $A\mu'(x)$ at $x = 0$ is the same as that used before. Thus, to compare with Fig. 2a, we set the field amplitude to 0.017 and obtain the corresponding response curve for the same γ value (0.001) in Fig. 2b. We see that the use of Eq. (3) has drastically reduced the size of the bistable region from 0.360–0.943 to 0.780–0.941. This can be attributed to the fact that $\mu'(x)$ decreases as the internuclear distance increases. Since $\mu(x)$ used here possesses some common features of realistic functions, namely, they go to maximum and then decay to zero when the atoms separate, the reduction of bistable region (and chaotic region) should be true in general. We note that a small kink appears at $\omega = 0.5$ for both cases (although it is not resolved in Fig. 2a). Associated with it is a change of shape of limit cycles: The acute side of the egg-shaped orbit faces left for $\omega < 0.5$ and faces right for $\omega > 0.5$. This kink may signal the birth of a second bistable region (due to 1:2 resonance between the oscillator and the field frequency), as discussed below. Finally, the response curves of Figs. 2a and b have another clear difference: The response curve in Fig. 2b crosses the natural frequency curve, while that for Fig. 2a does not.

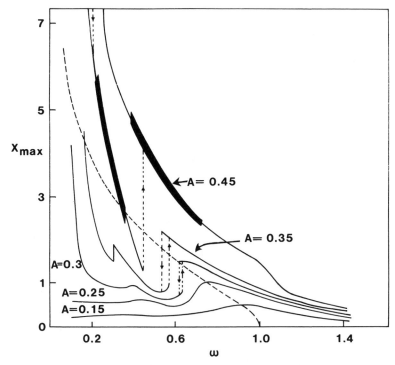

Figure 3. The response curves of a driven damped Morse oscillator for several field amplitudes (A). The symbols and curves are defined in Fig. 2a. The thick segments of the solid curves denote the chaotic response.

Bistable behavior still shows up as the field strength increases. However, when field strength increases, bond dissociation becomes probable and chaotic behavior often appears. For atomic and molecular dynamical processes chaos often emerges with ionization and dissociation. However, the converse does not hold; for instance, classical dynamics of atomic helium is chaotic even at low energies. (This may have something to do with the fact that the helium problem is scale-invariant with energy.)

Shown in Fig. 3 are response curves of a Morse oscillator at $\gamma = 0.4$ with several A values. At lower A values the peaks of the response curves nearly coincide with the intersections of the response and natural frequency curves. A narrow, but visible, bistable region appears near the peak ($\omega = 0.606$ to 0.626) at $A = 0.3$. This bistable domain grows as A increases, and at $A = 0.45$ a big bistable region is observed. Up until $A = 0.35$ all points on the curves represent the amplitudes of P1 orbits. The dynamical behavior for $A = 0.45$ is, however, much more complicated. An infinite cascade of period-doubling bifurcations occurs generating orbits of 2^n (where n is an integer) periods, that is, the following bifurcation sequence: P1 → P2 → P4 → P2^n → P2$^\infty$ = chaos. This scenario to chaos is quite common among driven systems and is epitomized by the well-known logistic map.[54] In the chaotic region many periodic windows have been found. The most pronounced are P3 and P6. The chaotic domain ends with the period-halving bifurcations. The above description applies to the chaotic domains on both the lower and upper branches, which do not overlap with each other. One clear way to present the above dynamical changes is to plot the maximal Lyapunov exponent as a function of ω. An example of this with more details is given in Ref. 49.

Two other features of Fig. 3 should be emphasized. First, response curves for $A > 0.25$ become singular at small ω values. This is because a dc field strength with $A > 0.25$ is already strong enough to pull the molecule apart, i.e., to cause it to dissociate. What is interesting is that an ac field at a frequency greater than about 0.2 actually stabilizes a molecule. This is similar to the stabilization of the upright position of a pendulum when its point of suspension is oscillating at a high frequency.[46] Here damping in our system further helps to stabilize the configuration. In Fig. 4, we show two stabilized P1 orbits, lying respectively on the upper and the lower branch in the bistable domain of the $A = 0.35$ curve. These two orbits plotted in phase space and on the potential curve are shown in Figs. 4a and b, respectively. We see that they are well separated both in phase space and energy space. An even more revealing example is shown in Fig. 5 where a strange attractor (chaos) on the upper branch is plotted both in phase space (Fig. 5a) and on the potential curve (Fig. 5b). In the latter case, the stabilization effect of the damping is clearly seen. The molecule may have energy higher than the dissociation energy in some outgoing parts of the trajectory, but its energy is quickly removed by the damping mechanism and the driving (retarding) force so that it eventually returns to the well region of the potential.

A second set of finite maxima can be seen on the response curves of Fig. 3. These maxima are located at frequencies ($\omega = 0.3$ to 0.5), which are about half of those of the primary bistable domains. These maxima eventually evolve into the secondary bistable zone as A increases, as shown in Fig. 6, where we have plotted the bistable domains in the $A\omega$ control parameter space. Two cusp-shaped zones of bistability are found. The primary bistable domain arises from the 1:1 resonance between the amplitude-dependent frequency of the oscillator and that of the external field, and the secondary domain occurs around the 1:2 resonance, that is, a first subharmonic resonance. In principle, higher-order subharmonic resonances may also occur.

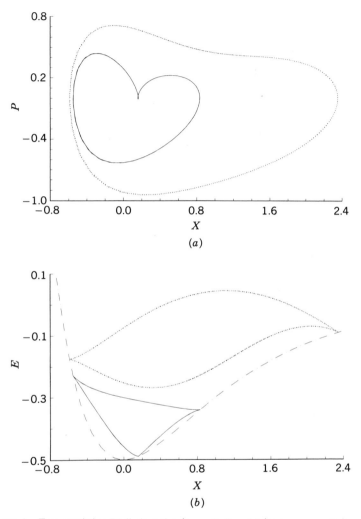

Figure 4. Two coexisting P1 limit cycles (or periodic orbits) lying, respectively, on the upper (dotted) and the lower (solid) branch in the bistable domain ($\omega = 0.5$) of the $A = 0.35$ curve of Fig. 3. (a) Limit cycles in phase space. (b) Limit cycles on the potential energy curve, shown as a dashed curve.

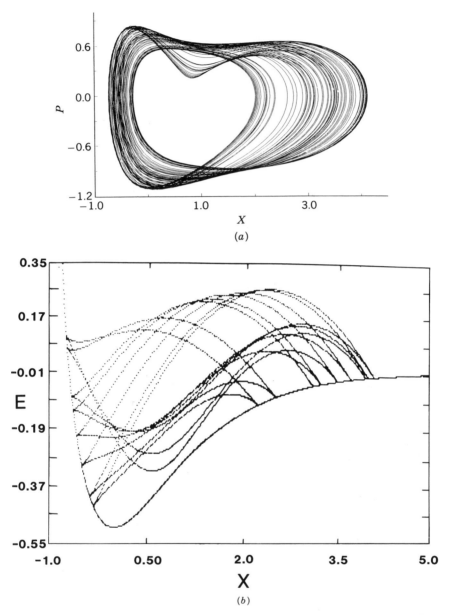

Figure 5. A chaotic attractor on the upper branch is plotted (a) in phase space and (b) on the potential energy curve. The parameters are set at $A = 0.45$, $\gamma = 0.4$, and $\omega = 0.45$.

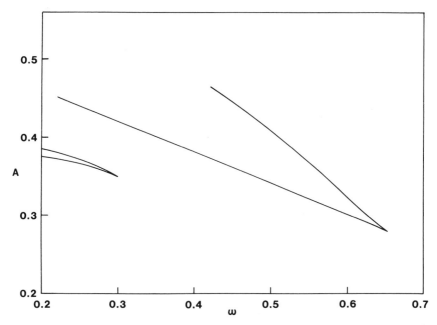

Figure 6. Two bistable domains in the control parameter space of the amplitude (A) and frequency (ω) of a driven damped Morse oscillator with $\gamma = 0.4$.

Finally, bistability also shows up when we fix the frequency ω and vary the amplitude A through a maximal value (A_0 in Eq. (4)). One way to facilitate such a calculation is to vary A adiabatically in a single run. Adiabaticity here means a rate much slower than the relaxation (damping) rate and the Morse frequency. In fact, these conditions are satisfied in a multiphoton absorption experiment using a nanosecond pulsed laser, such as a CO_2 laser. We show in Fig. 7 results of such a calculation, where the amplitude sweep is achieved by introducing a time-dependence to A according to

$$A(t) = A_0 \cos^2 \Omega t \qquad (4)$$

In Fig. 7, Ω used is equal to 10^{-4} of ω_0 and the parameters A_0, ω, and γ are set, respectively, to 0.4, 0.55, and 0.4. We see a big bistable region ($A(t) = 0.315$ to 0.37) and the response curve is roughly S-shaped with a missing middle branch connecting smoothly the upper and lower branches (it is not seen in the figure because it is unstable). The complication seen in the upper branch is due to the fact that a P1 orbit becomes unstable

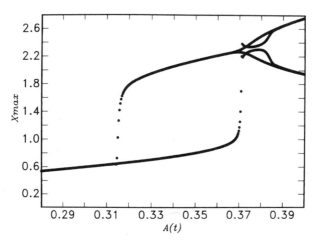

Figure 7. Vibrational amplitude (x_{max}) of a driven damped Morse oscillator plotted as a function of the slowly varying field amplitude $A(t)$ of Eq. (4), where $A_0 = 0.4$ and $\Omega = 10^{-4}$.

and bifurcates into a P2 as $A(t)$ becomes greater than about 0.37. When the lower branch ends, x_{max} jumps up to the upper branch and settles down to a P2 orbit after a transient.

Knop and Lauterborn[52] have studied extensively the bifurcation structures of the driven Morse oscillator system by calculating fixed-point bifurcation diagrams, generalization of normal bifurcation diagrams to include unstable as well as stable orbits, and bifurcation sets in the $A\omega$ parameter space. Dynamical behavior over a much larger parameter space than that of Fig. 6 is investigated for $\gamma = 0.4$ and 0.8. Knop and Lauterborn have found very rich saddle-node and period-doubling bifurcation structures (the only two types of bifurcations they have found in driven Morse oscillators) for larger A values (particularly A above 2.0), compared with what we have studied. They have further classified the bifurcations according to types, periods, stability, super- or sub-criticality, and torsion numbers. It would be interesting if this bifurcation complexity could be observed experimentally. Hysteresis and period-doubling bifurcations in anharmonic oscillators of the type of $V(x) = a_2 x^2 + a_4 x^4 + a_6 x^6$ have also been studied by Debnath and Chowdhury.[55] Since many parameters appear in the potential, possible variations increase drastically.

B. Semiclassical Approach

Since atoms and molecules obey quantum mechanics, the interesting phenomenon of bistability and chaos seen in the classical approach will

$\langle a^\dagger a \rangle$ by z, we obtain the equations of motion in the following form:

$$\dot{x} = -x + \delta y - \alpha yz$$
$$\dot{y} = -\delta x - y + \alpha xz - \Omega_R \qquad (11)$$
$$\dot{z} = -2\Omega_R y - \lambda z$$

where the dimensionless parameters α, λ, δ, and Ω_R are the scaled anharmonicity, longitudinal relaxation rate, detuning, and Rabi rate defined by

$$\alpha = \frac{2\varepsilon}{\gamma_\perp}$$

$$\lambda = \frac{\gamma_\parallel}{\gamma_\perp}$$

$$\delta = \frac{\omega_0 - 2\varepsilon - \omega}{\gamma_\perp} \qquad (12)$$

$$\Omega_R = \frac{\mu A}{\gamma_\perp \hbar}$$

with the transverse relaxation rate $\gamma_\perp$ given by

$$\gamma_\perp = \eta + \frac{\gamma_\parallel}{2} \qquad (13)$$

This set of equations, Eq. (11), was first derived by Narducci et al.[56]

Equation (11) looks simple, but like the Lorenz equations, it is nonlinear, thus dynamical behavior can be complicated. We begin by locating the steady states of Eq. (11) and analyzing their stability. By setting all velocities $(\dot{x}, \dot{y}, \dot{z})$ to zero, we can show[68] that the steady-state values of the average vibrational excitation z_0 are solutions of the following cubic equation:

$$z_0^3 - 2\frac{\delta}{\alpha}z_0^2 + \frac{1 + \delta^2}{\alpha^2}z_0 - 2\frac{\Omega_R^2}{\lambda \alpha^2} = 0 \qquad (14)$$

which can have either one or three real solutions. The case of three real solutions can exist only for $\delta > \sqrt{3}$ (scaled by the transverse relaxation time), i.e., only when the laser frequency is red-shifted from the fundamental frequency of the pumped mode by at least this amount can multiple steady states exist.

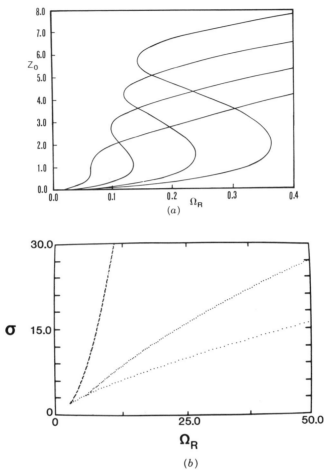

Figure 8. (a) Steady-state value of the average vibrational excitation (z_0) of Eq. (14) plotted against the Rabi frequency Ω_R for $\alpha = 1.152$ and $\lambda = 0.008$. The four S-shaped curves correspond to, from large to small, $\delta = 6.7, 5.0, 3.3$, and 1.6. (b) Turning points of the S-shaped curve as a function of the detuning (δ) and Ω_R for $\alpha = 1.152$ and $\lambda = 8$. The middle curve is a bifurcation set from which limit cycles bifurcate as described in the text.

Numerical results[68] show that in general z_0 is an S-shaped function of the Rabi rate Ω_R. An example is shown in Fig. 8a, in which we have plotted z_0 versus Ω_R for several δ values with $\alpha = 1.152$ and $\lambda = 0.008$. As shown in the figure the range of Ω_R where triple solutions exist (only two are stable) increases as the detuning δ increases. Figure 8b shows more clearly how this region of bistability changes with δ and Ω_R for

$\alpha = 1.152$ and $\lambda = 8$. It is bounded by two outer curves in the figure (formed by the turning points of the S-shaped curve) and is in the shape of a cusp. The S-shaped curves of Fig. 8a are very similar to the classical response curve of a Morse oscillator as a function of the field strength shown in Fig. 7. The cusp-shaped region of bistability (Fig. 8b), on the other hand, resembles those classical regions of Fig. 6. In fact, the figures suggest that z_0 is actually a cusp-shaped surface as a function of the laser frequency and amplitude. This is exactly what we would expect based on the catastrophe theory,[72, 73] which states generically that Eq. (14) can be represented by the gem, $z^3 - az + b = 0$, which is a cusp catastrophe. Thus, the response of a Morse oscillator to the control parameters ($A\omega$ or $\delta\Omega_R$) of a laser field is more adequately described by a cusp-catastrophe surface, as shown schematically in Fig. 9a. From a different viewing angle the cusp catastrophe looks like what is plotted in Fig. 9b, which also shows a smaller second bistable region, as suggested by the results presented in Fig. 6.

However, there are clear differences between the classical and semiclassical results.[20] One difference is that linear stability analysis of Eq. (11) shows that the upper and lower branches are always stable and the middle branch unstable for λ less than 2, whereas classically both upper and lower branches can lose their stability, as seen in Fig. 3. This could be a flaw of the semiclassical model, which reflects the inadequacy of the spectroscopic Hamiltonian, Eq. (5), to simulate the Morse oscillator near dissociation. Instability does show up on the lower branch if λ is allowed to increase above 2, which is unphysical but may be of mathematical interest. The lower branch loses its stability at a bifurcation point where the real part of a pair of complex conjugate roots goes through zero and becomes positive. According to the Hopf bifurcation theorem a family of limit cycles (periodic orbits) bifurcates from this point, which can be either supercritical or subcritical. Numerical results show that the Hopf bifurcation in our case is subcritical, that is, the family of limit cycles arises on the side of stable foci and is thus unstable. This is shown in Fig. 10 and the location (z_0) of the bifurcation point as a function of Ω_R is the middle curve appearing in Fig. 8b. Detailed analysis and tests of numerical results near the bifurcation point are carried out in Ref. 68.

Finally, searching over the control parameter space, we have not found any chaos in this semiclassical model. This is another striking difference between the classical and semiclassical predictions and may again be attributed to the inappropriateness of the semiclassical model in describing dissociation. However, there is an important extension of the model that may bring chaos back into the model. So far we have not considered explicitly the rotational degree of freedom of the molecule. In principle,

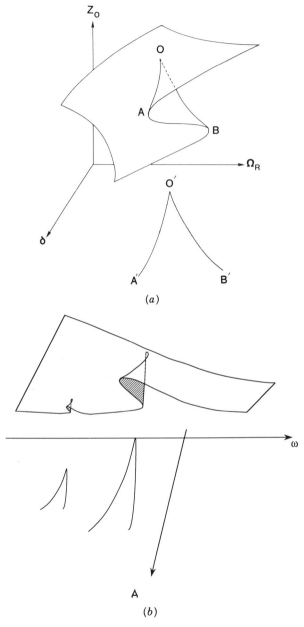

Figure 9. A schematic plot of the steady-state vibrational excitation (z_0) against the Rabi frequency Ω_R and detuning δ, as suggested by the semiclassical model, Eq. (11). It forms the shape of a cusp catastrophe. (a) The cusp, $A'O'B'$, is a projection of the bistable region onto the control parameter space of δ and Ω_R. (b) Viewing the cusp catastrophe from another angle. Also shown is a small second bistable region as suggested by Fig. 6.

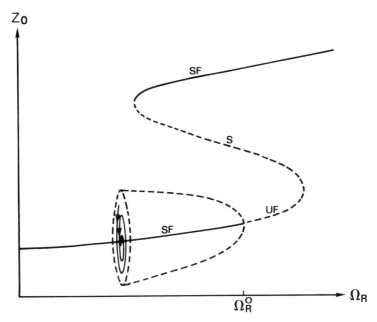

Figure 10. A schematic plot of z_0 as a function of Ω_R based on Eq. (11). Solid curves denote stable states and dashed curves represent unstable ones. Ω_R^0 marks the bifurcation point where unstable limit cycles bifurcate. SF, stable foci; UF, unstable foci; and S, saddle point.

some effects of rotational motion can be included as a part of the reservoir, thus contributing to the relaxation constants $\gamma_\parallel$ and $\gamma_\perp$. Another way to consider the dynamical effects of rotational motion is to introduce an amplitude modulation at the rotational frequency ω_m. This is done[74] by introducing a time-dependence to the Rabi rate; that is, we replace Ω_R everywhere in Eq. (11) by

$$\Omega_R(t) = \Omega_R^0 \cos^2 \omega_m t \tag{15}$$

Equation (11) thus becomes nonautonomous and no longer possesses fixed points or steady states, but it does have limit cycles. We can use the stroboscopic representation to study these limit cycles. That is, we record the position of the system trajectory in phase space, (x, y, z) in this case, once every period: A Pn periodic orbit would show up as n distinct points in the stroboscopic representation.

This modulation of field amplitude introduces instability to the upper branch. Here the upper and lower branches can be distinguished according to the average sizes of the coexisting attractors, but a better way is to convert the nonautonomous set of equations into an autonomous set by introducing two new variables, $u = \cos^2 \omega_m t$ and $v = \sin 2\omega_m t$. The steady states of the resulting five-variable equations, which are equivalent to those obtained by taking the time average of the nonautonomous ones, form again a surface of the shape of a cusp catastrophe. With reference to this surface we shall continue to use the terms, the upper and lower branches, below. As a control parameter (either A or ω) is varied, interesting behavior occurs on the upper surface. For example, if we set parameters at $\alpha = 1$, $\lambda = 0.1$, $\delta = 10$, $\Omega_R^0 = 1.7$, and decrease the value of ω_m down from 0.5 to 0.3, a stable P1 limit cycle may undergo an infinite sequence of period-doubling bifurcations leading to chaos. Then transition from chaotic to periodic behavior takes place via period-halving bifurcations. As shown in Table I, bifurcation thresholds up to the P16–P32 bifurcation are determined[74] from both sides of the chaotic zone, and the scaling factors δ_{calc} agree with Feigenbaum's universal value (4.669 2102) to within 1%. Chaotic attractors of this driven Morse system show interesting topological shape. An example at $\omega_m = 0.378$ is given in Fig. 11, along with a sequence of 12 Poincare surfaces of sections. They show clearly the stretching, twisting, folding, and even self-intersecting properties of the Morse strange attractor. Finally, for small ω_m values, such as $\omega_m = 0.01$ and $\Omega_R^0 = 4.0$, we have also found in this system relaxation oscillations (Fig. 3 of Ref. 74), in which a periodic orbit goes between the upper and lower branches, similar to the classical behavior of Fig. 7.

TABLE 1

Bifurcation Thresholds and the Geometrical Factors

	ω_k	δ_{calc}
1P–2P	0.414 85	
2P–4P	0.393 14	
4P–8P	0.389 756 3	6.4161
8P–16P	0.388 993	4.4330
16P–32P	0.388 828 3	4.6345
Chaos with periodic windows	0.388 783–0.364 7761	
32P–16P	0.364 685 2	
16P–8P	0.364 613 8	
8P–4P	0.364 282	4.6471
4P–2P	0.362 96	3.9843
2P–1P	0.359 0	2.9955

Note. δ_{calc} is obtained according to the formula δ_{calc}
$$= \frac{\omega_k - \omega_{k+1}}{\omega_{k+1} - \omega_{k+2}}.$$

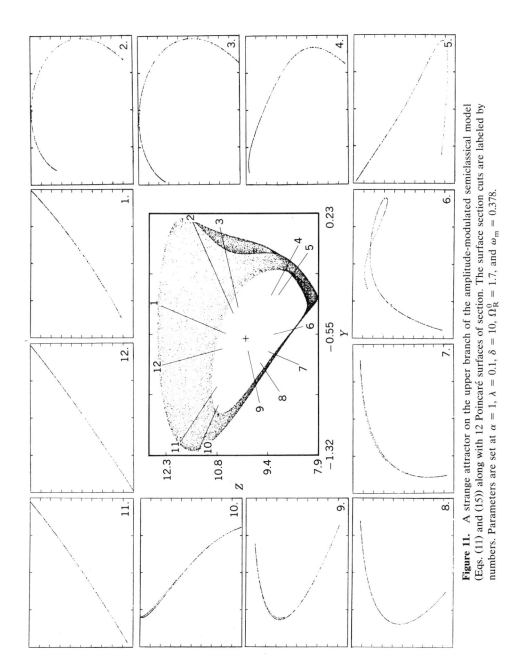

Figure 11. A strange attractor on the upper branch of the amplitude-modulated semiclassical model (Eqs. (11) and (15)) along with 12 Poincaré surfaces of section. The surface section cuts are labeled by numbers. Parameters are set at $\alpha = 1$, $\lambda = 0.1$, $\delta = 10$, $\Omega_R^0 = 1.7$, and $\omega_m = 0.378$.

23

As discussed above, Eq. (11) has been derived by truncating an infinite hierarchy of equations of motion for operators, we have also extended this treatment to the next higher level. This level yields seven equations,[75] one each for time evolution of the expectation values of a, $a^\dagger$, $a^\dagger a$, aa, $a^\dagger a^\dagger$, $a^\dagger aa$, and $a^\dagger a^\dagger a$. Complicated steady-state behavior has been found with this new set of equations at high Rabi rates (Ω_R), but for the range of the Rabi rate that is of interest here the steady-state curve has a similar shape to that of the previous set. The differences are that the upper branch is steeper than before and there are sets of parameters for which the upper part of the lower branch becomes unstable. The steady-state changes from a stable node to a saddle as the Rabi rate increases.[75]

C.   Quantum Approach

The semiclassical model shows some interesting nonlinear dynamical behavior of an IR multiphoton process. It has the merit that the equations are simple, and thus easy to analyze. But precise statements cannot be made about the quantum–classical correspondence based on these results alone, because the oscillator considered is not exactly a Morse oscillator and the relaxation terms included in the Liouville equation are not those for a Morse oscillator either. In this section we discuss a much more accurate quantum-mechanical treatment of the driven damped Morse oscillator problem.[76, 77]

Our starting point is an accurate description of damping in quantum mechanics. The safest way of doing it is to start with a bigger system in which the system of interest is coupled to a reservoir of harmonic oscillators.[78] The key here is a powerful representation of the Morse oscillator, such as by means of Lie algebraic method.[79–85] Particularly useful for us is an algebraic expression for the Hamiltonian of a one-dimensional Morse oscillator in terms of the generators of an su(2) algebra.[82, 83] It is given by

$$H_s = \hbar\omega_0\left(A^+A^- + \frac{I_0}{2}\right) \qquad (16)$$

where the generators A^+, A^-, and I_0 satisfy the following commutation relations:

$$[A^-, A^+] = I_0$$
$$[A^\pm, I_0] = \pm 2x_0 A^\pm \qquad (17)$$

In Eqs. (16) and (17), ω_0 and x_0 stand, respectively, for the Morse fundamental frequency and the anharmonic parameter, related to one-half of the total number of bound states supported by the Morse potential. A^+ and A^- are the raising and lowering operators for the Morse oscillator, and along with I_0, their operations on Morse eigenfunctions are given by

$$A^+|m\rangle = [(1 - x_0 m)(m + 1)]^{1/2}|m + 1\rangle$$
$$A^-|m\rangle = \{[(1 - x_0(m - 1)]m\}^{1/2}|m - 1\rangle \qquad (18)$$
$$I_0|m\rangle = (1 - 2x_0 m)|m\rangle$$

The Hamiltonian for the entire system, Morse oscillator + reservoir + driving field, becomes

$$H = H_s + H_r + H_{sr} + H_{sf} \qquad (19)$$

where H_r is the Hamiltonian of the reservoir defined by

$$H_r = \sum_{j=1}^{\infty} \hbar\omega_j b_j^{\dagger}b_j \qquad (20)$$

where $b_j^{\dagger}$ and b_j are the creation and annihilation operators, respectively, for the jth phonon mode. H_{sr} and H_{sf} are the interaction Hamiltonians given by

$$H_{sr} = \hbar(\Lambda_1 A^+ + \Lambda_1^{\dagger} A^-) + \hbar\Lambda_2 A^+ A^-$$
$$H_{sf} = \mu\varepsilon(A^+ + A^-)\cos\omega t \qquad (21)$$

Λ_1 and Λ_2 are operators in the space of bath modes; Λ_1 annihilates a phonon and Λ_2 is a phonon-conserved operator.

In what follows we skip details of the derivation and just outline the procedure of arriving at the equation of motion for the reduced density matrix of the Morse oscillator. We start with the Liouville equation,

$$\frac{dW(t)}{dt} = -i[L^s + L^r + L^{sr} + L^{sf}]W(t) \qquad (22)$$

for the statistical density matrix $W(t)$ for the whole system, where $L^i\rho$ is defined by $[H^i, \rho]/\hbar$. We assume that each bath mode is only weakly coupled to the anharmonic oscillator and that the reservoir is never far from thermal equilibrium and any deviations from equilibrium are rapidly eliminated. But this does not necessarily mean that the interaction between the anharmonic oscillator and the bath is weak, because the number of bath modes is very large.[78] Under these assumptions, a separable ansatz can be introduced in which $W(t)$ is written as

$$W(t) = \rho(t)W_r^\beta \qquad (23)$$

where $\rho(t)$ is the reduced density matrix for the system defined by

$$\rho(t) = \mathrm{Tr}_r\left[W(t)\right] \qquad (24)$$

and W_r^β by

$$W_r^\beta = \frac{e^{-\beta H_r}}{\mathrm{Tr}_r\left(e^{-\beta H_r}\right)} \qquad (25)$$

If in addition to the weak-coupling approximation we further invoke the Markoffian approximation that the relaxation time of the reservoir correlation functions is much shorter than the characteristic time of the system, the generalized master equation can then be written in the following form[76, 77]:

$$
\begin{aligned}
\frac{d\rho(t)}{dt} = {}& -iL^s\rho(t) - iL^{sf}\rho(t) \\
& + \int d\tau \exp(-i\omega_0 x_0\tau)\big\{\big\langle \Lambda_1(\tau)\Lambda_1^\dagger\big\rangle_0\left[\exp(i\omega_0 I_0\tau)A^-\rho(t), A^+\right] \\
& + \big\langle \Lambda_1^\dagger\Lambda_1(\tau)\big\rangle_0\left[A^+, \rho(t)\exp(i\omega_0 I_0\tau)A^-\right] \\
& + \big\langle \Lambda_1\Lambda_1^\dagger(\tau)\big\rangle_0\left[A^-, \rho(t)\exp(-i\omega_0 I_0\tau)A^+\right] \\
& + \big\langle \Lambda_1^\dagger(\tau)\Lambda_1\big\rangle_0\left[\exp(-i\omega_0 I_0\tau)A^+\rho(t), A^-\right]\big\} \\
& + \eta\big\{\left[A^+A^-\rho(t), A^+A^-\right] + \left[A^+A^-, \rho(t)A^+A^-\right]\big\}
\end{aligned}
\qquad (26)
$$

where $\langle\ \rangle_0$ denotes expectation value in a bath for which the bath reduced density matrix is represented by W_r^β and η is the scaled phase relaxation rate related to Λ_2. $\Lambda_1(\tau)$ denotes $\exp(iH_r\tau/\hbar)\Lambda_1\exp(-iH_r\tau/\hbar)$. Based on Eq. (26), the time evolution equation for the matrix element between the mth and the nth Morse eigenstate becomes[76, 77]

$$
\begin{aligned}
\frac{d\rho_{km}}{dt} = &-i\omega_{km}\rho_{k,m} + \left((\overline{k+1})(\overline{m+1})\right)^{1/2}[\gamma_{\downarrow k+1} + \gamma_{\downarrow m+1}]\rho_{k+1,m+1} \\
&-\left[\overline{k}\gamma_{\downarrow k} + \overline{m}\gamma_{\downarrow m} + (\overline{k+1})\gamma_{\uparrow k} + (\overline{m+1})\gamma_{\uparrow m} \right. \\
&\left. + \eta(\overline{k} - \overline{m})^2\right]\rho_{k,m} \\
&+ (\overline{km})^{1/2}[\gamma_{\uparrow k+1} + \gamma_{\uparrow m-1}]\rho_{k-1,m-1} \\
&- i\Omega \cos \omega t\left[\overline{k}^{1/2}\rho_{k-1,m}\right. \\
&\qquad\qquad + \overline{k+1}^{1/2}\rho_{k+1,m} - \overline{m}^{1/2}\rho_{k,m-1} \\
&\qquad\qquad \left. - \overline{m+1}^{1/2}\rho_{k,m+1}\right]
\end{aligned}
\tag{27}
$$

where ω_{km} = transition frequency between the kth and the mth
      level
      η = phase relaxation time
      $\gamma_\uparrow, \gamma_\downarrow$ = transition rates up and down the ladder due to the
interaction with the reservoir
      $\overline{m}$ = $m[1 - x_0(m - 1)]$
      Ω = Rabi rate = $\mu\varepsilon/\hbar$

We notice that $\gamma_\uparrow$ and $\gamma_\downarrow$ depend on the vibrational quantum number, but not η. Therefore, it is only an approximation that a unique longitudinal relaxation constant can be defined for an anharmonic oscillator. On the other hand, for the quadratic interaction assumed in H_{sr}, η indeed defines a level-independent phase relaxation time.

In what follows we present, by skipping the technical details, some of the results that we have obtained by integrating Eq. (27) numerically.[77] The parameters used are those for a hydrogen fluoride molecule: $\omega_0 = 4140$ cm^{-1} and $x_0 = 0.0211$. In Fig. 12 we show the time evolution of the average energy $\langle E(t)\rangle$ of the Morse oscillator for several phase relaxation times (η). As η increases, transient oscillations damp out increasingly faster, as expected. Furthermore, Fig. 12 shows that the asymptotic $\langle E(t)\rangle$ settles down into a period (or quasi-periodic) orbit; that is, it oscillates around a mean value (averaged over several periods), E_{av}. This asymptotic mean, E_{av}, on the other hand, is not a monotonically increasing function of η. It first increases with η, but eventually goes down as η becomes too large. This observation is consistent with the interpretation that η is

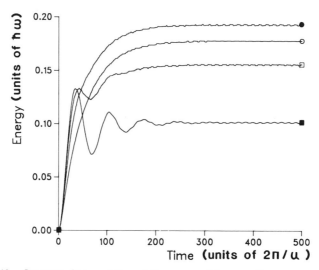

Figure 12. Quantum time evolution of the average Morse oscillator energy $\langle E(t) \rangle$ for several phase relaxation times (η) by solving Eq. (27). In the order of increasing asymptotic energy the curves correspond to $\eta = 0.05, 0.15, 0.80,$ and 0.35. Molecular constants for HF are used, $\omega = 4000$ cm^{-1}, the scaled Rabi rate $\alpha_f = (2\pi/\omega)\Omega_R = 0.22$, and the scaled relaxation time $\sigma = 0.05$.

related to the level width. As level width increases it compensates for the detuning caused by the anharmonic defect and thus facilitates the excitation. But as the level width becomes too large ($\eta > 0.45$), the oscillator strength spreads out so thin that excitation becomes less effective.

In Fig. 13 we show that E_{av} grows with the driving amplitude with a relatively sudden threshold. We have seen in the previous section that threshold behavior is associated with hysteretic bistability in the classical and semiclassical treatments. Hysteresis will not show up in the asymptotic density matrix in the quantum approach, because the equation is linear in ρ; thus if ρ_1 and ρ_2 are solutions of the Liouville equation so is any arbitrary combination $c_1\rho_1 + c_2\rho_2$. But the sudden increase here seems to indicate that the quantum results represent an average between two branches. This interpretation is supported by the asymptotic vibrational level population distributions P_n, plotted in Fig. 14 for several scaled field strength values as measured by $\alpha_f = (2\pi/\omega)\Omega$. For a very weak field (not shown in the figure) P_n peaks at the ground state and drops quickly to zero as n increases. As α_f increases P_n still peaks at the ground state, but a second peak grows at a finite quantum number. At high α_f we see that a second peak appears at $n = 3$ to form a bimodal distribution. Further

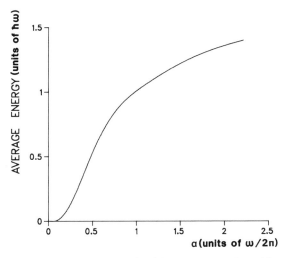

Figure 13. Average asymptotic energy (E_{av}) as a function of α with $\omega = 3960$ cm^{-1}, $\eta = 0.2$, and $\sigma = 0.05$. Symbols are defined in the caption of Fig. 12.

increase of α_f yields a distribution with a single maximum at $n = 3$. These results are very interesting, because the quantum picture supplied here gives a physical meaning to the lower and upper branches, seen so far only in classical and semiclassical approaches. That is, the lower branch corresponds to a level population distribution which peaks at the ground state and the upper branch a population distribution peaking at a finite quantum number. Therefore in quantum mechanics bistability shows up not as a hysteresis loop, but as a bimodal population distribution or a distribution representing a mixture of the two existing branches. The significance of this bimodal distribution is discussed in Subsection II.D.

To summarize, the quantum approach starts with an exact description of the Morse oscillator in terms of su(2) algebra. Based on the coupling of such an oscillator to a bath of infinite number of harmonic modes, we derive the relaxation master equation within the Born-Markoffian approximation. The external laser field has been treated as a classical field and its coherent interaction with the Morse oscillator has been derived without invoking the rotating wave approximation. In the final equation of motion derived for the density matrix, both the diagonal and off-diagonal elements are coupled together. Of course, unlike the semiclassical approach, no factorization ansat has been introduced in the derivation.

Ray[86, 87] has also studied the driven damped Morse oscillator problem, but with somewhat different starting point from Eq. (26). Instead of

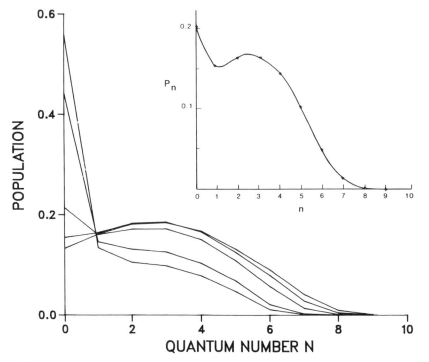

Figure 14. Asymptotic vibrational distributions as a function of the quantum number N for several driving amplitudes. Parameter values are the same as those of Fig. 13. From the top curve down on the left (or from the bottom curve up on the right) $\alpha_f = 1.9$, 2.2, 3.1, 3.8, and 4.4. The inset is a smooth curve connecting points of $\alpha_f = 3.1$.

deriving the generalized master equation from first principle, he adds in the Liouville equation a phenomenological term of the form of $\gamma\{[A^-, \rho A^+] + [A^- \rho, A^+]\}$ for the interaction between the oscillator and a bath. In contrast, two bath interaction terms are contained in Eq. (26): One represents energy relaxation, which is similar to the term discussed above, but includes the memory effect; the other is a phase relaxation term. Based on the su(2) coherent states, Ray derives a Fokker-Planck equation for the driven damped Morse oscillator from the Liouville equation. Furthermore, he has also written down the corresponding Langevin equation. Analyzing the latter equation in the semiclassical limit he has shown the existence of bistability, but without hysteresis. More recently, Gangopadhyay and Ray[88] have rederived the relaxation master equation by coupling the su(2) Morse Hamiltonian with a harmonic bath.

The equations they derived are in exact agreement with our earlier results, Eq. (34) of Ref. 76, or Eq. (2.17) of Ref. 77. They then consider broadband excitations of the oscillator in the quasicontinuum regime in which many photons can be effectively absorbed through a near-resonant incoherent process. In this regime rate equations[89, 90] have been used to treat the field-induced excitation process. Combining relaxation equations with the excitation equations, they have obtained a Pauli master equation, which involves only the diagonal elements of the reduced density matrix. Finally, by introducing a phenomenological decay term into the continuum,[88] they are able to discuss the roles of excitation, dissipation, and nonlinearity of the multiphoton dissociation process of a Morse oscillator. The last step is needed for the discussion of the dissociation process, for the su(2) representation of the Morse oscillator does not account for the continuum of a Morse potential.

D. Comparison with Experiments

We have found evidence of molecular vibrational bistability from all three approaches: hysteresis loops in the classical and semiclassical treatments and bimodal distributions in the asymptotic vibrational level populations. This bistable property should play a role in molecular processes, for instance, in effective infrared multiphoton excitation of a polyatomic molecule to high vibrational states, which can then be used for the study of state-selective chemical reactions. From a practical point of view, molecular bistability is potentially useful in the fabrication of mirrorless optical bistable devices, as discussed in the introduction. Unfortunately, so far no experiment has verified directly the bistability in a microscopic molecular systems. We discuss below some indirect evidence of vibrational bistability and experiments that may be carried out to verify the bistability.

A bimodal vibrational distribution can arise if two characteristic vibrational distributions coexist in which a fraction of molecules is found to be highly excited and the rest barely excited. Bimodal distributions have been observed in the experiments of infrared multiphoton excitation of SF_6 by Bagratashvili et al.[91, 92] and Mazur et al.[93] In the former experiment the spectra of the Stokes signals from Raman scattering were determined. It was found that they have two well-separated peaks: One ensemble contains molecules that are vibrationally hot and the other contains molecules that are vibrationally cold. This result is consistent with that observed by Mazur et al.[93] in the anti-Stokes signal with low fluence. Bagraashvli et al.[91, 92] also found that by adding buffer Xe gas to SF_6 the signal of the hot ensemble is enhanced. These results are consistent with the molecular bistable property discussed above, which predicts that for a certain range of parameter values a fraction of molecules can be effectively excited and the rest is still vibrationally cold. Furthermore, the effects of adding buffer gas can be simulated by increasing η in our model. Results of Fig. 12 then suggest that increasing the buffer gas pressure (within a certain limit) enhances the excitation efficiency, thus increasing the fraction of hot molecules. This is in agreement with the observation.[91, 92]

Another way to observe molecule bistability is to measure the shape change of a single infrared laser pulse through a nonlinear bistable medium. In fact, by plotting the output power of a CO_2 pulse passing through a cell containing SF_6 versus the input power, Galarneau et al.[94] obtained a hysteresis loop instead of a single retraced curve, as expected from linear response of a medium. This could be a direct measurement of the bistable behavior of SF_6, if one can eliminate the possibility of self-focusing that might be present in the experimental setup. More accurate measurements with fast IR detectors need be performed to verify that the hysteresis is indeed due to molecular bistability.

Finally, it has been suggested that the drastic growth observed in the Stokes wave intensity by three to six orders of magnitude in a stimulated Raman scattering (SRS)[95-97] may be due to molecular bistability, particularly in a medium for which self-focusing is absent. In ordinary treatments the molecular vibrational motion is described by a harmonic oscillator. In view of the fact that the input laser intensity can reach as high as GW/cm^2 and that of the Stokes wave reaches a few MW/cm^2, it is more proper to treat the vibrational motion as an anharmonic oscillator, such as a Morse oscillator. In such cases it has been suggested[98, 99] that the sudden increase of the bistable vibrational amplitude (see, for example, in Fig. 7) may then be responsible for the drastic growth of the Stokes wave intensity in a non-self-focusing medium. More calculations need to be done to confirm the role of molecular bistability and chaos in SRS.

III. BISTABILITY AND CHAOS ON A LENNARD-JONES POTENTIAL

The above study of the bistable property of a Morse oscillator raises several questions: First, how general is molecular bistability? Does it also show up in other anharmonic potentials? Second, as we discuss in Section VII, the required field strength to show molecular bistability and chaos is quite high among strongly bound molecules, when modeled by a Morse oscillator. Will other anharmonic potentials predict lower field strength requirement? To answer these questions, it is useful to study other anharmonic potentials.

Another commonly used molecular potential is the Lennard-Jones $(6, 12)$ potential,[100]

$$V(r) = 4\varepsilon \left[\left(\frac{\sigma}{r} \right)^{12} - \left(\frac{\sigma}{r} \right)^{6} \right] \tag{28}$$

where ε is the potential well depth and σ a measure of the atomic or molecular size. Note that many variations of the potential exist in the literature; most involve the use of different exponents and/or the truncation of the potential after a few σ's and smoothly connect to zero. The fundamental classical frequency of the potential is inversely proportional to σ and proportional to $\sqrt{\varepsilon}$, as given by

$$\omega_0 = \sqrt{\frac{c\varepsilon}{m\sigma^2}} \tag{29}$$

where $c = 36*3^{2/3}$. This property makes the potential very unsuitable for describing strong intra- or intermolecular interactions, which usually have a high ε and a small σ. For example, in the case of HF, Eq. (29) predicts the frequency to be about twice as large as the actual value. Thus, the potential usually finds its applications in systems with weak interactions, such as Ar-Ar and CH_4-CH_4.[100]

For a damped Lennard-Jones oscillator driven by an external field of $A_f \cos \omega_f t$, the equation of motion reads as

$$m\ddot{r} + \lambda\dot{r} - \frac{4\varepsilon}{r} \left[12 \left(\frac{\sigma}{r} \right)^{12} - 6 \left(\frac{\sigma}{r} \right)^{6} \right] = A_f \cos \omega_f t \tag{30}$$

which can be cast into a dimensionless equation,

$$\ddot{x} + \gamma\dot{x} - [ax^{-13} - bx^{-7}] = A \cos \omega t \tag{31}$$

by making the transformation

$$t \rightarrow \frac{t}{\omega_0} \quad \text{and} \quad r \rightarrow x\sigma \tag{32}$$

where $a = 48/c$ and $b = 24/c$. A, ω, and γ are defined in the same way as those given previously for the Morse potential.

As in the Morse case, Eq. (31) contains three external control parameters, γ, A, and ω. To facilitate the search of a bistable region in this three-dimensional parameter space, we need some strategy. We choose to fix the γ value and scan more thoroughly the $A\omega$ space. The value of γ is fixed in the following way: From Eq. (31), the maximal attractive force is about 0.041 93 occurring at $x = 1.2445$. This is equivalent to the minimal field strength of a static field required to pull apart a molecule, bounded by a Lennard-Jones (LJ) potential. We then let $A = 0.041\,93$ and vary γ until the response curve is just about to show bistable behavior. The value of γ thus found is 0.37. For a fixed γ, the bistable frequency domain usually increases with A. Keeping A equal to or less than 0.041 93 will not only prevent static field dissociation, guaranteeing the nonsingular behavior of the response curve at low frequency, but also insure that no bistable regions will appear. The following qualitative picture thus emerges from our procedure for the Lennard-Jones potential:

1. For $\gamma = 0.37$, bistability can be found for $A > 0.041\,93$ and none for $A < 0.041\,93$.

2. If we lower the γ value below 0.37, bistability starts to occur at an A value less than 0.041 93. The lower the γ value, the smaller is the minimal A value to observe bistability.

3. For $\gamma > 0.37$, the converse to (1) is true. That is, bistability starts to occur at an A value greater then 0.041 93. The greater the γ value, the larger is the minimal A value to observe bistability.

With this general picture in mind, we present in Fig. 15 several response curves at $\gamma = 0.37$ for several A values. We see striking similarities between Figs. 3 and 15, despite the drastic difference in the potential forms. We proceed to discuss some of the details of Fig. 15. First, for the specific choice of γ value, bistability is about to occur around $\omega = 0.7$ for $A = 0.04$. When A is increased to 0.06, a bistable region between $\omega = 0.417$ and 0.560 emerges. Examination of the phase space trajectories reveals that both the upper and lower branches represent P1 limit cycles, similar to the Morse response curves of $A = 0.3$ and 0.35 in Fig. 3. When A is increased to 0.08, the bistable frequency domain becomes larger, from $\omega = 0.220$ to 0.403. Furthermore, chaotic attractors are found over large ranges of ω on both branches, indicated by thicker curves in Fig. 15. Notice that over the bistable domain chaotic behavior is observed simultaneously on both the upper and lower branches. This is different from the Morse case, where the upper and lower chaotic zones do not exactly overlap (see Fig. 3). As usual, amid these chaotic regions are many windows of stable limit cycles. We have discovered a P16 limit cycle

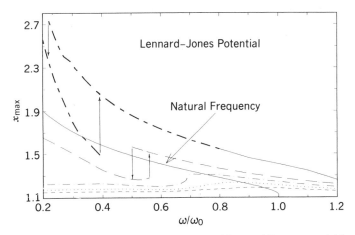

Figure 15. The response curves of a driven damped Lennard-Jones potential for several field amplitudes (A). In the increasing order of A, the curves correspond to $A = 0.02$ (dash), 0.03 (dot), 0.04 (dash-dot), 0.06 (long dash), and 0.08 (solid). The thick dashed segments denote the chaotic response. Vertical lines with arrows denote the hysteresis loop and the middle solid curve denotes the natural frequency curve.

around $\omega = 0.564$. What seems to be very different from the Morse case is the appearance of many P5 limit cycles in the upper chaotic region (In the Morse case P3 and P6 limit cycles dominate.). For example, all the stable limit cycles that we have examined in the windows between $\omega = 0.431$ and 0.500 are P5. Finer scan reveals that a P5 window occupies the frequency range from $\omega = 0.4304$ to 0.4319.

The above results seem to support the notion that molecular bistability is a general property of anharmonic potentials representing interactions between atoms. A question that remains is whether bistability also exists for a potential representing molecular bending motions. Some such potentials are harder than the harmonic one near the bottom of the well and become softer as energy increases. Going back to one of the motivations for studying the dynamical behavior of a Lennard-Jones potential, we discuss whether it is more likely to observe bistability in a LJ system. The real field strength (A_f) is related to the scaled one (A) through the following relation:

$$A_f = \frac{m\omega^2}{\sigma}A \qquad (33)$$

For about the same damping factor, the A value used here is about an order-of-magnitude smaller than that used for a Morse oscillator. If the frequency of a LJ system is also an order-of-magnitude smaller than that of a Morse system (in general, σ is larger for a LJ system too), then the A_f of a LJ system can be a thousand times smaller than that of a Morse system. This corresponds to a six order of magnitude decrease in the threshold field intensity required. However, such gain in the reduction of field intensity is at least partly compensated by the weakness of the transition dipole moments of the LJ systems which can be several orders of magnitude smaller than those of the Morse systems. Thus, it is inconclusive whether the LJ systems are more suitable for observing molecular bistability.

IV.   PHYSICAL PICTURE OF MOLECULAR BISTABILITY

With all the evidence that we have on molecular bistability we still feel that a physical intuitive picture about how bistability arises at the microscopic level is lacking. Such physical interpretation, if it exists, will be very useful for the prediction and understanding of bistability in other systems.

First we focus on Figs. 2 and 3 to see how we can understand the response of a Morse oscillator to a frequency sweep from a microscopic point of view. This frequency response is schematically reproduced at the top part of Fig. 16. First of all, the response curve tells us that if we can adiabatically vary the frequency of a pulsed field at will, the most efficient way of pumping energy into a molecule is to vary the frequency from high to low, instead of from low to high. But let us follow the arrows shown in the figure in which the frequency is swept from low to high and back down again. We shall discuss the four cases in the order of 1 to 4 as labeled in the figure. We assume that the molecule is in the ground state and the frequency increases from zero on the lower branch. At point 1, as labeled in Fig. 16, the frequency is too low compared to that of the $0 \rightarrow 1$ transition, meaning that the frequency detuning is much larger than the width of level 1, that the molecule is not effectively excited. This is true even if this frequency is in near resonance with a transition involving higher energy levels, such as the $4 \rightarrow 5$ transition, because these levels are not populated and all population essentially is and remains in the ground state of the oscillator. These facts are illustrated in the corresponding plot 1, which shows the level diagram and the vibrational population distribution.

At point 2 where the lower branch disappears, a sudden jump in the average energy of the oscillator occurs. At this point the frequency is still tuned to the red side of the fundamental frequency ω_0, but it falls inside the widths of the energy levels. Since the level spacings become smaller as

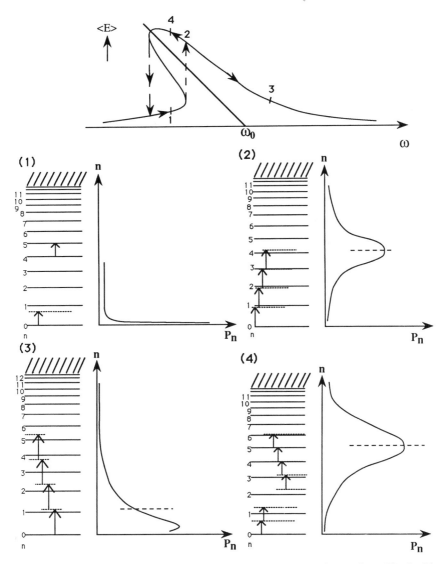

Figure 16. Physical interpretation of bistability. The top panel reproduces Fig. 1 with several points on the response curve labeled by numbers 1 to 4. In the four plots below (plots 1 through 4), we show the energy level diagrams and resonances with the photons as well as the vibrational distributions at these four points. More details are given in the text.

energy increases for molecular systems, several transitions involving low-energy levels will be in near resonance with the photon and are therefore excited in the photoabsorption process. The excitation process becomes efficient. There may even be an accidental n-photon resonance (a three-photon resonance is shown between levels 0 and 3 in plot 2) that helps pump energy into the system efficiently. Here the key to the sudden increase in energy is that the photon energy is still red-shifted, but falls inside the level widths of the lowest few levels.

When the photon energy further increases, a fewer number of levels will be involved in resonance excitation and the average energy drops. For example, at point 3 the photon energy is slightly greater than the fundamental frequency of the oscillator. The photon energy may still lie inside the width of level 1, but further excitation is not effective, because it quickly gets out of tune with higher energy levels.

As the photon frequency decreases after it reaches a maximal value beyond point 3, the response of the molecule retraces its path until point 2. Let us see what happens after point 2, for example, between points 2 and 4. As discussed above, at point 2 population has been built up among intermediate energy levels. Further decrease of frequency will bring about further transitions to higher levels. This is again due to the shrinkage of level spacings as energy increases. On the other hand, as in point 1 the detuning here is larger than the level widths of the lowest few levels to make downward transitions back to the ground state less efficient. The population is essentially locked in the high levels, thus high average energy of the system. This rise continues until the frequency is so small that the field strength, which has been held fixed, cannot maintain the upward transitions of population to even higher levels and everything collapses back down to the ground state.

This physical picture is simple and easy to understand. It seems that it has to be correct and can be directly verified in a quantum calculation. Such a calculation, however, has yet to be carried out.

Another way to understand molecular bistability is based on interpreting the response curve of Fig. 7. In this figure we vary the field amplitude adiabatically, while keeping the red-shifted (from the fundamental frequency ω_0) frequency constant. From Fig. 7 we see that as field amplitude $A(t)$ increases from zero, x_{max} barely increases until $A(t)$ reaches about 0.37. At that point x_{max} jumps up suddenly. The key to understanding molecular bistability from this viewpoint is to find out what mechanism is responsible for such a sudden surge of efficient photoexcitation. Since the frequency is held constant, the only variable is the field amplitude. The important factor that the field amplitude can bring into the picture is the power-broadening of energy levels. When the linewidth broadens to the extent that the effective linewidth is larger than the detuning, transitions to excited states suddenly become efficient. Since the frequency is tuned to the red side of the fundamental frequency, a sequence of transition to higher levels will follow immediately. This leads to the observed sudden increase of oscillator energy or amplitude.

Once on the upper branch, further increase of field amplitude will cause the oscillator energy to increase only slightly. When the field amplitude reaches a maximal value and sweeps back to a value slightly below 0.37, x_{max} does not drop back to the lower branch and retrace the original path, because the population has already been built up among the excited energy levels and the laser is still in resonance with transitions between them and its power is large enough to sustain the population among them. In addition, transitions back down to the ground states have been switched off due to weak power broadening. Thus, again the population has been locked up among the excited states. Further decrease of field amplitude only causes x_{max} to go down graduately, until laser intensity is so low that it can no longer sustain the resonant transitions among excited states. At this point, suddenly the entire population collapses down essentially to the ground state.

This interpretation has also yet to be verified directly in a calculation. We note that the rise and decay of field amplitude takes place naturally in a laser pulse. Thus, this second interpretation is much easier to realize experimentally than the adiabatical sweeping of frequency in the first interpretation.

V. MULTISTABILITY AND CRISES IN COUPLED SYSTEMS

So far we have focused only on the bistable behavior of molecules. In principle, multistability can occur in molecules too; that is, more than two attractors (neglecting dissociation) can coexist in the finite phase space. An example in the static case is a system having a multiwell potential,[101] such as molecules with several stable configurations or conformations. Again of interest here is dynamical multistability instead of static one. We do not know any concrete example of molecular dynamical multistability, but, as far as we know, no systematic investigation has been undertaken to search for them either.

One of the theoretical models that have been proposed for the study of population dynamics or physico-chemical processes is the coupled logistic map:[102–107]

$$x_{n+1} = 4\lambda_1 x_n(1 - x_n) + g_1(x_n, y_n)$$
$$y_{n+1} = 4\lambda_2 y_n(1 - y_n) + g_2(x_n, y_n) \tag{34}$$

where $g_i(x_n, y_n)$ is a polynomial in x_n and y_n. The most studied cases[102-106] are (1) symmetric bilinear coupling with $g_1(x_n, y_n) = g_2(x_n, y_n) = \gamma x_n y_n$; and (2) symmetric linear coupling with $g_1(x, y) = \gamma y$ and $g_2(x, y) = \gamma x$. We shall concentrate on case (1) for most of our discussion here. It is well known that the logistic map $x_{n+1} = 4\lambda_1 x_n(1 - x_n)$ is a paradigm for the period-doubling bifurcation route to chaos. In some cases a two-dimensional map is expected to behave just like a one-dimensional one, because the attractor that a trajectory is asymptotic to, as phase space shrinks, may be equivalent to the one-dimensional one. It is therefore not clear whether map (34) exhibits dynamical behavior that is intrinsically different from that of the one-dimensional logistic map.

To answer this question we[102, 103] and others[104-106] have carried out detailed dynamical studies of the coupled logistic map (34) or its equivalents. Three parameters exist in map (34): λ_1, λ_2, and γ. Chaos arises when the value of any one of these parameters increases, while the others are kept constant. We have calculated the chaotic thresholds on the $\lambda_1 \lambda_2$ plane when γ is kept constant. For small constant γ (e.g., less than 0.4), the scenario to chaos when one of the λ's increases is, in most cases, via the infinite cascade of the period-doubling bifurcations. But when γ is large, there exists a large region close to the $\lambda_1 \lambda_2$ diagonal in the parameter space where the dynamical behavior is very different, characterizing the competition between two attractors or oscillators in near resonance. In this region, map (34) has been shown to exhibit complicated behavior, including quasiperiodicity, phase locking, period adding, intermittency, and long-lived chaotic transient. The route to chaos is that generally characteristic of bifurcations of tori: First, Hopf bifurcations of periodic orbits into tori. Then, tori become increasingly unstable by developing "ears" on the attractor, i.e., the folding of a torus onto itself. This is followed by merging of separate parts into a single part and finally the breakup of the torus.

In what follows we concentrate only on one important aspect of the problem: the multistability associated with map (34) along the $\lambda_1 = \lambda_2$ diagonal of the parameter space with γ fixed at 0.1.[102] We gradually vary λ, which is defined by $\lambda_1 = \lambda_2$, and study the route to chaos of one attractor by selecting as the starting point of a given iteration the final point corresponding to the previous value of λ. We also trace an attractor by varying λ in the backward direction to reveal any hysteretic behavior.

The quantitative tool that we have used[103] is the Lyapunov exponents, which are measures of the exponential divergent rates of two neighboring points near an attractor. For example, the maximal Lyapunov exponent is negative for a stable period attractor, positive in a chaotic region, and zero at the bifurcation point and for a quasiperiodic interval. Using Lyapunov

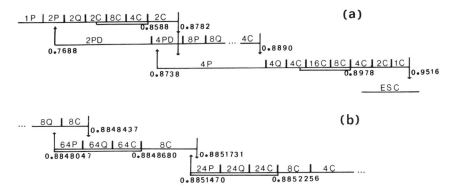

Figure 17. Bifurcation diagram of map (34) as a function of $\lambda = \lambda_1 = \lambda_2$ with $\gamma = 0.1$. The notations of symbols are given in the text. Multiple horizontal lines denote coexisting attractors. (b) An enlargement of a segment in (a), denoted by dots.

exponents as a tool, we have obtained a bifurcation diagram in the one-dimensional parameter space of λ, shown in Fig. 17. The symbols used in the figure are P (periodic), PD (periodic attractor located on the diagonal xy space), Q (quasiperiodic), C (chaotic), and ESC (escape to infinity). The integer n in front of the attractor denotes that the attractor can be separated into n parts, where each part belongs to an attractor of the nth convoluted map of map (34). The nth-convoluted map can be obtained iteratively from the $(n-1)$th map according to $F^{(n)}(x) = F(F^{(n-1)}(x))$. We use multiple horizontal lines to represent several coexisting attractors with their own basins. Double lines under an attractor symbol denote coexisting attractors that are mirror images of each other. There exist also many frequency lockings and windows within the quasiperiodic and chaotic regions that are not shown in the figure. As shown in Fig. 17a, the entire stability domain can be separated into three main branches, each with its own bifurcation route to chaos. The bifurcation routes of the upper and lower branches show strong similarities. The middle branches contain more complicated substructures, as shown in Fig. 17b. Even if we regard the coexisting mirror-imaged attractors as a single attractor, we still observe tristability in several parameter intervals of the bifurcation diagram, such as $\lambda = 0.8738$–0.8782, $0.884\,804\,7$–$0.884\,843\,7$, and $0.885\,147\,0$–$0885173\,1$.

Associated with Fig. 17 is another interesting phenomenon called crisis, which represents a sudden qualitative change in chaotic dynamical behavior induced by a collision of a strange attractor with an unstable fixed point. Crises in some cases signal the end of a branch in the multiple

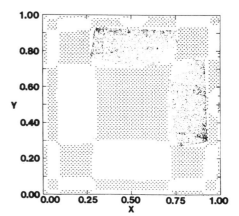

Figure 18. Basin structure and 2C attractor of map (34) for $\gamma = 0.1$ and $\lambda = 0.8782$. The blank region represents the basin of 2C; vertical bars denote the basin of 4P; horizontal bars the basin of 4PD.

stability zone. Two types of crises are observed in Fig. 17: boundary and cyclic. A boundary crisis occurs when a strange attractor collides with an unstable fixed point on the basin boundary, causing the disappearance of both the basin and the chaotic attractor. There are five cases of boundary crises, denoted in Fig. 17 by downward arrows, each corresponding to the end of a chaotic band of an upper branch. An example is shown in Fig. 18 at $\lambda_c = 0.8782$, where three attractors (2C, 4PD, and 4P) coexist. It is also clear from the figure that the phase portrait of the 2C attractor almost collides with its basin boundary. In fact, it seems that the 2C approaches the boundaries with the 1P and with the 4PD attractors simultaneously. Above λ_c, the 2C attractor and its basin disappear and the basins of the 4P and 4PD expand to fill the original 2C basin in a highly interlaced fashion. Notice also that hysteresis is observed for all five cases of the boundary crises shown in Fig. 17.

Four of the crises in Fig. 17 are cyclic crises, in which several coexisting chaotic attractors are involved in a simultaneous cyclic collision with the boundaries that separate the respective basins of attraction. After collision all the attractors that take part in the cyclic collision merge into a single large attractor. Its dynamical behavior is characterized by cyclic transitions of the phase space trajectory from one to the other of the original attractors in a fixed order, but with seemingly random residence times on each of them. This type of crises was first discovered[103] in the coupled logistic map and the four cases are located at the right ends of the four horizontal double lines of Fig. 17. Incidentally, no hysteresis exists with cyclic crises.

For many other interesting observations on coupled logistic maps, we refer readers to the extensive work done by Kaneko, as summarized in his book.[106] Ferretti and Rahman[107] have also studied a coupled logistic map and its potential applications in chemical physics. They have especially investigated the coupled map with nonsymmetric linear couplings: $g_1(x, y)$ = $\gamma_1 y$ and $g_2(x, y) = \gamma_2 x$, where in general γ_1 is not equal to γ_2. They have argued that the coupled logistic map can be used as a qualitative model for the study of bifurcations in oscillatory reactions, such as the Belousov-Zhabotinskii reaction, by monitoring two reactants, and for the simulation of two competing modes in the process of infrared multiphoton dissociation of polyatomic molecules.

Following the approaches presented in the previous sections a more realistic model for the study of dynamics involving two competing modes in the process of IR multiphoton dissociation of polyatomic molecules appears to be a (driven damped) coupled Morse oscillator system. It would be very interesting to find out whether tristability and crises exhibit themselves in this model.

VI. BASIN BOUNDARIES OF ATTRACTORS AND ESCAPE RATES

We report in this section some preliminary results on two important concepts concerning systems with multiple stable attractors, one of which may be located at infinity. The purpose of reporting these results here is to stimulate more work along these directions.

For a system that has several coexisting stable attractors an important concept that an experimentalist should worry about is the basin boundary of an attractor.[108] The set of initial points in phase space that approach an attractor is defined as the basin of attraction of that attractor. A basin boundary is a curve in phase space that separates one basin of attraction from the neighboring one(s). The importance of a basin boundary comes from the fact that all initial experimental conditions lying within the basin boundary will lead to the same attractor, thus leading to the robustness of the experimental results. In the 'bistable regime' of a driven damped Morse oscillator three or more attractors may actually coexist. For example, two attractors exist in phase space near the equilibrium configuration ($x = 0$, $p = 0$) of the ground-state molecule, and the others with the average interatomic distance so large that for all practical purpose it can be considered to correspond to a single dissociation channel. The entire noncompact phase space can therefore be partitioned into three basins of attraction. Under the experimental conditions that dissociation can take place only for those molecules with the initial energies close to the dissociation threshold, we can neglect dissociation and concentrate on the basin boundaries of the other two "finite" attractors. An example of basin boundaries is plotted in Fig. 19, which is obtained by solving Eq. (2) for

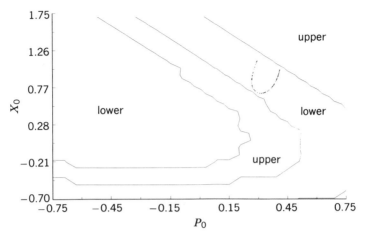

Figure 19. Basin boundaries of the upper and the lower branches for a driven damped Morse oscillator. The "lower" basin converges to a strange attractor, whose surface section (recorded every period) is also shown in the figure. The "upper" basin converges to a P1 orbit, not shown. The parameters used correspond to those of Fig. 3, where $A = 0.45$, $\omega = 0.3$, and $\gamma = 0.4$.

the following parameters: $A = 0.45$, $\omega = 0.3$, $\gamma = 0.4$, and the initial time $t_0 = 0$. The initial phase points of trajectories are chosen so that they are uniformly distributed over the phase space of the figure. The results presented in Fig. 19 show that the limited phase space considered can be partitioned into two basins of attraction: one for the upper branch, which is asymptotic to a periodic orbit, and the other for the lower branch, which asymptotic to a strange attractor. The basin boundary is found to be essentially smooth, rather than fractal. (It is not uncommon to find fractal basin boundaries.[108]) Also shown in the figure is the Poincaré surface of section (ωt mod by 2π) of the strange attractor residing on the lower branch. This strange attractor has been generated by the initial conditions $x_0 = 0.50$ and $p_0 = -0.15$. This figure is similar to Fig. 18, in which basin boundaries and a strange attractor of the coupled logistic map is shown.

In the above paragraph we mention that for a certain set of parameters some initial conditions may lead to "dissociation," that is, escape of trajectories from the finite phase space region near the origin ($x = 0$, $p = 0$) which defines the bound state domain. The escape rate is a problem of current interests.[33, 109–111] It is generally believed that for Hamiltonian systems algebraic decay marks phase space mixed with regular and chaotic regions, whereas hyperbolic systems decay exponentially.[112, 113] Exceptions to this empirical rule have been found.[111] It would

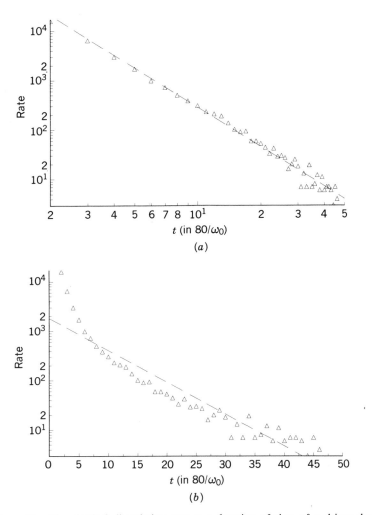

Figure 20. The classical dissociation rate as a function of time of a driven damped Morse oscillator for $A = 0.06$, $\omega = 0.9$, and $\gamma = 10^{-5}$. (a) A log–log plot and (b) a log–linear plot. More than 20 000 initial points uniformly distributed over the energy curve of the 10th Morse eigenstates of the HF molecule are used.

be of some interest to find out whether the same relationships apply to the dissociation rate of the present dissipative flow system. But here we are immediately faced with a numerical difficulty in defining dissociation rate. A trajectory is considered dissociated if it passes a critical distance, x_c, and the dissociation rate can be defined in terms of the decreasing rate of the surviving trajectories. Only when this rate converges as x_c increases can we obtain a well-defined dissociation rate. This criterion for the definition of a dissociation rate may already be too stringent for a Hamiltonian system and is particularly hard to satisfy for a dissipative system. This is because an outgoing trajectory at a large distance outside the potential well region can still lose its energy and stops moving outward due to damping. To avoid this difficulty in defining a dissociation rate, we study instead the escape rate of trajectories from a region of phase space, characterizing the bound states of the Morse oscillator. Thus, escape here means leaving an artificially defined bound-states region in phase space.

We have studied the escape rate of a driven damped Morse oscillator for $A = 0.06$, $\omega = 0.9$, and a very small damping constant, $\gamma = 10^{-5}$. These conditions mimic the Hamiltonian case (with $\gamma = 0$) studied in Ref. 33. We are not able to find any bound-state attractor near the potential minimum for this set of parameters. For the trajectory calculations, more than 20 000 points evenly distributed over the energy curve of the 10th Morse eigenstate of the HF molecule are used as initial conditions. A molecule is considered to have escaped from the bound state region once its trajectory reaches x_c. Two values are used for x_c: 8.0 and 10.0. The escape rate is found to be independent of the x_c chosen. Results for $x_c = 10.0$ are plotted in Fig. 20, which shows that the present escape rate is better fitted by a power law (Fig. 20a, a log–log plot) than by an exponential law (Fig. 20b, a log–linear plot). Both x_c's give the following time dependence: The survival probability is proportional to $t^{-1.65}$. However, we caution that more thorough studies are needed to reach any definitive conclusions about the systematics of the escape rate of a dissipative molecular system.

VII.   DISCUSSION AND CONCLUDING REMARKS

Most results reported in previous sections are based on calculations using dimensionless variables. For example, all the classical calculations reported are carried out by using a dimensionless Hamiltonian with a scaled potential valid for all diatomic systems with nonzero dipole moments. We give below some estimated values of the field strength, frequency, and damping constant required to observe the phenomena discussed in previous sections in terms of the molecular constants of some realistic molecules.

The field strength required to observe the interesting behavior predicted in previous sections for polyatomic molecules is quite reasonable. Semiclassical results presented in Fig. 8 are obtained by using parameters simulating the ν_3 mode of SF_6 (Refs. 114, 115): $\omega_0 = 948$ cm^{-1}, anharmonicity defect $= 2.88$ cm^{-1}, the transition dipole moment $\mu = 0.388$ D (Ref. 116), and $\gamma_{\parallel} = 5$ cm^{-1}. The curve in Fig. 8 with $\delta = 1.6$ corresponds to the experimental situation of the ν_3 mode pumped by a CO_2 laser. Figure 8a shows that effective excitation takes place at $\Omega_R = 0.069$, which yields a threshold field intensity of 6 MW/cm^2 (a factor of 4 smaller than the observed threshold). Furthermore, comparison between curves with $\delta = 1.6$ and 5.0 (corresponding to the isotope shift of 17 cm^{-1} between $^{32}SF_6$ and $^{34}SF_6$) shows that at $\Omega_R = 0.069$ only $^{23}SF_6$ is effectively excited; thus, it can be efficiently separated from the $^{34}SF_6$ isotopes.

Similar to the infrared multiphoton dissociation process, the field strengths required for diatomic molecules tend to be very high. For example, molecular constants for HF are $\omega_e = 4138.66$ cm^{-1}, $D = 6.125$ eV, $m = 0.95$ amu, and $\beta = 1.1741 a_0^{-1}$. The true field strength A_0 is related to the scaled value by $A_0 = A m \omega_0^2 / \beta \mu'(x)$ or $A_0 = 2A\beta D/\mu'(x)$. The parameter values $A = 0.01$, $\omega = 0.6$, and $\gamma = 0.001$ correspond, respectively, to a laser intensity of 220 GW/cm^2, a frequency around 2483 cm^{-1}, and a collisional deactivation rate attainable in a gas under high pressure or in liquids, if an effective charge $\mu'(x)$ equal to the electronic charge is used. A damping rate of $\gamma = 0.1$ roughly corresponds to the vibrational lifetime of the C-O stretching mode of the CO molecule absorbed on a Pt(11) surface. The damping value of $\gamma = 0.4$ used in many of the classical calculations could be too high to be realistic, but its usage can be justified by the motivation of speeding up the calculations without changing qualitatively the dynamics involved. The field strength required to see interesting behavior increases correspondingly.

From this review it should be clear that for molecular bistable and chaotic behavior what is still needed most is more quantum-mechanical calculations of some realistic molecules. One of the most urgent goals is to verify the physical pictures of bistability presented in Section IV. This will more firmly establish the fact that the response of the molecular systems can be described by a cusp catastrophe. Other objectives include establishing the classical–quantum correspondence of systems which exhibit bistable and chaotic behavior.

One of the most fundamental aspects of the problems that we are faced with in studying driven damped Morse oscillators concerns the treatment of dissipation in quantum mechanics. Dissipative quantum mechanics has a long history; important developments include the Feynman-Vernon influence functional theory[117] based on the path-integral formulation, the work on the damped harmonic oscillator problem as reviewed by Dekker,[118] treatments of dissipative quantum tunneling and coherence,[78, 119, 120] and the quantum-dynamical semigroup approach.[121, 122] Quite differently, non-linear time-dependent Schrödinger equation approaches have been proposed by Kostin,[123] Gisin,[124] and Schuh et al.,[125] in which a phenomenological dissipative (nonlinear) term is introduced into the Schrödinger equation. Korsch and Steffen,[126] on the other hand, have argued that a more correct approach should start with a nonlinear evolution equation for the density operator. In their approach again a nonlinear phenomenological term is added into the Liouville equation, which preserves the norm. The difficulty of this kind of approach resides in the solution of the nonlinear density matrix equations and the conceptual interpretation of the density operators involved. Progress has also been made in dissipative molecular processes involving highly excited states of polyatomic molecules.[127, 128] Classical and semiclassical calculations are compared with the quantum results in terms of master equations to study the classical–quantum correspondence of such processes.

A very interesting, but different, approach to classical atomistic dynamics involving a reservoir has been developed based on the molecular dynamics methods by Nosé,[129, 130] Hoover,[131] and others.[132] Nosé uses an extended system approach in which an additional degree of freedom s, which acts as a reservoir, is introduced to a physical system of N particles. By using a microcanonical ensemble of the extended system, a canonical ensemble statistics of the N particle system is derived. The resulting formulation is reversible, but not Hamiltonian, thus non-volume-preserving. This approach uses feedback to link thermodynamics to microscopic mechanics and could represent the simplest possible system that maintains both the reversibility and thermodynamical significance. The dynamics of reversible systems is in general quite different from that of dissipative systems, but it does exhibit similar interesting behavior, such as attracting set and chaos. However, the dynamics associated with the Nosé mechanics[131, 132] are only virtual and may not represent those of any real system. The latter, of course, can always be described by the statistical average values of the variables.

It will be interesting to apply this formalism to the molecular vibrational excitation and relaxation problem. In this problem one may introduce several Nosé degrees of freedom. Thus, as the characteristic frequency of the system changes during time evolution, different bath degrees of freedom will be in strong resonant interactions with the system. Furthermore, it is important to develop a corresponding quantum-mechanical approach.[133]

The importance of dissipative (or non-Hamiltonian reversible) quantum dynamics to the understanding of dynamical processes involving large systems, such as energy transfer and relaxation of polyatomic molecules, processes related to surfaces and condensed-phase materials, cannot be overemphasized. But it is fair to say that dissipative quantum dynamics is still far from being well understand and that much work remains to be done and many open questions remain to be answered.

Acknowledgments

We are grateful to Yan Gu, Mingwhei Tung, and Erning Liu for their collaboration of the investigations based on which this review has been written. We thank also Michael Klein for stimulating discussions on Nosé mechanics and Conan W. Yuan for making some of the drawings in this review. GCL expresses his gratitude for the support of Dow Corning Co. during this work. JMY acknowledges the Donors of Petroleum Research Fund, Administered by the American Chemical Society, for support of this research.

References

1. See, for example, Abstracts of Papers for the Third International Symposium on Molecular Electronics Devices, 1986.

2. Proceedings of the Molecular Electronic Devices Workshop, *NRL Memorandum Report 4662*, 1981.

3. E. Abraham and S. D. Smith, *Rep. Prog. Phys.* **45**, 815 (1982).

4. A. Aviram and P. E. Ratner, *Chem. Phys. Lett.* **29**, 277 (1974).

5. S. Yanchinski, *New Sci.* **93**, 68 (1982).

6. B. Macke, E. M. Pessina, B. Segard, and J. Zemmoari, *Phys. Rev. A* **44**, R5350 (1991).

7. C. M. Bowden, C. C. Sung, J. W. Haus, and J. M. Cook, *J. Opt. Soc. Am. B* **5**, 11 (1988).

8. F. Henneberger and H. Rossmann, *Phys. Status Solidi B* **121**, 685 (1984).

9. J. Guckenheimer and P. Holmes, *Nonlinear Oscillations, Dynamical Systems, and Bifurcations of Vector Fields*, Springer, New York, 1983.

10. D. K. Arrowsmith and C. M. Place, *An Introduction to Dynamical Systems*, Cambridge University Press, Cambridge, UK, 1990.

11. E. T. Jackson, *Perspectives of Nonlinear Dynamics*, Cambridge University Press, Cambridge, UK, 1989.

12. A. D. Morosov, *Differ. Equa.* **12**, 164 (1976).

13. P. J. Holmes, *Appl. Math. Mod.* **1**, 362 (1977).

14. B. A. Huberman and J. P. Crutchfield, *Phys. Rev. Lett.* **43**, 1743 (1979).

15. F. C. Moon and P. J. Holmes, *J. Sound Vib.* **65**, 285 (1979); **69**, 339 (1980).

16. P. J. Holmes, *Philos. Trans. R. Soc. A* **292**, 419 (1979).

17. Y. Ueda, *J. Stat. Phys.* **20**, 181 (1979); Y. Ueda, in R. H. G. Helleman (Ed.), *Nonlinear Dynamics*, New York Academy Sciences, New York, 1980, p. 422.

18. C. S. Wang, Y. H. Kao, J. C. Huang, and Y. S. Gou, *Phys. Rev. A* **45**, 3471 (1992).

19. I. S. Averbukh, V. A. Kovarsky, and N. F. Perelman, *Phys. Lett.* **24A**, 36 (1979); N. F. Perelman, V. A. Kovarskii, and I. S. Averbukh, *Sov. Phys.-JETP* **52**, 9 (1980).

20. J. M. Yuan, in Hao Bai-lin (Ed.), *Directions in Chaos*, Vol. 1, World Scientific, Singapore, 1987, p. 164.

21. P. M. Morse, *Phys. Rev.* **34**, 57 (1929).

22. D. ter Haar, *Phys. Rev.* **70**, 222 (1946).

23. K. S. Jhung, I. H. Kim, K. B. Hahn, and K. H. Oh, *Phys. Rev. A* **40**, 7409 (1989); **42**, 6497 (1990); **44**, 5611 (1991).

24. B. R. Henry and W. J. Siebrand, *J. Chem. Phys.* **49**, 5369 (1968).

25. B. R. Henry, *Acc. Chem. Res.* **10**, 207 (1977).

26. E. L. Sibert III, W. P. Reinhardt, and J. T. Hynes, *J. Chem. Phys.* **77**, 3538 (1982); **81**, 1115 (1984).

27. J. Kommandeur, W. A. Majeuski, W. L. Meerts, and D. W. Pratt, *Annu. Rev. Phys. Chem.* **38**, 433 (1987).

28. R. B. Walker and R. K. Preston, *J. Chem. Phys.* **67**, 2017 (1977).

29. S. K. Gray, *Chem. Phys* **75**, 67 (1983); **83**, 125 (1984).

30. M. Davis and R. E. Wyatt, *Chem. Phys. Lett.* **86**, 235 (1982).

31. R. C. Brown and R. E. Wyatt, *Phys. Rev. Lett.* **57**, 1 (1986); *J. Phys. Chem.* **90**, 3590 (1986).

32. R. B. Shirts and T. F. Davis, *J. Phys. Chem.* **88**, 4665 (1984).

33. Y. Gu and J. M. Yuan, *Phys. Rev. A* **36**, 3788 (1987).

34. D. Poppe and J. Korsch, *Physica* **24D**, 367 (1987).

35. M. E. Goggin and P. W. Milonni, *Phys. Rev. A* **37**, 796; **38**, 5174 (1988).

36. J. J. Tanner and M. M. Maricq, *Phys. Rev. A* **40**, 4054 (1989); *Chem. Phys. Lett.* **14**, 503 (1988).

37. P. Gaspard and S. A. Rice, *J. Phys. Chem.* **93**, 6947 (1989).

38. J. Heagy and J. M. Yuan, *Phys. Rev. A* **41**, 571 (1990).

39. R. Graham and M. Hohnerbach, *Phys. Rev. Lett.* **64**, 637 (1990).

40. C. Leforestier and R. E. Wyatt, *J. Chem. Phys.* **82**, 752 (1985).

41. K. F. Milfeld and R. E. Wyatt, *Phys. Rev. A* **27**, 72 (1983).

42. Z. M. Lu, J. F. Heagy, M. Vallieres, and J. M. Yuan, *Phys. Rev. A* **43**, 1118 (1991).

43. R. Heather and H. Metiu, *J. Chem. Phys.* **86**, 5009 (1987); **88**, 5496 (1988).

44. Z. M. Lu, M. Vallieres, J. M. Yuan, and J. F. Heagy, *Phys. Rev. A* **45**, 5512 (1992).

45. J. F. Heagy, Z. M. Lu, J. M. Yuan, and M. Vallieres, in D. H. Feng and J. M. Yuan (Eds.), *Directions in Chaos*, Vol. 4, World Scientific, Singapore, 1992, p. 322.

46. L. D. Landau and E. M. Lifshitz, *Mechanics*, Pergamon, Oxford, UK, 1976.

47. A. H. Nayfeh, and D. T. Mook, *Nonlinear Oscillator*, Wiley, New York, 1977.

48. J. B. Marion, *Classical Dynamics of Particles and Systems*, 2d ed., Academic, New York, 1970.

49. G. C. Lie and J. M. Yuan, *J. Chem. Phys.* **84**, 5486 (1986).

50. R. Kapral, M. Schell, and S. Fraser, *J. Phys. Chem.* **86**, 2205 (1982).

51. G. C. Lie, *J. Chem. Phys.* **60**, 2991 (1974).

52. W. Knop and W. Lauterborn, *J. Chem. Phys.* **93**, 3950 (1990).

53. See, for example, K. Ohwada, *Spectrochim. Acta A: Mol. Spectrosc.* (*UK*) **47A**, 1751 (1991).

54. M. J. Feigenbaum, *J. Stat. Phys.* **19**, 25 (1978); **21**, 669 (1979).

55. M. Debnath and A. R. Chowdhury, *Phys. Rev.* **44**, 1049 (1991).

56. L. M. Narducci, S. S. Mitra, R. A. Shatas, and C. A. Coulter, *Phys. Rev. A* **16**, 247 (1977).

57. L. M. Narducci and J. M. Yuan, *Phys. Rev. A* **22**, 261 (1980).

58. X. Y. Huang, L. M. Narducci, and J. M. Yuan, *Phys. Rev. A* **23**, 3084 (1981).

59. C. D. Cantrell, V. S. Letokhov, and A. A. Makarov, in M. S. Feld and V. S. Lefokhov (Eds.), *Coherent Nonlinear Optics: Recent Advances*, Springer, Berlin, 1980.

60. P. A. Schultz, A. S. Sudbo, D. J. Krajnovich, H. S. Kwok, Y. R. Shen, and Y. T. Lee, *Annu. Rev. Phys. Chem.* **30**, 379 (1979).

61. H. S. Kwok, E. Yablonovitch, and N. Bloembergen, *Phys. Rev. A* **23**, 3094 (1981).

62. H. W. Galbraith, J. R. Ackerhalt, and P. W. Milonni, *J. Chem. Phys.* **79**, 5345 (1983).

63. N. Bloembergen, *Opt. Commun.* **15**, 416 (1975).

64. Y. Weissman, and J. Jortner, *Chem. Phys.* **59**, 1 (1981).

65. Z. Z. Gan, G. Z. Yang, X. Y. Huang, and K. A. Feng, *Acta Phys. Sinica* **29**, 743 (1980); Z. Z. Gan, G. Z. Yang, K. A. Feng, and X. Y. Huang, *Acta Phys. Sinica* **27**, 664 (1978).

66. See, for example, H. Gzyl, *Phys. Rev. A* **27**, 2297 (1983) and references therein.

67. D. M. Greenberg, *J. Math. Phys.* **20**, 762 (1979).

68. J. M. Yuan, E. Liu, and M. Tung, *J. Chem. Phys.* **79**, 5034 (1983).

69. H. Haken, *Handbuch der Physik*, Springer, Berlin, 1970.

70. R. Bonifacio and F. Haake, *Z. Phys.* **200**, 526 (1967).

71. W. H. Louisell, *Quantum Statistical Properties of Radiation*, 2d ed., Wiley, New York, 1984.

72. T. Poston and I. Stewart, *Catastrophe Theory and Its Applications* Pitman, London 1978.

73. R. Gilmore, *Catastrophe Theory for Scientists and Engineers*, Wiley, New York, 1981.

74. E. Liu and J. M. Yuan, *Phys. Rev. A* **29**, 2257 (1984).

75. P. You and J. M. Yuan, unpublished.

76. M. Tung, E. Eschenazi, and J. M. Yuan, *Chem. Phys. Lett.* **115**, 405 (1985).

77. M. Tung and J. M. Yuan, *Phys. Rev. A* **36**, 4463 (1987).

78. A. O. Caldeira and A. J. Leggett, *Ann. Phys.* (*NY*) **149**, 374 (1983).

79. R. D. Levine and C. E. Wulfman, *Chem. Phys. Lett.* **60**, 372 (1979).

80. F. Iachello, *Chem. Phys. Lett.* **78**, 581 (1981).

81. Y. Alhassid, F. Gursey, and F. Iachello, *Phys. Rev. Lett.* **50**, 873 (1983); *Ann. Phys.* (*NY*) **148**, 346 (1983); *Chem. Phys. Lett.* **99**, 27 (1983).

82. R. D. Levine, *Chem. Phys. Lett.* **95**, 87 (1983).

83. R. D. Levine, in J. Jortner and B. Pullman (Eds.), *Intramolecular Dynamics*, Reidel, Dordrecht, 1982, p. 17.

84. Y. Alhassid, J. Engel, and F. Iachello, *Phys. Rev. Lett.* **57**, 9 (1986).

85. A. O. Barut, A. Inomata, and R. Wilson, *J. Math. Phys.* **28**, 605 (1987).

86. D. S. Ray, *J. Chem. Phys.* **92**, 1145 (1990).

87. D. S. Ray, *Phys. Lett. A* **122**, 479 (1987).

88. G. Gangopadhyay and D. S. Ray, *J. Chem. Phys.* **97**, 4104 (1992).

89. J. L. Lyman, *J. Chem. Phys.* **67**, 1868 (1977).

90. P. A. Schultz, A. S. Sudbo, E. A. Grant, Y. R. Shen, and Y. T. Lee, *Chem. Phys.* **72**, 4895 (1980).

91. V. N. Bagratashvili et al., *Sov. Phys.-JETP* **53**, 512 (1981).

92. V. N. Bagratashvili et al., *Opt. Lett.* **6**, 148 (1981).

93. E. Mazur, I. Burak, and N. Bloembergen, *Chem. Phys. Lett.* **105**, 258 (1984).

94. P. Galarneau, S. L. Chin, X. X. Ma, G. Farks, and F. Yergeau, *SPIE Proc.* **380**, 392 (1983).

95. Y. R. Shen, *The Principle of Nonlinear Optics*, Wiley-Interscience, 1984, Chapter 10.

96. W. Kaiser and M. Maier, in F. T. Arecchi and E. O. Schulz-Dubois (Eds.), *Laser Handbook*, Vol. 2, North-Holland, Amsterdam, 1972, p. 1077.

97. G. X. Jin, J. M. Yuan, L. M. Narducci, Y. S. Liu, and E. Seibert, *Opt. Commun.* **68**, 379 (1988).

98. G. P. Djotyan, and L. L. Minasyan, *Opt. Commun.* **49**, 117 (1984).

99. C. Flytzanis and C. L. Tang, *Phys. Rev. Lett.* **45**, 441 (1980).

100. G. C. Maitland, M. Rigby, E. B. Smith, *Intermolecular Forces*, Clarendon Press, Oxford, UK, 1987.

101. J. W. Gadzuk, *J. Opt. Soc. Soc. B* **4**, 201 (1987).

102. J. M. Yuan, M. Tung, D. H. Feng, and L. M. Narducci, *Phys. Rev. A* **28**, 1662 (1983).

103. Y. Gu, M. Tung, J. M. Yuan, D. H. Feng, and L. M. Narducci, *Phys. Rev. Lett.* **52**, 701 (1984).

104. K. Kaneko, *Prog. Theor. Phys.* **69**, 1427 (1983).

105. T. Hogg and B. A. Huberman, *Phys. Rev. A* **29**, 275 (1984).

106. K. Kaneko, *Collapse of Tori and Genesis of Chaos in Dissipative Systems*, World Scientific, Singapore, 1986.

107. A. Ferreti and N. K. Rahman, *Chem. Phys.* **119**, 275 (1988); *Chem. Phys. Lett.* **133**, 150 (1987); **140**, 71 (1987).

108. C. Grebogi, E. Ott, and J. A. Yorke, *Phys. Rev. Lett.* **50**, 935 (1983); S. W. McDonald, C. Grebogi, E. Ott, and J. A. Yorke, *Physica* **17D**, 125 (1985).

109. J. M. T. Thompson, *Proc. R. Soc. London Ser. A* **421**, 195 (1989).

110. M. Ding, T. Bountis, and E. Ott, *Phys. Lett. A* **151**, 400 (1990).

111. C. F. Hillermeier, R. Blumel, and U. Smilansky, *Phys. Rev. A* **45**, 3486 (1992).

112. T. Tel, *Phys. Rev. A* **36**, 1502 (1987); *J. Phys. A: Math. Gen.* **22**, L691 (1989).

113. Y. T. Lau, J. M. Finn, and E. Ott, *Phys. Rev. Lett.* **66**, 978 (1991).

114. R. V. Ambartzumian and V. S. Letokhov, in C. B. Moore (Ed.), *Chemical and Biological Applications of Lasers*, Academic, New York, 1977.

115. N. Bloembergen, *Opt. Commun.* **15**, 416 (1975).

116. K. Fox, *Opt. Commun.* **19**, 397 (1976).

117. R. P. Feynman and F. L. Vernon, Jr., *Ann. Phys.* (*NY*) **24**, 118 (1963); R. P. Feynman and A. R. Hibbs, *Quantum Mechanics and Path Integrals*, McGraw-Hill, New York, 1965.

118. H. Dekker, *Phys. Rep.* **80**, 1 (1981).

119. A. O. Caldeira and A. J. Leggett, *Ann. Phys.* (*NY*) **153**, 445 (1984).

120. A. J. Leggett, S. Chakravarty, A. T. Dorsey, M. P. A. Fisher, A. Garg, and W. Zwerger, *Rev. Mod. Phys.* **19**, 1 (1987).

121. G. Lindblad, *Commun. Math. Phys.* **48**, 119 (1976); *Non-Equilibrium Entropy and Irreversibility*, Reidel, Dordrecht, 1983.

122. R. Kosloff and S. A. Rice, *J. Chem. Phys.* **72**, 4591 (1980); R. Kosloff and M. A. Ratner, *J. Chem. Phys.* **80**, 2352 (1984).

123. M. O. Kostin, *J. Chem. Phys.* **57**, 3589 (1972); *J. Stat. Phys.* **12**, 145 (1975).

124. N. Gisin, *J. Phys. A: Math. Gen.* **14**, 2259 (1981); **19**, 205 (1986).

125. D. Schuh, K. M. Chung, and H. Hartmann, *J. Math. Phys.* **24**, 1652 (1983); **25**, 3086 (1984).

126. H. J. Korsch and H. Steffen, *J. Phys. A: Math. Gen.* **20**, 3787 (1987).

127. R. Parson and E. J. Heller, *J. Chem. Phys.* **85**, 2581, 2569 (1986).

128. R. Parson, *J. Chem. Phys.* **89**, 262 (1988); *Chem. Phys. Lett.* **129**, 87 (1986).

129. S. Nosé, *J. Chem. Phys.* **81**, 511 (1984).

130. S. Nosé, *Mol. Phys.* **52**, 255 (1984); **57**, 187 (1986).

131. W. G. Hoover, *Phys. Rev. A* **37**, 252 (1988); **31**, 1695 (1985).

132. D. Kusnezov, A. Bulgac, and W. Bauer, *Ann. Phys.* (*NY*) (1990).

133. M. Grilli and E. Tosatti, *Phys. Rev. Lett.* **62**, 2889 (1989).

MEASUREMENT OF FEMTOSECOND DYNAMICS OF NONLINEAR OPTICAL RESPONSES

TAKAYOSHI KOBAYASHI

Department of Physics, University of Tokyo, Bunkyo, Tokyo, Japan

CONTENTS

Modern Nonlinear Optics, Part 3, Edited by Myron Evans and Stanisław Kielich. Advances in Chemical Physics Series, Vol. LXXXV.
ISBN 0-471-30499-9   © 1994 John Wiley & Sons, Inc.

I. INTRODUCTION

A. Nonlinear Optical Phenomena

Recently nonlinear optical properties of materials have been widely studied from both the basic scientific and practical points of view. Nonlinear optical processes are essential in optoelectronics and "photonics," which will be developed in the 21st century. Nonlinear electronic processes such as modulation or frequency multiplication are key phenomena in advanced electronics. Ultrafast and large optical nonlinearity is needed to realize nonlinear optical devices, such as optical switches or optical information processors. From a fundamental scientific viewpoint, nonlinear optical processes can be used to study optical properties of materials. Nonlinear optical responses disclose the electronic structure and dynamics of elementary excitations in materials with high sensitivity and time resolution.

Nonlinear optical phenomena are accounted for in terms of induced nonlinear polarization $P^{(n)}$, which is related to the light field E through the relation $P^{(n)} = X^{(n)}E^n$, where $X^{(n)}$ is the nonlinear optical susceptibility and E^n represents the tensor products of applied field. In general, the lowest order of nonlinearity present in materials in any symmetry is the third-order nonlinearity, $n = 3$. To obtain complete information, we need to obtain both the real and imaginary parts of $X^{(n)}$, which are associated with the in-phase and quadrature-phase components, respectively, of the induced nonlinear polarization. The former corresponds to the absorbance change and the latter component is proportional to the induced phase change in the limit of the small change.

To elucidate the mechanism of the optical nonlinearity, it is important to obtain both the real and imaginary parts of the nonlinear susceptibility, including their optical frequency dependence, and to investigate the dynamics of the nonlinear response to clarify the origins of the nonlinearity. We must study the temporal response of the nonlinear processes to find the relaxation mechanisms of excitation in condensed matters relevant to the ultrafast nonlinear optical processes.

B. Femtosecond Nonlinear Optical Response

This chapter examines the relaxations of the prepared states in condensed-phase material by the interaction with laser light. There are two relaxation processes of the state prepared: depopulation and dephasing. Depopulation is the decay of the population change induced by the absorption of light to the thermal equilibrium state of the population distribution. The population decay time (T_1) is the decay time constant of the population difference and is sometimes called the energy relaxation time or longitudinal relaxation time. The latter is the disappearance of the macroscopic polarization, sometimes called polarization decay or transverse relaxation. The dephasing time (T_2) is the decay time constant of the macroscopic polarization. Coherently induced microscopic polarizations constructively interfere with each other, resulting in a finite macroscopic polarization. The phases of these microscopic polarizations become gradually random because of scattering between carriers or excitons, so that they destructively interfere and the macroscopic polarization decays.

Ultrafast spectroscopy has developed rapidly in the last decade. Improved time resolution has clarified the detailed mechanisms of the relaxation processes in various condensed-phase materials, including bulk and quantum-well structured semiconductors, metals, polymers, organic liquids, molecules dissolved in solutions, molecules absorbed on solid surfaces, and biological systems. Since the success of the picosecond light pulse generation using mode-locking,[1] extensive continuous efforts to generate shorter pulses have been made. The shortest pulse ever obtained has only 6-fs width, corresponding to three oscillation periods of an optical cycle.[2] It was obtained by nonlinear optical processes, i.e., by compressing the output of a group-velocity-dispersion-compensated colliding-pulse mode-locked (CPM) rhodamine 6G laser with DODCI saturable absorber using a prism pair and a grating pair for the compensation of both first- and second-order group velocity dispersions. By ultrashort light pulses with pulse durations between 6 and about 100 fs, time-resolved and conventional coherent spectroscopies have been applied to the condensed-phase materials. In conventional time-resolved spectroscopy the time resolution is limited by temporal width of the pulse used.

There are several difficulties and disadvantages to using ultrashort pulses for the study of the ultrafast phenomena:

1. Lasers for the generation of ultrashort pulses with enough intensity to be used in spectroscopies are often highly sophisticated and expensive.

2. The wavelength of a laser pulse shorter than 100 fs used to be extremely limited in the region around 615–625 nm from a rhodamine 6G colliding pulse mode-locked (CPM) laser. Recently the situation has become slightly better in that synchronously pumped dye lasers with about 100-fs pulse width tunable in the red spectral region (580–660 nm) are also available. Recent outstanding developments of the solid-state laser have enabled us to obtain sub-100-fs pulses in the spectral region of 800–900 nm from a Ti^{3+}-doped sapphire laser. However, it is still difficult to obtain stable and inexpensive femtosecond laser systems.

3. An ultrashort light pulse is easily broadened due to the inevitable dispersion of optical components because of its broad power spectrum. Even though recompression is possible by using negatively dispersive optical components, such as grating pairs, prism pairs, or both, perfect recompression is not easy and the laser intensity may be reduced substantially by passing through these components. Even after getting ultrashort pulses it is still difficult to maintain the pulse width because of the dispersion of optical components, through which the short pulses pass.

II. PICOSECOND AND FEMTOSECOND RESOLVED SPECTROSCOPY WITH THE USE OF TIME-INCOHERENT LASERS

A. Application of Time-Incoherent Light to Time-Resolved Measurement

To overcome the difficulties in ultrafast spectroscopy using a short pulse, a new spectroscopic technique utilizing transient coherent nonlinear optical effects with incoherent light was proposed and applied to nonlinear optical processes in various systems. In the transient coherent spectroscopy for the studies of the ultrafast dynamics, the signal light generated or modulated by nonlinear optical effects is detected, using the correlation between the excitation and the probe light beams, and the signal intensity is expressed by the time integration of functions of the field amplitude (or intensity) and the response function of the matter. Therefore, in such cases, the time resolution is expected to be determined not by the pulse duration but by the correlation time.

According to this principle, experiments with extremely high time resolution can be performed much more easily by using time-incoherent light with a short enough correlation time. The applicability of this principle to studies of ultrafast processes has been verified for the electronic dephasing-time measurement by degenerate four-wave mixing (DFWM) spectroscopy.[3-9]

Though transient DFWM spectroscopy, using either coherent short pulses[10] or incoherent light sources as mentioned above, is a powerful tool for studying dynamic processes in condensed matter, it cannot be applied to optically forbidden transition and the range of available wavelength is limited, as mentioned before. Dephasing of Raman active vibrational modes in molecules can be studied by so-called transient coherent Raman spectroscopy such as CARS (coherent anti-Stokes Raman scattering) or CSRS (coherent Stokes Raman scattering).[11-13] We studied theoretically and performed experiments to verify the principle that the correlation time determines the resolution time to the vibrational dephasing-time measurement by this method. [14]

The techniques can be applied not only to the dephasing-time measurement but also to the population lifetime or energy decay time of electronically excited states with time resolution much shorter than the pulse width by using temporally incoherent light. For the purpose of the population relaxation study two types of techniques have been reported. One is the pump-probe technique, in which the delay-time dependence of the change in the intensity of the transmitted probe beam is measured.[15] The other method is DFWM with three incident beams, in which a fourth signal beam is detected as a function of the delay time of one of the three incident beams.[16] Further, as an extended case of the longitudinal relaxation time measurement, we showed that the relaxation of the nonresonant third-order susceptibility in optical Kerr media can also be studied by DFWM and Kerr-shutter techniques[17] with temporally incoherent light.

In the remainder of Section II, we summarize the theoretical studies on this new technique using temporally incoherent light. In Section III, we describe the experimental studies performed by our group as the applications of this technique to the studies of the ultrafast relaxation dynamics of materials in condensed phase.

B. Assumptions for the Theoretical Description of the Nonlinear Optical Response Using Time-Incoherent Light

The following assumptions and symbols are used throughout this and following sections. The duration of the incoherent light is assumed to be much longer than the other characteristic times, such as the correlation time τ_c, the transverse relaxation time T_2, and the longitudinal relaxation time T_1. The notation $E(t)$ represents the total electric field; $\tilde{E}(t)$ denotes the envelope function of the electric field slowly varying compared with the optical frequency; $f(t)$ stands for the autocorrelation function of the incoherent light field normalized to unity at its peak, i.e., $f(0) = 1$; τ and τ' are the delay times between the incident beams; and angle brackets $\langle \ \rangle$ represent the statistical average. The rotating-wave and the electric-dipole approximations are also used throughout.

Here we would like to review the definition of the correlation time of incoherent light. The correlation time is an important value for characterizing the temporal coherence of the incoherent light. However, since there are several definitions of correlation time[14, 15, 18-22] and the value varies by an order of magnitude with the definition for the same autocorrelation function, we should make clear which definition is used when we refer to the correlation time. The four definitions of correlation time are listed in Table I, each giving a characteristic time width of the nonvanishing

TABLE I
Definitions of the Correlation Times $\tau_c^{(i)}$

Definition		$\tau_c^{(i)}/\Delta\tau$
$\tau_c^{(1)}$	$\mathrm{Re}\left\{ \int_0^\infty dt\, f(t) \right\}$	0.89
$\tau_c^{(2)}$	$\int_{-\infty}^\infty dt\, \lvert f(t) \rvert^2$	1.25
$\tau_c^{(3)}$	$\left[\int_0^\infty dt \int_0^\infty dt'\, f^*(t) f(t') f(t-t) \right]^{1/2}$	0.78
$\tau_c^{(4)}$	$\left[Re \int_0^\infty dt \int_0^\infty dt'\, f(t) f(t') f^*(t+t') \right]^{1/2}$	0.55

Note. Numerical calculations are performed for the Gaussian autocorrelation function defined by (1).

autocorrelation function $f(t)$. Note that the correlation time can vary with the shape of the autocorrelation function, even with the same spectral width. In numerical calculations we assume $f(t)$ to have a real Gaussian profile,

$$f(t) = \exp\left(-\frac{t^2}{\Delta\tau^2} \right) \qquad (1)$$

and make $\Delta\tau$ a unit of time. The relationship between $\tau_c^{(i)}$ and $\Delta\tau$ is calculated in Table I. τ_c represents an indefinite value of the order of $\tau_c^{(i)}$.

C. Two-Beam Degenerate Four-Wave Mixing for Electronic Dephasing Measurement

Transient DFWM spectroscopy with two incident beams with wave vectors $\mathbf{k}_1$ and $\mathbf{k}_2$ with common frequency from the same source is often adopted for electronic dephasing-time measurement because of the simple experimental arrangement. The configuration of the incident and the signal beams is schematically shown in Fig. 1. Two incident beams focused on a resonant nonlinear material generate two parametrically mixed waves by the third-order nonlinear optical effect.

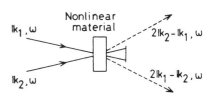

Figure 1.   Schematic of the two-beam DFWM experiment for the measurement of the dephasing process.

To derive the dependence of the signal intensity on the delay time τ of beam 2 relative to beam 1, the resonant nonlinear material was assumed to consist of two-level systems, to which the density matrix formalism was applied. It is assumed that the resonant material can be described by a two-level energy system, and that the grating amplitude induced by the incoherent beams is small, so that perturbation techniques can be applied. Using the rotating-wave approximation, the total electric field of the two beams with the two wave vectors involved in the degenerate four-wave mixing is given by

$$E(t) = \tilde{E}(t + \tau)\exp\left[i(\mathbf{k}_1\mathbf{r} - \omega(t + \tau))\right] + \tilde{E}(t)\exp\left[i(\mathbf{k}_2\mathbf{r} - \omega t)\right] \quad (2)$$

The third-order nonlinear polarization with the wave vector $2\mathbf{k}_2 - \mathbf{k}_1$ is calculated by the perturbation method:

$$
\begin{aligned}
P^{(3)}&(t, \tau; 2\mathbf{k}_2 - \mathbf{k}_1, \omega) \\
&= \exp\left[i((2\mathbf{k}_2 - \mathbf{k}_1)\mathbf{r} - \omega t)\right] \int_{-\infty}^{\infty} d(\Delta\omega) g(\Delta\omega) \\
&\quad \times \int_{-\infty}^{t} dt_3 \int_{-\infty}^{t_3} dt_2 \int_{-\infty}^{t_2} dt_1 \\
&\quad \times \exp\left(-\frac{t_3 - t_2}{T_1^{\text{ele}}}\right) \exp\left(-\frac{t - t_3 + t_2 - t_1}{T_2^{\text{ele}}}\right) \\
&\quad \times \left\{ \tilde{E}(t_3)\tilde{E}(t_2)\tilde{E}^*(t_1 + \tau)\exp\left[-i\Delta\omega(t - t_3 - t_2 + t_1)\right] \right. \\
&\quad \left. + \tilde{E}(t_3)\tilde{E}^*(t_2 + \tau)\tilde{E}(t_1)\exp\left[-i\Delta\omega(t - t_3 + t_2 - t_1)\right] \right\}
\end{aligned}
\quad (3)
$$

Here the proportionality coefficient is set at unity and $g(\Delta\omega)$ is the distribution function of the difference between the transition frequency of the material and the laser frequency ω; T_1^{ele} and T_2^{ele} denote the population relaxation time and the dephasing time of the relevant electronic transition, respectively. The superscript "ele" indicates an electronic transition to be discriminated from vibrational relaxation. The delay-time dependence of the signal intensity $I(\tau)$ is expressed simply by the statistical average of the squared absolute value of the macroscopic polarization using a unity proportionality coefficient as follows:

$$I(\tau) = \int_{-\infty}^{\infty} dt \left\langle \left| P^{(3)}(t, \tau; 2\mathbf{k}_2 - \mathbf{k}_1, \omega) \right|^2 \right\rangle \quad (4)$$

Now we assume the extremely broad inhomogeneous width of the transition frequency and $T_1^{\text{ele}} \gg T_2^{\text{ele}}, \tau_c$. In this case Eq. (4) is reduced to the following equation irrespective of the characteristics of the light[6, 10]

$$I(\tau) = \int_0^\infty dt \int_0^\infty dt' f(t' - t) f(t - \tau) f^*(t' - \tau) \exp\left[-\frac{2(t + t')}{T_2^{\text{ele}}}\right] \quad (5)$$

where $f(t)$ is the autocorrelation function of the electric field amplitude. When the dephasing time is much longer than the correlation time of the light field, the signal decays exponentially at the rate of $4/T_2$. In Fig. 2, the temporal shape of the signal intensity given by (5) is shown for several values of the dephasing time T_2^{ele}, where a Gaussian autocorrelation function (1) is assumed.

When the dephasing time is comparable with or shorter than the correlation time ($T_2^{\text{ele}}/\Delta\tau \lesssim 1$), the signal trace curves have no clear decaying tails. However, the peak of the signal intensity shifts toward the positive delay time depending on the dephasing time, while the peak of the other signal with the wave vector $2\mathbf{k}_1 - \mathbf{k}_2$ shifts toward the negative delay time, since the delay-time dependence of the latter signal is obtained by replacing τ with $-\tau$ in (5).[7, 10] Therefore, even such a short dephasing time can be estimated from the peak separation of the two signals detected simultaneously at the two directions $2\mathbf{k}_1 - \mathbf{k}_2$ and $2\mathbf{k}_2 - \mathbf{k}_1$. The relationship between the peak separation τ_s and the dephasing time T_2^{ele} is calculated by using a Gaussian autocorrelation function. The results are shown in Fig. 3.

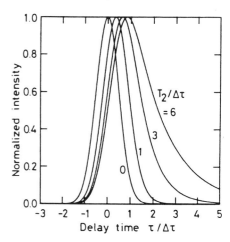

Figure 2. Calculated delay-time dependence of the signal intensity for the measurement of the electronic dephasing by the two-beam DFWM experiment. Extremely wide inhomogeneous broadening and Gaussian autocorrelation function are assumed. T_2 represents an electronic dephasing time T_2^{ele}

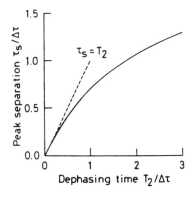

Figure 3. Calculated speak separation as a function of the electronic dephasing time. Extremely wide inhomogeneous broadening and a Gaussian autocorrelation function are assumed.

We would like to mention here the effect of the limited inhomogeneous width on the magnitude of the peak separation. Equation (5) is correct only for much broader inhomogeneous width than the homogeneous width. In actual materials, the inhomogeneous width is finite, and the effect of the width on the peak separation should be taken into account. For narrower inhomogeneous width the peak separation becomes smaller, and there is no separation for a completely homogeneous system. Hence, the dephasing time estimated by the observed peak separation using the result obtained with infinite inhomogeneous width may be shorter than the real value. The relation between the peak separation and the inhomogeneous width relative to the homogeneous width is calculated in the region where the peak separation is proportional to the dephasing time in Fig. 3. We assume that the inhomogeneous broadening has a Gaussian profile with width of $\Delta\Omega$, i.e.,

$$g(\Delta\omega) = \frac{1}{\sqrt{\pi}\,\Delta\Omega}\exp\left[-\left(\frac{\Delta\omega}{\Delta\Omega}\right)^2\right] \tag{6}$$

The following relation can be obtained using this distribution:

$$\frac{\tau_s}{T_2^{\mathrm{ele}}} = \frac{v\left[(2v + 1/v) - (4v^2 + 4 - 1/v^2)\exp(v^2)\phi(v)\right]}{1 - (2v - 1/v)\exp(v^2)\phi(v)} \tag{7}$$

where v represents the ratio of the homogeneous width $1/T_2^{\mathrm{ele}}$ to the inhomogeneous width $\Delta\Omega$.

$$v = \frac{1}{\Delta\Omega\,T_2^{\mathrm{ele}}} \tag{8}$$

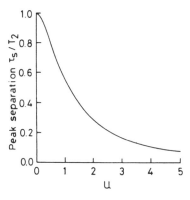

Figure 4. The relation between the peak separation and the inhomogeneous width of the transition, which is calculated in the region where the peak separation is proportional to the dephasing time. The inhomogeneous broadening is assumed to have a Gaussian profile defined by (6). The value v is the ratio of the homogeneous width to the inhomogeneous width, which is defined by (8).

and the error function $\phi(v)$ is defined by

$$\phi(v) = \int_v^\infty dt \, \exp[-t^2] \qquad (9)$$

The relation given by (7) is depicted in Fig. 4.

D.   Vibrational Dephasing

The method for observing electronic-dephasing dynamics discussed in the previous sections can be extended to the vibrational dephasing-time measurement. For this purpose two types of nondegenerate wave mixing, namely coherent Stokes Raman scattering (CSRS) and coherent anti-Stokes Raman scattering (CARS), with both coherent and incoherent light can be utilized. In the next subsection a technique based on the transient coherent Raman process is described.[14] However, since this process does not include a rephasing process, the results may have some ambiguities when the relevant transition is inhomogeneously broadened. To eliminate this ambiguity, a Raman echo process may be utilized, which is discussed in Section II.F.

E.   Coherent Raman Scattering

We calculate the time-dependent CSRS intensity by using a simple three-level system. Its derivation for CARS is essentially the same procedure. This technique is based on a transient coherent Raman process with three beams, two of them are incoherent light from a single broad-band laser; the delay time between the two is variable. A third beam is coherent in the measuring range of the delay time and has a higher (lower in case of CARS) frequency than the incoherent light by a vibrational energy in a

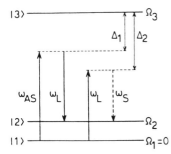

Figure 5. Energy diagram of the model system for measuring vibrational dephasing.

molecule of interest. The delay-time dependence of CSRS (or CARS) intensity offers information about the coherence dynamics of the vibrational transition and about electronic transition in case of resonant CSRS or CARS with a resolution time limited by the correlation time of the incoherent light.

Figure 5 shows a simple model of a three-level system, which is usually used for theoretical analysis of coherent Raman scattering.[23, 24] Vibrational levels $|1\rangle$ and $|2\rangle$ belong to the ground electronic state, whereas level $|3\rangle$ belongs to an electronically excited state. Levels $|1\rangle$ and $|3\rangle$ and levels $|2\rangle$ and $|3\rangle$ are coupled with each other by electronic transition dipoles. In most CSRS (or CARS) experiments two beams are used, but for the purpose of time-resolved measurement, a three-beam (BOXCARS) configuration is employed,[25] where two beams of frequency ω_L and one beam of frequency ω_{AS} are used, as shown in Fig. 6. The total electric field in the interaction region in a sample is given as

$$E(t,\mathbf{r}) = \tilde{E}_{AS}(t)\exp\left[i(\mathbf{k}_{AS}\mathbf{r} - \omega_{AS}t)\right]$$
$$+ \tilde{E}_{L}(t + \tau)\exp\left[i(\mathbf{k}_{L1}\mathbf{r} - \omega_{L}(t + \tau))\right] \quad (10)$$
$$+ \tilde{E}_{L}(t)\exp\left[i(\mathbf{k}_{L2}\mathbf{r} - \omega_{L}t)\right]$$

where τ is the delay time of beam L2 relative to beam L1. The intensity of

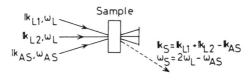

Figure 6. Schematic of the CSRS experiment for measuring vibrational dephasing.

the scattered light of frequency $\omega_S = 2\omega_L - \omega_{AS}$ and wave vector $\mathbf{k}_S = \mathbf{k}_{L1} + \mathbf{k}_{L2} - \mathbf{k}_{AS}$ was measured as a function of the delay time τ.

The macroscopic polarization of frequency ω_S and wave vector $\mathbf{k}_S$ is derived by a perturbation method under the following conditions: (1) the difference of light frequencies is tuned exactly to the vibrational energy, and (2) the relevant energy levels of the molecular system are homogeneously broadened. Under these conditions, the third-order polarization $P^{(3)}(\mathbf{k}_S, \omega_S)$ is obtained as

$$P^{(3)}(t, \tau; \mathbf{k}_S, \omega_S)$$

$$= \exp[i(\mathbf{k}_S\mathbf{r} - \omega_S t)] \int_{-\infty}^{t} dt_3 \int_{-\infty}^{t_3} dt_2 \int_{-\infty}^{t_2} dt_1$$

$$\times \exp\left[-(t - t_3)\left(\frac{1}{T_{2,32}^{\text{ele}}} + i\Delta_2\right)\right] \exp\left[-\frac{t_3 - t_2}{T_2^{\text{vib}}}\right] \quad (11)$$

$$\times \exp\left[-(t_2 - t_1)\left(\frac{1}{T_{2,31}^{\text{ele}}} - i\Delta_1\right)\right]$$

$$\times \left[\tilde{E}_L(t_3)\tilde{E}_L(t_2 + \tau) + \tilde{E}_L(t_3 + \tau)\tilde{E}_L(t_2)\right]\tilde{E}_{AS}^*$$

Here the proportionality coefficient is set at unity for simplicity and T_2^{vib} is the dephasing time of the vibrational transition, $T_{2,32}^{\text{ele}}$ and $T_{2,31}^{\text{ele}}$ are the dephasing times of the electronic transitions between $|3\rangle$ and $|2\rangle$ and between $|3\rangle$ and $|1\rangle$, respectively, and Δ_1 and Δ_2 are the frequency detunings shown in Fig. 5. The vibrational dephasing time is replaced with the electronic dephasing time when the electronic Raman is studied. The light of frequency ω_{AS} is assumed to be coherent and hence E_{AS} to be constant in the delay time region of observation. This expression has two terms, in which the roles of the two incoherent light beams are exchanged with each other.

When we assume that all the light beams are off-resonant with the electronic transitions, i.e., Δ_1 and Δ_2 are large enough, and that the stochastic property of the incoherent light is expressed in terms of a Gaussian random process, we can derive an expression for the signal intensity as follows, with a unity proportionality coefficient for simplicity:

$$I(\tau) = \int_0^\infty dt \int_0^\infty dt' \exp\left[-\frac{t + t'}{T_2^{\text{vib}}}\right]$$

$$\times [2f(0)f(t - t') + 2f(t)f^*(t') \quad (12)$$

$$+ 2\,\text{Re}[f(\tau)f(t - t' - \tau)]$$

$$+ f(t - \tau)f^*(t' - \tau) + f(t + \tau)f^*(t' + \tau)]$$

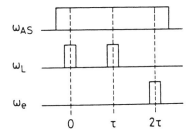

Figure 7. Schematic of the pulse sequence in the Raman echo experiment.

The integration of the first and the second terms in (12) gives a constant background, that of the third term forms a coherence spike, and that of the fourth and the last terms decays with the time constant of $T_2^{\text{vib}}/2$ in the positive and the negative delay-time region, respectively. The signal trace is symmetrical with respect to $\tau = 0$ and the peak intensity $I(0)$ is twice as large as the background intensity.

F. Raman Echo

The rephasing process included in photon echoes does not take place in CSRS or CARS. The value of the dephasing time is ambiguous when the relevant levels in the system are inhomogeneously broadened.[26, 27] By Raman echo experiment[23] we can clarify the dephasing dynamics and determine the dephasing time without ambiguities.[24] However, this is a higher-order (seventh-order) process, and signal detection is very difficult. Raman echo experiments have been carried out in solids,[28] gases,[29, 30] and liquid nitrogen,[31] but not yet in liquids at room temperature.

The Raman echo type experiment depicted in Fig. 7 is feasible with incoherent light. The molecules are brought in coherent superposition of the ground state and a vibrationally excited state by light of two frequencies, ω_{AS} and ω_{L} at $t = 0$. The difference between these two frequencies is matched to the vibrational transition. Rephasing is introduced by light of the same pair of the frequencies at $t = \tau$. The macroscopic coherence is recovered, if the vibrational transition is inhomogeneously broadened, at 2τ. It is probed by the light of frequency ω_{AS}, and the emitted light of frequency $\omega_e = 2\omega_{\text{AS}} - \omega_{\text{L}}$ is detected, as shown in Fig. 8. The two ω_{L}

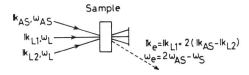

Figure 8. Schematic of the Raman echo experiment for measuring vibrational dephasing.

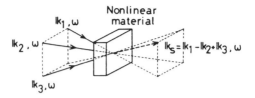

Figure 9. Schematic of the BOXCARS-type DFWM experiment for measuring population relaxation.

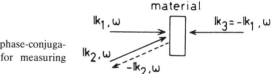

Figure 10. Schematic of the phase-conjugation-type DFWM experiment for measuring population relaxation.

pulses in Fig. 7 are replaced in the actual experiment by two incoherent light beams with a delay of τ obtained from the same light source.

G. Kerr Relaxation

For Kerr relaxation studies we utilize either of the configurations depicted in Figs. 9 and 10. They are the same as that of the DFWM with three incident beams. The third-order nonlinear polarization in a Kerr medium is given by

$$P_i^{(3)}(t) = E_j(t) \int_{-\infty}^{t} dt'\, r_{ijkl}(t - t') E_k(t') E_l(t') \tag{13}$$

where r_{ijkl} is a fourth-rank tensor representing the electronic and the nuclear contributions to the nonlinearity, which is assumed to relax exponentially with a time constant of T^{Kerr}, and $E_i(t)$ is the ith polarization direction component of total electric field.

If we neglect the contribution of the two-photon resonance, we find that (13) is equivalent to an expression for a limiting case of the measurement of T_1^{ele}, where T_2^{ele} is set at zero for the population measurement discussed below. The third-order nonlinear polarization with the wave vector $\mathbf{k}_s = \mathbf{k}_1 - \mathbf{k}_2 + \mathbf{k}_3$ is given as follows:

$$
P_i^{(3)}(t, \tau, \tau'; \mathbf{k}_s, \omega) = \left[r_{ijkl}(0) + r_{ijlk}(0) \right] \int_{-\infty}^{t} dt' \exp\left[-\frac{t - t'}{T^{\text{Kerr}}} \right]
$$
$$
\times \left[\tilde{E}_j(t - \tau) \tilde{E}_k^*(t' + \tau') \tilde{E}_l(t') \right. \tag{14}
$$
$$
\left. + \tilde{E}_j(t) E_k^*(t' + \tau') \tilde{E}_l(t' - \tau) \right]
$$

After summation over the polarization indices j, k, and l, Eq. (14) is reduced to the following with a unity proportionality coefficient:

$$
P^{(3)}(t, \tau, \tau'; \mathbf{k}_s, \omega) = \int_{-\infty}^{t} dt' \exp\left[-\frac{t - t'}{T^{\text{Kerr}}} \right]
$$
$$
\times \left[\tilde{E}(t - \tau) \tilde{E}^*(t' + \tau') \tilde{E}(t') \right. \tag{15}
$$
$$
\left. + g\tilde{E}(t) \tilde{E}^*(t' + \tau') \tilde{E}(t' - \tau) \right]
$$

where g is a constant coefficient determined by the polarization condition of the incident light beams. For example,

1. When all the polarizations are parallel, $g = 1$.
2. When the polarizations of beams 1 and 3 are parallel and that of beam 2 is perpendicular to them, $g = 1$.
3. When the polarizations of beams 1 and 2 are parallel (y) and that of beam 3 is perpendicular to them (x), $g = (r_{xyyx}(0) + r_{xyxy}(0)) / 2r_{xxyy}(0)$.
4. When the polarizations of beams 2 and 3 are parallel (y) and that of beam 1 is perpendicular to them (x), $g = 2r_{xxyy}(0)/(r_{xyyx}(0) + r_{xyxy}(0))$.

The calculated result of signal intensity with a unity proportionality coefficient is given as follows:

Case 1: Variable τ and fixed $|\tau'| \ll \tau_c$

$$
I(\tau) = B_1 + 2gB_2 + (1 + g^2)B_3 + (2g + g^2)B_4
$$
$$
+ (2g + g^2)S_1(\tau) + 2g^2 S_2(\tau) + 2g S_3(\tau) + 2g S_4(\tau) \tag{16}
$$
$$
+ 2D_1(\tau) + 2gD_2(\tau) + 2D_3(\tau) + g^2 D_4(\tau)
$$

Case 2: Variable τ' and fixed $|\tau| \ll \tau_c$

$$I(\tau') = (1 + g)^2 [B_3 + B_4 + S_1(\tau') + 2S_3(\tau') + D_3(\tau')] \quad (17)$$

Case 3: Variable τ' and fixed $|\tau| \gg \tau_c$; it is also assumed that $|\tau| \gg T^{\text{Kerr}}$

$$I(\tau') = (1 + g^2)B_3 + 2gB_4 + S_1(\tau') + 2gS_3(\tau') + g^2 D_3(\tau') \quad (18)$$

Case 4: Variable τ and fixed $|\tau'| \gg \tau_c$; it is also assumed that $|\tau'| \gg T^{\text{Kerr}}$

$$I(\tau) = (1 + g^2)B_3 + 2gB_4 + 2gS_4(\tau) + D_3(\tau) + g^2 D_4(\tau) \quad (19)$$

Here B_i, S_i, and D_i are given as follows:

$$\left.\begin{array}{r} B_i \\ S_i(\tau) \\ D_i(\tau) \end{array}\right\} = \int_0^\infty dt \int_0^\infty dt' \exp\left[-\frac{t + t'}{T_1^{\text{ele}}}\right] \left\{\begin{array}{l} b_i \\ s_i(\tau) \\ d_i(\tau) \end{array}\right. \quad (20)$$

$$\text{where} \quad b_1 = f(0)^3$$

$$b_2 = f(0)|f(t)|^2$$

$$b_3 = f(0)|f(t - t')|^2$$

$$b_4 = f^*(t)f(t')f(t - t')$$

$$s_1(\tau) = f(0)|f(\tau)|^2$$

$$s_2(\tau) = \text{Re}[f(\tau)f(t)f^*(t + \tau)]$$

$$s_3(\tau) = \text{Re}[f(\tau)f^*(t)f(t - \tau)] \quad (21)$$

$$s_4(\tau) = \text{Re}[f(\tau)f(t - t' - \tau)f^*(t - t')]$$

$$d_1(\tau) = f(0)|f(t - \tau)|^2$$

$$d_2(\tau) = \text{Re}[f^*(t - \tau)f(t')f(t - t' - \tau)]$$

$$d_3(\tau) = f^*(t - \tau)f(t' - \tau)f(t - t')$$

$$d_4(\tau) = f^*(t + \tau)f(t' + \tau)f(t - t')$$

In cases 2 and 3, τ is replaced by τ' in the above expressions. The terms B_i are constant background, the terms S_i correspond to a coherence spike,

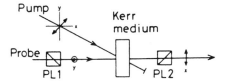

Figure 11. Schematic of the Kerr-shutter experiment for measuring Kerr relaxation. PL1 and PL2 are the polarizer and analyzer, respectively, with perpendicular polarization to each other. The polarization of the incoherent gate-opening pump beam is at 45° with respect to that of probe beam.

D_1 and D_2 increase with the time constant of τ_c and decay with T_1^{ele}, D_3 increases with τ_c and decays with $T_1^{\text{ele}}/2$, and D_4 increases with $T_1^{\text{ele}}/2$ and decays with τ_c.

The appearance of the coherence spike in Kerr media has been pointed out by Oudar in the case of a Kerr shutter.[32] As shown above, the response time of the optical Kerr effect can also be measured by using incoherent light in the same way as in the population-decay measurement by DFWM.

We also found that the relaxation times in optical Kerr media can be measured by the so-called Kerr-shutter configuration shown in Fig. 11, using incoherent light.[33] The pump beam is linearly polarized at 45° from the polarization direction of the probe beam y, i.e.,

$$\tilde{E}_x^{\text{pump}}(t) = \tilde{E}_y^{\text{pump}}(t) = \frac{\tilde{E}(t)}{\sqrt{2}} \tag{22}$$

$$\tilde{E}_x^{\text{probe}}(t) = 0 \qquad \tilde{E}_y^{\text{probe}}(t) = E(t - \tau)$$

where τ is the delay time of the probe from the pump beam. We note that this case is equivalent to case 1, where $\mathbf{k}_2 = \mathbf{k}_1$, under the condition that the two-photon contribution is negligible. By using (14) the x component of the nonlinear polarization in the direction of the transmitted probe beam is calculated to be (15) with $\tau' = 0$ and

$$g = \frac{1}{2} + \frac{r_{xxyy}(0)}{r_{xyyx}(0) + r_{xyxy}(0)} \tag{23}$$

Theoretical considerations predict $g = 1$ for the electronic contribution and $g = \frac{1}{6}$ for the nuclear contribution in isotropic media.[34] The delay-time dependence of the signal is obtained as (16). If the nonlinearity is due to the nuclear orientation and T^{Kerr} is much longer than τ_c, the expression is reduced to the following:

$$I(\tau) = 1 + \frac{1}{3}|f(\tau)|^2 + \frac{2\tau_c^{(2)}}{T^{\text{Kerr}}} \exp\left(-\frac{\tau}{T^{\text{Kerr}}}\right) \qquad (24)$$

where the "tail" term appears only in the positive delay time and $\tau_c^{(2)}$ is defined in Table I. This result shows that the relative intensity of the coherence spike is suppressed for a smaller g.

H. Population Relaxation

Like the Kerr relaxation measurement in the previous section, population lifetimes (longitudinal relaxation times) of electronically excited states can also be measured by using temporally incoherent light with the time resolution determined by the correlation time of the incident light instead of the pulse duration. Two methods have been reported: the pump-probe[15] and DFWM[16] techniques.

1. Pump–Probe Spectroscopy

In contrast to all the other methods described here, in which the signal beams to be detected are generated by the parametric wave-mixing process, in the pump-probe technique the transmission change of the probe beam is measured as the delay time between the pump and the probe beams being varied. The beam configuration is shown in Fig. 12.

The delay-time dependence of the change in the transmitted probe light intensity $\Delta T(\tau)$ is derived as follows.[15, 32] Pump and probe beams are obtained from a single light source. The total electric field is given by

$$E(t) = \tilde{E}(t)\exp[i(\mathbf{k}_1\mathbf{r} - \omega t)] + \tilde{E}(t - \tau)\exp[i(\mathbf{k}_2\mathbf{r} - \omega(t - \tau))] \quad (25)$$

In the small-signal regime the transmission change is written by the third-order nonlinear polarization. We assume that the material consists of two levels and that the dephasing time T_2^{ele} is much shorter than the correlation time of the incident light τ_c and the population lifetime T_1^{ele}.

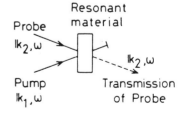

Figure 12. Schematic of the pump–probe experiment for measuring population relaxation.

The delay-time dependent transmission change of the probe beam is expressed by setting a proportionality coefficient to be unity as follows:

$$
\Delta T(\tau) = \int_0^\infty dt' \exp\left(-\frac{t'}{T_1^{\mathrm{ele}}}\right)
$$

$$
\times \int_{-\infty}^\infty dt \left[\tilde{E}^*(t-\tau)\tilde{E}(t-\tau)\tilde{E}^*(t-t')\tilde{E}(t-t') \right. \tag{26}
$$

$$
\left. + \tilde{E}^*(t-\tau)\tilde{E}(t)E^*(t-t')\tilde{E}(t-t'-\tau) \right]
$$

If we assume that the incoherent light field $E(t)$ follows a stationary complex Gaussian random process and if we take a statistical average of the right side of Eq. (26), then (26) is reduced to

$$
\Delta T(\tau) = \int_0^\infty dt \exp\left(-\frac{t}{T_1^{\mathrm{ele}}}\right)
$$

$$
\times \left[|f(0)|^2 + |f(t)|^2 + |f(\tau)|^2 + |f(t-\tau)|^2 \right] \tag{27}
$$

After integrating Eq. (27), the first and the second terms make a constant background, the third term forms a coherence spike, and the last term reflects the population-relaxation time. The ratio of the intensity at zero delay to the background intensity of the signal is two, regardless of T_1^{ele}.

In the limiting case where τ_c is much shorter than T_1^{ele}, (27) is reduced to the following for the positive delay time $\tau > 0$:

$$
\Delta T(\tau) = 1 + |f(\tau)|^2 + \frac{\tau_\mathrm{c}^{(2)}}{T_1^{\mathrm{ele}}} \exp\left(-\frac{\tau}{T_1^{\mathrm{ele}}}\right) \tag{28}
$$

where $\tau_\mathrm{c}^{(2)}$ is defined in Table I. No tail appears when the delay time is negative.

Since the change in the incident probe light intensity is to be detected in this method, a considerably stable light source is required to obtain data with high quality.

2. Degenerate Four-Wave Mixing

The population relaxation time can also be measured by the DFWM with three incoherent incident light beams. The BOXCARS or the phase-conjugation configurations, (Figs. 9 and 10) can be used. For the population-relaxation-time determination we measure the intensity of the signal in the direction of $\mathbf{k}_s = \mathbf{k}_1 - \mathbf{k}_2 + \mathbf{k}_3$ as varying the delay time of one of the three incident beams.

The delay-time dependence of the signal intensity is derived in the same way as in Section II.H.1 (Refs. 16, 32). Here we assume that beams 2 and 3 are retarded by $-\tau'$ and τ, respectively, from beam 1. The total electric field $E(t)$ is given by

$$
\begin{aligned}
E(t) = {} & \tilde{E}(t)\exp[i(\mathbf{k}_1\mathbf{r} - \omega t)] + \tilde{E}(t + \tau')\exp[i(\mathbf{k}_2\mathbf{r} - \omega(t + \tau'))] \\
& + \tilde{E}(t - \tau)\exp[i(\mathbf{k}_3\mathbf{r} - \omega(t - \tau))]
\end{aligned}
\tag{29}
$$

The third-order nonlinear polarization in the direction $\mathbf{k}_s$ is obtained by setting a proportionality coefficient to be unity as follows:

$$
\begin{aligned}
P^{(3)}(t, \tau, \tau'; \mathbf{k}_s, \omega) = {} & \exp[i(\mathbf{k}_s\mathbf{r} - \omega t)] \int_{-\infty}^{\infty} d(\Delta\omega)g(\Delta\omega) \\
& \times \int_{-\infty}^{t} dt_3 \int_{-\infty}^{t_3} dt_2 \int_{-\infty}^{t_2} dt_1 \exp\left(-\frac{t_3 - t_2}{T_1^{\text{ele}}}\right) \\
& \times \exp\left(-\frac{t - t_3 + t_2 - t_1}{T_2^{\text{ele}}}\right) \\
& \times \Big[\big[\tilde{E}(t_3 - \tau)\tilde{E}(t_2)\tilde{E}^*(t_1 + \tau') \\
& \qquad + \tilde{E}(t_3)\tilde{E}(t_2 - \tau)\tilde{E}^*(t_1 + \tau')\big] \\
& \qquad \times \exp[-i\Delta\omega(t - t_3 - t_2 + t_1)] \\
& \quad + \big[\tilde{E}(t_3 - \tau)\tilde{E}^*(t_2 + \tau')\tilde{E}(t_1) \\
& \qquad + \tilde{E}(t_3)\tilde{E}^*(t_2 + \tau')\tilde{E}(t_1 - \tau)\big] \\
& \qquad \times \exp[-i\Delta\omega(t - t_3 + t_2 - t_1)]\Big]
\end{aligned}
\tag{30}
$$

where $g(\Delta\omega)$ is the distribution function of the frequency difference between the transition frequency of the two-level system and the frequency of the light. For the sake of simplicity, we restrict our discussion to the case where T_2^{ele} is much shorter than τ_c and T_1^{ele}. Then (30) is reduced to the following in both homogeneous and inhomogeneous broadening cases:

$$
\begin{aligned}
P^{(3)}(t, \tau, \tau'; \mathbf{k}_s, \omega) = \int_{-\infty}^{t} dt' \exp\left(-\frac{t - t'}{T_1^{\text{ele}}}\right) \\
\times \Big[\tilde{E}(t - \tau)\tilde{E}^*(t' + \tau')\tilde{E}(t') \\
+ \tilde{E}(t)\tilde{E}^*(t' + \tau')\tilde{E}(t' - \tau)\Big]
\end{aligned}
\tag{31}
$$

The signal intensity, with the unity proportionality coefficient set for simplicity, is obtained as follows:

$$
I(\tau, \tau') = \int_{-\infty}^{\infty} dt \left\langle \left| P^{(3)}(t, \tau, \tau'; \mathbf{k}_s, \omega) \right|^2 \right\rangle
\tag{32}
$$

If the envelope function of the incoherent light field $E(t)$ is expressed by a stationary complex Gaussian random process, the right side of (32) can be given by 24 terms, shown below.

Four types of experiments can be performed. They are classified in terms of delay times τ and $-\tau'$ of beams 3 and 2, respectively, from beam 1:

Case 1: Variable τ, the small value of fixed τ', $|\tau'| \ll \tau_c$
Case 2: Variable τ', the small value of fixed τ, $|\tau| \ll \tau_c$
Case 3: Variable τ', the large value of fixed τ, $|\tau| \gg \tau_c$
Case 4: Variable τ, the large value of fixed τ', $|\tau'| \gg \tau_c$

Four cases are discussed when τ_c is much shorter than T_1^{ele} as follows.

Case 1. This is the case treated in Ref. 16. The signal trace is expressed as follows for the positive delay time:

$$
I(\tau) = 1 + 3|f(\tau)|^2 + \frac{2\tau_c^{(2)}}{T_1^{\text{ele}}} \exp\left(-\frac{\tau}{T_1^{\text{ele}}}\right)
\tag{33}
$$

where $\tau_c^{(2)}$ is defined in Table I.

Case 2. In this case the signal intensity for the positive delay time is reduced to the following:

$$
\begin{aligned}
I(\tau') = 1 + \frac{2T_1^{\text{ele}}}{\tau_c^{(2)}}|f(\tau')|^2 \\
+ \frac{4\left[\left(\tau_c^{(3)}\right)^2 + \left(\tau_c^{(4)}\right)^2\right]}{T_1^{\text{ele}}\tau_c^{(2)}} \exp\left[-\left(\frac{2}{T_1^{\text{ele}}}\right)'\right]
\end{aligned}
\tag{34}
$$

where $\tau_c^{(3)}$ and $\tau_c^{(4)}$ are defined in Table I. For the negative delay time the tail term vanishes.

Case 3. In this case the signal intensity for the positive delay time is given by

$$
\begin{aligned}
I(\tau') = 1 &+ \frac{T_1^{\text{ele}}}{\tau_c^{(2)}}|f(\tau')|^2 \\
&+ 2\frac{\left[\left(\tau_c^{(3)}\right)^2 + \left(\tau_c^{(4)}\right)^2\right]}{T_1^{\text{ele}}\tau_c^{(2)}}\exp\left[-\left(\frac{2}{T_1^{\text{ele}}}\right)\tau'\right]
\end{aligned} \tag{35}
$$

The tail term disappears when the delay time is negative.

Case 4. In this case the signal intensity is reduced to the following:

$$
\begin{aligned}
I(\tau) = 1 &+ \text{Re}\left[f(\tau)f^{(2)*}(\tau)\right] \\
&+ \frac{2\left[\left(\tau_c^{(3)}\right)^2 + \left(\tau_c^{(4)}\right)^2\right]}{T_1^{\text{ele}}\tau_c^{(2)}}\exp\left(-\frac{2}{T_1^{\text{ele}}}|\tau|\right)
\end{aligned} \tag{36}
$$

where $f^{(2)}(\tau)$ is defined by

$$
f^{(2)}(\tau) = \frac{\langle f(t + \tau)f^*(t)\rangle}{\langle|f(t)|^2\rangle} \tag{37}
$$

The above results show us that, in each case, the signal consists of three parts, i.e., a constant background, a coherence spike, and a tail which decays with the time constant of T_1^{ele} or $T_1^{\text{ele}}/2$, but the ratio of the intensities of these three components varies from one case to another. Because only the tail includes the information of the relaxation time, a larger fraction of the tail part gives a better signal to noise ratio for the determination of the relaxation time. Another factor that affects the signal to noise ratio is an absolute intensity of the tail part. It is proportional to $(T_1^{\text{ele}}\tau_c)$ in case 1, and proportional to $(\tau_c)^2$ in the other cases. Therefore, case 1 is more advantageous than the other three for improving the quality of the data.

However, when we adopt the phase-conjugation configuration, case 1 requires the delay time between the counterpropagating beams to be varied. In such a case the time resolution is determined not by the correlation time but by the interaction length of the beams, and it is difficult to determine the delay time unless the sample is thinner than the correlation length of the incoherent light, which is about 30 μm for a correlation time of 100 fs. To avoid this degradation of the time resolution resulting from the thickness of the sample, an experiment under the conditions of case 3 can be performed with the phase-conjugation configuration.

I. Advantages and Disadvantages

The study of ultrafast processes with the use of temporally incoherent light has several outstanding advantages. First, sophisticated and expensive femtosecond or picosecond pulsed lasers are not necessary. It is much easier to construct, operate, and maintain lasers with nanosecond or longer pulses than those with femtosecond or picosecond pulses. Second, the wavelength of the broadband nanosecond dye lasers can be changed easily by changing dye species, dye concentration, solvent, or pumping power. Therefore, the present method is suited for the studies of excitation-energy dependence of relaxation processes. And third, the time resolution is not degraded by dispersion, even when incoherent light propagates in optical media with long optical path length.

There are also a few disadvantages. Because of the appearance of a coherence spike and/or a constant background in the signal trace, high signal to noise ratio is sometimes difficult to obtain. It is therefore necessary to adjust the spectral width of the laser, or τ_c of incoherent light, to the optimum condition to enhance the signal to noise ratio. Also, since in the present method the energy density of the beams in the sample is necessarily higher than that in the method using femtosecond pulses with the same peak power, the thermal effect can be more serious in experiments using incoherent light than in those using femtosecond pulses. This fact is explained in the following.

The thermal effect can be represented by the following nonlinear refractive index:

$$n_2^{\text{th}} = \left(\frac{\mathrm{d}n}{\mathrm{d}T}\right)\left(\frac{\alpha}{\rho C_{\text{p}}}\right)\tau_{\text{d}} \tag{38}$$

where the temperature derivative of the refractive index $\mathrm{d}n/\mathrm{d}T$, specific heat capacity C_{p}, density ρ, absorption coefficient α of the sample, and decay rate $1/\tau_{\text{d}}$ of the temperature increase are the factors determining n_2^{th}. The heat-release time τ_{d} is roughly given by $1/(\alpha^2 D)$. Here D is the thermal diffusion constant in (or perpendicular to) the direction of light propagation when the size of the area, on which heat is deposited by light absorption, is much larger (or smaller) than the absorption depth $d = 1/\alpha$.

If d is of the order of 10^2–10^4 Å, τ_d is of the order of nanoseconds to picosections. When femtosecond pulses with the duration of t_p is used, the effective n_2^{th} due to temperature increase is reduced by a factor of t_p/τ_d. Therefore, the thermal effect is more serious when incoherent light with the duration longer than a few nanoseconds is used.

For example, in the case of the electronic dephasing-time measurement by the two-beam DFWM under the resonance condition, if the polarizations of the two incoherent light beams are parallel to each other, then the signal due to thermal grating can surpass the electronic contribution. In such a case, incident light beams must be used in perpendicular polarization. However, when the two light beams propagate in an anisotropic system, the crossed polarization condition is destroyed if one of the two polarization axes does not coincide with the optical axis. Therefore, it can be very difficult to eliminate the thermal effect from the signal obtained for anisotropic systems with larger n_2^{th} than electronic n_2 using incoherent light.

III.   EXAMPLES OF THE APPLICATIONS OF INCOHERENT LIGHT TO ULTRAFAST DYNAMIC STUDY

In this section we discuss the details of experimental apparatus utilized and results obtained by our group using the techniques described in the preceding section. Two types of experiments are described: (1) measurements of electronic dephasing in a semiconductor doped glass sample and (2) the Kerr relaxation in optical Kerr media CS_2 and mixtures of CS_2 and other liquids.

A.   Electronic Dephasing in a Semiconductor-Doped Glass Sample

The absorption spectrum of the sample of CdS colloids in Fig. 13 exhibits a shoulder around 420 nm, which corresponds to the exciton absorption. The width of the shoulder is estimated to be about 40 nm. If the size of the particles is uniform, exciton absorption should appear as a narrow peak. The particles in this sample may have a broad distribution of diameters.

Figure 14 shows the experimental apparatus. The incoherent light source was a broadband dye laser pumped by a nanosecond excimer laser. The laser dye was stilbene 420, selected because of the near resonance on the exciton absorption. The dye laser had no tuning elements, and τ_c was about 120 fs. The laser beam, which was linearly polarized by a Glan-Thompson prism, was divided into two beams with equal intensity. The probe beam was focused on the sample through a polarizer, and directed

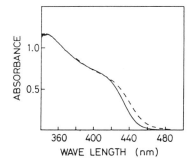

Figure 13. Absorption spectra of CdS microcrystallites doped in a polymer film: solid line, at 2 K; dashed line, at room temperature.

to a photomultiplier through an analyzer. A Babinet compensator and a pinhole were used to eliminate the background. The pump beam was polarized 45° with respect to the probe beam, variably delayed with a translational stage, and focused at the same spot. Samples were immersed directly in superfluid helium. The signal transmitted through a crossed analyzer was detected as a function of the delay of the probe with respect to the pump.

We could successfully detect the signal by the new method using the Kerr-shutter configuration. We failed in the ordinary two-beam DFWM configuration, probably because precise alignment can be performed more easily in the former method. That is why we called the new method "signal-selective." Two decay components can be seen in the obtained data from CdS microcrystallites with a mean diameter less than 50 Å (Fig. 15). From this trace we estimate the decay time constant to be 200 fs for

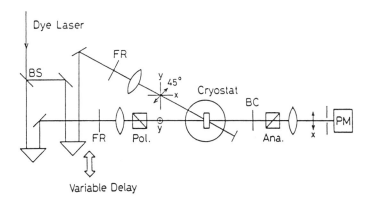

Figure 14. Experimental setup for T_2 measurement in the Kerr-shutter configuration. PM, photomultiplier; BS, beamsplitter; FR, Fresnel rhomb; BC, Babinet-Soleil compensator.

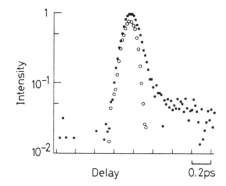

Figure 15. Semilogarithmic plots of the delay time dependence of signal intensity at 2 K obtained from CdS microcrystallites. Signal and autocorrelation traces are shown by closed and open circles, respectively.

the faster component. Since the exciton absorption of our sample is inhomogeneously broadened, higher levels of excitons are excited and the dephasing of the higher levels may be faster. On the other hand, the homogeneous linewidth reported in Ref. 35 represents a dephasing time of 80 fs. Dephasing time obtained indirectly by hole burning is shorter than that obtained by DFWM. Systematic errors are expected to increase the hole width. This means estimated value is a lower limit of the dephasing time. For a more detailed study the particle size of the sample must be controlled to have a much narrower size distribution.

B. Kerr Relaxation in Neat and Mixed Liquids

Measurement of the dynamics of the optical Kerr effect (OKE) was performed with the Kerr shutter configuration. An excimer laser (Hamamatsu) pumped dye laser (Molectron DL-14) with a single-stage amplifier was used as an incoherent light source by eliminating a frequency tuning element from the dye laser system. The peak wavelength and a pulse width at half-maximum of the nearly Gaussian laser spectrum were 476 and 8 nm, respectively. About 10 to 20% output light of a 30-μJ laser pulse with 5-ns width was employed. The laser dye used for this specific experiment was coumarin 102. Autocorrelation width measured using benzene solution of 4-diethyl-4'-aminonitrostilbene (DEANS) was determined to be 61 $\pm$ 4 fs, giving a corresponding coherent pulse width of 43 $\pm$ 3 fs of a Fourier transform limited pulse. The measurement temperature was 22 $\pm$ 1°C.

Figures 16, 17, and 18 show the delay time dependence of the Kerr signal intensity from neat CS_2 liquid, the mixed liquids of CS_2 and hexane, and the mixed liquids of CS_2 and methanol, respectively, after the constant background contributions have been subtracted. At first sight of the data of neat CS_2 shown in Fig. 18, the signal seems to be composed mainly

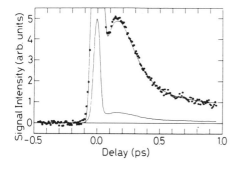

Figure 16. The delay-time dependence of the OKE signal intensity obtained from carbon disulfide. The lower line is the experimental data. Circles are the data expanded by a factor of 10. The upper line is the fitting curve.

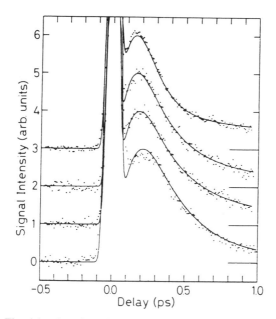

Figure 17. The delay-time dependence of the signal intensity obtained from a binary mixture of CS_2 and hexane. The signal intensity, after the background intensity was subtracted, was multiplied by a factor of 6–12 to normalize the intensity of the shoulder part of the signal. The volume fractions of CS_2 are 100, 50, 25, and 10% from top to bottom. Dots are the experimental data points and the lines are the fitting curves. The fitting functions are described in the text and the obtained fitting parameters are presented in Table II.

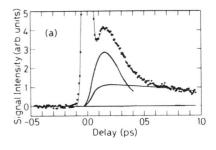

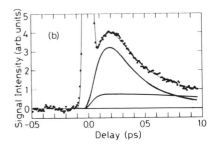

Figure 18. The delay-time dependence of the signal intensity from binary mixtures of CS_2 and methanol. Closed circles are experimental data. Fitting curves together with two components—I-I and orientational contributions—are displayed. The volume fractions of CS_2 are (a) 100%; (b) 33%; and (c) 10%. The fitting functions are described in the text and the obtained fitting parameters are presented in Table II.

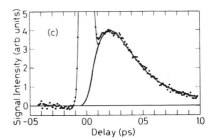

of two parts: a very intense peak at zero delay time and a component with finite rise and slow decay times. The former can be attributed to the coherent spike term and also to a contribution from the instantaneous response. The latter was analyzed in terms of two contributions: intermolecular interaction-induced polarizability changes (I-I) and anisotropic polarizability introduced by the orientational distribution change (OR).[33] The former is faster in the solvent studied.

As shown in Fig. 17, both the rise and fall times of the fast I-I component become longer and the relative amplitude with respect to the slow OR contribution increases upon dilation. The enhancement of the amplitude ratio of the fast component to the reorientational contribution is attributed mainly to enhancement of the amplitude of the fast component, as shown later.

Analysis of these results, including those obtained from other mixtures, is performed by fitting them using two functions, the reorientational response $G_{OR}(t)$ and the I-I response $G_{II}(t)$, as expressed in the following equations for the nuclear response:

$$G_{OR}(t) = C_{OR}\theta(t)\exp\left(-\frac{t}{\tau_{col}}\right)\left[1 - \exp\left(-\frac{t}{\tau_v}\right)\right] \tag{39}$$

$$G_{II}(t) = C_{II}\theta(t)\text{Im}\left[1 - i\omega_0 t\right]^{-q-2} \tag{40}$$

Here C_{OR} and C_{II} are constants, $\theta(t)$ is the normalized step function, and τ_{col} is the collective orientational relaxation time, which describes the decay of OKE. The initial rising behavior in (39) is a consequence of a finite decay rate of the molecular angular momentum and the rise time τ_v is the relaxation time of the angular momentum. This OKE response function $G(t)$ can be related to the following I-I dynamic light scattering (DLS) spectrum using the fluctuation-dissipation theorem.[33, 36] An expression for the DLS intensity as a function of the angular frequency of shift ω is given with a unity proportionality coefficient for simplicity as follows to describe the II-DLS spectrum of spherical molecules induced by binary collisions[37]:

$$I_{DLS}(\omega) = \omega^q \exp\left(-\frac{\omega}{\omega_0}\right) \tag{41}$$

This equation has also been used to interpret DLS spectra from nonspherical molecular systems.[37-41] In the derivation of this equation, an effective polarizability change $\Delta\alpha$ is assumed to be proportional to a power of distance between the interacting molecules r as follows:

$$\Delta\alpha = ar^{-m} \tag{42}$$

Where a is an appropriate proportionality coefficient. The relation between exponent q in (41) and m in (42)

$$q = 2\frac{(m-7)}{7} \tag{43}$$

has been derived[37] and the frequency parameter ω_0 corresponds to the time duration in which the interaction occurs. Although in dense media the assumption of the binary collision does not hold and cancellation occurs, this relation has often been used for the analysis of I-I DLS spectra of liquids.[37-41] In dense media, ω_0 has been suggested to be correlated to the decay time of the translational motion autocorrelation[42] and q can be regarded as the exponent in an effective function for the intermolecular I-I polarization change. In this study of the orientational component, if τ_{col} is determined from the long time OKE relaxation, τ_r can be derived using the following equation, and the value of g_2/j_2 (Ref. 43):

$$\tau_{col} = \left(\frac{g_2}{j_2}\right)\tau_r \qquad (44)$$

where g_2 and j_2 are the static and the dynamic orientational correlation coefficients of the liquid, respectively.

We could not determine precisely the relaxation time of the long-time OKE response owing to a large background signal, which is inherent in the method using incoherent light. To circumvent this difficulty, we determined the relaxation time of the orientational component by measuring the viscosity of the mixture. The results of the data analysis for the different mixtures are presented in Table II. In this table, $M(2)$ is the

TABLE II
Parameters Obtained from the Fitting of Optical Kerr Effect Signal Intensities.

Volume(molar) Fraction of CS_2	τ_{col} (ps)	q	m	ω_0^{-1} (fs)	$\bar{\nu}_0 = \dfrac{\omega_0}{(cm^{-1})2\pi c}$	$\dfrac{M(2)}{(cm^{-2})(2\pi c)^2}$	Peak Ra
1.00(1.00)	1.70	0.580	9.03	400	13.3	721	2.6
Hexane							
0.50(0.68)	1.56	−0.759	4.43	265	20.0	65	11.4
0.25(0.42)	1.44	−0.756	4.35	290	18.3	102	8.7
0.10(0.19)	1.35	−0.622	4.82	357	14.9	115	Infinity
Methanol							
0.33(0.25)	2.17	−0.768	4.31	300	17.7	90	4.2
0.10(0.07)	2.25	−0.591	4.93	361	14.7	125	Infinity
Ethanol							
0.50(0.49)	1.92	0.052	7.18	410	13.0	363	3.0
0.10(0.10)	2.06	−0.801	4.20	310	17.1	68	Infinit

Note. τ_{col} values are calculated values using viscosities. Peak ratios are the ratios of the peak valu the I-I contribution to the orientational contribution before convolution with the coherent s "Infinity" means that the best fit was obtained without the orientational contribution.

second-order moment of the I-I contribution to the DLS spectrum, which is defined by

$$M(2) = \frac{\int_0^\infty d\omega \, \omega^2 I_{DLS}(\omega)}{\int_0^\infty d\omega \, I_{DLS}(\omega)} \qquad (45)$$

By using (41), $M(2)$ can be calculated to be

$$M(2) = (q + 2)(q + 1)\omega_0^2 \qquad (46)$$

This second-order moment is a measure of the broadening of the DLS spectrum, or the relaxation rate of the Kerr response.

From the data shown in Figs. 16–18 and Table II, several trends are found: (1) q decreases upon dilution down to -0.6 to -0.8; (2) ω_0 does not vary essentially; (3) $M(2)$ decreases upon dilution; (4) the peak ratio of the I-I to the orientational component increases upon dilution and the amplitude of the orientational component becomes negligible in case of 10% CS_2 volume fraction. This dependence on density has been observed in the DLS studies of rare gases and by theoretical and simulation studies[44-51] and has been explained in terms of the three- and four-body correlations which cancel the two-body correlation effect on the I-I polarizability fluctuation. In particular, low-frequency polarizability fluctuation is reduced effectively, resulting in the spectral broadening, or increase in decay rate, at higher densities.

On the other hand, the description of the I-I polarization change, or the I-I optical Kerr effect, by binary interactions is appropriate at low densities. The I-I polarization change in CS_2 has been attributed primarily to dipole–induced-dipole interaction proportional to the cubic inverse of the intermolecular distance.[52] When the binary collision model, in which the molecules are assumed to be in free space, is applied,[37] the exponent q in (41) is derived to be $-8/7$ at low volume fractions of CS_2, in which the binary collision approximation is appropriate, by using (42) and (43). The experimentally determined values, as shown in Table II, of -0.6 to -0.8 are close to this theoretical value. This shows that molecular motions of CS_2 in dilute solutions in these solvents are fairly free even with the presence of the solvent. The reasonably good fit and the consistent change of parameters in a series of binary mixtures with different mixing ratios support the attribution of the fast optical Kerr effect component mainly to the I-I effect even in neat CS_2.

IV. NEW METHODS FOR THE STUDY OF FEMTOSECOND
NONLINEAR RESPONSE

A. Femtosecond Nonlinear Optical Measurement Using Short Pulses

To find materials with ultrafast and large optical nonlinearity to be used in optical devices it is necessary to develop reliable methods of determining nonlinear optical susceptibility. In searching for such materials, we must have guiding principles to obtain higher nonlinearity and faster response by clarifying the mechanism of optical nonlinearity. To elucidate the mechanism, it is important to obtain both the real and imaginary parts of the nonlinear susceptibility with frequency and time resolution. Recently we proposed a new highly sensitive method of the time-resolved Kerr measurement using heterodyne detection.[53] We now propose two new methods for separating the real and imaginary parts of the nonlinear susceptibility by a single measurement with femtosecond time resolution, which is based on time-resolved interferometry.[54]

B. Optical Heterodyne Detection of Induced Phase Modulation
for the Study of Femtosecond Optical Kerr Effect

An intense light pulse propagating through an optical medium introduces rapid change of refractive index with time, resulting in a phase modulation in proportion to the changing rate of the refractive index. This induced phase modulation (IPM) process causes a shift of the probing light spectrum, which is proportional to the time derivative of the refractive index. By the index change of the medium caused by an ultra-short pump pulse and using another probe pulse, the time derivative of the optical Kerr effect (OKE) response in the medium can be measured.[55-57]

In this section, we develop a new method by applying the polarization selective optical heterodyne-detection (OHD) scheme,[53] resulting in great enhancement in sensitivity. In our new technique, optical-heterodyne-detected induced phase modulation (OHD-IPM), the nonlinear polarization is heterodyne detected by a weak local oscillator. Here the term heterodyne is used only to meet convention.[57, 53] The detection technique is more accurately described as homodyne detection since oscillation frequencies of the nonlinear polarization signal and the local oscillator field are the same.

The experimental schematic is shown in Fig. 19. A standard combination of a colliding-pulse mode-locked (CPM) ring dye laser and a multiple-pass amplifier pumped by a copper vapor laser was employed.[59, 60] A 3-mm flow cell was used for the amplification, and the output of the CPM laser was amplified six times. The amplifying medium was a 3 ×

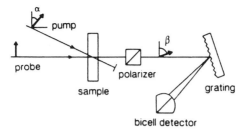

Figure 19. The schematic of the experiment for measuring optical heterodyne-detected induced phase modulation.

10^{-4}-mol/l solution of rhodamine 640 in a 1:3 v/v mixture of methanol and water with 4% Ammonyx LO. The output was passed through a sequence of four prisms to minimize the pulse width at the sample by compensating for the group-velocity dispersion of optical components. The repetition rate, the energy, and the wavelength of the amplified pulses were 10 kHz, $2\mu J$, and 620 nm, respectively. The output pulses were attenuated for the measurement to about 100 nJ to avoid any higher-order effects. The pulse width was obtained as 55 fs from the second-harmonic autocorrelation measurement, assuming a squared hyperbolic secant shape. The pump pulse is linearly polarized at angle α, usually 45°, from the linear polarization of the probe pulse. A fraction of the probe light passes through a polarization analyzer placed after the sample, which is polarized at an angle β nearly equal to 90°, from the original polarization of the probe light. Thus, the component of the nonlinear polarization orthogonal to the incident probe light is heterodyne (homodyne) detected with the fraction of the probe light, which serves as the local oscillator. The intensity of the local oscillator was adjusted so that it was greater than about 20 times of that of the nonlinear signal field in order to ensure linear detection of the OKE response. The polarization angle of the analyzer was typically 1 to 2° from orthogonal to the input polarization direction. A grating dispersed the light onto a bicell (photodiode) detector (BiPD). Spectral shift of the light was detected by taking the difference of the two outputs of the BiPD. The pump beam was chopped at 2.3 kHz, and the difference of the two outputs of the BiPD was fed to a lock-in amplifier. The BiPD was positioned so that the two outputs balanced when the pump beam was intercepted. The spectral shift against the delay time of the probe pulse with regard to the pump pulses gave the signal trace of the OHD-IPM measurements.

Conventional OHD-OKE measurements were also performed for the comparison with the OHD-IPM measurements. For the OHD-OKE mea-

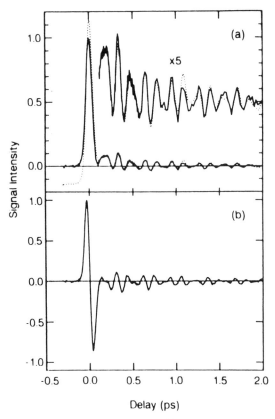

Figure 20. Signal traces obtained from CCl_4. (a) the OHD-OKE data (solid line) are plotted with the time integral of the OHD-IPM signal trace (dotted line). Also displayed are the oscillating parts, which are expanded fivefold after being shifted vertically. (b) Original signal trace of OHD-IPM measurements.

surements, a quartz quarter-wave plate was inserted in the probe beam after the sample cell. One of the axes of the quarter-wave plate was fixed to be parallel to the polarization direction of the probe light. The analyzer was oriented by a small angle from an orthogonal direction to the input probe light polarization. The intensity of the output of the analyzer was detected by a photodiode. All the measurements were performed at $22 \pm 1°C$ The sample liquids of CCl_4, benzene, and CS_2 were in a 1-mm-thick glass cell.

The results of a conventional OHD-OKE measurement and an OHD-IPM measurement from liquid CCl_4 are shown in Fig. 20. The oscillations in the data are due to intramolecular vibrational modes of 218 and 314

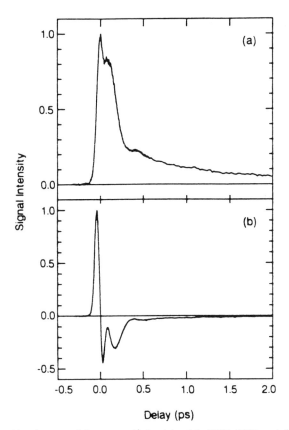

Figure 21. Signal trace of benzene obtained by (a) OHD-OKE and (b) OHD-IPM measurements.

cm^{-1}. The time integral of the OHD-IPM data is also shown in the figure to confirm the validity of OHD-IPM measurement method. The integral is expanded and vertically shifted to compare with the OHD-OKE data. Except for the misfit near zero delay time and the baseline shift at negative delay times, the agreement between the OHD-OKE and the time integral of the OHD-IPM signal is remarkably good. The disagreements are due to the contribution of the coherent coupling, since that to the OHD-IPM signal trace is not the time derivative of that to the OHD-OKE signal trace, as shown later.

The OHD-OKE and OHD-IPM data obtained from liquid benzene are shown in Fig. 21. Overdamped oscillations are apparent in the OHD-IPM data. The dynamics of the OKE response of benzene have been studied by a conventional transient-grating measurement with femtosecond pulses[61] and by a Kerr-shutter experiment with temporally incoherent light.[17, 33, 62, 63] Although the oscillations are visible in the previously reported data[61-64] and in the present OHD-OKE data, the advantage of the time-derivative technique by OHD-IPM measurements for the study of vibrational dynamics of molecular systems is clear. The OHD-IPM signal trace can be divided into three contributions: (1) the nearly antisymmetric signal with a large amplitude around time zero due to the electronic and the coherent coupling contributions, (2) damped cosine oscillations with a period of about 400 fs, and (3) slowly decaying negative signal. The oscillations in the OKE and the IPM data are attributed to an intermolecular vibrational mode, since benzene has no low-frequency intramolecular modes in the wave-number region below 400 cm^{-1} (Ref. 65). The character of the intermolecular mode, however, is not clear.

Molecular dynamics in liquid benzene has also been studied by means of the measurement of the DLS spectrum. The angular velocity correlation function (AVCF) of benzene molecules in liquids has been calculated from the DLS spectrum by several groups.[66-68] They separated the DLS spectrum into the I-I contribution, $I_{int}(\omega)$, and the orientational contribution, $I_{OR}(\omega)$, by assuming the functional form of the I-I contribution, and calculated the AVCF.

We reconstructed the time derivative of the nuclear contribution to the OKE response function $\dot{r}(t)$ for comparison with those results, since it should be nearly equal to the AVCF if the I-I contribution can be neglected. The reconstructed IPM response function $\dot{r}(t)$ for benzene is shown in Fig. 22. The reduced Raman spectrum $R_2(\omega)$ up to 200 cm^{-1}

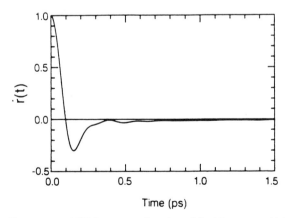

Figure 22. Reconstructed IPM response function $r(t)$ of benzene, which is due to the nuclear response.

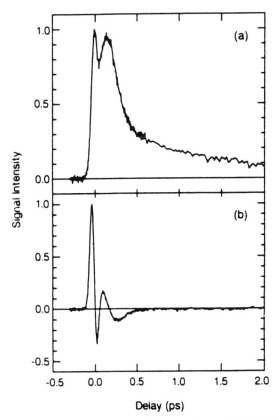

Figure 23. Signal trace of CS_2 obtained by (a) OHD-OKE and (b) OHD-IPM measurements.

was used for the Fourier inverse transformation. It clearly shows damped oscillations with a period of about 400 fs.

The humps around 250 and 550 fs not seen in the original experimental data are probably artifacts. The reconstructed IPM response agrees well with the AVCFs obtained from the DLS spectrum,[41, 42, 66] except for those humps due to the truncation of the Fourier spectrum at 200 cm^{-1} or due to imperfect separation of coherent-coupling and electronic contributions and a slowly decaying negative feature.

The conventional OHD-OKE data and the OHD-IPM data obtained from CS_2 are shown in Fig. 23. In the IPM signal trace of CS_2 the oscillatory character is not so apparent as in that of benzene. The IPM response function of CS_2 reconstructed from the IPM signal trace is shown in Fig. 24. A Fourier transform up to 100 cm^{-1} was used. The

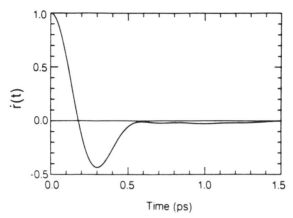

Figure 24. Reconstructed IPM response function $r(t)$ of CS_2 due to the nuclear response.

oscillation almost damps in a single period, which is about 600 fs. Although the femtosecond OKE response of CS_2 has already been reported by several groups,[33, 67-69] the present IPM data, being closely related to the AVCF, exhibit more directly the features of the coherent molecular motions in CS_2 liquids.

The present experimental results are consistent with those of the simulation studies.[52, 72] Although the dipole–induced-dipole effect on the susceptibility is found to be the main origin of the I-I contribution by theoretical and simulation studies,[52, 72] the molecular motions that induce the susceptibility change are not well characterized yet. The I-I effect due to translational intermolecular motion was predicted by calculation to dominate the high-frequency region of the DLS spectrum of CS_2 (Ref. 73). However, since the time scales of the I-I and the orientational contributions are comparable, these two contributions can interfere with each other in the subpicosecond time region.[71] Librational motions of molecules, therefore, may affect the I-I response of CS_2 in this time region, as suggested by Ruhman et al.[64] Detailed simulation studies are needed to clarify the character of the molecular motions that induce the I-I response.

C. Femtosecond Interferometer with Reference Arms Using Fast Fourier Transform of Interference Fringe

In this section a sensitive interferometric method is demonstrated that allows the separation of the real and imaginary parts of the nonlinear susceptibility with a femtosecond time resolution by a single set of measurement.[54] A special reference interferometer compensates for any fluc-

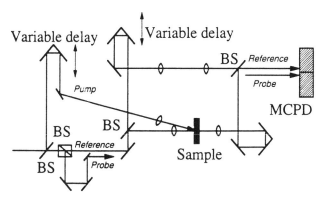

Figure 25. Block diagram of the femtosecond time-resolved Mach-Zehnder interferometer. BS, beamsplitters; MCPD, multichannel photodiode.

tuations of the fringe pattern and enhances detection sensitivity of a fringe shift as small as 0.025 rad ($\lambda/250$) by averaging only 100 shots using a low-repetition-rate laser. This method can be applied with high sensitivity to materials even with optical anisotropy or absorption.

Figure 25 shows the apparatus. The light source is a colliding-pulse mode-locked (CPM) ring dye laser amplified by a four-stage dye amplifier pumped by a 10-Hz Q-switched Nd:YAG laser. The typical time-width, energy, and peak wavelength of the amplified pulses are 100 fs, 200 μJ, and 630 nm, respectively. One of the two beams divided by a beamsplitter was led to a variable optical delay and used as a pump. The other was divided again into parallelly propagating probe and reference beams, with an appropriate delay and a small separation, which permitted selective detection.

Both the probe and the reference beams were divided again and formed two Mach-Zehnder interferometers using common optical components, such as mirrors and beamsplitters. The sample was placed in one of the two arms of each interferometer. Both the probe and the reference beams were focused onto the same position of the sample. The two interference fringes of equal inclination of the probe and reference beams were detected simultaneously by a multichannel photodiode (Hamamatsu S2301-512Q). The reference pulse reached the sample before the pump pulse at any delay time; therefore, the reference interference fringe, which is not influenced by the pump, can be used to compensate for fluctuations of the fringe. There are two origins of fluctuations: (1) optical path changes due to mechanical vibration and distortion of the optical components, and (2) the change in intensity and position and direction of the incident beam, which is difficult to eliminate. Both can be eliminated by the present method.

Similar reference interferometry was performed by Olbright and Peyghambarian.[74] Two beams were spatially separated with a double-image prism, and one beam did not pass through the sample, so as to form a reference interferometer. Therefore, the intensities of probe and reference beams can be much different in the case of samples with strong absorption, resulting in lower visibility and hence lower sensitivity than our method.

The interference fringe pattern was then fast Fourier transformed to determine the real and imaginary parts of nonlinear susceptibility. They were separately obtained by a single set measurement, whereas in most previous studies two sets of experiments were needed[75, 76] or experiments were performed on materials of negligibly small imaginary parts.[77, 78] The Fourier transform also has the advantage of improving accuracy by its use of the complete fringe data for analysis, in comparison with the case in which only the peak shift of the fringe is used to obtain real parts.[74] The principle of the measurement is described as follows.

The intensity I of the interference fringe can be ideally described by the following function of the position (x) on the detection plane of the multichannel photodiode plate, using a single wave number k_0 and assuming a plane-wave approximation:

$$I(x) = \left[E_1^2 + E_2^2 + 2E_1E_2 \cos(k_0x - \phi_0) \right] \exp\left(-\frac{x^2}{\alpha^2} \right) \qquad (47)$$

Here E_1 and E_2 are the field amplitudes of the light in each arm of the interferometer, and ϕ_0 is the phase shift at $x = 0$. The Fourier transform F of $I(x)$ is

$$\begin{aligned}
I(k) = \alpha\sqrt{2} \Bigg\{ &\left(E_1^2 + E_2^2 \right) \exp\left(-\frac{\alpha^2k^2}{4} \right) \\
&+ E_1E_2 \exp\left[-\frac{\alpha^2(k - k_0)^2}{4} \right] \exp(-i\phi_0) \qquad (48) \\
&+ E_1E_2 \exp\left[-\frac{\alpha^2(k + k_0)^2}{4} \right] \exp(i\phi_0) \Bigg\}
\end{aligned}$$

The spectrum of $I(k)$ has three peaks at $k = 0, \pm k_0$, which correspond to the three terms of (48). The shift and distortion of the fringe due to fluctuations were compensated for in every shot and then averaged to reduce the finesse in our measurement. We obtained the change in phase and transmittance due to excitation of the sample as

$$\Delta\phi = \left(\phi_{\text{pro}}^{\text{ex}} - \phi_{\text{ref}}^{\text{ex}}\right) - \left(\phi_{\text{pro}}^{\text{ne}} - \phi_{\text{ref}}^{\text{ne}}\right) \tag{49}$$

$$\frac{\Delta T}{T} = \frac{(E_1 E_2)_{\text{pro, ex}}^2 / (E_1 E_2)_{\text{ref, ex}}^2 - (E_1 E_2)_{\text{pro, ne}}^2 / (E_1 E_2)_{\text{ref, ne}}^2}{(E_1 E_2)_{\text{pro, ne}}^2 / (E_1 E_2)_{\text{ref, ne}}^2} \tag{50}$$

where $\Delta T/T$ is the relative transmittance change of the sample owing to excitation, ex and ne represent with and without pump, respectively, and pro and ref correspond to the probe and reference interferometers, respectively.

When the reference interferometer was used, the standard deviation was reduced to 0.25 rad, four times smaller than that in the case without a reference, resulting in high sensitivity that can detect a fringe shift as small as 0.025 rad ($\lambda/250$) by averaging over only 100 shots using a low-repetition-rate laser.

Figure 26 shows the phase change of CS_2 without absorption at 630 nm due to excitation at 297 K. The polarization of the pump and probe light are parallel (case I) (Fig. 26a) and perpendicular (case II) (Fig. 26b) to each other. We obtained $\chi_{zzzz}^{(3)}$ and $\chi_{zzxx}^{(3)}$ in cases I and II, respectively, whereas with the Kerr shutter method, only the difference between two tensor components can be determined. The time dependence of the phase change in CS_2 was composed of the instantaneous response and the two exponential decays. The time constants of exponential decays under parallel polarization were determined to be 0.30 ± 0.08 and 1.6 ± 0.9 ps, using the convolution with the pulse-shape function.

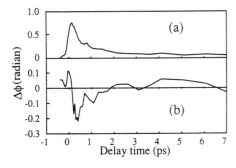

Figure 26. Transient phase change ($\Delta\phi$) in CS_2. The polarizations of the pump and probe light were (a) parallel and (b) perpendicular to each other.

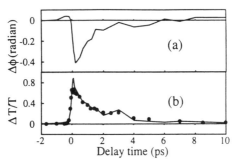

Figure 27. (a) Transient phase change $\Delta\phi$ and (b) the normalized transmittance change $\Delta T/T$ (solid curve) obtained by the present method in CdS_xSe_{1-x}-doped glass. The filled circles in (b) show $\Delta T/T$ by a conventional pump–probe absorbance change measurement using only the sample arm of the interferometer.

Figure 26 shows that case II has positive phase change just after excitation, and then becomes negative; in case I it was positive at all delay times. In case I the contributions of electronic and molecular origin are known to be positive. Therefore, in case II the electronic contribution is considered to be positive and the molecular contribution is negative from the tensor analysis. Thus, the obtained signal in case II has both of these components.

Figure 27 shows the phase change (Fig. 27a) and the transmittance change (Fig. 27b, solid curve) obtained by our method for the CdS_xSe_{1-x}-doped glass. The transmittance change by conventional absorbance change measurement is also shown in Fig. 27b (filled circles) for comparison. It was performed by blocking the arm without a sample. The transmittance change in Fig. 27b (solid curve) agrees well with data obtained by the conventional absorption measurement. This result verifies proper separation of the real and the imaginary parts by our method. The third-order nonlinear susceptibility of CdS_xSE_{1-x}-doped glass was determined to be $(5 + 3i) \times 10^{-14}$ esu at the peak, which is several orders of magnitude smaller than the results obtained with much longer pulses, such as with a duration of a nanosecond,[79] which includes the effect of accumulation over the pulse widths.

D. Frequency Domain Interferometer for the Determination of Time-Resolved Complex Nonlinear Susceptibility

A new femtosecond time-resolved interferometer is described here that utilizes interference fringes in the frequency domain to obtain simultaneously difference phase spectra (DPS) and difference transmission spectra

(DTS) with a multichannel spectrometer.[80] By analogy with the beat, which is interference in the time domain caused by two frequency components, two components in the time domain (the reference and probe pulses) interfere in the frequency domain, as readily derived by the Fourier transform. Spectral interference (frequency-domain interference) can therefore be observed even when the two pulses are displaced by more than the pulse duration, in contrast to spatial interference. For the first time to the author's knowledge, transient oscillations were observed in DPS and the spectral shift of a probe pulse was time resolved together with the rise in DPS, which is clear evidence for induced phase modulation in absorptive materials.

The pump–probe technique with a white-light continuum pulse has played a major role in the progress of femtosecond absorption spectroscopy during the past decade, because difference transmission spectra can be obtained by using a multichannel spectrometer without scanning a probe wavelength. On the other hand, femtosecond phase spectroscopy, which is complementary to femtosecond absorption spectroscopy, is still in a very primitive stage. This is mainly because there have been no methods by which difference phase spectra (DPS) can be obtained with a multichannel spectrometer as in the case of DTS.

To obtain a phase change on excitation, various time-resolved interferometers have been developed.[54, 75–79] They utilize spatial interference fringes, which require the time coincidence of the reference and the probe pulses from the same light source within the coherence time. To obtain DPS using this interference, one needs a tunable laser or spectral filtering of a continuum. Such a measurement is difficult, however, because of the complicated experimental setup. Our new frequency domain interferometeric method enables us to obtain DPS and DTS simultaneously with a multichannel spectrometer as explained in the following.

Two identical pulses displaced temporally by T are considered, a probe pulse and a reference pulse, which are expressed by

$$E_{pr}(t) = E(t)\exp(i\omega_0 t) \tag{51}$$

$$E_{ref}(t) = E(t - T)\exp[i\omega_0(t - T)] \tag{52}$$

where $E(t)$ is an envelope function, which can be complex. The Fourier transform of the two pulses is given by

$$F[E_{pr}(t) + E_{ref}(t)] = E(\omega - \omega_0)[1 + \exp(-i\omega T)] \tag{53}$$

where $E(\omega) = F[E(t)]$. A grating in a spectrometer Fourier transforms the time-dependent signals, and power spectra of the signal are detected experimentally. The detected signal intensity of (53) is then given by

$$\left| F\left[E_{pr}(t) + E_{ref}(t) \right] \right|^2 = \left| E(\omega - \omega_0) \right|^2 (2 + 2\cos \omega T)$$

which represents the frequency-domain interference with the periodical spacing of $2\pi/T$.

A general expression for the electric field of a probe pulse propagating in a medium is

$$
\begin{aligned}
E'_{pr}(t) = (2\pi)^{-1} \int d\omega \, E(\omega - \omega_0) \\
\times \exp\{i\omega[t - n(\omega)x/c + ik(\omega)x/c]\}
\end{aligned}
\tag{54}
$$

where x is the medium thickness, c is the light velocity, and $n_c(\omega) = n(\omega) - i\kappa(\omega)$ is the complex refractive index of the medium. The Fourier transform of (64) is given as

$$
\begin{aligned}
E'_{pr}(\omega) &= F\left[E'_{pr}(t) \right] \\
&= E(\omega - \omega_0)\exp\{[-in(\omega) - \kappa(\omega)]\omega x/c\}
\end{aligned}
\tag{55}
$$

In the experimental configuration both the reference and the probe pulses are set to be transmitted through the same medium, if the effect of pump pulse disappears after a repetition period, only the probe pulse undergoes a change in the complex refractive index, $\Delta n_c(\omega, \tau) - i\Delta\kappa(\omega, \tau)$, introduced by the pump pulse preceding the probe pulse by τ such that

$$
\begin{aligned}
E'_{pr}(\omega, \tau) &= F\left[E'_{pr}(t, \tau) \right] \\
&= E'_{pr}(\omega)\exp\{[-i\Delta n(\omega, \tau) - \Delta\kappa(\omega, \tau)]\omega x/c\}
\end{aligned}
\tag{56}
$$

The intensities of interference signals without and with excitation are respectively given by

$$\left| E'_{pr}(\omega) + E'_{ref}(\omega) \right|^2 = \left| E'_{pr}(\omega) \right|^2 (2 + 2\cos \omega T) \tag{57}$$

and

$$
\begin{aligned}
\left| E'_{pr}(\omega, \tau) + E_{ref}(\omega) \right|^2 \\
= \left| E'_{pr}(\omega) \right|^2 \{ 1 + \exp[-2\Delta\kappa(\omega, \tau)\omega x/c] \\
+ 2\exp[-\Delta\kappa(\omega, \tau)\omega x/c] \\
\times \cos \omega[T - \Delta n(\omega, \tau)x/c] \}
\end{aligned}
\tag{58}
$$

Comparing (58) with (57), $\Delta n(\omega, \tau)$ and $\Delta \kappa(\omega, \tau)$ can be obtained simultaneously from the peak shifts and fringe amplitude changes, respectively. The phase change is given by $\Delta \Phi(\omega, \tau) = -\Delta n(\omega, \tau)\omega x/c$.

Temporally separated pulses experience broadening in a spectrometer due to the linear dispersion of a grating and interferes with each other. For example, when light is normally incident upon a grating, it is diffracted at an angle θ, and optical path differences are made between different transverse components of the light. The components diffracted from neighboring grating grooves interfere constructively when $d \sin \theta = m\lambda$ is satisfied, where d is the period of the grating grooves, λ is the light wavelength, and m is an integer representing the order of interference. In the case of a short light pulse, the pulse is broadened up to $D_0 \sin \theta/c$, where D_0 is the cross-sectional size of the pulse. If two pulses are displaced by time T, the components from the two pulses transversely separated by D can interfere with each other when $D \sin \theta = cT$. In most experiments the condition $D \gg d$ is satisfied, and this frequency-domain interference causes a shorter period of oscillation as a function of θ than in the interference by $d \sin \theta = m\lambda$.

The experiments were performed as follows. A homemade colliding-pulse mode-locked ring dye laser was amplified by a six-pass amplifier pumped by a copper-vapor laser.[60] The wavelength, duration, energy, and repetition rate of the amplified pulses were 620 nm, 60 fs, 2 μJ, and 10 kHz, respectively. Figure 28 shows the frequency-domain interferometer apparatus, where the time-division technique of Ref. 75 is employed to

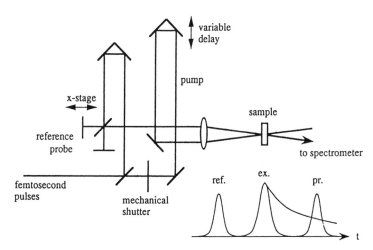

Figure 28. Experimental setup of the frequency-domain interferometer and the time sequence of the pump (ex.), probe (pr.), and reference (ref.) pulses with separation τ and T.

make the separation simpler. The pulse was divided into the pump and the probe pulses. The probe pulse was further divided into two arms of a Michelson interferometer-type optical system for the reference and probe pulses, which were displaced temporally by T adjusting one of the arm lengths. Any optical system can be used to create are optical delay of T, but the Michelson interferometer is one of the simplest to be modified from the ordinary pump–probe experiment configuration.

The setup in Fig. 28 was obtained by adding only two more optical parts, a beamsplitter and a mirror. This is the simplest configuration of all the time-resolved interferometers developed so far and is more stable than a standard two-arm interferometer against several fluctuations because the path difference between the reference and the probe is only the two short arms of the Michelson interferometer.

The reference and probe pulses then propagated through the common path and were focused by a lens onto a sample. Both the transmitted pulses were detected by a spectrometer coupled with a multichannel photodiode. The pump passed through a variable-delay line and was focused onto the sample at the same angle from the reference and probe beams. The pump was blocked at 5 Hz by a mechanical shutter to get signals with and without excitation alternately. Difference spectra were obtained as a function of the delay time τ between the probe and the pump. The reference was not affected by the pump. In the present experiment the displacement T was fixed at 410 fs. The polarizations of all the pulses were parallel. All data were taken at room temperature. By simply blocking the reference beam, the ordinary pump-probe measurement was also performed to obtain DTS.

To demonstrate the validity of the method, we measured a commercially available Toshiba R63 glass filter containing CdS_xSe_{1-x} microcrystallites of a few weight percent, which is expected to have a large nonlinearity and is suited for excitation at 620 nm. Figure 29 shows signals observed by the frequency-domain interferometer with delay time at $\tau = 20$ fs and $T = 410$ fs. The excitation density was approximately 3.8×10^{-3} J/cm^2. The open circles give the DPS. The spectra are calculated from the ith fringe-valley wavelengths with and without excitation λ_i^{ex} and λ_i as $2\pi(\lambda_{i+1}^{ex} - \lambda_i^{ex})/(\lambda_{i+1} - \lambda_i)$.

Figure 30 shows DTS, DPS, and the transmitted probe spectra with and without excitation at several delay times. The DPS shows a positive phase change (negative refractive-index change) on average and changes the sign at the absorption saturation peak, as expected from the Kramers-Kronig relations.[81] There are oscillatory structures in both DTS and DPS at -100 fs. These are known as transient oscillations,[82] observed in DTS, and were observed in DPS for the first time to our knowledge.[80] From -50 to 0 fs,

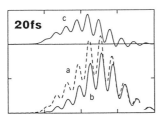

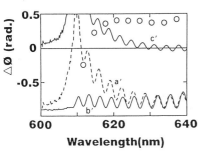

Figure 29. Signals for the R63 filter obtained by the frequency-domain interferometer at $\tau = 20$ fs and $T = 410$ fs. *Upper*: Directly observed interference spectra with excitation (curve a) and without excitation (curve b); the difference spectrum (curve c) is also shown. *Lower*: Curves a', b', and c', respectively. The open circles (DPS) are calculated from the fringe-valley shifts between curves a' and b'.

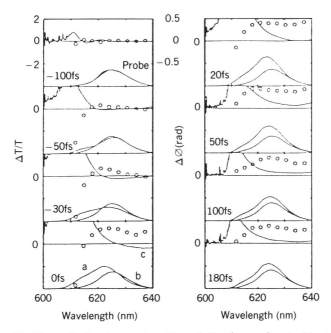

Figure 30. Transmitted probe spectra with excitation (curve a) and without excitation (curve b) and the DTS (curve c) and the DPS (open circles) for the R63 filter.

the transmitted probe spectra are blue-shifted upon excitation. Even though real absorbance change also contributes to the signal, the rapid rise in the DPS from negative to zero delay evidences a major role of induced phase modulation in the spectral shift. The shift of the probe due to induced phase modulation and the corresponding rise in DPS were detected with time resolution simultaneously for the first time to our knowledge.[80] This gives clear evidence for induced phase modulation in absorptive materials, because without DPS data a spectral shift may be assigned to absorption change (absorption increase and decrease) as well as to induced phase modulation, in contrast to the case with transparent materials.

Owing to the simple setup, the frequency-domain interferometer can be readily performed by using a femtosecond white-light continuum to obtain DPS over the whole visible region of the spectrum. It thereby will open a new field, femtosecond phase spectroscopy.

Acknowledgments

The studies described in the present paper was performed in collaboration with T. Hattori, A. Terasaki, E. Tokunaga, K. Minoshima, and M. Taiji. A part of the works was performed at Riken when the author was involved in the Advanced Material Research (A. F. Garito, A. Yamada, and H. Sasabe) of the Frontier Project. He would like to acknowledge these colleagues.

References

1. H. W. Mocker and R. J. Collins, *Appl. Phys. Lett.* **7**, 270 (1965).

2. C. H. Brito-Cruz, R. L. Fork, and C. V. Shank, *Opt. Lett.* **12**, 483 (1987).

3. N. Morita and T. Yajima, *Phys. Rev. A* **30**, 2525 (1984).

4. S. Asaka, H. Nakatsuka, M. Fujiwara, and M. Matsuoka, *Phys. Rev. A* **29**, 2286 (1984).

5. R. Beach and S. R. Hartmann, *Phys. Rev. Lett.* **53**, 663 (1984).

6. H. Nakatsuka, M. Tomita, M. Fujiwara, and S. Asaka, *Opt. Commun.* **52**, 150 (1984).

7. M. Fujiwara, R. Kuroda, and H. Nakatsuka, *J. Opt. Soc. Am. B* **2**, 1634 (1985).

8. S. R. Meech, A. J. Hoff, and D. A. Wiersma, *Chem. Phys. Lett.* **121**, 287 (1985).

9. T. Hattori and T. Kobayashi, *Chem. Phys. Lett.* **133**, 230 (1987).

10. A. M. Weiner, S. De Silvestri, and E. P. Ippen, *J. Opt. Soc. Am.* **B2**, 654 (1985).

11. A. Laubereau, and W. Kaiser, *Rev. Mod. Phys.* **50**, 607 (1978).

12. S. M. George, H. Auwester, and C. B. Harris, *J. Chem. Phys.* **73**, 5573 (1980).

13. S. M. George, A. L. Harris, M. Berg, and C. B. Harris, *J. Chem. Phys.* **80**, 83 (1984).

14. T. Hattori, A. Terasaki, and T. Kobayashi, *Phys. Rev.* **A35**, 715 (1987).

15. M. Tomita and M. Matsuoka, *J. Opt. Soc. Am.* **B3**, 560 (1986).

16. N. Morita, T. Tokizaki, and T. Yajima, *J. Opt. Soc. Am.*, 1269 (1987).

17. K. Kurokawa, T. Hattori, and T. Kobayashi, *Phys. Rev.* **A36**, 1298 (1987).

18. N. Morita, K. Torizuka, and T. Yajima, *J. Opt. Soc. Am.* **B3**, 548 (1986).

19. E. Wolf, *Proc. Phys. Soc.* (London) **71**, 257 (1958).

20. L. Mandel, *Proc. Phys. Soc.* (*London*) **74**, 223 (1959).

21. T. Yajima and N. Morita, in Y. Prior (Ed.), *Methods of Laser Spectroscopy*, Plenum, New York, 1986, p. 75.

22. R. Beach, D. DeBeer, and S. R. Hartmann, *Phys. Rev. A* **32**, 3467 (1985).

23. S. R. Hartmann, *IEEE J. Quantum Electron.* **QE-4**, 802 (1968).

24. R. F. Loring and S. Mukamel, *J. Chem. Phys.* **83**, 2116 (1985).

25. A. C. Eckbreth, *Appl. Phys. Lett.* **32**, 421 (1978).

26. W. Zinth, H.-J. Polland, A. Laubereau, and W. Kaiser, *Appl. Phys. B* **26**, 77 (1981).

27. S. M. George and C. B. Harris, *Phys. Rev. A* **28**, 863 (1983).

28. P. Hu, S. Geschwind, and T. M. Jedju, *Phys. Rev. Lett.* **37**, 1357 (1976).

29. K. P. Leung, T. W. Mossberg, and S. R. Hartmann, *Phys. Rev. A* **25**, 3097 (1982).

30. V. Brückner, E. A. J. M. Bente, J. Langelaar, D. Bebelaar, and J. D. W. van Voorst, *Opt. Commun.* **51**, 49 (1984).

31. J. D. W. van Voorst, D. Brandt, and B. L. van Hensbergen, in *Technical Digest of Topical Meeting on Ultrafast Phenomena*, (1986).

32. J. L. Oudar, *IEEE J. Quantum Electron.* **QE-19**, 713 (1983).

33. T. Hattori and T. Kobayashi, *J. Chem. Phys.* **94**, 3332 (1991).

34. R. W. Hellwarth, *Prog. Quantum Electron.* **5**, 1 (1977).

35. A. P. Alivisatos, A. L. Harris, N. J. Levinos, M. L. Steigerwald, and L. E. Brus, *J. Chem. Phys.* **89**, 4001 (1988).

36. Y.-X. Yan and K. A. Nelson, *J. Chem. Phys.* **87**, 6240, 6257 (1987).

37. J. A. Bucaro and T. A. Litovitz, *J. Chem. Phys.* **54**, 3846 (1971).

38. H. B. Levine and G. Birnbaum, *Phys. Rev. Lett.* **20**, 439 (1968).

39. G. C. Tabisz, W. R. Wall, and D. P. Shelton, *Chem. Phys. Lett.* **15**, 387 (1972).

40. J. F. Dill, T. A. Litovitz, and J. A. Bucaro, *J. Chem. Phys.* **62**, 3839 (1975).

41. W. Danninger and G. Zundel, *Chem. Phys. Lett.* **90**, 69 (1982).

42. J. P. McTague, P. A. Fluery, and D. B. DuPre, *Phys. Rev.* **188**, 303 (1969).

43. T. Keyes and D. Kivelson, *J. Chem. Phys.* **56**, 1057 (1972).

44. J. van der Elsken and R. A. Huijts, *J. Chem. Phys.* **88**, 3007 (1988).

45. S. C. An, L. Fishman, T. A. Litovitz, C. J. Montrose, and H. A. Posch, *J. Chem. Phys.* **70**, 4626 (1979).

46. M. Zoppi, F. Barocchi, D. Varshneya, M. Neumann, and T. JA. Litovitz, *Can. J. Phys.* **59**, 1475 (1981).

47. P. A. Fleury, W. B. Daniels, and J. M. Worlock, *Phys. Rev. Lett.* **27**, 1493 (1971).

48. S.-C. An, C. J. Montrose, and T. A. Litovitz, *J. Chem. Phys.* **64**, 3717 (1976).

49. B. J. Alder, J. J. Weis, and H. L. Strauss, *Phys. Rev. A* **7**, 281 (1973).

50. A. J. C. Ladd, T. A. Litovitz, and C. J. Montrose, *J. Chem. Phys.* **71**, 4242 (1979).

51. B. Guillot, S. Bratos, and G. Birnbaum, *Phys. Rev. A* **22**, 2230 (1980).

52. T. I. Cox and P. A. Madden, *Mol. Phys.* **39**, 1487 (1980).

53. T. Hattori, A. Terasaki, T. Kobayashi, T. Wada, A. Yamada, and H. Sasabe, *J. Chem. Phys.* **95**, 937 (1991).

54. K. Minoshima, M. Taiji, and T. Kobayashi, *Opt. Lett.* **16**, 1683 (1991).

55. R. L. Fork, C. V. Shank, C. Hirlimann, R. Yen, and W. J. Tomlinson, *Opt.. Lett.* **8**, 1 (1983).

56. S. Ruhman, A. G. Joly, and K. A. Nelson, *J. Chem. Phys.* **86**, 6563 (1987).

57. J. Chesnoy and A. Mokhtari, *Phys. Rev. A* **38**, 3566 (1988).

58. G. L. Eesley, M. D. Levenson, and W. M. Tolles, *IEEE J. Quantum Electron.* **QE-14**, 45 (1978).

59. W. H. Knox, M. C. Downer, R. L. Fork, and C. V. Shank, *Opt. Lett.* **9**, 552 (1984).

60. A. Terasaki, M. Hosoda, T. Wada, H. Tada, A. Koma, A. Yamada, H. Sasabe, A. F. Garito, and T. Kobayashi, *J. Phys. Chem.*, **96**, 10534 (1992).

61. J. Etchepare, G. Grillon, G. Hamoniaux, A. Antonetti, and A. Orszag, *Rev. Phys. Appl.* **22**, 1749 (1987).

62. T. Kobayashi, T. Hattori, A. Terasaki, and K. Kurokawa, *Rev. Phys. Appl.* **22**, 1773 (1987).

63. T. Kobayashi, T. Hattori, A. Terasaki, and K. Kurokawa, in R. J. H. Clark and R. E. Hester (Eds.), *Time Resolved Spectroscopy*, Wiley, New York, 1989, pp. 113–156.

64. S. Ruhman, B. Kohler, A. G. Joly, and K. A. Nelson, *Chem. Phys. Lett.* **141**, 16 (1987).

65. G. Herzberg, *Molecular Spectra and Molecular Structure*, van Nostrand, New York, 1945.

66. H. D. Dardy, V. Volterra, and T. A. Litovitz, *J. Chem. Phys.* **59**, 4491 (1973).

67. J. Etchepare, G. Grillon, J. P. Chambaret, G. Hamoniaux, and A. Orszag, *Opt. Commun.* **63**, 329 (1987).

68. D. McMorrow, W. T. Lotshaw, and G. A. Kenney-Wallace, *IEEE J. Quantum Electron.* **QE-24**, 443 (1988).

69. C. Kalpouzos, D. McMorrow, W. T. Lotshaw, and G. A. Kenney-Wallace, *Chem. Phys. Lett.* **150**, 138 (1988); Comment in *Chem. Phys. Lett.* **155**, 240 (1989).

70. D. J. Tildesley and P. A. Madden, *Mol. Phys.* **48**, 129 (1983).

71. P. A. Madden, in D. H. Auston and K. B. Eisenthal (Eds.), *Ultrafast Phenomena IV*, Springer, Berlin, 1985, p. 244.

72. L. C. Geiger and B. M. Ladanyi, *Chem. Phys. Lett.* **159**, 413 (1989).

73. P. A. Madden and T. I. Cox, *Mol. Phys.* **43**, 287 (1981).

74. G. R. Olbright and N. Peyghambarian, *Appl. Phys. Lett.* **48**, 1184 (1986).

75. M. J. LaGasse, K. K. Anderson, H. A. Haus, and J. G. Fujimoto, *Appl. Phys. Lett.* **54**, 2068 (1989).

76. D. Cotter, C. N. Ironside, B. J. Ainslie, and H. P. Girdlestone, *Opt. Lett.* **14**, 317 (1989).

77. J.-M. Halbout and C. L. Tang, *Appl. Phys. Lett.* **40**, 765 (1982).

78. M. J. LaGasse, D. Liu-Wong, J. G. Fujimoto, and H. A. Haus, *Opt. Lett.* **14**, 311 (1989).

79. R. K. Jain and R. C. Lind, *J. Opt. Soc. Am.* **73**, 647 (1983).

80. E. Tokunaga, A. Terasaki, and T. Kobayashi, *Opt. Lett.* **17**, (1992).

81. See, for example, A. Yariv, *Quantum Electronics*, 3d ed. Wiley, New York, 1988.

82. C. H. Brito-Cruz, J. P. Gordon, P. C. Becker, R. L. Fork, and C. V. Shank, *IEEE J. Quantum Electron.* **24**, 261 (1988).

THEORY AND MEASUREMENT OF RAMAN OPTICAL ACTIVITY

LAURENCE A. NAFIE AND DIPING CHE

*Department of Chemistry and Center for Molecular Electronics,
Syracuse University, Syracuse, New York*

CONTENTS

Modern Nonlinear Optics, Part 3, Edited by Myron Evans and Stanisław Kielich. Advances in Chemical Physics Series, Vol. LXXXV.
ISBN 0-471-30499-9 © 1994 John Wiley & Sons, Inc.

I. INTRODUCTION

With the past decade, both the theory and measurement of natural Raman optical activity (ROA)[1] have been extended in ways not previously imagined. Some of these advances have been conceptual while others have been technical in nature. The result is that ROA has emerged as a technique with remarkably enhanced potential for stereochemical studies. The purpose of this Chapter is to review these recent advances from the perspective of theoretical formalism. Advances in instrumental aspects of ROA will be described from the perspective of how elements of the theory can be isolated and measured. Reviews of recent progress in ROA from a broader viewpoint, including applications to organic and biological molecules, have appeared elsewhere.[2-8]

Raman optical activity is a form of natural optical activity. ROA spectra can be measured in samples that possess a net chirality at the level of individual molecules. ROA is closely related to its companion effect, vibrational circular dichroism (VCD),[9-11] and these two spectroscopies comprise the wider spectroscopic area known vibrational optical activity (VOA). Although VCD is the direct extension of electronic circular dichroism into the infrared vibrational region of the spectrum, ROA is not the analog of any previously known form of natural optical activity. Both VCD and vibrational ROA involve the difference in the vibrational spectral intensity of a molecule with respect to the use of left versus right circularly polarized radiation.

The theoretical basis for ROA emerged from a study of the polarization dependence of interference scattering between the ordinary polarizability tensor and what are now referred to as the magnetic dipole and electric quadrupole optical activity tensors.[12] The original theory of ROA, developed two years later, was formulated as the difference in Raman scattering intensity for right minus left circularly polarized *incident* radiation.[13] For reasons that will be clear below, we now refer to this form of ROA as *incident* circular polarization (ICP) ROA. The original theory is based on the far-from-resonance approximation in which the incident and scattered light frequencies are regarded as equally distant from any electronic excited-state resonances in the scattering molecule. This frequency equivalence results in many simplifications of the theory, simplifications that have subsequently been removed to reveal a rich and complex texture upon which the more general theory of ROA rests. The evolution of new aspects of the general theory of ROA is the principal focus of this review.

II. HISTORICAL BACKGROUND

In this section, we review briefly important historical developments related to the theory and measurement of Raman optical activity. The motivation is to provide the reader with a sense of perspective for the material in the sections that follow. The review is not meant to be exhaustive, but rather selective of key conceptual advances that have shaped our present understanding of this rapidly progressing field.

A. Early Developments

The experimental observation of genuine ROA was first reported in 1973.[14] The measurement was carried out with a right-angle scattering geometry using the ICP-ROA approach. For the remainder of the decade, this form of ROA was nearly the only one practiced or described theoretically. Further restrictions for most of these measurements were that they corresponded to the depolarized form of Raman scattering in which only the component of the scattered light polarized parallel to the scattering plane was measured and that only Stokes scattering transitions were measured. In 1976, anti-Stokes ROA was reported[15] and shown to confirm the far-from-resonance theoretical prediction that, aside from the Boltzmann population difference, there is exact symmetry between the Stokes and anti-Stokes forms of ROA. No further experimental attention has since been given to anti-Stokes ROA. In 1979, polarized ICP ROA was successfully measured[16]; however, this form of ROA remains much more difficult to measure, without interfering artifacts, than the depolarized form.

In 1980, a paper by Andrews described theoretically the dependence of polarized and depolarized ICP ROA on scattering angle, again in the far-from-resonance approximation.[17] He demonstrated that all three ROA tensor invariants could be isolated by procedures involving measurement of the angular dependence of polarized and depolarized ROA. Hug advanced this line of investigation further by constructing an ROA spectrometer capable of backscattering measurements, and demonstrating that, together with right-angle scattering, three ROA tensor invariants could be isolated.[18] Unfortunately, his spectrometer was destroyed by fire before he could carry out backscattering ROA measurements.

B. New Forms of ROA

In 1982, Barron published a book on optical activity and light scattering that contained, among other things, elegant formalism related to the theoretical foundations of ROA and several significant extensions of the far-from-resonance theory of ROA.[1] The most important of these, for the purposes of this review, was the definition of the ROA degree of circular polarization. In essence, this is a second form of ROA in which one measures the degree to which right or left circularly polarized light predominates in the scattering beam. Barron,[1] and previously Barron and Buckingham,[19] predicted that the spectrum of the degree of circular polarization would be the same as the conventional (ICP) form of ROA

(in the far-from-resonance approximation). In his book, Barron also presented theoretical expressions for Raman and ROA in forward scattering
($0°$) and backscattering ($180°$), as well as the conventional polarized and
depolarized right-angle scattering geometry.

In 1985, Barron and Escribano published a generalization of previous
formulations of ROA[20] in which they dropped certain features of the
far-from-resonance approximation. The optical activity tensors they employed distinguished the differing frequencies of the incident and scattered photons, although complete provision for strong vibronic resonance
was not included due to the absence of imaginary damping factors in the
resonance energy denominators. Nevertheless, a number of key features of
general ROA theory emerged for the first time from this groundbreaking
work. First, they identified two forms of magnetic dipole optical activity
tensors (distinguished by roman and script typefaces) and correspondingly
two forms of electric quadrupole optical activity tensors. Using the complete complex forms of these tensors, they recognized that the conventional (normalized) circular intensity difference, CID,

$$\Delta\alpha = \frac{I_\alpha^R - I_\alpha^L}{I_\alpha^R + I_\alpha^L} \tag{1}$$

differed from the degree of circular polarization $P_c(\alpha)$, defined as the
intensity of the circular polarized component for the scattered light S_3
divided by the total scattered intensity S_0 for a fixed incident polarization
state α,

$$P_c(\alpha) = \frac{S_3}{S_0} \tag{2}$$

Further, they found that the symmetry between Stokes and anti-Stokes
ROA did not hold in this more general formulation. They also found that,
to a very good approximation, Δ_α for Stokes scattering equaled $P_c(\alpha)$ for
anti-Stokes scattering, and that $P_c(\alpha)$ for Stokes scattering corresponds to
Δ_α for anti-Stokes scattering. In the original far-from-resonance theory[1]
all four of these quantities are equal to one another.

Several years later, Nafie and coworkers succeeded in measuring the
$P_c(\alpha)$ form of ROA, which they referred to as *scattered* circular polarization (SCP) ROA.[21] They recognized that $P_c(\alpha)$ could be measured in a
manner directly analogous to that of Δ_α, namely by a difference measurement between the intensity of scattered light with right versus left circular

polarization. The normalized CID for SCP ROA can then be written as

$$\Delta^\alpha = \frac{I_R^\alpha - I_L^\alpha}{I_R^\alpha + I_L^\alpha} \tag{3}$$

where the superscripts correspond to the polarization state of the incident light and the subscripts correspond to those of the scattered light. This work confirmed, to within experimental uncertainty, the far-from-resonance prediction[1] that the ICP and SCP ($P_c(\alpha)$) forms of ROA are equal to one another. The key experimental innovation that made SCP ROA measurements possible was the use of a zeroth-order quarter waveplate to distinguish the left and right circular polarization components of the scattered light across a wide range of optical frequencies. For reason of angular aperture acceptance, such measurements are not possible with an electro-optical modulator, the previous standard for the control of (incident) polarization states in ROA measurements.

With the success in the measurement of SCP ROA came the realization that simultaneous, synchronous circular polarization modulation of the incident and the scattered light beams would lead to new and possibly improved approaches to the measurement of ROA.[22] Two new forms of ROA were identified and referred to as in-phase dual circular polarization (DCP$_I$) ROA, with a normalized CID defined as

$$\Delta_I = \frac{I_R^R - I_L^L}{I_R^R + I_L^L} \tag{4}$$

and out-of-phase dual circular polarization (DCP$_{II}$) ROA given by

$$\Delta_{II} = \frac{I_L^R - I_R^L}{I_L^R + I_R^L} \tag{5}$$

In the far-from-resonance approximation DCP$_{II}$ ROA is predicted to vanish. For right-angle scattering DCP$_I$ and DCP$_{II}$ ROA correspond to half the sum and difference, respectively, of unpolarized ICP and SCP ROA in single measurements. For forward and backward scattering, in the far-from-resonance approximation, DCP$_I$ ROA spectra are equal to unpolarized ICP ROA spectra, although fewer total Raman photons (depolarized versus polarized) are required for the measurement of DCP$_I$ ROA.

C. Recent Advances

In 1989, a period of rapid experimental advance for ROA was initiated by the use of backscattering as a more efficient approach to ROA measurements[23] and shortly thereafter by the availability of charge coupled detectors (CCDs).[24, 25] Two designs for backscattering ROA measurements have emerged. In one developed at the University of Glasgow, a collection mirror, with a hole to pass the incident laser radiation to the sample, is used to reflect the backscattered light at a right angle to the spectrograph. This design distorts the polarization of the scattered light and requires a Lyot depolarizer to eliminate large polarization artifacts induced by the collection mirror.[8, 23, 24, 26] In the second design, initially proposed by Hecht while at Syracuse University, a small prism acting as a mirror reflects the incident laser radiation at right angles to the sample.[5-7, 27] The backscattered light then passes by the prism to the polarization analysis and collection optics. The advantage of this approach is the preservation of the polarization of the scattered light which permits SCP, DCP_I, or DCP_{II} ROA measurements in the backscattering configuration. In fact, the first report of the Syracuse backscattering ROA spectrometer included the first example of a DCP_I ROA spectrum[27] and its comparison to a corresponding unpolarized ICP backscattering ROA spectrum.

A number of studies have been carried out with multichannel ROA spectrometers to ascertain the mechanistic origin (magnetic dipole or electric quadrupole) of observed intensities. Four of these involve direct comparisons of right-angle polarized and depolarized ICP ROA.[28-31] Within the bond polarizability model of ROA, and under the simplifying assumption of axially symmetric bond polarizabilities, one of three ROA invariants vanishes, the isotropic magnetic dipole invariant, and the other two, the magnetic dipole and the electric quadrupole anisotropic invariants, equal one another. These studies compare experimental ROA spectra with the theoretical prediction that the polarized ROA spectrum should be just twice the depolarized ROA spectrum if these theoretical assumptions prevail. Further insight was gained with right-angle ICP ROA with the analyzing polarizer set at the so-called magic angle of 35.26° (Ref. 32). With this analyzer setting, all of the electric quadrupole contributions vanish from the ROA spectrum. Finally, the first measurements of ROA in the forward scattering direction were carried out.[33] In this setup, ICP ROA consists predominantly of isotropic magnetic dipole contributions, allowing further insight into scattering mechanisms.

More recently, a scheme has been devised for unambiguously isolating the three far-from-resonance ROA invariants identified above through a series of four experimental Raman and ROA measurements: right-angle polarized and depolarized ICP, and backscattering ICP and DCP_I.[34] The practicality of this scheme was then demonstrated experimentally by isolating, unequivocally for the first time, the anisotropic magnetic dipole and electric quadrupole invariant spectra for two molecules, *trans*-pinane and α-pinene. The approach was extended to consider wider apertures for the collection optics,[35] although sufficiently small apertures were maintained in the original experiments to keep such corrections to a few percent or less, well below the level of uncertainty due to noise. A slightly improved variation on the general scheme for ROA invariant isolation is presented below.

Several sets of experiments have been carried out that test the validity of the far-from-resonance approximation. Two of these involve comparing ICP and SCP ROA using a right-angle scattering geometry.[36, 37] In the third, ICP and DCP_I backscattering ROA are compared.[27] In all cases these pairs of ROA spectra are predicted to give the same ROA spectrum in the far-from-resonance approximation. In the most recent case,[37] ROA difference spectra are presented for the first time and compared with the pure noise spectrum associated with the difference spectrum. In this way, very sensitive measures of the breakdown of the far-from-resonance approximations were obtained.

We conclude this section with a discussion of two formal extensions of the theory of ROA. In the first, a complete rederivation of the theory of natural circular polarization (CP) ROA was undertaken several years ago.[38] The rationale for this work was to provide a coherent, general formalism (including the strong resonance case) for all the new forms of ROA that had been developed and implemented since the last formal version of the theory, namely, that of Barron and Escribano in 1985.[20] In particular, general expressions for ICP, SCP, DCP_I, and DCP_{II} ROA for arbitrary scattering angles were needed. A new derivation of the ROA equations was provided that featured explicit dependence on the photon polarization and propagation vectors of the experiment. The resulting expressions can easily be adapted to any experimental scattering configuration. It was found that there are ten CP ROA invariants; however, they appear in the theory in only six linearly independent combinations. As a result, not all ten invariants can be isolated.

The other theoretical development involves a new form of ROA in which the Raman intensity difference is developed between two orthogonal linear polarized, rather than circularly polarized, states.[39-41] This form of ROA is referred to as linearly polarized (LP) ROA, and as with the CP forms, there are ILP, SLP, DLP_I, and DLP_{II} modulation schemes. In the far-from-resonance approximation, all forms of LP ROA vanish. As a new form of natural optical activity, LP ROA became apparent only as the most general levels of the ROA formalism came under close scrutiny,[38] although its existence in principle had been recognized in an earlier theoretical study on the polarization properties of scattered light.[40]

In the following section, a complete, unified formalism for both CP and LP ROA will be presented. A version of this unified formalism including polarization and propagation vectors has already appeared in the Ph.D.

thesis of one of the authors.[42] Our aim below is to present a complete statement of the theory in a form that is both compact and efficient for use and application. We have chosen the Stokes-Mueller formalism as a convenient vehicle for this objective.[39, 42–44]

III. INTENSITIES AND SCATTERING TENSORS

In this section we provide theoretical expressions that are needed for the description of the various forms of ROA. For the most part we are concerned with defining the polarizability and optical activity tensors that are used in the subsequent sections. The starting point is the Raman intensity expression for an ensemble of molecules whose reference frame is averaged relative to that of the laboratory frame. The polarization vectors for the incident and scattered light, denoted by $\mathbf{e}^i$ and $\mathbf{e}^d$, respectively, are described in the laboratory frame. We have

$$I(\tilde{e}^d, \tilde{e}^i) = 90K \left\langle \left| \tilde{e}^{d*}_\alpha \tilde{a}_{\alpha\beta} \tilde{e}^i_\beta \right|^2 \right\rangle \tag{6}$$

Here, Greek subscripts refer to Cartesian tensor or vector components; repeated Greek subscripts for each term are to be summed over all three Cartesian directions; the angular brackets represent orientation averaging of the molecular frame relative to the laboratory frame; and K is a constant given by

$$K = \frac{1}{90} \left(\frac{\omega^2 \mu_0 \tilde{E}^{(0)}}{4\pi R} \right)^2 \tag{7}$$

where ω is the angular frequency of the scattered light, μ_0 is the magnetic permeability, $\tilde{E}^{(0)}$ is the electric field strength of the incident radiation, and R is the distance from the point of scattering to the detector. The complex conjugate is taken for the polarization vector of the scattered (detected) photon in Eq. (6) because it represents emission and not absorption.[38, 45] The complex, general Raman scattering tensor $\tilde{a}_{\alpha\beta}$ consists of the ordinary Raman polarizability at the lowest order and the Raman optical activity tensors at the next order of the interaction of radiation with matter. Given that a tilde over a symbol indicates a complex quantity, that ω_0 is the angular frequency of the incident radiation, and that $\mathbf{n}^i$ and $\mathbf{n}^d$ are the propagation vectors of the incident and scattered photons, respectively, the general Raman scattering tensor is given by[1, 20, 38]

$$\tilde{a}_{\alpha\beta} = \tilde{\alpha}_{\alpha\beta} + \frac{1}{c}\left[\varepsilon_{\gamma\delta\beta}n_{\delta}^{i}\tilde{G}_{\alpha\gamma} + \varepsilon_{\gamma\delta\alpha}n_{\delta}^{d}\tilde{\mathscr{G}}_{\gamma\beta} + \frac{1}{3}\left(\omega_{0}n_{\gamma}^{i}\tilde{A}_{\alpha,\gamma\beta} - \omega n_{\gamma}^{d}\tilde{\mathscr{A}}_{\beta,\gamma\alpha}\right)\right]$$

$$(8)$$

where $\varepsilon_{\gamma\delta\alpha}$ is the alternating tensor equal to $+1$ or -1 depending on whether the Cartesian components are either even or odd permutations, respectively, of xyz. Previously published expressions for the general scattering tensor[1, 20, 38] have not distinguished the roles of the incident and scattered photon frequencies, and we do so here to provide greater accuracy of ROA expressions in the case of resonance and preresonance Raman scattering. The Raman tensors that appear in Eq. (8) are, in the order of their appearance, the electric dipole polarizability, the electric dipole–magnetic dipole optical activity tensor, the magnetic dipole–electric dipole optical activity tensor, the electric dipole–electric quadrupole optical activity tensor, and the electric quadrupole–electric dipole optical activity tensor. They are given by

$$\tilde{\alpha}_{\alpha\beta} = \frac{1}{\hbar}\sum_{j\neq m,n}\left[\frac{\langle m|\hat{\mu}_{\alpha}|j\rangle\langle j|\hat{\mu}_{\beta}|n\rangle}{\omega_{jn} - \omega_{0} + i\Gamma_{J}} + \frac{\langle m|\hat{\mu}_{\beta}|j\rangle\langle j|\hat{\mu}_{\alpha}|n\rangle}{\omega_{jm} + \omega_{0} + i\Gamma_{J}}\right] \quad (9a)$$

$$\tilde{G}_{\alpha\beta} = \frac{1}{\hbar}\sum_{j\neq m,n}\left[\frac{\langle m|\hat{\mu}_{\alpha}|j\rangle\langle j|\hat{m}_{\beta}|n\rangle}{\omega_{jn} - \omega_{0} + i\Gamma_{j}} + \frac{\langle m|\hat{m}_{\beta}|j\rangle\langle j|\hat{\mu}_{\alpha}|n\rangle}{\omega_{jm} + \omega_{0} + i\Gamma_{j}}\right] \quad (9b)$$

$$\tilde{\mathscr{G}}_{\alpha\beta} = \frac{1}{\hbar}\sum_{j\neq m,n}\left[\frac{\langle m|\hat{m}_{\alpha}|j\rangle\langle j|\hat{\mu}_{\beta}|n\rangle}{\omega_{jn} - \omega_{0} + i\Gamma_{j}} + \frac{\langle m|\hat{\mu}_{\beta}|j\rangle\langle j|\hat{m}_{\alpha}|n\rangle}{\omega_{jm} + \omega_{0} + i\Gamma_{j}}\right] \quad (9c)$$

$$\tilde{A}_{\alpha,\beta\gamma} = \frac{1}{\hbar}\sum_{j\neq m,n}\left[\frac{\langle m|\hat{\mu}_{\alpha}|j\rangle\langle j|\hat{\Theta}_{\beta\gamma}|n\rangle}{\omega_{jn} - \omega_{0} + i\Gamma_{j}} + \frac{\langle m|\hat{\Theta}_{\beta\gamma}|j\rangle\langle j|\hat{\mu}_{\alpha}|n\rangle}{\omega_{jm} + \omega_{0} + i\Gamma_{j}}\right] \quad (9d)$$

$$\tilde{\mathscr{A}}_{\alpha,\beta\gamma} = \frac{1}{\hbar}\sum_{j\neq m,n}\left[\frac{\langle m|\hat{\Theta}_{\beta\gamma}|j\rangle\langle j|\hat{\mu}_{\alpha}|n\rangle}{\omega_{jn} - \omega_{0} + i\Gamma_{j}} + \frac{\langle m|\hat{\mu}_{\alpha}|j\rangle\langle j|\hat{\Theta}_{\beta\gamma}|n\rangle}{\omega_{jm} + \omega_{0} + i\Gamma_{j}}\right] \quad (9e)$$

where $\omega_{jn} = \omega_{j} - \omega_{n}$ and $\omega_{jm} = \omega_{j} - \omega_{m}$ are the angular transition frequencies between states j and either n or m, respectively, and Γ_{j} is the bandwidth of electronic state j, which is inversely proportional to its excited state lifetime. The initial state is n and the final state is m, with $\omega_{mn} = \omega_{m} - \omega_{n}$ representing the Raman angular frequency shift. The

electric dipole, magnetic dipole, and electric quadrupole moment operators are given in that order by

$$\hat{\mu}_\alpha = \sum_k e_k r_{k\alpha} \tag{10a}$$

$$\hat{m}_\alpha = \frac{1}{2} \sum_k \frac{e_k}{m_k} \varepsilon_{\alpha\beta\gamma} r_{k\beta} p_{k\gamma} \tag{10b}$$

$$\hat{\Theta}_{\alpha\beta} = \frac{1}{2} \sum_k e_k \left(3 r_{k\alpha} r_{k\beta} - r_k^2 \delta_{\alpha\beta} \right) \tag{10c}$$

where $\delta_{\alpha\beta}$ is the Kronecker delta symbol and the sum is over all the
particles in the molecule that have charge e_k, mass m_k, position $\mathbf{r}_k$, and
momentum $\mathbf{p}_k$. If Cartesian components in the laboratory frame are
indicated by capital Roman letters and those in the molecular frame by
Greek letters, the expression for the averaging over all orientations of
molecules in the molecular frame is given by

$$
\begin{aligned}
I(\tilde{e}^{d*}, e^i) = 90K \Bigg\{ &\mathrm{Re}\left[\tilde{e}_A^{d*} \tilde{e}_B^i \tilde{e}_C^d \tilde{e}_D^{i*} \tilde{\alpha}_{\alpha\beta} \tilde{\alpha}_{\gamma\delta}^* \right] \langle l_{A\alpha} l_{B\beta} l_{C\gamma} l_{D\delta} \rangle \\
&+ \frac{2}{c} \mathrm{Im}\Bigg\{ \tilde{e}_A^{d*} \tilde{e}_B^i \tilde{e}_C^d \tilde{e}_E^{i*} \tilde{\alpha}_{\alpha\beta} \Bigg[n_D^i \left(i\varepsilon_{\rho\delta\varepsilon} \tilde{G}_{\gamma\rho}^* + \frac{1}{3} \omega_0 \tilde{A}_{\gamma,\delta\varepsilon}^* \right) \\
&+ n_D^d \left(i\varepsilon_{\rho\delta\gamma} \tilde{\mathscr{G}}_{\rho\varepsilon}^* - \frac{1}{3} \omega \tilde{\mathscr{A}}_{\varepsilon,\gamma\delta}^* \right) \Bigg] \Bigg\} \langle l_{A\alpha} l_{B\beta} l_{C\gamma} l_{D\delta} l_{E\varepsilon} \rangle \Bigg\}
\end{aligned}
\tag{11}
$$

where the symbol $l_{A\alpha}$ is the direction cosine between the laboratory
direction A and molecular direction α, and Im indicates the imaginary
part of the complex expression.

IV.   COMPLETE ROA THEORY

In this section we present a comprehensive set of theoretical expressions
for the intensities of CP and LP ROA. From these expressions it is
straightforward to determine the appropriate ROA intensity expression
for any type of experiment that is presently foreseeable. The complete
theory of ROA can be cast in a variety of forms. Some choices are (1) a
large set of equations of representative intensities,[1, 20] (2) a more general
set of expressions into which the appropriate propagation and polarization

vectors must be inserted,[38, 39] and (3) a compact Mueller matrix expression for which the appropriate Stokes vectors must be supplied.[39] Further, the intensities themselves can be expressed directly in Cartesian tensor components,[20] as standard tensor products,[22, 38] or as tensor invariants.[38, 39] For the sake of simplicity and the isolation of key conceptual elements, we present the theory in terms of a Raman/ROA Mueller matrix using tensor invariants to express the intensity.

A.   General Expressions for Raman and ROA Intensities

We begin with expression of the Raman intensity I associated with incident and scattered (detected) polarization states specified by superscript i and subscript d, respectively. For elliptically (or linearly) polarized radiation, the angle of the major polarization axis relative to the vertical direction (0°) is represented by θ for the incident light and ϕ for the scattered light, and the scattering angle is ξ. A diagram illustrating these definitions is given in Fig. 1. We have then for the scattered intensity,

$$I_d^i(\theta, \phi, \xi) = \frac{K}{2}\left[(\mathbf{S}^d)^t \cdot \mathbf{M} \cdot \mathbf{S}^i\right] = \frac{K}{2} \sum_{I, J = 0}^{3} S_I^d M_{IJ} S_J^i. \quad (12)$$

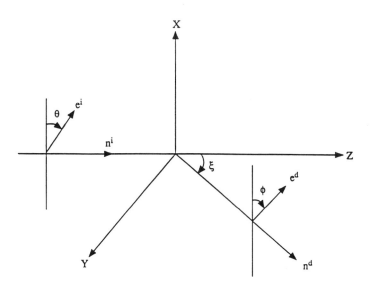

Figure 1.   Diagram of the scattering geometry where ξ is the scattering angle from the forward direction, and θ and ϕ are the angles of the planes of polarization (major polarization ellipse axes) of the incident and scattered light beams, respectively.

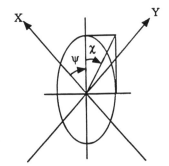

Figure 2. Diagram illustrating the polarization ellipse for a light beam traveling the positive Z direction. The angle ψ is the azimuthal angle and χ is the ellipticity angle.

The Raman/ROA Mueller matrix is represented by $\mathbf{M}$ and the Stokes vectors are represented by $\mathbf{S}$, where the transpose of the Stokes vector for the detected light beam is required for the matrix multiplication. The equation is continued with an explicit expression for the matrix multiplication of the (4×4) Mueller matrix and the (1×4) Stokes vectors. The four components of the Stokes vector are given by

$$
\mathbf{S} = \begin{pmatrix} S_0 \\ S_1 \\ S_2 \\ S_3 \end{pmatrix} = \begin{pmatrix} 1 \\ P \cos 2\chi \cos 2\psi \\ P \cos 2\chi \sin 2\psi \\ P \sin 2\chi \sin 2\psi \end{pmatrix}
\tag{13}
$$

where S_0, the total intensity of the light beam, is unity, P is the degree of polarization that may vary from 0 to 1, and ψ and χ are the angles associated with the azimuthal and ellipticity of the polarization ellipse, as illustrated in Fig. 2. The role of S_0 is to represent the total intensity. S_1 is the difference in the intensity of vertically minus horizontally polarized intensity components of the beam, S_2 is the corresponding difference for light polarized along the $+45°$ axis minus that along the $-45°$ axis, and S_3 is the difference of right minus left circularly polarized components. Using these definitions we identify the Stokes vector (transpose, for convenience) for six key polarization states that are associated with the last three Stokes vector components:

$$S_X = (1 \quad 1 \quad 0 \quad 0) \qquad S_Z = (1 \quad -1 \quad 0 \quad 0) \tag{14a}$$

$$S_+ = (1 \quad 0 \quad 1 \quad 0) \qquad S_- = (1 \quad 0 \quad -1 \quad 0) \tag{14b}$$

$$S_R = (1 \quad 0 \quad 0 \quad 1) \qquad S_L = (1 \quad 0 \quad 0 \quad -1) \tag{14c}$$

These will be useful in specifying the intensities associated with certain polarization sum and difference spectra to be identified below. The elements of the Raman/ROA Mueller matrix of Eq. (12) are given more explicitly below as

$$\mathbf{M} = \begin{pmatrix} M_{00} & M_{01} & M_{02} & M_{03} \\ M_{10} & M_{11} & M_{12} & M_{13} \\ M_{20} & M_{21} & M_{22} & M_{23} \\ M_{30} & M_{31} & M_{32} & M_{33} \end{pmatrix} \tag{15}$$

where the individual elements are given in terms of scattering angle and combinations of invariants by

$$M_{00} = \left(2R_1 - R_3 \sin^2 \xi \right) \tag{16a}$$

$$M_{01} = M_{10} = R_3 \sin^2 \xi \tag{16b}$$

$$M_{02} = \frac{2}{c} I_4 \sin^2 \xi \tag{16c}$$

$$M_{03} = \frac{2}{c} \left(2I_1 + 2\mathscr{I}_2 \cos \xi - I_3 \sin^2 \xi \right) \tag{16d}$$

$$M_{11} = R_3 \left(1 + \cos^2 \xi \right) \tag{16e}$$

$$M_{12} = \frac{2}{c} \left(I_4 + I_4 \cos^2 \xi - 2\mathscr{I}_4 \cos \xi \right) \tag{16f}$$

$$M_{13} = \frac{2}{c} I_3 \sin^2 \xi \tag{16g}$$

$$M_{20} = \frac{2}{c} \mathscr{I}_4 \sin^2 \xi \tag{16h}$$

$$M_{21} = \frac{2}{c} \left(\mathscr{I}_4 + \mathscr{I}_4 \cos^2 \xi - 2I_4 \cos \xi \right) \tag{16i}$$

$$M_{22} = 2R_3 \cos \xi \tag{16j}$$

$$M_{23} = M_{32} = 0 \tag{16k}$$

$$M_{30} = \frac{2}{c} \left(2\mathscr{I}_1 + 2I_2 \cos \xi - \mathscr{I}_3 \sin^2 \xi \right) \tag{16l}$$

$$M_{31} = \frac{2}{c} \mathscr{I}_3 \sin^2 \xi \tag{16m}$$

$$M_{33} = 2R_2 \cos \xi \tag{16n}$$

From these expressions, it is clear that the diagonal elements, as well as elements M_{10} and M_{01}, which equal one another, are associated with ordinary Raman intensity, while the remaining off-diagonal elements, except for M_{23} and M_{32} which vanish, are associated with ROA intensity. It can also be noticed that the ROA off-diagonal pairs are related to one another by exchange of roman and script invariant combinations. The Raman invariant combinations are given by

$$R_1 = 45\alpha^2 + 7\beta_s(\tilde{\alpha})^2 + 5\beta_a(\tilde{\alpha})^2 \tag{17a}$$

$$R_2 = 45\alpha^2 - 5\beta_s(\tilde{\alpha})^2 + 5\beta_a(\tilde{\alpha})^2 \tag{17b}$$

$$R_3 = 45\alpha^2 + \beta_s(\tilde{\alpha})^2 - 5\beta_a(\tilde{\alpha})^2 \tag{17c}$$

where the tensor invariants are the standard expressions encountered in the theory of Raman scattering[1] and are given by

$$\alpha^2 = \tfrac{1}{9}\operatorname{Re}\left[(\tilde{\alpha}_{\alpha\alpha})^s(\tilde{\alpha}_{\beta\beta})^{s*}\right] \tag{18a}$$

$$\beta_s(\tilde{\alpha})^2 = \tfrac{1}{2}\operatorname{Re}\left[3(\tilde{\alpha}_{\alpha\beta})^s(\tilde{\alpha}_{\alpha\beta})^{s*} - (\tilde{\alpha}_{\alpha\alpha})^s(\tilde{\alpha}_{\beta\beta})^{s*}\right] \tag{18b}$$

$$\beta_a(\tilde{\alpha})^2 = \tfrac{3}{2}\operatorname{Re}\left[(\tilde{\alpha}_{\alpha\beta})^a(\tilde{\alpha}_{\alpha\beta})^{a*}\right] \tag{18c}$$

and where the symmetric and antisymmetric parts of complex tensor components are obtained using the general relations

$$\left(\tilde{T}_{\alpha\beta}\right)^s = \tfrac{1}{2}\left[\tilde{T}_{\alpha\beta} + \tilde{T}_{\beta\alpha}\right] \tag{19a}$$

$$\left(\tilde{T}_{\alpha\beta}\right)^a = \tfrac{1}{2}\left[\tilde{T}_{\alpha\beta} - \tilde{T}_{\beta\alpha}\right] \tag{19b}$$

which hold for both Raman and ROA tensors. The sense in which the quantities defined in Eq. (18) are invariant, is with respect to the choice of the molecular Cartesian coordinate frame. These invariants, as well as those defined below for ROA, maintain the same values as the molecule is rotated with respect to the Cartesian frame, or vice versa.

The corresponding tensor invariant combinations for ROA are given by[38, 39, 42]

$$I_1 = 45\alpha G + 7\beta_s(\tilde{G})^2 + 5\beta_a(\tilde{G})^2 + \beta_s(\tilde{A})^2 - \beta_a(\tilde{A})^2 \tag{20a}$$

$$I_2 = 45\alpha G - 5\beta_s(\tilde{G})^2 + 5\beta_a(\tilde{G})^2 - 3\beta_s(\tilde{A})^2 - \beta_a(\tilde{A})^2 \tag{20b}$$

$$I_3 = 45\alpha G + \beta_s(\tilde{G})^2 - 5\beta_a(\tilde{G})^2 + 3\beta_s(\tilde{A})^2 - 3\beta_a(\tilde{A})^2 \tag{20c}$$

$$I_4 = 45(\alpha G)' + \beta_s'(\tilde{G})^2 - 5\beta_a'(\tilde{G})^2 - \beta_s'(\tilde{A})^2 + \beta_a'(\tilde{A})^2 \tag{20d}$$

and

$$\mathscr{I}_1 = -45\alpha\mathscr{G} - 7\beta_s(\tilde{\mathscr{G}})^2 - 5\beta_a(\tilde{\mathscr{G}})^2 + \beta_s(\tilde{\mathscr{A}})^2 + \beta_a(\tilde{\mathscr{A}})^2 \quad (21a)$$

$$\mathscr{I}_2 = -45\alpha\mathscr{G} + 5\beta_s(\tilde{\mathscr{G}})^2 - 5\beta_a(\tilde{\mathscr{G}})^2 - 3\beta_s(\tilde{\mathscr{A}})^2 + \beta_a(\tilde{\mathscr{A}})^2 \quad (21b)$$

$$\mathscr{I}_3 = -45\alpha\mathscr{G} - \beta_s(\tilde{\mathscr{G}})^2 + 5\beta_a(\tilde{\mathscr{G}})^2 + 3\beta_s(\tilde{\mathscr{A}})^2 + 3\beta_a(\tilde{\mathscr{A}})^2 \quad (21c)$$

$$\mathscr{I}_4 = 45(\alpha\mathscr{G})' + \beta_s'(\tilde{\mathscr{G}})^2 - 5\beta_a'(\tilde{\mathscr{G}})^2 + \beta_s'(\tilde{\mathscr{A}})^2 + \beta_a'(\tilde{\mathscr{A}})^2 \quad (21d)$$

The roman CP ROA tensor invariants appearing in these expressions are given by[38]

$$\alpha G = \tfrac{1}{9} \operatorname{Im}\left[(\tilde{\alpha}_{\alpha\alpha})^s(\tilde{G}_{\beta\beta})^{s*}\right] \quad (22a)$$

$$\beta_s(\tilde{G})^2 = \tfrac{1}{2} \operatorname{Im}\left[3(\tilde{\alpha}_{\alpha\beta})^s(\tilde{G}_{\alpha\beta})^{s*} - (\tilde{\alpha}_{\alpha\alpha})^s(\tilde{G}_{\beta\beta})^{s*}\right] \quad (22b)$$

$$\beta_a(\tilde{G})^2 = \tfrac{3}{2} \operatorname{Im}\left[(\tilde{\alpha}_{\alpha\beta})^a(\tilde{G}_{\alpha\beta})^{a*}\right] \quad (22c)$$

$$\beta_s(\tilde{A})^2 = \tfrac{1}{2}\omega_0 \operatorname{Im}\left\{i(\tilde{\alpha}_{\alpha\beta})^s\left[\varepsilon_{\alpha\gamma\delta}(\tilde{A}_{\gamma,\delta\beta})\right]^{s*}\right\} \quad (22d)$$

$$\beta_a(\tilde{A})^2 = \tfrac{1}{2}\omega_0 \operatorname{Im}\left\{i(\tilde{\alpha}_{\alpha\beta})^a\left\{\left[\varepsilon_{\alpha\gamma\delta}(\tilde{A}_{\gamma,\delta\beta})\right]^{a*} + \left[\varepsilon_{\alpha\beta\gamma}(\tilde{A}_{\delta,\gamma\delta})\right]^{a*}\right\}\right\} \quad (22e)$$

and the corresponding script CP ROA invariants are defined in precisely the same form. The roman LP ROA invariants are designated by primes and are given by the real parts, rather than the imaginary parts, of the tensor combinations that appear for the CP ROA invariants[39]:

$$\alpha G' = \tfrac{1}{9} \operatorname{Re}\left[(\tilde{\alpha}_{\alpha\alpha})^s(\tilde{G}_{\beta\beta})^{s*}\right] \quad (23a)$$

$$\beta_s'(\tilde{G})^2 = \tfrac{1}{2} \operatorname{Re}\left[3(\tilde{\alpha}_{\alpha\beta})^s(\tilde{G}_{\alpha\beta})^{s*} - (\tilde{\alpha}_{\alpha\alpha})^s(\tilde{G}_{\beta\beta})^{s*}\right] \quad (23b)$$

$$\beta_a'(\tilde{G})^2 = \tfrac{3}{2} \operatorname{Re}\left[(\tilde{\alpha}_{\alpha\beta})^a(\tilde{G}_{\alpha\beta})^{a*}\right] \quad (23c)$$

$$\beta_s'(\tilde{A})^2 = \tfrac{1}{2}\omega_0 \operatorname{Re}\left\{i(\tilde{\alpha}_{\alpha\beta})^s\left[\varepsilon_{\alpha\gamma\delta}(\tilde{A}_{\gamma,\delta\beta})\right]^{s*}\right\} \quad (23d)$$

$$\beta_a'(\tilde{A})^2 = \tfrac{1}{2}\omega_0 \operatorname{Re}\left\{i(\tilde{\alpha}_{\alpha\beta})^a\left\{\left[\varepsilon_{\alpha\gamma\delta}(\tilde{A}_{\gamma,\delta\beta})\right]^{a*} + \left[\varepsilon_{\alpha\beta\gamma}(\tilde{A}_{\delta,\gamma\delta})\right]^{a*}\right\}\right\} \quad (23e)$$

and again, the script LP ROA invariants are defined in exactly the same way as those in Eq. (23). At this point, we note that the five LP ROA invariants given in Eq. (23) and the five other script LP ROA invariants, as well as the invariant combinations I_4 and $\mathscr{I}_4$ in Eqs. (20d) and (21d), vanish in the far-from-resonance approximation. All of these invariants will be able to contribute, and LP ROA should become observable, when the imaginary damping terms in the denominators of the optical activity tensors, given in Eqs. (9b–e), become comparable to the contributions of the real frequency terms in the denominators, i.e., close to resonance when the frequency differences become small.

B. General Intensities for Particular Classes of Polarization Modulation

In this section, we reduce the generality of Eq. (12) by specifying the Stokes vectors for particular classes of polarization modulation ROA experiments. In each subsection we identify the Stokes vector and give the corresponding intensity expression in terms of Mueller matrix elements. We then give the expressions for the corresponding circular intensity difference (CID) and circular intensity sum (CIS) in terms of the invariant combinations given in Eqs. (17), (20), and (21). These expressions still retain a high degree of generality in that the scattering angle and the angle of any linear polarization analyzer remain to be specified and may assume any possible values.

1. Incident Circular Polarization (ICP) ROA

For ICP ROA we select Stokes vectors for right and left circularly polarized incident radiation and a Stokes vector for the linear polarization at the angle ϕ of the scattered radiation. The Stokes vectors and resulting Mueller matrix intensity expression from Eq. (12) are

$$S^i = (1, 0, 0, \pm 1) \qquad S^d = (1, \cos 2\phi, 0, 0) \qquad (24a)$$

$$I_\phi^{R,L}(\xi) = \frac{K}{2}\left[M_{00} \pm M_{03} + (M_{10} \pm M_{13})\cos 2\phi \right] \qquad (24b)$$

where the first superscript (R) is associated the upper sign choice and the second (L) with the lower sign. Using Eq. (16) for the Mueller matrix elements yields

$$I_\phi^R(\xi) - I_\phi^L(\xi) = \frac{4K}{c}\left[I_1 + \mathscr{I}_2 \cos \xi - I_3 \sin^2 \xi \sin^2 \phi \right] \qquad (24c)$$

$$I_\phi^R(\xi) + I_\phi^L(\xi) = 2K\left[R_1 - R_3 \sin^2 \xi \sin^2 \phi \right] \qquad (24d)$$

A common form of ICP ROA is the absence of polarization discrimination in the scattered beam, the sum of polarization parallel and perpendicular to the scattering plane, referred to as unpolarized ICP. In this case we have

$$S^i = (1, 0, 0, \pm 1) \qquad S^d = (2, 0, 0, 0) \tag{25a}$$

$$I_u^{R,L}(\xi) = \frac{K}{2}[M_{00} \pm M_{03}] \tag{25b}$$

and the corresponding CID and CIS expressions are

$$I_u^R(\xi) - I_u^L(\xi) = \frac{4K}{c}\left[2I_1 + 2\mathscr{I}_2 \cos \xi - I_3 \sin^2 \xi\right] \tag{25c}$$

$$I_u^R(\xi) + I_u^L(\xi) = 2K\left[2R_1 - R_3 \sin^2 \xi\right] \tag{25d}$$

2.   Scattered Circular Polarization (SCP) ROA

The Stokes vectors for SCP ROA are reversed from those of ICP ROA, and the general intensity expression takes a similar form. We have

$$S^i = (1, \cos 2\theta, 0, 0) \qquad S^d = (1, 0, 0, \pm 1) \tag{26a}$$

$$I_{R,L}^\theta(\xi) = \frac{K}{2}\left[M_{00} \pm M_{30} + (M_{01} \pm M_{31})\cos 2\theta\right] \tag{26b}$$

and the corresponding CID and CIS intensity expressions are given by

$$I_R^\theta(\xi) - I_L^\theta(\xi) = \frac{4K}{c}\left[\mathscr{I}_1 + I_2 \cos \xi - \mathscr{I}_3 \sin^2 \xi \sin^2 \theta\right] \tag{26c}$$

$$I_R^\theta(\xi) + I_L^\theta(\xi) = 2K\left[R_1 - R_3 \sin^2 \xi \sin^2 \theta\right] \tag{26d}$$

The CIS expression is the same as that for ICP Raman intensity in Eq. (24d) and the CID expression is obtained from the ICP ROA intensity in Eq. (24c) by exchanging the roman invariant combinations for the script ones and vice versa. Unpolarized SCP ROA is not generally measured since laser sources are polarized, and the overall efficiency is lower, compared to ICP ROA with constant laser power at the sample, due to the need for a polarization analyzer in the scattered beam.

3.   In-Phase Dual Circular Polarization (DCP$_I$) ROA

The Stokes vectors needed for DCP ROA are simply the circular polarization Stokes vectors from ICP and SCP ROA. For DCP$_I$ ROA we have

$$S^i = (1, 0, 0, \pm 1) \qquad S^d = (1, 0, 0, \pm 1) \tag{27a}$$

$$I_{R,L}^{R,L}(\xi) = \frac{K}{2}\left[M_{00} + M_{33} \pm (M_{03} + M_{30})\right] \tag{27b}$$

and the CID and CIS are obtained from the association of the first and second super(sub)scripts with the upper and lower sign choices, respectively:

$$I_R^R(\xi) - I_L^L(\xi) = \frac{4K}{c}\left[I_1 + \mathscr{I}_1 + (I_2 + \mathscr{I}_2)\cos\xi - \frac{1}{2}(I_3 + \mathscr{I}_3)\sin^2\xi\right]$$

(27c)

$$I_R^R(\xi) + I_L^L(\xi) = 2K\left[R_1 + R_2\cos\xi - \frac{1}{2}R_3\sin^2\xi\right]$$

(27d)

4.   Out-of-Phase Dual Circular Polarization (DCP_{II}) ROA

In DCP_{II} ROA the Stokes vectors are the same as those in DCP_I ROA, except that the phase of the detected Stokes vector is opposite with respect to right and left circular polarization states. The Stokes vectors and the intensity expressed in Mueller matrices are given by

$$S^i = (1,0,0,\pm 1) \qquad S^d = (1,0,0,\mp 1)$$

(28a)

$$I_{L,R}^{R,L}(\xi) = \frac{K}{2}\left[M_{00} - M_{33} \pm (M_{03} - M_{30})\right]$$

(28b)

We can then work out the DCP_{II} CID and CIS expressions

$$I_L^R(\xi) - I_R^L(\xi) = \frac{4K}{c}\left[I_1 - \mathscr{I}_1 + (I_2 - \mathscr{I}_2)\cos\xi - \frac{1}{2}(I_3 - \mathscr{I}_3)\sin^2\xi\right]$$

(28c)

$$I_L^R(\xi) + I_R^L(\xi) = 2K\left[R_1 - R_2\cos\xi - \frac{1}{2}R_3\sin^2\xi\right]$$

(28d)

The DCP_I and DCP_{II} CIS intensities differ in the sign of the $\cos\xi$ term, and the CID intensities differ in the relative signs of the roman and script invariant combinations, being positive for DCP_I ROA and negative for DCP_{II} ROA.

5.   Incident Linear Polarization (ILP) ROA

The Stokes vectors for ILP ROA differ from those of ICP ROA in that the modulation of the extremes of the S_2 component of the Stokes vector for the incident light rather than the corresponding modulation of the S_3 component are needed. Thus, the Stokes vectors and basic intensity expression are

$$S^i = (1, 0, \pm 1, 0) \qquad S^d = (1, \cos 2\phi, 0, 0) \qquad (29a)$$

$$I_\phi^{+,-}(\xi) = \frac{K}{2}[M_{00} \pm M_{02} + (M_{10} \pm M_{12})\cos 2\phi] \qquad (29b)$$

where again ϕ is the angle of the polarizer in the scattered beam and ξ is the scattering angle as defined above. The linear intensity difference (LID) and linear intensity sum (LIS) expressions are

$$I_\phi^+(\xi) - I_\phi^-(\xi) = \frac{4K}{c}\{I_4 - \mathscr{I}_4 \cos \xi$$
$$- [I_4(1 + \cos^2 \xi) - 2\mathscr{I}_4 \cos \xi]\sin^2 \phi\} \qquad (29c)$$

$$I_\phi^+(\xi) + I_\phi^-(\xi) = 2K[R_1 - R_3 \sin^2 \xi \sin^2 \phi] \qquad (29d)$$

6. Scattered Linear Polarization (SLP) ROA

The intensity expressions SLP ROA follow from those of ILP and SCP ROA. The Stokes vectors and general intensity expressions are

$$S^i = (1, \cos \theta, 0, 0) \qquad S^d = (1, 0, \pm 1, 0) \qquad (30a)$$

$$I_{+,-}^\theta(\xi) = \frac{K}{2}[M_{00} \pm M_{20} + (M_{01} \pm M_{21})\cos 2\theta] \qquad (30b)$$

and the corresponding LID and LIS expressions are

$$I_+^\theta(\xi) - I_-^\theta(\xi) = \frac{4K}{c}\{\mathscr{I}_4 - I_4 \cos \xi - [\mathscr{I}_4(1 + \cos^2 \xi)$$
$$- 2I_4 \cos \xi]\sin^2 \theta\} \qquad (30c)$$

$$I_+^\theta(\xi) + I_-^\theta(\xi) = 2K[R_1 - R_3 \sin^2 \xi \sin^2 \theta] \qquad (30d)$$

As in the case of CP ROA, the LIS expressions for the ILP and SLP ROA are the same, aside from the definition of the polarizer angle, and the LID expressions differ only by the interchange roman and script invariant combinations.

7. In-Phase Dual Linear Polarization (DLP_I) ROA

By analogy with previous expressions it is straightforward to write the Stokes vectors and general intensity expression for DLP_I ROA as

$$S^i = (1, 0, \pm 1, 0) \qquad S^d = (1, 0, \pm 1, 0) \tag{31a}$$

$$I^{+, -}_{+, -}(\xi) = \frac{K}{2}[M_{00} + M_{22} \pm (M_{02} + M_{20})] \tag{31b}$$

and the associated LID and LIS expressions are

$$I^+_+(\xi) - I^-_-(\xi) = \frac{2K}{c}[I_4 + \mathscr{I}_4]\sin^2 \xi \tag{31c}$$

$$I^+_+(\xi) + I^-_-(\xi) = K[2R_1 + R_3(2\cos\xi - \sin^2 \xi)] \tag{31d}$$

8. Out-of-Phase Dual Linear Polarization (DLP_II) ROA

Finally, the Stokes vectors and Mueller matrix intensity expressions for DLP_{II} ROA are given by

$$S^i = (1, 0, \pm 1, 0) \qquad S^d = (1, 0, \mp 1, 0) \tag{32a}$$

$$I^{+, -}_{-, +}(\xi) = \frac{K}{2}[M_{00} - M_{22} \pm (M_{02} - M_{20})] \tag{32b}$$

and the LID and LIS expressions are

$$I^+_-(\xi) - I^-_+(\xi) = \frac{2K}{c}[I_4 - \mathscr{I}_4]\sin^2 \xi \tag{32c}$$

$$I^+_-(\xi) + I^-_+(\xi) = K[2R_1 - R_3(2\cos\xi + \sin^2 \xi)] \tag{32d}$$

Here again, the DLP LIS expressions differ by the sign of the $\cos\xi$ term, and the LID expressions for DLP_I and DLP_{II} ROA differ only by the relative sign for the roman and script invariant combinations.

C. Isolation of ROA Invariant Combinations

It is well known that the three invariants associated with ordinary Raman intensities (CIS) given in Eq. (18) can be isolated by an appropriate choice of experimental measurements.[46, 47] On the other hand, for the complete theory of CP and LP ROA this is not possible. There are ten CP ROA invariants, five roman invariants given in Eq. (22) and five script invariants, not explicitly given, and there are also ten LP ROA invariants, five roman invariants in Eq. (23) and another five script invariants not given. For CP ROA there are six linearly independent invariant combinations that can be extracted from experiment. Those combinations that carry a distinct dependence on the scattering angle, either none, $\cos \xi$, or $\cos^2 \xi$ (or $\sin^2 \xi$), are presented in Eqs. (20a)–(20c) and (21a)–(21c), but there are obviously other sets of linearly independent invariant combinations that could be formed. In the case of LP ROA, there are only two linearly independent combinations of invariants that appear in LP ROA intensities. A natural pair of combinations are those that appear in Eqs. (20d) and (21d) in which there is a separation of roman and script invariants, but not more. Due to the limited potential for reducing the number of invariants that appear in the LP ROA invariant combinations, we will focus the remainder of our attention in this section on optimizing the number of invariants in the six invariant combinations of CP ROA, and in this way we seek to reveal as much as possible about the relative contributions of the ten invariants that contribute to CP ROA intensity.

We begin by specifying in Table I the general CIS and CID expressions for the principal modulation forms of CP ROA that were presented in the previous subsections for three limiting sets of scattering angles: forward, right angle, and backward scattering. Further, we present these intensity expressions in terms of the ten CP ROA invariants so a clearer understanding of the relationship between the experimental form of ROA and the appearance of specific invariants may be obtained. More specifically, Table I is based on the following generic CID and CIS expressions:

$$
\begin{aligned}
I(R) - I(L) = \frac{8K}{c} \Big[& C_1 \alpha G + C_2 \beta_s(\tilde{G})^2 + C_3 \beta_a(\tilde{G})^2 + C_4 \beta_s(\tilde{A})^2 \\
& + C_5 \beta_a(\tilde{A})^2 + C_6 \alpha \mathscr{G} + C_7 \beta_s(\tilde{\mathscr{G}})^2 + C_8 \beta_a(\tilde{\mathscr{G}})^2 \\
& + C_9 \beta_s(\tilde{\mathscr{A}})^2 + C_{10} \beta_a(\tilde{\mathscr{A}})^2 \Big]
\end{aligned}
\tag{33a}
$$

$$
I(R) + I(L) = 4K \Big[C_{11} \alpha^2 + C_{12} \beta_s(\tilde{\alpha})^2 + C_{13} \beta_a(\tilde{\alpha})^2 \Big]
\tag{33b}
$$

in which a particular Raman and ROA setup for polarization modulation, linear polarization angle (when needed), and scattering angle leads to a specification of the ten ROA invariant coefficients C_1–C_{10} and three Raman invariant coefficients C_{11}–C_{13}. These coefficients are given in Table I.

In the case of forward scattering $(0°)$ in Table I, there are four sets of CIS and CID coefficients corresponding to the four modulation experiments: ICP, SCP, DCP_I, and DCP_{II}. All thirteen invariants make nonzero

TABLE I

Values of Raman and ROA Invariant Coefficients for the General ROA and Raman Intensity Expressions in Eqs. (33a) and (33b)

ξ	Form	Raman $4K$			ROA $\dfrac{8K}{c}$									
		α^2	$\beta_s(\tilde{\alpha})^2$	$\beta_a(\tilde{\alpha})^2$	αG	$\beta_s(\tilde{G})^2$	$\beta_a(\tilde{G})^2$	$\beta_s(\tilde{A})^2$	$\beta_a(\tilde{A})^2$	$\alpha\mathcal{G}$	$\beta_s(\tilde{\mathcal{G}})^2$	$\beta_a(\tilde{\mathcal{G}})^2$	$\beta_s(\tilde{\mathcal{A}})^2$	$\beta_a(\tilde{\mathcal{A}})^2$
0°	$\mathrm{ICP_u}$	45	7	5	$+45$	$+7$	$+5$	$+1$	-1	-45	$+5$	-5	-3	$+1$
	$\mathrm{SCP_u}$	45	7	5	$+45$	-5	$+5$	-3	-1	-45	-7	-5	$+1$	$+1$
	$\mathrm{DCP_I}$	45	1	5	$+45$	$+1$	$+5$	-1	-1	-45	-1	-5	-1	$+1$
	$\mathrm{DCP_{II}}$		6		$+6$	$+6$		$+2$			$+6$		-2	
90°	$\mathrm{ICP_p}$	$\frac{45}{2}$	$\frac{7}{2}$	$\frac{5}{2}$	$+\frac{45}{2}$	$+\frac{7}{2}$	$+\frac{5}{2}$	$+\frac{1}{2}$	$-\frac{1}{2}$					
	$\mathrm{ICP_d}$		3	5		$+3$	$+5$	-1	$+1$					
	$\mathrm{ICP*}$	$\frac{45}{3}$	$\frac{10}{3}$	$\frac{10}{3}$	$+\frac{45}{3}$	$+\frac{10}{3}$	$+\frac{10}{3}$							
	$\mathrm{SCP_p}$	$\frac{45}{2}$	$\frac{7}{2}$	$\frac{5}{2}$						$-\frac{45}{2}$	$-\frac{7}{2}$	$-\frac{5}{2}$	$+\frac{1}{2}$	$+\frac{1}{2}$
	$\mathrm{SCP_d}$		3	5							-3	-5	-1	-1
	$\mathrm{SCP*}$	$\frac{45}{3}$	$\frac{10}{3}$	$\frac{10}{3}$						$-\frac{45}{3}$	$-\frac{10}{3}$	$-\frac{10}{3}$		
	$\mathrm{DCP_I}$	$\frac{45}{4}$	$\frac{13}{4}$	$\frac{15}{4}$	$+\frac{45}{4}$	$+\frac{13}{4}$	$+\frac{15}{4}$	$-\frac{1}{4}$	$+\frac{1}{4}$	$+\frac{45}{4}$	$+\frac{13}{4}$	$+\frac{15}{4}$	$-\frac{1}{4}$	$-\frac{1}{4}$
	$\mathrm{DCP_{II}}$	$\frac{45}{4}$	$\frac{13}{4}$	$\frac{15}{4}$	$+\frac{45}{4}$	$+\frac{13}{4}$	$+\frac{15}{4}$	$-\frac{1}{4}$	$+\frac{1}{4}$	$+\frac{45}{4}$	$+\frac{13}{4}$	$+\frac{15}{4}$	$+\frac{1}{4}$	$+\frac{1}{4}$
180°	$\mathrm{ICP_u}$	45	7	5	$+45$	$+7$	$+5$	$+1$	-1	$+45$	-5	$+5$	$+3$	-1
	$\mathrm{SCP_u}$	45	7	5	-45	$+5$	-5	$+3$	$+1$	-45	-7	-5	$+1$	$+1$
	$\mathrm{DCP_I}$		6			$+6$		$+2$			-6		$+2$	
	$\mathrm{DCP_{II}}$	45	1	5	$+45$	$+1$	$+5$	-1	-1	$+45$	$+1$	$+5$	$+1$	-1

contributions to the first three modulation forms, but only the symmetric anisotropic invariants, i.e., the β_s invariants, contribute in the case of DCP_{II}. On the other hand, we note that for DCP_I, the intensities have very weak contributions from the β_s invariants. Other patterns can be observed for forward scattering. The CIS values for ICP_u or SCP_u are simply the sum of those for DCP_I and DCP_{II}. The CIDs for ICP_u and SCP_u are the sum and difference, respectively, of the CIDs for DCP_I and DCP_{II}. The CID for DCP_I is the mean of the CIDs for ICP_u and SCP_u, and the DCP_{II} CID is the difference of the CIDs for ICP_u and SCP_u.

For backscattering, a similar set of patterns is observed, with the intensity roles of DCP_I and DCP_{II} reversed. Further, the relative signs of the roman and script invariants have changed for each of the four forms of ROA, and the relative signs between the invariants for ICP_u and SCP_u have changed.

The intensity expressions for right-angle scattering are more numerous. For ICP we list the polarized (linearly polarized component perpendicular to the plane of the scattering) intensities (ICP_p), the depolarized (linearly polarized component in the plane of the scattering) intensities (ICP_d), and the magic-angle intensities ($ICP*$).[32] A similar listing is given for SCP intensities. We can see for this scattering geometry that the CIDs for ICP involve only the roman invariants, whereas those for SCP involve only the script invariants. Further, for the depolarized intensities, only anisotropic invariants contribute, and for the magic-angle CIDs only the magnetic dipole invariants contribute. The magic-angle experiments are simply linear combinations of polarized and depolarized experiments that lead to a cancellation of the electric quadrupole invariants. The CIS values for DCP_I and DCP_{II} are unpolarized right-angle intensities and are one-half the sum of polarized and depolarized CIS values. The CID for DCP_I is the average of the $ICP_p + ICP_d$ and $SCP_p + SCP_d$ CIDs, and the CID for DCP_{II} is one-half the difference of the $ICP_p + ICP_d$ and $SCP_p + SCP_d$ CIDs. Note that $ICP_p + ICP_d$ equals ICP_u and $SCP_p + SCP_d$ equals SCP_u for right-angle scattering.

The three CIS invariant combinations in Eq. (18) and the six CID invariant combinations for CP ROA given in Eqs. (20a–20c) and (21a)–(21c) provide rather poor discrimination among the various tensor invariants that contribute to these combinations. As noted above, although the ten CP ROA invariants cannot be isolated, the three Raman invariants can be. Further, from Table I, it can be seen the the six magnetic dipole CID invariants (three roman and three script) follow the relative intensity pattern of the three Raman CIS invariants. We show below a method for segregating the six magnetic dipole CID invariants among six CID invariant combinations in which only one or two of the four electric quadrupole invariants also appears. This achieves a complete natural separation of the six magnetic dipole CID invariants, and perhaps a maximum degree of isolation of all ten CP ROA invariants.

We show first the expressions for the isotropic invariants, complete isolation for Raman and partial isolation for CP ROA:

$$4K[45\alpha^2] = \left[I_L^R(180°) + I_R^L(180°) \right] + \frac{1}{3}\left[I_R^R(180°) + I_L^L(180°) \right]$$

$$- \left[I_z^R(90°) + I_z^L(90°) \right] \tag{34a}$$

$$\frac{8K}{c}\left[45\alpha G + \frac{2}{3}\beta_s(\tilde{A})^2 - 2\beta_a(\tilde{A})^2 \right]$$

$$= \frac{1}{2}\left\{ \left[I_R^R(0°) - I_L^L(0°) \right] + \frac{1}{3}\left[I_L^R(0°) - I_R^L(0°) \right] \right.$$

$$\left. + \left[I_L^R(180°) - I_R^L(180°) \right] + \frac{1}{3}\left[I_R^R(180°) - I_L^L(180°) \right] \right\}$$

$$- \left[I_z^R(90°) - I_z^L(90°) \right] \tag{34b}$$

$$\frac{8K}{c}\left[-45\alpha\mathscr{G} + \frac{2}{3}\beta_s(\tilde{\mathscr{A}})^2 + 2\beta_a(\tilde{\mathscr{A}})^2 \right]$$

$$= \frac{1}{2}\left\{ \left[I_R^R(0°) - I_L^L(0°) \right] - \frac{1}{3}\left[I_L^R(0°) - I_R^L(0°) \right] \right.$$

$$\left. - \left[I_L^R(180°) - I_R^L(180°) \right] + \frac{1}{3}\left[I_R^R(180°) - I_L^L(180°) \right] \right\}$$

$$- \left[I_R^y(90°) - I_L^y(90°) \right] \tag{34c}$$

The Raman measurements are specified assuming the availability of the RCP and LCP intensities needed for ROA spectra. Outside the context of ROA measurements, a somewhat simpler, although somewhat less precise, representation of ordinary Raman intensities would be sufficient to achieve the desired invariant solution. The subscript z refers to ICP$_d$ scattering where $\phi = 90°$, and the superscript y for SCP$_d$ scattering, $\theta = 90°$.

The isolation of the CIS symmetric anisotropy and the CID invariant combinations is achieved as follows:

$$4K\left[6\beta_s(\tilde{\alpha})^2\right] = \left[I_R^R(180°) + I_L^L(180°)\right] \tag{35a}$$

$$\frac{8K}{c}\left[6\beta_s(\tilde{G})^2 + 2\beta_s(\tilde{A})^2\right] = \frac{1}{2}\{[I_R^R(180°) - I_L^L(180°)]$$
$$+ [I_L^R(0°) - I_R^L(0°)]\} \tag{35b}$$

$$\frac{8K}{c}\left[-6\beta_s(\tilde{\mathscr{G}})^2 + 2\beta_s(\tilde{\mathscr{A}})^2\right] = \frac{1}{2}\{[I_R^R(180°) - I_L^L(180°)]$$
$$- [I_L^R(0°) - I_R^L(0°)]\} \tag{35c}$$

Finally, we provide expressions for the isolation of the CIS antisymmetric anisotropy and the separation of the antisymmetric magnetic dipole anisotropic invariant from the other magnetic dipole CID invariants. These expressions are

$$4K\left[5\beta_a(\tilde{\alpha})^2\right] = \left[I_z^R(90°) + I_z^L(90°)\right] - \frac{1}{2}\left[I_R^R(180°) + I_L^L(180°)\right] \tag{36a}$$

$$\frac{8K}{c}\left[5\beta_a(\tilde{G})^2 - 2\beta_s(\tilde{A})^2 + \beta_a(\tilde{A})^2\right]$$
$$= \left[I_z^R(90°) - I_z^L(90°)\right] - \frac{1}{4}\{[I_R^R(180°) - I_L^L(180°)]$$
$$+ [I_L^R(0°) - I_R^L(0°)]\} \tag{36b}$$

$$\frac{8K}{c}\left[-5\beta_a(\tilde{\mathscr{G}})^2 - 2\beta_s(\tilde{\mathscr{A}})^2 - \beta_a(\tilde{\mathscr{A}})^2\right]$$
$$= \left[I_R^y(90°) - I_L^y(90°)\right] - \frac{1}{4}\{[I_R^R(180°) - I_L^L(180°)]$$
$$- [I_L^R(0°) - I_R^L(0°)]\} \tag{36c}$$

In the limit that the magnetic dipole invariants dominate the electric quadrupole invariants as the source of CP ROA intensity, the preceding intensity expressions would afford a method for isolating all six contributing magnetic dipole CID invariants.

V. THE FAR-FROM-RESONANCE THEORY OF ROA

The original theory of ROA was cast in the far-from-resonance approximation for right-angle scattering,[1, 13] and hence the expression to be presented here for ICP, as well as the same expressions for SCP ROA, have been available in the literature for slightly more than two decades and approximately one decade for the forward and backward scattering cases. In this section, we present, for the first time, all the far-from-resonance intensity expressions for the principal forms of ROA in forward, right-angle, and backward scattering geometries. We first obtain these intensity expressions by relaxing the constraints of resonance of the CIS and CID expressions given in the previous sections. We then present an efficient approach to the isolation of all three far-from-resonance CID tensor invariants.

A. Reduction of the General Theory

Each of the Raman transition tensors in Eq. (9) is complex, and this is designated by a tilde above the tensor symbol. Far from resonance, the polarizability tensor and the two electric quadrupole optical activity tensors become real, while the two magnetic dipole optical activity tensors become pure imaginary. In general, these complex tensors can be separated into real and imaginary parts by writing

$$\tilde{T} = T - iT' \tag{37}$$

where the unprimed quantity refers to the real part and the primed quantity to the imaginary part. In the far-from-resonance approximation, all of the antisymmetric invariants vanish. In particular, we have

$$\beta_a(\tilde{\alpha})^2, \beta_a(\tilde{G})^2, \beta_a(\tilde{A})^2, \beta_a(\tilde{\mathscr{G}})^2, \beta_a(\tilde{\mathscr{A}})^2 = 0 \tag{38}$$

Further, consideration of the Hermitian character of the operators in the polarizability and optical activity tensors leads, in the far-from-resonance approximation, to the relations

$$\beta_s(\tilde{\alpha})^2 = \beta(\alpha)^2 \tag{39a}$$

$$\beta_s(\tilde{G})^2 = -\beta_s(\tilde{\mathscr{G}})^2 = \beta(G')^2 \tag{39b}$$

$$\beta_s(\tilde{A})^2 = \beta_s(\tilde{\mathscr{A}})^2 = \beta(A)^2 \tag{39c}$$

$$\alpha G = -\alpha\mathscr{G} = \alpha G' \tag{39d}$$

The definitions of the far-from-resonance tensor invariants are given by

$$\beta(\alpha)^2 = \tfrac{1}{2}(3\alpha_{\alpha\beta}\alpha_{\alpha\beta} - \alpha_{\alpha\alpha}\alpha_{\beta\beta}) \tag{40a}$$

$$\beta(G')^2 = \tfrac{1}{2}(3\alpha_{\alpha\beta}G'_{\alpha\beta} - \alpha_{\alpha\alpha}G'_{\beta\beta}) \tag{40b}$$

$$\beta(A)^2 = \tfrac{1}{2}\omega_0\alpha_{\alpha\beta}\varepsilon_{\alpha\gamma\delta}A_{\gamma,\delta\beta} \tag{40c}$$

$$\alpha = \tfrac{1}{3}\alpha_{\alpha\alpha} \quad \text{and} \quad G' = \tfrac{1}{3}G'_{\alpha\alpha} \tag{40d}$$

where as before repeated Greek subscripts are to be summed over the three Cartesian directions. If the simplifying relations in Eqs. (38) and (39) are valid, the general CID and CIS expressions in Eqs. (33) reduce to

$$I(R) - I(L) = \frac{8K}{c}\left[D_1\alpha G' + D_2\beta(G')^2 + D_3\beta(A)^2\right] \tag{41a}$$

$$I(R) + I(L) = 4K\left[D_4\alpha^2 + D_5\beta(\alpha)^2\right] \tag{41b}$$

·It is clear from these equations that there are now only three CID invariants and two CIS invariants, and the values corresponding to the coefficients D_1–D_5 are given in Table II.

The CIS contributions in Table II are the same as those in Table I for the isotropic and symmetric anisotropic invariants. The CIDs vanish in all cases for DCP$_{II}$ ROA. In forward and in backward scattering, the CID contributions are the same for ICP$_u$, SCP$_u$, and DCP$_I$ ROA. In forward scattering there is a strong contribution for the isotropic ROA invariant with relatively small contributions for the two anisotropic ROA invariants, whereas for backward scattering, the magnetic dipole anisotropic ROA invariant makes a very large contribution with no contribution from the isotropic ROA invariant. In both forward and backward scattering the ROA spectrum can be obtained with the least amount of Raman intensity using the DCP$_I$ setup, and hence the greatest intrinsic efficiency is available with this technique. In right-angle DCP$_I$ scattering, the CIS is one-half the sum of the CISs for polarized and depolarized ICP (or SCP), but the CID is equal to the sum of the polarized and depolarized CIDs for ICP (or SCP). As before, for magic-angle ROA scattering, the contribution of the electric quadrupole invariant is zero. Finally, we note that the CIS for backscattering DCP$_I$ is twice that for right-angle depolarized ICP, and for the CIDs, the absolute values of the coefficients for DCP$_I$ are four times those for right-angle depolarized ROA. The fact that the relative signs of the invariants are opposite in these two ROA experiments will

TABLE II

Values of Raman and ROA Invariant Coefficients for the Far-from-Resonance ROA
and Raman Intensity Expressions in Eqs. (41a) and (41b)

ξ	Form	Raman $4K$		ROA $\dfrac{8K}{c}$		
		α^2	$\beta(\alpha)^2$	αG	$\beta(G')^2$	$\beta(A)^2$
0°	ICP_u	45	7	90	2	-2
	SCP_u	45	7	90	2	-2
	DCP_I	45	1	90	2	-2
	DCP_{II}		6			
90°	ICP_p	$\frac{45}{2}$	$\frac{7}{2}$	$+\frac{45}{2}$	$+\frac{7}{2}$	$+\frac{1}{2}$
	ICP_d		3		$+3$	-1
	ICP^*	$\frac{45}{3}$	$\frac{10}{3}$	$+\frac{45}{3}$	$+\frac{10}{3}$	
	SCP_p	$\frac{45}{2}$	$\frac{7}{2}$	$+\frac{45}{2}$	$+\frac{7}{2}$	$+\frac{1}{2}$
	SCP_d		3		$+3$	-1
	SCP^*	$\frac{45}{3}$	$\frac{10}{3}$	$+\frac{45}{3}$	$+\frac{10}{3}$	
	DCP_I	$\frac{45}{4}$	$\frac{13}{4}$	$+\frac{45}{2}$	$+\frac{13}{2}$	$-\frac{1}{2}$
	DCP_{II}	$\frac{45}{4}$	$\frac{13}{4}$			
180°	ICP_u	45	7		$+12$	$+4$
	SCP_u	45	7		$+12$	$+4$
	DCP_I		6		$+12$	$+4$
	DCP_{II}	45	1			

prove useful in the isolation of ROA invariants, as described in the
following section.

B.   Complete Isolation of ROA Invariants

We present here a procedure for isolating all three ROA invariants in the
far-from-resonance approximation. In Raman scattering, absolute intensities are rarely measured due to the difficulty of taking into account the
solid collection angle of the detected radiation and all of the instrumental
response factors. The relative values of the two CIS invariants are easily
obtained by measuring the standard depolarization ratio in right-angle
scattering. The magnitudes of ROA spectra are measured relative to the
parent Raman spectrum, and hence one can consider obtaining magnitudes of all three ROA invariants relative to a parent Raman spectrum.

The isolation method for ROA invariants presented here is a slight
variant of one presented previously.[34] As recognized before, the simplicity
of the method relies on keeping the solid angle of collection sufficiently
small. We start with depolarized right-angle scattering with CID and CIS

intensities given by

$$I_z^R(90°) - I_z^L(90°) = 8\frac{K_d}{c}\left[3\beta(G')^2 - \beta(A)^2\right] \tag{42a}$$

$$I_z^R(90°) + I_z^L(90°) = 4K_d\left[3\beta(\alpha)^2\right] \tag{42b}$$

where the constant K_d has a subscript to reflect the instrumental response and collection efficiency associated with the particular depolarized setup used to obtain the CIS and CID spectra. If we now measure backscattering DCP$_I$ ROA spectra with a different instrumental setup, a different proportionality constant will enter the expression, denoted K'_d, and we have

$$I_R^R(180°) - I_L^L(180°) = 8\frac{K'_d}{c}\left[12\beta(G')^2 + 4\beta(A)^2\right] \tag{42c}$$

$$I_R^R(180°) + I_L^L(180°) = 4K'_d\left[6\beta(\alpha)^2\right] \tag{42d}$$

where the depolarized nature of this form of ROA is clear from its CIS spectrum. We can now scale Eqs. (42c) and (42d) by the same factor such that the DCP$_I$ CIS spectrum is exactly twice that of the depolarized ICP CIS spectrum, as required theoretically if the two constants were identical. Once the spectra are scaled in this way, the difference between these two proportionality constants is effectively eliminated, and hence the corresponding two ROA spectra can be combined in a manner that provides isolated spectra for $\beta(G')^2$ and $\beta(A)^2$.

In Fig. 3 we illustrate this isolation method for (+)-*trans*-pinane, where a common CIS spectrum is shown at the bottom of the figure. Above this is first the CID spectrum for DCP$_I$ ROA, and above that the CID spectrum for depolarized ICP ROA. The large difference in relative intensity of the two depolarized ROA spectra, for an identical parent Raman spectrum, is easily seen. Finally, these two CID spectra are added and subtracted in a manner that allows isolation of the spectral contributions of $\beta(G')^2$ and $\beta(A)^2$ to the backscattering DCP$_I$ ROA spectrum.

We can proceed with the isolation procedure by making further measurements of a polarized nature in order to isolate the isotropic magnetic dipole ROA invariant. However, we first have to find a way to relate the instrumental setup for polarized scattering back to the setups above used for depolarized scattering. This can be accomplished by measuring unpolarized ICP in backscattering, where the ROA spectrum is predicted

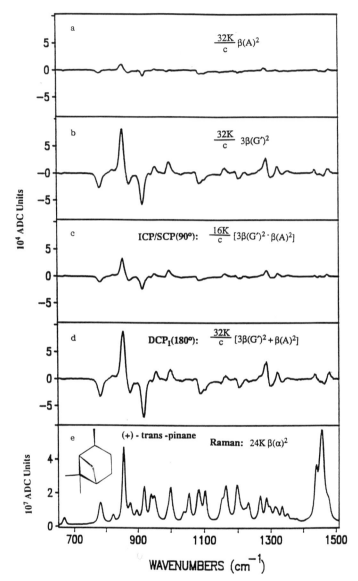

Figure 3. ROA and Raman spectra for (+)-*trans*-pinane using 488-nm laser excitation. The (d) DCP$_I$ (180°) and (c) ICP/SCP (90°) ROA spectra are associated with the same (e) depolarized parent Raman spectrum. The isolated anisotropic (a) electric quadrupole and (b) magnetic dipole optical activity invariant spectra are shown. The sum of the two ROA invariant spectra equals the DCP$_I$ (180°) ROA spectrum, and half of their difference equals the ICP/SCP (90°) ROA spectrum. The notation ICP/SCP refers to the experimental average of the ICP and SCP ROA spectra, which in the far-from-resonance theoretical limit are equal to one another. (Reproduced with permission from Ref. 34.)

theoretically to be identical, for the same scaling constant, as that for DCP$_I$ ROA above, even though the parent Raman spectra are different. The CIS and CID intensities are

$$I_u^R(180°) - I_u^L(180°) = 8\frac{K'_p}{c}\left[12\beta(G')^2 + 4\beta(A)^2\right] \tag{42e}$$

$$I_u^R(180°) + I_u^L(180°) = 4K'_p\left[45\alpha^2 + 7\beta(\alpha)^2\right] \tag{42f}$$

If these initially measured CID and CIS spectra are both scaled by the same factor such that the CIDs for Eqs. (42c) and (42e) become equal to one another, the polarized proportionality constant K'_p will become equal to the two depolarized proportionality constants above (which have already been scaled to be equal to each other). Finally, we can measure a different polarized Raman spectrum in which the isotropic ROA invariant appears, such as the CIS and CID spectra for unpolarized ICP in forward scattering,

$$I_u^R(0°) - I_u^L(0°) = 8\frac{K''_p}{c}\left[90\alpha G' + 2\beta(G')^2 - 2\beta(A)^2\right] \tag{42g}$$

$$I_u^R(0°) + I_u^L(0°) = 4K''_p\left[45\alpha^2 + 7\beta(\alpha)^2\right] \tag{42h}$$

Previously, we had suggested right-angle polarized ICP scattering for this purpose[34]; however, it is well known that polarized right-angle ROA spectra are difficult to measure. Forward scattering spectra should be more easily measured and the desired isotropic ROA invariant is present with high intensity and little interference from the anisotropic ROA invariants. To complete the analysis, the final set of spectra are scaled such that the forward scattering ICP CIS spectrum matches that of the scaled ICP backscattering CIS spectrum. The previous isolated anisotropic invariant spectra can then be combined with the forward scattering CID spectrum to isolate the isotropic ROA invariant spectrum.

It has been pointed out that if the effects of finite collection angle are taken into account, the uniform scaling procedure cannot be rigorously followed, since the correct scaling factor has a dependence on the depolarization ratios of the parent Raman bands.[35] The analysis confirms our empirical finding that if the collection angle is kept at small but operationally feasible sizes, the variations of solid angle collection can be kept below 2%, well below the noise level of approximately 5%.

VI. AB INITIO THEORETICAL CALCULATIONS OF ROA

The interpretation of ROA spectra has been of considerable interest ever since the first ROA spectra were measured. Until recently, the only route to spectral interpretation has been through simplifying models,[1-4, 8] of which there have been many. In general, it is found that ROA theory is complex, and good examples illustrative of simple intensity mechanisms have not yet been found. Recently, advances in the calculation of molecular properties using ab initio molecular orbital theory have made possible the direct calculation of ROA observables for a variety of molecules.[48-56]

In this section we discuss the nature of these calculations relative to the theoretical and experimental advances that have taken place in ROA in the past several years. The theoretical basis of these calculations has been described in detail.[51] The calculations have been carried out using the Cambridge Analytical Derivatives Package (CADPAC),[57] and this package has also been used by Stephens and coworkers to carry out ab initio calculations of VCD intensities.[58] The central idea for ROA studies is to calculate the polarizability, the magnetic dipole optical activity tensor, and the electric quadrupole optical activity tensor using the first derivatives of field-perturbed wave functions. In the far-from-resonance approximation, we can write the expression for the electric dipole polarizability without the imaginary damping term, and without the need to distinguish between the energies or wave functions of states m and n. Under these conditions Eq. (15a) becomes

$$\tilde{\alpha}_{\alpha\beta} = \frac{1}{\hbar} \sum_{j \neq n} \left[\frac{\langle n|\hat{\mu}_\alpha|j\rangle\langle j|\hat{\mu}_\beta|n\rangle}{\omega_{nj} - \omega_0} + \frac{\langle n|\hat{\mu}_\beta|j\rangle\langle j|\hat{\mu}_\alpha|n\rangle}{\omega_{nj} + \omega_0} \right] \quad (43)$$

Using the general definition for the real and the imaginary parts of these complex tensors given in Eq. (37) we can further simplify Eq. (43) to be

$$\alpha_{\alpha\beta} = \frac{2}{\hbar} \sum_{j \neq n} \frac{\omega_{jn}}{\omega_{jn}^2 - \omega_0^2} \, \text{Re}\left[\langle n|\hat{\mu}_\alpha|j\rangle\langle j|\hat{\mu}_\beta|n\rangle\right] \quad (44a)$$

With the same approximations the optical activity tensors in the far-from-resonance approximation are given by

$$G'_{\alpha\beta} = \frac{-2}{\hbar} \sum_{j \neq n} \frac{\omega_0}{\omega_{jn}^2 - \omega_0^2} \, \text{Im}\left[\langle n|\hat{\mu}_\alpha|j\rangle\langle j|\hat{m}_\beta|n\rangle\right] \quad (44b)$$

$$A_{\alpha\beta\gamma} = \frac{2}{\hbar} \sum_{j \neq n} \frac{\omega_{jn}}{\omega_{jn}^2 - \omega_0^2} \, \text{Re}\left[\langle n|\hat{\mu}_\alpha|j\rangle\langle j|\hat{\theta}_{\beta\gamma}|n\rangle\right] \quad (44c)$$

These tensors have essentially no vibrational or vibronic detail. They can be used as they are for Rayleigh scattering and Rayleigh optical activity. Their application to Raman and ROA is developed by finding their variation with nuclear motion, i.e., their first derivatives with respect to vibrational coordinates. These far-from-resonance tensors will describe preresonance Raman behavior through their dependence on the incident laser frequency ω_0.

To calculate ROA intensities with the CADPAC program, a further approximation is needed. This can be referred to as the zero-frequency limit, and corresponds to the response of the molecule to a static perturbing field at zero frequency. In this approximation, we write $\omega_{jn}^2 - \omega_0^2$ simply as ω_{jn}^2. The tensors in Eq. (44) can then be written as

$$\alpha_{\alpha\beta} = 2 \sum_{j \neq n} \frac{\text{Re}\left[\langle n|\hat{\mu}_\alpha|j\rangle\langle j|\hat{\mu}_\beta|n\rangle\right]}{E_j - E_n} \tag{45a}$$

$$G'_{\alpha\beta} = -2\hbar\omega_0 \sum_{j \neq n} \frac{\text{Im}\left[\langle n|\hat{\mu}_\alpha|j\rangle\langle j|\hat{m}_\beta|n\rangle\right]}{\left(E_j - E_n\right)^2} \tag{45b}$$

$$A_{\alpha\beta\gamma} = 2 \sum_{j \neq n} \frac{\text{Re}\left[\langle n|\hat{\mu}_\alpha|j\rangle\langle j|\hat{\theta}_{\beta\gamma}|n\rangle\right]}{E_j - E_n} \tag{45c}$$

and these expressions can in turn be evaluated by a sum over molecular orbital contributions using unperturbed and field-perturbed molecular orbitals. Under these circumstances, the tensor expressions in Eq. (45) become[51]

$$\alpha_{\alpha\beta} = 4 \sum_k \left\langle \phi_k^0|\hat{\mu}_\alpha|\phi_k'(F_\beta)\right\rangle \tag{46a}$$

$$G'_{\alpha\beta} = -4\hbar\omega_0 \sum_k \text{Im}\left[\left\langle \phi_k'(F_\alpha)\big|\phi_k'(B_\beta)\right\rangle\right] \tag{46b}$$

$$A_{\alpha\beta\gamma} = 4 \sum_k \left\langle \phi_k^0|\theta_{\beta\gamma}|\phi_k'(F_\alpha)\right\rangle \tag{46c}$$

where $\mathbf{F}$ and $\mathbf{B}$ are perturbing electric and magnetic fields, respectively, and the prime represents a derivative with respect to the perturbing field. The expressions in Eqs. (46a) and (46c) are pure real, and hence the real part does not need to be taken.

Using these expressions, ROA intensity calculations have been performed for several three-membered-ring molecules,[4-50, 56] alanine,[53, 54] and tartaric acid.[55] The comparison between theory and experiment is encouraging, particularly relative to earlier model calculations; however, the degree of agreement is not as high as what has been accomplished with ab initio VCD calculations.[59] If the relative contributions of the three ROA invariants to the observed intensities are isolated, which can be accomplished far more easily theoretically than experimentally, no particular pattern emerges in favor of one ROA intensity source over another. Unfortunately, there is a gauge origin dependence problem that arises for the magnetic dipole optical activity tensor which leads to some variation in the observed ROA intensity with the choice of molecular origin for the calculations. This disturbs the relative intensity balance among the three invariants, making a unique determination of their relative contributions somewhat problematic. Nevertheless, the magnitude of the gauge origin dependence does not appear to be very large in the calculations carried out to date. For example, the relative sign pattern of the individual bands is not changed by the relocation of the origin from one part of the molecule to another. The gauge origin problem is related to the shortcoming of the wave functions used for the calculations. In the limit of exact wave functions, the origin dependence of the calculations vanishes. Thus, the experimental determination of the ROA invariant spectra, as outlined in the previous section, will provide valuable data for the comparison of theory to experiment.

Despite the progress achieved to date in the calculation of ROA, there is a clear need for further improvements. The first step is to carry out direct calculations of the far-from-resonance theory using a sum-over-states approach, or the equivalent, so that the preresonance dependence of the polarizability and the optical activity tensors of the exciting laser frequency can be studied. The necessary expressions for such calculations are given in Eq. (44). Second, such calculations need to be extended to the more general tensor definitions in Eq. (9) so that differences in the roman and script tensors as a function of the exciting laser frequency can be determined. It would be useful to find theoretically the conditions under which differences between ICP and SCP ROA can be expected to be observed or, similarly, under what conditions measurable LP ROA spectra could be observed. Many interesting questions remain to be answered regarding the origin of ROA intensities, and the answers to these questions should soon become available through the application of ab initio molecular orbital techniques.

VII.   EXPERIMENTAL ASPECTS OF ROA MEASUREMENTS

Parallel to the recent theoretical development of ROA has been an equally significant advance in instrumental technique. The instrumental advances have not only made possible the accurate measurement of many new forms of ROA, but they have also opened the way for reliable routine measurement of samples that previously were regarded as too difficult to measure. In this section we focus on the new instrumental techniques that have led to a dramatic improvement in ROA measurement capability. This is followed by considerations of artifact reduction, and finally a brief overview of the recent applications of ROA to problems involving the stereochemistry of chiral molecules.

A. Instrumental Configurations

Presently, there are two ROA instruments with advanced capabilities, one at the University of Glasgow,[8, 26] and a second at Syracuse University.[5-7, 25, 27, 37, 42] They share two critical features that are responsible in part for their advanced performance: backscattering geometry and charge-coupled device (CCD) detection. Beyond that the instruments are quite different and illustrate alternative aspects of improved ROA instrumentation.

The Glasgow instrument in its most recent configuration[26] (Fig. 4) features polarization modulation of the incident laser beam with a longitudinal electro-optic modulator (EOM). The beam is focused at the sample through small holes in a collection mirror, a collimating lens, and a Lyot

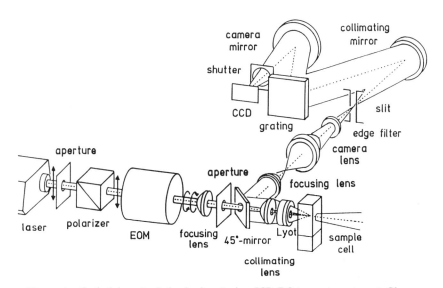

Figure 4. Optical layout of the backscattering ICP ROA spectrometer at Glasgow University. (Reproduced with permission from Ref. 26.)

depolarizer. The backward scattered light is then depolarized, collimated, and reflected at by the collection mirror, which is oriented at 45° with respect to the direction of the incident laser beam and the backscattered light. The scattered light is then passed through a holographic edge filter and is focused by a camera lens onto the entrance slit of fast single-stage monochromator. The spectrograph is equipped with a holographic grating and a thinned back-illuminated CCD detector. The Lyot depolarizer is required to remove optical artifacts induced by the right-angle reflection of the scattered light by the collection mirror. The depolarizer precludes further analysis of the polarization states of the scattered light associated with the measurement of the SCP, DCP_I, and DCP_{II} forms of ROA. The spectrograph, as described,[26] is capable of measuring only backscattered unpolarized ICP ROA, although an earlier version was adapted for forward scattering ICP ROA.[33] The principal advantage of this spectrograph is its extraordinary efficiency in the measurement of difficult samples such as biological molecules. An upgraded right-angle ICP ROA instrument at the University of Glasgow has recently been reported.[56]

The Syracuse ROA instrument employs one or two zeroth-order quarter waveplates in digital, computer-controlled rotation stages for the control of polarized states. The instrument is capable of measuring all forms of CP or LP ROA, in either backward, right-angle, or forward scattering geometry. The instrument has been described in detail in its right-angle SCP mode of operation,[25] and further improvements have been described separately, such as its backscattering configuration in which DCP_I and unpolarized ICP have been compared in detail with a single optical alignment,[27] and its ability to compare ICP and SCP forms of ROA in right-angle scattering without relocating any of the optical components.[37] The key to its backscattering configuration is a small prism that reflects the incident laser beam through a right angle and around which the scattered light passes unreflected on its way to the spectrograph. Also in backscattering, a single quarter waveplate is responsible for the modulation of the polarization of both the incident and the scattered light. A second quarter waveplate may be used to fine tune the polarization of the incident light prior to its reaching the main quarter waveplate. The optical diagram for this instrument in its backscattering configuration is given in Fig. 5.

The efficiency of the Syracuse instrument, although more than two orders of magnitude better than the same instrument of a few years ago, is still nearly an order of magnitude below that of the Glasgow instrument. Improvements are available in several areas, such as the use of a back-thinned CCD having quantum efficiency near 80% rather than our more

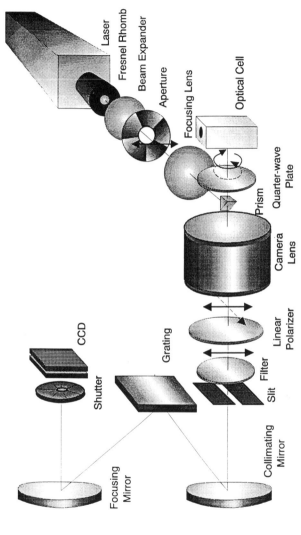

Figure 5. Optical layout for the ROA spectrometer at Syracuse University in its backscattering mode for ICP, SCP DCP$_I$, or DCP$_{II}$ ROA measurements.

141

convention front-side illuminated design which has a quantum efficiency near 25%. The principal advantage of our instrument is flexibility and stability. The quarter waveplates are essentially passive rather than active optical elements that are much less sensitive to variations in optcial alignment and temperature. The motorized stages that control the angular position of the quarter waveplate has a reproducible step size of 0.05°. The instrument can maintain a very precise alignment for weeks at a time, and we have successfully averaged data sets without change in the ROA spectrum for over 60 h. Hence, our instrument is capable of producing ROA of extremely high signal quality, even though our efficiency is not at the state-of-the-art level enjoyed by the Glasgow instrument.

B. Elimination of Artifacts

The principal impediment to the widespread application of ROA has been the occurrence of optical artifacts in the observed spectra. The origin of artifacts has been traced to imperfections in the balance of the polarization modulation cycle between right and left circularly polarized states.[60, 61] Hug demonstrated the relationship between observed artifacts and the ordinary Raman depolarization ratio.[60] His analysis explained the tendency for highly polarized Raman bands to exhibit large artifacts. Hug's approach to artifact reduction was to employ a dual optical collection system which effectively canceled artifacts due to the depolarization ratio. Barron and Vrbancich extended this analysis to include solid angle collection and a description of the ellipticities of the incident circular polarization states.[61] An interesting aspect of this work is the prediction that artifacts are suppressed if the ellipticities of the right and left circularly polarized states are simply balanced, rather than perfectly circular. This principle is used in the design of SCP and DCP ROA experiments where perfectly circularly polarized *scattered* light cannot be simultaneously maintained over a wide range of wavelengths.

Further refinements in the understanding of artifacts of ROA has involved the use of Mueller matrix theory.[43] In this approach, the contributions of all the optical elements in the optical train can be determined, which leads to a completely quantitative specification of artifact intensities. Using this approach, artifact analyses have been carried out for all four forms of ROA for various scattering angles.[42, 44, 62] From the analysis carried out for the Syracuse ROA instrument[42, 62] has emerged specific procedures for reducing artifacts through fine adjustments of the zeroth-order quarter waveplates employed in the optical train.

Based on the artifact analyses carried out to date, an in-depth understanding of artifact origin and control has been achieved. As a result, ROA spectra can now be measured routinely with little or no contribution from artifacts above the instrument noise level.

C. Applications to Molecules of Biological Interest

The remarkable improvements in ROA instrumentation that have occurred over the last several years have sparked a flurry of activity to measure, for the first time, ROA in biological molecules. The earliest of these came from the Glasgow laboratory,[63–66] where unpolarized ICP backscattering measurements were carried out for amino acids,[63] peptides and proteins,[64] cyclodextrins,[65] and a variety of carbohydrates, including simple mono- and disaccharides.[66] This progress was reviewed,[4] and soon thereafter DCP$_I$ backscattering measurements, exhibiting high signal-to-noise ratios from long collection times, were reported for amino acids, peptides, and monosaccharides.[5] A more detailed analysis of the backscattering ICP ROA spectra for alanine at several pH levels was carried out using ab initio molecular orbital calculations, as mentioned above,[53] where some encouraging degree of agreement was achieved between theory and experiment. This work was refined experimentally based on upgrades in the ROA instrument, extension of the pH ranges to isolate pure cationic and anionic species, and a revised interpretation.[54] The backscattering ICP spectra of tartaric acid in H_2O and D_2O have been presented and analyzed using ab initio methods[55], and the subject of backscattering ICP ROA in proteins has been revisited[67] with improved instrumentation compared with the earlier spectra.[64] Additional aspects of the application of ROA to biological molecules have appeared in several additional reviews.[6–8, 68] These include the observation of ICP ROA in nucleosides[8, 26] and DCP$_I$ ROA in ephedrine molecules.[6, 42, 69]

From these studies it has become clear that backscattering ROA, in either the ICP or DCP$_I$ modes, can be used to study biological molecules in aqueous solution. Further improvements in instrumentation and technique will permit spectra to be observed in lower concentrations. Background fluorescence is a problem for most samples, although purification procedures reduce this interference, as well as the observation that the fluorescence decreases monotonically with time if the sample is placed in the laser beam. Hence, for some samples, fluorescence can be reduced to acceptable levels by this form of laser pretreatment, although care needs to be taken that the sample itself has not been compromised.

VIII.   RESONANCE ENHANCED ROA

Although, ROA has matured to a point where the routine measurement of biological samples is now at hand, the technique has not yet been extended to the successful measurement of resonance ROA. In this section we explore some of the reasons for the situation, some of the theoretical consequences of the resonance limit for ROA, and some of the goals for future use of resonance ROA in the study of molecules of chemical and biological interest.

The successful measurement of resonance ROA will require surmounting a number of experimental obstacles. Because resonance enhancement requires laser excitation near or within electronic absorption bands in the sample, beam attenuation and interfering fluorescence are more of a problem than in ordinary Raman scattering. For chiral samples (required for ROA), optical activity associated with electronic absorption and fluorescence can interfere with ROA. The largest effects will arise from a fluorescence optical activity background. If circular dichroism were present at the wavelength of the incident radiation, a parental Raman bias would be imposed on an ICP or DCP ROA spectrum because the intensity left and right circularly polarized incident radiation reaching the point of scattering would not be the same. Further, if fluorescence were present, a fluorescence background would appear in the ROA due to unequal amounts of fluorescence associated with the intensity modulations of the right and left circularly polarized incident laser radiation. A similar, but more compolex, situation arises for SCP and the DCP ROA experiments due to circular dichroism affecting the amounts of left and right circularly polarized scattered radiation emerging from the sample cell across the entire range of scattered wavelengths, and the spectrum of circularly polarized emission associated with the excited state fluorescence emission spectrum of the sample.

A second set of difficulties arises from sample heating, degradation, and photochemistry. All of these problems are associated with exposing an absorbing sample to intense laser radiation for long periods of time, times that are long relative to the measurement of an ordinary resonance Raman spectrum due to the requirement of sufficient signal-to-noise ratio. To observe a typical ROA spectrum, the signal-to-noise ratio of the parent Raman spectrum must be 10^4 or greater, a requirement that has probably not been met by any resonance Raman spectrum published to date. Prospects for overcoming this difficulty involve using a spinning cell or a flow cell of some kind so that a given volume of sample is in the beam for only a brief moment, and any heating is gradual and nonlocal.

When these spectral and experimental difficulties have been overcome, a rich variety of fascinating ROA spectra will await the investigator. As resonance is approached, differences will emerge in a variety of ROA spectra, as illustrated by the differences in intensities given in Table I for the general case compared to Table II for the far-from-resonance case. In the limit of strong resonance enhancement, the summation over excited states j in Eq. (9) reduces to just those vibronic components that are in direct resonance with the incident laser radiation. In this case the tensors in Eqs. (9a)–(9e) reduce to

$$\tilde{\alpha}_{\alpha\beta} = \frac{1}{\hbar} \frac{\langle m|\hat{\mu}_\alpha|j\rangle\langle j|\hat{\mu}_\beta|n\rangle}{(\omega_{jn} - \omega_0 + i\Gamma_j)} \tag{47a}$$

$$\tilde{G}_{\alpha\beta} = \frac{1}{\hbar} \frac{\langle m|\hat{\mu}_\alpha|j\rangle\langle j|\hat{m}_\beta|n\rangle}{(\omega_{jn} - \omega_0 + i\Gamma_j)} \tag{47b}$$

$$\tilde{\mathscr{G}}_{\alpha\beta} = \frac{1}{\hbar} \frac{\langle m|\hat{m}_\alpha|j\rangle\langle j|\hat{\mu}_\beta|n\rangle}{(\omega_{jn} - \omega_0 + i\Gamma_j)} \tag{47c}$$

$$\tilde{A}_{\alpha,\beta\gamma} = \frac{1}{\hbar} \frac{\langle m|\hat{\mu}_\alpha|j\rangle\langle j|\hat{\Theta}_{\beta\gamma}|n\rangle}{(\omega_{jn} - \omega_0 + i\Gamma_j)} \tag{47d}$$

$$\tilde{\mathscr{A}}_{\alpha,\beta\gamma} = \frac{1}{\hbar} \frac{\langle m|\hat{\Theta}_{\beta\gamma}|j\rangle\langle j|\hat{\mu}_\alpha|n\rangle}{(\omega_{jn} - \omega_0 + i\Gamma_j)} \tag{47e}$$

where the so-called nonresonance term is dropped as negligible compared to the resonance term. In the limit of exact resonance, the frequency difference in the denominator vanishes and the denominator is governed by the pure imaginary bandwidth term. In this limit, which can probably be satisfied for only a few vibronic band components at a time, all complex tensors that are predominantly real will become predominantly imaginary and vice versa. This is illustrated for the roman magnetic dipole optical activity tensor in Eq. (47b), which can be rewritten as

$$\tilde{G}_{\alpha\beta} = \frac{1}{\hbar} \left[\frac{\langle m|\hat{\mu}_\alpha|j\rangle\langle j|\hat{m}_\beta|n\rangle}{(\omega_{jn} - \omega_0)^2 + \Gamma_j^2} \right] [(\omega_{jn} - \omega_0) - i\Gamma_j] \tag{48}$$

Far from resonance the $\omega_{jn} - \omega_0$ terms dominate the Γ_j terms, but as $\omega_{jn} - \omega_0$ becomes smaller than Γ_j and approaches zero, the numerator is governed by the $i\Gamma_j$ term and the components in the matrix elements that made imaginary contributions would now make a real ones to the overall complex tensor and vice versa. It follows that as the limit $\omega_{jn} - \omega_0 = 0$ is approached for an ideal isolated excited-state resonance, the LP ROA invariants in Eq. (23) may become large and the CP ROA invariants in Eq. (22) may become small and vanish. By investigating the CP and LP ROA spectra in the regions of strong laser resonance, these interesting effects should be observable. It is also clear by considering the pairs of roman and script optical activity tensors in Eq. (47) that in the limit of one resonant vibronic excited state, there is a greater likelihood for large differences to arise, and hence for large differences in right-angle ICP and SCP ROA spectra, or in backscattering ICP and DCP_I ROA, for example.

The principal attraction of resonance ROA is its potential for biological applications. Ordinary resonance Raman scattering is already established as a powerful tool in the study of biological chromophores. The extremely local character of resonance Raman spectra and the need to carry out measurements at very low concentrations to avoid absorption losses of the excited and scattered light are the most important factors in the success of this technique. By adding the dimension of optical activity to resonance Raman scattering, it will be possible to probe the stereochemistry of biological chromophores at a level of sensitivity and specificity that is not available by other means. It is interesting to note that metallic species with variable oxidation states are often the site of both catalytic biological activity and resonance Raman chromophoric enhancement. It is an exciting prospect to consider adding the stereosensitivity of ROA to biological studies of this kind.

As a final note to this section, we mention that the related phenomenon of magnetic Raman optical activity has been observed, but only in the case of resonance enhancement.[70] In magnetic ROA, however, the spectra are more intense than they are in natural ROA. Also, as with other forms of magnetic optical activity, there is no requirement that the sample molecules be chiral, and in the absence of natural ROA, a simple reversal of the magnetic field yields a reversal in sign of the entire magnetic ROA spectrum. The most recent progress in this area has involved studies of the anion of irrdium hexabromide,[71] and the protein ferrocytochrome c.[72]

IX. CONCLUSIONS

In this review we have traced the principal conceptual and technical advances that have brought ROA from its initial discovery to its present state of theoretical and experimental development. A secondary aim has been to present the fundamental equations associated with the complete theory of CP and LP ROA in a compact and coherent notation. From considerations of both the theory and measurement of ROA, this area of spectroscopy has undergone a renaissance in the last several years and is now poised to make significant contributions to our understanding of the structural and conformational details of chiral molecules in solution. Particularly promising are studies of biological molecules in aqueous solutions. Ab initio molecular orbital calculations of ROA show promise and are crucial to the detailed interpretation of ROA spectra and the translation of these spectra into new information about the observed conformational states of chiral molecules. The field of resonance ROA represents an exciting new frontier for ROA from the standpoints of understanding more about the basic theory of CP and LP ROA and of probing sites of catalytic activity in biological molecules.

Acknowledgments

Support of this work from a grant from the National Institute of Health (GM-23567) is gratefully acknowledged.

References

1. L. D. Barron, *Molecular Light Scattering and Optical Activity*, Cambridge University Press, Cambridge, UK, 1982.

2. L. A. Nafie and C. G. Zimba, in T. A. Spiro (Ed.), *Biological Applications of Raman Spectroscopy*, Vol. 1, Wiley, New York, 1987, p. 307.

3. L. D. Barron, in H. D. Bist, J. R. Durig, and J. F. Sullivan (Eds.), *Vibrational Spectra and Structure*, Vol. 17B, Elsevier, Amsterdam, 1989, p. 343.

4. L. D. Barron, A. R. Gargaro, L. Hecht, Z. Q. Wen, and W. Hug, in S. A. Akhmanov, M. Y. Poroshino, N. I. Koroteev and B. N. Toleutaev (Eds.), *Laser Applications in the Life Sciences*, Proc. *SPIE* **1403**, 66 (1991).

5. L. A. Nafie, D. Che, G.-S. Yu, and T. B. Freedman, in R. R. Birge and L. A. Nafie (Eds.), *Biomolecular Spectroscopy II*, Proc. *SPIE* **1432**, 37 (1991).

6. L. A. Nafie, in H. Klein and G. Snatzke (Eds.), *Lectures and Posters of the Fourth International Conference on Circular Dichroism*, *Bochum*, *Germany*, Ruhrgebiet, Essen, 1991, p. 101.

7. L. A. Nafie, in J. M. Lerner (Ed.), *Optically Based Methods for Process Analysis*, Proc. *SPIE* **1681**, 29 (1992).

8. L. D. Barron and L. Hecht, in R. E. Hester and R. J. H. Clark (Eds.), *Advances in Spectroscopy*, Vol. 20, in press.

9. T. B. Freedman and L. A. Nafie, in E. Eliel and S. Wilen (Eds.), *Topics in Stereochemistry*, Vol. 17, Wiley, New York, 1987 pp. 113–206.

10. T. A. Keiderling in J. R. Ferraro and K. Krishman (Eds.), *Practical Fourier Transform Infrared Spectroscopy*, Academic, San Diego, 1990, pp. 203–284.

11. T. A. Keiderling and P. Pancoska, in R. E. Hester and R. J. H. Clark (Eds.), *Advances in Spectroscopy*, Vol. 20, in press.

12. P. W. Atkins and L. D. Barron, *Mol. Phys.* **16**, 453 (1969).

13. L. D. Barron and A. D. Buckingham, *Mol. Phys.* **20**, 1111 (1971).

14. L. D. Barron, M. P. Bogaard, and A. D. Buckingham, *Nature* **241**, 113 (1973).

15. L. D. Barron, *Mol. Phys.* **31**, 1929 (1976).

16. W. Hug and H. Surbeck, *Chem. Phys. Lett.* **60**, 186 (1979).

17. D. L. Andrews, *J. Chem. Phys.* **72**, 4141 (1980).

18. W. Hug, in J. Lascombe and P. V. Huong, (Eds.) *Raman Spectroscopy*, Wiley-Heyden, Chichester, UK, 1982, p. 3.

19. L. D. Barron and A. D. Buckingham, *Annu. Rev. Phys. Chem.* **26**, 381 (1975).

20. L. D. Barron and J. R. Escribano, *Chem. Phys.* **98**, 437 (1985).

21. K. M. Spencer, T. B. Freedman, and L. A. Nafie, *Chem. Phys. Lett.* **149**, 367 (1988).

22. L. A. Nafie and T. B. Freedman, *Chem. Phys. Lett.* **154**, 260 (1989).

23. L. Hecht, L. D. Barron, and W. Hug, *Chem. Phys. Lett.* **158**, 341 (1989).

24. L. D. Barron, L. Hecht, W. Hug, and M. J. MacIntosh, *J. Am. Chem. Soc.* **111**, 8731 (1989).

25. L. Hecht, D. Che and L. A. Nafie, *Appl. Spectrosc.* **45**, 18 (1991).

26. L. Hecht, L. D. Barron, A. R. Gargaro, Z. Q. Wen, and W. Hug, *J. Raman Spectrosc.* **23**, 401 (1992).

27. D. Che, L. Hecht, and L. A. Nafie, *Chem. Phys. Lett.* **180**, 182 (1991).

28. L. D. Barron, J. R. Escribano, and J. F. Torrance, *Mol. Phys.* **57**, 653 (1986).

29. L. D. Barron and J. R. Escribano, *Chem. Phys. Lett.* **126**, 461 (1986).

30. L. D. Barron, L. Hecht and P. L. Polavarapu, *Chem. Phys. Lett.* **154**, 251 (1989).

31. L. D. Barron, L. Hecht, and S. M. Blyth, *Spectromchim. Acta* **45A**, 375 (1989).

32. L. Hecht and L. D. Barron, *Spectromchim. Acta* **45A**, 671 (1989).

33. L. D. Barron, L. Hecht, A. R. Gargaro, and W. Hug, *J. Raman Spectrosc.* **21**, 375 (1990).

34. D. Che and L. A. Nafie, *Chem. Phys. Lett.* **189**, 35 (1992).

35. L. Hecht, *Chem. Phys. Lett.* **195**, 518 (1992).

36. D. Che, L. Hecht, and L. A. Nafie, in J. R. Durig and J. F. Sullivan (Eds.), *Proceedings of the 12th International Conference on Raman Spectroscopy*, Wiley, New York, 1990, p. 846.

37. L. Hecht, D. Che, and L. A. Nafie, *J. Phys. Chem.* **96**, 4266 (1992).

38. L. Hecht and L. A. Nafie, *Mol. Phys.* **72**, 441 (1991).

39. L. Hecht and L. A. Nafie, *Chem. Phys. Lett.* **174**, 575 (1990).

40. R. A. Harris and W. M. McClain, *Chem. Phys. Lett.* **195**, 633 (1992).

41. L. Hecht and L. A. Nafie, *Chem. Phys. Lett.* **195**, 637 (1992).

42. D. Che, Thesis, Syracuse University, Syracuse, NY, 1992.

43. L. Hecht, B. Jordanov, and B. Schrader, *Appl. Spectrosc.* **41**, 295 (1987).

44. L. Hecht and L. D. Barron, *Appl. Spectrosc.* **44**, 483 (1990).

45. J. Michl and E. W. Thulstrup, *Spectroscopy with Polarized Light*, VCH, New York, 1986, p. 25.

46. L. A. Nafie, M. Pezolet, and W. L. Peticolas, *J. Raman Spectrosc.* **1**, 455 (1973).

47. J. Nestor and T. G. Spiro, *J. Raman Spectrosc.* **1**, 539 (1973).

48. P. K. Bose, L. D. Barron, and P. L. Polavarapu, *Chem. Phys. Lett.* **155**, 423 (1989).

49. P. K. Bose, P. L. Polavarapu, L. D. Barron, and L. Hecht, *J. Phys. Chem.* **94**, 1734 (1990).

50. T. M. Black, P. K. Bose, P. L. Polavarapu, L. D. Barron, and L. Hecht, *J. Am. Chem. Soc.* **112**, 1479 (1990).

51. P. L. Polavarapu, *J. Phys. Chem.* **94**, 8106 (1990).

52. P. L. Polavarapu, *Chem. Phys. Lett.* **173**, 511 (1990).

53. L. D. Barron, A. R. Gargaro, L. Hecht, and P. L. Polavarapu, *Spectrochim. Acta* **47A**, 1001 (1991).

54. L. D. Barron, A. R. Gargaro, L. Hecht, and P. L. Polavarapu, *Spectrochim. Acta* **48A**, 261 (1992).

55. L. D. Barron, A. R. Gargaro, L. Hecht, P. L. Polavarapu, and H. Sugeta, *Spectrochim. Acta* **48A**, 1051 (1992).

56. L. D. Barron, L. Hecht, and P. L. Polavarapu, *Spectrochim. Acta* **48A**, 1993 (1992).

57. R. D. Amos and J. E. Rice, CADPAC: The Cambridge Analytic Derivatives Package, issue 4.0, Cambridge University, Cambridge, UK, 1987.

58. R. D. Amos, N. C. Handy, K. J. Jalkanen, and P. J. Stephens, *Chem. Phys. Lett.* **133**, 21 (1987).

59. K. J. Jalkanen, R. W. Kawiecki, P. J. Stephens, and R. D. Amos, *J. Phys. Chem.* **94**, 7040 (1990).

60. W. Hug, *Appl. Spectrosc.* **35**, 115 (1981).

61. L. D. Barron and J. Vrbancich, *J. Raman Spectrosc.* **15**, 47 (1984).

62. D. Che and L. A. Nafie, *Appl. Spectrosc.* **47**, 544 (1993).

63. L. D. Barron, A. R. Gargaro, and L. Hecht, in J. R. Durig and J. F. Sullivan (Eds.), *Proceedings of the 12th International Conference on Raman Spectroscopy*, Wiley, New York, 1990, p. 834.

64. L. D. Barron, A. R. Gargaro, and Z. Q. Wen, *J. Chem. Soc. Chem. Commun.* 1034 (1990).

65. L. D. Barron, A. R. Gargaro, Z. Q. Wen, D. D. MacNicol, and C. Butters, *Tetrahedron: Asymmetry* **1**, 513 (1990).

66. L. D. Barron, A. R. Gargaro, and Z. Q. Wen, *Carb. Res.* **210**, 39 (1991).

67. L. D. Barron, Z. W. Wen, and L. Hecht, *J. Am. Chem. Soc.* **114**, 784–786 (1992).

68. L. A. Nafie and T. B. Freedman, in J. F. Riordan and B. L. Vallee (Eds.), *Metallochemistry*, *Methods in Enzymology*, Part C, Academic, Orlando, FL, 1993, in press.

69. G.-S. Yu, D. Che, T. B. Freedman, and L. A. Nafie, submitted for publication.

70. L. D. Barron, *Nature* **275**, 372 (1975).

71. L. D. Barron, J. Vrbancich, and R. S. Watts, *Chem. Phys. Lett.* **89**, 71 (1982).

72. L. D. Barron, C. Meehan, and J. Vrbancich, *J. Raman Spectrosc.* **12**, 251 (1982).

QUANTUM OPTICS AND THE CLASSICAL LIMIT OF QUANTUM MECHANICS

LUCA BONCI AND PAOLO GRIGOLINI*

Dipartimento di Fisica dell'Università di Pisa, Pisa, Italy

CONTENTS

I. INTRODUCTION

According to the orthodox view quantum mechanics is a satisfactory universal theory, from which classical mechanics is naturally derived in the proper limit. However, quantum mechanics does not completely fulfill this promise and since its inception has been fraught with paradoxical aspects. As shown by von Neumann,[1] the process of measurement itself, even in a

---

*Paolo Grigolini is also affiliated with the Istituto di Biofisica del CNR, Pisa, Italy, and the Department of Physics, University of North Texas, Denton, Texas.

Modern Nonlinear Optics, Part 3, Edited by Myron Evans and Stanisław Kielich. Advances in Chemical Physics Series, Vol. LXXXV.
ISBN 0-471-30499-9 © 1994 John Wiley & Sons, Inc.

total idealized version, has the unavoidable effect of making the microscopic quantum uncertainty grow and become macroscopic. Let us review
this basic aspect of quantum mechanics, so that the reader who is not
acquainted with these problems may understand the essence of this
chapter.

Let us consider an observable A of a quantum system S, with two
eigenstates $|a\rangle$ and $|b\rangle$. To measure A we must set the quantum system
into contact with the macroscopic measurement apparatus M. Von
Neumann[1] imagined the measurement act as a collision-like process
between A and M, and he used a unitary transformation to describe the
effect of this collision. If prior to collision the system is placed in a state
$|a\rangle$ and the measurement device in state $|\mu\rangle$, then the measurement act
produces the following transition:

$$|a\rangle|\mu\rangle \rightarrow |a\rangle|+\rangle \tag{1}$$

If the system S is originally placed in state $|b\rangle$ then the collision-like
process of interaction between the system S and the measuring apparatus
M results in

$$|b\rangle|\mu\rangle \rightarrow |b\rangle|-\rangle \tag{2}$$

$|+\rangle$ and $|-\rangle$ are quantum-mechanical states corresponding, for instance,
to distinct positions of a pointer. This is a nice theory showing that it is
possible for us to assess whether the quantum system is in state $|a\rangle$ or $|b\rangle$.

However, there are deep problems involved by this way of modeling the
measurement act. Quantum mechanics would have no dynamical relevance without the existence of states that are linear combinations of
distinct states. Thus, we have also to answer the question of what the
result of a measurement would be in the general case when the system is
placed in the state

$$|\Psi_0\rangle = \alpha|a\rangle + \beta|b\rangle \tag{3}$$

Since quantum mechanics is linear, from (1) and (2) we immediately have
that the measurement act results in the transition

$$|\psi_0\rangle|\mu\rangle \rightarrow \alpha|a\rangle|+\rangle + \beta|b\rangle|-\rangle \tag{4}$$

This is a paradoxical condition equivalent to the well-known Schrödinger's
cat.[2] Indeed, states $|+\rangle$ and $|-\rangle$ are macroscopically distinct, and (4)
would be equivalent to stating that it is possible to see a macroscopic
object in a superposition of two distinct macroscopic states. Schrödinger[2]

made this paradox still more striking by imagining that the object to measure is an atom which can either stay in an excited state, $|a\rangle$, or decay into the ground state, $|b\rangle$. If it decays, it triggers a device resulting in the death of a cat, $|-\rangle$. If it does not, the cat is still alive, $|+\rangle$. Thus, Eq. (4) in the language of Schrödinger would represent a cat in a strange state of suspended animation, neither alive nor dead.

There have been many attempts to amend quantum mechanics from this paradoxical aspect. In recent years, especially thanks to the proposal made by Leggett,[3] a new approach has emerged in this field of research. The supposedly unlimited validity of quantum mechanics is no longer accepted and the reductionist point of view according to which the macroscopic world is still quantum mechanical is questioned.[3] It is thought that it might be beneficial to purposely look for experimental manifestation of quantum mechanics at a macroscopic level. A benefit of this research work might be the disclosure of regions where reality departs from quantum mechanics.

The purpose of this review is to illustrate some recent results in the field of quantum optics which are claimed by the authors to bear a closed relation with the historic paradoxes of quantum mechanics. Although we are not totally convinced that these claims are proper, we do believe that these interesting results have to do with the intriguing problem of recovering classical from quantum physics. The latter is supposed to be more fundamental than the former, and the former is expected to derive from the latter when large quantum numbers or small de Broglie wave lengths are considered. The invention of coherent states by Glauber[4] is considered to be a key ingredient for this reconciliation to be operated. On the contrary, we shall see that quantum optics affords interesting theoretical conditions where this is not possible. We also devote some efforts to defining the physical conditions that might make it easier to detect these interesting quantum-mechanical effects.

II. THE SPIN-BOSON MODEL AND THE ROTATING WAVE APPROXIMATION: THE JAYNES-CUMMING MODEL

The simplest possible way of modeling the interaction between radiation and matter rests on approximating an atom with a two-level quantum system and the electromagnetic field with a single mode, namely a boson. The two-level system is described by using the Pauli matrices and the boson field is represented by a quantum harmonic oscillator, thereby leading to the following Hamiltonian:

$$H = \tfrac{1}{2}\omega_0\sigma_z + g\sigma_x x + \tfrac{1}{2}\Omega^2 x^2 + \tfrac{1}{2}v^2 \qquad (5)$$

This is called the spin-boson Hamiltonian and is widely used in the literature, for a large variety of applications ranging from quantum optics[5-8] and quantum dissipation[9] to chemical reaction theory.[10] In spite of the heavy simplifications on which Eq. (5) is rooted, the calculations necessary to evaluate the time evolution of the system are still beyond the range of an analytical treatment. For this reason researchers in related areas frequently replace the Hamiltonian of (5) with one easier to handle and leading to exact results, one resting on the so-called rotating wave approximation (RWA).[11] In quantum optics the Hamiltonian of (5) supplemented by the RWA is also known as the Jaynes-Cummings model (JCM)[12] and in recent years real experiments have been carried out in physical conditions closer and closer to that supposed by the RWA.[13] This explains the increasing popularity of the JCM, and its wide use in quantum optics. The JCM has also been used to deal with the controversial quantum-mechanical aspects mentioned in the Introduction, namely measurement theory and the search for macroscopic quantum superpositions (MQS),[14, 15] and we plan to discuss these interesting applications of the JCM in detail.

A proper discussion of these problems cannot be made without a deep understanding of the physical significance of the RWA. For this reason we devote this section to reviewing the RWA, and especially those aspects of this approximation that we do not find conveniently illustrated in the current literature. A popular way of illustrating the rotating wave approximation is derived from the theory of magnetic resonance.[16] Let us consider a magnetic dipole **M** interacting with a constant magnetic field H_0, directed along the z axis and with the magnetic field $H_1(t)$ of a radiation polarized along the x direction. The classical Hamiltonian describing this system reads

$$\mathscr{H} = -g_M \mathbf{M} \cdot \mathbf{H} = -g_M M_z H_0 - g_M M_x H_1 \cos(\Omega t) \qquad (6)$$

where g_M is the gyromagnetic ratio and Ω is the frequency of the radiation field. The connection between this expression and the Hamiltonian of Eq. (5) is straightforward: The components M_z and M_x of the magnetic dipole correspond to the mean value of the Pauli operators σ_z and σ_x, respectively, and frequency ω_0 appearing in (5) is identified as the Larmor angular velocity, $g_M H_0$.

Note that the Hamiltonian (5) is a quantum-mechanical operator, whereas (6) describes the energy of a classical dipole interacting with a classical radiation field. A still more significant difference between the quantum Hamiltonian (5) and the classical Hamiltonian (6) lies in the fact that the time evolution of the classical radiation field is unaffected by the dynamics of the magnetic dipole. The corresponding condition does not apply to the system described by the Hamiltonian (5). In such a case the time evolution of the quantum harmonic operator is also affected by its interaction with the two-level system. When we focus our attention on the two-level system and on the action exerted on its dynamics by the radiation, we are tempted to disregard the action that, in turn, the two-level system exerts on the radiation. Throughout this paper we denote this latter action as *reaction*, to stress its feedback nature.

In spite of the assumption that the reaction field can be neglected, Eq. (6) still results in a time evolution for the components of the magnetic dipole incompatible with an easy analytical treatment. We can make the Hamiltonian (6) tractable by adopting the following approximation. The unperturbed motion of the dipole in the presence of the static field H_0 is a precession around the z direction with a frequency equal to the Larmor frequency. In a reference frame rotating with the magnetic dipole unperturbed by the radiation field, the total magnetic field changes into

$$\mathbf{H}' \equiv \left(H_1 \cos(\Omega t + \varphi)\cos(\omega_0 t), \; -H_1 \cos(\Omega t + \varphi)\sin(\omega_0 t), \; H_0 \right) \quad (7)$$

where φ is the initial angle between the direction of the magnetic dipole and the abscissa axis of the new reference frame. With a proper choice of the phase φ the magnetic field reduces to

$$\mathbf{H}'' \equiv \left(H_1 \cos(\Omega - \omega_0)t, \; H_1 \sin(\Omega + \omega_0)t, \; H_0 \right) \quad (8)$$

which means a frequency for the x component of the magnetic field different from that of the y component. At resonance,

$$\Omega = \omega_0 \equiv g_{\mathrm{M}} H_0 \quad (9)$$

we see that the x component becomes time independent and that the y component oscillates with a frequency that is twice as large as the Larmor frequency. If the intensity of the perturbing field H_1 is small we can safely neglect the fast oscillating component, since the magnetic dipole moves with a much slower rate and perceives its (vanishing) mean value rather than the instantaneous one. In conclusion, in the rotating frame of reference the precession of the spin takes place about the constant effective field

$$\mathbf{H}'' \equiv \left(H_1, 0, H_0 \right) \quad (10)$$

In other words, in the original reference system, the linearly polarized magnetic field can be expressed as the superposition of two circularly polarized fields, lying on the xy plane and counterrotating with the frequency Ω. If Ω is very close to the Larmor frequency, the rotation of one of these two fields follows closely that of the dipole, while the other component of the field rotates in the opposite direction. The former component is perceived by the magnetic dipole as almost time indepen-dent, whereas the latter component undergoes extremely fast oscillations with respect to the magnetic dipole. Under the assumption

$$H_1 \ll H_0 \tag{11}$$

the Larmor frequency, and so the frequency of the counterrotating compo-nent, turns out to be much larger than the frequency of the dipole precession around the effective field, especially if we consider a condition close to the resonance condition (9). Indeed, the precession frequency around the effective field is

$$\omega_{\mathrm{eff}} = \left[\left(\omega_0 - \Omega \right)^2 + g_{\mathrm{M}} H_1 \right]^{1/2} \tag{12}$$

At resonance we get the largest time-scale separation between the slow precession about the effective field and the fast oscillations of the counter-rotating component of the magnetic field. Thus, the motion of the mag-netic dipole is not affected by the counterrotating component, since again the magnetic dipole does not "feel" its instantaneous value, but the (vanishing) time average of it. All this is a sound justification for the replacement of the Hamiltonian (6) with the following one:

$$\mathscr{H}_{\mathrm{RWA}} = -g_{\mathrm{M}} M_z H_0 - g_{\mathrm{M}} \left(M_x H_1 \cos(\Omega t) - M_y H_1 \sin(\Omega t) \right) \tag{13}$$

This intuitive illustration of the RWA suggests that in some cases the RWA Hamiltonian might be an exact representation of the system rather than an approximation. This is because in principle it is possible to adopt a field that is circularly polarized rather than linearly polarized. However, if we establish the RWA within the quantum-mechanical formalism as is shown in the quantum optics literature, we must conclude that the quantum version of the RWA is never exact, not even if we use a circularly polarized rather than a linearly polarized one. Let us see this aspect in detail by rewriting the quantum-mechanical Hamiltonian of Eq. (5) as

$$\mathscr{H} = \frac{1}{2}\omega_0 \sigma_z + \frac{g}{\sqrt{2\Omega}}(\sigma_+ + \sigma_-)(a + a^+) + \Omega a^+ a \tag{14}$$

In writing Eq. (14) we used the explicit expressions for the operators position and momentum of the oscillator in terms of the creation and destruction operator a^+ and a,

$$x = \frac{1}{\sqrt{2\Omega}}(a^+ + a) \qquad v = i\sqrt{\frac{\Omega}{2}}(a^+ - a) \qquad (15)$$

and we replaced σ_x with its expression in terms of the operator σ_+ and σ_- defined by

$$\sigma_{\pm} \equiv \frac{(\sigma_x \pm i\sigma_y)}{2} \qquad (16)$$

The formal counterpart of this intuitive illustration is obtained by writing the Schrödinger equation

$$i\frac{\partial}{\partial t}|\psi(t)\rangle = \mathcal{H}|\psi(t)\rangle \qquad (17)$$

in the interaction picture. This implies that the Hamiltonian $\mathcal{H}$ is divided into an unperturbed and a perturbation part as follows

$$\mathcal{H} = \mathcal{H}_0 + \mathcal{H}_1 \qquad (18)$$

Then we make the transformation

$$|\tilde{\psi}(t)\rangle \equiv \exp(-i\mathcal{H}_0 t)|\psi(t)\rangle \qquad (19)$$

With this transformation the Schrödinger equation reads

$$i\frac{\partial}{\partial t}|\tilde{\psi}(t)\rangle = \tilde{\mathcal{H}}_1(t)|\tilde{\psi}(t)\rangle \qquad (20)$$

where

$$\tilde{\mathcal{H}}_1(t) \equiv \exp(-i\mathcal{H}_0 t)\mathcal{H}_1 \exp(i\mathcal{H}_0 t) \qquad (21)$$

Equations (19) and (21) define the interaction picture. In general, this transformation makes an observable A change in time with the law

$$\tilde{A}(t) \equiv \exp(-i\mathcal{H}_0 t) A \exp(i\mathcal{H}_0 t) \qquad (22)$$

which is precisely the Heisenberg dynamics corresponding to the unperturbed Hamiltonian. The obvious definition of unperturbed Hamiltonian is

$$\mathscr{H}_0 = \tfrac{1}{2}\omega_0\sigma_z + \Omega a^+ a \tag{23}$$

which implies

$$\mathscr{H}_1 = \frac{g}{\sqrt{2\Omega}}(\sigma_+ + \sigma_-)(a + a^+) \tag{24}$$

We note that the unperturbed dynamics of the operators a, a^+ and σ_+ and σ_- are given by

$$\begin{aligned}
\tilde{a}(t) &= a(0)e^{-i\Omega t} & \tilde{\sigma}_+(t) &= \sigma_+(0)e^{-i\omega_0 t} \\
\tilde{a}^+(t) &= a^+(0)e^{i\Omega t} & \tilde{\sigma}_-(t) &= \sigma_-(0)e^{i\omega_0 t}
\end{aligned} \tag{25}$$

Consequently, at resonance the Hamiltonian of Eq. (21) consists of a fast oscillating term plus a time-independent contribution. By averaging over the fast oscillations, we obtain the following expression for the interaction Hamiltonian:

$$\overline{\mathscr{H}}_1 = \frac{g}{\sqrt{2\Omega}}(\sigma_+ a + \sigma_- a^+) \tag{26}$$

namely, the time-independent contribution to (20). The RWA of quantum optics[5-8] rests on assuming that (21) can be replaced with (26). Coming back to the original representation we get the following expression for the approximated RWA Hamiltonian:

$$\mathscr{H}_{\mathrm{RWA}} = \frac{1}{2}\omega_0\sigma_z + \frac{g}{\sqrt{2\Omega}}(\sigma_+ a + \sigma_- a^+) + \Omega a^+ a \tag{27}$$

This approximation is equivalent to considering only the processes corresponding to the absorption (emission) of one photon and to the concurrent transition from the ground (excited) to the excited (ground) state of the atomic system.[17] In other words, the RWA Hamiltonian leaves room only for those processes that maintain the total energy of the system unchanged. Adopting the conventional perturbation approach we see that the virtual processes corresponding to an energy change would produce energy corrections proportional to λ^2, where

$$\lambda \equiv \frac{g}{\sqrt{2\Omega}}\sqrt{\bar{n}} \tag{28}$$

is the effective strength of the interaction. This effective strength depends on both the value of the coupling g and the intensity of the field, namely, the square root of the mean number of photons in the field, $\sqrt{\bar{n}}$. It is evident that for the RWA to work, it is necessary to keep small the value of the key parameter λ.

By applying the Heisenberg prescriptions, we obtain from the RWA Hamiltonian of (27) the following equations of motion:

$$\dot{\sigma}_x = -\omega_0\sigma_y - \frac{g}{\Omega}\upsilon\sigma_z$$

$$\dot{\sigma}_y = \omega_0\sigma_x - gx\sigma_z$$

$$\dot{\sigma}_z = gx\sigma_y + \frac{g}{\Omega}\upsilon\sigma_x \tag{29}$$

$$\dot{x} = \upsilon - \frac{g}{2\Omega}\sigma_y$$

$$\dot{\upsilon} = -\Omega^2 x - \frac{g}{2}\sigma_x$$

Note that the Heisenberg picture applied to the exact Hamiltonian (Eq. (5)) results in

$$\dot{\sigma}_x = -\omega_0\sigma_y$$

$$\dot{\sigma}_y = \omega_0\sigma_x - 2gx\sigma_z$$

$$\dot{\sigma}_z = 2gx\sigma_y \tag{30}$$

$$\dot{x} = \upsilon$$

$$\dot{\upsilon} = -\Omega^2 x - g\sigma_x$$

Let us disregard the reaction field, namely, according to our earlier definition, the contributions to the time evolution of x and y stemming from the action of the $\frac{1}{2}$-spin system on the harmonic oscillator. We see that even under this approximation the RWA Hamiltonian leads to a dynamical process distinct from that resulting from the full Hamiltonian, since the RWA Hamiltonian makes the motion of the $\frac{1}{2}$-spin system also depend on the velocity υ of the oscillator. This is correct, because according to the earlier interpretation of the RWA, this is equivalent to replacing the linearly polarized field with one circularly polarized, which, in turn, is equivalent to the linear superposition of two orthogonal one-dimensional oscillations with the phase difference $\delta\varphi = \pi/2$. Let us see this in detail and study the Hamiltonian describing the interaction between the $\frac{1}{2}$-spin system and two orthogonal oscillators, one oscillating along the x axis and one along the y axis. We get

$$H = \tfrac{1}{2}\omega_0\sigma_z + g(\sigma_x x + \sigma_y y) + \tfrac{1}{2}\Omega^2(x^2 + y^2) + \tfrac{1}{2}(v^2 + w^2) \quad (31)$$

The corresponding Heisenberg dynamics lead to the following set of equations:

$$\dot{\sigma}_x = -\omega_0\sigma_y + 2gy\sigma_z$$
$$\dot{\sigma}_y = \omega_0\sigma_x - 2gx\sigma_z$$
$$\dot{\sigma}_z = 2g(x\sigma_y - y\sigma_x)$$
$$\dot{x} = v \qquad\qquad\qquad (32)$$
$$\dot{v} = -\Omega^2 x - g\sigma_x$$
$$\dot{y} = w$$
$$\dot{w} = -\Omega^2 y - g\sigma_y$$

Let us now compare the set (32) to the set (29) and let us disregard again the reaction field in both sets. We see that the two sets of equations give rise to the same dynamics if we choose the initial conditions of the two oscillators, x and y, in such a way as to produce a circular polarization. The exact equivalence between the two pictures, with the reaction fields neglected, also implies the parameter g appearing in (29) to be identified with the quantity $2g$ appearing in (32). Under this condition $-v/\Omega$ can be identified with y.

It seems, in other words, that the RWA Hamiltonian of Eq. (27) is the quantum counterpart of the Hamiltonian describing the interaction between a classical dipole and a circularly polarized classical radiation field. However, by comparing (32) to (29) we see that the reaction field makes the RWA Heisenberg dynamics distinct from the dynamics of a $\frac{1}{2}$-spin system interacting with a circularly polarized field. Since the RWA Hamiltonian results in many interesting properties and there is an increasing search for the experimental realization of these effects (a part of this experimental and theoretical research work is reviewed here), we must preliminarily assess how accurate the representation of the physical reality the RWA is. Here we show that, due to the action of the reaction field, not even in the case of an interaction between atom and a circularly polarized radiation field does this approximation become exact.

We expect that Eqs. (29) and (32) lead to the same result in the limiting condition of very intense radiation field, since in this case the role of the reaction fields that make them different is strongly reduced. To quantitatively assess the discrepancy between the RWA, which is an approximation, and the interaction with a circularly polarized light, which in principle is a realizable physical condition, we must find a convenient approximation scheme for the calculation of the $\frac{1}{2}$-spin system dynamics to be easily carried out. Let us make the assumption that the oscillator is classical. We shall refer to this as the *semiclassical* approximation, rather than merely classical, since the quantum-mechanical nature of the $\frac{1}{2}$-spin system is fully retained. Note that the semiclassical approximation is equivalent to replacing terms such as $\langle y(t)\sigma_z(t)\rangle$ with $\langle y(t)\rangle\langle\sigma_z(t)\rangle$, thereby making it possible to truncate a hierarchy of equations which would become infinite. We are thus in a position to carry out an easy numerical calculation to compare, for instance, the time evolution $\langle\sigma_x(t)\rangle$ as resulting from the semiclassical approximation applied to (29) (note that this results in the set (33) discussed at the end of this section) and to (32), with equivalent initial conditions. The comparison is made by means of the results illustrated in Fig. 1. We see that in a case of a radiation field of weak intensity the two Hamiltonians result in different predictions. The validity of the semiclassical approximation is precisely the object of scrutiny of this paper. We show that it cannot be trusted in general, not even in the case of fields of large intensity if circularly polarized fields are used. It certainly cannot be trusted in the weak field case of Fig. 1. However, the results of Fig. 1 are a nice indication of the fact that at radiation fields of weak intensity the RWA Hamiltonian and the Hamiltonian of (31) are expected to lead to different results.

In conclusion, the RWA Hamiltonian and the spin-boson Hamiltonian with a circularly polarized field are expected to lead to identical predictions only in the limiting case of very large radiation field. In this case, the RWA Hamiltonian can no longer be regarded as a good approximation to the full Hamiltonian of (5). However, circularly polarized radiation fields should produce the same effects as those predicted by the RWA Hamiltonian. Thus, all the interesting predictions of the RWA Hamiltonian with fields of large intensity must be compared to experimental realizations of the processes of radiation–matter interaction with circularly polarized rather than linearly polarized radiation fields.

Before concluding this section, let us make some final remarks on the physical consequences of the semiclassical approximation. As mentioned

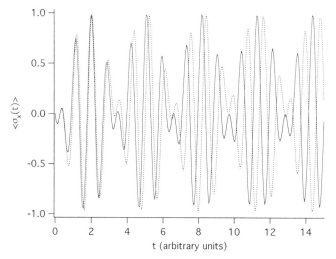

Figure 1. Time evolution for $\langle \sigma_x(t) \rangle$, as resulting from the semiclassical approximation on (29), equivalent to using (33) (solid line) and on (32) (dashed line). The parameters of the two systems are given the following values: $\omega_0 = 1$, $\Omega = 1$, $g = 3$, $\bar{n} = 5$.

earlier, the semiclassical approximation is equivalent to a factorization such as that of $\langle y(t)\sigma_z(t) \rangle$ into $\langle y(t) \rangle \langle \sigma_z(t) \rangle$. This factorization, in turn, generates a nonlinear structure that produces physical effects, which turn out to be dependent on whether we use the RWA Hamiltonian or the full Hamiltonian. Let us discuss some of these interesting properties in detail. As a consequence of the semiclassical approximation the two sets of equations (29) (RWA Hamiltonian) and (30) (full Hamiltonian) are changed into

$$\langle \dot{\sigma}_x \rangle = -\omega_0 \langle \sigma_y \rangle - \frac{g}{\Omega} v \langle \sigma_z \rangle$$

$$\langle \dot{\sigma}_y \rangle = \omega_0 \langle \sigma_x \rangle - gx \langle \sigma_z \rangle$$

$$\langle \dot{\sigma}_z \rangle = gx \langle \sigma_y \rangle + \frac{g}{\Omega} v \langle \sigma_x \rangle \qquad (33)$$

$$\dot{x} = v - \frac{g}{2\Omega} \langle \sigma_y \rangle$$

$$\dot{v} = -\Omega^2 x - \frac{g}{2} \langle \sigma_x \rangle$$

and

$$\langle \dot{\sigma}_x \rangle = -\omega_0 \langle \sigma_y \rangle$$

$$\langle \dot{\sigma}_y \rangle = \omega_0 \langle \sigma_x \rangle - 2gx \langle \sigma_z \rangle$$

$$\langle \dot{\sigma}_z \rangle = 2gx \langle \sigma_y \rangle \tag{34}$$

$$\dot{x} = v$$

$$\dot{v} = -\Omega^2 x - g \langle \sigma_x \rangle$$

respectively. Notice that $x(t)$ and $v(t)$ now denote c numbers rather than operators. These two sets of equations are expected to provide coinciding predictions under those conditions where the RWA is a good approximation, namely, at resonance with fields of weak intensity. However, their nonlinear structure leads to remarkable differences. The most striking difference between these two sets of equations concerns their integrability. In both systems the semiclassical approximation generates nonlinearity. However, in addition to the energy, the RWA admits a further constant of motion, thereby becoming integrable. The set of equations (34), on the contrary, is not integrable and in specific conditions is proven to be characterized by positive Lyapunov coefficients.[18-22]

Another interesting aspect of the semiclassical treatment of (29) and (30) concerns their stationary properties.[23] Setting equal to zero the left-side everywhere, we obtain (from the RWA set of (33))

$$\langle \sigma_z \rangle_{eq} = -\frac{2\omega_0 \Omega^2}{g^2} \equiv -\eta_{RWA}$$

$$\langle \sigma_x \rangle_{eq}^2 + \langle \sigma_y \rangle_{eq}^2 = 1 - \eta_{RWA}^2 \tag{35}$$

$$\Omega^2 x_{eq}^2 + v_{eq}^2 = \frac{g^2}{4\Omega^2} \left(1 - \eta_{RWA}^2\right)$$

and (from the set (34) corresponding to the Hamiltonian (5) with no approximation)

$$\langle \sigma_z \rangle_{eq} = -\frac{\omega_0 \Omega^2}{2g^2} \equiv -\eta$$

$$\langle \sigma_x \rangle_{eq}^2 = 1 - \eta^2 \qquad \langle \sigma_y \rangle_{eq} = 0 \tag{36}$$

$$x_{eq} = -\frac{g}{\Omega^2} \langle \sigma_x \rangle_{eq} \qquad v_{eq} = 0$$

The existence of the stationary states of (35) and (36) implies $\eta_{RWA} < 1$ and $\eta < 1$, respectively.

This analysis of the stationary states leads to the following scenario. With weak interactions $\eta_{RWA} > 1$ and $\eta > 1$, and only the conventional equilibrium condition, with the $\frac{1}{2}$-spin dipole directed along the z axis and with $\langle \sigma_x \rangle_{eq} = \langle \sigma_y \rangle_{eq} = x_{eq} = v_{eq} = 0$, applies. Upon increasing the coupling between atom and radiation field, the conditions $\eta_{RWA} < 1$ and $\eta < 1$ are reached and this conventional stationary condition becomes unstable and is replaced by the high-coupling stationary condition illustrated by Eqs. (35) and (36).

We see that in the high-coupling limit the RWA leads to an infinite number of fixed points corresponding to all the possible orientation of the spin in the xy plane with the z component of the $\frac{1}{2}$-spin dipole kept fixed. Correspondingly, the oscillator is characterized by a continuum of stationary conditions, and precisely all the points lying on the circle determined by the last equation of (35). When the RWA Hamiltonian is replaced by the full Hamiltonian of (5) the high-coupling limit yields only two possible stationary conditions for the $\frac{1}{2}$-spin system, and correspondingly only two possible stationary states for the oscillator. This will have remarkable consequence on the problem of "quantum measurement" which is discussed later in this chapter.

III. THE OPTICAL SCHRÖDINGER CAT

Some recent papers[15, 24] stress an interesting aspect of the time evolution of the boson field of the Jaynes-Cummings model: the splitting of the oscillator distribution into two distinct states. This splitting can be observed within the classical phase space of the oscillator, and this observation, in turn, is carried out without adopting the assumption that the oscillator is classical. This seemingly contradictory operation is made possible by the adoption of the quasi-probability distribution formalism.[25, 26] This method originates from the earlier work of Wigner[25] and makes it possible to associate a classical-like phase space to a quantum system, thereby shedding insights into the quantum–classical correspondence. In the next section we illustrate a generalization of this method making it possible to study the dynamics of the whole spin-boson system as well as that of the boson field.

Eiselt and Risken[24] applied the conventional Wigner technique to study the dynamics of the oscillator of the Jaynes-Cummings model, following an initial condition with the oscillator in a coherent state and the $\frac{1}{2}$-spin system in an eigenstate of σ_z. They found that the process of collapse of $\langle \sigma_z(t) \rangle$ corresponds to the splitting of the oscillator distribution into two distinct states. These two distinct components of the oscillator state counterrotate on the phase space of the oscillator and, after a time corresponding to the time interval between a collapse and a revival, merge again in a single state. Thus, the process of revival of $\langle \sigma_z(t) \rangle$ corresponds to the reunion of the two distinct components of the oscillator.

In a more recent paper Gea-Banacloche[15] established an interesting connection between the interesting phenomenon discovered by Eiselt and Risken[24] and a new basis set of states of the $\frac{1}{2}$-spin system, $|+\rangle$ and $|-\rangle$. They are defined as follows:

$$| \pm \rangle \equiv \frac{1}{\sqrt{2}} \left(e^{-i\varphi} |+\rangle_z \pm |-\rangle_z \right) \tag{37}$$

where $| \pm \rangle_z$ indicates the eigenstate of σ_z, φ is the initial phase of the coherent state

$$|\alpha\rangle \equiv e^{-(1/2)|\alpha|^2} \sum_{n=0}^{\infty} \frac{|\alpha|^n}{\sqrt{n!}} e^{-in\varphi} |n\rangle \tag{38}$$

and, according to the prescriptions of Glauber,[4] $|\alpha|^2 = \bar{n}$, with $\bar{n}$ denoting the average number of photons in the field. In the semiclassical case, the operators a and a^+ appearing in (27) are replaced by the c numbers α and α^*, thereby resulting in a new Hamiltonian acting only on the space of the $\frac{1}{2}$-spin system. Then the two states $|+\rangle$ and $|-\rangle$ become the eigenstates of this resulting Hamiltonian. This means that in the semiclassical limit, the two states are stationary. In the quantum case, on the contrary, states $|+\rangle|\alpha\rangle$ and $|-\rangle|\alpha\rangle$ are not eigenstates of the corresponding RWA Hamiltonian, and therefore they are not stationary. Their time evolution can be described by means of a simple analytical expression in the limiting condition of a very large number of photons. In such a condition, the time evolution of the whole system, evolving from an initial condition with the $\frac{1}{2}$-spin system either in state $|+\rangle$ or state $|-\rangle$ and the field in state $|\alpha\rangle$, gets the following simple expression for the wave function:

$$|\psi(t)\rangle = \frac{1}{\sqrt{2}} \left(e^{-i\varphi \mp (i/2)gt\sqrt{\bar{n}}} |+\rangle_z \pm |-\rangle_z \right)$$

$$\times e^{-\bar{n}/2} \sum_{n=0}^{\infty} \frac{\bar{n}^{n/2}}{\sqrt{n!}} e^{-in\varphi \mp igt\sqrt{n}} |n\rangle \tag{39}$$

Here the upper and lower signs correspond to an initial condition with the $\frac{1}{2}$-spin dipole in state $|+\rangle$ or $|-\rangle$, respectively. Gea-Banacloche adopts a notation slightly different from ours: for a proper comparison between the result of this subsection and the rest of the paper we should replace the coupling strength g with the expression $g/\sqrt{2\Omega}$.

Gea-Banacloche[15] proves that the motion of the radiation field component corresponds to a rotation in the phase space of the oscillator, which is clockwise or counterclockwise according to whether the $\frac{1}{2}$-spin system is initially in the $|-\rangle$ or $|+\rangle$ state, respectively. The time period for a complete rotation is proven to be $T = 4\pi\sqrt{\bar{n}}/g$ (Refs. 15 and 24). Let us now discuss the physical meaning of these states. According to definition (37) the eigenstates of the $\frac{1}{2}$-spin system along the z direction are a linear combination of these two states, equally weighted. Consequently, the system departing from such initial conditions will evolve in time with the oscillator in a linear superposition of a clockwise and a counterclockwise rotation. It is well known that this initial condition accompanies the collapse of $\langle \sigma_z(t) \rangle$ (Refs. 5–8). Consequently, it is evident that the onset of the collapse process corresponds to the splitting of the oscillator probability distribution. According to Gea-Banacloche[15] the revival time is $t_R \equiv 2\pi\sqrt{\bar{n}}/g$, namely half of the full rotation time T of the two oscillator components. This means that the revival takes place when the two oscillator components are at half of their complete counterrotations, thus being temporarily reunited. It is also evident that the maximum departure of the two oscillator components from one another is reached at half revival time, namely at one-fourth of their complete rotations. Similar results have been obtained by other authors.[14]

We are now in a position to explain why many authors[14] believe that there is a close connection between this interesting quantum optics effect and the theory of measurement. According to the traditional wisdom,[4] classical electrodynamics is recovered from quantum physics by making the number of photons tend to infinity. If the term "classical" and the term "macroscopic" were equivalent and interchangeable at will, then the phenomenon here under discussion would be as paradoxical as the paradox of Schrödinger's cat described in Section I. We would be in the presence of a classical system in a linear superposition of distinct macroscopic states.

We discuss this important issue in Section VI. Here we limit ourselves to noticing that the experimental observation of this interesting effect is fraught with technical and conceptual difficulties, which might be settled using the theoretical results of Section II. On the one hand, Eq. (39) implies that a very large number of photons have to be considered. On the other hand, radiation fields of extremely large intensity might provoke the breakdown of the RWA. Gea-Banacloche[15] points out that the condition of Eq. (39) can be satisfactorily approximated with a moderately large number of photons, i.e., $\bar{n} = 25$. Of course, the statement this is a moderately strong radiation field should be supplemented by a detailed discussion of the energy spectrum of the atomic system and of the transitions among the corresponding states caused by interaction with the radiation field. One way out could be to adopt the remarks of Section II, where we show that for a very large number of photons the RWA Hamiltonian is very well approximated by the Hamiltonian of a $\frac{1}{2}$-spin system interacting with a circularly polarized radiation.

IV. THE RWA HAMILTONIAN AND THE BREAKDOWN OF THE SEMICLASSICAL APPROXIMATION

As we discuss in Section VI, we are inclined to think that the so-called optical Schrödinger cat is not a genuine macroscopic quantum effect. Rather it is an expression of the fact that the way from quantum to classical physics is not as easily practicable as usually supposed. Here, we want to discuss another example of the same kind, concerning the discrepancy between the semiclassical approximation of Eqs. (33) and (34) and the exact quantum prediction in the limit of a large number of photons.[27] According to the traditional wisdom[4] the classical radiation field is the natural limit of a coherent state of the quantum radiation field when $\bar{n}$ is very large.

Thus, at first sight one would be tempted to say that the time evolution described by Eqs. (33) and (34) coincides with the exact predictions of quantum mechanics for very large values of $\bar{n}$. On the contrary, if the RWA Hamiltonian is used, the agreement between the semiclassical approximation and the fully quantum-mechanical calculation is lost, regardless the intensity of the radiation field. Indeed, the splitting of the oscillator distribution is a quantum-mechanical property that is lost once the classical approximation on the boson field is made. From the analysis of Phoenix and Knight[28] it is also evident that the phenomenon of collapses and revivals is based on the quantum-mechanical properties of the radiation field. Consequently, both processes are incompatible with the semiclassical approximation behind Eqs. (33) and (34). This is confirmed either by the analytical solution of the set (33), which can be obtained by disregarding the reaction field on the right side of the last two equations, and by the numerical calculation. Both approaches show that $\langle \sigma_z(t) \rangle$ is characterized by harmonic oscillations. The set of Eqs. (34)

results in a time evolution for $\langle \sigma_z(t) \rangle$ with nonharmonic oscillations. However, even in this case the time evolution of $\langle \sigma_z(t) \rangle$ exhibits oscillations with no sign of either collapses or revivals.

In this section we show that the breakdown of the classical prediction is provoked by the adoption of the RWA Hamiltonian. The spin-boson Hamiltonian of Eq. (5) leads to a satisfactory agreement with the predictions of the classical approximation in a region of the parameters where the discrepancy between quantum and classical physics is still strong if the RWA Hamiltonian is adopted. We reach this important conclusion by adopting a variety of analytical and numerical techniques. Prior to the illustration of the theoretical result we illustrate these techniques in detail.

A. The Wigner Method

By using the Wigner prescription it is possible to express quantum-mechanical averages in the same form one writes for classical averages. This might generate confusion between operators and c-number variables, so in this subsection we denote the operators A that might be confused with a corresponding c-number variable, with the symbol $\hat{A}$. Within the Wigner formalism every operator of the Hilbert space is linked[25] to a corresponding c-number by means of the Weyl transformation.[29] As an illustrative example, let us consider a particle moving in one-dimension space. The position and momentum of this particle are characterized by the operators $\hat{q}$ and $\hat{p}$, respectively. Consequently, the most general quantum-mechanical observable of this physical system is given by the operator $\hat{A}(\hat{q}, \hat{p})$. The corresponding c-number variable, $A_W(q, p)$, is given the following from:

$$A_{\mathrm{W}}(q, p) = \int_{-\infty}^{+\infty} \mathrm{d}z \, \exp\left(i\frac{p}{\hbar}z\right)\left\langle q - \frac{z}{2} |\hat{A}| q + \frac{z}{2} \right\rangle \qquad (40)$$

Equation (40) can be used to express the expectation value of the operator $\hat{A}$ as

$$\left\langle \hat{A}(t) \right\rangle = \mathrm{Tr}\left[\hat{A}\hat{\rho}(t) \right] = \int \mathrm{d}q \, \mathrm{d}p \, A_{\mathrm{W}}(q, p)\rho_{\mathrm{W}}(q, p; t) \qquad (41)$$

where $\rho_{\mathrm{W}}(q, p; t)$ is the Wigner pseudo-probability function associated with the density matrix $\hat{\rho}(t)$

$$\rho_{\mathrm{W}}(q, p; t) = \frac{1}{\pi\hbar} \int_{-\infty}^{\infty} \mathrm{d}z \, \exp\left(2iz\frac{p}{\hbar}\right)\left\langle q - z|\hat{\rho}(t)|q + z \right\rangle \qquad (42)$$

Note the similarity between (41) and the classical definition of an average. For convenience, we shall refer to ρ_W as the Wigner distribution or the Wigner density.

The most remarkable property of the Wigner distribution is its dynamical evolution, which also closely resembles the classical situation. Let us suppose that the quantum system is described by the following Hamiltonian:

$$\mathscr{H} = \frac{\hat{p}^2}{2m} + V(\hat{q}) \tag{43}$$

The evolution of $\rho_W(q, p; t)$ is then determined by[25]

$$\frac{\partial}{\partial t}\rho_W(q, p; t) = \mathscr{L}\rho_W(q, p; t)$$

$$= -\{H, \rho_W\} + \sum_n \frac{1}{(2n + 1)!} \left(\frac{\hbar}{2_i}\right)^{2n} V^{(2n+1)}(q) \tag{44}$$

$$\times \frac{\partial^{2n+1}}{\partial p^{2n+1}}\rho_W(q, p; t)$$

where the curly brackets denote the Poisson brackets, H is the c-number Hamiltonian, $V(q)$ is the c-number potential, $V^{(l)}(q)$ is the lth derivative of $V(q)$, and $\mathscr{L}$ is the Liouville operator implicitly defined by (44). The main properties of the Wigner distribution are reviewed by Hillery et al.[25]

The most important features of the above Wigner formalism are summarized as follows.

1. Equation (44) introduces a new kind of dynamical evolution for quantum systems. Following the standard treatment of statistical mechanics we define the dynamical operator for the function $A_W(q, p; t)$ as

$$\frac{\partial A_W(q, p; t)}{\partial t} = \Gamma A_W(p, q; t) \tag{45}$$

The operator Γ must satisfy the equation

$$\langle \hat{A}(t) \rangle = \int dq \int dp A_W[\exp(\mathscr{L}_t)\rho_W] = \int dq \int dp[\exp(\Gamma_t)A_W]\rho_W \tag{46}$$

where we have used (41). Thus, the quantum average for the observable $\hat{A}(t)$ can be found, at least in principle, by obtaining the time evolution of $A_W(q, p)$ determined by the operator Γ and averaging over the ensemble of initial conditions represented by the Wigner distribution $\rho_W(q, p; 0)$. Note that in the limit of vanishing $\hbar$ the operator $\mathscr{L}$ recovers the exact classical result, i.e., it becomes the Poisson bracket with the Hamiltonian (cf. (44)) which is just the classical Liouvillian. This led us to rewrite the Liouvillian of Eq. (44) as $\mathscr{L} = \mathscr{L}_{cl} + \mathscr{L}_Q$, where

$$\mathscr{L}_{cl} = -\{H, \ldots\} \tag{47}$$

is formally indistinguishable from the classical Liouvillian, while the operator

$$\mathscr{L}_Q = \sum_{n=1} \frac{1}{(2n+1)!} \left(\frac{\hbar}{2i}\right)^{2n} V^{(2n+1)}(q) \frac{\partial^{2n+1}}{\partial p^{2n+1}} \tag{48}$$

accounts for the quantum corrections.

2. The concept of a quantum phase space, in apparent violation of the uncertainty principle, is introduced in a natural way as the corresponding classical phase space. On the other hand, the state of a quantum system is not defined by a precise position in this phase space. In fact, due to the intrinsically statistical nature of quantum mechanics, even for a single quantum state, i.e., a pure state, the Wigner distribution cannot be a delta function. Thus, a pure quantum state is not described by a point in this phase space, but rather is described by an ensemble of systems distributed according to the Wigner density. For this reason, even if the quantum correction $\mathscr{L}_Q$ (58) is neglected, some important quantum properties can be retained by the resulting Wigner picture.

From the preceding discussion we can derive an alternative way to solve the dynamical evolution of a quantum system. In principle, according to feature 1 we should solve the equation of motion (45) and derive from it the quantum evolution (46) by averaging over the initial Wigner distribution. However, in the nonlinear case Eq. (45) does not result in a close set of equations, and we would be obliged to solve a infinite number of coupled differential equations. This might turn out to be troublesome even if we adopt a numerical procedure and use a computer calculation. One way out rests on disregarding the term $\mathscr{L}_Q$. In this case procedure 1 coincides with the evaluation of the classical trajectories (which can be easily carried out numerically) and the quantum effects are obtained by averaging over an ensemble of initial conditions distributed according to the Wigner prescription. Throughout this paper we refer to this approximated procedure as the *Wigner recipe*. The inconvenience of the Wigner recipe is that it rests on the uncontrolled assumption that the term $\mathscr{L}_Q$ does not produce significant corrections. However, by application of it to the spin-boson Hamiltonian we shall show that there are physical conditions where the Wigner recipe works well.

The application of the Wigner recipe to the spin-boson Hamiltonian requires the Wigner method to be suitable extended to the the spin-boson Hamiltonian. Let us show how to do that. The Wigner density can also be defined as the Fourier transform of the characteristic function. Then we can extend its definition to the spin-boson Hamiltonian and write

$$\rho_W(\underline{x}, q, p; t) = \int d\underline{k} \frac{e^{-i\underline{k} \cdot \underline{x}}}{(2\pi)^3} \int d\alpha \, d\tau \frac{e^{-i(q\alpha + p\tau)}}{(2\pi)^2} F(\underline{k}, \alpha, \tau; t) \quad (49)$$

Here $F(\underline{k}, \alpha, \tau; t)$ is the quantum characteristic function defined by

$$F(\underline{k}, a, \tau; t) = \text{Tr}\{\exp[i(k_1\sigma_x + k_2\sigma_y + k_3\sigma_z)] \exp[i(\alpha\hat{q} + \tau\hat{p})]\hat{\rho}(t)\} \quad (50)$$

and $\hat{\rho}(t)$ is the density operator for the complete quantum system. The variables $\underline{x} \equiv (x_1, x_2, x_3)$, p and q are the phase-space variables associated, via a generalized Weyl rule of the form (40), to the spin-boson operators: $\hat{\sigma}_j \to x_j, \hat{q} \to q, \hat{p} \to p$. The quantum average values of these operators are given by

$$\langle \hat{\sigma}_j(t) \rangle = \int d\underline{x} \, dq \, dp \, x_j \rho_W(\underline{x}, q, p; t) \quad (51)$$

$$\langle \hat{q}(t) \rangle = \int d\underline{x} \, dq \, dp \, q \rho_W(\underline{x}, q, p; t) \quad (52)$$

After some algebra, we obtain that the corresponding equation of motion for the Wigner distribution is given by[30, 31]

$$\frac{\partial}{\partial t}\rho_W(\underline{x}, q, p; t) = (\mathscr{L}_{cl} + \mathscr{L}_{QGD})\rho_W(\underline{x}, q, p; t) \quad (53)$$

where

$$
\mathscr{L}_{cl} \equiv \omega_0 \left(x_2 \frac{\partial}{\partial x_1} - x_1 \frac{\partial}{\partial x_2} \right) + 2gq \left(x_3 \frac{\partial}{\partial x_2} - x_2 \frac{\partial}{\partial x_3} \right)
$$
$$
+ \Omega^2 q \frac{\partial}{\partial p} - p \frac{\partial}{\partial q} + g x_1 \frac{\partial}{\partial p} \tag{54}
$$

and

$$
\mathscr{L}_{QGD} \equiv g \frac{\partial}{\partial p} \left[\frac{\partial}{\partial x_1} - \frac{\partial}{\partial x_1} x_1^2 - \frac{\partial}{\partial x_2} x_1 x_2 - \frac{\partial}{\partial x_3} x_1 x_3 \right] \tag{55}
$$

Before proceeding we have to explain the subdivision of the Wigner Liouvillian established with Eqs. (54) and (55). There is a subtle difference between this case and that of Eqs. (47) and (48). In fact, in the latter case the subdivision of the Wigner Liouvillian was based on the expansion of the dynamical operator $\mathscr{L}$ into a series expansion in powers of $\hbar$, thereby making natural the choice of $\mathscr{L}_{cl}$ as the zero-order term in this expansion. In the spin-boson case both the last term in (54) and the operator $\mathscr{L}_{QGD}$ are proportional to $\hbar$ (set for notational convenience equal to the unity). Thus, in this case we adopt a different criterion to define the classical part of the Liouvillian, and we select the operator $\mathscr{L}_{cl}$ as that part of the Liouvillian fully equivalent to a classical magnetic dipole interacting with a constant magnetic field and a classical oscillator. This led us to define $\mathscr{L}_{cl}$ by (54) and, consequently, the quantum part as in (55). Note that the quantum term of (55) has a diffusion-like structure, i.e., it is a sum of second-order differential operators, with a resulting diffusion coefficient that is not positive definite. This means that it can act as an antidiffusional mechanism, competing against fluctuations and constraining the dipole, which otherwise would freely diffuse over all possible orientations, to vacillate between two possible orientations, thereby remembering its quantum nature, and, consequently, the discrete nature of its state. For this reason this process was called the quantization generating diffusion (QGD) by Roncaglia et al.[31]

In conclusion, the operator $\mathscr{L}_{cl}$ is identical to the Liouvillian of a classical dipole interacting with a classical oscillator; i.e., this term alone corresponds to the semiclassical set of equations discussed by various authors[18-22] and shown in Eq. (34). We refer to the calculations based on the study of the single trajectory solutions of the nonlinear dynamical equations as the semiclassical predictions. Note that in the present case the Wigner recipe is obtained making averages over the semiclassical trajectories distributed according to the prescription of the Wigner method.

B. A Perturbative Approach

The perturbation method that we describe here is suggested by the remark that in the limiting condition of a vanishing Larmor frequency ω_0, the problem admits an exact analytical solution. This leads us to write $\langle \sigma_z(t) \rangle$ as

$$\langle \sigma_z(t) \rangle = \mathrm{Tr} \left\{ \rho(0) e^{\mathscr{L}_0 t} \, T \exp \int_0^t \mathrm{d}s \, e^{-\mathscr{L}_0 s} \, \mathscr{L}_a \, e^{\mathscr{L}_0 s} \, \sigma_z \right\} \qquad (56)$$

with

$$\mathscr{L}_a \equiv -\mathrm{i}[\mathscr{H}_a, \dots] \quad \text{and} \quad \mathscr{L}_0 \equiv -\mathrm{i}[\mathscr{H}_0, \dots] \qquad (57)$$

and the Hamiltonian terms $\mathscr{H}_a$ and $\mathscr{H}_0$ defined by

$$\mathscr{H}_a \equiv \frac{\omega_0}{2} \sigma_x \qquad (58)$$

and

$$\mathscr{H}_0 \equiv \Omega a^+ a + g \Gamma (a + a^+) \sigma_x \qquad (59)$$

where

$$\Gamma \equiv \frac{1}{\sqrt{2\Omega}} \qquad (60)$$

This subdivision means that we consider ω_0 to be small compared to the intensity of the interaction so as to result in a weak perturbation condition admitting an exact solution. We must remark again that the actual intensity of the spin-boson interaction not only depends on the coupling strength g but also on the field intensity. More precisely the interaction is proportional to λ as defined in Eq. (28) and we shall show that the semiclassical approximation is recovered by making $\bar{n}$ tend to infinity. In conclusion we will study the case where the condition

$$g \Gamma \bar{n}^{1/2} \gg \omega_0 \qquad (61)$$

applies. We shall show that in this limiting condition also the results obtained disregarding the terms of first-order in ω_0 are reliable. Replacing

Eq. (56) with its approximation at the zeroth order in $\mathscr{L}_a$ (and consequently in ω_0), we obtain the following expression:

$$\langle \sigma_z(t) \rangle = \text{Tr}\{\rho(0)e^{\mathscr{L}_0 t} \sigma_z\} \qquad (62)$$

Note that, surprisingly enough, the resulting analytical expression will be proved by the numerical calculation of Section IV.C to lead to reliable predictions also in the time region $\omega_0 t \cong 1$. Throughout the remaining part of this paper we shall study the dynamics of the spin-boson system subsequent the initial condition with the $\frac{1}{2}$-spin system polarized along the positive direction of the z axis and the field in a coherent state, i.e.,

$$\rho(0) = \rho_B | + \rangle_z \langle + |_z \qquad (63)$$

where $| + \rangle_z$ is the eigenstate of σ_z and ρ_B is the initial density matrix of the field mode.

We plan to use (62) to derive the time evolution of $\langle \sigma_z(t) \rangle$ subsequent to the initial condition (63). To make this procedure straightforward, let us express $| + \rangle_z$ on the basis set of the eigenstates of σ_x:

$$\rho(0) = \rho_B[| + \rangle_x \langle + |_x + | - \rangle_x \langle - |_x - i| + \rangle_x \langle - |_x + i| - \rangle_x \langle + |_x] \qquad (64)$$

Using this statistical weight, we obtain from (62) the following expression

$$\langle \sigma_z(t) \rangle = \frac{1}{2} \text{Tr}_B \{\rho_B[\exp[i(\Omega a^+ a - gx)t] \exp[-i(\Omega a^+ a + gx)t] + \text{h.c.}]\} \qquad (65)$$

with the symbol x denoting the space coordinate

$$x \equiv \Gamma(a + a^+) \qquad (66)$$

To make (65) workable we have to adopt a proper form for the initial distribution ρ_B. According to a point of view widely shared in quantum optics,[4] for a correspondence with the predictions of classical mechanics to be established the radiation field must be assumed to be in the state of coherent oscillation $|\alpha\rangle$. It is well known[11] that a coherent oscillation can be seen as being the motion of a ground-state oscillator, the center of oscillation of which is shifted by a suitable quantity. In other words, the coherent states are displaced forms of the ground state of the oscillator and we can write

$$|\alpha\rangle = D(\alpha)|0\rangle \tag{67}$$

where

$$D(\alpha) \equiv \exp(\alpha b^+ - \alpha^* b) \tag{68}$$

is the Glauber displacement operator. By applying this displacement operator to a, a^+ and the space coordinate x, we obtain

$$
\begin{aligned}
\tilde{a} &= a + \alpha \\
\tilde{a}^+ &= a^+ + \alpha^* \\
\tilde{x} &= x + 2\Gamma \operatorname{Re} \alpha
\end{aligned}
\tag{69}
$$

This means that if we adopt the displaced coordinates of (69), rather than the conventional ones, then the initial distribution of the oscillator reduces to

$$\rho_{Bg} = |0\rangle\langle 0| \tag{70}$$

i.e., the oscillator in its ground state. In other words, we have

$$
\begin{aligned}
\operatorname*{Tr}_{B} &\{\rho_B F(\Omega, \Gamma, a, a^+)\} \\
&= \operatorname*{Tr}_{B}\{D(\alpha)|0\rangle\langle 0|D(\alpha)^{-1}F(\Omega, \Gamma, a, a^+)\} \\
&= \operatorname*{Tr}_{B}\{|0\rangle\langle 0|D(\alpha)^{-1}F(\Omega, \Gamma, a, a^+)D(\alpha)\} \\
&= \operatorname*{Tr}_{B}\{|0\rangle\langle 0|F(\Omega, \Gamma, \tilde{a}, \tilde{a}^+)\}
\end{aligned}
\tag{71}
$$

By applying this property to (65) we obtain

$$
\begin{aligned}
\langle \sigma_z(t) \rangle = \tfrac{1}{2}\langle 0| \{ & \exp[-4ig\Gamma \operatorname{Re}(\alpha)t] \\
& \times \exp[i(\Omega a^+ \tilde{a} - Aa - A^* a^+)t] \\
& \times \exp[-i(\Omega a^+ a + Ba + B^* a^+)t] + \text{h.c.}\}|0\rangle
\end{aligned}
\tag{72}
$$

where

$$A \equiv g\Gamma - \Omega\alpha^* \tag{73}$$

$$B \equiv g\Gamma + \Omega\alpha^* \tag{74}$$

At this stage, we make use of a general result by Weiss and Maradudin[32] which allows us to rewrite Eq. (72) as follows (for details see Appendix A of Ref. 33):

$$
\begin{aligned}
\langle \sigma_z(t) \rangle = \tfrac{1}{2} \langle 0 | & \exp[-\delta^*(t)a^+ + \delta(t)a] \\
& \times \exp[-i\Omega a^+ at]\exp[-\gamma^*(t)a^+ + \gamma(t)a]|0\rangle \\
& \times \exp\left\{[-4ig\Gamma\,\mathrm{Re}(\alpha)t] - i\left(\frac{|A|^2 - |B|^2}{\Omega^2}\right) \right. \\
& \left. \times (\Omega t - \sin\Omega t)\right\} + \text{c.c.}
\end{aligned}
\tag{75}
$$

where

$$
\delta(t) \equiv \frac{A}{\Omega}(\exp(i\Omega t) - 1)
\tag{76}
$$

$$
\gamma(t) \equiv -\frac{B}{\Omega}(\exp(-i\Omega t) - 1)
\tag{77}
$$

Using now the well-known property[4]

$$
\exp[-\xi^*a^+ + \xi a] = \exp[-\xi^*a^+]\exp[\xi a]\exp\left\{-\tfrac{1}{2}|\xi|^2[a, a^+]\right\}
\tag{78}
$$

we rewrite Eq. (75) as follows:

$$
\begin{aligned}
\langle \sigma_z(t) \rangle = \tfrac{1}{2} \langle 0 | & \exp[\delta(t)a]\exp[-i\Omega a^+ at]\exp[-\gamma^*(t)a^+]|0\rangle \\
& \times \exp\left[-4ig\Gamma\,\mathrm{Re}(\alpha)t - i\left(\frac{|A|^2 - |B|^2}{\Omega^2}\right)[\Omega t - \sin(\Omega t)]\right] \\
& \times \exp\left\{-\tfrac{1}{2}\left(|\delta(t)|^2 + |\gamma(t)|^2\right)\right\}
\end{aligned}
\tag{79}
$$

Expanding the exponentials and applying the creation and destruction operators we show that

$$
\begin{aligned}
\langle 0 | & \exp[\delta(t)a]\exp[-i\Omega a^+ at]\exp[-\gamma^*(t)a^+]|0\rangle \\
& = \exp\left[\frac{2g^2\Gamma^2}{\Omega^2}\left(1 - \frac{\Omega^2|\alpha|^2}{g^2\Gamma^2} + \frac{2\Omega\,\mathrm{Im}(\alpha)}{g\Gamma}\right)(\cos(\Omega t) - 1)\right]
\end{aligned}
\tag{80}
$$

By replacing Eq. (70) into (71) and using Eqs. (73)–(77), we finally obtain

$$
\langle \sigma_z(t) \rangle = \cos\left[4g\Gamma \operatorname{Re}(\alpha)\frac{\sin(\Omega t)}{\Omega} \right]
$$

$$
\times \exp\left\{ [-(1 - \cos \Omega t)]\left(\frac{4g^2\Gamma^2}{\Omega^2} + \frac{4g\Gamma}{\Omega}\operatorname{Im}(\alpha) \right) \right\} \quad (81)
$$

This analytical expression is the central result of the perturbation method used here and is used later to make more transparent our analysis on the breakdown of the semiclassical approximation with the RWA Hamiltonian.

C. The Numerical Calculation

Here we give details on the numerical calculation we adopt as a useful check of our theoretical arguments and analytical predictions. This method rests on the numerical integration of the Liouville-von Neumann equation

$$
\dot{\rho}(t) = -i[\mathscr{H}, \rho(t)] \quad (82)
$$

where the density matrix operator $\rho(t)$ describes completely the quantum system. The numerical solution is obtained by expanding the operator equation (72) over the basis set of the direct products $| \pm \rangle_z |n\rangle$. Here $| +\rangle_z$ and $| -\rangle_z$ denotes the eigenstates of the operator σ_z, and the states $|n\rangle$ denote the eigenstates of the operator $n = a^+ a$. This results in a set of coupled linear differential equations, which is then solved numerically.

It must be stressed that also the numerical result rests on approximations, since to make the computer calculation possible it is necessary to truncate the basis set, thereby using a finite dimension Hilbert space rather than the infinite dimension one, which would make the solution exact. For this reason the long-time limit of the system and its behavior under the action of field of large intensity cannot be carried out with this numerical method.

D. Quantum Behavior Versus Semiclassical Approximation

We now compare the two models of Eqs. (5) and (27) to their respective semiclassical approximations of Eqs. (34) and (33). All the results herein correspond to the same initial condition, with the spin in the positive eigenstate of σ_z and the field in the coherent state

$$|\alpha\rangle \equiv e^{-(1/2)|\alpha|^2} \sum_{n=0}^{\infty} \frac{|\alpha|^n}{\sqrt{n!}} |n\rangle \tag{83}$$

where $\bar{n} = |\alpha|^2$ is the average value of photons in the field. According to the traditional wisdom, this initial condition should make it possible to recover the semiclassical prediction upon increasing the value of $\bar{n}$.

Before addressing this key issue, let us compare the time evolution for the population difference predicted by the Wigner recipe of Section IV.A and the analytical result of Section IV.B with the numerical exact result stemming from Eq. (82). We study both the spin-boson Hamiltonian (5) and its RWA counterpart, Eq. (27). In principle both cases could be addressed by using the numerical treatment of Section IV.C. However, as far as the RWA Hamiltonian is concerned, we have available an exact solution[12] and we would find it more convenient to use it. The exact solution reads

$$\langle \sigma_z(t) \rangle_{\mathrm{RWA}} = \sum_{n=0}^{\infty} P(n)\cos(2gn^{1/2}t) \tag{84}$$

where $P(n)$ represents the distribution of the diagonal elements of the initial density matrix of the field. In the limiting case of a very large number of photons, this series becomes slowly convergent and, as pointed out by various authors[5-8], it is convenient to replace this series with an analytical approximation. Thus, for very large field, instead of using the exact but troublesome result of (84), we adopt the analytical expression suggested by Narhozny et al.[5b]

Figures 2 and 3 refer to a case of a radiation field of moderate intensity. In these figures we compare the semiclassical approximation of Eqs. (33) and (34) and the related predictions of the Wigner recipe to the corresponding exact results. By exact result we mean, in the case of the Hamiltonian of Eq. (5), the numerical integration of Eq. (82) illustrated in Section IV.C, and in the case of the RWA Hamiltonian of Eq. (27) the analytical expression of Eq. (84). We see that the Wigner recipe is in satisfactory agreement with the exact calculations. Furthermore, we see that the quantum mechanical calculations (namely both the exact approach and the Wigner recipe) result in a marked relaxation behavior, whereas the semiclassical approximation results in regular oscillations.

In this case, that of radiation fields of moderate intensity, the disagreement between semiclassical prediction and fully quantum calculation is not

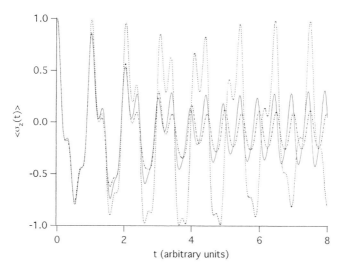

Figure 2. Dynamics of the full Hamiltonian (5). We compare the exact result (full line) and the Wigner recipe (dot-dashed line) to the numerical solution of the corresponding set of semiclassical equations, namely (34) (dashed line). We consider the case with a field of moderate intensity, namely $\bar{n} = 30$, while the values of the other parameters are $\omega_0 = \Omega = 2\pi$ and $g = 2$.

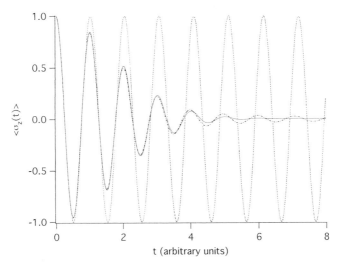

Figure 3. Dynamics of the RWA Hamiltonian (27). We compare the exact result of Eq. (84) (full line) and the Wigner recipe (dot-dashed line) to the numerical solution of the corresponding set of semiclassical equations, namely (33) (dashed line). We consider the case with a field of moderate intensity, namely $\bar{n} = 30$, while the values of the other parameters are $\omega_0 = \Omega = 2\pi$ and $g = 2$.

yet perceived as being in conflict with the traditional wisdom, according to which classical is derived from quantum physics upon increase of the quantum numbers. We plan to show that upon increase of the radiation field intensity, the full Hamiltonian (5) leads to a satisfactory agreement with the corresponding semiclassical approximation, whereas the RWA Hamiltonian does not. Unfortunately, for the reasons given in Section IV.C, the numerical treatment of the full Hamiltonian is not reliable in the case of radiation fields of very large intensity. In the case of fields of very large intensity we refer to the prediction of the Wigner recipe as the "exact result." In truth, the prediction of the Wigner recipe is not exact and its accuracy decreases with time. This can be observed by applying the Wigner recipe to the RWA case. We assessed in that case that the Wigner recipe does not satisfactorily reproduce large time processes such as the revivals, which are an exact prediction of the RWA Hamiltonian. However, the precision of the Wigner recipe does not depend on the radiation field intensity and thus we expect that in the case of radiation fields of very large intensity the disagreement between the prediction of the Wigner recipe and the exact behavior is not worse than that illustrated in Fig. 2. In conclusion, the prediction of the Wigner recipe is reliable for fields of any intensity, at least in the short-time region where the first collapse process takes place. We expect that the joint use of the Wigner recipe and of the analytical treatment of Section IV.B might provide a reliable prediction. This is so because the analytical treatment of Section IV.B rests on the assumption that the frequency ω_0 is very small. This is an approximation completely different from that behind the Wigner recipe, which is based on neglecting the QGD term. Thus, if the predictions of these approximated but distinct procedures are in a good agreement with each other, we must refer to the their common prediction as an "exact result."

The effects of the interaction between the $\frac{1}{2}$-spin system and radiation fields of very large intensity, namely $\bar{n} = 300$, are illustrated in Figs. 4 and 5. We split Fig. 4 into two parts, so as to stress that both the Wigner recipe and the analytical result of Eq. (81) are in remarkable agreement with the semiclassical assumption. Note that this, in turn, is indistinguishable from the conventional semiclassical prediction resting on the replacement of the full Hamiltonian of Eq. (5) with the time-dependent Hamiltonian

$$\mathcal{H}_{\text{sc}} = \omega_0 \sigma_z + \lambda \sigma_x \cos(\Omega t) \qquad (85)$$

where

$$\lambda \equiv 2g\Gamma\sqrt{\bar{n}} \qquad (86)$$

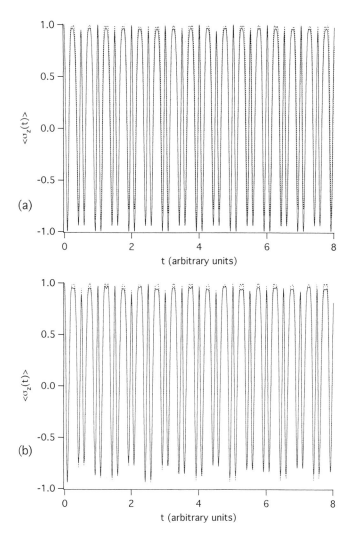

Figure 4. Dynamics of the full Hamiltonian (5). (a) Comparison between the analytical prediction of (81) (solid line) and the semiclassical approximation of Eq. (34) (dashed line); (b) comparison between the Wigner recipe (solid line) and the semiclassical approximation of (34) (dashed line). The values of the parameters are $\omega_0 = \Omega = 2\pi$, $g = 2$, and $\bar{n} = 300$, which correspond to a high field intensity.

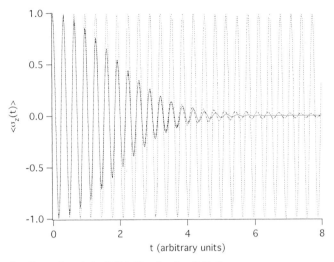

Figure 5. Dynamics of the RWA Hamiltonian (27). We compare the exact result of Eq. (84) calculated via the analytical approximation of Ref. 5b (full line) and the Wigner recipe (dot-dashed line) to the numerical solution of the corresponding set of semiclassical equations, namely (33) (dashed line). We consider the case with a field of high intensity, namely $\bar{n} = 300$, while the values of the other parameters are $\omega_0 = \Omega = 2\pi$ and $g = 2$.

This can be easily explained by remarking that the reaction term is independent of the field intensity, whereas the influence exerted by the radiation field on the system increases upon increase of the radiation field intensity. Thus, in the limiting cases of radiation fields of very large intensity, the reaction field becomes negligible compared to the action exerted by the radiation field on the system. It must be remarked that in the case of a radiation field of very large intensity, the exact numerical methods cannot be used. We are led to conclude that for $\bar{n} = 300$ the accordance between a quantum prediction, "exact" in the sense discussed above, and the semiclassical approximation is remarkably good.

The RWA Hamiltonian results in a quite different scenario, illustrated by Fig. 5. In fact, while the numerical solution of the semiclassical equations of the set (33) leads to the regular Rabi oscillation, the quantum-mechanical treatment results in a collapse, depending on the value g of the coupling strength and independent of the radiation field intensity, namely $t_c \approx \sqrt{\Omega}/g$ (Ref. 5b). Note that Fig. 5 refers to a case of a radiation field of extremely large intensity. Thus the RWA quantum prediction is not obtained from a direct numerical evaluation of the sum appearing in Eq. (84). Instead, to avoid numerical divergences, this sum is evaluated by adopting the approximation illustrated in Ref. 5b.

On the basis of the results illustrated in this and the previous sections, we are now in a position to stress some striking differences between the Jaynes-Cummings model and the spin-boson system without RWA. The most striking aspect of the Jaynes-Cummings evolution is represented by the collapse and the revivals of the atomic population inversion. Both the collapse time $t_c \approx \sqrt{\Omega}/g$ and the revival time $t_r \approx 2\pi\sqrt{2\Omega\bar{n}}/g$ are inversely proportional to the coupling, and the revival is also proportional to the square root of the strength of the field $\bar{n}$. Therefore, both are made to decrease by increasing the coupling strength, but the effect of increasing the radiation field intensity is that of making the revival time larger (see also the remarks following Eq. (39)). As shown earlier in this section, the collapse seems to be a robust property and it is not weakened by increasing the radiation field intensity, thereby making it impossible to recover the classical approximation within the RWA.

The splitting of the pseudo-probability distribution of the boson is a genuinely quantum-mechanical property, which has to do with the linear superposition of two distinct quantum states (see Eq. (37)). However, its time scale is the same as that of the phenomenon of revivals, because the maximum distance between the two peaks of the boson distribution is reached at half the revival time t_r. Thus, increasing the mean photon number has the effect of slowing down this process, and the project of an hypothetical experiment aiming at revealing the existence of the optical Schrödinger cat should take this property into account.

On the other hand, it must be pointed out that the current quantum optics experiments[13] are carried out using a linear polarized field. Thus, the corresponding theoretical treatment should rest on the proper Hamiltonian of Eq. (5) rather than on its RWA version. The real experiment of Ref. 13 led to a satisfactory agreement with the RWA prediction because it rested on physical conditions compatible with the RWA, and consequently on radiation fields of moderate intensity. To further stress the breakdown of the classical approximation one might be tempted to increase the intensity of the radiation field (this, according to the traditional wisdom, should have the effect of making the field closer to the classical approximation[4]). At the same time the coupling g should be increased if possible, so as to compensate the slowing down process triggered by the augmented intensity of the radiation field. Both effects would contribute to breaking the RWA.

Which kind of effects would one obtain in such a case? Would the interesting process of boson probability splitting into two distinct states still occur? To answer this question we use the full spin-boson Hamiltonian of Eq. (5), and we turn again to a numerical calculation, i.e., the numerical treatment of the Liouville-von Neumann equation (82). We choose a region of physical parameters where the field intensity is large enough to make the full Hamiltonian case depart significantly from the RWA predictions, but not so large as to generate the numerical difficulties mentioned in Section IV.C. The results are illustrated in Fig. 6, and the reader should compare them to the corresponding RWA results of Fig. 7. The difference between the prediction of the RWA Hamiltonian and that of the full Hamiltonian, namely the Hamiltonian referring to linearly polarized field without approximation, is striking: In the latter case there is no splitting of the boson pseudo-probability distribution!

For graphical purposes, Figs. 6 and 7 refer to the Husimi distribution[34] rather than to the Wigner distribution. The Husimi distribution can be easily derived from the statistical density matrix and has the nice property of being always positive.[34] Note also that in these two figures the motion of the oscillator is described within the interaction picture, i.e., in a reference system undergoing a clockwise rotation with angular velocity Ω.

In addition to the disappearance of the splitting of the boson pseudo-probability distribution, the full Hamiltonian results in other interesting properties. The dynamics of the boson pseudo-probability distribution follows a rather complex behavior, and it spreads asymmetrically with respect to the x direction (see Fig. 6).

Before ending this section we illustrate another aspect of the boson field dynamics, the time evolution of the mean values $\langle x(t) \rangle$ and $\langle v(t) \rangle$ (Figs. 8 and 9). First, we note that the RWA splitting of the distribution function corresponds to a collapse of the amplitude of the motion of the coordinate $\langle x(t) \rangle$. This happens at half the revival time, when the two counterrotating peaks of the distribution have a relative phase of π, so that the average values of both $\langle x(t) \rangle$ and $\langle v(t) \rangle$ are equal to zero. Of course, we are considering the envelope of the oscillations, since, as we remarked earlier, the splitting of the probability function is illustrated within the interaction representation. Thus, the motion within the conventional laboratory frame of reference would also be characterized by the fast oscillations accompanying the unperturbed dynamics of the boson Hamiltonian. In conclusion, the time evolution of $\langle x(t) \rangle$ is a harmonic oscillation with the frequency Ω (the frequency of the radiation field) modulated by the process of splitting and recombination of the pseudo-distribution. All this is illustrated in Fig. 8, where, for the sake of comparison, we also show the corresponding time evolution of the population inversion, i.e., the time evolution of $\langle \sigma_z(t) \rangle$.

As shown in Fig. 9, these properties are lost if the full Hamiltonian of Eq. (5) is adopted. The relaxation of the envelope of boson field oscillations is determined by the spreading of the corresponding distribution function and can probably be related to the effect of the underlying

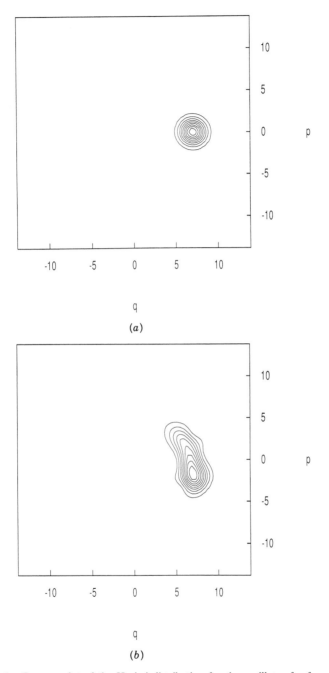

Figure 6. Contour plot of the Husimi distribution for the oscillator for four different times up to the revival time. The dynamics is driven by the full Hamiltonian of Eq. (5) and the values of the parameter are $\omega_0 = \Omega = 1$, $g = 0.5$, and $\bar{n} = 25$. The different times are $t = 0$, $t = 22.5$, $t = 45$, and $t = 90$.

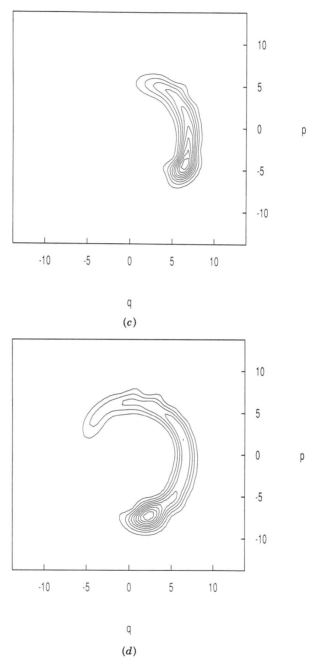

Figure 6b.   *(Continued)*

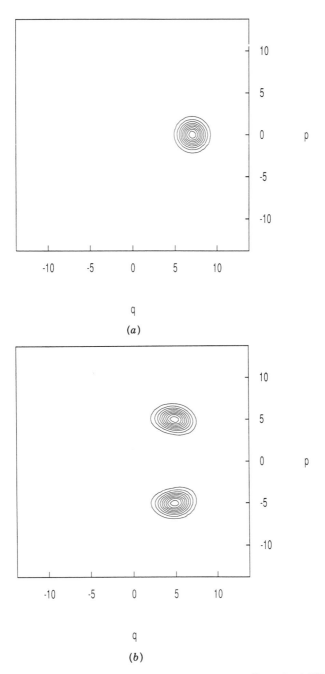

Figure 7. Contour plot of the Husimi distribution for the oscillator for 4 different times up to the revival time. The dynamics are driven by the RWA Hamiltonian of Eq. (27) and the values of the parameter are $\omega_0 = \Omega = 1$, $g = 0.5$, and $\bar{n} = 25$. The different times are $t = 0$, $t = 22.5$, $t = 45$, $t = 90$.

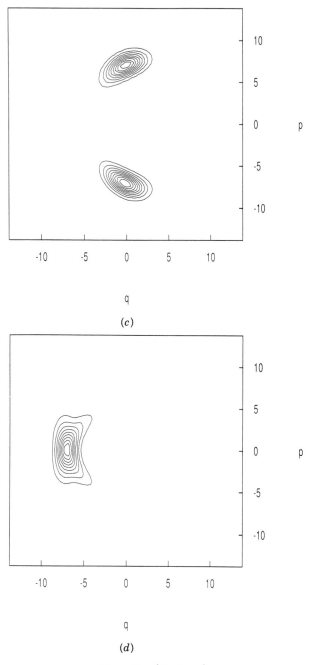

Figure 7b. *(Continued)*

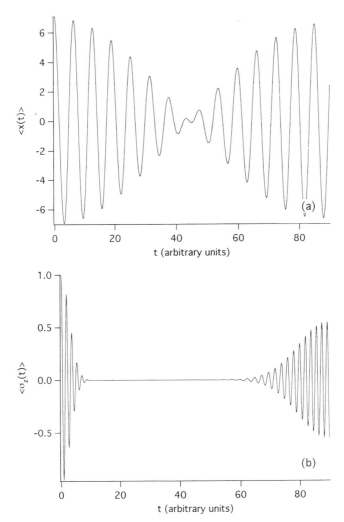

Figure 8. Time evolution of (a) $\langle x(t) \rangle$ and of (b) $\langle \sigma_z(t) \rangle$ in the RWA case of Hamiltonian (27) up to the first revival. The values of the parameters are $\omega_0 = \Omega = 1$, $g = 0.5$, and $\bar{n} = 25$.

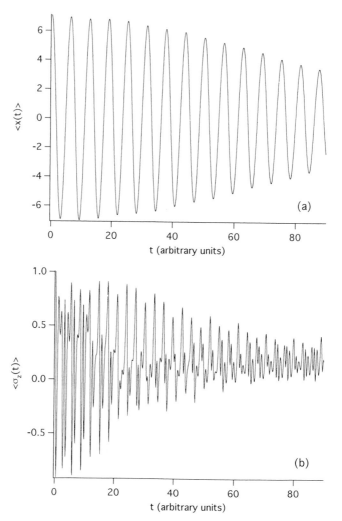

Figure 9. Time evolution of (a) $\langle x(t) \rangle$ and of (b) $\langle \sigma_z(t) \rangle$ in the case of Hamiltonian (5) in the same condition of Fig. 8.

semiclassical chaos (see the discussion of Refs. 30 and 31). Following this interpretation we predict that in the limit of infinitely large values of $\bar{n}$, the time scale of this relaxation process behavior should be the same as that of the population inversion discussed earlier (see Fig. 4). This is because the fields of very large intensity destroy the chaotic properties of the semiclassical approximation.

V. FURTHER DISCUSSION ON THE OPTICAL SCHRÖDINGER CAT: THE JOINT ROLE OF DISSIPATION AND SEMICLASSICAL APPROXIMATION

The purpose of this section is to discuss further the Schrödinger cat and the optical Schrödinger cat. From a formal point of view, the optical Schrödinger cat stems from the linear nature of quantum mechanics. This is why many authors propose it as a paradigm of the theory of quantum measurement.[14, 15, 35] The similarity with the measurement process illustrated in Section I is great. We prepare the $\frac{1}{2}$-spin system in a linear superposition of states $|+\rangle$ and $|-\rangle$ defined in Eq. (37) and we set the oscillator, imagined as the "measurement device," in a coherent state $|\alpha\rangle$, thereby realizing this initial state:

$$|\psi_0\rangle = (\gamma|+\rangle + \delta|-\rangle)|\alpha\rangle \tag{87}$$

As a result of the interaction with the quantum system under measurement the "pointer" splits into two distinct states, each moving with a law of motion corresponding to the respective state of the $\frac{1}{2}$-spin system. This is proven[15] to lead to

$$|\psi(t)\rangle = \gamma\, e^{-igt/2\sqrt{\bar{n}}}|+\rangle|\alpha\, e^{-igt/2\sqrt{\bar{n}}}\rangle + \delta\, e^{igt/2\sqrt{\bar{n}}}|-\rangle|\alpha\, e^{igt/2\sqrt{\bar{n}}}\rangle \tag{88}$$

This is also reminiscent of the interpretation scheme used by Leggett within his theoretical approach to the macroscopic effects of quantum mechanics.[3]

If we interpret the two states of the oscillator as macroscopic distinct states, corresponding to two different positions of the pointer of the measurement apparatus, then we recover in full the paradoxical aspect illustrated in Section I. Moreover, the quantum-optical model would seem to suggest also the resolution of this paradox. If we evaluate the reduced density matrix for the $\frac{1}{2}$-spin system, we get

$$\rho_{\text{at}}(t) = |\gamma|^2|+\rangle\langle +| + |\delta|^2|-\rangle\langle -| + \\ + \left(\gamma\delta^*\, e^{-ig\sqrt{\bar{n}}}\langle \alpha\, e^{igt/2\sqrt{\bar{n}}}|\alpha\, e^{-igt/2\sqrt{\bar{n}}}\rangle|+\rangle\langle -| + \text{h.c.}\right) \tag{89}$$

We see that in this reduced representation the coherence between state $|+\rangle$ and state $|-\rangle$ is not completely lost, due to the fact that states

$|\alpha\,e^{-igt/2\sqrt{n}}\rangle$ and $|\alpha\,e^{igt/2\sqrt{n}}\rangle$ are not exactly orthogonal. They correspond to the two counterrotating terms in the oscillator phase space, and in a time of the order of the collapse time t_c their scalar products vanish.

It must be stressed that at times of the order of the revival time these two distinct components are reunited, thereby resulting in the original strong correlation between states $|+\rangle$ and $|-\rangle$. However, the revival times tend to infinity, which would suggest a nice solution of the paradox, along the lines suggested by Glauber,[36] who pointed out that amplification is necessary to produce a macroscopic signal; this by itself is sufficient to produce a breakdown of the quantum-mechanical coherence responsible for the paradoxical manifestations of quantum mechanics.

In conclusion, as a result of the amplification process, in a time of the order of the collapse time the density matrix (89) would reduce to a diagonal form, and the process of optical collapse should be identified with the collapse process assumed by the orthodox interpretation of quantum mechanics. Furthermore, the statistical weight of the resulting uncorrelated states of the $\frac{1}{2}$-spin system would be proportional to $|\gamma|^2$ and $|\delta|^2$. This is precisely what, according to the orthodox interpretation of quantum mechanics, an ideal measurement apparatus is supposed to do.

This nice scenario is fraught with problems. The suppression of the phenomenon of the revivals, which would be a clear manifestation of the quantum-mechanical nature of the process, is expected to be caused by radiation fields of large intensity, but the use of such intense fields might invalidate the RWA, which lies behind the expression (89). In principle, the suppression of the quantum revivals can be produced by the influence of the environment, which might destroy the quantum-mechanical coherence, and no fields of extremely large intensity should be used. This is reminiscent of the theoretical scheme proposed by Zurek,[37] known as the *environment-induced superselection rule*. We shall come back to this issue in Section V.A, where we illustrate in more detail the proposal of Zurek. However, even if we do succeed in getting rid of the off-diagonal elements of the density matrix, it is still true that the whole system is placed in a state with the structure of Eq. (88), and the role of the environment on the whole system is not different from that exerted by the radiation field on the $\frac{1}{2}$-spin system. We have to admit that within the reduced picture disregarding the degrees of freedom of the environment, the system would really relax to a state described by a diagonal density matrix. However, the total space would be found in a linear superposition of the same kind as (87), with "generalized" states $|+\rangle$ and $|-\rangle$, namely, states including also the infinite degrees of freedom of the environment as well as those of the radiation field, thereby making the situation worse. In conclusion, the price we have to pay for an apparent destruction of quantum correlation on the space of interest, is to establish quantum-mechanical correlations over systems of still larger dimensions.

A. The Zurek Proposal

Let us go back to Eq. (4), here renumbered for the reader's convenience:

$$|\psi_0\rangle|\mu\rangle \rightarrow \alpha|a\rangle|+\rangle + \beta|b\rangle|-\rangle \qquad (90)$$

This is the result of the measurement process that, according to von Neumann,[1] is imagined as an abrupt collision between the quantum system S and the measurement apparatus M. At the end of the collision process we are in the presence of a paradoxical condition because $|+\rangle$ and $|-\rangle$ refer to macroscopically distinct states. Within von Neumann process, these are the states of a pointer thereby involving a macroscopic object, the measurement apparatus M, with an infinite number of degrees of freedom. This establishes a crucial difference with the optical model, where $|+\rangle$ and $|-\rangle$ refer to large quantum numbers of an individual oscillator, rather than to a collection of many oscillators (this latter condition might be regarded as a genuinely macroscopic system).

A more complete description should supplement Eq. (90) by including within the theoretical picture the environment of the system M. Prior to the interaction with the quantum system, the apparatus M and its environment are supposed to be in a state $|\mu\rangle|E_0\rangle$, where $|\mu\rangle$ is the unperturbed position of the pointer (see Section I) and $|E_0\rangle$ is the corresponding equilibrium state of the environment. Let us make the plausible assumption that throughout the extremely fast collision-like process of von Neumann the environment is left in its original state. Then, the von Neumann collision process creates the following state:

$$|\psi\rangle = (\alpha|a\rangle|+\rangle + \beta|b\rangle|-\rangle)|E_0\rangle \qquad (91)$$

where $|E_0\rangle$ denotes the state of the environment. We assume this instant as our origin of time. After this initial time the whole system, namely the quantum system S, the measurement apparatus M, and the environment, will move under the action of the interaction between M and the environment. Since the pointer is macroscopic, its environment is characterized by a virtually infinite number of degrees of freedom and the interaction between apparatus and environment is perceived by the apparatus as a fluctuation–dissipation process. Prior to the experimental observation of the pointer position, the irreversible process resulting from the interaction with the environment, slow compared to the collision-like stage, but fast compared to the macroscopic time scale of the experimenter, produces significant effects on the state of the pointer, and has the effect of dissolving the ghost of the Schrödinger cat.

This point of view, originally expressed by van Kampen[38] and Daneri et al.[39] has been recently reformulated by Zurek.[37] Herein, we illustrate this viewpoint adopting the procedure of Zurek. If a contraction over the degrees of freedom of the environment is carried out, the density matrix of the system $S + M$ reads

$$
\begin{aligned}
\rho(t) = {} & |\alpha|^2 |+\rangle\langle +| + |\beta|^2 |-\rangle\langle -| \\
& + z(t)\alpha\beta^* |+\rangle\langle -| \, |a\rangle\langle b| + z^*(t)\alpha^*\beta |+\rangle\langle -| \, |a\rangle\langle b|
\end{aligned}
\tag{92}
$$

where $z(t)$ is a function of time, which is expressed by Zurek in terms of the interaction between apparatus and environment. Zurek shows that for t tending to infinity the function $z(t)$ tends to vanish.

The similarity of the optical model result (89) with the key result of Zurek, (92), is striking. Here recurrences are eliminated as a consequence of the fact that the interaction with the infinite degree of freedom of the environment is responsible for the taking over of a genuinely irreversible process. However, the amplification process necessary to establish a connection with the macroscopic observer might also open up the avenue to an interaction between the optical Schrödinger cat and the environment, thereby making it equivalent to the Zurek model. Thus, we would be tempted to refer to the optical Schrödinger cat as a genuine Schrödinger cat, since the genuinely macroscopic environment is now a part of it. We would also be inclined to agree with Glauber[36] that the amplification dissolves this ghost.

Unfortunately, this happy state of affairs is fraught with problems, since the Zurek proposal, in our opinion, is not a satisfactory settlement of the quantum-mechanical paradoxes. Let us see why. The result of Zurek is interesting, because it shows that the quantum-mechanical coherence is destroyed by the influence of the environment, an effect also taken into account by Leggett and coworkers[40] in their study of the quantum-mechanical behavior of a macroscopic variable. However, this theory does not have anything to do with the occurrence of a real collapse, although it is indistinguishable from the consequences produced by the occurrence of real collapses. A real collapse would make the state (90) randomly decay into state $|a\rangle|+\rangle$ or state $|b\rangle|-\rangle$, with probabilities proportional to $|\alpha|^2$ and $|\beta|^2$, respectively, thereby resulting in the density matrix

$$
\rho(\infty) = |\alpha|^2 |+\rangle\langle +| + |\beta|^2 |-\rangle\langle -|
\tag{93}
$$

which is indistinguishable from the stationary regime of Zurek. However, the reader familiar with magnetic resonance relaxations[41] will immediately realize that the state (93) is reached by (92) as a result of a transverse relaxation. The dephasing process responsible for the attainment of the stationary state (93) implies the adoption of a Gibbs ensemble, and each system of the Gibbs ensemble is still a Schrödinger cat, in a state of suspended animation, even if the statistical average over the infinite systems makes the condition "cat alive" statistically independent of the condition "cat dead." This is an unsatisfactory state of affairs indeed!

B.   A Last Resort: The Nonlinear Schrödinger Equation

The paradoxical aspects of quantum mechanics stem from the linear nature of the Schrödinger equation. If it were possible to replace it with a nonlinear one, then it might be possible to get out of it a new quantum mechanics with no paradoxes.[42] We expect it, however, to be a hard job indeed to derive from a nonlinear Schrödinger equation a picture of the microscopic world as satisfactory as ordinary quantum mechanics. If, on the contrary, the nonlinear Schrödinger equation were to be interpreted in an effective sense, then the avenue to the solution of quantum-mechanical paradoxes would become practicable, because it would simply the adoption of ordinary quantum mechanics.

What do we mean by effective nonlinear Schrödinger equation? This meaning can be conveniently illustrated by using the full Hamiltonian of Eq. (5). Let us assume that the space of interest is that of the $\frac{1}{2}$-spin system. Then we are interested in building up an equation of motion concerning only this space and we might do that via a contraction procedure on the irrelevant space. By irrelevant space we mean either that of the oscillator or, if the case applies, that of the oscillator plus the environment of the oscillator. It has been widely remarked that the contraction procedure makes irreversibility enter into play.[43] If in addition to the action that the irrelevant space exerts on the system of interest we also consider the action that the system of interest exerts on the irrelevant system, then the resulting equation of motion for the system of interest has a nonlinear structure.[43] This nonlinear structure can be imagined as being produced by a nonlinear Schrödinger equation. This is what we mean by effective nonlinear Schrödinger equation. Although the Schrödinger equation driving the motion of the whole system is linear, the system of interest has dynamical properties that are equivalent to those that would be generated by a nonlinear Schrödinger equation.

This is a problem much more general than those concerning quantum optics. However, since all the simple models for the nonlinear Schrödinger equation have been based on Hamiltonians with the same structure as the spin-boson Hamiltonian of this paper, we think we should discuss it.

In a recent paper Bonilla and Guinea[44] illustrated a further version of this physical model, and showed that under the assumption that the oscillator is classical, a real collapse occurs. Since they mean a real collapse rather than a dephasing process à la Zurek, this would be a satisfactory settlement of the quantum-mechanical paradoxes, or, at least, a promising direction to investigate.

The model of Ref. 44 is essentially equivalent to the semiclassical version of our spin-boson system of Eq. (5). Thus it is immediately clear to the reader that the collapses discovered by these authors correspond to an irreversible decay into one of the two stable states discussed in Section II. In the case of very large couplings these two equilibrium states correspond to the $\frac{1}{2}$-spin system in states $|+\rangle_x$ and $|-\rangle_x$, and the irreversible decay into one or the other of these two states from the initial state $(|+\rangle_x + |-\rangle_x)/\sqrt{2}$ would be equivalent to a real collapse of this state into one of its two component states. Bonilla and Guinea[44] adopt a dissipation process acting directly on the $\frac{1}{2}$-spin system. This serves the purpose of making irreversible the process of collapse. The same task is accomplished by assuming the oscillator to be dissipative.[45] In our opinion the second condition is more natural, since the oscillator plays the role of a "pointer" and since this is macroscopic, it is plausible to imagine it to interact with a bath making it dissipative. Therefore, we make this latter choice and we replace the last equation of set (34) with

$$\dot{v} = -\Omega^2 x - g\langle \sigma_x \rangle - \gamma v \tag{94}$$

The numerical treatment of the collapse process by means of the semiclassical approximation carried out on the full Hamiltonian corresponds to the numerical solution of the set of differential equations (34) supplemented by (94). What about carrying out the semiclassical approximation on the RWA Hamiltonian (27)? This is equivalent to a numerical solution of the set of differential equations (33). In this case, to make the classical oscillator dissipative we must replace the last equation of this set with

$$\dot{v} = -\Omega^2 x - \frac{g}{2}\langle \sigma_x \rangle - \gamma v \tag{95}$$

However, according to the theoretical remarks of Section II, we expect that the semiclassical approximation on the RWA Hamiltonian does not bring about the collapses. Indeed, we have seen that the semiclassical approximation carried on the RWA Hamiltonian leads to a continuum of stationary states rather than the two equilibrium states of the full Hamiltonian.

In Figs. (10) and (11) we show the results of numerical calculations carried out using the semiclassical approximation to the full and the RWA Hamiltonian.

The parameters are chosen so as to satisfy the strong coupling condition yielding the nonordinary fixed points of Eqs. (35) and (36). This means that $\eta < 1$ and $\eta_{\text{RWA}} < 1$. The inclusion of dissipation does not affect (36). Thus, the position of the two points (36) is unchanged. The fixed points (35) on the contrary, are modified as follows:

$$\langle \sigma_x \rangle_{\text{eq}}^2 + \langle \sigma_y \rangle_{\text{eq}}^2 = 1 - \eta_{\text{RWA}}^2$$

$$\times \Omega^2 x_{\text{eq}}^2 + \left(1 + \frac{\gamma^2}{\Omega^2}\right) v_{\text{eq}}^2 + 2\gamma x_{\text{eq}} v_{\text{eq}} = \frac{g^2}{4\Omega^2}\left(1 - \eta_{\text{RWA}}^2\right) \qquad (96)$$

We see that the action of friction does not much change the situation, since it still admits an infinite number of equilibrium states.

We set the system into an initial condition with the $\frac{1}{2}$-spin system in the positive eigenstate of the Pauli matrix σ_z and the oscillator in the classical ground state, i.e., in the unperturbed equilibrium position. Notice that according to the analysis of Section II this initial condition is a unstable equilibrium state. Consequently, to provoke a collapse, if there is any, it is necessary to add to this initial condition small biases by displacing the oscillator or the spin from their equilibrium positions. From Fig. 10 we see that even an extremely small bias is amplified and turned into a real collapse. This fits our expectation on the effect of the semiclassical approximation on the full Hamiltonian. In the case of the RWA Hamiltonian, on the contrary, we derive from Fig. 11 a quite different behavior. In the early part of the relaxation process the system seems to regress to a collapsed state, namely $|+\rangle_x$ or $|-\rangle_x$. However, rather than settling in the collapsed state, as the full Hamiltonian system does, it departs from the collapsed state and undergoes a steady regression to its "true" stationary condition, with a vanishing mean value for σ_x and consequently for the oscillator position. Thus, the numerical analysis shows that the effect of damping is not limited to affecting the stationary states as shown in Eq. (96), but it makes all of them unstable.

In conclusion, our semiclassical analysis seems to support the results of Bonilla and Guinea. Furthermore, it is also evident that an experimental verification of this collapse process should be carried out adopting a linear

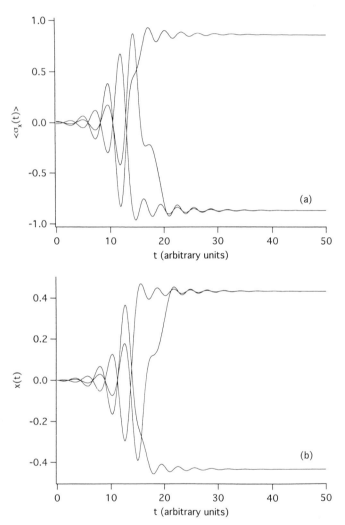

Figure 10. Time evolution for (a) $\langle \sigma_x(t) \rangle$ and (b) $x(t)$ according to Eqs. (34) supplemented by a dissipative term (see Eq. (94)). We show three trajectories starting from slightly different initial conditions for the spin close to the σ_z positive eigenstate, i.e., $\langle \sigma_z(0) \rangle \approx 1$ and $\langle \sigma_x(0) \rangle \approx 0$. Initially the oscillator is in the fixed point $x = 0$, $v = 0$. The parameters defining the system are given the following values: $\omega_0 = 1$, $\Omega = 2$, $g = 2$, $\gamma = 3$.

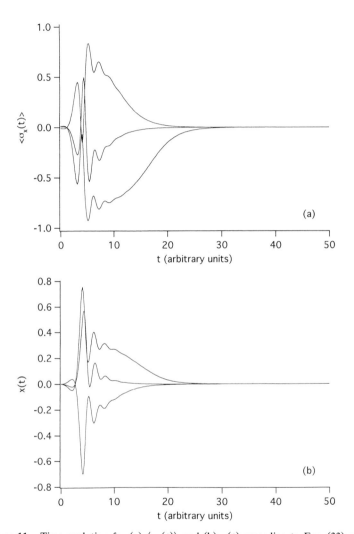

Figure 11. Time evolution for (a) $\langle \sigma_x(t) \rangle$ and (b) $x(t)$ according to Eqs. (33) supplemented by a dissipative term (see Eq. (95)). We show three trajectories starting from slightly different initial conditions for the spin close to the σ_z positive eigenstate, i.e., $\langle \sigma_z(0) \rangle \approx 1$ and $\langle \sigma_x(0) \rangle \approx 0$. Initially the oscillator is in the fixed point $x = 0$, $v = 0$. The parameters defining the system are given the following values: $\omega_0 = 1$, $\Omega = 2$, $g = 4$ (see remark on page 12), $\gamma = 3$.

polarized field, rather than the circularly polarized radiation. The reader should contrast this aspect to the fact that the optical Schrödinger cat[14, 15, 24] rests on the RWA Hamiltonian, and thus, according to the results of Section II, on the adoption of a circularly polarized radiation field.

We do believe, however, that if the ordinary quantum mechanics holds true in the specific physical conditions behind the results illustrated in Fig. 10, then no collapse can occur. We believe that the collapse process is determined by the classical approximation on the harmonic oscillator, and that this approximation neglects quantum-mechanical fluctuations as significant as the nonlinear terms responsible for the collapse process. This is made transparent if we use the Wigner analysis of Section IV.A. The semiclassical approximation behind our results, and behind those of Bonilla and Guinea as well, is equivalent to disregarding the QGD term of (55). The system is driven only by the classical-like Liouvillian of (54). Thus, if only this classical-like term is used, the resulting dynamical process must coincide with that illustrated by the authors of Ref. 46. Adopting a contraction procedure on the irrelevant space, namely the dissipative oscillator, these authors have shown that it is possible to express the behavior of the $\frac{1}{2}$-spin system by means of a nonlinear Fokker-Planck equation. In the adiabatic limit of an oscillator instantaneously regressing to equilibrium, this Fokker-Planck equation is shown[46] to be equivalent to the prediction of the two-site discrete nonlinear Schrödinger equation.

This is what the effective nonlinear Schrödinger equation is all about. According to the careful analysis of these authors, fitting the point of view of many others,[47-49] the key ingredient of this effective nonlinear behavior is the feedback operator

$$\mathscr{L}_R \equiv gx_1 \frac{\partial}{\partial p} \tag{97}$$

(last term on the right side of Eq. (54)), namely the term responsible for the action exerted by the $\frac{1}{2}$-spin system on its own bath. However, if we compare this key term to the neglected QGD term of (55), we find that the latter contains a contribution as intense as (97). Indeed, from the second term in the square brackets of (55) we obtain

$$g\frac{\partial}{\partial p}\frac{\partial}{\partial x_1}x_1^2 = g\frac{\partial}{\partial p}x_1^2\frac{\partial}{\partial x_1} - 2g\frac{\partial}{\partial p}x_1 = g\frac{\partial}{\partial p}x_1^2\frac{\partial}{\partial x_1} - 2\mathscr{L}_R \tag{98}$$

This raises strong doubts on the assumption that the QGD term can be

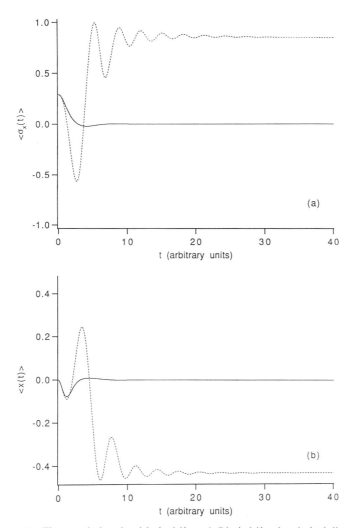

Figure 12. Time evolution for (a) $\langle \sigma_x(t) \rangle$ and (b) $\langle x(t) \rangle$; the dashed line is the prediction of the semiclassical set of Eqs. (34), and the solid line is obtained from the numerical solution of the quantum Liouville for the spin-boson model of Eq. (5) supplemented by dissipation (see Ref. 44). The initial state for the spin has a strong bias along the x direction ($\langle \sigma_x(0) \rangle = 0.5$). The parameters defining the system are given the following values: $\omega_0 = 1$, $\Omega = 2$, $g = 2$, $\gamma = 3$.

neglected. To justify this assumption one might be tempted to invoke the limit $\hbar \to 0$. However, whatever one does, it is not possible to prevent the reaction term, responsible for the nonlinear nature of the reduced motion of the system, and the QGD contribution, expressing the joint quantum fluctuations of system and oscillator, from scaling in the same way. In conclusion, there is no convincing derivation of the high-coupling nonlinearity that is necessary for the collapse to occur.

Of course, the fact that the neglected term of Eq. (55) is as intense as the reaction term does not rule out the possibility that nonlinearity might result in significant effects. For this reason we now adopt a numerical calculation, aiming at establishing the role of the QGD term on the collapse process. To do that we solve again the Liouville-von Neumann equation (82), and we illustrate the corresponding results in Fig. 12. In this case, we set a strong bias on the $\frac{1}{2}$-spin system, which should make it easier to provoke the onset of the process of wave-function collapse. However, from Fig. 12 we see that, in spite of the significant bias the "exact" quantum evolution of $\langle \sigma_x(t) \rangle$ always regresses to the equilibrium condition $\langle \sigma_x \rangle = 0$, corresponding to the equally weighted superposition of the two states $|+\rangle_x$ and $|-\rangle_x$ of the $\frac{1}{2}$-spin system. Note that the exact quantum-mechanical evolution of $\langle \sigma_x(t) \rangle$ with the same weak biases as those used in Fig. 10 would be indistinguishable from the abscissa axis of Fig. 12. In conclusion, the possibility of deriving collapses from the ordinary quantum mechanics by means of an effective nonlinear Schrödinger equation is jeopardized by the ill-founded microscopic derivation of this equation.

VI. CONCLUSIONS

We have seen that an important aspect of the JCM is its connection with the intriguing problem of the semiclassical limit. Many authors[14, 15, 36] stress the relevance with the problems raised by the theory of quantum measurement and quantum-mechanical paradoxes, and many refer to some aspects of the rich dynamics of this system as a quantum optical Schrödinger cat. We believe that this is probably an arbitrary extension to optics of the problems connected with the tendency for quantum mechanics to become macroscopic.[50] It has been pointed out[50] that the historic paradoxes of quantum mechanics are a small class of a much more general family of systems, where the quantum-mechanical uncertainty is made to grow up very quickly, to reach macroscopic dimensions within a time comparable to the time duration of the experiment under consideration. This must be perceived as a paradox only when by "macroscopic" property is meant something related with our daily experience of world. For a rigorous definition of a macroscopic variable we refer the reader to the remarkable work of Leggett and coworkers[40] and to their search for macroscopic quantum manifestation.

However, we must recognize that the JCM system well serves the purpose of stressing the difficulties associated with the derivation of classical physics from quantum physics. Our theoretical analysis confirms this aspect of the RWA model, and shows that the quantum-mechanical properties are present in the RWA model, at any value of the radiation field intensity. This is in a great contrast with the behavior of the full Hamiltonian, which is proven to lead to an agreement with the classical prediction in the condition where the RWA Hamiltonian fails to do that.

An experimental search of these effects is certainly welcome, especially because the experiments carried out so far are limited to a region where the replacement of the full Hamiltonian with the RWA Hamiltonian is still valid.[13] The main result of this paper is that with a real system driven by a full Hamiltonian the traditional semiclassical predictions are recovered at fields of extremely large intensity. A circularly polarized field of extremely large intensity, on the contrary, would lead to effects indistinguishable from those predicted by the RWA quantum Hamiltonian, and these would be in great conflict with the expectation of the corresponding semiclassical approximation. Therefore, an experimental search is highly welcome with physical conditions where the RWA Hamiltonian, although in conflict with the traditional viewpoints on the semiclassical approximation, would be a genuine representation of reality rather than an approximation.

Acknowledgments

This review would not have been possible without the invaluable help we have had from Roberto Roncaglia and David Vitali, to whom we are indebted for many illuminating discussions. We thank the Texas Higher Education Coordinating board for partial support of this research (Texas Advanced Research Project 0035494-038).

References

1. J. von Neumann, *Mathematical Foundation of Quantum Mechanics*, Princeton University Press, Princeton, NJ, 1955.

2. E. Schrödinger, *Naturwissenschaften* **23**, 807, 823, 844 (1935); translation in English in *Proc. AM. Philos. Soc.* **124**, 323 (1980).

3. A. J. Leggett, in J. Souletie, J. Vannimenus, and R. Stora (Eds.), *Chance and Matter*, Proceedings, 1986 Les Houches Summer School, North Holland, Amsterdam, 1987, p. 395.

4. R. J. Glauber, *Phys. Rev.* **131**, 2766 (1963).

5. (a) J. H. Eberly, N. B. Narozhny, and J. J. Sánchez-Mondragón, *Phys. Rev. Lett.* **44**, 1323 (1980); (b) N. B. Narozhny, J. J. Sánchez-Mondragón, and J. H. Eberly, *Phys. Rev. A* **23**,

236 (1981); (c) H. I. Yoo, J. J. Sánchez-Mondragón, and J. H. Eberly, *J. Phys. A* **14**, 1383 (1981); (d) H. I. Yoo and J. H. Eberly, *Phys. Rep.* **118**, 239 (1985).

6. S. M. Barnett and P. L. Knight, *Phys. Rev. A* **33**, 2444 (1986); R. R. Puri and G. S. Agarwal, *Phys. Rev. A* **33**, 3610 (1986).

7. X.-S. Li, D. L. Lin, and C.-D. Gong, *Phys. Rev. A* **36**, 5209 (1987); F.-L. Li, D. L. Lin, T. F. George, and X.-S. Li, *Phys. Rev. A* **40**, 1394 (1989); F.-L. Li, X.-S. Li, D. L. Lin, and T. F. George, *Phys. Rev. A* **40**, 5129 (1989).

8. G. M. Palma and P. L. Knight, *Phys. Rev. A* **39**, 1962, 1989; S.-C. Gou, *Phys. Rev. A* **40**, 5116 (1989); M. Venkata Satyanarayana, P. Rice, R. Vyas, and H. J. Carmichael, *J. Opt. Soc. Am. B* **6**, 228 (1989); J. I. Cirac and L. L. Sanchez-Soto, *Phys. Rev. A* **40**, 3743 (1989).

9. C. Aslangul, N. Pottier, and D. Saint James, *J. Phys. (Paris)* **47**, 1657 (1986); S. Chakravarty, *Phys. Rev. Lett.* **49**, 681 (1982); A. J. Bray and M. A. Moore, *Phys. Rev. Lett.* **49**, 1545 (1982); V. Cápek and P. Chvosta, *Phys. Rev. A* **43**, 2819 (1991).

10. C. H. Mak and D. Chandler, *Phys. Rev. A* **32**, 2462 (1991).

11. W. H. Louisell, *Quantum Statistical Properties of Radiation*, Wiley, New York, 1973.

12. E. T. Jaynes and F. W. Cummings, *Proc. IEEE* **51**, 89 (1963).

13. G. Rempe, H. Walther, and N. Klein, *Phys Rev. Lett.* **58**, 353 (1987).

14. B. Sherman and G. Kurizki, *Phys. Rev. A* **45**, R7674 (1992); I. Sh. Averbukh, *Phys. Rev. A* **46**, R2205 (1992); V. Buzek, H. Moya-Cessa, P. L. Knight, and S. J. D. Phoenix, *Phys. Rev. A* **45**, 8190 (1992).

15. J. Gea-Banacloche, *Phys. Rev. A* **44**, 5913 (1991).

16. A. Abragam, *The Principles of Nuclear Magnetism*, Clarendon, Oxford, UK, 1961.

17. M. D. Crisp, *Phys. Rev. A* **43**, 2430 (1991).

18. P. I. Belobrov, G. M. Zaslavskii, and G. Kh. Tartakovskii, *Sov. Phys.-JETP* **44**, 945 (1977).

19. P. W. Milonni, J. R. Ackerhalt, and H. W. Galbraith, *Phys. Rev. Lett.* **50**, 966 (1983); **51**, 1108 (E) (1983).

20. R. Graham and M. Höhnerbach, *Z. Phys.-Cond. Matt.* **57**, 233 (1984).

21. R. F. Fox and J. Eidson, *Phys. Rev. A* **34**, 482 (1986).

22. A. Nath and D. S. Ray, *Phys. Lett. A* **117**, 341 (1986); G. Gangopadhyay and D. S. Ray, *Phys. Rev. A* **40**, 3750 (1989).

23. M. A. M. de Aguiar and K. Furuya, *Ann. Phys.* **216**, 291 (1992).

24. J. Eiselt and H. Risken, *Phys. Rev. A* **43**, 346 (1991).

25. E. P. Wigner, *Phys. Rev.* **40**, 749 (1932); M. Hillery, R. F. O'Connell, M. O. Scully, and E. P. Wigner, *Phys. Rept.* **106**, 122 (1984).

26. K. Takahashi, *Prog. Theor. Phys. Suppl.* **98**, 109, (1989).

27. L. Bonci and P. Grigolini, *Phys. Rev. A* **46**, 4445 (1992).

28. S. J. D. Phoenix and P. L. Knight, *Phys. Rev. A* **44**, 6023 (1991).

29. H. Weyl, *Z. Phys.* **46**, 1 (1927).

30. L. Bonci, R. Roncaglia, B. J. West, and P. Grigolini, *Phys. Rev. Lett.* **67**, 2593 (1991).

31. R. Roncaglia, L. Bonci, B. J. West, and P. Grigolini, *J. Stat. Phys.* **68**, 321 (1992); L. Bonci, R. Roncaglia, P. Grigolini, and B. J. West, *Phys. Rev. A* **45**, 8490 (1992).

32. G. H. Weiss and A. A. Maradudin, *J. Math. Phys.* **3**, 771, (1962).

33. L. Bonci, P. Grigolini, and D. Vitali, *Phys. Rev. A* **42**, 4452 (1990).

34. K. Husimi, *Proc. Phys. Math. Soc. Japan* **22**, 264 (1940).

35. G. J. Milburn and C. A. Holmes, *Phys. Rev. Lett.* **56**, 2237 (1986); B. Yurke and D. Stoler, *Phys. Rev. Lett.* **57**, 13 (1986); A. Mecozzi and P. Tombesi, *Phys. Rev. Lett.* **58**, 1055 (1987).

36. R. J. Glauber, in E. R. Pike and S. Sarkar (Eds.), *Frontiers in Quantum Optics*, Adam Hilger, Bristol, UK, 1986, p. 534.

37. W. G. Unruh, W. H. Zurek, *Phys. Rev. D* **40**, 1071 (1989).

38. N. G. van Kampen, *Physica* **20**, 603 (1954).

39. A. Daneri, A. Loinger, and G. M. Prosperi, *Nucl. Phys.* **33**, 297 (1962).

40. A. J. Leggett. S. Chakravarty, A. T. Dorsey, M. P. A. Fisher, A. Garg, and W. Zwerger, *Rev. Mod. Phys.* **59**, 1 (1987).

41. C. P. Poole, Jr., and H. A. Farach, *Relaxation in Magnetic Resonance*, Academic, New York, 1971; S. Stenholm, in P. Meystre and M. O. Scully (Eds.), *Quantum Optics Experimental Gravitation and Measurement Theory*, Plenum, New York, 1982; G. Emch, *Helv. Phys. Acta* **45**, 1409 (1972); R. A. Harris and L. Stodolsky, *J. Chem. Phys.* **74**, 2145 (1981).

42. S. Weinberg, *Phys. Rev. Lett.* **62**, 885 (1989).

43. P. Grigolini, in F. Moss and P. V. E. McClintock (Eds.), *Noise in Nonlinear Dynamical System*, Cambridge University Press, Cambridge, UK, 1989, Chapter 5, p. 161.

44. L. L. Bonilla and F. Guinea, *Phys. Rev. A* **45**, 7718 (1992).

45. L. Bonci, P. Grigolini, R. Roncaglia, and D. Vitali, *Phys. Rev. A* **47**, 3538.

46. L. Bonci, P. Grigolini, G. Trefan, R. Manella, and D. Vitali, *Phys. Rev. A* **43**, 2694 (1991); L. Bonci, P. Grigolini, R. Mannella, and D. Vitali, *Phys. Rev. A* **44**, 876. (1990).

47. P. Grigolini, *Int. J. Mod. Phys. B* **6**, 171 (1992).

48. D. Feinberg and J. Ranninger, *Phys. Rev. A* **30**, 3339 (1984); D. Feinberg and J. Ranninger, *Physica D* **14**, 29 (1984); D. Feinberg and J. Ranninger, *Phys. Rev. A* **33**, 3466 (1986).

49. P. Grigolini, V. M. Kenkre, and D. Vitali, *Phys. Rev. A* **44**, 1015 (1991).

50. P. Grigolini, *Quantum Mechanical Irreversibility and Measurement*, World Scientific, Teaneck, NJ, 1993.

THEORETICAL FORMALISM AND MODELS FOR VIBRATIONAL CIRCULAR DICHROISM INTENSITY

TERESA B. FREEDMAN

and

LAURENCE A. NAFIE

Department of Chemistry and Center for Molecular Electronics
Syracuse University, Syracuse, New York

CONTENTS

Modern Nonlinear Optics, Part 3, Edited by Myron Evans and Stanisław Kielich. Advances in Chemical Physics Series, Vol. LXXXV.
ISBN 0-471-30499-9 © 1994 John Wiley & Sons, Inc.

I. INTRODUCTION

Vibrational circular dichroism (VCD),[1–6] the differential absorbance of left and right circularly polarized infrared radiation by a chiral molecule during vibrational excitation, arises from simultaneous changes in the electric and magnetic dipole moments of the molecule due to the nuclear motion. VCD can thus be considered to be a nonlinear interaction of matter and radiation due to the contribution from the magnetic dipole interaction. Although VCD was first measured nearly twenty years ago,[7,8] a priori formulations of the theoretical basis of the effect have been developed only within the past decade.[9–22] Implementation of several of these formalisms as ab initio molecular orbital calculations has been carried out by a number of laboratories.[2, 23–33] It is currently possible to calculate VCD intensities in good agreement with experiment for molecules of moderate size. For large molecules, approximate models,[34–51] whose form is motivated by descriptive mechanisms for VCD intensity, continue to be employed for spectral interpretation.

The a priori formalisms for VCD intensity have been developed independently by a number of investigators; unfortunately, the definition of terms and the symbols used are not mutually consistent. Although there has been some effort to relate various theoretical approaches, there remains a need for a unifying presentation that clearly defines the relationships between all the a priori formalisms. We provide in Section II a systematic presentation of the tensor elements required in VCD intensity expressions; these tensor elements are formulated in terms of energy gradients and perturbation expansions, as well as expansions over LCAO molecular orbitals for SCF wavefunctions.

The models for VCD intensity are derived in Section III from the a priori expressions in order to evaluate the relationship of the models to rigorous theory and to assess the approximations involved. The implementations of both the a priori formalisms and approximate models for the calculation of VCD intensity are described in Section IV. Examples are provided where calculated intensities from several approaches can be compared for the same molecule and vibrational force field.

II.   A PRIORI THEORETICAL FORMALISM

A.   Atomic Polar Tensors and Atomic Axial Tensors

Infrared absorption intensity for a fundamental vibrational transition at angular frequency ω_a is proportional to the dipole strength[52] in the position form,

$$D_r^a(g0 \to g1) = \left| \left\langle \tilde{\Psi}_{g1}^a \middle| \hat{\mu}_r \middle| \tilde{\Psi}_{g0}^a \right\rangle \right|^2 \tag{1}$$

or velocity form,

$$D_v^a(g0 \to g1) = \omega_a^{-2} \left| \left\langle \tilde{\Psi}_{g1}^a \middle| \hat{\mu}_v \middle| \tilde{\Psi}_{g0}^a \right\rangle \right|^2 \tag{2}$$

where $\tilde{\Psi}_{g0}^a$ and $\hat{\Psi}_{g1}^a$ are the ground and first excited vibrational states, respectively, of normal mode Q_a (a mass weighted normal coordinate) in the ground electronic state,

$$\hat{\mu}_r = \hat{\mu}_r^E + \hat{\mu}_r^N = -\sum_j e \mathbf{r}_j + \sum_J Z_J e \mathbf{R}_J \tag{3}$$

is the position form of the electric dipole moment operator, and

$$\hat{\mu}_v = \hat{\mu}_v^E + \hat{\mu}_v^N = -\sum_j e \dot{\mathbf{r}}_j + \sum_J Z_J e \dot{\mathbf{R}}_J = \sum_j \frac{ie\hbar}{m} \frac{\partial}{\partial \mathbf{r}_j} - \sum_J \frac{ie\hbar Z_J}{M_J} \frac{\partial}{\partial \mathbf{R}_J} \tag{4}$$

is the velocity form of the electric dipole moment operator, broken down into an electronic contribution summed over electrons j with mass m and charge $-e$, and a nuclear contribution summed over nuclei J with mass M_J and charge $Z_J e$. In the expressions to follow, Cartesian component notation is employed, wherein summation over repeated Greek subscripts is implied; $\varepsilon_{\alpha\beta\gamma}$ is the unit alternating tensor. A tilde denotes a complex quantity.

In the harmonic approximation, the electric dipole transition moments can be expressed in terms of the atomic polar tensor (APT)[53, 54] for nucleus A with position $\mathbf{R}_A$ and velocity $\dot{\mathbf{R}}_A$,

$$\langle \tilde{\Psi}_{g1}^a | \hat{\mu}_{r,\beta} | \tilde{\Psi}_{g0}^a \rangle = \left(\frac{\hbar}{2\omega_a} \right)^{1/2} \sum_A P_{r,\alpha\beta}^A s_{A\alpha,a} \tag{5}$$

and

$$\langle \tilde{\Psi}_{g1}^a | \hat{\mu}_{v,\beta} | \tilde{\Psi}_{g0}^a \rangle = i \left(\frac{\hbar\omega_a}{2} \right)^{1/2} \sum_A P_{v,\alpha\beta}^A s_{A\alpha,a} \tag{6}$$

The atomic polar tensors are defined in the position form

$$P_{r,\alpha\beta}^A = E_{r,\alpha\beta}^A + N_{\alpha\beta}^A \tag{7}$$

or the velocity form

$$P_{v,\alpha\beta}^A = E_{v,\alpha\beta}^A + N_{\alpha\beta}^A \tag{8}$$

where

$$N_{\alpha\beta}^A = Z_A e \delta_{\alpha\beta} \tag{9}$$

is the nuclear contribution. For the ground state electronic wave function $\tilde{\Psi}_g$

$$E_{r,\alpha\beta}^A = \left(\frac{\partial}{\partial R_{A\alpha}} \langle \tilde{\Psi}_g | \hat{\mu}_{r,\beta}^E | \tilde{\Psi}_g \rangle \right)_{R=0,\, \dot{R}=0} \tag{10}$$

is the electronic contribution to the position-form APT, and

$$E_{v,\alpha\beta}^A = \left(\frac{\partial}{\partial \dot{R}_{A\alpha}} \langle \tilde{\Psi}_g | \hat{\mu}_{v,\beta}^E | \tilde{\Psi}_g \rangle \right)_{R=0,\, \dot{R}=0} \tag{11}$$

is the electronic contribution to the velocity-form APT. The s-vector (displacement vector) for nucleus A in normal mode Q_a is defined in terms of either nuclear position or nuclear velocity,

$$s_{A\alpha, a} = \left(\frac{\partial R_{A\alpha}}{\partial Q_a}\right)_{Q=0} = \left(\frac{\partial \dot{R}_{A\alpha}}{\partial \dot{Q}_a}\right)_{\dot{Q}=0} \tag{12}$$

where $\dot{Q}_a = P_a$ is the conjugate momentum for the normal mode.

The vibrational circular dichroism intensity for a fundamental vibrational transition is proportional to the rotational strength,[9, 55]

$$R_r^a(g0 \rightarrow g1) = \mathrm{Im}\left(\left\langle \tilde{\Psi}_{g0}^a | \hat{\mu}_r | \tilde{\Psi}_{g1}^a \right\rangle \left\langle \tilde{\Psi}_{g1}^a | \hat{m} | \tilde{\Psi}_{g0}^a \right\rangle\right) \tag{13}$$

or

$$R_\nu^a(g0 \rightarrow g1) = \omega_a^{-1} \mathrm{Re}\left(\left\langle \tilde{\Psi}_{g0}^a | \hat{\mu}_\nu | \tilde{\Psi}_{g1}^a \right\rangle \left\langle \tilde{\Psi}_{g1}^a | \hat{m} | \tilde{\Psi}_{g0}^a \right\rangle\right) \tag{14}$$

which involve transition moments of both the electric dipole moment operator (Eq. (3) or (4)) and the magnetic dipole moment operator

$$\hat{m} = \hat{m}^E + \hat{m}^N = -\sum_j \frac{e}{2c}\mathbf{r}_j \times \dot{\mathbf{r}}_j + \sum_J \frac{Z_J e}{2c}\mathbf{R}_J \times \dot{\mathbf{R}}_J \tag{15}$$

In the harmonic approximation, the magnetic dipole transition moment

$$\left\langle \tilde{\Psi}_{g1}^a | \hat{m}_\beta | \tilde{\Psi}_{g0}^a \right\rangle = i\left(\frac{\hbar \omega_a}{2}\right)^{1/2} \sum_A M_{\alpha\beta}^A s_{A\alpha, a} \tag{16}$$

is expressed in terms of the atomic axial tensor (AAT)[11, 12, 16, 17, 22]

$$M_{\alpha\beta}^A = I_{\alpha\beta}^A + J_{\alpha\beta}^A \tag{17}$$

with nuclear contribution

$$J_{\alpha\beta}^A = \frac{Z_A e}{2c}\varepsilon_{\alpha\beta\gamma}R_{A\gamma}^0 \tag{18}$$

and electronic contribution

$$I_{\alpha\beta}^A = \left(\frac{\partial}{\partial \dot{R}_{A\alpha}}\left\langle \tilde{\Psi}_g | \hat{m}_\beta^E | \tilde{\Psi}_g \right\rangle\right)_{R=0, \dot{R}=0} \tag{19}$$

These definitions for the AAT and its electronic and nuclear contributions differ by a factor of $i/2\hbar$ from those introduced by Stephens.[11, 12] With our definition, the APT and AAT are defined in a parallel fashion as direct dipole–moment derivatives.

Insight into the nature of the atomic polar and axial tensors can also be gained by expressing them as derivatives involving the ground-state charge density $\rho_g(\mathbf{r}, \mathbf{R}, \dot{\mathbf{R}})$, current density $\mathbf{j}_g(\mathbf{r}, \mathbf{R}, \dot{\mathbf{R}})$, and angular current density $\mathbf{r} \times \mathbf{j}_g(\mathbf{r}, \mathbf{R}, \dot{\mathbf{R}})$ of the molecule (including both electronic and nuclear contributions), which are treated as functions of the electron position and nuclear position and velocity[48, 51, 56]:

$$P_{r,\alpha\beta}^A = \left(\frac{\partial}{\partial R_{A\alpha}} \int r_\beta \rho_g(\mathbf{r}, \mathbf{R}, \dot{\mathbf{R}})\, d\tau \right)_{R=0,\, \dot{R}=0} \tag{20}$$

$$P_{v,\alpha\beta}^A = \left(\frac{\partial}{\partial \dot{R}_{A\alpha}} \int j_{g,\beta}(\mathbf{r}, \mathbf{R}, \dot{\mathbf{R}})\, d\tau \right)_{R=0,\, \dot{R}=0} \tag{21}$$

$$M_{\alpha\beta}^A = \left(\frac{\partial}{\partial \dot{R}_{A\alpha}} \int \varepsilon_{\beta\gamma\delta} r_\gamma j_{g,\delta}(\mathbf{r}, \mathbf{R}, \dot{\mathbf{R}})\, d\tau \right)_{R=0,\, \dot{R}=0} \tag{22}$$

where $\rho_g(\mathbf{r}, \mathbf{R}, \dot{\mathbf{R}})$ and $\mathbf{j}_g(\mathbf{r}, \mathbf{R}, \dot{\mathbf{R}})$ are expressed in terms of complete adiabatic ground electronic state wave functions, $\tilde{\Psi}_g(\mathbf{r}, \mathbf{R}, \dot{\mathbf{R}})$, as[22, 57]

$$\rho_g(\mathbf{r}, \mathbf{R}, \dot{\mathbf{R}}) = -e \tilde{\Psi}_g^*(\mathbf{r}, \mathbf{R}, \dot{\mathbf{R}}) \tilde{\Psi}_g(\mathbf{r}, \mathbf{R}, \dot{\mathbf{R}}) + \sum_J Z_J e \delta(\mathbf{r} - \mathbf{R}_j) \tag{23}$$

$$\mathbf{j}_g(\mathbf{r}, \mathbf{R}, \dot{\mathbf{R}}) = \frac{-e\hbar}{2im} \left[\tilde{\Psi}_g^*(\mathbf{r}, \mathbf{R}, \dot{\mathbf{R}}) \frac{\partial}{\partial \mathbf{r}} \tilde{\Psi}_g(\mathbf{r}, \mathbf{R}, \dot{\mathbf{R}}) - \tilde{\Psi}_g(\mathbf{r}, \mathbf{R}, \dot{\mathbf{R}}) \frac{\partial}{\partial \mathbf{r}} \tilde{\Psi}_g^*(\mathbf{r}, \mathbf{R}, \dot{\mathbf{R}}) \right]$$
$$+ \sum_j Z_J e \dot{\mathbf{R}}_j \delta(\mathbf{r} - \mathbf{R}_J) \tag{24}$$

The AAT in Eq. (22) is described by a moment arm $\mathbf{r}$ directed to an element of current density $\mathbf{j}_g$ located at position $\mathbf{r}$; this formulation demonstrates that the contributions to the AAT are local in nature.

We note that in terms of derivatives of the electric and magnetic moments of the molecule with respect to the normal coordinate Q_a and conjugate momentum P_a for normal mode a, the dipole and rotational strengths in the harmonic approximation are expressed as[48]

$$D_r^a(g0 \rightarrow g1) = \frac{\hbar}{2\omega_a} \left| \left(\frac{\partial \boldsymbol{\mu}_r}{\partial Q_a} \right)_{0,0} \right|^2 \tag{25}$$

$$D_v^a(g0 \rightarrow g1) = \frac{\hbar}{2\omega_a} \left| \left(\frac{\partial \boldsymbol{\mu}_v}{\partial P_a} \right)_{0,0} \right|^2 \tag{26}$$

$$R_r^a(g0 \rightarrow g1) = \frac{\hbar}{2} \left(\frac{\partial \boldsymbol{\mu}_r}{\partial Q_a} \right)_{0,0} \cdot \left(\frac{\partial \boldsymbol{m}}{\partial P_a} \right)_{0,0} \tag{27}$$

$$R_v^a(g0 \rightarrow g1) = \frac{\hbar}{2} \left(\frac{\partial \boldsymbol{\mu}_v}{\partial P_a} \right)_{0,0} \cdot \left(\frac{\partial \boldsymbol{m}}{\partial P_a} \right)_{0,0} \tag{28}$$

B. Perturbation Expressions

Expressions for the electronic contributions to the atomic polar tensors and atomic axial tensor can be derived by considering a perturbed adiabatic wave function that is factorable into electronic and vibrational terms of the form (for the zeroth vibrational level of the ground electronic state)[22, 57]

$$\tilde{\Psi}_{g0}^a(\mathbf{r}, \mathbf{R}, \mathbf{X}) = \tilde{\Psi}_g(\mathbf{r}, \mathbf{R}, \mathbf{X}) \chi_{g0}^a(\mathbf{R}) \tag{29}$$

where $\mathbf{X}$ is a perturbation variable associated with a perturbed electronic Hamiltonian

$$H_E = H_E^0 + H_E'(\mathbf{X}) \tag{30}$$

The relevant perturbations are[17, 22]

$$H_E'(\mathscr{E}) = -\hat{\boldsymbol{\mu}}_r \cdot \mathscr{E} = e \sum_j \mathbf{r}_j \cdot \mathscr{E} \qquad \text{[electric field ($\mathscr{E}$) perturbation]}$$

$$\tag{31}$$

$$H_E'(\mathscr{A}) = -\left(\frac{1}{c} \right) \hat{\boldsymbol{\mu}}_v \cdot \mathscr{A} = -\left(\frac{ie\hbar}{mc} \right) \sum_j \mathscr{A} \cdot \nabla_j$$

$$\text{[electric-dipole vector potential ($\mathscr{A}$) perturbation]} \quad (32)$$

$$H_E'(\mathscr{H}) = -\hat{m} \cdot \mathscr{H} = -\left(\frac{ie\hbar}{2mc} \right) \sum_j \mathbf{r}_j \times \nabla_j \cdot \mathscr{H}$$

$$= -\left(\frac{ie\hbar}{2mc} \right) \sum_j \mathscr{H} \times \mathbf{r}_j \cdot \nabla_j \tag{33}$$

$$\text{[magnetic field ($\mathscr{H}$) perturbation]}$$

$$H_E'(\dot{\mathbf{R}}) = -i\hbar \sum_J \dot{\mathbf{R}} \cdot \nabla_J \qquad \text{[nuclear velocity ($\dot{\mathbf{R}}$) perturbation]} \quad (34)$$

For the electric-dipole vector potential, magnetic field, and nuclear velocity perturbations, $H'_E(\mathbf{X})$ causes the wave function to be complex. The magnetic field perturbation depends on the gauge origin or reference point for the electron position in the angular momentum operator.

The electronic contributions to the atomic polar and axial tensors can be expressed as second-order properties arising from the simultaneous perturbation by a field and either nuclear position or nuclear momentum. Expressed as derivatives of the electronic energy of the molecule, these tensor contributions are[17, 22, 58]

$$E^A_{r,\alpha\beta} = -\left(\frac{\partial^2 E_{el}}{\partial \mathscr{E}_\beta \partial R_{A\alpha}} \right)_{0,0} \tag{35}$$

$$E^A_{v,\alpha\beta} = -c\left(\frac{\partial^2 E_{el}}{\partial \mathscr{A}_\beta \partial \dot{R}_{A\alpha}} \right)_{0,0} \tag{36}$$

$$I^A_{\alpha\beta} = -\left(\frac{\partial^2 E_{el}}{\partial \mathscr{H}_\beta \partial \dot{R}_{A\alpha}} \right)_{0,0} \tag{37}$$

From second-order perturbation theory, for the general case in which the Hellman-Feynman theorem[59] is satisfied for perturbation λ_1, we have[60]

$$\left(\frac{\partial^2 E_{el}}{\partial \lambda_1 \partial \lambda_2} \right)_{0,0} = \left(\frac{\partial^2 E_{el}}{\partial \lambda_2 \partial \lambda_1} \right)_{0,0} = \left[\frac{\partial}{\partial \lambda_2} \left\langle \tilde{\Psi}_g \left| \frac{\partial H_E}{\partial \lambda_1} \right| \tilde{\Psi}_g \right\rangle \right]_{0,0} \tag{38}$$

$$= \left\langle \Psi^0_g \left| \left(\frac{\partial^2 H_E}{\partial \lambda_1 \partial \lambda_2} \right)_{0,0} \right| \Psi^0_g \right\rangle + 2\left[\left\langle \tilde{\Psi}_g \left| \frac{\partial H_E}{\partial \lambda_1} \right| \frac{\partial \tilde{\Psi}_g}{\partial \lambda_2} \right\rangle \right]_{0,0} \tag{39}$$

$$= \left\langle \Psi^0_g \left| \left(\frac{\partial^2 H_E}{\partial \lambda_1 \partial \lambda_2} \right)_{0,0} \right| \Psi^0_g \right\rangle$$
$$- 2 \sum_{e \neq g} \left[\frac{\langle \tilde{\Psi}_g | \partial H_E/\partial \lambda_1 | \tilde{\Psi}_e \rangle \langle \tilde{\Psi}_e | \partial H_E/\partial \lambda_2 | \tilde{\Psi}_g \rangle}{E_e - E_g} \right]_{0,0} \tag{40}$$

For most types of basis functions, the first term on the right in Eqs. (39) and (40) does not contribute to the polar or axial tensor.

C. A Priori Formalisms

The various a priori theoretical formulations for VCD intensity differ primarily in their approach to the calculation of the atomic axial tensor. Each approach also leads to a corresponding expression for the velocity form of the APT.

1. Vibronic Coupling Theory

In the vibronic coupling theory (VCT) formulation of VCD intensity developed by Nafie and Freedman,[9] $I_{\alpha\beta}^A$, given in Eq. (37), is expressed as a sum over excited electronic states

$$I_{\alpha\beta}^A = \frac{-e\hbar^2}{mc} \sum_{e \neq g} \varepsilon_{\beta\gamma\delta} \left\{ \left\langle \Psi_g \left| \sum_j r_{j\gamma} \right| \frac{\partial \Psi_e}{\partial r_{j\delta}} \right\rangle \left\langle \Psi_e \left| \frac{\partial \Psi_g}{\partial R_{A\alpha}} \right\rangle E_{eg}^{-1} \right\} \right._{R=0} \tag{41}$$

where E_{eg} is the energy difference $(E_e - E_g)$ between the ground and excited electronic states and Ψ_g and Ψ_e are (real) ground and excited state Born-Oppenheimer adiabatic electronic wave functions. This expression arises from the second term in Eq. (40) with $\lambda_1 = \mathscr{H}_\beta$ (magnetic field) and $\lambda_2 = \dot{R}_{A\alpha}$ (nuclear velocity), and where H_E in Eq. (40) includes the magnetic field and nuclear velocity perturbation as operators in Eqs. (33) and (34), respectively.

The velocity form APT, given in Eq. (36) can also be formulated[9] as a sum-over-states from Eq. (40), with $\lambda_1 = \mathscr{A}_\beta$ (electric-dipole vector potential) and $\lambda_2 = \dot{R}_{A\alpha}$ (nuclear velocity), and the appropriate perturbation operators in H_E,

$$E_{\nu,\alpha\beta}^A = \frac{-2e\hbar^2}{m} \sum_{e \neq g} \left\{ \left\langle \Psi_g \left| \sum_j \frac{\partial \Psi_e}{\partial r_{j\beta}} \right\rangle \left\langle \Psi_e \left| \frac{\partial \Psi_g}{\partial R_{A\alpha}} \right\rangle E_{eg}^{-1} \right\} \right._{R=0} \tag{42}$$

2. Magnetic Field Perturbation

In the magnetic field perturbation (MFP) formulation of VCD intensity, developed independently by Stephens[10-12] and by Buckingham and Galwas,[13, 14] $I_{\alpha\beta}^A$ is expressed in terms of the overlap of derivatives of the electronic wave function with respect to an external magnetic field, $\mathscr{H}_\beta$, and nuclear position, $R_{A\alpha}$,

$$I_{\alpha\beta}^A = -2 \left[\left\langle \tilde{\Psi}_g \left| -i\hbar \frac{\partial}{\partial R_{A\alpha}} \right| \frac{\partial \tilde{\Psi}_g}{\partial \mathscr{H}_\beta} \right\rangle \right]_{\mathscr{H}=0, R=0}$$

$$= -2 \left[\left\langle \frac{\partial \tilde{\Psi}_g}{\partial \mathscr{H}_\beta} \left| \left(-i\hbar \frac{\partial}{\partial R_{A\alpha}} \right) \right| \tilde{\Psi}_g \right\rangle^* \right]_{\mathscr{H}=0, R=0} \tag{43a}$$

$$= 2i\hbar \left(\left\langle \frac{\partial \tilde{\Psi}_g}{\partial \mathscr{H}_\beta} \left| \frac{\partial \tilde{\Psi}_g}{\partial R_{A\alpha}} \right\rangle \right)_{\mathscr{H}=0, R=0} = -2i\hbar \left(\left\langle \frac{\partial \tilde{\Psi}_g}{\partial R_{A\alpha}} \left| \frac{\partial \tilde{\Psi}_g}{\partial \mathscr{H}_\beta} \right\rangle \right)_{\mathscr{H}=0, R=0}$$

$$\tag{43b}$$

These expressions can be deduced from the second term in Eq. (39) with $\lambda_1 = \dot{R}_{A\alpha}$ (nuclear velocity) and $\lambda_2 = \mathcal{H}_\beta$ (magnetic field), and with the Hermitian property of the operator $-i\hbar\nabla_{A\alpha} = -i\hbar\,\partial/\partial R_{A\alpha}$. The AAT expressions in Eqs. (41) and (43) are thus related by an interchange of the order of the perturbation operators. A similar relation involving interchange of perturbation operators has been identified with respect to the magnetic field perturbation and nuclear velocity perturbation approaches to the AAT in the context of the velocity gauge formalism[22] (see Section II.C.3). As noted above, Stephens' equations[12] employ a definition of the AAT that differs by a factor of $i/2\hbar$ from Eq. (43). A matrix formulation of the derivation Eq. (43) has also been described.[26]

For LCAO molecular orbitals $\tilde{\psi}_j = \sum_\mu \tilde{c}_{\mu j}\varphi_\mu \equiv \sum_\mu \tilde{c}_{\mu j}(\mathbf{R}, \mathcal{H})\phi_\mu(\mathbf{r} - \mathbf{R})$, the MFP electronic AAT given in the second equality in Eq. (43), with a common magnetic field origin (CO) for all nuclei, is expanded as[24]

$$\left(I_{\alpha\beta}^A\right)^{\mathrm{CO}}_{\mathcal{H}} = 4\hbar\,\mathrm{Im}\,\sum_j^{occ}\sum_B{}^B\sum_\mu\sum_D{}^D\sum_\nu \left(\frac{\partial c_{\mu j}}{\partial R_{A\alpha}}\frac{\partial \tilde{c}_{\nu j}}{\partial \mathcal{H}_\beta}\langle \phi_\mu^B | \phi_\nu^D \rangle \right.$$
$$\left. + c_{\mu j}\frac{\partial \tilde{c}_{\nu j}}{\partial \mathcal{H}_\beta}\left\langle \frac{\partial \phi_\mu^B}{\partial R_{A\alpha}}\middle| \phi_\nu^D \right\rangle \right)_{R=0,\,\mathcal{H}=0} \tag{44}$$

When the origin for the magnetic field perturbation is placed at the equilibrium position of the nucleus, A, corresponding to that AAT, a distributed origin (DO) expression is derived,[12] which involves the position form of the electronic APT for that nucleus,

$$\left(I_{\alpha\beta}^A\right)^{\mathrm{DO}} = \frac{1}{2c}\varepsilon_{\beta\gamma\delta}R_{A\gamma}^0 E_{r,\alpha\delta}^A + \left(I_{\alpha\beta}^A\right)^A \tag{45}$$

It can be shown that the electronic contributions to the distributed origin AAT can be expressed in terms of the common origin electronic AAT and both the position and velocity APTs,[18]

$$\left(I_{\alpha\beta}^A\right)^{\mathrm{DO}} = \frac{1}{2c}\varepsilon_{\beta\gamma\delta}R_{A\gamma}^0\left(E_{r,\alpha\delta}^A - E_{v,\alpha\delta}^A\right) + \left(I_{\alpha\beta}^A\right)^{\mathrm{CO}} \tag{46}$$

With Eq. (46), the derivatives of the LCAO coefficients with respect to magnetic field can be calculated for all the AATs for the molecule at once, requiring one calculation for each Cartesian direction, whereas with Eq. (45), these coefficient derivatives must be calculated separately for the three Cartesian directions for each AAT. We note that Eqs. (45) and (46) hold for all the formulations of the AAT described here; in the limit of exact wave functions, the common origin and distributed origin AATs will be identical.

The position form atomic polar tensor element for LCAO molecular orbitals can be formulated as a ground-state expression[9] (second term in Eq. (39), $\lambda_1 = \mathscr{E}_\beta$ (electric field) and $\lambda_2 = R_{A\alpha}$ (nuclear position)),

$$
E^A_{r,\alpha\beta} = -4e \sum_j^{occ} \sum_B \sum_\mu^B \sum_D \sum_\nu^D \left(\frac{\partial c_{\mu j}}{\partial R_{A\alpha}} c_{\nu j} \left\langle \phi^B_\mu | r_{1\beta} | \phi^D_\nu \right\rangle \right.
$$
$$
\left. + c_{\mu j} c_{\nu j} \left\langle \frac{\partial \phi^B_\mu}{\partial R_{A\alpha}} | r_{1\beta} | \phi^D_\nu \right\rangle \right) \Bigg|_{R=0}
$$

(47)

In the field perturbation approach,[18] the velocity form of the APT is formulated as an overlap of electronic wave-function derivatives from the second term in Eq. (39) with $\lambda_1 = \dot{R}_{A\alpha}$ (nuclear velocity) and $\lambda_2 = \mathscr{A}_\beta$ (electric-dipole vector potential),

$$
E^A_{v,\alpha\beta} = -2i\hbar c \left(\left\langle \frac{\partial \tilde{\Psi}_g}{\partial R_{A\alpha}} \Bigg| \frac{\partial \tilde{\Psi}_g}{\partial \mathscr{A}_\beta} \right\rangle \right)_{\mathscr{A}=0,\, R=0}
$$

(48)

which is expanded in atomic orbitals as

$$
E^A_{v,\alpha\beta} = 4\hbar c \, \mathrm{Im} \sum_j^{occ} \sum_B \sum_\mu^B \sum_D \sum_\nu^D \left(\frac{\partial c_{\mu j}}{\partial R_{A\alpha}} \frac{\partial \tilde{c}_{\nu j}}{\partial \mathscr{A}_\beta} \left\langle \phi^B_\mu | \phi^D_\nu \right\rangle \right.
$$
$$
\left. + c_{\mu j} \frac{\partial \tilde{c}_{\nu j}}{\partial \mathscr{A}_\beta} \left\langle \frac{\partial \phi^B_\mu}{\partial R_{A\alpha}} | \phi^D_\nu \right\rangle \right) \Bigg|_{R=0,\, \mathscr{A}=0}
$$

(49)

We note that Stephens et al.[18] have defined a perturbation $H'_E(A_\beta) = A_\beta \sum_i p_i$ (where p_i is the electronic momentum operator for the ith electron) in the derivation of an equation that is similar in form to Eq. (48) for $E^A_{v,\alpha\beta}$. The equivalence of Eq. (48) and the formulation of Stephens et al. is obtained from the relationship $A_\beta = (e/mc)\mathscr{A}_\beta$, for the electric-dipole vector potential component $\mathscr{A}_\beta$.

3. Nuclear Velocity Perturbation–Velocity-Gauge Formalism

Two approaches have been proposed that introduce nuclear velocity dependence into the calculation of $E^A_{v,\alpha\beta}$ and $I^A_{\alpha\beta}$.[17, 22] The nuclear velocity perturbation (NVP) velocity-gauge formalism, developed by Nafie,[22] employs a complete adiabatic[57] (CA) nuclear-velocity gauge molecular orbital, which has the form

$$
\tilde{\psi}_j^{CA}\left(\mathbf{r}_1, \mathbf{R}, \dot{\mathbf{R}}\right) = \sum_B \sum_\mu {}^B \tilde{c}_{\mu j}(\mathbf{R}, \dot{\mathbf{R}}) \phi_\mu(\mathbf{r}_1 - \mathbf{R}_B) \exp\left(\frac{i m \dot{\mathbf{R}}_B \cdot \mathbf{r}_1}{\hbar}\right) \quad (50)
$$

Equation (50) can be generalized to include the electron velocity gauge due to a magnetic field or an electric-dipole vector potential. The exponential factor in Eq. (50) serves explicity to set in motion the atomic orbital centered on atom B with the velocity $\dot{\mathbf{R}}_B$, and permits the direct differentiation of $\tilde{\psi}_j^{CA}$ with respect to the orbital (nuclear) velocity, a non-Born-Oppenheimer property.

The NVP velocity-gauge formalism may be developed through the second term in Eq. (39) with $\lambda_1 = \mathcal{H}_\beta$ (magnetic field) and $\lambda_2 = \dot{R}_{A\alpha}$ (nuclear velocity), an interchange of the λ_1 and λ_2 perturbation operators compared to the MFP formalism. The molecular orbital expression for $I_{\alpha\beta}^A$ in the common origin gauge is given by[22]

$$
\left(I_{\alpha\beta}^A\right)_{\dot{R}}^{CO} = -\frac{e}{c} \sum_j^{occ} \sum_B \sum_\mu {}^B \varepsilon_{\beta\gamma\delta}\left\{ \sum_D \sum_\nu {}^D \frac{\hbar}{im} \frac{\partial\left(\tilde{c}_{\mu j}^* \tilde{c}_{\nu j}\right)}{\partial \dot{R}_{A\alpha}} \left\langle \phi_\mu^B | r_{1\gamma} | \frac{\partial \phi_\nu^D}{\partial r_{1\delta}} \right\rangle \right.
$$

$$
+ \sum_\nu^A c_{\mu j}^* c_{\nu j}\left(\left\langle \phi_\mu^B | r_{1\gamma} | \phi_\nu^A \right\rangle \delta_{\delta\alpha}\right. \quad (51)
$$

$$
\left.\left. + \left\langle \phi_\mu^B | r_{1\alpha} r_{1\gamma} | \frac{\partial \phi_\nu^A}{\partial r_{1\delta}} \right\rangle - \left\langle \varphi_\nu^A | r_{1\alpha} r_{1\gamma} | \frac{\partial \varphi_\mu^B}{\partial r_{1\delta}} \right\rangle\right)\right\}_{R=0,\dot{R}=0}
$$

The atomic polar tensor $E_{\nu,\alpha\beta}^A$ in the nuclear-velocity gauge formalism is derived from the second term in Eq. (39) with $\lambda_1 = \mathcal{A}_\beta$ (electric-dipole vector potential) and $\lambda_2 = \dot{R}_{A\alpha}$ (nuclear velocity),

$$
E_{\nu,\alpha\beta}^A = -2e \sum_i^{occ} \sum_B \sum_\mu {}^B \left\{ \sum_D \sum_\nu {}^D \frac{\hbar}{im} \frac{\partial\left(\tilde{c}_{\mu j}^* \tilde{c}_{\nu j}\right)}{\partial \dot{R}_{A\alpha}} \left\langle \phi_\mu^B \left| \frac{\partial \phi_\nu^D}{\partial r_{1\beta}} \right.\right\rangle \right.
$$

$$
\left. + \sum_\nu^A c_{\mu j} c_{\nu j}\left(\left\langle \phi_\mu^B | \phi_\nu^A \right\rangle \delta_{\alpha\beta} + \left\langle \phi_\mu^B | r_{1\alpha} | \frac{\partial \phi_\nu^A}{\partial r_{1\beta}} \right\rangle - \left\langle \phi_\nu^A | r_{1\alpha} | \frac{\partial \phi_\mu^B}{\partial r_{1\beta}} \right\rangle\right)\right\}_{R=0,\dot{R}=0}
$$

$$
\quad (52)
$$

4. Floating Basis Set Formalism

A earlier method of introducing nuclear velocity dependence into the calculation of the AAT or velocity APT, proposed by Freedman and Nafie,[17] employs atomic orbital basis functions that float[61, 62] with the nuclear velocities and the nuclear positions. With this method, the form of the atomic orbital basis function does not change with nuclear displacement or velocity, and the LCAO coefficients carry all the nuclear displacement dependence. The electronic position and velocity operators depend on the *nuclear* position and velocity, respectively, when they act on the atomic orbitals of a moving nucleus. The expressions for $E^A_{\nu,\alpha\beta}$ and $I^A_{\alpha\beta}$ involve a ground-state, perfect-nuclear-following contribution arising from the first term in Eq. (40) and a sum-over-states non-perfect-nuclear-following contribution arising from the second term in Eq. (40), where $\lambda_1 = \mathscr{A}_\beta$ (electric-dipole vector potential) or $\mathscr{H}_\beta$ (magnetic field) and $\lambda_2 = \dot{R}_{A\alpha}$ (nuclear velocity), as was also the case for the VCT theory.[17] The floating basis set perfect-following term is identical to the second term in Eq. (52) or (51) for the APT and AAT, respectively.

5. Nuclear Shielding Tensor Expressions

Two additional methods that employ sum-over-states expressions have also been proposed for the calculation of the atomic polar and axial tensors.[13–16, 19, 20] From the relationship

$$\left(\frac{\partial H_E}{\partial R_{A\alpha}}\right)_{R=0} = \sum_i \frac{Z_A e^2 (r_{i\alpha} - R^0_{A\alpha})}{|\mathbf{r}_i - \mathbf{R}^0_A|^3} + \left(\frac{\partial V_{nn}}{\partial R_{A\alpha}}\right)_{R=0}$$

$$= -Z_A e F^A_\alpha + \left(\frac{\partial V_{nn}}{\partial R_{A\alpha}}\right)_{R=0} \tag{53}$$

where V_{nn} is the nuclear–nuclear repulsion energy and F^A_α is the electric field due to the electrons at nucleus A, and the expansion

$$\tilde{\Psi}_g(\mathbf{r}, \mathbf{R}, \mathbf{X}) = \Psi^0_g - \sum_{\substack{e \neq g \\ A, \alpha}} \Psi^0_e \frac{\left\langle \Psi^0_e \left| (\partial H_E / \partial X_{A\alpha})_{X=0} \right| \Psi^0_g \right\rangle}{E^0_e - E^0_g} X_{A\alpha} + \cdots \tag{54}$$

where $X_{A\alpha}$ is a perturbation (including nuclear displacement), the formulations for the AAT and APT in Eqs. (43b) and (48) in terms of wave-function derivatives can be expanded as [13-6, 19]

$$
\begin{aligned}
I_{\alpha\beta}^{A} &= -2i\hbar \sum_{e \neq g} \left\langle \Psi_g^0 \left| \left(\frac{\partial H_E}{\partial R_{A\alpha}} \right)_{R=0} \right| \Psi_e^0 \right\rangle \left\langle \Psi_e^0 \left| \left(\frac{\partial H_E}{\partial \mathcal{H}_\beta} \right)_{\mathcal{H}=0} \right| \Psi_g^0 \right\rangle \left(E_{eg}^0 \right)^{-2} \\
&= \frac{Z_A e^2 \hbar^2}{mc} \sum_{e \neq g} \varepsilon_{\beta\gamma\delta} \left\langle \Psi_g^0 \left| F_\alpha^A \right| \Psi_e^0 \right\rangle \left\langle \Psi_e^0 \left| \sum_i r_{i\gamma} \left| \left(\frac{\partial \Psi_g}{\partial r_{i\delta}} \right)_{R=0} \right\rangle \right(E_{eg}^0 \right)^{-2} (55)
\end{aligned}
$$

and

$$
\begin{aligned}
E_{\nu,\alpha\beta}^{A} &= -2i\hbar c \sum_{e \neq g} \left\langle \Psi_g^0 \left| \left(\frac{\partial H_E}{\partial R_{A\alpha}} \right)_{R=0} \right| \Psi_e^0 \right\rangle \left\langle \Psi_e^0 \left| \left(\frac{\partial H_E}{\partial \mathcal{A}_\beta} \right)_{\mathcal{A}=0} \right| \Psi_g^0 \right\rangle \left(E_{eg}^0 \right)^{-2} \\
&= \frac{2 Z_A e^2 \hbar^2}{m} \sum_{e \neq g} \left\langle \Psi_g^0 \left| F_\alpha^A \right| \Psi_e^0 \right\rangle \left\langle \Psi_e^0 \left| \sum_i \left(\frac{\partial \Psi_g}{\partial r_{i\beta}} \right)_{R=0} \right\rangle \left(E_{eg}^0 \right)^{-2} (56)
\end{aligned}
$$

In Eqs. (55) and (56), the unperturbed electronic wave functions at the equilibrium nuclear positions are employed. Alternatively, Eqs. (55) and (56) may be regarded as a further sum-over-states perturbation expansion of $|\partial H_E/\partial \dot{R}_{A\alpha}|\Psi_g\rangle = -i\hbar|\partial \Psi_g/\partial R_{A\alpha}\rangle$ in Eq. (40) for nuclear velocity perturbation. The electronic contribution to the position form APT can be expressed as the summation (Eq. (40), $\lambda_1 = R_{A\alpha}$ (nuclear position) and $\lambda_2 = \mathcal{E}_\beta$ (electric field))

$$
\begin{aligned}
E_{r,\alpha\beta}^{A} &= -2 \sum_{e \neq g} \left\langle \Psi_g^0 \left| \left(\frac{\partial H_E}{\partial R_{A\alpha}} \right)_{R=0} \right| \Psi_e^0 \right\rangle \left\langle \Psi_e^0 \left| \left(\frac{\partial H_E}{\partial \mathcal{E}_\beta} \right)_{\mathcal{E}=0} \right| \Psi_g^0 \right\rangle \left(E_{eg}^0 \right)^{-1} \\
&= 2 Z_A e^2 \sum_{e \neq g} \left\langle \Psi_g^0 \left| F_\alpha^A \right| \Psi_e^0 \right\rangle \left\langle \Psi_e^0 \left| \sum_i r_{i\beta} \right| \Psi_g^0 \right\rangle \left(E_{eg}^0 \right)^{-1} (57)
\end{aligned}
$$

The sum-over states expressions (55), (56), and (57) are related to nuclear shielding tensors,[63-67] that is, the electric field induced at a nucleus by an external magnetic or electric field. Hunt and Harris[20] have developed a nonlocal susceptibility density theory for VCD to describe the electric field shielding at the nucleus; their expressions for the AAT are shown to be equivalent to Eq. (55).

III.   MODELS FOR CALCULATING VCD INTENSITY

Although numerous a priori formulations for VCD intensity, as summarized above, have been developed and in many cases implemented (Section IV.A) for a variety of ab initio molecular orbital packages, in practice interpretation of VCD spectra still relies on approximate models, both in applications to larger molecules for which the ab initio calculations are not practical, and for their mechanistic insight.[34-51] The models were introduced before the a priori methods were proposed or fully implemented, in order to circumvent the non-Born-Oppenheimer nature of the problem, that is, the dependence of the electronic contribution to VCD intensity on nuclear velocity, in addition to dependence on nuclear position. It was recognized that the derivative of the magnetic dipole moment with respect to nuclear position vanished within the Born-Oppenheimer approximation.[68] The property that the a priori formalisms have made clear is that in addition to requiring nuclear velocity dependent electronic wave functions, the rotational strength and velocity form of the dipole strength involve nuclear *velocity* derivatives, rather than nuclear *position* derivatives. The models for VCD intensity for the most part seek to approximate nuclear velocity derivatives with nuclear displacement derivatives. It is valuable at this stage in the evolution of formalisms for VCD intensities to analyze how the models can be derived in the context of the more complete expressions.

A.   Locally Distributed Origin Gauge Expressions

Models for VCD intensity include the fixed partial charge (FPC) model,[36, 37] the coupled oscillator (CO) model,[34, 35] the APT model,[43] the localized molecular orbital (LMO) model[38, 39] the dynamic polarization model,[40] and various charge-flow models.[41, 42, 44-48] With the exception of the APT model, these models for VCD intensity consider localized contributions. These models can be all deduced from the a priori formulations by first writing the velocity form APT and the AAT as a sum of atomic contributions, denoted $[P_{v,\alpha\beta}^A]^B$ and $[M_{\alpha\beta}^A]^B$ for the contribution from atom B to the $\alpha\beta$ component of the tensor for atom A; that is, we consider the local effects that moving atom A has on the contributions to the dipole moments from each of the atoms B in the molecule.[56] For a priori formulations expressed explicitly in terms of atomic orbitals and electronic momentum or angular momentum operators acting on those orbitals (e.g., Eq. (51) and (52) for the velocity gauge expressions,[22] the floating basis set expressions[17] and the vibronic coupling expressions[9]), the electronic contribution from an atom B is uniquely defined[56] by summing the terms for which these operators act on an orbital belonging to atom B.

For the magnetic field perturbation and vector potential perturbation expressions, Eqs. (43) and (49), a contribution from atom B can be defined by summing over the terms involving a field perturbation derivative of a coefficient of an orbital on atom B.

The APT, which represents the electric dipole moment derivative of the *entire* molecule with respect to $\mathbf{R}_A$, is simply the sum over all of the atomic contributions.

$$P^A_{\nu,\alpha\beta} = \sum_B \left[P^A_{\nu,\alpha\beta} \right]^B \tag{58}$$

The form of the AAT depends on the choice of gauge, that is, the choice of the origin for the magnetic dipole moment operator. Two such gauges are the common origin (CO) gauge and the distributed origin (DO) gauge introduced in Section II.C. The gauge most useful in deriving the models for VCD intensity is the locally distributed origin (LDO) gauge, in which the origin of the magnetic dipole moment operator is placed at atom B for the contribution from that atom.[56] In this case, the AAT reduces to

$$\left(M^A_{\alpha\beta} \right)^{\text{LDO}} = \sum_B \frac{1}{2c} \varepsilon_{\beta\gamma\delta} R^0_{B\gamma} \left[P^A_{\nu,\alpha\delta} \right]^B + \sum_B \left[\left(M^A_{\alpha\beta} \right)^B \right]^B \tag{59}$$

The first term in this expression involves a moment arm to each atom B, crossed into the APT contribution from that atom. The second term is a sum over the intrinsic magnetic dipole moment contribution for each atom B with origin at B, that is, terms involving angular momentum and position operators located at atom B. For example, these contributions are readily deduced for the velocity-gauge formulation by substituting $r_{1\gamma} = R^0_{B,\gamma} + r^B_{\mu,\gamma}$ in Eq. (51). Only atom B and nearest-neighbor atoms with orbitals that have significant overlap with the orbitals on atom B make large contributions to the intrinsic term. With locally distributed origins, the rotational strength expression becomes

$$(R^a_\nu)^{\text{LDO}} = \frac{\hbar}{4c} \left(\sum_A \sum_B \left[P^A_{\nu,\alpha\beta} \right]^B s_{A\alpha,a} \right)$$

$$\cdot \left(\sum_{A'} \sum_{B'} \left\{ \varepsilon_{\beta\gamma\delta} R^0_{B'\gamma} \left[P^{A'}_{\nu,\alpha\delta} \right]^{B'} + \left[\left(I^{A'}_{\alpha\beta} \right)^{B'} \right]^{B'} \right\} s_{A'\alpha,a} \right) \tag{60}$$

$$= \frac{\hbar}{4c} \left[\sum_B \left(\frac{\partial \boldsymbol{\mu}^B_\nu}{\partial P_a} \right)_0 \right] \cdot \left[\sum_{B'} \left\{ \mathbf{R}^0_{B'} \times \left(\frac{\partial \boldsymbol{\mu}^{B'}_\nu}{\partial P_a} \right)_0 + \left(\frac{\partial \mathbf{m}^{B'}_{\text{loc}}}{\partial P_a} \right)_0 \right\} \right]$$

where the second equality is in vector form and the summations over A and A' have been carried out. We stress that no approximations are involved in obtaining the form of the rotational strength in Eq. (60). This expression retains the local nature of the AAT expressed explicity by Eq. (22). Most of the models for VCD can be derived from the form of Eq. (60). In general, the models ignore the intrinsic term; the triple product in Eq. (60) can then be rearranged to obtain the following form for the rotational strength:

$$(R_\nu^a)^{\mathrm{LDO}} \approx \frac{\hbar}{4c} \sum_B \sum_{B<B'} (\mathbf{R}_B^0 - \mathbf{R}_{B'}^0) \cdot \left(\frac{\partial \boldsymbol{\mu}_\nu^B}{\partial P_a}\right)_0 \times \left(\frac{\partial \boldsymbol{\mu}_\nu^{B'}}{\partial P_a}\right)_0 \qquad (61)$$

The molecular origin independence of the rotational strength is clear from Eq. (61). The intrinsic term is by nature independent of the molecular origin.

B.   Fixed Partial Charge Model

The simplest model for VCD intensity, the fixed partial charge (FPC) model,[36, 37] treats the electronic contribution in terms of a fixed partial electronic charge located at the nuclear position, which shields the nuclear charge and perfectly follows the nuclear motion. This type of contribution is explicitly present in the second term of the velocity-gauge expression, Eq. (51), as seen by employing $r_{1\gamma} = R_{B,\gamma}^0 + r_{\mu,\gamma}^B$ and setting $B = A$,

$$I_{\alpha\beta}^A \approx -\frac{e}{c} \sum_j^{occ} \sum_B \sum_\mu^B \sum_A \sum_\nu^A \varepsilon_{\alpha\beta\gamma} c_{\mu j}^0 c_{\nu j}^0 \langle \phi_\mu^B | R_{B\gamma}^0 + r_{\mu,\gamma}^B | \phi_\nu^A \rangle \qquad (62)$$

$$\approx -\frac{e}{2c} \varepsilon_{\alpha\beta\gamma} P_{AA}^0 R_{A\gamma}^0 \qquad (63)$$

where in Eq. (63), the terms involving the local position operator for the orbital are ignored as small and P_{AA}^0 is the Mulliken population for atom A at equilibrium. This FPC contribution is also present explicitly in the floating basis set expressions.[17] In practice, the magnitude of the FPCs are sometimes adjusted to obtain a better fit to the observed dipole strengths. From Eq. (60), the FPC model is obtained by ignoring the intrinsic contribution, and setting

$$\left[P_{\nu,\alpha\beta}^A\right]^B = (Z_A e - \xi_A) \equiv q_A \qquad \text{(for } B = A) \qquad (64a)$$

$$= 0 \qquad \text{(for } B \neq A) \qquad (64b)$$

where q_A is the fixed partial charge on the atom and ξ_A is the fixed electronic screening charge. From Eq. (61), the FPC model assumes $(\partial \boldsymbol{\mu}_\nu^B / \partial P_a)_0 = q_B s_{B,a}$.

C. Coupled Oscillator Models

It has long been recognized that bisignate VCD features are frequently observed for systems in which chirally oriented, local identical oscillators (primarily stretching oscillators) can couple in and out of phase.[34, 35] We consider here the more general case of a mode in which N identical, achiral isolated local oscillators, j, are coupled.[56] We restrict each oscillator to the stretch of two bonded atoms with an electric-dipole transition moment along the bond axis, and write a (normalized) normal mode as $Q_i = \sum_j a_{ij} \Delta \ell_j$. Local oscillator contributions to the electric-dipole moment derivative can be obtained by regrouping the atomic contributions as

$$\sum_B \left(\frac{\partial \boldsymbol{\mu}_\nu^B}{\partial P_i} \right)_0 = \sum_j \sum_{B_j} \left(\frac{\partial \boldsymbol{\mu}_\nu^{B_j, j}}{\partial P_i} \right)_0 = \sum_j \left(\frac{\partial \boldsymbol{\mu}_\nu^j}{\partial P_i} \right)_0 \qquad (65)$$

where B_j is the sum over the two atoms in oscillator j. By ignoring the intrinsic contribution since the oscillators are achiral, and choosing a moment arm directed to any point along the bond axis (see below), we can write the coupled oscillator (CO, not to be confused with common origin) rotational strength as

$$(R_i)^{\mathrm{co}} = \frac{\hbar}{4c} \left[\sum_j \left(\frac{\partial \boldsymbol{\mu}_\nu^j}{\partial P_i} \right)_0 \right] \cdot \left[\sum_{j'} \mathbf{R}_{j'}^0 \times \left(\frac{\partial \boldsymbol{\mu}_\nu^{j'}}{\partial P_a} \right)_0 \right] \qquad (66)$$

For identical, isolated oscillators, the electric dipole moment derivative for the jth oscillator can be expressed as

$$\left(\frac{\partial \boldsymbol{\mu}_\nu^j}{\partial P_i} \right)_0 = \left| \frac{\partial \boldsymbol{\mu}^0}{\partial S} \right| \hat{\mathbf{u}}_j a_{ij} \qquad (67)$$

where we denote the magnitude of the local electric dipole moment derivative as $|\partial \boldsymbol{\mu}^0 / \partial S|$, and the unit vector along the oscillator axis in the direction of positive dipole moment change as $\hat{\mathbf{u}}_j$.

The dipole and rotational strengths for the ith coupled oscillator mode become

$$(R_i)^{\text{co}} = \frac{\hbar}{4c} \left| \frac{\partial \mathbf{\mu}^0}{\partial S} \right|^2 \sum_{j,j'}^{N} a_{ij} a_{ij'} \hat{\mathbf{u}}_j \cdot \mathbf{R}_{j'}^0 \times \hat{\mathbf{u}}_{j'} \tag{68}$$

$$(D_i)^{\text{co}} = \frac{\hbar}{2\omega_i} \left| \frac{\partial \mathbf{\mu}^0}{\partial S} \right|^2 \sum_{j,j'}^{N} a_{ij} a_{ij'} \hat{\mathbf{u}}_j \cdot \hat{\mathbf{u}}_{j'} \tag{69}$$

We can then substitute the total dipole strength for the N-coupled oscillator modes,

$$D_T = \sum_{i=1}^{N} (D_i)^{\text{co}} = \frac{\hbar N}{2\omega_{\text{av}}} \left| \frac{\partial \mathbf{\mu}^0}{\partial S} \right|^2 \tag{70}$$

into Eq. (68) and rearrange the triple product to obtain

$$(R_i)^{\text{co}} = -\pi \nu_{\text{av}} \frac{D_T}{N} \sum_{j>j'}^{N} a_{ij} a_{ij'} \mathbf{R}_{jj'} \cdot \hat{\mathbf{u}}_j \times \hat{\mathbf{u}}_{j'} \tag{71}$$

where $\nu_{\text{av}}(= \omega_{\text{av}}/2\pi c)$ is the average frequency (cm^{-1}) of the coupled modes and $\mathbf{R}_{jj'} = (\mathbf{R}_{j'}^0 - \mathbf{R}_j^0)$ is the separation vector from oscillator j to oscillator j'. Equation (71) is the generalized coupled oscillator expression for N nonchiral identical oscillators. Only chirally oriented oscillator pairs will contribute to the rotational strength. From Eq. (71), it is clear that the moment arm in Eq. (66) can be located at any position along the bond axis of the oscillator and can be the same for both oscillators, since the rotational strength is not affected by moving $\mathbf{R}_j^0$ to $\mathbf{R}_j^0 + \lambda\hat{\mathbf{u}}_j$. The more familiar coupled oscillator expression for two identical oscillators vibrating in and out of phase is a special case of Eq. (71).[34, 35] If the local oscillators are themselves chiral, or are strongly affected by a local chiral environment, the intrinsic rotational strength due to the inherent chirality of the oscillator should be included as an additional term.[35]

D. Localized Molecular Orbital Model

In the localized molecular orbital (LMO) model,[38, 39] the molecular orbitals ψ_k are first localized. The summation over B in Eq. (60) for the electronic contributions is regrouped into a summation over the localized orbitals; the locally distributed origins are the centroids $\mathbf{r}_k^0$ of the localized molecular orbitals when the nuclei are at their equilibrium positions.

$$
(R_\nu^a)^{lmo} = \frac{\hbar}{4c} \sum_A \left\{ Z_A e \delta_{\alpha\beta} + \sum_k \left[E_{\nu,\alpha\beta}^A \right]_{loc}^k \right\} s_{A\alpha,a}
$$

$$
\cdot \sum_{A'} \left\{ \varepsilon_{\alpha\beta\gamma} Z_{A'} e R_{A'\gamma}^0 + \sum_{k'} \varepsilon_{\beta\gamma\delta} r_{k'\gamma}^0 \left[E_{\nu,\alpha\delta}^{A'} \right]_{loc}^{k'} \right.
$$

$$
\left. + \sum_{k'} \left[\left(I_{\alpha\beta}^{A'} \right)^{k'} \right]_{loc}^{k'} \right\} s_{A'\alpha,a} \tag{72}
$$

Equation (72) is still an exact a priori rotational strength expression. The LMO model results from ignoring the intrinsic term and approximating the velocity dipole moment derivative contribution for the kth localized orbital with the position dipole moment derivative contribution,[38]

$$
\left[E_{\nu,\alpha\beta}^A \right]_{loc}^k \approx \left[E_{r,\alpha\beta}^A \right]_{loc}^k = \left(\frac{\partial}{\partial R_{A\alpha}} \langle \psi_k^{loc} | e r_{k\beta} | \psi_k^{loc} \rangle \right)_0
$$

$$
= e \left(\frac{\partial \langle r_{k\beta} \rangle}{\partial R_{A\alpha}} \right)_0 \equiv e \sigma_{k,\alpha\beta}^A \tag{73}
$$

where $\sigma_{k,\alpha\beta}^A$ is the β-displacement of the orbital centroid when atom A is displaced in the α direction. The LMO model expression for the rotational strength is

$$
(R_r^a)^{lmo} = \frac{\hbar e^2}{4c} \left[\left\{ \sum_A Z_A \delta_{\alpha\beta} + \sum_k \sigma_{\alpha\beta}^k \right\} s_{A\alpha,a} \right]
$$

$$
\cdot \left[\left\{ \sum_{A'} Z_{A'} \varepsilon_{\alpha\beta\gamma} R_{A'\gamma}^0 + \sum_{k'} \varepsilon_{\beta\gamma\delta} r_{k'\gamma}^0 \sigma_{k,\alpha\delta}^A \right\} s_{A'\alpha,a} \right] \tag{74}
$$

The intrinsic term in Eq. (72) has been described as an orbital rocking term,[38] arising from the rotations of nonspherical LMOs at their fixed equilibrium centroid positions.

We note that at all levels of consideration, the LMO formalism is origin independent. When applied at the ab initio level, the LMO model is the only naturally origin-independent formalism using R_r^a (Eq. (23)), whereas the a priori formulations can only achieve this through R_ν^a (Eq. (14)).

E.   Charge Flow Models

The charge flow models augment the FPC approximation to allow for changes in the electronic charge contribution due to the nuclear motion, most often introduced as charge flow between bonded atoms. The two original charge flow formulations[41, 42] can be shown to be equivalent. The bond-dipole model for VCD[45-47] can also be formulated[47] to be equivalent to these charge flow models. The nonlocalized molecular orbital model[44] is a molecular orbital version of the charge flow model that also includes local rehybridization effects. Special cases of charge flow have been classified for a single driving oscillator.[48, 51] To motivate the derivation of the charge flow models from the a priori theories, we first examine the consequences of directly substituting the position form of the atomic contribution from atom B to the APT for the velocity form of that contribution in Eq. (60). The electric dipole moment in the position form (Eqs. (7), (9), (10), and (47)) can be broken down into the atomic contributions

$$\left[P^A_{r,\alpha\beta}\right]^B = \left(\frac{\partial}{\partial R_{A\alpha}}\left\{Z_B e R_{B\beta} - e\sum_D \sum_\mu^D \sum_\nu^B P_{\mu\nu}\langle\phi^D_\mu|r_{1\beta}|\phi^B_\nu\rangle\right\}\right)_0 \quad (75)$$

$$\left[P^A_{r,\alpha\beta}\right]^B = Z_A e \delta_{\alpha\beta}\delta_{AB}$$

$$-\left(\frac{\partial}{\partial R_{A\alpha}}\left\{e\sum_D \sum_\mu^D \sum_\nu^B P_{\mu\nu}\left(R_{B\beta}S_{\mu\nu} + \langle\phi^D_\mu|r^B_{\nu,\beta}|\phi^B_\nu\rangle\right)\right\}\right)_0 \quad (76)$$

$$= Z_A e \delta_{\alpha\beta}\delta_{AB}$$

$$-e\sum_D \sum_\mu^D \sum_\nu^B \left\{P^0_{\mu\nu}S^0_{\mu\nu}\delta_{\alpha\beta}\delta_{AB} + P^0_{\mu\nu}\left(\frac{\partial}{\partial R_{A\alpha}}\langle\phi^D_\mu|r^B_{\nu,\beta}|\phi^B_\nu\rangle\right)_0\right.$$

$$\left.+\left(\frac{\partial P_{\mu\nu}}{\partial R_{A\alpha}}\right)_0 \langle\phi^D_\mu|r^B_{\nu,\beta}|\phi^B_\nu\rangle_0\right\}$$

$$-e\sum_D \sum_\mu^D \sum_\nu^B R^0_{B\beta}\left\{\left(\frac{\partial P_{\mu\nu}}{\partial R_{A\alpha}}\right)_0 S^0_{\mu\nu} + P^0_{\mu\nu}\left(\frac{\partial S_{\mu\nu}}{\partial R_{A\alpha}}\right)_0\right\} \quad (77)$$

where $P_{\mu\nu} = 2\sum_i^{occ}c_{\mu i}c_{\nu i}$ and $S_{\mu\nu} = \langle\phi^D_\mu|\phi^B_\nu\rangle$.

We note that three terms in Eq. (77), which can be rearranged as

$$
-e \sum_D \sum_\mu^D \sum_\nu^B P^0_{\mu\nu} \left\{ S^0_{\mu\nu} \delta_{\alpha\beta} \delta_{AB} + R^0_{B\beta} \left(\frac{\partial S_{\mu\nu}}{\partial R_{A\alpha}} \right)_0 + \left(\frac{\partial}{\partial R_{A\alpha}} \langle \phi^D_\mu | r^B_{\nu,\beta} | \phi^B_\nu \rangle \right)_0 \right\}
$$

$$
\equiv -2e \sum_i^{occ} \sum_D \sum_\mu^D \sum_\nu^A c^0_{\mu i} c^0_{\nu i} \left(\langle \phi^D_\mu | \phi^A_\nu \rangle \delta_{\alpha\beta} + 2 \langle \phi^D_\mu | r_{1\beta} | \frac{\partial \phi^A_\nu}{\partial R_{A\alpha}} \rangle \right)_{R=0}
$$

$$(78)$$

correspond to the contribution from atom B to the last three terms in Eq. (52) for the velocity gauge APT, but with the α and β indices interchanged, if the equivalence $(\partial \phi^D_\nu / \partial R_{A\alpha})_0 = -(\partial \phi^D_\nu / \partial r_\alpha)_0 \delta_{AD}$ is introduced. The first four terms in Eq. (77) can reasonably be located at atom B in a locally distributed origin AAT expression in the form of Eq. (60), since these are the nuclear term, an FPC-like term, and two terms involving a local position operator at atom B. These latter two terms have been described as a rehybridization or polarization contribution.[69, 70] However, the last two terms involving $R^0_{B\beta}$ in Eq. (77) are charge flux terms that will vanish in the vector cross product in the rotational strength expression if they are associated with a moment arm to atom B. In the charge flow models, the charge flux at atom B due to the motion of atom A is described in terms of currents between atom B and the other atoms in the molecule:

$$
\left(\frac{\partial \xi_B}{\partial R_{A\alpha}} \right)_0 = \sum_D \sum_\mu^D \sum_\nu^B \left[\left(\frac{\partial P_{\mu\nu}}{\partial R_{A\alpha}} \right)_{R=0} S_{\mu\nu} + P_{\mu\nu} \left(\frac{\partial S_{\mu\nu}}{\partial R_{A\alpha}} \right)_{R=0} \right]
$$

$$
= \sum_K \left(\frac{\partial I_{BK}}{\partial \dot{R}_{A\alpha}} \right)_{\dot{R}=0}
$$

$$(79)$$

where I_{BK} is the current, defined as positive for positive charge flow from atom B to atom K. The derivative of the current is taken with respect to the nuclear velocity. The total charge flux contribution to the total electric-dipole moment derivative can be rearranged as

$$
\sum_B \sum_D \sum_\mu^D \sum_\nu^B R^0_{B\beta} \left[\left(\frac{\partial P_{\mu\nu}}{\partial R_{A\alpha}} \right)_{R=0} S_{\mu\nu} + P_{\mu\nu} \left(\frac{\partial S_{\mu\nu}}{\partial R_{A\alpha}} \right)_{R=0} \right]
$$

$$
= \sum_B \sum_K R^0_{B\beta} \left(\frac{\partial I_{BK}}{\partial \dot{R}_{A\alpha}} \right)_{\dot{R}=0}
$$

$$
= \sum_B \left[\sum_K \frac{1}{2} (R^0_{B\beta} - R^0_{K\beta}) \left(\frac{\partial I_{BK}}{\partial \dot{R}_{A\alpha}} \right)_{\dot{R}=0} \right]
$$

$$(80)$$

The last expression in Eq. (80) suggests an alternative way to describe a contribution from atom B, in which the charge flux is expressed in terms of current between two atoms multiplied by the vector directed between the atoms. The earliest charge flow models added such a charge flow contribution to the FPC contribution to obtain an improved description of VCD intensity.[41, 42] In the nonlocalized molecular orbital (NMO) model,[44] a molecular orbital version of the charge flow model, a more complete atomic contribution to the position form APT is employed,

$$
(R_\nu^a)^{\text{NMO}} = \frac{\hbar}{4c} \left(\sum_A \sum_B \left[P_{r,\alpha\beta}^A \right]_{\text{NMO}}^B S_{A\alpha,a} \right)
$$
$$
\cdot \left(\sum_{A'} \sum_{B'} \left\{ \varepsilon_{\beta\gamma\delta} R_{B'\gamma}^0 \left[P_{r,\alpha\delta}^{A'} \right]_{\text{NMO}}^{B'} \right\} S_{A'\alpha,a} \right) \tag{81}
$$

where

$$
\left[P_{r,\alpha\beta}^A \right]_{\text{NMO}}^B = Z_A e \delta_{\alpha\beta} \delta_{AB} - e \sum_D^D \sum_\mu \sum_\nu^B \left\{ P_{\mu\nu}^0 S_{\mu\nu}^0 \delta_{\alpha\beta} \delta_{AB} \right.
$$
$$
+ P_{\mu\nu}^0 \left(\frac{\partial}{\partial R_{A\alpha}} \left\langle \phi_\mu^D \middle| r_{\nu,\beta}^B \middle| \phi_\nu^B \right\rangle \right)_0 + \left(\frac{\partial P_{\mu\nu}}{\partial R_{A\alpha}} \right)_0 \left\langle \phi_\mu^D \middle| r_{\nu,\beta}^B \middle| \phi_\nu^B \right\rangle_0 \right\}
$$
$$
+ \sum_K \frac{1}{2} \left(R_{B\beta}^0 - R_{K\beta}^0 \right) \left(\frac{\partial I_{BK}}{\partial \dot{R}_{A\alpha}} \right)_0 \tag{82}
$$

The expressions in Eqs. (81) and (82) extend the original CNDO formulation[44] of the NMO model to ab initio wave functions. These expressions are equivalent to the formulation of this model recently described by Maurer and Wieser.[71]

Although the NMO expression includes current between any pair of atoms, in practice the current is restricted to flow along bonds. When no rings are present in the molecule, the vibrationally generated current along bonds can be uniquely deduced from the charge fluxes, from Eq. (79). However, the current along the bonds in a closed molecular ring can be determined only from the charge fluxes to within a constant value of charge flow around the ring, termed the ring current.[42, 48] In such a case, the contribution due to vibrationally generated ring current[48]

$$
\sum_{A'} \left(\frac{\partial I_{\text{ring}}}{\partial \dot{R}_{A'\alpha}} \right) \sum_{K \to L}^{\text{ring}} \varepsilon_{\beta\gamma\delta} R_{K\gamma} R_{L\delta} S_{A'\alpha,a} \tag{83}
$$

must be added to the magnetic dipole moment part of Eq. (81), where the second summation in Eq. (83) is over all adjacent pairs of atoms around the ring.

F. Atomic Polar Tensor Model

Finally, we develop the atomic polar tensor (APT) model[43] for VCD intensity from the LDO expression in Eq. (60). The APT model can be considered to arise[12] from ignoring the second term in the distributed origin gauge expression for the electronic contribution to the AAT in Eq. (45). Rupprecht has also demonstrated the relationship between the APT model and the more complete a priori formalism in a form equivalent to the DO gauge expression.[72] However, the limitations of the APT model are best assessed by a derivation via Eq. (60). We first substitute $R_{B'\delta}^0 = R_{B'\delta}^0 + R_{A'\delta}^0 - R_{A'\delta}^0$, carry out the summation over B or B' where possible, and then convert the resulting velocity form APTs to the position form,

$$
\begin{aligned}
R^a &= \frac{\hbar}{4c} \left(\sum_A P_{r,\alpha\beta}^A S_{A\alpha,a} \right) \left(\sum_{A'} \varepsilon_{\beta\gamma\delta} R_{A'\gamma}^0 P_{r,\alpha\delta}^{A'} S_{A'\alpha,a} \right. \\
&\quad + \left. \sum_{A'} \sum_{B'} \left\{ \varepsilon_{\beta\gamma\delta} \left(R_{B'\gamma}^0 - R_{A'\gamma}^0 \right) \left[P_{\nu,\alpha\delta}^{A'} \right]^{B'} + \left[\left(I_{\alpha\beta}^{A'} \right)^{B'} \right]^{B'} \right\} S_{A\alpha,a} \right) \\
&= \frac{\hbar}{4c} (\mathbf{P}_a \cdot \mathbf{L}_a + \mathbf{P}_a \cdot \mathbf{M}_a)
\end{aligned}
\tag{84}
$$

where the last line represents the notation introduced by Stephens[12] to decompose the distributed origin rotational strength into $\mathbf{P} \cdot \mathbf{L}$ and $\mathbf{P} \cdot \mathbf{M}$ contributions. The APT model rotational strength expression retains only the first term in the magnetic dipole transition moment portion of Eq. (84) (the $\mathbf{P} \cdot \mathbf{L}$ term):

$$
\left(R_r^a \right)^{\mathrm{APT}} = \frac{\hbar}{4c} \left(\sum_A P_{r,\alpha\beta}^A S_{A\alpha,a} \right) \left(\sum_{A'} \varepsilon_{\beta\gamma\delta} R_{A'\gamma}^0 P_{r,\alpha\delta}^{A'} S_{A'\alpha,a} \right)
\tag{85}
$$

A comparison of Eqs. (84) and (85) reveals that the APT model is highly nonlocal in character; there is no reason that the terms left out of the model (the $\mathbf{P} \cdot \mathbf{M}$ term) should in general be small. Although the *form* of some of the models for VCD can be deduced from the APT model expression,[12, 73] it is not valid to evaluate the models for VCD intensity based on how well the APT model reproduces VCD intensities. The derivation of the models from the LDO expression allows a better assessment of the specific terms that are being ignored or approximated in each model.

IV. CALCULATIONS OF VCD INTENSITY

A. Implementation of a Priori Formalisms

The magnetic field perturbation, vibronic coupling theory, and nuclear shielding tensor a priori formalisms for calculating vibrational circular dichroism intensities have been implemented and applied by several laboratories for ab initio molecular orbital wave functions, either by utilizing the output of commercial molecular orbital packages or as integrated parts of these programs. The velocity gauge formalism has not yet been implemented. The general procedure for the VCD intensity calculations has been to use standard ab initio molecular orbital programs to obtain an optimized geometry and s-vectors from the calculated vibrational force field, with the largest basis set practical. The APTs and AATs are then calculated with that same geometry, by employing the largest basis set that is practical for that a priori method; this basis set may be smaller than that employed for the geometry and frequency calculations. In most cases, the best agreement with experimental VCD intensities has been achieved with force fields that have been scaled to produce better agreement with the experimental frequencies. In some cases an experimental geometry and empirical force field have been used.

1. Magnetic Field Perturbation

The magnetic field perturbation expression for the atomic axial tensor, Eq. (43b), was first implemented by Stephens and coworkers[23] as a modification to the *Gaussian 80* program,[74] with finite nuclear displacement and finite field perturbation introduced to obtain the wave function derivatives. The position form of the APT was employed. Calculations on (S,S)-cyclopropane-1,2-$^2H_2$ were carried out with a 4-31G basis set for both experimental and ab initio force fields and geometries.[23] Although experimental data were not available at the time the calculations were obtained, subsequent comparison with experiment[75, 76] reveals good agreement for the calculations obtained with the experimental geometry and the empirical force field of Duncan and Burns.[77] Calculations on (R,R)-cyclobutane-1,2-$^2H_2$ and (S)-propylene oxide also compared favorably to experiment.[78] Subsequent calculations by Lowe and Alper,[27] which examined the force field and geometry dependence of the calculated VCD intensities in (S)-propylene oxide, revealed considerable sensitivity of the calculated intensities to the force field. Although enough of the major VCD bands were reproduced in the calculations to allow confident prediction of the absolute configuration, there were notable features that were not calculated with the correct sign or relative magnitude even with presumably high-quality force fields. Lowe and coworkers[28] also adapted the finite perturbation calculations to examine the effect of gauge origin. For (R)-methylthiirane, use of the common origin gauge with the origin at the center of mass was found to give the best agreement with experiment.[28] Calculations on deuteriated isotopomers of ethanol were consistent with an excess of gauche conformers; the distributed origin gauge produced the better agreement with experiment in this case.[79]

The MFP calculations were made more efficient by employing Eq. (44) for the AAT and using coupled Hartree-Fock (CHF) methods[58, 80-82] to obtain the LCAO coefficient derivatives.[24] This implementation was incorporated into the CADPAC (Cambridge Analytical Derivatives Package) ab initio program,[83] which already had the CHF procedures in place. Morokuma and Sugeta[25] also implemented a calculation via Eq. (44) with CHF evaluation of the position derivatives of the LCAO coefficients and finite displacement evaluation of the magnetic field derivatives. Stephens and coworkers[84] explored the basis set dependence of the rotational strengths of NHDT at the SCF level. Modes with large calculated rotational strengths were found to have the same signs and comparable magnitudes for all the basis sets tested, from STO-3G to VD/3P. Both the signs and magnitudes of the calculated rotational strengths for the modes predicted to have weak VCD showed much more variation with the size of the basis set. The values of the rotational strengths stabilized at approximately the DZ or 6-31G* basis set size. Favorable agreement with experiment was found for a 6-31G** (125 basis functions) MFP-VCD calculation on $trans$-$(1S,2S)$-dicyanocyclopropane.[85]

The distributed origin gauge expression, Eq. (45), was presented initially as a way to relate the a priori formalism to the models for VCD intensity.[12] The form of the FPC and coupled oscillator models was deduced from Eq. (85) (the APT model expression) by employing only the first term in Eq. (45). If the position form of the APT is employed, the CO gauge AAT will lead to a rotational strength that is molecular origin dependent, as demonstrated for NHDT, whereas the DO gauge AAT yields rotational strengths that are independent of molecular origin.[33] The signs and magnitudes of the rotational strengths were also much more stable for the DO gauge as the size of the basis set was increased. We note, however, that the DO and CO gauge results for DZ basis sets and larger are comparable if the common origin is the center of mass. Subsequent MFP VCD calculations on molecules for which experimental data are available have generally employed the DO gauge, with very good overall agreement to experiment. These include studies of $(3R,4R)$-dideuteriocyclobutane-1,2-dione,[86] $(2R,3R)$-dimethylcyclopropane[26] (CH-stretching region, CO gauge), $(2S,3S)$-oxirane-$^2H_2$,[73, 87] (S,S)-cyclopropane-1,2-$^2H_2$,[73, 88] (S)-propylene oxide,[89] $(2S,3S)$-dideuteriobutyrolactone,[90] and (R)-methyl lactate.[91] The VCD spectra of several molecules for which no experimental data are available have also been calculated for comparison to other methods of calculation or for assessment of gauge and basis set dependence; these include (R)-methyl glycolate and (R)-methyl glycolate-d_4,[92] (S)-allene-1,3-$^2H_2$,[93] propane-1,1,1-$^2H_3$-2-$^2H_1$,[73, 88] and chiral conformers of hydrazine,[73, 88] oxaziridine,[88] fluorohydroxylamine,[73, 88] and carbodiimide.[73] Bour and Keiderling[94] have carried out MFP calculations of the amide I and amide II VCD intensities in model peptides in conformations that mimic various peptide secondary structures. Simulations of the calculated VCD spectra were comparable to observed peptide amide I and II VCD features.

An alternative to employing an empirically scaled force field for improving the description of the normal modes is to introduce electron correlation in the calculation of the vibrational force field. Amos and coworkers have demonstrated that the use of MP2 force fields and MP2 electric-dipole moment derivatives improves the agreement between VCD intensities calculated with the MFP formalism and the observed VCD intensities for (R)-methylthiirane.[95] MP2 procedures have not been implemented for the magnetic-dipole moment derivative. Stephens has also discussed the effects of using post-SCF force fields on the accuracy of MFP VCD calculations.[96]

Calculations of the velocity form of the APT with Eqs. (48) and (49) have also been implemented[18] with the CADPAC package. The electronic contribution $E^A_{\nu,\alpha\beta}$ is employed via Eq. (46) as an efficient way to determine the DO gauge AAT. For NHDT, only at the largest basis sets were the position and velocity form APTs comparable.[18]

2. Nuclear Shielding Tensor and Random Phase Approximation Calculations

Calculations of the atomic polar and atomic axial tensor expressions in Eqs. (55)–(57) have been implemented by using the random phase approximation (RPA)[97] to evaluate the sum-over-states. Lazzeretti and Zanasi had previously calculated the APT with nuclear shielding tensor expressions and determined that good results could be obtained with large basis sets including polarization functions, when the RPA method was used to evaluate the sum-over-states.[63, 65, 66] The single transition approximation (STA; excited state is represented by a single particle transition) and

Tamm-Dancoff approximation (TDA; a representation of the excited state by a linear combination of single-particle transitions) methods[97] were less reliable than the RPA method (excited state represented by linear combination of single-particle transitions and inclusion of other configurations in the ground state). The first RPA calculation of both dipole and rotational strengths with nuclear shielding tensors was for NHDT.[32] The position form for the APT (Eq. (57)) gave more reliable results than the velocity-form (Eq. (56)) or acceleration-form[32] nuclear-shielding tensor expressions; polarized basis sets were required. The most reliable rotational strengths were obtained with the DO gauge, where the position form atomic polar tensor used in Eq. (45) was calculated with Eq. (57). Calculations of APTs and AATs for small molecules such as HF, H_2O, NH_3, and CH_4 with a variety of formalisms and basis sets at the SCF level have also been carried out to compare the CHF analytical derivative and RPA nuclear shielding tensor approaches.[98] RPA calculations of the CH-stretching VCD intensities in $(2S,3S)$-oxirane-$^2H_2$ exhibited good agreement with experiment.[99] The "localized orbital/local origin" (LORG) method of Hansen and Bouman, in which localized molecular orbitals, each with it own origin, are employed, has been combined with the RPA approach and the nuclear shielding tensor expressions to develop a formalism for APTs and AATs using noncanonical orbitals.[19] The implementation of these expressions will employ the RPAC program of Hansen and Bouman,[100] which is interfaced to *Gaussian 90*.[101]

3. Vibronic Coupling Theory

Calculations employing the vibronic coupling theory sum-over-states expressions for $E_{v,\alpha\beta}^A$, Eq. (42), and $I_{\alpha\beta}^A$, Eq. (41) were first implemented in our laboratory with CNDO wave functions.[54, 102] Subsequently, Dutler and Rauk[29, 103] developed an implementation of the VCT equations with ab initio wave functions by creating a modified version of *Gaussian 82*[104] and software (named FREQ85)[103] that uses the read-write-file from the modified *Gaussian 82* execution to evaluate the VCT dipole and rotational strengths. This program employed the single transition approximation (STA) for the sum-over-states, wherein each excited state is a single configuration that differs from the ground state by one singly excited molecular orbital. The summation is over all the singly excited states defined by the basis set. The calculations were carried out with a basis set designated 6-31G$\sim$, which is constructed from the 6-31G basis by the addition of all nonredundant first derivatives with respect to nuclear displacement of the 6-31G basis functions. With this basis set, the effects of truncation of the infinite sum over excited states in Eq. (42) does not lead to serious errors,[103] as defined by the "completeness" criterion that the two formulations for the position form APT,

$$E_{r,\alpha\beta}^{a} = -2e \sum_{j} \left\{ \left\langle \Psi_{g} \left| r_{j\beta} \frac{\partial}{\partial R_{A\alpha}} \right| \Psi_{g} \right\rangle \right\}_{R=0} \qquad (86)$$

and

$$E_{r,\alpha\beta}^{a} = -2e \sum_{e \neq g} \left\{ \left\langle \Psi_{g} \left| \sum_{j} r_{j\beta} \right| \Psi_{e} \right\rangle \left\langle \Psi_{e} \left| \frac{\partial}{\partial R_{A\alpha}} \right| \Psi_{g} \right\rangle \right\}_{R=0} \qquad (87)$$

give nearly the same results for the basis set employed.[103] The best rotational strength results were obtained with the position form APT.[103] The calculated VCD spectra[29] of (2S,3S)-oxirane-^{2}H$_2$ agreed well with experiment.[100] For (S)-ethanol-1-^{2}H, the calculated VCD spectrum for the equilibrium mixture of OH rotamers was in excellent agreement with experiment.[106] A series of calculations on model systems such as chiral conformations of hydrazine and hydrogen peroxide were also carried out with the 6-31G ~ basis set.[30]

The FREQ85 program has been subsequently upgraded to a form, FREQ86, compatible with *Gaussian 86*[107] and to its most recent version, VCT90,[108] which employs read-write-file output from a slightly modified version of *Gaussian 90*.[101] The incorporation of electron correlated excited states at the level of all singles CI was found to have little effect.[31] For hydrazine, the VCD patterns calculated with the MFP and VCT formalisms were in very good agreement with the exception of one band.[31] To permit the VCT calculation of VCD intensities on larger molecules, other basis sets that do not include all the nuclear derivative functions were explored. A basis set denoted 6-31G$^{*(0.3)}$, which is derived from the 6-31G* basis set by changing the d-orbital exponent from 0.8 to 0.3, was found to produce calculated VCD intensities in good agreement with experiment even though the "completeness" criterion is not met.[31, 108] VCT calculations have been carried out with this basis set on (S)-2-methyloxirane,[31] (R,R)-2,3-dimethyloxirane,[31, 109] (R)-2-methylthiirane,[110] and (R,R)-2,3-dimethylthiirane,[110] which show very good agreement with experiment in the mid-infrared range. For the CH-stretching regions, use of the 6-31G$^{*(0.3)}$ basis set may introduce errors in sign and magnitude.[31]

The distributed origin gauge expressions have also been implemented for the VCT formalism.[108, 111] In contrast to the MFP results, the VCT calculations with the DO and CO gauges do not converge at the largest basis sets for all modes, but the basis set dependence was reduced with the DO gauge, consistent with a similar observation for the MFP calculations.

VCT calculations with the DO gauge were carried out for the oxiranes and thiiranes listed previously in this section, and for NHDT, H_2O_2, and S-oxaziridine.[111] For (2R)-2-methylaziridine, which exists in *cis* and *trans* forms, and (S,S)-2,3-dimethylaziridine, the calculated VCD spectra show slightly better agreement with the observed spectra when the DO gauge was employed.[108]

4. The Floating Basis Set Formulation

The floating basis set formulation has been implemented at the CNDO level with finite nuclear displacement.[17] Since the LCAO coefficients carry all the nuclear displacement dependence in the floating basis molecular orbitals, the derivatives of the floating basis set LCAO coefficients differ from those calculated from standard CHF methods with non-floating basis functions. This difference arises from the additional nuclear displacement derivatives of the potential energy that are required in the CHF equations for the floating basis set,[17, 22] which do not enter the non-floating basis set CHF equations. Although it is probable that the large polarized basis sets required for the VCT and RPA calculations will not be needed with the floating basis set formulation since the perfect-following contributions enter as ground-state terms, modification of the CHF procedures to incorporate the additional terms is required before this formalism can be implemented with ab initio molecular orbital programs.

5. Nuclear Velocity Perturbation–Velocity-Gauge Formalism

The NVP velocity gauge-formalism[22] avoids the sum-over-states needed in the VCT, nuclear shielding tensor, and floating basis set methods, and no magnetic field coefficient derivatives are required. The NVP velocity-gauge method also allows explicit evaluation of the relative importance of terms that are directly related to simplified models for VCD intensity. Implementation of the NVP velocity-gauge formalism requires CHF calculations of the nuclear *velocity* derivatives of the LCAO coefficients. Evaluation of these derivatives requires additional terms in the matrices used in the CHF procedure, including the nuclear displacement coefficient derivatives and additional overlap integral and one- and two-electron integral derivatives that vanish in the evaluation of coefficient derivatives with respect to $\mathscr{A}$ and $\mathscr{H}$.[22] Implementation of the NVP velocity-gauge formalism for VCD intensities is currently underway.

B. Implementation of VCD Model Calculations

1. Coupled Oscillator Models

The coupled oscillator model[34, 35] has been used to explain bisignate VCD features observed for modes with large contributions from the coupling of local oscillators that are chirally oriented. Two approaches can be taken to implement the coupled oscillator model. Since the sign of a VCD band arising from the coupled motion of local oscillators is defined by the chiral orientation and the relative phases of the oscillators, the sense, $(+, -)$ or $(-, +)$, of a coupled oscillator VCD couplet will be determined by the frequency ordering of the coupled modes. The generalized coupled oscillator model,[56] Eq. (71), described in Section III.C, uses the vibrational force field to define the frequencies and sense of coupling. A second approach is to view the coupling as arising from the electrostatic dipole–dipole potential energy,[112]

$$V_{i,j} = \frac{1}{|R_{i,j}|^3}\left(\mathbf{\mu}_i \cdot \mathbf{\mu}_j - \frac{3(\mathbf{R}_{i,j} \cdot \mathbf{\mu}_i)(\mathbf{R}_{i,j} \cdot \mathbf{\mu}_j)}{|\mathbf{R}_{i,j}|^2}\right) \tag{88}$$

where $\mathbf{R}_{i,j} = \mathbf{R}_j - \mathbf{R}_i$ is the separation vector between oscillators i and j, which have local electric dipole transition moments $\mathbf{\mu}_i$ and $\mathbf{\mu}_j$. For N coupled degenerate local oscillators $\Delta \ell_i$, the frequencies (cm^{-1}) of the coupled modes $Q_k = \Sigma_i C_{ik} \Delta \ell_i$ are given by

$$\bar{\nu}_k = \bar{\nu}_0 + \frac{1}{hc} \sum_{i=1}^{N} \sum_{j=1}^{N} C_{ik} C_{jk} V_{ij} \tag{89}$$

where $\bar{\nu}_0$ is the unperturbed frequency of a local oscillator, and the C_{ik}, which are defined by the interaction potential, can also be determined by using group theory for a coupled system with symmetry. For N-coupled degenerate oscillators, the rotational strength expression

$$R_k = -\pi \bar{\nu}_0 \sum_{i=1}^{N} \sum_{j=1}^{N} C_{ik} C_{jk} \mathbf{R}_{ij} \cdot \mathbf{\mu}_i \times \mathbf{\mu}_j \tag{90}$$

and dipole strength expression

$$D_k = \sum_{i=1}^{N} \sum_{j=1}^{N} (C_{ik}\mathbf{\mu}_i) \cdot (C_{jk}\mathbf{\mu}_j) \tag{91}$$

are special cases of the generalized coupled oscillator expressions Eqs. (71) and (72) for modes coupled via a dipole–dipole interaction potential. The sum of the integrated absorption intensities for the set of coupled modes is used to obtain the magnitude of the local electric-dipole transition moments (Eq. (70)).

Interpretation of VCD couplets with this model provides information on the chiral orientation of the oscillators and thus on the conformation of the molecule containing the oscillators. This form of the coupled oscillator model has been used successfully in our laboratory to interpret VCD couplets observed for the chiral metal complex bis(acetylacetonato)(L-alaninato)cobalt(III),[113] for N-urethanyl-L-amino acid derivatives,[114] and for interchain hydrogen-bonded tripodal peptides.[115] Keiderling and coworkers have interpreted VCD couplets for $C = O$ stretching modes in dicarbonyl containing steroids,[116] for base vibrations in poly(ribonucleic acids),[117] and for phenyl modes in (S)-1,3-diphenylallene[118] in terms of degenerate oscillators with dipolar coupling. Diem and coworkers have formulated the coupled oscillator equations for coupled n-mers and have implemented their calculations in a FORTRAN program to compute VCD intensities for up to 30 dipoles interacting via a dipolar potential.[119] They have successfully modeled VCD features in model deoxyoligo-nucleotides,[119] models for DNA in the B-conformation,[120] alanyl tripep-tides,[121] and prototypical peptide conformations[122] as extended coupled oscillators.

These implementations of the coupled oscillator model are successful for interpreting VCD couplets arising from local oscillators with large local electric-dipole transition moments, such as $C = O$ stretches, which are separated by several bonds in a molecule, but are spatially close enough for dipolar coupling to occur. In the NH- and OH-stretching regions, the assumption of a dipolar interaction potential leads to the wrong VCD sign patterns for coupled modes.[115, 123] In such cases, it is still likely that the coupled oscillator mechanism is valid, that is, the bisignate character of the VCD features does arise from the coupling of chirally oriented oscillators, but that the coupling is dominated by potential or kinetic energy effects other than dipolar coupling.

In the CH-stretching region, bisignate VCD features are common for molecules with chirally oriented CH oscillators, when functional groups with lone pairs are not adjacent to the CH groups. In such cases, the sign pattern of the VCD spectrum has been reproduced by using the general-ized coupled oscillator expression in conjunction with a normal mode calculation.[56, 76]

The coupled oscillators need not be degenerate or identical for VCD intensity to be generated. In the application of the generalized coupled oscillator formula, Eq. (71), to $(2S,3S)$-cyclopropane-1-$^{13}C,^2H$-2,3-$^2H_2$,[56] the individual CH oscillators have different frequencies due the ^{13}C-sub-stitution, and in the N-urethanyl-L-amino acid derivatives,[114] the urethane and acid $C = O$ stretches couple when formation of an acid dimer shifts the acid $C = O$ stretching frequency close to the urethane $C = O$

stretching band. The generalized coupled oscillator equations and Eqs. (89)–(91) are readily extended to nonidentical oscillator situations.[56, 124] However, the need to ascribe individual electric-dipole moment derivatives to each oscillator then introduces added complexity to the application of the model. Similarly, use of the coupled oscillator model for deformation modes is less justified, since determination of the location of the local electric-dipole transition moments is less precise. The coupled oscillator model must be viewed as only one mechanistic contribution to VCD intensity. Coupled oscillator effects may not be observed if the splitting between the modes is too small. In addition, an expected VCD couplet may not be observed or may be observed as a biased rather than conservative couplet if an individual local oscillator is in a chiral environment that gives rise to intrinsic monosignate VCD intensity for that oscillator.[35] Finally, it should be noted that the term "coupled oscillator" is used by some investigators to imply coupling only via a dipolar potential. However, local achiral oscillators that are coupled and are chirally oriented generate bisignate VCD features independent of the source of coupling, and this more general definition is preferred.

2. Fixed Partial Charge Model

The fixed partial charge model[36, 37] is readily implemented once a vibrational force field for a molecule is obtained and the s-vectors are calculated. The simplest way to use the FPC model is to employ the Mulliken populations as in Eq. (63). In some implementations, the fixed charges are scaled to produce better agreement with the dipole strengths. Despite its simplicity and the severe approximations involved, the FPC model is reasonably successful in reproducing VCD patterns in rigid hydrocarbon systems with no heteroatoms or double bonds, such as (S,S)-cyclopropane-1,2-$^2H_2$[23] and (R,R)-cyclobutane-1,2-$^2H_2$,[125, 126] and the CH-stretches of $(3R)$-methylcyclohexanone.[127] The FPC model also reproduces the VCD sign pattern for most of the midinfrared modes in (R)-2-methyloxetane.[128] The VCD intensities predicted from the FPC model with Mulliken charges are considerably smaller than observed, which may be due largely to the inability of the FPC model to reproduce the dipole strengths. The FPC model by nature must give rise to conservative VCD spectra (equal positive and negative integrated VCD intensity) over a region of coupled modes, and the model fails to reproduce the monosignate or biased VCD features observed in molecules with heteroatoms and unsaturated groups. The cases in which the FPC model is successful suggest an underlying mechanism for those modes, where charge flows distant from a moving nucleus are very small and the VCD intensity arises largely from perfect nuclear following contributions.

3. Localized Molecular Orbital Model

The localized molecular orbital model, Eq. (74), was first implemented with CNDO wave functions.[39] Although this implementation provides calculated VCD spectra in fairly good agreement with experiment in the CH-stretching region,[125, 129] spectral features in the midinfrared are not reproduced.[93, 126]

Recently, Polavarapu has implemented the LMO model with ab initio wave functions.[49, 50] The GAMESS program[130] is used with the Boys localization method[131] to obtain the localized orbital centroids. A finite displacement method is employed to obtain the nuclear displacement derivatives of the orbital centroids. Polavarapu and coworkers have used this implementation of the LMO model to calculate the VCD spectra of $(2S,3S)$-oxirane-$^2H_2$,[50] (S,S)-cyclopropane-1,2-$^2H_2$,[49] (R,R)-2,3-dimethylthiirane,[110] methylthiirane,[132] and (R,R)-2,3-dimethyloxirane and its 2-$^2H_1$ and $2,3$-$^2H_2$ isotopomers.[109] Satisfactory agreement with the experimental spectra and with corresponding a priori VCD intensity calculations was found, although the magnitudes of the rotational strength differed for the various calculations and some differences in sign also occur. The good agreement with the observed VCD sign patterns for molecules with known absolute configuration has led to the application of LMO VCD calculations to the determination of absolute configuration from observed VCD spectra; assignments of absolute configuration have been made for $(+)$-2-methylthiirane-3,3-$^2H_2$,[133] and the chiral anesthetics isoflurane[134] and desflurane.[135]

4. Charge Flow Models

The simplest charge flow model[41, 42] has been implemented by two laboratories. Moscowitz and coworkers applied the model to the CH- and CD-stretching modes of chirally deuteriated phenylethanes.[136] With principal charge flows along the CH bonds obtained from the infrared intensities, the rotational strengths calculated with the charge flow model underestimated the observed VCD magnitudes. Introduction of polarization terms for the phenyl group led to the correct order of magnitude for the calculated rotational strengths. Moscowitz and coworkers have also proposed, but not implemented, a version of the charge flow model for oriented species, which includes electric quandrupole contributions.[137]

The most recent implementations of the charge flow models have been presented by Wieser and coworkers. In a study of 2-methyloxetane, augmentation of the FPC calculation with a single charge flow parameter for the C–O bond stretch greatly improved the fit to the experimental

VCD spectrum.[128] For six methyl substituted 6,8-dioxabicyclo[3.2.1]octanes, a single charge flow parameter for the C–O stretch in addition to fixed partial charges from an STO-3G basis set MO calculation provided good agreement with experiment for the calculated midinfrared VCD intensites.[138] Introduction of charge-flow parameters also provided fairly good agreement with experiment for calculated spectra of camphor, camphene, and fenchone,[139] and a series of analogues of α-pinene.[140] In all these fairly large molecules, scaled harmonic 3-21G or higher force fields were employed, following the method of Pulay.[141] It is clear that the fit of the calculations to experiment depends on the quality of the force field. Since the sign patterns are reproduced fairly well with the FPC model, much of the VCD intensity in these large molecules must be determined by the chiral arrangement and coupling of the oscillators contributing to each mode, and accurate mode descriptions are crucial. The absorption intensities as well as the VCD intensity magnitudes are improved with the introduction of the charge flow parameters, but the latter do not significantly alter the calculated VCD signs. One can conclude that the major improvement in the rotational strength provided by the C–O charge flow parameter lies in the electric dipole transition moment contribution.

The molecular orbital charge flow model, the NMO model, was first implemented with CNDO wave functions.[44] For the CH-stretching region in L-alanine and deuteriated isotopomers, the NMO rotational strengths showed considerable improvement over the FPC and APT models, but the large positive VCD intensity bias in the CH-stretching region was not reproduced.[44] Recently, Maurer and Wieser[71] have implemented an ab initio version of the NMO model in a program that uses the output of a slightly modified version of *Gaussian 90*.[101] Calculated rotational strengths for propane-1,1,1-$^2H_3$-2-$^2H_1$ agreed fairly well with corresponding calculations with the MFP and VCT approaches.

5. Ring Current Model

The ring current model[3, 48] was proposed as an extension of the charge flow models to explain large monosignate VCD intensity due to local oscillators that are external or internal to hydrogen-bonded or covalently bonded rings. For closed rings, vibrationally generated current that flows around the ring at constant electron density does not contribute to the dipole strength, but can make large contributions to the rotational strength due to the magnetic dipole transition moment arising from the current in the ring. The impetus for the development of the ring current model was the observation that the large monosignate VCD intensity for the stretch of NH and OH bonds involved in intramolecular hydrogen bonding and for the stretch of a C^*H bond adjacent to the heteroatom in a hydrogen-bonded ring was not observed for similar modes involving free NH and OH stretches and methine bonds not associated with such rings.[113, 115, 142–147] Empirical rules were proposed for determining the direction of the postulated vibrationally generated current flow in the ring for different types of oscillators.[3, 143, 148] These rules were an attempt to generalize a large body of empirical observations that categorized specific types of environments for local oscillators as leading to large monosignate VCD intensity of a specific sign.[113–115, 142–151]

Stephens and coworkers have called into question the validity of the ring current model based on MFP VCD intensity calculations on (R)-methyl lactate[91] and (R)-methyl glycolate-d_1,[92] wherein the calculated VCD intensities for conformations with and without an intramolecular hydrogen bond were compared. The VCD intensity for the isolated C^*H stretch in (R)-methyl lactate-d_6 and (R)-methyl glycolate-d_4 was calculated to be fairly large and of the same sign for both types of conformation, which appears to contradict the premise of the ring current mechanism that attributes the enhanced VCD intensity to current around a closed ring.

The MFP VCD calculations for these two molecules do not in fact rule out the possibility of vibrationally generated ring current. For (R)-methyl lactate-d_6, the calculated rotational strength for the uncoupled C^*H stretch decreases by a factor of two in going from the hydrogen-bonded to the non-hydrogen-bonded conformer.[91] In (R)-methyl glycolate-d_4, the calculated anisotropy ratio for the uncoupled C^*H stretch decreases from 1.2×10^{-4} for the most stable hydrogen-bonded conformer to 2.6×10^{-5} for the conformer without a hydrogen bond.[92] The latter value is considerably below what is considered to be a large or enhanced VCD signal. Conformers of (R)-methyl glycolate-d_4 with weaker $OH \cdots OCH_3$ hydrogen-bonding were calculated to have smaller methine stretching rotational strengths than the most stable $OH \cdots O=C$ conformer. The observation that the calculated anisotropy ratios decrease by a factor of two to four for the non-hydrogen-bonded and more weakly hydrogen-bonded conformers is in complete agreement with the hypothesis of the ring current mechanism that the presence of a closed ring can lead to enhanced signals, and that the magnitude of vibrationally generated current in a ring is related to the strength of the hydrogen bond.

The magnitude of the calculated anisotropy ratio ($g = \Delta\varepsilon/\varepsilon = 4R/D$) for the C^*H stretch in (R)-methyl lactate[91] (-1.1×10^{-4} for the hydrogen-bonded conformer) is considerably smaller than our recent measurement of this parameter for (S)-(methyl-d_3)lactate (2.9×10^{-4}).[152] Recent experiments in our laboratory for a variety of α-hydroxy ester derivatives[152] indicates that the anisotropy ratios for this type of C^*H stretch can be as large as 5.5×10^{-4}. It is thus possible that the MFP calculations or possibly all the present a priori formalisms may be missing some of the magnetic dipole transition moment contribution. Calculation of VCD intensity of the observed sign is not sufficient evidence to rule out the possibility of an additional mechanism that leads to enhanced signals. In addition, we find experimentally that a conventional hydrogen bond is not required for the generation of large monosignate VCD intensity in α-hydroxy ester derivatives, since the methine stretch in (S)-methyl-$d_3$2-(methoxy-d_3)-proprionate, a derivative of methyl lactate with $O–CD_3$ substituted for the OH group, exhibits an intense monosignate methine stretching VCD band with anisotropy ratio 2.8×10^{-4} (Ref. 152). Our empirical evidence reveals that orbital overlap between the heteroatom adjacent to the methine bond and a second group with lone pairs or π-orbitals is necessary for enhanced methine stretching VCD to be observed. In this context, the definition of a "ring" that may be responsible for enhanced VCD intensity must be expanded to include such orbital overlap.

Stephens and coworkers[91, 92] also criticize the ring current mechanism based on their observation that the largest contribution to the calculated rotational strength scalar product was found to be a $\mu_\alpha m_\alpha$ term, where α is in the plane of the ring, not perpendicular to it as predicted for enhancement due to ring current. However, there is a fallacy in the latter argument, since although the rotational strength itself is not origin dependent, the individual terms in the scalar product summation do depend on the origin for the magnetic moment terms; the magnitude of the individual terms in the summation has no real significance. Furthermore, there are numerous contributions to the rotational strength in addition to a postulated ring current, which may occur along any direction.

The major difficulty with the ring current mechanism is that there is at present no independent method for calculating such a contribution and thus no definitive way to confirm or invalidate the proposed mechanism. The primary utility of the ring current mechanism for the interpretation of VCD spectra lay in the empirical association of large monosignate VCD intensity of a specific sign with a specific chiral environment for a local oscillator.[113–115, 142–152] This aspect of the empirical evidence remains unquestioned and provides a context in which characteristic VCD signals can be associated with stereo-specific local chiral structures, without the need to postulate a specific descriptive mechanism for the large signal. The more recent studies from our laboratory have emphasized correlations of VCD intensity with the local chiral environment of the oscillator in order to obtain evidence for the solution conformations from the VCD spectra.[115, 153, 154]

6. Atomic Polar Tensor Model

The atomic polar tensor model, Eq. (85), has been implemented with ab initio wave functions in conjunction with MFP, VCT, and NMO calculations. This term is a part of the distributed origin gauge formulation for the rotational strength, Eq. (45). As noted above, the APT model is nonlocal in nature. The changes in the dipole moment of the entire molecule arising from the displacement of a single nucleus are placed at the moving nucleus, and the terms that are left out of the model (see Eq. (84)) in part correct for the misplacement of local contributions. In general, the agreement of APT model rotational strengths with experiment is poor, particularly in the midinfrared region and for small molecules. Exceptions are the calculations on the series of methyl substituted 6,8-dioxabicyclo[3.2.1]octanes[138] for which the APT model rotational strengths were in fairly good agreement with experiment. Stephens and coworkers[73] have compared the APT model, the $\mathbf{P} \cdot \mathbf{L}$ term, with the $\mathbf{P} \cdot \mathbf{M}$ term (Eq. (84)) for MFP calculations of rotational strengths. For most cases, the $\mathbf{P} \cdot \mathbf{L}$ term was not the dominant contribution to the total rotational strength, and for a number of modes the signs of the two contributions differed.

C. Comparison of Calculational Methods

The quality of calculations of VCD intensity may be assessed by comparison with the observed VCD spectra of small rigid "benchmark" molecules. Two such examples are highlighted below, for $(2S,3S)$-oxirane-$^2H_2$, and $(2S,3S)$-cyclopropane-1-$^{13}C,^2H$-2,3-$^2H_2$. Several additional examples of comparisons among calculational approaches have been published. The results of VCT and MFP calculations have also been compared for hydrazine,[31] (S)-2-methyloxirane,[111] NHDT,[111] (S)-oxaziridine,[111] and (R)-2-methylthiirane.[110, 111] The LMO and VCT approaches were found to yield calculated spectra of comparable agreement with experiment for (R)-2-methylthiirane, (R,R)-2,3-dimethylthiirane, and (R,R)-2,3-dimethylthiirane-2,3-d_2.[110] The coupled oscillator model with dipolar coupling has been compared to MFP VCD calculations in order to define the limits of applicability of the model.[155] We also provide below examples of utilizing the form of the VCD intensity expressions to understand the origin of VCD intensity for molecules that are chiral due to isotopic substitution.

TABLE I

Comparison of Observed and Calculated Frequencies and Rotational and Dipole Strengths for (S,S)-[2,3-^{2}H$_2$]Oxirane

	Frequencies (cm^{-1})			Rotational Strengths (esu^{2}cm$^2 \times 10^{44}$)					Dipole Strengths (esu^{2}cm$^2 \times 10^{39}$)				
	Observed					Calculated[c]					Calculated[c,d]		
	Gas	Solution[e]	Calculated[a]	Observed[b]	MFP	LMO(I)	LMO(II)	VCT	Observed[b]	MFP	LMO(I)	LMO(II)	VCT
B	3028	3027	3040	+11.4	+30.7	+10.6	+10.4	+14.4	5.3	5.65	5.00	4.52	5.06
A	3015	3014	3036	−8.9	−26.7	−18.2	−13.2	−11.8		0.89	0.90	0.49	0.61
A	2254	2252	2239	+12.1	+14.4	+7.9	+6.0	+13.2	0.57	1.06	1.05	0.56	0.72
B	2240	2232	2224	−10.4	−14.8	−2.9	−3.4	−12.0	2.7	5.11	4.55	4.01	4.48
A	1397		1397	(−15)	−10.0	−9.8	−9.5	−18.5	(1.2)	1.48	1.08	1.02	1.08
B	1339		1328	(−2.5)	−2.2	−2.6	−1.4	−0.9	0.23	0.19	0.12	0.08	0.12
A	1235	1226	1235	+24.1	+13.6	+17.4	+17.7	+19.2	2.96	3.26	2.87	2.59	2.87
A	1112	1109	1123	−4.9	+4.3	+3.3	+1.5	+3.3		0.10	0.07	0.06	0.07
B	1106	1102	1117	+11.1	+9.3	+7.8	+4.2	+14.1	0.86	1.09	0.93	0.90	0.93
A	961	948	947	−29.0	−40.8	−35.7	−34.3	−18.8	5.4	12.98	9.08	9.19	10.28
B	914	914	918	−6.2	−1.0	−4.1	−2.9	+5.7	0.63	2.42	2.30	1.88	2.09
A	885		871	(∼ +5)	+5.8	+30.5	+31.6	−5.5		16.29	14.53	11.71	12.99
B	817		820	(+)	+1.3	+2.7	+2.9	−2.1		6.09	5.75	4.17	4.52
A	754		751		+15.0	−6.7	−8.0	+2.1		15.68	19.00	12.76	14.03
B	673		646		+0.7	+0.2	+0.5	+3.4		0.12	0.11	0.27	0.18

[a] Calculated with ab initio force field of Lowe et al.[157] and experimental geometry of Hirose.[156]

[b] Measured from integrated intensities for solution phase sample for all bands except for values in parentheses, which were estimated from the relative intensities in the gas phase spectrum.

[c] MFP, magnetic field perturbation, from Ref. 73, with 6–31G(ext) basis; LMO(I) and LMO(II), localized molecular orbital model, from Ref. 50, with 6–31G(ext) and 6–31$\bar{\text{G}}$(-2$\bar{S}$,2$\bar{P}$,2$\bar{P}_H$) basis sets, respectively; VCT, vibronic coupling, sum-over-states method, from Ref. 29, 6–31$\bar{\text{G}}$(-2$\bar{S}$,2$\bar{P}$,2$\bar{P}_H$) basis.

[d] The dipole strength calculations are identified by the rotational strength method to which they correspond.

[e] C$_2$Cl$_4$ solution for CH- and CD-stretching modes; CS$_2$ solution for mid-infrared modes.

245

1.   (2S,3S)-Oxirane-$^2H_2$

Experimental VCD spectra have been obtained[105] for (2S,3S)-oxirane-$^2H_2$ in the gas and solution phases in the CH- and CD-stretching regions, in the gas phase from 800 to 1500 cm^{-1}, and in CS$_2$ solution from 900 to 1350 cm^{-1}. The results of VCD intensity calculations are compared with the experimental values in Table I for three formalisms, MFP[73] (6–31G(ext) basis set, experimental geometry[156]), VCT[29] (6-31G$\sim$ basis set, optimized geometry), and LMO[50] (6-31G(ext) basis set, experimental geometry, and 6-31G$\sim$ basis set, optimized geometry). Simulations of the calculated spectra with uniformly scaled frequencies are compared with the observed solution phase data in Fig. 1. For four bands in the midinfrared that were obscured by solvent absorption, the data for the gas phase

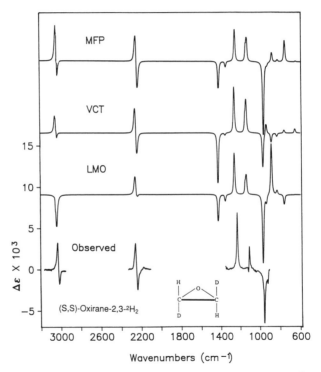

Figure 1. Comparison of the observed VCD spectra of (2S,3S)-oxirane-$^2H_2$ (Ref. 105, 0.019 M in C$_2$Cl$_4$ solution, 2000–3200 cm^{-1}; 0.052 M in CS$_2$ solution, 900–1350 cm^{-1}) with simulations (10 cm^{-1} half-width for CH and CD stretches, 6 cm^{-1} half-width for 600–1600 cm^{-1}, Lorentzian bands) of the VCD spectra calculated with the MFP formalism (Ref. 73, 6–31G(ext) basis), the VCT formalism (Ref. 29, 6-31G$\sim$ basis), and the LMO model (Ref. 50, 6-31G(ext) basis).

sample (Table I) can be compared to the calculations. All three calculations reproduce the signs of the major VCD bands. The negative VCD band observed at 1109 cm^{-1} is calculated as positive with all three methods. For modes below 950 cm^{-1}, several differences in sign and VCD intensity occur among the calculations. Unfortunately, most of these bands could not be measured experimentally with our present instrumentation. The calculated rotational strengths are within a factor of two of those obtained from the integrated VCD intensities for most of the modes. The MFP calculation yields the best overall agreement with experiment. However, any of the methods could be used unambiguously for a definitive identification of the absolute configuration based on an overall fit to the VCD data.

2.   Isotopically Substituted Cyclopropanes

$(2S,3S)$-Cyclopropane-1-$^{13}C,^{2}H$-2,3-$^{2}H_2$, **1**, is chiral due to ^{13}C-substitution at C-1 in achiral *anti*-cyclopropane-1,2,3-$^2H_2$, **2** (see figure). This

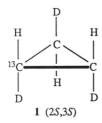 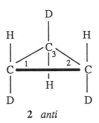

1 (2S,3S)   **2** *anti*

small difference in mass introduces chirality into the CH- and CD-stretching modes due to altered contributions to the modes from the CHD group at C-1 compared to those from the CHD groups at C-2 and C-3. The VCD and absorbance spectra[56] in these regions are shown in Fig. 2. In the midinfrared region, no measurable VCD intensity was observed. We have carried out MFP VCD calculations on **1** with a 6–31G(ext) basis set, by employing the experimental geometry and the empirical force field of Duncan and Burns,[77] which we have slightly modified[56] by changing the interaction force constant for *anti* CH stretches from -0.0033 to -0.015 mdyne/A. This modification was introduced to obtain the correct frequency order[76] of the in- and out-of-phase CH stretches at C-1 and C-2 in (S,S)-cyclopropane-1,2-$^2H_2$ and to obtain the correct relative VCD intensities in the CH-stretching region compared to the CD-stretching region in

$(2S,3S)$-cyclopropane-1-^{13}C,^{2}H-2,3-^{2}H$_2$ (Ref. 56). Similar MFP VCD calculations[73] on (S,S)-cyclopropane-1,2-^{2}H$_2$ with the unmodified Duncan and Burns force field and the 6–31G(ext) basis set exhibited good overall agreement with the observed VCD spectra[76] in the 950- to 1400-cm^{-1} region. We have also carried out generalized coupled oscillator calculations, Eq. (71) for the CH- and CD-stretching modes of **1**.[56] The observed and calculated frequencies and rotational strengths for these modes are listed in Table II. The observed rotational strengths are determined from a curve fit to the experimental data; these represent a lower limit to the actual rotational strengths, since somewhat more intense bands with larger cancellation due to the overlap of positive and negative features could also reproduce the observed VCD pattern. The simulated absorbance and VCD spectra derived from the generalized coupled oscillator calculations are compared to the experimental data in Fig. 2. The absorbance and

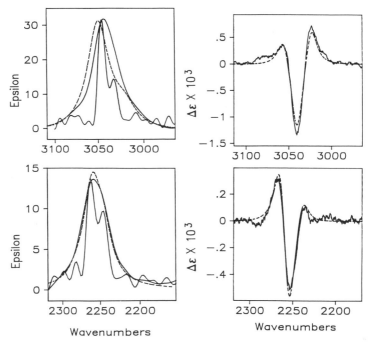

Figure 2. Comparison of observed absorbance (left) and VCD (right) spectra (—) of $(2S,3S)$-cyclopropane-1-^{13}C,^{2}H-2,3-^{2}H$_2$ in the CH- and CD-stretching regions with simulations of calculated spectra (---) obtained with the generalized coupled oscillator model. Fourier deconvolutions of the observed absorbance spectra are also shown.

TABLE II

Observed and Calculated Absorption and VCD spectra of (2S,3S)-Cyclopropane-1-^{13}C,^{2}H-2,3-^{2}H$_2$

	Observed[a]			Calculated			
Freq (cm^{-1})	ε (10^3 cm^2 mol^{-1})	$10^3\,\Delta\varepsilon$ (10^3 cm^2 mol^{-1})	$10^{44}R$ (esu^2 cm^2)	Freq (cm^{-1})	Coup. Oscillator[b] ($10^{44}R$, esu^2 cm^2)	MFP (DO)[c] ($10^{44}R$, esu^2 cm^2)	MFP (CO)[d] ($10^{44}R$, esu^2 cm^2)
3054		0.35	1.5	3085	1.3	8.0	6.3
3041	31.7	−1.30	−3.2	3071	−3.0	−18.1	−20.8
3025		0.70	1.8	3060	1.7	10.2	11.0
2271		0.13	1.3	2270	1.7	7.0	11.3
2258	14.2	−0.51	−2.1	2263	−2.4	−9.4	−11.6
2245		0.11	0.7	2251	0.7	2.3	6.9

[a] Reference 56.
[b] Generalized coupled oscillator model.
[c] Magnetic field perturbation, distributed origin gauge.
[d] Magnetic field perturbation, common origin gauge.

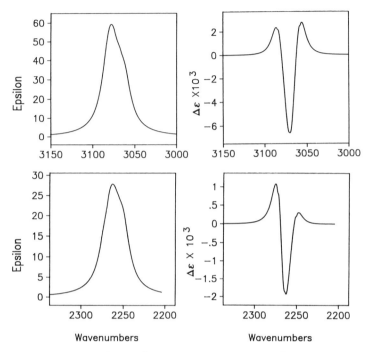

Figure 3. Simulations (10 cm^{-1} half-width, Lorentzian bands) of the VCD spectra of (2S,3S)-cyclopropane-1-^{13}C,^{2}H-2,3-^{2}H$_2$ calculated with the MFP formalism, 6–31G(ext) basis set.

VCD spectra simulated from the MFP calculation with the distributed origin gauge are depicted in Fig. 3.

For this simple molecule, the generalized coupled oscillator model provides an excellent fit to the experimental VCD data. The absorption band shape is also reasonably well reproduced by describing the modes as linear combinations of local isolated oscillators as required by the coupled oscillator model. With the MFP calculation, the overall absorption and VCD band envelopes in both regions are closely reproduced with the band widths (10 cm^{-1} half-width) used in the spectral simulations. The calculated absorption intensity is a factor of two larger than experiment, which is a typical discrepancy for this type of ab initio calculation. In contrast, the calculated MFP VCD intensity is approximately a factor of five too large in both regions. Increasing the band widths or decreasing the band separation to obtain more cancellation in the simulated calculated spectrum results in band shapes that no longer agree with experiment. If the *anti* CH/CH force constant is decreased, the same $(+, -, +)$ VCD patterns are maintained, but the calculated magnitudes of the CD-stretching rotational strengths increase while those of the CH-stretching rotational strengths decrease. The VCD intensities calculated with the MFP formalism and the common origin gauge (Table II) are also too large by a similar factor. In the CH-stretching region, the simulated CO-gauge overall band envelope is in better agreement with experiment compared to the DO-gauge results; however, in the CD-stretching region, the CO-gauge results have an overall positive intensity bias, in contrast to the experimental results, which are slightly negatively biased. We note that in the MFP calculations for (S,S)-cyclopropane-1,2-$^2H_2$, the CH-stretching VCD intensities are also larger than experiment by a factor of three to six, and the VCD intensities calculated for two modes in the midinfrared are smaller than experiment by a similar factor.[73, 76]

3. Descriptive Interpretations

A difficulty with all the a priori approaches to calculating VCD intensity is that in achieving the ability to readily calculate an entire absorbance and VCD spectrum starting only with an assumed molecular conformation, there is a tendency merely to compare the overall calculated spectrum with experiment. There is less concern with the precise nuclear motion that is generating the VCD intensity, and there is little effort or possibly little need seen to seek an underlying mechanism that may be responsible for an intense VCD feature. With the present emphasis on a priori calculations, when only very general assignments are provided for vibrational modes, interpretational insight is lost, and with it some of the capability to use VCD to obtain information on solution conformations in molecules that are too large for current ab initio calculations. For molecules that have numerous possible conformations, it is important to retain the capacity to use characteristic observed VCD features to eliminate unreasonable starting conformations in proceeding with a priori calculations.

As discussed in Section IV.B.4, certain types of localized vibrational modes can give rise to large monosignate VCD intensity that is characteristic of the chiral environment of the oscillator. Such VCD signals can be used as marker bands for a specific conformation, without the necessity for detailed intensity calculations. Coupled oscillator effects can also dominate the VCD spectrum in a local region; interpretation of the spectra with the coupled oscillator model can provide an estimate of the geometry or a starting conformation for more detailed calculations.

For simple molecules that are chiral due to isotopic substitution, the form of the locally distributed origin gauge expression, Eq. (60), can be used to provide rather extensive insight into the origin of the VCD intensity. We consider here the CH-stretching modes in $(2S,3S)$-

cyclopropane-1-^{13}C,^{2}H-2,3-^{2}H$_2$ and the "methylene scissors" modes in (S,S)-cyclopropane-1,2-^{2}H$_2$ and (2S,3S)-oxirane-^{2}H$_2$.

The calculated CH-stretching modes[56] in (2S,3S)-cyclopropane-1-^{13}C,^{2}H-2,3-^{2}H$_2$, **1**, and *anti*-cyclopropane-1,2,3-^{2}H$_2$, **2**, are shown schematically in Fig. 4. We approximate the $(\partial\boldsymbol{\mu}_\nu^B/\partial P_a)_0$ contributions in Eq. (60) as vectors, represented by the arrows, that are located at the hydrogens and proportional to and in the opposite direction to the hydrogen displacement (since positive electric dipole moment change is directed H $\longrightarrow$ C for CH elongation[158]). The origin of the coordinate

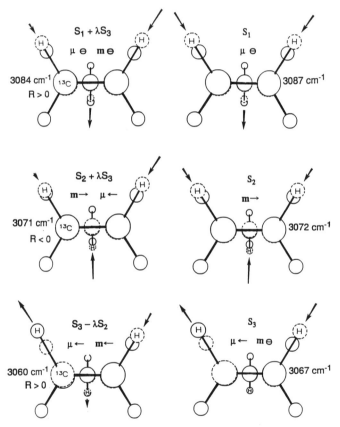

Figure 4. Schematic representations of the calculated CH-stretching modes of (2S,3S)-cyclopropane-1-^{13}C,^{2}H-2,3-^{2}H$_2$, left, and the parent achiral molecule, *anti*-cyclopropane-1,2,3-^{2}H$_2$, right. The local electric dipole transition moment contributions are indicated as solid arrows, along with the directions of the net electric and magnetic dipole transition moments for each mode for the phase of vibration going from the solid to dotted circles.

system is at the center of mass, and the intrinsic contribution in Eq. (60) is ignored. The direction of the electric and magnetic dipole transition moments (denoted μ and $\mathbf{m}$) for the modes of the parent achiral molecule **2** are readily deduced from the diagrams in Fig. 4 (see Eqs. (60) and (14)) as follows: The vector sum of the $(\partial \mu_\nu^B / \partial P_a)_0$ arrows gives the direction of μ and the sense of circulation of the arrows about the center of mass is used in conjunction with the right-hand-rule relating current flow and magnetic dipole moment to give the direction of $\mathbf{m}$. Mode S_1 is only electric dipole active, S_2 is only magnetic dipole active, and for S_3 the electric and magnetic dipole transition moments are orthogonal, as expected for an achiral molecule. The ^{13}C substitution destroys the symmetry, allowing the CH-stretching modes of **2** with A' symmetry (S_1 or S_2) to mix with the mode of **2** with A'' symmetry (S_3) to form the modes of **1** shown on the left; these modes are decomposed by inspection into the linear combinations of S_1 to S_3 shown, where λ is a small mixing parameter. This admixture results in electric- and magnetic-dipole transition moments along the same Cartesian axis. The parallel or antiparallel alignment of these transition moments determines the sign of the rotational strength for each mode of **1**. This approach results in predicted VCD signs in agreement with experiment and provides mechanistic insight into the origin of the VCD intensity.

As second example, this descriptive interpretational approach can be applied to the pair of modes of A and B symmetry near 1350 cm^{-1} in (S,S)-cyclopropane-1,2-$^2H_2$, **3**, and $(2S,3S)$-oxirane-$^2H_2$, **4**, shown schematically in Fig. 5 from the results of normal coordinate calculations.[76, 105] These modes are primarily in- and out-of-phase combinations of the trans CCH deformations in the planes of the methylene groups, and since **3** and **4** are mirror images in terms of deuterium substitution, one might expect that the VCD spectra for the pair of modes will be couplets of opposite sense for the two molecules. The modes in **3** do in fact give rise to a $(-, +)$ VCD couplet, but the modes of **4** both exhibit negative VCD intensity. The modes can again be decomposed into linear combinations of symmetry coordinates of **A** or **B** symmetry of the parent achiral nondeuteriated molecules, in this case methylene scissors and rocking modes. For the modes of A symmetry, the symmetric ring stretch also contributes. For the scissors and rocking coordinates, the primary $(\partial \mu_\nu^B / \partial P_a)_0$ contributions to Eq. (60) are assumed to coincide with the hydrogen displacements,[156] resulting in the net electric- and magnetic-dipole transition moment directions for each mode of the parent molecule as shown in the figures. For oxirane, the ring contraction contribution results in a large $(\partial \mu_\nu^B / \partial P_a)_0$ vector in the opposite direction to the oxygen displacement. The symmetric ring stretch in cyclopropane does not lead to a change in electric or

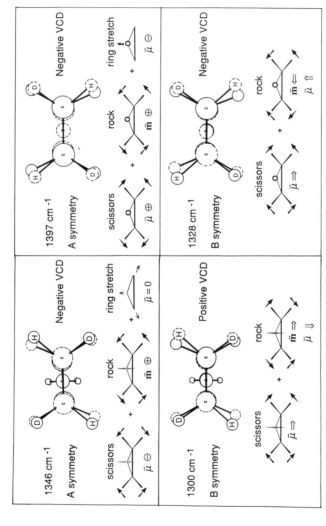

Figure 5. Schematic representations of the modes of $(2S,3S)$-oxirane-$^2H_2$ and (S,S)-cyclopropane-1,2-$^2H_2$ near 1350 cm^{-1} showing a decomposition into contributions from scissors, rock, and symmetric ring stretch symmetry coordinates of the parent nondeuteriated molecules, for the phase of vibration going from the solid to the dotted circles. Also shown are the local electric dipole transition moment contributions (solid arrows) and the directions of the net electric and magnetic dipole transition moments.

magnetic dipole moment; a small contribution from scissors motion at C-3 in the cyclopropane has been ignored here for simplicity.

If only the CHD displacements are considered, the pair of modes in **3** and **4** are indeed predicted to give rise to VCD couplets of opposite sense, based on the relative directions of the electric-dipole transition moment from the methylene scissors contribution and the magnetic dipole transition moment from the methylene rock contribution. However, for the A-symmetry mode in **4**, the electric dipole transition moment from the ring-stretching contribution is opposite in direction and dominates that from the scissors contribution, resulting in antiparallel net electric- and magnetic-dipole transition moments and negative VCD for the mode. In **3**, the ring stretch does not contribute to the electric-dipole transition moment for the A-symmetry mode. In the absorption spectra, the contribution from the ring stretches is reflected in a larger intensity for the A-symmetry mode in **4** compared to the B-symmetry mode, whereas for **3**, the opposite intensity pattern (B > A) occurs. The different absorbance and VCD intensity patterns for these two molecules in this region are thus readily understood from a descriptive interpretation that is motivated by the form of the a priori VCD formalism.

This descriptive mechanistic interpretation also points out a deficiency in generic assignments of modes as, for example, "methylene scissors modes" or "ring deformations," with which VCD intensity may be associated and then compared to a similar generic assignment in another molecule. The sign of VCD intensity is dependent on the phasing of any coupled chiral nuclear motion and on local intrinsic effects; VCD intensity may arise largely from contributions to the mode that do not appear as the major contributions in a potential energy distribution. For example, a mode may be largely ring stretch that is not chiral, but the VCD intensity may be determined by a small contribution from chiral methylene motions. Certainly, an understanding of VCD spectra, as opposed to merely comparing observed and calculated VCD patterns, still requires precise knowledge of the nuclear vibrational motion.

V. CONCLUSIONS

Vibrational circular dichroism spectra can now be calculated with reasonable accuracy with a variety of calculational approaches. Major VCD features are calculated to be large and of the observed sign with the magnetic field perturbation and vibronic coupling theory a priori approaches and with the approximate, localized molecular orbital model, all of which have been implemented for ab initio wave functions. For weaker VCD features, agreement with experiment is less favorable with all three approaches; the MFP calculations appear to yield the best results for the weak modes in some comparisons,[105] whereas the VCT calculations exhibit the better agreement in others.[108] The approximations inherent in the LMO model do result in less accurate calculated VCD magnitudes in some cases, but do not appear to preclude application of the LMO model for the determination of absolute configuration. The VCT implementation has demonstrated that the single transition approximation is sufficient for over-states calculation; excited-state electron correlation does not seem to have a large effect on the calculated VCD intensities. Basis sets with first-derivative functions included are required for the best VCT calculations, but introduction of the 6-31G$^{(0.3)}$ basis set seems promising for application of the VCT approach to larger molecules. The nuclear shielding tensor approach has not been extensively compared to experiment, but these calculations do appear to require both polarized basis sets and the random phase approximation for inclusion of some electron correlation effects in the sum-over-states, possibly because of the larger influence from the squared energy denominator compared to the VCT expression. The MFP approach avoids a sum-over-states, but requires complex wave functions and magnetic field coefficient derivatives obtained via a coupled-Hartree-Fock method, and the results depend on the choice of gauge. The nuclear velocity perturbation velocity-gauge formalism, which is not yet implemented, avoids a magnetic field derivative of the wave function or a sum-over-states; however, the wave functions are also complex and the required atomic orbital coefficient derivatives with respect to nuclear velocity, as derived with coupled-Hartree-Fock methods, involve additional overlap integral and one- and two-electron integral derivatives that vanish in the evaluation of coefficient derivatives with respect to $\mathscr{A}$ and $\mathscr{H}$.

The field perturbation, vibronic coupling, and nuclear shielding calculations all demonstrate that the position form electric dipole transition moments are calculated with much higher accuracy than the velocity form electric-dipole transition moments. Since the position form rotational strength, Eq. (13), is thus preferred, the common origin gauge rotational strengths (determined with Eq. (13)) for any of the a priori methods will be molecular origin dependent for nonexact wave functions. The center of mass remains a good origin choice for such calculations. With the distributed origin gauge, a portion of the rotational strength is calculated with the more accurate position form atomic polar tensor. The distributed origin gauge appears to yield more stable results as the size of the basis set is changed, and in most cases, the rotational strengths calculated with the distributed origin gauge yield the best agreement with experiment. For both MFP and VCT calculations, the magnitudes of the rotational strengths are in general smaller when the DO gauge is employed, compared to the CO gauge results.

Application of the approximate models for VCD has been criticized based on the argument that the models all derive from the APT model expression and it is not theoretically justified to neglect the $\mathbf{P} \cdot \mathbf{M}$ contribution to the rotational strength in Eq. (84).[73] However, as pointed out in Section III, the models for VCD are better viewed in terms of approximations to the locally distributed origin gauge expression, Eq. (60) and the approximations leading to each model are reasonable ones for certain types of vibrational modes. Clearly, application of the APT model yields poor agreement with experiment in most cases, but the LMO model appears to be a good approximation. The danger with using models for VCD intensity is their indiscriminant application. Certainly, not every VCD couplet is due to a coupled oscillator. However, observed VCD couplets are reasonably interpreted in terms of a coupled oscillator mechanism when such coupled motion might be expected based on the most probable molecular conformations and where either dipolar coupling should be strong enough to allow assignment of the phasing of the coupled modes or normal mode calculations support the presence of coupling and provide information on relative phase. Conversely, the absence of an expected couplet does not rule out a contribution from a coupled oscillator mechanism, since a chiral environment for the oscillators can introduce monosignate VCD intensity that can obscure any bisignate character, or the splitting between the modes may be insufficient for a couplet to be observed. Reasonable agreement of FPC model calculations with experiment remains valuable in suggesting an underlying mechanism wherein the dominant vibrationally generated charge flow follows and is localized near the nuclear motion.

There is still room for improvement in the agreement of calculated VCD intensities with experiment for any of the existing calculational methods. It is not yet clear whether the a priori formalisms include all the contributions to VCD intensity, since the largest anisotropy ratios observed for some localized vibrational motions have not yet been reproduced with the current calculational methods. Effort aimed toward efficient applications of all the a priori approaches to larger molecules is still needed. Certainly, none of the present implementations of a priori formalisms need to be discontinued due to a demonstrated inferior agreement with experiment. Similarly, models for VCD intensity remain valuable in understanding the origin of VCD intensity.

Future applications of VCD to larger molecules with multiple solution conformations has clearly been strengthened by the availability and the demonstrated reliability of calculational methods reviewed here.

Acknowledgments

Support of this work through National Institutes of Health Grant GM-23567 (to T. B. Freedman and L. A. Nafie) is gratefully acknowledged. We thank the Northeast Parallel Architectures Center at Syracuse University for access to an IBM RS/6000 workstation for the Gaussian 90 calculations. This work was partially supported by the National Center for Supercomputing Applications under Grant TRA920315N and utilized the CRAY Y-MP4/464 at the National Center for Supercomputing Applications, University of Illinois at Urbana-Champaign. We thank Denise Gigante for carrying out some of the calculations.

References

1. L. A. Nafie, *Adv. Infrared Raman Spectrosc.* **11**, 49 (1984).

2. P. J. Stephens and M. A. Lowe, *Annu. Rev. Phys. Chem.* **36**, 213 (1985).

3. T. B. Freedman and L. A. Nafie, in E. L. Eliel and S. H. Wilen (Eds.) *Topics in Stereochemistry*, Vol. 17, Wiley, New York, 1987, p. 113.

4. T. M. Keiderling, in J. R. Ferraro and K. Krishnan (Eds.), *Practical Fourier Transform Spectroscopy*, Academic, San Diego, 1990, p. 203.

5. P. L. Polavarapu, in H. D. Bist, J. R. Durig and J. F. Sullivan (Eds.), *Vibrational Spectra and Structure*, Vol. 17B, Elsevier, Amsterdam, 1989, p. 319.

6. A. Rauk, in P. B. Mezey (Ed.), *New Developments in Molecular Chirality*, Kluwer Academic, Dordrecht, The Netherlands, 1991, p. 57.

7. E. C. Hsu and G. Holzwarth, *J. Chem. Phys.* **59**, 4678 (1973).

8. G. Holzwarth, E. C. Hsu, H. S. Mosher, T. R. Faulkner, and A. Moskowitz, *J. Am. Chem. Soc.* **96**, 251 (1974).

9. L. A. Nafie and T. B. Freedman, *J. Chem. Phys.* **78**, 7108 (1983).

10. P. J. Stephens, American Chemical Society Meeting, San Francisco, Nov. 1983, Paper GC1.

11. P. J. Stephens, *J. Phys. Chem.* **89** 784 (1985).

12. P. J. Stephens, *J. Phys. Chem.* **91**, 1712 (1987).

13. P. A. Galwas, Thesis, University of Cambridge, Cambridge, UK, 1983.

14. A. D. Buckingham, P. W. Fowler, and P. A. Galwas, *Chem. Phys.* **112**, 1 (1987).

15. P. Lazzeretti, R. Zanasi, and P. J. Stephens, *J. Phys. Chem.* **90**, 6761 (1986).

16. L. A. Nafie and T. B. Freedman, *Chem. Phys. Lett.* **134**, 225 (1987).

17. T. B. Freedman and L. A. Nafie, *J. Chem. Phys.* **89**, 374 (1988).

18. R. D. Amos, K. J. Jalkanen, and P. J. Stephens, *J. Phys. Chem.* **92**, 5571 (1988).

19. Aa. E. Hansen, P. J. Stephens, and T. D. Bouman, *J. Phys. Chem.* **95**, 4255 (1991).

20. K. L. C. Hunt and R. A. Harris, *J. Chem. Phys.* **94**, 6995 (1991).

21. D. P. Craig and T. Thirunamachandran, *Can. J. Chem.* **63**, 1773 (1985).

22. L. A. Nafie, *J. Chem. Phys.* **96**, 5687 (1992).

23. M. A. Lowe, G. A. Segal, and P. J. Stephens, *J. Am. Chem. Soc.* **108**, 248 (1986).

24. R. D. Amos, N. C. Handy, K. J. Jalkanen, and P. J. Stephens, *Chem. Phys. Lett.* **133**, 21 (1987).

25. K. Morokuma and H. Sugeta, *Chem. Phys. Lett.* **134**, 23 (1987).

26. R. D. Amos, N. C. Handy, A. F. Drake, and P. Palmieri, *J. Chem. Phys.* **89**, 7287 (1988).

27. M. A. Lowe and J. S. Alper, *J. Phys. Chem.* **92**, 4035 (1988).

28. H. Dothe, M. A. Lowe, and J. S. Alper, *J. Phys. Chem.* **92**, 6246 (1988).

29. R. Dutler and A. Rauk, *J. Am. Chem. Soc.* **111**, 6957 (1989).

30. A. Rauk, R. Dutler, and D. Yang, *Can. J. Chem.* **68**, 258 (1990).

31. A. Rauk and D. Yang, *J. Phys. Chem.* **96**, 437 (1992).

32. K. J. Jalkanen, P. J. Stephens, P. Lazzeretti, and R. Zanasi, *J. Chem. Phys.* **90**, 3204 (1989).

33. K. J. Jalkanen, P. J. Stephens, R. D. Amos, and N. C. Handy, *J. Phys. Chem.* **92**, 1781 (1988).

34. G. Holzwarth and I. Chabay, *J. Chem. Phys.* **57**, 1632 (1972).

35. H. Sugeta, C. Marcott, T. R. Faulkner, J. Overend, and A. Moscowitz, *Chem. Phys. Lett.* **40**, 397 (1976).

36. C. W. Deutsche and A. Moscowitz, *J. Chem. Phys.* **49**, 3257 (1968); **53**, 2630 (1970).

37. J. A. Schellman, *J. Chem. Phys.* **58**, 2882, (1973); **60**, 343 (1974).

38. L. A. Nafie and T. H. Walnut, *Chem. Phys. Lett.* **49**, 441 (1977); T. H. Walnut and L. A. Nafie, *J. Chem. Phys.* **67**, 1501 (1977).

39. L. A. Nafie and P. L. Polavarapu, *J. Chem. Phys.* **75**, 2935 (1981).

40. C. J. Barnett, A. F. Drake, R. Kuroda, and S. F. Mason, *Mol. Phys.* **41**, 455 (1980).

41. S. Abbate, L. Laux, J. Overend, and A. Moscowitz, *J. Chem. Phys.* **75**, 3161 (1981).

42. M. Moskovits and A. Gohin, *J. Phys. Chem.* **86**, 3947 (1982).

43. T. B. Freedman and L. A. Nafie, *J. Chem. Phys.* **78**, 27 (1983); **79**, 1104 (1983).

44. T. B. Freedman and L. A. Nafie, *J. Phys. Chem.* **88**, 496 (1984).

45. L. D. Barron, *Molecular Light Scattering and Optical Activity*, Cambridge University Press, Cambridge, UK, 1982, p. 317.

46. P. L. Polavarapu, *Mol. Phys.* **49**, 645 (1983).

47. J. R. Escribano, T. B. Freedman, and L. A. Nafie, *J. Phys. Chem.* **91**, 46 (1987).

48. L. A. Nafie and T. B. Freedman, *J. Phys. Chem.* **90**, 763 (1986).

49. P. L. Polavarapu and P. K. Bose, *J. Phys. Chem.* **95**, 1606 (1991).

50. P. L. Polavarapu and P. K. Bose, *J. Chem. Phys.* **93**, 7524 (1990).

51. L. A. Nafie and T. B. Freedman, *J. Mol. Struct.* **224**, 121 (1990).

52. P. A. M. Dirac, *Quantum Mechanics*, 4th ed., Clarendon, Oxford, 1958, p. 244.

53. W. B. Person and J. H. Newton, *J. Chem. Phys.* **61**, 1040 (1974).

54. T. B. Freedman and L. A. Nafie, *Chem. Phys. Lett.* **126**, 441 (1986).

55. Ae. E. Hansen and T. D. Bouman, in I. Prigogine and S. A. Rice (Eds.), *Advances in Chemical Physics*, Vol. 44, Wiley, New York, 1980, p. 545.

56. T. B. Freedman, S. J. Cianciosi, N. Ragunathan, J. E. Baldwin, and L. A. Nafie, *J. Am. Chem. Soc.* **113**, 8298 (1991).

57. L. A. Nafie, *J. Chem. Phys.* **79**, 4950 (1983).

58. R. D. Amos, *Adv. Chem. Phys.* **67**, 99 (1987).

59. H. Hellmann, *Einfuhring in Die Quantenchemie*, Franz Deuticke, Leipzig, 1937, p. 285; R. P. Feynman, *Phys. Rev.* **56**, 340 (1939).

60. J. A. Pople, J. W. McIver, Jr., and N. S. Ostlund, *J. Chem. Phys.* **49**, 2960 (1968).

61. (a) A. D. Liehr, *Z. Naturforsch. Teil A* **13**, 311 (1958); (b) *Can. J. Phys.* **35**, 1123 (1957); **36**, 1588 (1958).

62. W. C. Johnson, Jr., and O. E. Weigang, Jr., *J. Chem. Phys.* **63**, 2135 (1975).

63. P. Lazzeretti and R. Zanasi, *Chem. Phys. Lett.* **112**, 103 (1984).

64. P. W. Fowler and A. D. Buckingham, *Chem. Phys.* **98**, 167 (1985).

65. P. Lazzeretti and R. Zanasi, *J. Chem. Phys.* **83**, 1218 (1985).

66. P. Lazzeretti and R. Zanasi, *J. Chem. Phys.* **84**, 3916 (1986).

67. R. Zanasi and P. Lazzeretti, *J. Chem. Phys.* **85**, 5932 (1986).

68. N. V. Cohan and H. F. Hameka, *J. Am. Chem. Soc.* **88**, 2136 (1966).

69. W. T. King and G. B. Mast, *J. Chem. Phys.* **80**, 2521 (1976).

70. R. F. W. Bader, A. Larouche, C. Gatti, M. T. Carroll, P. J. MacDougall, and K. B. Wiberg, *J. Chem. Phys.* **87**, 1142 (1987).

71. F. Maurer and H. Wieser, *J. Phys. Chem.* (in press).

72. A. Rupprecht, *Mol. Phys.* **63**, 951 (1988).

73. P. J. Stephens, K. J. Jalkanen, and R. W. Kawiecki, *J. Am. Chem. Soc.* **112**, 6518 (1990).

74. J. S. Binkley, R. A. Whiteside, R. Krishnan, R. Seeger, D. J. Defrees, J. B. Schlegel, S. Topiol, L. R. Kahn, and J. A. Pople, *Gaussian 80*, Quantum Chemistry Program Exchange, Indiana University, Bloomington, IN.

75. S. J. Cianciosi, K. M. Spencer, T. B. Freedman, L. A. Nafie, and J. E. Baldwin, *J. Am. Chem. Soc.* **111**, 1913 (1989).

76. T. B. Freedman, N. Ragunathan, K. M. Spencer, S. J. Cianciosi, and J. E. Baldwin (in preparation).

77. J. L. Duncan and G. R. Burns, *J. Mol. Spectrosc.* **30**, 253 (1969).

78. M. A. Lowe, P. J. Stephens, and G. A. Segal, *Chem. Phys. Lett.* **123**, 108 (1986).

79. H. Dothe, M. A. Lowe, and J. S. Alper, *J. Phys. Chem.* **93**, 6632 (1989).

80. R. M. Stevens, R. M. Pitzer, and W. N. Lipscomb, *J. Chem. Phys.* **38**, 550 (1963).

81. J. Gerratt and I. M. Mills, *J. Chem. Phys.* **49**, 1719 (1968).

82. J. A. Pople, R. Krishnan, H. B. Schlegel, and J. S. Binkley, *Int. J. Quantum Chem. Symp.* **13**, 225 (1979).

83. R. D. Amos, *The Cambridge Analytic Derivatives Package*, Publication CCP 1/84/4, SERC, Daresbury Laboratory, Daresbury, Warrington, UK, 1984.

84. K. J. Jalkanen, P. J. Stephens, R. D. Amos, and N. C. Handy, *Chem. Phys. Lett.* **142**, 153 (1987).

85. K. J. Jalkanen, P. J. Stephens, R. D. Amos, and N. C. Handy, *J. Am. Chem. Soc.* **109**, 7193 (1987).

86. P. Malon, T. A. Keiderling, J.-Y. Uang, and J. S. Chickos, *Chem. Phys. Lett.* **179**, 282 (1991).

87. K. J. Jalkanen, P. J. Stephens, R. D. Amos, and N. C. Handy, *J. Am. Chem. Soc.* **110**, 2012 (1988).

88. K. J. Jalkanen, R. W. Kawiecki, P. J. Stephens, and R. D. Amos, *J. Phys. Chem.* **94**, 7040 (1990).

89. R. W. Kawiecki, F. J. Devlin, P. J. Stephens, R. D. Amos, and N. C. Handy, *Chem. Phys. Lett.* **145**, 411 (1988).

90. P. Malon, L. J. Mickley, K. M. Sluis, C. N. Tam, T. A. Keiderling, S. Kamath, J. Uang, and J. S. Chickos, *J. Phys. Chem.* (in press).

91. R. Bursi, F. J. Devlin, and P. J. Stephens, *J. Am. Chem. Soc.* **112**, 9430 (1990).

92. R. Bursi and P. J. Stephens, *J. Phys. Chem.* **95**, 6447 (1991).

93. A. Annamalai, K. J. Jalkanen, U. Narayanan, M.-C. Tissot, T. A. Keiderling, and P. J. Stephens, *J. Phys. Chem.* **94**, 194 (1990).

94. P. Bour and T. A. Keiderling, personal communication.

95. R. D. Amos, N. C. Handy, and P. Palmieri, *J. Chem. Phys.* **93**, 5796 (1990).

96. P. J. Stephens, *Biomolecular Spectroscopy II Symposium*, SPIE, Los Angles, January 1991, paper 1432-05.

97. C. W. McCurdy, Jr., T. N. Rescigno, D. L. Yeager, and V. McKoy, in H. F. Schaefer III (Ed.), *Methods of Electron Structure Theory*, Plenum, New York, 1977, p. 339.

98. P. J. Stephens, K. J. Jalkanen, R. D. Amos, P. Lazzeretti, and R. Zanasi, *J. Phys. Chem.* **94**, 1811 (1990).

99. K. J. Jalkanen, P. J. Stephens, P. Lazzeretti, and R. Zanasi, *J. Phys. Chem.* **93**, 6583 (1989).

100. T. D. Bouman and Aa. E. Hansen, *Int. J. Quantum Chem. Symp.* **23**, 381 (1989).

101. M. J. Frisch, M. Head-Gordon, B. W. Trucks, J. B. Foresman, H. B. Schlegel, K. Raghavachari, M. A. Robb, J. S. Binkley, C. Gonzalez, D. J. Defrees, D. J. Fox, R. A. Whiteside, R. Seeger, L. R. Martin, J. Baker, L. R. Kahn, J. J. Stewart, S. Topiol, and J. A. Pople, *Gaussian 90*, Gaussian, Inc., Pittsburgh, PA, 1990.

102. T. B. Freedman and L. A. Nafie, unpublished results.

103. R. Dutler, Thesis, The University of Calgary, 1988.

104. J. S. Binkley, M. J. Frisch, D. J. Defrees, K. Raghavachari, R. A. Whiteside, H. B. Schlegel, E. M. Fleuder, and J. A. Pople, *Gaussian 82*, Carnegie-Mellon Publishing Unit, Pittsburgh, PA, 1982.

105. T. B. Freedman, K. M. Spencer, N. Ragunathan, L. A. Nafie, J. A. Moore, and J. M. Schwab, *Can. J. Chem.* **69**, 1619 (1991).

106. R. A. Shaw, H. Wieser, R. Dutler, and A. Rauk, *J. Am. Chem. Soc.* **112**, 5401 (1990).

107. M. J. Frisch, J. S. Binkley, H. B. Schlegel, K. Raghavachari, C. F. Melius, L. R. Martin, J. J. Stewart, F. W. Bobrowicz, C. M. Rohlfing, L. R. Kahn, D. J. Defrees, R. Seeger, R. A. Whiteside, D. J. Fox, E. M. Fleuder, and J. A. Pople, *Gaussian 86*, Carnegie-Mellon Publishing Unit, Pittsburgh, PA, 1986.

108. D. Yang, Thesis, The University of Calgary (1992).

109. S. T. Pickard, H. E. Smith, P. L. Polavarapu, T. M. Black, A. Rauk, and D. Yang, *J. Am. Chem. Soc.* **114**, 6850 (1992).

110. P. L. Polavarapu, S. T. Pickard, H. E. Smith, T. M. Black, A. Rauk, and D. Yang, *J. Am. Chem. Soc.* **113**, 9747 (1991).

111. D. Yang and A. Rauk, *J. Chem. Phys.* **97**, 6517 (1992).

112. I. Tinoco, *Radiation Res.* **20**, 133 (1963).

113. D. A. Young, E. D. Lipp, and L. A. Nafie, *J. Am. Chem. Soc.* **107**, 6205 (1985).

114. A. C. Chernovitz, T. B. Freedman, and L. A. Nafie, *Biopolymers* **26**, 1879 (1987).

115. M. G. Paterlini, T. B. Freedman, L. A. Nafie, Y. Tor, and A. Shanzer, *Biopolymers* **32**, 765 (1992).

116. U. Narayanan and T. A. Keiderling, *J. Am. Chem. Soc.* **105**, 6406 (1983).

117. A. Annamalai and T. A. Keiderling, *J. Am. Chem. Soc.* **109**, 3125 (1987).

118. U. Narayanan, T. A. Keiderling, C. J. Elsevier, P. Vermeer, and W. Runge, *J. Am. Chem. Soc.* **110**, 4133 (1988).

119. M. Gulotta, D. J. Goss, and M. Diem, *Biopolymers* **28**, 2047 (1989).

120. W. Zhong, M. Gulotta, D. J. Goss, and M. Diem, *Biochemistry* **29**, 7485 (1990).

121. O. Lee, G. M. Roberts, and M. Diem, *Biopolymers* **28**, 1759 (1989).

122. S. S. Birke, I. Agbaje, and M. Diem, *Biochemistry* **31**, 450 (1992).

123. P. L. Polavarapu, C. S. Ewig, and T. Chandramouly, *J. Am. Chem. Soc.* **109**, 7382 (1987).

124. T. R. Faulkner, Thesis, University of Minnesota, 1976.

125. A. Annamalai, T. A. Keiderling, and J. S. Chickos, *J. Am. Chem. Soc.* **106**, 6254 (1984).

126. A. Annamalai, T. A. Keiderling, and J. S. Chickos, *J. Am. Chem. Soc.* **107**, 2285 (1985).

127. T. B. Freedman, J. Kallmerten, E. D. Lipp, D. A. Young, and L. A. Nafie, *J. Am. Chem. Soc.* **110**, 689 (1988).

128. R. A. Shaw, N. Ibrahim, and H. Wieser, *J. Phys. Chem.* **94**, 125 (1990).

129. T. B. Freedman, M. Diem. P. L. Polavarapu, and L. A. Nafie, *J. Am. Chem. Soc.* **104**, 3343 (1982).

130. M. W. Schmidt, J. A. Boatz, K. K. Baldridge, S. Koseki, M. S. Gordon, S. T. Elbert, and B. Lam, *QCPE Bull.* **7**, 115 (1987); M. Dupuis, D. Spangler, and J. J. Wendoloski, NRCC Program QG01, University of California, Berkeley, CA, 1980.

131. J. M. Foster and S. F. Boys, *Rev. Mod. Phys.* **32**, 300 (1960).

132. P. L. Polavarapu, P. K. Bose, and S. T. Pickard, *J. Am. Chem. Soc.* **113**, 43 (1991).

133. P. L. Polavarapu, S. T. Pickard, H. E. Smith, and R. S. Pandurangi, *Talanta*, **40**, 687 (1993).

134. P. L. Polavarapu, A. L. Cholli, and G. Vernice, *J. Am. Chem. Soc.*, **114**, 10953 (1992).

135. P. L. Polavarapu, A. L. Cholli, and G. Vernice, *J. Pharm. Sci*, in press.

136. S. Abbate, H. A. Havel, L. Laux, V. Pultz, and A. Moscowitz, *J. Phys. Chem.* **92**, 3302 (1988).

137. S. Abbate, L. Laux, V. Pultz, H. A. Havel, J. Overend, and A. Moscowitz, *Chem. Phys. Lett.* **113**, 202 (1985).

138. T. Eggimann, Thesis, The University of Calgary, 1991.

139. T. Eggimann and H. Wieser, 75th Canadian Chemical Conference, Edmonton Alberta, 1992, poster 775.

140. F. Maurer and H. Wieser, 75th Canadian Chemical Conference, Edmonton Alberta, 1992, poster 776.

141. P. Pulay, G. Fogarasi, F. Pang, and J. Boggs, *J. Am. Chem. Soc.* **101**, 2550 (1979).

142. L. A. Nafie, M. R. Oboodi, and T. B. Freedman, *J. Am. Chem. Soc.* **105**, 7449 (1983).

143. T. B. Freedman, G. A. Balukjian, and L. A. Nafie, *J. Am. Chem. Soc.* **107**, 6213 (1985).

144. M. G. Paterlini, T. B. Freedman, and L. A. Nafie, *J. Am. Chem. Soc.* **108**, 1389 (1986).

145. W. M. Zuk, T. B. Freedman, and L. A. Nafie, *J. Phys. Chem.* **93**, 1171 (1989).

146. W. M. Zuk, T. B. Freedman, and L. A. Nafie, *Biopolymers* **28**, 2025 (1989).

147. N. Ragunathan, Thesis, Syracuse University, 1991.

148. D. A. Young, T. B. Freedman, and L. A. Nafie, *J. Am. Chem. Soc.* **109**, 7674 (1987).

149. D. A. Young, T. B. Freedman, E. D. Lipp, and L. A. Nafie, *J. Am. Chem. Soc.* **108**, 7255 (1986).

150. T. B. Freedman, D. A. Young, M. R. Oboodi, and L. A. Nafie, *J. Am. Chem. Soc.* **109**, 1551 (1987).

151. T. B. Freedman, A. C. Chernovitz, W. M. Zuk, M. G. Paterlini, and L. A. Nafie, *J. Am. Chem. Soc.* **119**, 6970 (1988).

152. D. M. Gigante, J. E. Evans, T. B. Freedman, and L. A. Nafie, 46th Ohio State University International Symposium on Molecular Spectroscopy, June 1991, paper WG3; D. M. Gigante, Thesis, Syracuse University, 1993.

153. R. W. Bormett, S. A. Asher, P. J. Larkin, W. G. Gustafson, N. Ragunathan, T. B. Freedman, L. A. Nafie, S. Balasubramanian, S. G. Boxer, N. Yu, K. Gersonde, R. W. Noble, B. A. Springer, and S. G. Sligar, *J. Am. Chem. Soc.* **114**, 6864 (1992).

154. D. M. Gigante, T. B. Freedman, C. S. Swindell, L. E. Chirlian, M. M. Francl, J. M. Heerding, and N. E. Krauss, submitted.

155. P. Bour and T. A. Keiderling, *J. Am. Chem. Soc.* **114**, 9100 (1992).

156. C. Hirose, *Bull. Chem. Soc. Japan* **47**, 1311 (1974).

157. M. A. Lowe, J. S. Alper, R. Kawiecki, and P. J. Stephens, *J. Phys. Chem.* **90**, 41 (1986).

158. K. B. Wiberg and J. J. Wendoloski, *J. Phys. Chem.* **88**, 586 (1984).

SIMULATION OF CRITICAL PHENOMENA
IN NONLINEAR OPTICAL SYSTEMS

M. I. DYKMAN

Department of Physics, Stanford University, Stanford, California

D. G. LUCHINSKY

*All Russian Research Institute for Metrological Service,
Moscow, Russia*

R. MANNELLA

Dipartimento di Fisica, Università di Pisa, Pisa, Italy

and

P. V. E. McCLINTOCK, N. D. STEIN AND N. G. STOCKS*

*School of Physics and Materials, Lancaster University,
Lancaster, UK*

CONTENTS

*Present address: Department of Engineering, University of Warwick, Coventry, UK.

Modern Nonlinear Optics, Part 3, Edited by Myron Evans and Stanisław Kielich. Advances in Chemical Physics Series, Vol. LXXXV.
ISBN 0-471-30499-9 © 1994 John Wiley & Sons, Inc.

I. INTRODUCTION

A characteristic feature of nonlinear science generally, and of nonlinear optics in particular, is the relative difficulty of constructing a satisfactory theoretical model of any given phenomenon. The aim, of course, is to create a model that captures the essential physics of the problem, incorporating the most important mechanisms while ignoring the less significant ones. Even when this step has been accomplished, the existence of the nonlinearity usually necessitates the introduction of additional approximations and simplifications in order for one to be able to solve the equations even of the simplified model. These considerations apply a fortiori to the study of fluctuation phenomena in nonlinear systems, and thus account for the increasing role being played by analogue and digital simulations, which enable the behavior of the model systems to be investigated in considerable detail.

In this chapter, we discuss the application of simulation techniques to the study of *critical phenomena* in nonlinear optical systems: that is, situations where a very marked change occurs in the state of the system as the result of a relatively small change in one of the parameters that describes it. The topics to be discussed have been selected mainly for their own intrinsic scientific interest, but also to provide an indication of the power and utility of the simulation approach as a means of focusing on, and reaching an understanding of, the essential physics underlying the phenomena under investigation; they also provide examples of different theoretical approaches. Although the different sections all share the same general theme of critical phenomena in model nonlinear optical systems, they deal with quite different aspects of the subject. Thus, each section is to a considerable extent self-contained and can be read almost independently of the others. Before considering particular systems, we review briefly the scientific context of the work and discuss in a general way the significance of critical phenomena in nonlinear optics.

The investigation of fluctuations by means of analogue or digital simulation is usually found most useful for those systems where the fluctuations of the quantities of immediate physical interest can be assumed to be due to noise. The latter perception of fluctuations goes back to Einstein,[1] Smoluchowski,[2] and Langevin[3] and has been often used in optics (cf. Refs. 4–7). In nonlinear optics, the noise can be regarded as arising from two main sources. First there are internal fluctuations in the macroscopic system itself. These arise because spontaneous emission of light by individual atoms occurs at random, e.g., because of fluctuations of the populations of atomic energy levels. The physical characteristics of such noise are usually immediately related to the physical characteristics of the model that describes the "regular" dynamics of the system, i.e., in the absence of noise. In particular, the power spectrum of thermal noise and its intensity can be expressed in terms of the dissipation characteristics via the fluctuation–dissipation relations (cf. Ref. 8) and, if the dissipation is nonretarded so that the corresponding dissipative forces (e.g., the friction force) depend only on the instantaneous values of dynamical variables, the noise power spectrum is independent of frequency, i.e., the noise is white. The model of noise as being white and Gaussian is one of the most commonly used in optics because the quantities of physical interest often vary slowly compared with the fast random processes that give rise to the noise, like emission or absorption of a photon.[4-7] The second important source of noise is external: for example, fluctuations of the pump power in a laser. The physical characteristics of such noise naturally vary from one system to another; its correlation time is often much longer than that of the internal noise, and its effects can be large and sometimes unexpected (cf. Ref. 9).

In general, the fluctuations observed in nonlinear optics are both spatial and temporal; i.e., the variations of the quantities of interest occur to a large extent independently in time and in space. However, in many cases the spatial modes in a system are well separated and the dynamics of interest is then just that of a few dominant modes. The appearance of such modes is typical for high-Q active and passive optical cavities. In view of recent progress in microelectronics (quantum-dot technology, semiconductor-laser arrays, etc.), the investigation of systems with a discrete set of spatial structures (modes) is particularly interesting and important.[10] The amplitudes and phases of the actual modes (or other appropriate characteristics of a system that do not depend on coordinates) make a set of purely *dynamical* variables, and the analysis of fluctuations in a system reduces to the investigation of the kinetics of a *dynamical system*.

One of the hot topics of nonlinear optics over the last decade has been optical bistability (OB) (cf. Refs. 11–14). The phenomenon consists in systems possessing two (or more, in the case of optical multistability) coexisting stable states for given values of the control parameters. These states can be different modes of lasing (e.g., clockwise and counterclockwise propagating modes in a symmetrical ring laser) or different states of transmission in a nonlinear passive optical cavity (e.g., in a Fabry-Pérot cavity filled with a medium displaying nonlinear refraction so that, for a given intensity of the incident radiation, the intracavity radiation can be nonresonant and have low intensity or, alternatively, can be of high intensity and nearly resonant because of the appropriate self-consistent value taken by the intensity-dependent refractive index of the medium). The dependence of the intensity I_T of the transmitted radiation as a function of the intensity I of the incident radiation for a bistable Fabry-Pérot cavity is shown schematically in Fig. 1. In the absence of fluctuations I_T moves along the lower branch in Fig. 1a when I increases slowly from small values. At $I \doteq I_{B2}$ this branch ends, and the system jumps to the upper branch. If, on the contrary, I is decreasing from high values

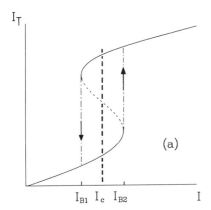

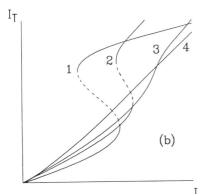

Figure 1. Sketches of transmitted light intensity I_T against input light intensity I, to illustrate optical bistability. (a) The system exhibits hysteresis, jumping up the dash-dot line at I_{B2} as I is increased, and jumping down the left hand one at I_{B1} as I is decreased. (b) As a control parameter is varied, the system can move from a situation where there is a range of bistability (curves 1 and 2) to one where there is not (curves 3 and 4). In both cases, the dashed curves represent regions of instability. The heavy dashed line in (a) indicates the position of the kinetic phase transition where the populations of the upper and lower branches are equal.

the system moves along the upper branch until it ends at $I = I_{B1}$, and then the system jumps down to the lower branch. Thus, the dependence of I_T on I displays hysteresis. Similar hysteretic behavior is also found to occur in optically bistable systems of other kinds. Figure 1b sketches a typical way in which monostability can give way to bistability as a control parameter is varied.

It is evident that fluctuations will change the above picture dramatically because they can give rise to fluctuational transitions between coexisting stable states on the upper and lower branches. As a result of these transitions the system will eventually "forget" its prehistory and a distribution over the stable states will be formed (for sufficiently weak fluctuations where the average fluctuational displacement from a stable state is small). It is important to note, however, that the probability of switching usually depends exponentially strongly on the number of degrees of freedom involved (cf. Ref. 8 where thermal equilibrium systems are considered).

One would therefore expect two-mode lasers[14] to provide prime candidate systems in which to observe fluctuational transitions and to investigate the distribution over the stationary states. Transitions of this kind, and the onset of a stationary distribution over the stable states, have indeed been observed in such lasers.[15, 16] On the other hand, in all-optical passive bistable systems with plane-parallel nonlinear Fabry-Pérot cavities, the probabilities of transitions are extremely small because of the in-plane spatial degrees of freedom. The first observations for passive bistable systems refer to hybrid devices,[17] and the initial results for all-optical systems[18] refer to the transient regime where the incident-light intensity I was driven through the bifurcational value I_{B1} and, as a result of fluctuations, the system was observed to occupy within a certain period of time both the low- and high-transmission states. This interesting phenomenon of *transient bimodality*[19] has also been observed in a laser with a saturable absorber.[20] A stationary distribution over the stable states of a passive, all-optical multistable system and fluctuational transitions between those states, has also been observed in Ref. 21. The system was a double-cavity: The first cavity was a semiconductor membrane, and the second was an air-spaced gap between this membrane and a dielectric mirror. The nonlinearity was due basically to bending of the membrane as a result of heating by radiation. The extent of the heating depended on the resonance conditions in the second cavity,[22] and the bending of the membrane was therefore the only degree of freedom of the system under the conditions studied.

One would expect that in a fluctuating bistable system, because of the fluctuational transitions between the states, there might arise some sort of critical phenomena similar to those in thermal equilibrium systems that experience a first-order phase transition and where there are also two coexisting stable states (phases). The reason is that the difference between the values taken by a physical quantity in the stable states is generally very much larger, for weak noise, than the fluctuational smearing of this quantity in the stable states. Therefore, for those parameter values of the system where the populations w_1, w_2 of both stable states 1, 2 are of the same order of magnitude (close to $I = I_c$ in Fig. 1a, for example), the fluctuations increase enormously. A strong increase of fluctuations of I_T in optically bistable systems for the corresponding range of parameters (the range of a *kinetic phase transition*[23]) was considered in Ref. 24.

The behavior of a system in the range of a kinetic phase transition (KPT) is determined by the following considerations. First, the probabilities W_{mn} of the fluctuational transitions ($n, m = 1, 2$ enumerate the stable states) are much smaller (exponentially smaller for small noise intensities) than the reciprocal relaxation time of the system t_r^{-1} and the reciprocal correlation time of the noise t_c^{-1}. Therefore, the kinetics of fluctuations related to the transitions between the states is *slow*. This means in particular that supernarrow spectral peaks will arise both in the spectral density of the fluctuations and in the susceptibility of the system, i.e., in the frequency dependence of the response of the system to an external force.[23, 25-28] In both cases, the characteristic width of these peaks is of order $W_{mn} \ll t_r^{-1}$. The second point to be borne in mind is that W_{12} and W_{21} differ strongly from each other for nearly all values of the parameters of the system. It is only within the KPT range that they become comparable and the aforementioned peaks would be expected to occur.

An interesting and, at first sight, highly counterintuitive phenomenon that arises in its most pronounced form in the KPT range is that of *stochastic resonance*: the noise-induced enhancement of the signal induced by a periodic force; it is closely associated with the onset of the supernarrow peaks. It has been observed in both active[29] and passive[30] optical systems, as well as many other areas of physics, engineering and biology.[31] It arises because, when the populations of coexisting states are comparable (as is the case in the KPT range), a relatively weak periodic force can substantially unbalance them, periodically. The net effect can thus be a relatively large signal, because its amplitude is then related, not to the amplitude of forced vibrations about one or other of the stable states, but to the *half-separation* of the stable states. The effect clearly comes about directly through the agency of noise: In the absence of noise there would be no transitions, and the concept of a modulation of the state populations would consequently be meaningless.

It is well known in statistical physics that systems that undergo first-order phase transitions exhibit special behavior at the critical point on the phase diagram where the coexisting phases merge. In the case of the Van der Waals type family of curves in Fig. 1b, the critical point would be represented by the point of inflection on the critical $I_T(I)$ curve that would fall between curves 2 and 3. The appearance of such a point of inflection can be regarded as precursor of the onset of bistability for other parameter values (as, for example, curves 1 and 2). It is to be expected that, just as for thermal equilibrium systems, large fluctuations will occur in the vicinity of the critical point because the effective "stiffness" of the system is then small (cf. critical opalescence). Such fluctuations would be expected to be strongly dependent on the type of noise that gives rise to them, and here an interesting question arises. On one hand, the system is slow (since, formally, its relaxation time diverges at the critical point), suggesting that almost any noise can reasonably be regarded as white. On the other hand, the system is also extremely "soft" so that it is easily driven away from the actual critical point toward a parameter range where its relaxation time is finite. Therefore, in principle (and also in practice, as we shall see), interesting effects can arise when a system in its critical state is driven by colored noise, effects that depend on an interrelation between the noise intensity and its correlation time.

Specific transient phenomena, which can also be regarded as critical, arise in systems where a control parameter is switched through some critical value on the opposite sides of which the behavior of the system is completely different (cf. thermal systems that experience a second-order phase transition). The problem is a very real one for lasers, for example, where, depending on the pump power, there either is, or is not, coherent light generation. The control parameter can either be swept, usually at a constant rate, or it can be switched extremely rapidly (almost discontinuously) from being below to being above a critical value. In cases where the system is initially stable (as a dynamical system at a potential minimum) it may end up in an unstable position exactly on a potential maximum after the relevant parameter has been swept or switched. The role of fluctuations in such a situation is clearly crucial because, in the absence of fluctuations, nothing would happen and the system would remain indefinitely in its unstable state so that, for example, a laser would never start up. The time taken for the system to leave that state, perhaps to attain stability in some new configuration, is bound to depend on the intensity and also, as we shall see, on the color of the external noise.

The plan of the rest of the chapter is as follows. In Section II we review briefly the basis of the simulation techniques themselves and discuss some of their advantages and disadvantages. Section III describes an investigation of the effect of colored (exponentially correlated) noise on stationary distributions in a model of dispersive optical bistability. It is shown in particular that, close to a critical point, a noise-induced transition from a monomodal to a bimodal distribution can result from a change of noise color alone. Some examples of time-evolving stochastic systems, relevant to the laser start-up problem, are then treated. Section IV discusses the decay of unstable states in terms of the nonlinear relaxation time (NLRT). Section V discusses swept parameter systems in terms of the mean first passage time and of the evolution of moments. In each case, the role played by noise color is explored both by simulation and in theory. Simulations of conventional stochastic resonance, in which a weak periodic signal in a bistable optical system can be amplified by the addition of external noise, are discussed in Section VI. In Section VII, results for the nearly resonantly driven single-well nonlinear oscillator are reviewed. It is a system that serves as an archetype for a variety of fluctuational phenomena in nonlinear optics, displays a striking example of a kinetic phase transition, and has made it possible to demonstrate the occurrence of stochastic resonance for periodic attractors. Section VIII describes simulations of an optical system with a multivalued potential, of relevance to phenomena occurring in the double optical cavity mentioned above. Sections III–VIII provide a range of examples in which the "tripod" of analytic approximation/analogue experiment/digital simulation enables progress to be made, with reliable results, in a relatively difficult field of study. Finally, in Section IX, we summarize the results and consider future perspectives.

II. ANALOGUE AND DIGITAL SIMULATION TECHNIQUES

In this section we review the techniques available for modeling nonlinear optical systems by analogue and digital simulation. Because full accounts have already been given elsewhere,[32-34] the discussion will be relatively brief. It is intended mainly to set the scene and to introduce the ideas in question to readers not yet familiar with them. Recent developments in the area are also summarized. We deal first with analogue simulation, then with digital simulation, and finally we offer a few thoughts, based on experience concerning the relative advantages and disadvantages of these two very different approaches.

A. Analogue Electronic Simulation

The use of electrical circuits to study stochastic differential equations of the kind that often occur in nonlinear optics, for example, Langevin equations, is well established. This approach was employed, for example, by Landauer[35] and by Stratonovich.[36, 37] It was subsequently developed and used by many other workers: We would mention particularly Morton and Corrsin,[38] Kawakubo et al.,[39] Arecchi and his coworkers,[25, 40] Sancho et al.,[41] Lange and his coworkers (see Ref. 42 and references therein) and Fauve and Heslot,[43] as well as the work by the Pisa[32] and Lancaster/Missouri[33] groups already cited above.

The procedure used for electronic analogue simulation is extremely simple. Suppose, for example, one wishes to simulate the simple Langevin equation

$$\dot{x} = h(x) + V_N(t)g(x) \tag{1}$$

where h and g are functions of x and $V_N(t)$ represents noise. It is straightforward to build an electronic model of Eq. (1) using standard circuit elements to perform the required mathematical operations. With the use of modern analogue components, this procedure is relatively straightforward and can usually be effected in a few hours at most.

Addition, subtraction, multiplication by a constant factor, integration, differentiation, etc., can be provided by suitably connected[44] operational amplifiers; more specialized integrated circuits can be used where appropriate (e.g., the AD534[45] or the MPY534[46] for general multiplication and division, or the AD639[45] for trigonometric functions).

The noise term $V_N(t)$ is obtained from an external broadband noise generator, whose output is usually first passed through an RC filter to produce a fluctuating voltage that is exponentially correlated such that

$$\langle V_N(t)V_N(t')\rangle = \frac{D}{\tau}e^{-|t-t'|/\tau} \tag{2}$$

where τ is the correlation time and D is the noise intensity. In practice, it is conventional to solve equations like Eq. (1) in integral form. The quantities appearing in Eq. (1) are normally scaled by numerical constants in the electronic model. This is done partly to speed up the rate of data acquisition (by scaling the time t) and partly to exploit to best advantage the necessarily limited dynamic range of the analogue components (scaling the coordinate x). The signal must be much larger than the internal noise ($\sim$ mV) of the active components, but it must also always be smaller than a maximum value ($\sim$ 10 V) tolerated by them. The connections between the scaled and unscaled quantities are readily established and are discussed in detail elsewhere.[32, 33, 47]

The noise generators used to provide the fluctuating voltage V_N have been of two main types. Either a commercial instrument such as the Quan-Tech model 420, or the Wandel and Goltermann model RG1, may be used; or a pseudo-random dichotomous noise generator may be constructed from a pair of digital shift registers.[32, 48, 49] In the latter case (assuming that dichotomous noise is not the type required), the distribution of the noise can be made Gaussian to an excellent approximation by passage through a low-pass filter. Although the noise is only pseudo-random, and repeats precisely after a finite period, it is straightforward to ensure that the repeat time is at least a few days, so that the output can be regarded as random for practical purposes.

The commonest type of noise to be investigated to date is Gaussian quasi-white noise. It is readily obtained by ensuring that the output of the noise generator, after filtration, is such that its correlation time is very small, which, in practice, means that it must be much smaller than the time constant of the integrator in the electronic circuit model to which V_N will be applied. The noise will then be perceived by the model as being white. Exponentially correlated colored noise is obtained simply by use of a longer constant in the filter. Quasimonochromatic noise (QMN), which provides a more accurate model of the noise occurring in many real systems than either white or exponentially correlated noise, may be obtained by "filtering" the white noise from the noise generator through a damped *harmonic* oscillator[50]; in fact, earlier work on this type of noise[51, 52] referred to it as harmonic noise. Its intensity and the shape of its spectral density are determined by the intensity of the white noise and the natural frequency and damping constant of the harmonic oscillator. Kai[53] has described the production of other types of noise (uniform noise, binomial noise) that could doubtless be applied successfully to electronic circuit models.

Once the electronic model has been built and tested, it is treated as an entity in its own right, on which experiments can be performed. The circuit variables, for example, $x(t)$ and $\dot{x}(t)$, will fluctuate in response to the noisy driving voltage. Their fluctuations are usually analyzed by means of a digital computer. The analogue signals from the circuit model are digitized, typically with 12-bit precision, and then processed to yield the statistical quantity of interest, e.g., a probability distribution, a spectral density, or the mean time to attain a threshold. The acquisition/analysis is performed on blocks of typically 1K–4K samples and the results of perhaps 10^2–10^5 such sweeps are usually averaged together to provide improved statistical quality in the quantity being sought.

The analogue electronic approach to stochastic nonlinear problems has enjoyed a number of successes, not only in models relevant to optics, but also more generally. It has been used to confirm some theoretical predictions, and to rebut others; it has also led to new discoveries in its own right. A full review would be inappropriate here, but highlights of the last decade might include the following: the first demonstrations of noise-induced phase transitions in the Stratonovich model[41] and the genetic model[54] equations; experimental clarification of the white noise Ito/Stratonovich controversy[55, 56]; the first observation[57] of stochastic postponements of critical onsets; measurements of stochastic phase portraits[58] for the double-well Duffing oscillator, leading to the unexpected discovery that these become skewed with increasing correlation time of colored noise; observation of a similar skewing effect with noise color in the ring-laser gyroscope equation[59]; demonstration of a noise-induced phase transition with additive noise, due to noise color[60] (see Section III); investigations of swept parameter or time-evolving systems related to chiral symmetry-breaking in prebiotic evolution,[61] transient optical bi-modality,[42] postponement of Hopf bifurcations in the Brusselator,[62] and studies[49, 63, 64] of the role of fluctuations in the laser start-up problem (see Sections IV and V); the first convincing demonstrations of stochastic resonance (SR)[43] and of its occurrence in underdamped systems[65]; vindication[66, 67] of the proposal that SR should be treated in terms of linear response theory and the fluctuation dissipation theorem (see Section VI), and the discovery of three new forms of SR, in systems with periodically modulated noise,[68] in periodically driven oscillators[69] (see Section VII) and in monostable systems[70]; a demonstration of the relevance of SR to sensory neurons[71]; rebuttal[72] of a theoretical prediction[73] of a noise-induced divergence in the relaxation time of the Verhulst model; the discovery of modulation-induced negative differential resistance (MINDR)[74] (a deterministic phenomenon); observation of supernarrow spectral peaks near a kinetic phase transition[75] (see Section VII); pioneering experiments with quasimonochromatic noise (QMN)[50]; studies of fluctuation phenomena associated with a multibranch potential[76] (see Section VIII); and a demonstration[77] of the physical reality of optimal paths, leading to the unexpected discovery of a marked increase in the dispersion of the prehistory probability distribution at intermediate times.

In a significant minority of these investigations, digital simulation of the stochastic/nonlinear equations was also used. The question arises, therefore, as to what may be the relative merits of these two very difficult techniques. Before attempting to address this, in Section II.C, we review briefly the basis and characteristics of the digital approach.

B. Digital Simulation

It is not possible to review here the whole field of digital simulation, or even all of the algorithms proposed for the integration of stochastic differential equations. We refer the interested reader to Ref. 34 for a survey of some of the existing literature. It must be said that digital simulations in optics are so ubiquitous that it is impossible to do justice to all the published material. Not only have digital simulations of stochastic differential equations been used to integrate the different (semiclassical) dynamical models studied by the various authors but, via suitable manipulations, even problems in nonlinear quantum optics have been mapped onto the integration of stochastic differential equations (among others, see Ref. 78 and references therein). Here, we are content with describing briefly a standard algorithm, pointing out problems, and referring to some recent relevant works. A word of warning is in order: Contrary to the case of ordinary differential equations, it is not straightforward to derive higher-order algorithms in the case of stochastic differential equations. Although it is trivial to derive integration schemes that intuitively (i.e., as far as the deterministic force is concerned) are correct, it is in general not trivial to prove that these integration schemes actually generate a trajectory close to the real one in the statistical sense (see Ref. 79 for a way to analyze algorithms for stochastic differential equations). It should also be borne in mind that, because of the discretization procedure, in the case of stochastic differential equations the various schemes will always spoil the final result (see, for instance, Refs. 80–82 and references therein). Typically, the spurious terms will be proportional to some power of the integration time step, a power that will not simply increase as higher-order algorithms are considered (this is in contrast with ordinary differential equations), but will also (amazingly) depend on the observable one is interested in. So, in many cases one is much better off considering a lower-order (and relatively faster) algorithm with a smaller integration time step than a higher-order algorithm with a larger time step.[81]

As for the analogue simulation case, in general one has to integrate the stochastic differential equation (sde)

$$\dot{x} = f(x) + g(x)\xi(t) \tag{3}$$

The most common $\xi(t)$ is given by a Gaussian fluctuation, with zero average and correlator

$$\langle \xi(s)\xi(t) \rangle = 2\delta(t - s) \tag{4}$$

We are also assuming that the noise has no space dependence; suitable generalizations can be readily obtained. Following Refs. 34 and 83, the basic and simplest algorithm (Euler scheme) to integrate Eq. (3) transforms it into the discrete recurrence relation (h will be the integration time step)

$$
\begin{aligned}
x(t + h) = x(t) &+ Z_1(h)g(x(t)) + hf(x(t)) \\
&+ \frac{1}{2}g(x(t))\frac{dg(x)}{dx}(t)[Z_1(h)]^2
\end{aligned} \tag{5}
$$

with

$$
\begin{aligned}
Z_1(h) &\equiv \int_0^h \xi(s)\, ds \\
&= \sqrt{2h}\, Y_1
\end{aligned} \tag{6}
$$

where Y_1 is a Gaussian random deviate, with average zero and standard deviation one. Incidentally, we are also assuming that Eq. (3) is to be integrated according to the Stratonovich prescription[55, 56]; otherwise, the last term in Eq. (5) should be programmed differently (see also next section, and Refs. 34 and 83). The algorithm given by Eq. (5) is a so-called one-step collocation scheme: Given the point $x(t)$ this scheme will "collocate" the solution of Eq. (3) to a "position" $x(t + h)$ according to the statistics implicitly given by Eq. (3). It is very simple to improve this algorithm, obtaining the so called Heun scheme.[84] One needs only to compute

$$x_1 = x(t) + Z_1(h)g(x(t)) + hf(x(t))$$

$$+ \frac{1}{2}g(x(t))\frac{dg(x)}{dx}(x(t))[Z_1(h)]^2$$

$$x_2 = x(t) + Z_1(h)g(x_1) + hf(x_1) \tag{7}$$

$$+ \frac{1}{2}g(x_1)\frac{dg(x)}{dx}(x_1)[Z_1(h)]^2$$

$$x(t + h) = \frac{1}{2}\{x_1 + x_2\}$$

with the same $Z_1(h)$ for both steps. This straightforward generalization (see Refs. 78, 80, and 85) can be shown[78, 80] to have the least possible error in the mean square at the order $O(h^3)$. We should mention that there is yet another recent integration scheme,[81] useful for white noise forcings, which, via a suitable discretization of Eq. (3) in both time and space, can achieve extremely high integration speeds.

In practice, however, one has often to deal with random noises that are neither delta correlated nor Gaussian. Also, very often the dynamical system is described by a set of sde with a number of stochastic sources. Algorithms suitable for the integration of systems of sde will be found in Refs. 34 and 83. We only want to point out, in passing, that the algorithms in this case are not a simple generalization of Eq. (7), but that the different stochastic drivings give rise to correlations in the discretized solution. In case of non-Gaussian random forces, clearly Eqs. (6) and (7) can be easily modified (though care must be taken with the term $[Z_1(h)]^2$ in Eq. (7)). The case of noise that is not delta correlated is, in some respects, the toughest to deal with as far as digital simulations are concerned, and we now focus on it at some length.

We have seen for the analogue simulation case that, very generally, it is possible to generate correlated noise by filtering the stochastic term prior to adding it to the system of interest. This is possible also in the case of digital simulations. However, it is clear that here this approach is very far from being an efficient one. Suppose that the system of interest has a typical frequency ν_s and we want to drive it with a single pole (an exponentially correlated, see Eq. (2)) noise. If, after filtering, the noise has a cutoff frequency $\nu = 1/\tau$ that is large compared to ν_s we would be forced to use a very small integration time step for the integration, even if, as far as the system of interest is concerned, the driving stochastic term is perceived as almost white. Clearly, this would imply an extremely large central processor unit (CPU) integration time, although we are simply wasting it in the (irrelevant) integration of the noise filter. To overcome this problem, and considering that the most interesting (from a physical point of view) correlation functions for the noise can be obtained from suitable linear filters, algorithms have been proposed to integrate implicitly the evolution equations of nonwhite noises. Here, we will understand that an implicit algorithm is an algorithm that is unconditionally stable whatever the relative values of h and $1/\nu_n$. In particular, an implicit algorithm should be able to integrate the filter equation of motion even in the case $h\nu_n \gg 1$. The idea is to generate the Gaussian variables necessary for each integration time step with the correct time correlation. The case of exponentially correlated noise has been discussed in Refs. 34, 83, and 85–92. The case of quasimonochromatic noise, i.e., noise with correlation

$$\langle \xi(t)\xi(0)\rangle = e^{-|t/\tau|}\cos(\omega t + \phi)$$

has been considered in Refs. 50 and 52. Several possibilities are available for the implicit integration of the noise. Given its simplicity and clarity, we will briefly outline here, as an example, a simple low-order, one-step collocation scheme algorithm for the more commonly studied exponentially correlated noise. Incidentally, in the light of Refs. 78 and 80, it is doubtful whether higher-order schemes (like the one considered in Refs. 87 and 88), although extremely interesting but also CPU time-consuming, would really introduce substantial improvements. One starts writing the sde

$$\dot{x} = f(x) + y \tag{8}$$

$$\dot{y} = -\frac{1}{\tau}y + \frac{\sqrt{2D}}{\tau}\xi(t) \tag{9}$$

where Eq. (9) (the linear filter) implies that the correlation function of $y(t)$ is an exponential function (cf. Eq. (2)); y is in turn used to drive the system of interest (Eq. (8)). It is straightforward to solve Eq. (9). The solution is given by

$$y(t) = e^{-t/\tau}y(0) + \frac{\sqrt{2D}}{\tau}\int_0^t e^{(s-t)/\tau}\xi(s)\,ds \tag{10}$$

The last term is a linear combination of Gaussian variables, and hence is itself a Gaussian variable with given (unknown) average and standard deviation. Some tedious algebra leads to

$$y(t + h) = e^{-h/\tau} y(t) + Z_0(h) \tag{11}$$

$$x(t + h) = x(t) + hf(x(t)) + \tau[1 - e^{-h/\tau}]y(t) + Z_1(h) \tag{12}$$

The explicit expressions for $Z_0(h)$ and $Z_1(h)$ are very cumbersome, and can be found in Refs. 83 and 85. For the present discussion, however, it is necessary only to know that they can be obtained simply by generating two Gaussian deviates and performing a very small number of elementary operations (multiplications and additions). But, most important, the scheme given by Eq. (12) is stable *no matter what the relative values of h and τ are.* In particular, the case $\tau \ll h$ can be handled by Eq. (12) with no problem,[85] and for $\tau \to 0$ Eq. (12) goes smoothly into Eq. (5). In practice, we have, as desired, that the integration time step h will be dictated only by the typical frequencies of the system of interest and not by the integration of the noisy driving. A refined scheme is obtained by rewriting Eq. (7) using the terms of Eq. (12) as building blocks. An equivalent algorithm for quasimonochromatic noise can be found in Refs. 50 and 52.

We should add that although similar schemes can be derived even for extremely complex noise correlation functions, a very elegant and efficient scheme for these cases has been proposed in Refs. 90 and 91. The idea is to start from

$$\dot{x} = f(x) + y \tag{13}$$

with y Gaussian and correlated as per

$$\langle y(t)y(s) \rangle = F(t - s)$$

Clearly, we have through a Fourier transform

$$\langle y(\omega)y(\omega') \rangle = G(\omega - \omega') = H(\omega)\delta(\omega - \omega')$$

Now, write (central limit theorem)

$$y(jh) = \frac{1}{\sqrt{h}} \sum_{i=0}^{N} a_i \cos(i\, \Delta\nu jh + \phi_i) \tag{14}$$

with ϕ_i uniformly distributed over $[0, 2\pi]$, $a_i^2 = H(i\, \Delta\nu)\, \Delta\nu$, and suitably chosen N and $\Delta\nu$, such that the conditions

$$\Delta\nu \ll \nu_s$$

and

$$N\Delta\nu \gg \nu_s$$

are satisfied. It is then possible to integrate (any integration method will do) Eq. (13). The key point (see Ref. 90) is that the sum appearing in Eq. (14) can be evaluated with a fast Fourier transform, which implies that to generate, say $y(ih)$ for times from 0 up to Nh, only $N \log N$ operations are required. Although for exponential or quasimonochromatic noise the direct simulation would be faster, for computationally expensive $H(\omega)$ this method is much preferable.[91, 92] For vector and parallel machines, furthermore, the whole algorithm can be made to run very efficiently.

For all schemes used in the digital simulations it is very important to have good algorithms to generate the necessary random deviates. In most cases, the generation of these random deviates is the slowest step of the whole simulation, and hence it is vital to have efficient random generators. General purpose routines, written in high-level languages, can be found, for instance in Ref. 93 (and references therein). On the other hand, to achieve very high speeds, it is necessary to have dedicated routines, which will vary from platform to platform. As a rule of thumb, routines from widely used libraries are reasonably fast and reliable, although our personal experience teaches that it is good practice to test the spectral characteristics[94] of a routine that has never been used before. A must for all algorithms considered here is also a good Gaussian generator. A safe (but slow) workhorse is the Box-Müller formula[93]; again all scientific libraries will have efficient (machine dependent) routines. It is worth mentioning, however, that the so-called Ziggurat algorithm,[95] which is easily portable in high-level languages, is still reasonably efficient.

A brief word about future directions of digital computing. The phenomenal increase in power of modern (often vector or parallel) machines is widening the field of application of digital simulations. Modern computers are so efficient and fast that they can simulate in real time quite complex dynamical systems. In fact, it is the possibility of fast interfacing between real systems and digital computers that has recently made it possible to control[96] chaotic dynamics. Also, modern parallel architectures make it possible to integrate, simultaneously, samples of thousands of particles, taking no longer than the time needed to integrate a single trajectory. The resulting numerical experiments enjoy an extremely good statistical quality. Suitable communications software can distribute particularly heavy calculations over geographically extended networks of computers, with added efficiency and speed.

C. Analogue Versus Digital Simulation

The analogue and digital simulation techniques are obviously very different, yet they are frequently applied to similar problems, and often simultaneously to the same problem. The question is frequently asked, therefore, Which of the two techniques is the better? Although no precise, universally applicable answer can be given (because their relative advantages and disadvantages naturally depend on the nature of the particular system under study), we can nonetheless offer the following general remarks. First, digital simulation is in general the more accurate of the two approaches, provided, of course, that the relevant algorithm has been correctly programmed. Digital simulations are not subject to the parameter drifts, or the effects of internal noise in active components, that can sometimes plague the analogue approach. Second, however, for many types of systems (especially, but not exclusively, those involving coupled equations, or the combined effects of more than one noise source), digital simulations can become very greedy (often requiring hours) of CPU time, if the results are to be of reasonable statistical quality. Analogue simulations, on the other hand, can produce results of excellent statistical quality relatively quickly (in minutes) to an accuracy typically of a few percent. Third, the analogue approach is in many ways closer to conventional experimental science than the digital one. In studying the Ito/ Stratonovich degeneracy,[55, 56] for example, an experimental resolution of matter at issue (i.e., the question of which of these two white noise stochastic calculi is applicable to the real world) would have been impossible on the basis of digital simulation, because of the need to program either an Ito or a Stratonovich algorithm into the simulation software. Thus, the results emerging from the simulation would in practice have been predetermined by the programmer. In the analogue case, on the other hand, the electronic model was obliged to respond as prescribed by nature, so that meaningful and scientifically illuminating results were obtained. Fourth, partly on account of its relative speed, analogue simulation readily enables large volumes of parameter space to be surveyed for interesting phenomena, often by turning knobs to adjust the relevant parameters while examining changes in the resultant distribution on a visual display; the equivalent procedure for a digital simulation is usually much slower and more ponderous.

In view of the above considerations, most scientists experienced in both analogue and digital simulation seem to regard them as *complementary* techniques for the study of stochastic nonlinear problems. Each has its own advantages and disadvantages; which of these is emphasized or de-emphasized will depend on the properties of the equation under study. As in any experimental study, it is possible to make mistakes and generate artifacts in either form of simulation. Consequently, in tackling really awkward problems for which there is no existing theory, or where the theory appears to be suspect, it is prudent to use both the analogue and the digital simulation techniques, with the one acting as a check on the other.

III. OPTICAL BISTABILITY WITH COLORED NOISE

The main purpose of this section is to discuss the characteristic behavior of a noise-driven system at the critical point that arises when two stable and one unstable state of the system merge together. Just such a situation would occur for the general system whose behavior is sketched in Fig. 1b if the control parameter were adjusted such that the resultant $I_T(I)$ curve (falling in between curves 2 and 3) acquires a vertical point of inflection. We shall see that the single-peaked (monomodal) probability density distribution found for white noise can then be split by noise color to become twin-peaked (bimodal), a phenomenon similar to the noise-induced transitions discussed in Ref. 97. The model that we shall consider comes from optics and describes a passive optical system adjusted to be just on the verge of bistability and driven by noise. The dynamics of this model are described by a stochastic differential equation.

We first remind the reader that given the stochastic differential equation

$$\dot{x} = \phi(x) + \psi(x)\xi(t) \tag{15}$$

where $\xi(t)$ is a random Gaussian noise, delta correlated and of intensity D, there is a partial differential equation (called the Fokker-Planck equation or FPE) for the probability distribution of the variable x associated with Eq. (15), which has the form

$$\frac{\partial}{\partial t}P(x,t) = \frac{\partial}{\partial x}\left\{-\phi + D\psi\frac{\partial}{\partial x}\psi\right\}P(x,t) \tag{16}$$

Let us now consider a general first-order differential equation in the absence of noise:

$$\frac{dx}{dt} = \phi(x,\lambda) \tag{17}$$

where we have introduced explicitly a control parameter λ. We suppose also that, for all values of λ this equation has only one stationary solution, $\bar{x}(\lambda)$. In the presence of noise $\xi(t)$, if this is white and additive, the stationary probability density will be given by the solution of the associated FPE (see Eq. (16)):

$$\sigma_s(x) = N \exp\left\{-\frac{1}{D} \int \phi(x, \lambda) \, dx\right\} \tag{18}$$

where N is a normalization constant. Equation (18) implies the existence of a single peak in the density located at exactly $x = \bar{x}(\lambda)$.

When the noise is still white but multiplicative, on the other hand, the behavior of the system is strikingly different. For appropriate choices of $\phi(x, \lambda)$ and $\psi(x)$, and in a suitable interval of the parameter λ, an increase of D can alter the stationary distribution in a continuous fashion from one that has a single peak to one that has two peaks.[54, 57, 97] It is this phenomenon that is referred to as a *noise-induced transition*.

A rather similar behavior is found in the steady-state probability distribution in a nonequilibrium phase transition of second order.[98] It must be noted, however, that the noise in the latter case is additive and that the behavior of the stationary distribution follows strictly that of the steady-state solution of Eq. (17), which exhibits a pitchfork bifurcation when plotted as a function of λ. This is in contrast to the transitions considered in Refs. 54 and 97 where the bimodal structure of the stationary distribution is produced exclusively by the multiplicative character of the noise, whence the appellation of the phenomenon.

The phenomenon of noise-induced transitions persists when the noise is changed from white to colored.[99, 100] By colored noise we mean that ξ is no longer delta correlated but has a correlation function of the form

$$\langle \xi(t)\xi(s)\rangle = \frac{D}{\tau} \exp\left\{-\frac{|t - s|}{\tau}\right\} \tag{19}$$

This change may even produce profound qualitative consequences in the behavior of the stationary probability distribution, leading, for example, from the scenario of a second-order phase transition to one of the first order.[16, 101, 102] We note in passing that the correlation function described by Eq. (19) can be obtained assuming that the variable ξ is driven by the stochastic differential equation

$$\dot{\xi} = -\frac{1}{\tau}\xi + \frac{\sqrt{2D}}{\tau}\eta(t) \tag{20}$$

where η is a stochastic Gaussian variable, delta correlated.

In the above references[16, 99, 100, 101, 102] the noise is multiplicative in nature. We show below that noise-induced transitions can arise, not only from multiplicative noise, but also from additive noise, provided, however, that it is colored rather than white. For the purpose of our demonstration we focus our attention on a model that is usually studied in connection with optical bistability,[11, 103] but is considered here for a value of the control parameter that corresponds, not to bistability, but to the critical point. For more details, the interested reader can also refer to Ref. 60.

A. The Model System

The model studied here describes a bistable system, which, in the last decade, has been the object of extensive investigations motivated both by its intrinsic theoretical interest and by practical considerations. The phenomenon of interest, passive optical bistability,[11] alluded to in Section I, occurs in an optical cavity filled with a material that has an intensity-dependent refractive index. The experimental arrangement is as shown in Fig. 2. A stationary, coherent beam, near to resonance both with the cavity and with the material, is injected into the cavity.

The steady-state intensity of the beam transmitted by this system is a nonlinear function of the input intensity. By suitably adjusting the frequency of the incident field, the stationary curve of transmitted vs. incident intensity becomes S-shaped, as shown in Fig. 1a. The part with negative slope (dashed) is unstable, so that there is an interval of the input intensity within which the system is bistable. If the incident power is slowly increased from zero to beyond the bistable region, and then decreased back to zero, one obtains a hysteresis cycle. By varying continuously the input frequency, the size of the hysteresis cycle can be reduced to zero and the bistability disappears, as in Fig. 1b. Just at the boundary between bistability and monostability there is a "critical" situation, in effect between curves 2 and 3, on which we focus in this section.

We describe this phenomenon by means of a very simple model, the derivation of which is discussed in Ref. 104. We indicate by x the normalized, intensity-dependent part of the refractive index; it obeys the

Figure 2. Fabry-Pérot cavity: M_1 and M_2 are partially transmitting mirrors; I, I_T, and I_R are the incident, transmitted, and reflected intensities, respectively. (After Ref. 60.)

dynamical equation

$$\frac{dx}{dt} = -x + \frac{I}{1 + (x - \theta)^2} \tag{21}$$

where the time t is normalized by the response rate γ of the material, I is the normalized input intensity, and θ is the detuning parameter, which arises both from the mismatch between the incident field frequency and the nearest cavity frequency and from the constant (intensity independent) part of the refractive index. The control parameters in this model are I and θ; in particular, θ plays the role of the control parameter λ introduced above.

The normalized transmitted intensity I_T is given by

$$I_T = \frac{I}{1 + (I_T - \theta)^2} \tag{22}$$

The model given by Eqs. (21) and (22) holds in the limit of dispersion dominant over absorption, and requires that the cavity relaxation time is much smaller than the atomic time γ^{-1}; if any noise is present, the cavity relaxation time must be much smaller also than the correlation time of the noise. At steady state, Eq. (21) gives the cubic stationary equation

$$I = x\left[1 + (x - \theta)^2\right] \tag{23}$$

and I_T coincides with x. The plot of x as obtained from Eq. (23) exhibits bistability[103] for $\theta > \sqrt{3}$ (Fig. 3) very similar in form to the type of behavior sketched for the $I_T(I)$ curve of Fig. 1a. The critical situation (Fig. 4) for which the curve displays a point of inflection with a vertical tangent, corresponds to $\theta = \sqrt{3}$; the coordinates of the critical point are $x_c = 2\sqrt{3}/3$ and $I_c = 8\sqrt{3}/9$. We consider now the case when there is thermal noise in the material, which means that Eq. (21) is replaced by a stochastic equation with additive noise:

$$\frac{dx}{dt} = -x + \frac{I}{1 + (x - \theta)^2} + \xi(t) \tag{24}$$

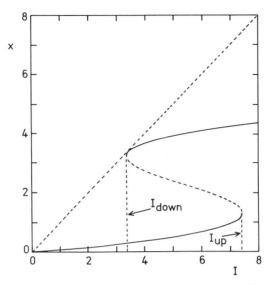

Figure 3. Steady-state curve representing Eq. (23) with $\theta = 2\sqrt{3}$; the part with a negative slope is unstable. (After Ref. 60.)

where we assume that the noise in Gaussian and that the variable ξ is correlated as per Eq. (19).

B. Theory

We now show that the system described by Eq. (24) will indeed become bimodal as the correlation time τ of the noise is increased, even if the system would be monostable if the noise were white. A similar phenomenon, for a prototype nonlinear system, has also been discussed in Ref. 105. We do not discuss in detail all the different effective FPEs proposed to treat the problem of a system subject to colored noise. The interested reader can consult, for instance, Refs. 60 and 106–108.

We start then from

$$\dot{x} = f(x) + \xi(t)$$

$$f(x) = -x + \frac{I}{1 + (x - \theta)^2}$$

$$\dot{\xi} = -\frac{1}{\tau}\xi + \frac{\sqrt{2D}}{\tau}\eta$$

$$(25)$$

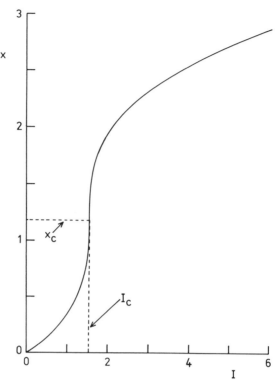

Figure 4. Critical steady-state curve representing Eq. (23) with $\theta = \sqrt{3}$. (After Ref. 60.)

where η is a Gaussian variable, delta correlated and with a standard deviation of unity.

We can easily show that in the limit of very large τ's the probability distribution becomes bistable, even if the parameters in Eq. (25) are chosen in such a way that in the white noise case the probability distribution is monostable.[107] Consider the equilibrium distribution of the variable ξ. Given the equation of motion Eq. (25) it is trivial to show that

$$P_{\text{st}}(\xi) = N \exp\left\{-\frac{\tau \xi^2}{2D}\right\} \tag{26}$$

where N is a normalization constant. Now, in the case of very large τ's we can assume that really the system moves much faster than the noise. This implies that essentially we will find the variable x almost always at the "equilibrium" solution (for the given noise value) obtained by setting $\dot{x} = 0$ in the first equation of Eq. (25). This yields, for the equilibrium distribution of the variable x

$$P_{\text{st}}(x) = N \frac{d\xi}{dx} \exp\left\{ -\frac{\tau \xi(x)^2}{2D} \right\} \tag{27}$$

At the critical value of the parameter $\theta = \sqrt{3}$, $I = \frac{8}{9}\sqrt{3}$ we can (approximately) write

$$\dot{x} = -x^3 + \xi \tag{28}$$

which yields, substituting ξ into Eq. (27),

$$P_{\text{st}}(x) = 3Nx^2 \exp\left\{ -\frac{\tau x^6}{2D} \right\} \tag{29}$$

which is clearly bimodal (with maxima located at $x = \pm(2D/3\tau)^{1/5}$).

In the general case, we start by saying that there is indeed an exact FPE corresponding to the evolution of Eq. (25), namely (we write for the deterministic right side the compact expression $f(x)$)

$$\frac{\partial}{\partial t} P(x, \xi, t) = \left\{ \frac{\partial}{\partial x}(-f(x) - \xi) + \frac{\partial}{\partial \xi}\left(\frac{\xi}{\tau} + \frac{D}{\tau^2} \frac{1}{\partial \xi} \right) \right\} P(x, \xi, t) \tag{30}$$

Equation (30) is exact and describes the problem completely. Unfortunately, for the $f(x)$ we are studying here, no solution is known. A typical approach, then, is to reduce the complexity of the problem by deriving an approximate (effective) one-dimensional FPE describing the evolution of the variable x alone (among others, see Refs. 60 and 106–108 and references therein). In the very general case, the one-dimensional FPE obtained will be in the form

$$\frac{\partial}{\partial t}\rho(x, t) = \frac{\partial}{\partial x}\left\{ -f(x) + D\frac{\partial}{\partial x}\Phi(x) \right\}\rho(x, t) \tag{31}$$

where the exact form of $\Phi(x)$ will depend on the approach followed to derive Eq. (31). Different forms of $\Phi(x)$ will obviously lead to different equilibrium distributions for the variable x.

In the literature there are basically three different prescriptions for $\Phi(x)$:

1. The so-called Best-Fokker-Planck equation (BFPE),[101, 109, 110] which yields

$$\Phi(x) = 1 + \tau f'(x) + \tau^2(f'(x)f'(x) - f(x)f''(x)) \quad (32)$$

obtained from Eq. (30) by expanding around the white noise $\tau = 0$ limit. Equation (32) was originally believed to be correct up to terms $O(\tau^2)$ included.

2. The Fox approach,[111, 112] based on a functional integral to derive $\Phi(x)$, which yields

$$\Phi(x) = \frac{1}{1 - \tau f'(x)} \quad (33)$$

which expanded to second order in τ reads

$$\Phi(x) = 1 + \tau f'(x) - [\tau f'(x)]^2 \quad (34)$$

which differs from the BFPE at second order in τ. A similar reduced FPE can be obtained with the local linearization theory (LLT).[113, 114] Within this approach one obtains the same equilibrium distribution, although the dynamics are different (see Refs. 113–115 for more details), given that the FPE obtained with the LLT reads

$$\frac{\partial}{\partial t}\rho(x, t) = \frac{\partial}{\partial x}\left\{-f(x) + \frac{D}{\tau}\frac{\partial}{\partial x}\int_0^t e^{-\Phi(x, t')/\tau}\,dt'\right\}\rho(x, t) \quad (35)$$

Closely related to the Fox approach is the so called unified colored noise approximation (UCNA),[118] which starts from a slightly different, contracted, one-dimensional FPE, namely

$$\frac{\partial}{\partial t}\rho(x, t) = \frac{\partial}{\partial x}\frac{1}{1 - \tau f'(x)}\left\{-f(x) + D\frac{\partial}{\partial x}\Phi(x)\right\}\rho(x, t) \quad (36)$$

and is characterized by the same $\Phi(x)$ and the same equilibrium distribution as the Fox approach (they also differ in the dynamics). The same FPE is obtained if one considers the uniform expansion (valid for small τ and D) of Refs. 116 and 117. The Fox approach is thought to be correct only up to first order in τ, the LLT should be correct for nearly harmonic systems, and the UCNA is putatively correct for both small and large τ's (but see Ref. 115). Incidentally, all three approaches yield for the equilibrium distribution of the variable x in the limit of very large τ the exact Eq. (29), a most puzzling result in the light of the above comment. Intuitively, then, one may expect that this approach should work well even in the intermediate region.[118, 119]

3. A fourth-order projection operation expansion. In all the approaches mentioned there is, more or less explicitly, a small parameter of expansion (τ for the BFPE and the Fox approach, $\tau\Phi'(x)$ for the LLT). Now, there is also another small parameter in the system, namely the noise intensity D. It was first argued in Ref. 120 that unless orders above the first in D are considered (thence, breaking the FPE structure), it may be meaningless to consider orders in τ beyond the first. This conjecture has been considered further in Refs. 107 and 108, where the term $D^2\tau^3$ in a projection operator expansion has been explicitly evaluated. Using a renormalization procedure it was then shown that indeed the above-mentioned term contributes also to order $D\tau^2$ in the small D, small τ limit. The final FPE written in Refs. 107 and 108 is still in the form of Eq. (31) with

$$\Phi = 1 + \tau f'(x) + \tau^2(f'(x))^2 + \tfrac{1}{2}\tau^2 f(x)f''(x) \qquad (37)$$

We can write Φ in the more compact form

$$\Phi = 1 + \tau f'(x) + \tau^2(f'(x))^2 + P_{\exp}\tau^2 f(x)f''(x) \qquad (38)$$

with $P_{\exp}$ equal to $-1, 0$ or $\tfrac{1}{2}$ for the BFPE, Fox approach, and fourth-order expansion, respectively. This leads to the following expression for the equilibrium distribution of the variable $\rho(x)$

$$\rho(x) = N \left| \frac{1 - \tau f'(x)}{1 + \tau^2 P_{\exp} f(x) f''(x)} \right| \exp\left\{ \int^x \frac{1}{D} \frac{f(y)(1 - \tau f'(y))}{1 + \tau^2 P_{\exp} f(y) f''(y)}\,dy \right\}$$
$$(39)$$

We should say that although in practice (see below) the correct prescription for $P_{\exp}$ seems to lie between 0 and $\tfrac{1}{2}$, remarkably, all three theoretical approaches lead to an equilibrium distribution which becomes bimodal for increasing τ.

C. Simulations and Comparison with the Theory

Equation (21) has been modeled using both analogue and digital techniques. In Section II.B we described the digital algorithm used. The analogue technique, as applied to this system, is particularly interesting and we now concentrate on describing the details. The direct simulation of Eq. (21) proved not entirely straightforward. We found that it was extremely difficult to perform a satisfactory analogue simulation for the required critical values of $\theta = \sqrt{3}$, $I = \frac{8}{9}\sqrt{3}$. The reason for this becomes clear immediately on inspection of Fig. 4, and is associated with the extreme sensitivity of x to small changes in I when x is close to x_c. In electronic circuits of the kind under discussion, parameters can be set typically to an accuracy of ± 2 mV; but they are liable to drift by up to about 10 mV during the several minutes usually needed to complete the measurement of a statistical density. This implied in practice that the deterministic ($D = 0$) value of x could not be relied upon to be closer than $\pm 5\%$ to x_c, on average, during a measurement. Consequently, although the results of the direct simulation of Eq. (25) were in good qualitative agreement with the theoretical predictions, they tended to be irreproducible because of parameter drift, and the quality of the quantitative agreement with theory was only fair. In part, the stability problem arose from the existence of the quotient in Eq. (21), which exacerbated the tendency of the parameter I to drift.

In an attempt to improve the quality of the simulation, a second electronic circuit was built that contrived to model the system under investigation without any need explicitly to calculate a quotient. The possibility of being able to do so depended on the realization that the equation under discussion (Eq. (21)) was originally obtained, by use of the technique of adiabatic elimination,[98] from the more complicated set of coupled differential equations normally used to model optical bistability,[11] namely

$$
\frac{dz}{dt} = -K\left[z - \sqrt{I} + iz(\theta - x)\right]
$$
$$
\frac{dx}{dt} = -\gamma\left[x - |z|^2\right] + \gamma\xi(t)
$$

(40)

where z is the transmitted field of intensity $|z|^2 = I_T$, K^{-1} is the cavity relaxation time, and the variable $\xi(t)$ obeys Eq. (20). In the limit $K \gg \gamma$, which usually obtains in practice, it can be shown that Eq. (40) yields Eq. (21). Although Eqs. (40) are, of course, very much more difficult to solve theoretically than Eq. (21), the corresponding analogue circuit is only slightly more complicated to connect up; and, as already mentioned, it has the considerable advantage in terms of stability that it does not include a quotient term.

A diagram of the second circuit, that is, of the circuit modeling Eqs. (40), is shown in Fig. 5. The design is of the minimum component type[32] in order to reduce to the lowest possible levels nonidealities, such as drift and internal noise; the necessary trimmers and offset adjustments are not shown. As usual with this particular kind of circuit, the system actually simulated differed from Eq. (40) to the extent that some of the quantities appearing there had to be scaled by suitable factors. For the final results given below, however, the appropriate normalization factors have been taken into account.

An interesting feature of the use of this circuit (in effect) to model Eq. (21) is that the adiabatic elimination is being made to occur in practice in the circuit itself by setting $K \gg \gamma$. Thus, it is possible to check that this procedure works in reality. The ratio K/γ for the circuit was actually equal to 2000 (with $K = 2 \times 10^5$ Hz, $\gamma = 10^2$ Hz); it was found that, for the noise of correlation time of order γ^{-1}, the distributions obtained were almost indistinguishable from those for $K/\gamma = 500$. This result can be

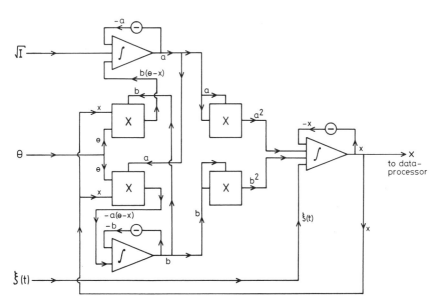

Figure 5. Block diagram of the analogue electronic circuit used to model Eq. (40), with $z = a + ib$. The multipliers form the product of the inputs shown at the top and at the left side of the square in all cases, each input being differential. (After Ref. 60.)

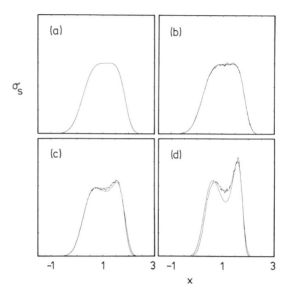

Figure 6. Probability distributions of x for a range of different noise correlation times τ: (a) $\tau = 0$; (b) 0.1; (c) 0.5; (d) 2.0. The jagged curves are the experimental results from the electronic circuit of Fig. 5. The smooth curve represents theory (see text). (After Ref. 60.)

construed as a direct and convincing demonstration that adiabatic elimination does work.

The circuit parameters were set so as to operate the system at its critical point $\theta_c = \sqrt{3}$, $I_c = \frac{8}{9}\sqrt{3}$ (see Fig. 4), and the noise intensity was kept fixed at $D = 0.1$. Several equilibrium distributions of x were then acquired for different values of τ. Some examples are shown by the jagged lines in Fig. 6, where τ is stated in each case in units of γ^{-1}. It is

TABLE I
Parameters Used in the Electronic Circuit and in Generating the Theoretical Curves of Fig. 6b. Obtained from a fit for $\tau = 0.1$

Parameter	Value in Circuit	Value from Fit
θ	1.733	1.734
I	1.541	1.543
D	1.000	1.000
τ	0.100	0.100
P_{exp}		2.8823

Note. The only adjustable parameter for all figures has been the quantity P_{exp}, setting τ to the experimental value.

immediately evident that, when the noise is effectively white (small τ), the distribution is monomodal (has a single peak); but as τ is increased, the distribution broadens and becomes bimodal (acquires two peaks), and the cleavage between the corresponding twin maxima steadily deepens with increasing τ. The change from a monomodal to a bimodal distribution represents a noise-induced transition, and it is important to note that, unlike the noise-induced transitions previously reported,[54, 57, 97] this transition is one that can be effected with *additive* noise as a result of changing only the noise *color*, with all other parameters held constant.

As already mentioned, given the extreme sensitivity of the circuit to shifts from the preset values and owing to the nonidealities that are in any case present in an analogue simulation, a problem is the identification of the actual parameters in the circuits for a fairer comparison with the theory. We answered this problem in a different way: Given that it is virtually impossible to adjust the circuit to give exactly some preset value for the various constants, is there any set of parameters that, within

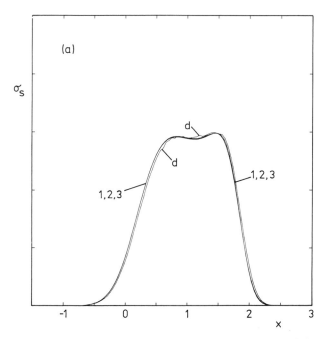

Figure 7. Comparisons between a digital simulation (labeled d) of Eq. (24) and theories for which $P_{exp} = 0$ (Fox theory, curve 1), $P_{exp} = 0.5$ (projection-operator method to fourth order, curve 2), and $P_{exp} = -1$ (best Fokker-Planck approximation, curve 3), for different values of the noise correlation time τ: (a) $\tau = 0.2$; (b) $\tau = 0.7$; (c) $\tau = 1.0$. (After Ref. 60.)

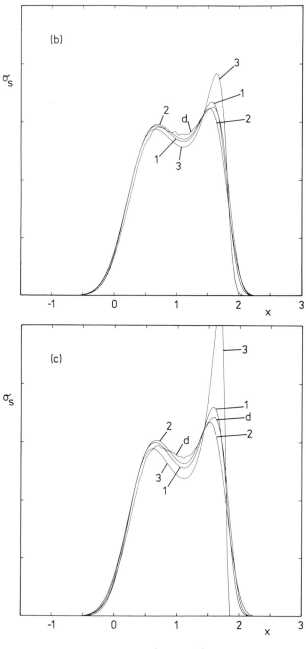

Figure 7. *(Continued)*

experimental error to those initially set in the circuit, can give a good agreement between experimental data and Eq. (39)? We have then assumed that the quantities θ, I, D, and τ were all affected by some experimental error (to fit, we also assumed that there could be a small arbitrary constant on the right side of Eq. (21)), and looked for the minimum of the function

$$\int \left| \rho_{\text{theoretical}}(x) - \rho_{\text{experimental}}(x) \right|^2 dx \qquad (41)$$

Also, note that a quantity that is unknown (theoretically) is P_{exp}; we treated it, too, as a fitting parameter. Not all parameters proved to be

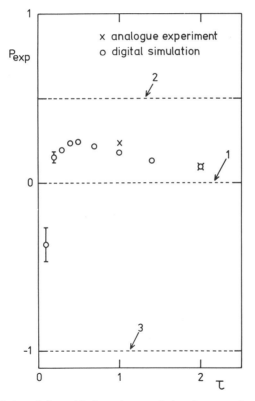

Figure 8. Variation of P_{exp} with the noise correlation time τ as determined from the analogue experiment and by digital simulation. The dashed lines indicate the values of P_{exp} corresponding to the different theories: line 1, Fox theory; line 2, projection-operator method to fourth order; line 3, best Fokker-Planck approximation. (After Ref. 60.)

necessary. Obviously, given that we have no prior knowledge of P_{exp}, we minimized the above function for the smallest possible value of τ. We used the routine MINUIT,[121] and varied the number of free parameters until the global correlation coefficient was judged to be acceptable. We found that only θ and I were actually necessary (see Table I), beside P_{exp}.

The full line in the curves for $\tau \neq 0.1$ is the theoretical prediction using P_{exp} as the only free fitting parameter. The agreement between simulations and theory is reasonable. In particular, the onset of bimodality induced by noise color is very clear.

As we mentioned, however, different theoretical approaches yield different values for the parameter P_{exp}. It is then most interesting to compare the result of our fits to the different theoretical prescriptions. To carry out a more refined comparison (in particular, to be able to control more exactly the various parameters), we also carried out a digital simulation of the system of Eq. (21) (i.e., after adiabatic elimination), and we used a procedure similar to the one described above to extract the best fitting P_{exp}.

In Fig. 7 we have plotted digital simulations and different theories. Again, the bimodality induced by noise color is very clear. Finally, in Fig. 8 we compare the values obtained from simulations for P_{exp} to the different theoretical prescriptions. It is remarkable that, apart from the lowest τ value considered, for which the error on P_{exp} is probably fairly large, the simulations give values in between 0 (Fox approach) and $\frac{1}{2}$ (fourth-order projector). Also, notice that after a rise for smaller τ values, P_{exp} smoothly decreases toward zero for larger τ values, in agreement with what is expected theoretically.

IV. DECAY OF UNSTABLE STATES

In both this section and Section V, the problem of how to quantify the dynamics of a stochastic nonlinear system that undergoes an instability (bifurcation) is addressed. The instability is assumed to occur as some appropriate control parameter is swept through the bifurcation point. Although this problem is of general interest, it has received most attention in the context of the laser switch-on, where the control parameter is taken to be the optical pump parameter and the bifurcation point is the first lasing threshold.

To date, two different cases have been treated. Historically, the first to be considered, which is also the subject of this section, was the case in which the optical pump parameter is changed discontinuously through the first lasing threshold and the transient statistics of the laser's intensity is monitored.[122–125] It was these early studies that first identified the important role that noise played in this process, with "anomalously large fluctuations" being reported. This line of research has been and continues to be the subject of intensive activity.[126–128] The second case to be considered, in which the control parameter is instead swept *slowly* through the bifurcation point, is the subject of Section V.

The majority of this work has focused on situations where the fluctuations can be modeled by white noise. Given that all physically occurring noise processes must have a finite correlation time, i.e., be colored, it is desirable to supplement these studies with studies of systems driven by colored noise.

The last few years have seen an increasing interest in the properties of systems driven by colored noise.[106] One of the most studied aspects has been the dynamics in the steady state,[129, 130] addressing for instance the study of correlation functions and relaxation times. However, some systematic work is still to be done on the more varied context of transient dynamics, that is, on the study of the relaxation of initial conditions. Most of the efforts in that direction have concerned the problem of time scales. The success of the first passage time (FPT) techniques in the context of white noise has motivated the study of the mean first passage time (MFPT) problem in the presence of colored noise. However, the difficulties involved in the mathematical treatment of this problem for non-Markovian processes have been a great handicap.[131] A possible alternative to these techniques has been proposed recently in relation to other definitions of characteristic times which in certain cases are expected to circumvent some of the difficulties of the non-Markovian FPT theory. These are the so-called nonlinear relaxation times (NLRT), which are defined in terms of integral expressions involving the time evolution of the transient moments. For a process $x(t)$ defined by a Langevin-like equation of the general form

$$\dot{x} = v(x, t) + g(x, t)\xi(t) \tag{42}$$

the NLRT associated with the average (over realizations of the noise $\xi(t)$ and over initial conditions) of any arbitrary quantity $\phi(x)$ is defined by[133]

$$T_\phi = \int_0^\infty \frac{\langle \phi(t) \rangle - \langle \phi \rangle_{st}}{\langle \phi \rangle_i - \langle \phi \rangle_{st}} \, dt \tag{43}$$

Here, $\langle \phi \rangle_i$ is the average value of ϕ in the initial state, and $\langle \phi(t) \rangle$ is the average value of ϕ during the evolution from a given initial distribution to the final stationary distribution, where the average value of ϕ is given by $\langle \phi \rangle_{st}$. Depending on the particular problem and for appropriate choices of the quantity $\phi(x)$, this may be a good definition of a global time scale for the relaxation of initial conditions toward the steady state. The interest of this definition lies in the fact that the NLRT can usually be calculated via very general techniques based on knowledge of the evolution operator of the probability densities, and with no use of any explicit form for the time dependence of the functions $\langle \phi(x(t)) \rangle$, which for nonlinear problems is usually unknown. The formalism can be systematized in a way that parallels the calculation of the FPT moments and has turned out to have some practical and theoretical advantages over them. Its usefulness has been checked in the context of white noise,[133, 134] but the advantages are particularly relevant in colored-noise problems.[135]

The fundamental difficulty in the study of the transient dynamics of processes driven by colored noise is that, due to the non-Markovian character of the process $x(t)$, the evolution equation of $P(x, t)$ depends on the preparation of the initial condition; that is, the evolution of $P(x, t)$ depends not only on its initial condition $P_i(x)$ but also on the state of the noise and its coupling to the variable x. The usual point of view is to assume that the noise variable is in its steady state and that it is initially decoupled from the system variable. This means that their joint probability density factorizes. One of the points we want to emphasize here is that this assumption, made for the sake of mathematical simplicity, is not necessarily the most interesting from the physical point of view. A more natural way to prepare the variable x with a given probability distribution could be as the steady state of an appropriate auxiliary model. A typical example would be the instantaneous change of a control parameter at $t = 0$, which leaves the system in an unstable situation but with a probability distribution that corresponds to a previously stable state. The point is that for any such steady state the joint probability density of the system variable and the noise variable cannot in general be factorized. In this section we discuss dynamical consequences of this initial coupling on transient dynamics, and particularly on the relaxation time.

The NLRT approach provides a particularly appropriate framework for this discussion since it permits a natural treatment of the case of random initial conditions (initially distributed states). Other advantages come from the absence of the absorbing boundary conditions that cause most of the troubles in the MFPT calculations if one uses the backward Kolmogorov equation. This makes it possible to obtain very general expressions that are relatively simple and provide a clear interpretation of the different transient and preparation effects, and particularly on the distinction between the purely non-Markovian ones and those that can be implemented in a quasi-Markovian description of the problem.

The theory of the decay of unstable states in terms of an NLRT was developed by J. M. Sancho and his colleagues in Barcelona.[136] In Section IV.A we outline their approach and explain how the main results were obtained. In Sections IV.B and IV.C we describe in turn analogue and digital simulations undertaken[49] to model the decay processes, and in Section IV.D we compare the results with predictions of the Barcelona theory. In Section IV.E we summarize the results and draw conclusions.

A. Theoretical Results for Unstable States

We now outline the Barcelona theory of the NLRT for the decay of unstable states; for further details, reference should be made to the original paper.[136]

1. Definitions and Models

The definition of an unstable state x_0 for a general model Eq. (42) is

$$v(x_0) = 0 \qquad v'(x_0) > 0 \tag{44}$$

If $v'(x_0) = 0$ then we have a marginal state, which is not the subject of this work. Nevertheless, our approach can be extended to that situation.[137] A very common model one can find in the literature as a prototype for the study of unstable states is the mean field Ginzburg-Landau model defined by the equation

$$\dot{x} = ax - bx^3 + \xi(t) \qquad a, b > 0 \tag{45}$$

$$\langle \xi(t)\xi(t') \rangle = \frac{D}{\tau} e^{-|t-t'|/\tau} \tag{46}$$

where $x_0 = 0$ is the unstable state. This model has two stable states at $|x| = \sqrt{a/b}$. However, even for such a simple model few results can be obtained exactly. To find analytical and explicit results and on the basis of the universal properties of the unstable states, we define a "representative model" which is mathematically simpler than Eq. (45) but captures its essential features. This model is defined by a linear equation

$$\dot{x} = ax + \xi(t) \qquad x \in [-R, R] \tag{47}$$

subject to two reflecting boundaries at $|x| = R$. The effect of these boundaries is essentially that of the nonlinearities in the model Eq. (45), that is, to provide a saturation regime that stabilizes the system at a typical distance R from the unstable state. The model Eq. (47) has thus the same parameters as the original model Eq. (45), with the equivalence $R = \sqrt{a/b}$ (a comparative study of these models can be found in Ref. 134). The models Eqs. (45) and (47) define the dynamics of the system for $t > 0$. Since we are interested in the preparation of steady-state-like initial conditions at $t = 0$, we have to fix now the preparation models which define the evolution of the system for $t < 0$, from an arbitrary state at $t = -\infty$. In our case, the preparation of the system at $t = 0$ for the models Eqs. (45) and (47) will be given respectively as the steady state of

$$\dot{x} = -a_0 x - bx^3 + \xi(t) \qquad a_0, b > 0, \quad -\infty < t < 0 \qquad (48)$$

and

$$\dot{x} = -a_0 x + \xi(t) \qquad -\infty < t < 0 \qquad (49)$$

Equation (49) is the linear approximation of Eq. (48). This approximation is justified when the intensity of the noise is very small, which is the limit we are interested in. In the case of large intensity of the noise, the dynamics of the relaxation are completely different and do not admit the usual picture of unstable states.[134] The steady-state probabilities for the Gaussian white noise case are respectively

$$P_{st}(x) \sim \exp\left[-\frac{a_0 x^2}{2D'} - \frac{bx^4}{4D'}\right] \qquad (50)$$

$$P_{st} \sim \exp\left[-\frac{a_0 x^2}{2D'}\right] \qquad (51)$$

where we admit in general a different noise intensity D' for $t < 0$. In the case of Gaussian exponentially correlated noise the steady distribution is exactly known for Eq. (49). The solution is the same as for Eq. (51), but with the substitution

$$D' \to \frac{D'}{1 + a_0\tau} \qquad (52)$$

The preparation models Eqs. (48) and (49) describe the typical situation in which the system suffers an instantaneous quench of the coefficient of the linear term $-a_0 \to a$, which leaves the system initially located around an unstable state. Up to very recently the usual initial condition considered was

$$P_i(x) = \delta(x) \tag{53}$$

From the experimental point of view, this means that the spread or uncertainty of the initial condition is much smaller than the actual intensity of the noise that will trigger the decay. However, in many situations both quantities will essentially originate from a unique noise source so that they will be of comparable size. If the noise is colored, particular attention has to be paid to the coupling of the system variable and the noise variable. When the system is prepared as a real steady state, the joint probability density of the variables x and ξ never factorizes; this means that due to the history of the system during the preparation at $t < 0$ there is a nonvanishing correlation between the noise and the system variable. Only if the noise source is not the same for $t < 0$ and $t > 0$ (or there is no correlation and the system starts from a point distributed at random) can the two variables be considered as statistically independent. Some recent theoretical results have considered the study of unstable states with the latter assumption. However, here we claim that this approach is not justified for simulations of quenched unstable states like Eq. (45) with Eq. (48) or those of Ref. 127, since, as we will show, the net effects of the initial coupling are nonnegligible.

For the sake of a better comprehension of the colored noise effects and for further reference we will start analyzing the white noise case with distributed initial conditions. In the rest of this section we always refer to the NLRT associated with the second moment $[\phi(x) = x^2]$.

2. White Noise

The NLRT problem for the decay of an unstable state with fixed initial conditions Eq. (53) and white noise has been solved.[133, 134] The conclusion was that for a (symmetric) unstable state, characterized by the parameter a and the length R separating the stable states from the unstable one, the NLRT, for small intensity of the noise is given by the general law

$$T^F(D, \tau = 0) = \frac{1}{a}\left[\frac{1}{2}\ln\left[\frac{R^2 a}{2D}\right] + C\right] + O(D) \tag{54}$$

where the logarithmic term is universal and the numeric constant C is characteristic of each model and accounts for the details of the deterministic relaxation in the nonlinear and saturation regimes. In Ref. 136, more details about the meaning and the calculation of that constant for fixed initial conditions and white noise can be found. For the linear model Eq. (47) we have

$$C_L = \tfrac{1}{2}(\gamma + 2\ln 2 - 1) = 0.482\ldots \tag{55}$$

where γ is the Euler constant. The value for the Ginzburg-Landau model Eq. (45) is

$$C_{NL} = C_L + \tfrac{1}{2} = 0.982\ldots \tag{56}$$

In this section we are interested in the corrections to that constant due to nonfixed initial conditions and to colored noise. Despite the fact that the value of the constant is model dependent, those corrections are expected to be universal. Let us see first the effects of the distributed initial conditions Eq. (51) in the model Eq. (47). The explicit calculation for the model Eq. (47) is given in Ref. 136. The final result in the limits of both $D, D' \to 0$ with D/D' finite reads

$$T_L^D(\tau = 0) = \frac{1}{a}\left[\frac{1}{2}\ln\left[\frac{R^2}{2(D/a + D'/a_0)} \right] + C_L \right] \tag{57}$$

We see that the effect of the distributed initial condition defines a larger effective diffusion coefficient. The difference between Eq. (57) and Eq. (54) is

$$T^D(\tau = 0) - T^F(\tau = 0) = -\frac{1}{2a}\ln\left[1 + \frac{D'a}{Da_0} \right] + O(D, D') \tag{58}$$

so the presence of distributed initial conditions speeds the decay of the unstable state. Its effect is important when the width of the initial distribution D' and the diffusion coefficient D are of comparable size, even if both are very small. Although Eq. (58) was calculated for the representative case of Eqs. (47), (49), and (51), that result has to be general ($R^2 = a/b$ in model Eq. (45)) since in the limits $D, D' \to 0$ only the linear part of the dynamics is involved. This result could also be used in real experiments to estimate the actual uncertainty on the initial condition.

3. Colored Noise

Now we will consider the case $\tau \neq 0$. To first order in τ the contribution to the NLRT is exactly that of the effective white-noise problem, so we can use the result Eq. (57) with the substitutions

$$D \to \frac{D}{1 - a_0\tau} \qquad D' \to \frac{D}{1 + a_0\tau} \tag{59}$$

And we get the final result

$$T^{\mathrm{D}}(\tau) = T^{\mathrm{D}}(\tau = 0) + O(\tau^2, D) \tag{60}$$

which has to hold also for the nonlinear model Eqs. (45) and (48). For decoupled (fixed) initial conditions we find,

$$T^{\mathrm{F}}(\tau) = T^{\mathrm{F}}(\tau = 0) + \frac{\tau}{2} + O(D, \tau^2)$$

$$= \frac{1}{2a}\left[\ln \frac{aR^2}{2D} + a\tau\right] + C + O(D, \tau^2) \tag{61}$$

This result coincides to first order in τ with the prediction of the quasideterministic theory (QDT).[138, 139] Assuming a distributed initial condition for the variable x given by Eqs. (51) and (52), but decoupled from the steady-state distribution for the noise, one gets

$$\hat{T}^{\mathrm{D}}(\tau) = T^{\mathrm{D}} + \frac{\tau}{1 + a_0} + O(\tau^2, D, D') \tag{62}$$

where the caret means uncoupled initial conditions. As in the previous cases, these results are claimed to be universal, so they have to hold also for the Ginzburg-Landau model.

B. The Analogue Experiment

An electronic circuit model of Eq. (45) was constructed.[49] It was driven by external noise and the time evolution of its response from given starting conditions was analyzed by means of a digital data processor. The operating principles of such circuits, and the general philosophy and practice of their application to problems in stochastic nonlinear dynamics, have been discussed in Section II.A and in more detail elsewhere.[32, 33] In this section, we describe the design and operation of the particular circuit used for the work, and we present the results that were obtained from it.

1. The Electronic Circuit

The circuit was based on two analogue multipliers (Analogue Devices type AD534) and a Miller integrator, connected as shown in Fig. 9. In terms of

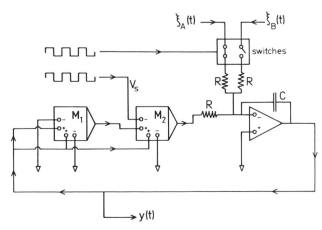

Figure 9. Block diagram of the analogue electronic circuit used to model the time evolution of the system from specified starting conditions. (After Ref. 49.)

the scaled variables y and t_s (see below) that were used to optimize its performance, the equation describing the operation of the circuit is readily shown to be

$$\tau_i \dot{y} = \left[V_s y - y^3 - \xi(t_s) \right] \tag{63}$$

where $\tau_i = RC$ is the integrator time constant, $\xi(t_s)$ is exponentially correlated Gaussian noise with correlation function

$$\langle \xi(t_s)\xi(t_s') \rangle = V_{rms}^2 \exp\left(-\frac{|t_s - t_s'|}{\tau_n} \right) \tag{64}$$

V_{rms}^2 is the variance of the noise voltage and τ_n is its correlation time. The sign and magnitude of the constant voltage V_s can be set externally; in practice the sign of $|V_s|$ is alternated periodically, while keeping $|V_s|$ constant, by application of a square wave of amplitude V_s, as shown in the figure. A second square wave of the same frequency and phase is used to operate a pair of solid-state switches through which different forms of external noise $\xi(t_s)$ can be applied to the circuit.

The operation of the circuit for fixed initial conditions was as follows. With V_s negative, so that the corresponding potential has a single minimum at $y = 0$, and with no noise applied (switches both open), the system was allowed to settle at $y = 0$. The sign of V_s was then suddenly made positive by the next half-cycle of the square wave. In terms of the corresponding double-well potential, the system was thereby prepared such that the representative particle was initially in unstable equilibrium on top of the central maximum at $y = 0$. Simultaneously, noise was applied to the circuit (by closing one of the switches) and a sweep of the Nicolet 1080 data processor was triggered. As the system subsequently relaxed toward one or other of the potential minima, its $y(t_s)$ trajectory was digitized (1024 points, 12 bits precision) and the 1st–4th moments were computed and added to their ensemble-averages. The process was then repeated. The number of blocks in the averages needed to provide acceptable statistical quality depended on the intensity of the applied noise, but was typically in the range 300–3000.

For operation with *random* but *coupled* initial conditions, the sequence of operations was exactly the same except that one of the noise switches was kept permanently closed. For *random* but *decoupled* initial conditions, two noise generators were used and the two switches were operated in phase opposition. Thus, no correlation existed between the noise producing random fluctuations about the equilibrium position ($y = 0$) at the bottom of the single-well potential prior to the transition in the sign of V_s, and the noise responsible for the relaxation toward one of the two minima in the double-well potential after the transition.

Particular care was taken to eliminate small dc offsets arising in the multipliers and integrator. The trimming circuitry that was used was of the conventional kind and is not shown in Fig. 9. The scaling factor in the transfer function of the multipliers was reduced below its default value of 10 in order to match the dynamic limits of the circuit more closely to the simulated model. The resultant loss of bandwidth (reduced to 80 kHz) was taken account of, first, by an appropriate choice of integrator time constant (typically $R = 10$ kΩ, $C = 100$ nF, hence $\tau = RC = 10^{-3}$ s) and, second, by an arrangement of the circuit (Fig. 9) such that the noise was injected directly into the summation input of the integrator.

2.   Scaling Laws for the Analogue Experiment

Although the system to be simulated was the Ginzburg-Landau model with unit coefficients

$$\dot{x} = x - x^3 + f(t) \tag{65}$$

it was often necessary in practice to use a value for the amplitude V_s of the square wave in Eq. (63) that differed from unity. This scaling was to ensure that the voltage swings in the circuit were as large as possible in comparison to its own internal noise while, at the same time, ensuring that "clipping" did not occur in any of the components, i.e., that their maximum voltage limits were not exceeded. To match the dynamical behavior of the system to the characteristics of the noise generators, and to enable the measurements to be completed reasonably quickly, the integrator time constant (see above) was chosen to be considerably less than unity. Consequently, the system Eq. (65) was also being modeled in terms of scaled time.[33] Assuming that the Gaussian random noise $f(t)$ in Eq. (65) has the correlation function

$$\langle f(t)f(t')\rangle = \frac{D}{\tau}e^{-|t-t'|/\tau} \qquad (66)$$

it is straightforward to demonstrate that, if we want to model Eq. (65) by means of data acquired from Eq. (63) with an integrator time constant τ_i, the proper scaling relations are

$$D = \langle \xi^2 \rangle \frac{\tau_n}{\tau_i} \frac{1}{V_s^2} = \frac{V_{rms}^2}{V_s^2}\tau \qquad (67)$$

$$\tau = \frac{\tau_n}{\tau_i} \qquad (68)$$

$$t = \frac{t_s V_s}{\tau_i} \qquad (69)$$

$$x = \frac{y}{V_s^{1/2}} \qquad (70)$$

3. Noise Generators for the Analogue Experiment

For most of the experiments with fixed initial conditions, a standard Wandel and Goltermann model RG2 noise generator was used. For the experiments with random but decoupled initial conditions, however, it was necessary to use a pair of noise generators whose outputs were uncorrelated: a twin-output pseudo-random noise generator was constructed specially for this purpose. It was of the type already described by Martano[140] and used successfully in analogue experiments.[48, 141] This kind of noise generator is based on the filtering of pseudo-random length sequences of dichotomous pulses to obtain Ornstein-Uhlenbeck noise according to the theorem of Rice.[142] The pulses are generated by a closed-loop-feedback shift register in which the ex-or feedback function is randomly inverted[140, 141] to ex-nor in order to eliminate skewness and thus obtain a virtually Gaussian distribution.

In the present version of the noise generator, two different stages of the same 18-stage feedback shift register are used as feedback inverters for two quite separate 41-stage shift registers. In this way, two independent random pulse sequences are obtained, which, after filtering, can serve as independent (uncorrelated) noise sources. The clock frequency is 4 MHz. With the filter time constants set to give a frequency cutoff above 40 kHz, the distribution functions of the noise at the outputs were found to be Gaussian to more than ± 4 standard deviations. The period of the pseudo-random sequences is about 6.5 days. This time exceeded by a large factor the characteristic times both for an input sweep to the Nicolet data processor (typically 10 ms) and also for the completion of an ensemble average of the statistical moments of several hundred such sweeps (typically 15 min).

4. Analogue Experimental Results

On completion of the ensemble averages of moments for any given set of conditions, the nonlinear relaxation time T_ϕ defined by Eq. (43) was computed, usually that from the second moment T_2. It was obviously impossible for the upper limit of integration to be ∞ in practice; rather, the data-processor sweep time was adjusted such that $\langle x^2(t) \rangle$ had effectively settled at its final value $\langle x^2 \rangle_{st}$ well before the end of the sweep. Some results of these procedures are presented below. In the interests of clarity, the raw values measured in terms of y and t_s, for various values of V_s, have all been scaled to the corresponding values of x and t, so as to be consistent with Eq. (65).

In Fig. 10a is shown a typical $x(t)$ input sweep immediately after being digitized by the Nicolet data processor; part (b) shows the same signal after being squared. A succession of such signals are ensemble-averaged to form the moments $\langle x^n(t) \rangle$. Figure 11a shows the measured evolution of the second moment under weak noise forcing. The shape of the curve is quite different from that of Fig. 11b, corresponding to relatively strong noise, where the system starts to rise almost immediately from its initial unstable equilibrium position at $\langle x^2 \rangle = 0$. The values of T_2 extracted from data such as those of Fig. 11 for various values of τ and D and for different initial condition are presented and discussed below.

C. The Digital Simulation

The digital simulation[49] was performed in Barcelona on an IBM-9375/60 computer using standard algorithms for white and colored noise.[101] Although further removed from being a real physical system, this type of simulation is closer to the ideal theoretical model and it has fewer sources

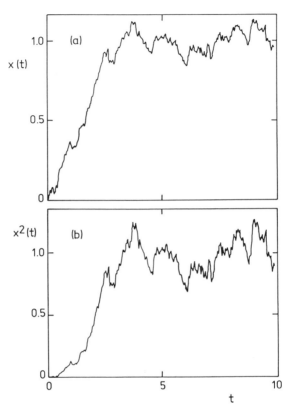

Figure 10. (a) An example realization of $x(t)$, starting from initial condition $x = 0$ at $t = 0$, with $\log(1/D) = 2$ and $\tau = 0.010$. (b) The same realization as in (a), but after being squared, ready to be added into the memory block in which $\langle x^2(t) \rangle$ is being ensemble-averaged. (After Ref. 49.)

of error than the analogue experiment. Hence, it can be used as a more precise test of the theoretical predictions.

An ensemble of typically 1000 or 2000 realizations is made for each point. The step of integration of the algorithm lies usually between 0.005 and 0.0005. The simulation is carried up to a time typically of the order of $4T_0$, where $T_0 \approx \frac{1}{2}\ln(1/D)$ is an estimate of the time scale of the process. The steady-state value of $\langle x^2 \rangle$ is determined using the data in the time interval $(3T_0, 4T_0)$. The NLRT is obtained from simple numerical integration (trapezoidal rule) of Eq. (43), from the time discretized evolution of $\langle x^2 \rangle$ between $t = 0$ and $t = 3T_0$.

The simulation of the initial conditions needs a bit more attention. For the case of uncoupled but distributed initial conditions we generate the

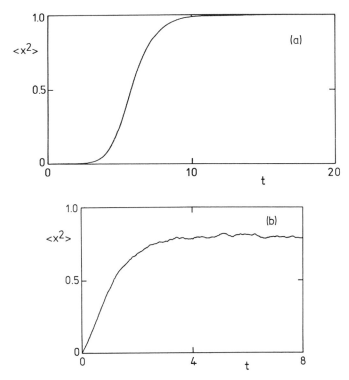

Figure 11. Measurements of the relaxation of the second moment $\langle x^2(t) \rangle$ from the fixed initial condition $x = 0$ at $t = 0$ with a noise correlation time $\tau = 0.010$, for two different noise intensities D: (a) an average of 300 realizations with $\log(1/D) = 4.69$; (b) 3000 realizations with $\log(1/D) = 0.80$. (After Ref. 49.)

initial values of x_0 according to the Gaussian distribution Eq. (51), and the values of the noise $\mu_0 = \sqrt{\tau}\,\xi(0)$ with another independent Gaussian number of variance D. For the case of coupled initial conditions, $\tau \neq 0$, we generate the initial values of x_0 and μ_0 using their joint probability density

$$P_{\mathrm{st}}(x_0, \mu_0) \sim \exp\left[\frac{1 + a_0\tau}{2D}\left(\mu_0 - a_0\sqrt{\tau}\,x_0\right)^2 - \frac{a_0}{2D}(1 + a_0\tau)x_0^2\right] \quad (71)$$

which is the steady-state solution of the linear model Eq. (49) with colored noise. To generate the pair of variables (x_0, μ_0) according to the statistics of Eq. (71) we use two independent Gaussian random numbers of zero mean and unit variance y_1 and y_2, so that

$$x_0 = \frac{\sqrt{D\tau}}{1 + a_0\tau} y_1 + \frac{\sqrt{D/a_0}}{1 + a_0\tau} y_2 \tag{72}$$

$$\mu_0 = \sqrt{D}\, y_1 \tag{73}$$

For the preparation with the nonlinear model Eq. (45), the statistics of (x_0, μ_0) are not strictly Gaussian. However, for the small intensities of the noise involved, the linear approximation (which we use also in the theoretical predictions) is completely justified since the corrections from the nonlinearity are of higher order than the theoretical predictions.

D. Results and Comparison with Theory

Given that the theoretical predictions are asymptotic (small τ, small D), the analogue and digital experiments[49] were useful not only as a confirmation of them, but also in determining their range of validity. The results we present below correspond to the Ginzburg-Landau model Eq. (45) with $a = a_0 = b = 1$.

The concrete predictions that we want to check are Eqs. (58) and (60)–(62). The reference framework will be the white noise case with fixed initial condition Eq. (64). This result was tested in Refs. 133 and 134 and it is valid for $D < 0.01$.

Our first result, Eq. (58), has been checked for $D = D'$, where D' is the intensity of the noise prior to the transition. The difference predicted in this case is

$$T^D(\tau = 0) - T^F(\tau = 0) = -\tfrac{1}{2}\ln 2 \tag{74}$$

In Fig. 12 both analogue and digital simulations support the predicted shift for all $D < 0.01$. In this regime, it is clear that the same net effect of speeding the decay is always present, irrespective of how small D is.

Our second prediction, Eq. (60), establishes that in the case of coupled initial conditions (the "quench" model Eqs. (45) and (48)) the effect of colored noise relative to that of white noise is of higher order than τ. We have checked the importance of these higher-order corrections with digital and analogue simulations. In Fig. 13 we see that the relative deviation between the two cases is practically zero even for relatively large values of $\tau \approx 1$. The conclusion is that, as far as the decay time is concerned, for a quench experiment the colored noise case is not essentially different from the white noise case, so in this kind of problem the white noise assumption is justified.

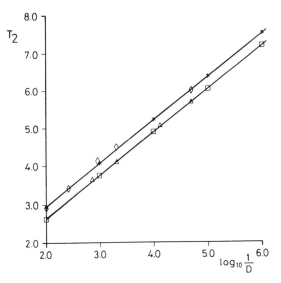

Figure 12. The NRLT T_2 plotted as a function of $\log(1/D)$ for white noise for fixed and distributed initial conditions $(D' = D)$. The straight lines are the theoretical predictions. Stars $(D' = 0)$ and squares $(D' = D)$ correspond to digital simulation, and rhombics $(D' = 0)$ and triangles $(D' = D)$ to analogue simulation. (After Ref. 49.)

The third prediction, Eq. (61), refers to the NLRT for fixed initial conditions and colored noise. For the model Eq. (45) the prediction is

$$T^{\mathrm{F}}_{\mathrm{NL}}(\tau) = \frac{1}{2}\left(\ln\frac{1}{2D} + \tau\right) + C_{\mathrm{NL}} \qquad (75)$$

In Fig. 14 we can see the excellent agreement between Eq. (75) and both kinds of simulation, covering a wide range of validity for different values of $D < 0.01$ and $\tau < 1$.

The last prediction, Eq. (62), refers to the case of distributed initial conditions in presence of colored noise. For the same distribution of the system variable x, the difference between being and not being coupled to the colored noise in our case, from Eq. (62) reads

$$\hat{T}^{\mathrm{D}}(\tau) - T^{\mathrm{D}}(\tau) = \frac{\tau}{2} + O(\tau^2, D, D') \qquad (76)$$

In Fig. 15a we see that this prediction is confirmed in a range of $\tau < 0.5$. The NLRT results for T_2 in theory, digital simulation and analogue

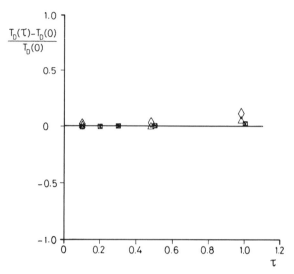

Figure 13. Relative deviation of the NLRT with coupled initial conditions with respect to the white-noise case vs. τ. Stars ($D = 10^{-6}$) and squares ($D = 10^{-5}$) correspond to digital simulations, and triangles ($D = 2.0 \times 10^{-5}$) and rhombics ($D = 4.7 \times 10^{-4}$) to analogue simulations. (After Ref. 49.)

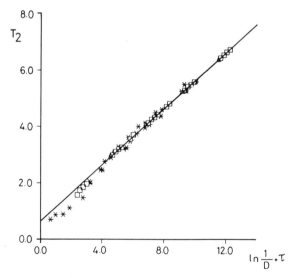

Figure 14. The NLRT T_2 with fixed initial conditions plotted against the scaling variable $\ln(1/D) + \tau$. The straight line is the theoretical prediction. Triangles and squares correspond to digital simulation with white noise and colored noise ($\tau < 1$), respectively. Stars correspond to analogue simulations for colored noise ($\tau < 1$). (After Ref. 49.)

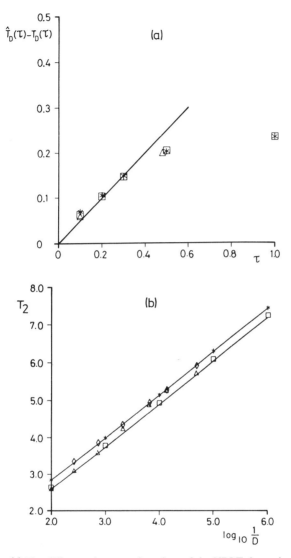

Figure 15. (a) The difference between the values of the NRLT determined for coupled and uncoupled initial conditions, as a function of the noise correlation time τ. The straight line corresponds to the theoretical prediction of Eq. (76). Stars represent ($D = D' = 10^{-6}$) and squares ($D = D' = 10^{-5}$). (b) The NRLT T_2 plotted as a function of $\log(1/D)$ for $\tau = 0.5$. The upper line and data refer to uncoupled initial conditions, and the lower line and data refer to coupled initial conditions. Stars and squares correspond to digital simulations, and rhombics and triangles to analogue experiments. (After Ref. 49.)

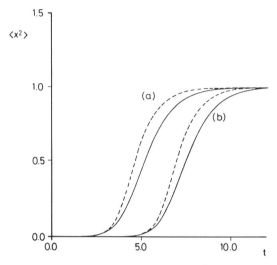

Figure 16. Digital simulation of the second moment $\langle x^2(t) \rangle$ for fixed initial conditions. The dashed curves correspond to the linear model with reflecting boundaries, $a = R = 1$, and the solid curves correspond to the Landau model ($a = b = 1$). (a) $D = 10^{-3}$; (b) $D = 10^{-6}$. (After Ref. 49.)

experiment are plotted as functions of $\log_{10}\left(\frac{1}{D}\right)$ in Fig. 15b; the agreement is excellent.

Finally, we have plotted in Fig. 16 the evolution of the second moment for the Ginzburg-Landau model Eq. (45) and the linear model Eq. (47) for fixed initial conditions and for different intensities of the noise (digital simulation), in order to illustrate the universal law Eq. (54). On one hand, it is clear that the effect of decreasing the noise intensity is just to shift the entire curve. On the other hand, the details of the shape in the nonlinear and saturation regimes are characteristic of each model and independent of D. Therefore, the constant C contains essentially the information on these last stages of the relaxation. The difference ΔC between the C of a given model and the reference value of the model Eq. (47) is essentially the area enclosed between the respective curves. This quantity is a measure of how fast the relaxation is, once the system has escaped from the unstable region.

E. Conclusions

The study of the decay of unstable states[136] has been described, showing how theoretical predictions for the NLRT may be obtained in different situations. These results were deduced using asymptotic methods, and the analogue experiments and digital simulations have demonstrated[49] their wide range of validity.

The NLRT provides a global characterization of the decay of unstable states including the final stages of the relaxation. The general picture establishes a universal logarithmic term plus a model-dependent constant which accounts for the nonlinear and saturation regime. This picture is valid for small noise intensities (far above threshold condition). For the Ginzburg-Landau model this condition turns out to be valid for $bD/a^2 < 10^{-2}$. In the large noise intensity ($bD/a^2 > 1$) domain, the system becomes marginally stable (near threshold).[134]

In this study, interest centered on the corrections introduced to this general picture by the presence of colored noise and random initial conditions. In all the cases studied, these effects can be taken into account as corrections to the model-dependent constant. Nevertheless, due to the universal character of the responsible mechanisms, which involve only the linear regime, these effects are claimed to be universal they are also present when other definitions of characteristic times like MFPT are used.

The prediction for the case of distributed initial conditions with white noise has been checked in the case $D' = D$ and indirectly in other cases in the calculations involving the Markovian contributions in the colored noise case. The domain of validity is that of the unstable state picture we commented on above. It is remarkable that the effect depends on the quotient D/D', so there may be a finite contribution irrespective of how weak the noise is. This could have some relevance in practice in determining the actual uncertainty on the initial conditions in switch-on problems or the effects of other internal noise sources in a "quenching" experiment,[127] for instance.

The colored noise predictions were carried out up to first order in τ. The simulations for finite values of τ have shown that the predictions are quite good even for moderate values of τ. The prediction of independence of τ for the coupled case has turned out to be very good up to $\tau \approx 1$. This remarkable result means that in a "quenched" experiment, it is almost irrelevant whether one has colored or white noise, as far as the decay time is concerned. The prediction for the uncoupled initial conditions, which is the most artificial case, has a shorter domain of validity. The first-order contribution on τ is valid up to $\tau < 0.5$. In the particular case of fixed initial condition with colored noise, the predicted law has been shown to be valid for $\tau < 1$.

In general, it was found[49] that digital simulation mimics the uncoupled case better than the coupled one. The contrary happens in analogue experiments that model better the coupled case. This is not unreasonable, given that the coupled case is closer to any real physical situation.

V. SWEPT PARAMETER SYSTEMS

We now consider the situations that can arise when, instead of being changed almost discontinuously, the control parameter is swept *slowly* through the bifurcation point. In this case it is known that the observation of the dynamical instability is delayed with respect to the static bifurcation point.[146, 147] However, due to the stochastic nature of the systems, there is no unique way of actually defining the time at which the instability is judged to have occurred. To date, two alternative characterizations have been proposed: a moments approach[154] and an MFPT approach.[63, 64, 155] A comparison of these two different characterizations is presented.

It is well known that most nonlinear systems can display a number of different stable attractors in different regions of parameter space. The transition from one of these stable attractors to another occurs under the influence of a change in the control parameter(s) of the system. The actual critical or bifurcation point is defined to be the point at which one stable attractor loses stability in favor of another when the control parameter is increased by an infinitesimal amount; this is typically associated with one or more of the eigenvalues of the system going to zero, which can, in itself, be used to define the bifurcation point.

A property common to all critical points is critical slowing down, which is a consequence of the restoring forces in the system diminishing in magnitude as the critical point is approached. This, combined with the fact that any internal noise inherent in the system is effectively amplified, makes experimental measurements in the vicinity of a critical point extremely difficult.[143] To try to overcome this problem, experimentalists have swept the control parameter through the bifurcation point, the assumption being that as long as the sweeping rate is small enough the system will remain close to equilibrium.[144, 145] This is often termed an adiabatic approximation. We will see later that the adiabatic approximation is generally valid, except in the region of the bifurcation point itself.

Although swept parameter systems are of general interest, the majority of work has been carried out in connection with the laser transition. This is largely because to switch on a laser, the pump (control) parameter must be switched through the lasing threshold (bifurcation point). Initial theoretical studies concentrated on a deterministic investigation of different laser models. These resulted in it being demonstrated, quite generally, that the dynamical bifurcation point is delayed with respect to that of the static one.[146, 147] A consequence of this is that the bifurcation diagram obtained using a swept parameter method can be greatly modified due to the nonzero sweeping rate. In addition to this it was predicted that systems could show spurious broken symmetries and hysteresis effects.[148]

However, subsequent experimental investigations of these predictions met with only partial success[149]; this was largely due to the effects of the noise, which is inherent to all physical systems and cannot be neglected in the region of an instability point. It thus became apparent that a stochastic description of a dynamical bifurcation point was necessary.[150]

A general approach in addressing the problem of a stochastic description is to identify it as a stability problem, with an associated lifetime. However, the question still remains of how to actually define the lifetime of the state. Obviously, the definition must involve some averaged quantity due to the randomness imposed by the noise, but in practice there is no unique way of formulating this average. At present two alternative approaches have emerged, each of which employs different averaging procedures. The first of these involves calculating the first moment of the stochastic trajectories $x^2(t)$ (Refs. 150 and 154), where $x(t)$ represents a single trajectory of the state variable as the control parameter is swept through its critical value. The dynamical bifurcation point is then defined to have been reached when $\langle x^2(t = t^*_{\text{mom}}) \rangle = x^2_{\text{th}}$, where x^2_{th} is an arbitrary threshold value and t^*_{mom}, which is identified with the lifetime of the state, is the time at which this condition is satisfied; hereafter this approach will be referred to as the moments approach. The second approach considers the mean first passage time (MFPT) to the threshold x^2_{th} (Refs. 155 and 156). In this case one measures the first time at which the condition $x^2(t = t') = x^2_{\text{th}}$ is satisfied for each individual trajectory; the lifetime or bifurcation time is then defined to be $t^*_{\text{MFPT}} = \langle t' \rangle$. Due to the different averaging procedures employed, these two methods should yield different results, that is, $t^*_{\text{MFPT}} \neq t^*_{\text{mom}}$. In this section, we show that this is indeed the case and we describe a detailed comparison between the two approaches. The theory for the moments approach is taken from the paper by Zeghlache et al.,[154] which hereafter will be referred to as ZMVB; much of the discussion centers on a new theory of the MFPT approach presented in Refs. 63 and 64, to which reference may be made for further details.

A. A Model for a Dynamical Bifurcation

1. A Deterministic Description

In the study of critical phenomena it is common to linearize equations describing the system about the stable state under investigation. The loss of stability of this state is then determined by a linear stability analysis. This technique can also be applied if we are interested in the dynamics of the system around the instability point itself. This is because the nonlinear effects govern only the final relaxation to the new stable state. For this reason the simplest linear differential equation possessing a critical point will be considered here and in the remainder of the study; that is,

$$\frac{dx(t)}{dt} = \mu x(t) \tag{77}$$

where μ is the control parameter and $x(t)$ is the state variable. The solution of Eq. (77) reads

$$x(t) = x(0)\, e^{\int_0^t \mu\, dt} = x(0)\, e^{\mu t} \tag{78}$$

It is obvious from Eq. (78) that $x(t) \to 0$ for $\mu < 0$; that is, the system approaches its stable state as $t \to \infty$. However, for $\mu > 0$ the variable $x(t)$ diverges; therefore, $\mu = 0$ is identified as the bifurcation point. In a multidimensional system this is equivalent to one or more of the eigenvalues resulting from a linear stability analysis of the system going to zero. This simple model also exhibits the critical slowing down that is common to all critical points, and has important consequences for the dynamics of the system in the vicinity of the instability. The relaxation time of the system is identified with μ^{-1}, a quantity that obviously diverges as $\mu \to 0$. One can now easily see why the adiabatic approximation fails as μ is swept through the bifurcation point. This is because, no matter how slowly the sweeping is carried out, as the bifurcation point is approached a point will always be reached when the rate of relaxation of the system will be much smaller than the sweeping rate.

A dynamical bifurcation may be defined in a similar fashion but in this case for a finite t; thus this is essentially a nonequilibrium description. Consider a time-dependent control parameter $\mu(t)$ with the following properties:

$$\mu(t) = \begin{cases} \mu(t) < 0: t < \bar{t} \\ \mu(t) = 0: t = \bar{t} \\ \mu(t) > 0: t > \bar{t} \end{cases} \tag{79}$$

The point of a dynamical bifurcation could now be defined by

$$\int_0^{t^*} \mu(t)\, dt = 0 \qquad t^* > 0 \tag{80}$$

It is interesting to note that because this definition involves time, the point of a dynamical bifurcation can be described uniquely by either the bifurcation time t^* or the value of the control parameter at the bifurcation point $\mu(t^*)$. In general $t^* > \bar{t}$ and hence the point of a dynamical bifurcation is said to be delayed with respect to the static bifurcation point, which is still identified with $\mu(t = \bar{t}) = 0$. This point has received much attention in the literature. However, this definition of a dynamical bifurcation is in itself quite arbitrary and the delay described above is simply a consequence of our adopted definition. For example, one could equally argue that, because the $x = 0$ solution is an asymptotically stable state, the instability is first manifest in the system by an increase in $x(t)$. This would suggest an alternative definition of a dynamical bifurcation, that is,

$$\left.\frac{dx(t)}{dt}\right|_{t^*} = 0 \tag{81}$$

With this definition it can be seen from Eqs. (78) and (79) that we obtain a bifurcation point given by

$$\mu(\bar{t}) = 0 \tag{82}$$

In this case there is no delay with respect to the static bifurcation point since $t^* = \bar{t}$. However, the main reason for adopting definition Eq. (80) is by analogy to an equilibrium system where the instability occurs as an exponent (eigenvalue) of the system goes to zero. In addition to this, an important feature that a definition for a dynamical bifurcation must possess is that it should be experimentally operational. The definition Eq. (80) gives a simple test to locate the bifurcation point because, when the condition Eq. (80) is satisfied, we have

$$x(t) = x(0) \tag{83}$$

whereas with the second definition we have to detect an increase in $x(t)$, which is arguably more difficult. However, in a stochastic description of a dynamical bifurcation even the definition Eq. (80) has certain drawbacks.

2. A Stochastic Description

Although noise in physical systems can have many different origins, it is usually incorporated in a model as either parametric fluctuations (multiplicative noise) or additive noise. A complete description of a dynamical bifurcation should take both of these types of noise into account; on doing this the following stochastic differential equation is arrived at:

$$\dot{x} = (\mu(t) + \eta(t))x + \xi(t) \tag{84}$$

$$\langle \eta(t) \rangle = 0; \qquad \langle \eta(t)\eta(t') \rangle = C_\eta(t - t')$$

$$\langle \xi(t) \rangle = 0; \qquad \langle \xi(t)\xi(t') \rangle = C_\xi(t - t')$$

Here, $\xi(t)$ represents the additive noise and $\eta(t)$ the multiplicative noise; we also assume that the independent stochastic processes $\eta(t)$ and $\xi(t)$ are Gaussianly distributed with noise intensities given by D_η and D_ξ respectively (C_η and C_ξ are arbitrary correlation functions). To illustrate possible physical origins for these two noise sources we will consider the example of the laser in the good cavity limit. In this case $\mu(t)$ is related to the pump parameter and $\eta(t)$ represents fluctuations in the pump parameter. The stochastic variable $\xi(t)$ models the spontaneously emitted photons, or quantum noise as it is commonly called. The state variable $x(t)$ represents the slowly varying amplitude of the electric field in the optical cavity. It is interesting to note that if $x(0) = 0$ prior to the pump parameter being swept through the lasing threshold (the bifurcation point)—which requires that D_ξ be zero during the preparation of the system (external noise case)—then the electrical field would remain zero even for positive $\mu(t)$; that is, in the absence of noise, a laser would not work. This is easily understood since even in the presence of a population inversion, photons must be present in the laser cavity to initiate stimulated emission.

Returning to the problem of defining a dynamical bifurcation point, the above example highlights a problem with the definition Eq. (80). In many physical situations the noise is internal to the system; that is, we cannot control exactly the initial conditions and the noise is ever present. This is the case in the above example, where $x(0) \neq 0$ below the bifurcation point but is actually determined by the spontaneous emission rate (additive noise). Therefore, the identity Eq. (83) no longer offers an experimental criterion for determining the bifurcation point because $x(0)$ is unknown. There is in fact an additional problem, even if $x(0)$ can be determined. This is because $x(0)$ is now itself a stochastic variable with an associated probability density. Therefore, condition Eq. (83) can be satisfied while $\mu(t)$ is well below the static bifurcation point. Consequently, this definition does not take into account the dynamics of the system itself. For this reason a dynamical bifurcation is usually judged to have occurred for an individual realization when the system reaches an arbitrary threshold value x_{th}, where $x_{th} > x(0)$. When the system is subject to internal noise the condition $x_{th} \gg \langle x^2(0) \rangle^{1/2}$ is usually considered.

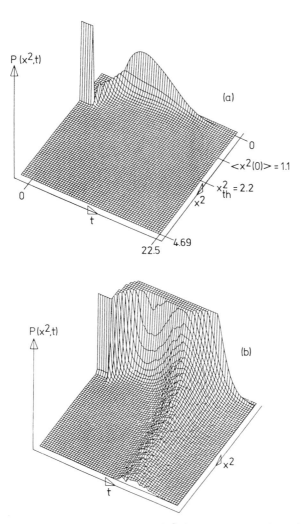

Figure 17. The time evolving distribution $P(x^2, t)$, as measured in the analogue simulation for the case of external noise. The system is held initially at $x = x(0)$ until $t = 0$, at which time noise is applied and the sweep of μ is simultaneously initiated. Both (a) and (b) represent the same set of data, but the vertical scale in (b) has been magnified compared to that in (a) to reveal more clearly the evolution at later times where $P(x^2, t)$ has become small because of the relatively high velocity. (After Ref. 63.)

The average lifetime (bifurcation time) of the system can be formulated by two different methods, each with its own averaging procedure. The first of these methods involves calculating the second moment $\langle x^2(t) \rangle$ of the realizations and finding the time at which this averaged trajectory reaches x_{th}^2. We will denote the bifurcation time obtained using this method by t_{mom}^*. The second method obtains the first passage time (FPT) for each individual trajectory to reach x_{th}^2, and then averages these FPTs to obtain a mean first passage time (MFPT) t_{MFPT}^*; this is then taken to be the bifurcation time. [The reason that the square of the state variable is

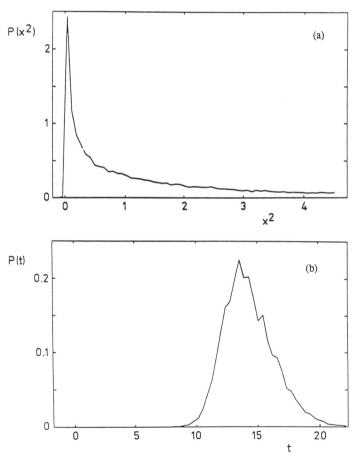

Figure 18. Sections taken through the evolving distribution of Fig. 17 to clarify the differences between the ZMVB and MFPT definitions of t^*. (a) Section giving the $P(x^2)$ distribution at the t^* defined by ZMVB. (b) Section at x_{th}^2 giving the $P(t)$ distribution from which the MFPT definition of t^* is derived. (After Ref. 63.)

considered is because in some cases, for example, with internal noise and where the system is symmetric, $\langle x(t) \rangle = 0$. In addition to this, if one is making a connection to laser applications the relevant quantity is the *intensity* of the electric field, which is proportional to x^2.] The difference in these two averaging procedures can be illustrated by considering Figs. 17 and 18. Figure 17 shows a time-evolving probability density for the case of external noise measured by the analogue simulator to be described below. The ZMVB (moments) approach corresponds to taking the average $\int_0^\infty x^2 P(x^2, t) \, dx$ at time t^*_{mom} (Fig. 18a), whereas the MFPT approach corresponds to taking the average $\int_0^\infty t P(x^2, t) \, dt$ at the threshold x^2_{th} (Fig. 18b). These two averaging procedures are obviously greatly different and, therefore, there is no reason to expect them to give the same result.

Each approach has advantages and disadvantages. In the case of the moments approach it is possible, for the linear equation considered here, to obtain exact analytical results and it was in this spirit that this procedure was first used. On the other hand, an MFPT approach is necessarily approximate, although it does have the advantage that once the FPT distribution is obtained, higher-order moments can also be obtained (for example, the variance of the bifurcation times) and it is thus arguably a more complete description of the process. The MFPT approach is also the one most commonly used by experimentalists. Having said this, there is, in principle, no reason to prefer one definition over the other.

B. Theory of the Mean First Passage Time

The theory for obtaining the FPT distribution for a system described by a single-variable Langevin equation subject to white noise forcing is well established. The procedure is to solve the associated time-dependent Fokker-Planck equation in the presence of an absorbing boundary placed at the threshold value of interest. This would yield the probability density $P'(x, t|x_0, 0)$, that is, the probability of the system being between x and $x + dx$ at time t, given that it was at x_0 at a time $t = 0$, subject to the condition that the "particle" had not crossed x at a previous time $t' < t$. The FPT distribution $F(t, x_{th}|x_0)$ is then given by

$$F(t, x_{th}|x_0) = -\frac{\partial}{\partial t} \int_{-\infty}^{x_{th}} P'(x, t|x_0) \, dx \tag{85}$$

The MFPT and higher moments of the FPT distribution can then be extracted in the normal manner:

$$\langle t^n \rangle = \int_0^\infty t^n F(t, x_{th}|x_0) \, dt \tag{86}$$

Alternatively, if one is interested only in the MFPT or higher moments, the Kolmogorov backward equation can be used to obtain a recurrence relationship for the moments and then, in principle, all the moments can be obtained.

Although this theory is well established, it is far from straightforward to apply in practice. In fact, time-dependent solutions of the Fokker-Planck equation are known in only a few special cases and these are in the absence of an absorbing boundary condition. In the case of colored noise the situation is even more difficult. For exponentially correlated noise, which is analyzed in this section, one has to consider a two-variable Markovian process and consequently the now two-dimensional Fokker-Planck equation involves the noise variable directly. As a result, most theories of the MFPT are necessarily approximate.

The MFPT has previously been calculated, for a similar system to the one studied here, for the cases of internal white noise[156] and internal colored noise.[155] The method used to calculate the MFPT was based on the idea of averaging over an "effective" initial condition. Essentially, this approach treats the evolution of the system as deterministic at times greater than $t = 0$, and averages over the stochastic initial conditions. This technique has been successfully employed in the study of the decay of an unstable state.[128, 157, 158] However, the problem with this approach is that it is difficult to interpret in the framework of the standard MFPT described above. In addition to this, the calculations of Torrent et al.[155, 156] considered the case of the weak noise limit only. The theory presented below extends these results to the cases of external and multiplicative noise and also extends them to encompass the whole parameter range where a MFPT approach is meaningful. In addition to this the theory is developed within the standard MFPT framework and as a consequence the assumptions used are easily interpretable.

The theory[63, 64] outlined here makes use of a characteristic that is particular to a swept parameter system. Figure 20b in Section V.C shows four typical trajectories obtained by analogue simulation of Eq. (84) with $D_\xi = 0$. After an initial decrease from their starting point, the trajectories can be seen to increase rapidly as they begin to diverge to infinity. If we now consider their FPTs to a threshold value $x_{th}^2 > x^2(0)$ it can be seen that, after attaining the value x_{th}^2, the probability of returning to this value some time later is extremely small. More precisely, considering the joint probability density $P(x_{th}, t; x_{th}, s)$, where $s > t$, the assumption is that this joint probability density is a highly peaked and narrow function centered about t. This is a tacit assumption that is implicit in all current theories for the MFPT in swept parameter systems. The advantage of making this assumption is that, in the parameter range where this assumption holds, we can replace the absorbing boundaries with natural boundary conditions. In other words, the probability density $P'(x, t|x_0)$ can be approximated by the standard transitional probability $P(x, t|x_0)$. The FPT distribution can then be obtained by using Eq. (85). This is the only assumption that will be employed in the derivation of the MFPT presented below. Following ZMVB, Eq. (84) will be considered in three different situations: (1) internal noise and $D_\eta = 0$; (2) external noise and $D_\eta = 0$; and (3) external noise and $D_\xi = 0$. External noise refers to the fact that we are free to prepare the system at some initial value $x(0)$ and then apply the noise (which is itself prepared at $t = -\infty$) at $t = 0$; while internal noise means that the system is prepared with the noise at $t = -\infty$.

First we consider the derivation of the FPT distribution with $D_\eta = 0$, that is, for the internal and external noise cases. The starting point is the solution of Eq. (84) which reads

$$x(t) = x(0)e^{\int_0^t \mu(t)\,dt} + e^{\int_0^t \mu(t)\,dt}\int_0^t \xi(s)e^{-\int_0^s \mu(r)\,dr}\,ds \tag{87}$$

This can be rewritten:

$$y(t) = x(0) + \int_0^t \xi(s)e^{-\int_0^s \mu(r)\,dr}\,ds \tag{88}$$

$$y(t) = x(t)e^{-\int_0^t \mu(t)\,dt} \tag{89}$$

The stochastic variable $y(t)$ is a linear combination of Gaussianly distributed stochastic variables (given by Eq. (88)) and hence is itself a Gaussian variable. Therefore, its probability density is completely characterized by its mean $\langle y \rangle$ and variance $\sigma^2 = \langle y^2 \rangle - \langle y \rangle^2$, so

$$P(y) = \frac{1}{\sqrt{2\pi\sigma^2}}e^{-[(y-\langle y\rangle)^2]/2\sigma^2} \tag{90}$$

Performing a transformation of variable of Eq. (90) from y to x using (89), the time-dependent probability density for the process $x(t)$ can be obtained. Upon doing this we obtain

$$P(x, t|x_0) = \frac{e^{-\int_0^t \mu(t)\,dt}}{\sqrt{2\pi\sigma^2}}\exp - \left\{\frac{\left[x\,e^{-\int_0^t \mu(t)\,dt} - \langle y\rangle\right]^2}{2\sigma^2}\right\} \tag{91}$$

where

$$\langle y \rangle = \langle x(0) \rangle \tag{92}$$

$$\sigma^2 = \langle x^2(0) \rangle - \langle x(0) \rangle^2$$

$$+ 2 \int_0^t \langle x(0)\xi(s) \rangle e^{-\int_0^s \mu(r)\,dr}\,ds \tag{93}$$

$$+ \int_0^t \int_0^t \langle \xi(p)\xi(q) \rangle e^{-\int_0^p \mu(r)\,dr - \int_0^q \mu(s)\,ds}\,dp\,dq$$

This result is exact, the only restriction that has been imposed being the Gaussian nature of $\xi(t)$. There is no restriction on the form of $\mu(t)$ or the correlation function $\langle \xi(p)\xi(q) \rangle$. The FPT distribution to the threshold x_{th}^2 is now given by

$$F(t, x_{th}^2 | x_0) = -\frac{\partial}{\partial t} \int_{-x_{th}}^{x_{th}} P(x, t | x_0)\,dx \tag{94}$$

Introducing the change of variable,

$$W = \frac{x\,e^{-\int_0^t \mu(r)\,dr} - \langle y \rangle}{\sqrt{2\sigma^2}} \tag{95}$$

Eq. (94) can be rewritten:

$$F(t, x_{th}^2 | x_0) = -\frac{1}{\sqrt{\pi}} \left[\frac{\partial W_+}{\partial t} e^{-W_+^2} - \frac{\partial W_-}{\partial t} e^{-W_-^2} \right] \tag{96}$$

$$W_{\pm} = \frac{\pm x_{th}\,e^{-\int_0^t \mu(r)\,dr} - \langle y \rangle}{\sqrt{2\sigma^2}} \tag{97}$$

For the case of external noise $x(0)$ is a constant and hence,

$$\langle y \rangle = x(0) \tag{98}$$

$$\sigma^2 = \int_0^t \int_0^t \langle \xi(p)\xi(q) \rangle e^{-\int_0^p \mu(r)\,dr - \int_0^q \mu(s)\,ds}\,dp\,dq \tag{99}$$

For internal noise the system is allowed to reach equilibrium with the noise before the sweep $\mu(t)$ is applied. Consequently, $x(0)$ is a stochastic variable which is determined by the noise; in this case we have

$$\langle y \rangle = 0 \tag{100}$$

$$\sigma^2 = \langle x^2(0) \rangle + 2 \int_0^t \langle x(0)\xi(s) \rangle e^{-\int_0^s \mu(r)\,dr}\,ds \tag{101}$$

$$+ \int_0^t \int_0^t \langle \xi(p)\xi(q) \rangle e^{-\int_0^p \mu(r)\,dr - \int_0^q \mu(s)\,ds}\,dp\,dq$$

and Eqs. (96) and (97) simplify to

$$F(t, x_{\text{th}}^2 | x_0) = -\frac{2}{\sqrt{\pi}}\frac{\partial W}{\partial t}e^{-W^2} \tag{102}$$

$$W = \frac{x_{\text{th}}}{\sqrt{2\sigma^2}}e^{-\int_0^t \mu(r)\,dr} \tag{103}$$

For the multiplicative noise case ($D_\xi = 0$), the starting point is again the solution of Eq. (84), which reads

$$x(t) = x(0)e^{\int_0^t \mu(r)\,dr + \int_0^t \eta(s)\,ds} \tag{104}$$

This can be rewritten:

$$y(t) = -\int_0^t \mu(r)\,dr + \ln\left(\frac{x(t)}{x(0)}\right) \tag{105}$$

$$y(t) = \int_0^t \eta(s)\,ds \tag{106}$$

The problem is now reduced to finding the probability density for the stochastic process $y(t)$ given by Eq. (106). Again we will assume that $\eta(s)$ is Gaussianly distributed with zero mean. Following the same procedure as in the additive noise case, the probability density for $x(t)$ can be written as

$$P(x, t | x_0) = \frac{1}{|x|\sqrt{2\pi\sigma^2}}\exp -\left\{\frac{\left[\ln(x/x(0)) - \int_0^t \mu(r)\,dr\right]^2}{2\sigma^2}\right\} \tag{107}$$

$$\sigma^2 = \int_0^t \int_0^t \langle \eta(p)\eta(q) \rangle\,dp\,dq \tag{108}$$

The FPT distribution is then given by

$$F\left(t, x_{\text{th}}^2 | x_0\right) = \frac{-1}{\sqrt{\pi}} \frac{\partial W}{\partial t} e^{-W^2} \tag{109}$$

$$W = \frac{\ln\left(x_{\text{th}}/x(0)\right) - \int_0^t \mu(r)\, dr}{\sqrt{2\sigma^2}} \tag{110}$$

To proceed further, the explicit forms of $\mu(t)$, $\langle \eta(t)\eta(t')\rangle$, and $\langle \xi(t)\xi(t')\rangle$ must now be specified. Again, following ZMVB, the case of a linear sweep will be studied:

$$\mu(t) = \mu(0) + vt \tag{111}$$

where $\mu(0) < 0, v > 0$. The noise will be taken to be an Ornstein-Uhlenbeck process; therefore, the correlation functions are given by

$$\langle \eta(t)\eta(t')\rangle = \frac{D_\eta}{\tau_\eta} \exp\left(-\frac{|t - t'|}{\tau_\eta}\right) \tag{112}$$

$$\langle \xi(t)\xi(t')\rangle = \frac{D_\xi}{\tau_\xi} \exp\left(-\frac{|t - t'|}{\tau_\xi}\right) \tag{113}$$

where D is the noise intensity and τ is the correlation time.

The three cases of external, internal, and multiplicative noise will now be studied. For the sweeping defined by Eq. (111) it is useful to introduce the following quantities:

$$
\begin{aligned}
\alpha &= \frac{x_{\text{th}}^2}{x^2(0)} \\[6pt]
b &= \frac{D_\xi}{x^2(0)} \left(\frac{\pi}{v}\right)^{1/2} \\[6pt]
\beta &= \frac{D_\xi}{x_{\text{th}}^2} \left(\frac{\pi}{v}\right)^{1/2} \\[6pt]
a &= \frac{\mu(0)}{v^{1/2}} \\[6pt]
z(t) &= \frac{\mu(0) + vt}{v^{1/2}} \\[6pt]
c &= \frac{1}{\tau_\xi v^{1/2}}
\end{aligned}
\tag{114}
$$

1. External Noise, $D_\eta = 0$

As previously mentioned a dynamical bifurcation point can be uniquely specified either by the bifurcation time t^* or by the value of the control parameter $\mu(t^*)$. For the case of additive noise it is convenient to use the scaled control parameter z, the bifurcation point then being defined as $z = z(t^*)$. The FPT distribution is now transformed to $F(z, \alpha)$, that is, the probability that the scaled control parameter will equal z when the system first reaches the scaled threshold value α. Using Eqs. (96)–(99), (111), and (113) the FPT distribution is calculated to be

$$F(z, \alpha) = -\frac{1}{\sqrt{\pi}} \left\{ \frac{\partial W_+}{\partial z} e^{-W_+^2} - \frac{\partial W_-}{\partial z} e^{-W_-^2} \right\} \tag{115}$$

$$\sigma'^2 = \sqrt{2}\, bc\, e^{a^2 + c^2} \int_a^z e^{-(1/2)(s+c)^2}$$

$$\times \left\{ \text{erf}\left[\frac{s-c}{\sqrt{2}} \right] - \text{erf}\left[\frac{a-c}{\sqrt{2}} \right] \right\} ds \tag{116}$$

$$W^\pm = \frac{\pm a^{1/2}\, e^{(1/2)(a^2 - z^2)} - 1}{\sqrt{2\sigma'^2}} \tag{117}$$

where we have written $\sigma^2 = x_0^2 \sigma'^2$ and erf is the standard error function.[159] In the white noise limit $c \to \infty$, σ'^2 simplifies to

$$\sigma'^2 = b\, e^{a^2}\{\text{erf}[z] - \text{erf}[a]\} \tag{118}$$

2. Internal Noise, $D_\eta = 0$

Using Eqs. (100)–(103), (111), and (113) and noting,

$$\langle x(0)\xi(t) \rangle = \frac{D}{1 - \mu(0)\tau_\xi} e^{-|t|/\tau_\xi} \tag{119}$$

$$\langle x(0)^2 \rangle = \frac{D}{-\mu(0)\big(1 - \mu(0)\tau_\xi\big)} \tag{120}$$

then the FPT distribution is calculated to be

$$F(z) = -\frac{2}{\sqrt{\pi}} \frac{\partial W}{\partial z} e^{-W^2} \tag{121}$$

$$W = \frac{e^{-(z^2-a^2)}}{\sqrt{2\sigma'^2}} \tag{122}$$

$$\sigma'^2 = \frac{\beta}{\pi^{1/2}|a|\left(1 + \dfrac{|a|}{c}\right)} + \frac{\sqrt{2}\,\beta\, e^{(1/2)(a+c)^2}}{1 + \dfrac{|a|}{c}}\left\{\mathrm{erf}\left[\frac{z+c}{\sqrt{2}}\right] - \mathrm{erf}\left[\frac{a+c}{\sqrt{2}}\right]\right\}$$

$$+ \sqrt{2}\,\beta c\, e^{a^2+c^2}\int_a^z e^{-(1/2)(s+c)^2}\left\{\mathrm{erf}\left[\frac{s-c}{\sqrt{2}}\right] - \mathrm{erf}\left[\frac{a-c}{\sqrt{2}}\right]\right\} ds \tag{123}$$

where $\sigma^2 = x_{\mathrm{th}}^2\sigma'^2$. Again taking the white noise limit the variance simplifies to

$$\sigma'^2 = \frac{\beta}{\pi^{1/2}|a|} + \beta\, e^{a^2}\{\mathrm{erf}[z] - \mathrm{erf}[a]\} \tag{124}$$

The second term in the expression for the variance in the colored noise case disappears in the white noise limit. Hence, this term is a direct consequence of the non-Markovian dynamics of the system. This theory reduces to the results of Ref. 155 in the limit of very small D_ξ.

3. Multiplicative Noise, $D_\xi = 0$

Taking Eqs. (108)–(112), the FPT distribution is calculated to be

$$F(t, \alpha) = \frac{\mu(t) + \sqrt{2}\,\dot{\sigma}W}{\sqrt{4\pi D_\eta t'}} \exp - \left\{\frac{\left[\frac{1}{2}\ln\alpha - \mu(0)t - \frac{1}{2}vt^2\right]^2}{4D_\eta t'}\right\} \tag{125}$$

$$t' = t + \tau_\eta\left[\exp\left(-\frac{t}{\tau_\eta}\right) - 1\right] \tag{126}$$

In the white noise limit $\tau_\eta \to 0$, $t' \to t$.

C. Analogue and Digital Simulations

To make a comparison between the MFPT and moments approach to the characterization of a dynamical bifurcation point, and to test the validity of the theory presented in Section V.B, both an analogue and a digital simulation were carried out. The analogue circuit used to simulate Eq. (84) is shown in Fig. 19a. The Gaussian noise sources $\eta(t), \xi(t)$ were passed through low-pass filters before being applied to the circuit to ensure that they were exponentially correlated with correlation times τ_η

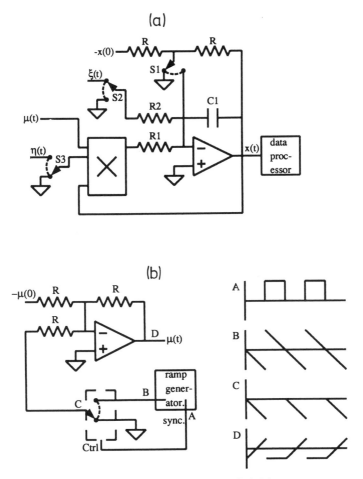

Figure 19. (a) Analogue circuit used to simulate Eqs. (84); (b) circuitry used to produce the linear ramp.

and τ_ξ, respectively, and correlation functions

$$\langle \eta(t)\eta(t') \rangle = \frac{D_\eta}{\tau_\eta} e^{-|t-t'|/\tau_\eta} \tag{127}$$

$$\langle \xi(t)\xi(t') \rangle = \frac{D_\xi}{\tau_\xi} e^{-|t-t'|/\tau_\xi} \tag{128}$$

A simple analysis of the circuit yields

$$R_1 C_1 \dot{x} = (\mu(t) + \eta(t))x + \frac{R_1}{R_2}\xi(t) \tag{129}$$

Setting $R_1 = R_2$, $R_1 C_1 = \tau_1$ gives the time-scaled version of Eq. (84). The circuitry used to obtain the sweeping function $\mu(t)$ is shown in Fig. 19b. A frequency synthesizer was used to obtain a ramp linear in time, which, using the control switch C and an adder (which initializes the ramp to the value $\mu(0)$), produced the sweeping function:

$$\mu(t) = \mu(0) + vt \tag{130}$$

The velocity v could be controlled by altering the frequency of the ramp.

The synchronization output from the frequency synthesizer, which produced a positive going pulse at the zero crossing point of the sweep, was used to control the switches S1, S2, S3 and also to trigger the data processor at time $t = 0$. The operation of the circuit for external noise is as follows. While the sync. output is low, switch S1 initiates the output to $x(0)$ and $\mu(t)$ is held constant at $\mu(0)$; switches S2, S3 disconnect the noise sources from the circuit [during the additive noise experiments the noise source $\eta(t)$ was removed, while $\xi(t)$ was absent during the multiplicative noise experiments]. On a positive going synchronization pulse the ramp was applied to the input, S1 was switched to earth and the noise was applied via S3 or S4. A triggering pulse was also sent to the data processor to start data acquisition. The operation of the circuit for internal noise was essentially the same, the only difference being that the $x(0)$ input was constantly grounded and the noise was present during the initialization stage (synchronization output low). Figure 20a shows the ramp and Fig. 20b four trajectories for external noise; the latter were used to obtain the MFPT and $\langle x^2 \rangle$, which is shown in Fig. 20c.

Data acquisition was carried out using a Nicolet 1080 data processor. The second moment was constructed by summation-averaging some 2000

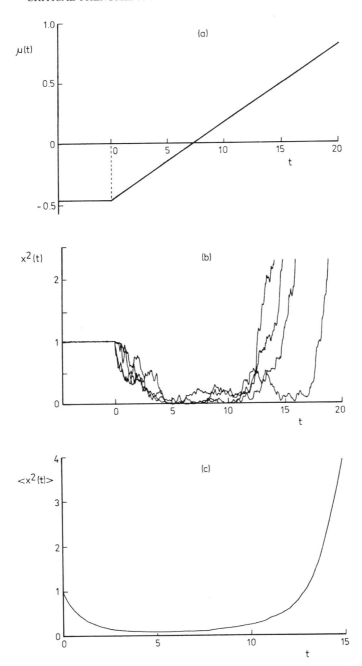

Figure 20. The sweep of $\mu(t)$. Until $t = 0$, μ is held at a fixed negative value; for $t > 0$, it is swept linearly toward higher values. (b) Corresponding typical trajectories of $x^2(t)$ for the case of external noise. Until $t = 0$, the circuit is held at a fixed value $x^2(0)$, at $t = 0$, the system is released and the noise is applied simultaneously. (c) Ensemble average $\langle x^2(t) \rangle$ for a set of $x^2(t)$ trajectories like those shown in (b). Note that the abscissa scale differs from those of (a) and (b). The bifurcation time t^* for the ZMVB theory is given by the point at which $\langle x^2(t) \rangle$ crosses x_{th}^2. (After Ref. 63.)

realizations, while the FPT distribution, used to obtain the MFPT, consisted of typically 2000 monitored crossings of x_{th}^2.

The algorithm used for the digital simulation is an algorithm specifically designed for colored noise, which allows for integration time steps much larger than the correlation time of the noise. For more details see Section II.B. The values used in the simulations were the same as the scaled quantities used in the analogue experiments. The integration time step was varied between 0.1 and 0.01 and the number of realizations incorporated in each average was typically 10 000.

D.   Results and Discussion

1.   External Noise

a.   White Noise. Analogue results for small noise intensity ($b = 0.01$) are shown in Fig. 21. The dot-dashed line corresponds to the deterministic result $z = \sqrt{a^2 + \ln \alpha}$ and the dashed line corresponds to the theory of ZMVB. For the range of a shown the influence of the noise is extremely small and both definitions (the circles are the analogue MFPT measurements and the squares are the analogue results corresponding to the moments definition) yield results close to the deterministic values. This confirms that both definitions tend to the deterministic results as $b \propto D_\xi \rightarrow 0$. However, for $a \sim z$ it can be seen that the results start to differ; if a

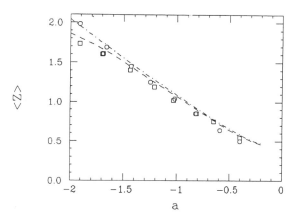

Figure 21.   Experiment and theory for external white noise, $b = 0.01$ and $\alpha = 1.2$. The scaled bifurcation point $\langle z \rangle$ is plotted against the scaled initial curvature a. The dot-dashed line corresponds to the deterministic (noise free) result, the circles represent the analogue simulation results of the MFPT, the squares are the analogue simulation results using the moments definition and the dashed curve is the ZMVB theory based on the moments definition. (After Ref. 47.)

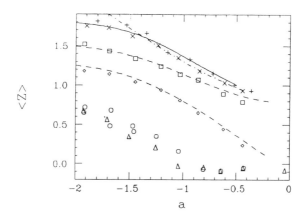

Figure 22. Results for external white noise, $b = 0.1, \alpha = 1.03$. The dashed lines correspond to the ZMVB theory, the top one for $\alpha = 2.0$ and the bottom one for $\alpha = 1.03$; the squares and diamonds are the corresponding analogue simulation results. The full time corresponds to the MFPT for $\alpha = 2.0$ calculated using Eqs. (86), (114)–(117); the crosses (analogue) and the pluses (digital) are the associated simulation results. The dot-dashed line represents the deterministic result for $\alpha = 2.0$; the circles (analogue) and the triangles (digital) are the simulation results for $\alpha = 1.03$. (After Ref. 47.)

was decreased further the results would flatten to a plateau, the value of which would differ for the two definitions. This effect can be seen better in Fig. 22, where the noise has been increased to $b = 0.1$. The dashed lines correspond to the theory of ZMVB, with the upper curve for $\alpha = 2$ and the lower one for $\alpha = 1.03$; the squares and the diamonds are the corresponding analogue simulation results. The solid line (theory), pluses (digital), and crosses (analogue) correspond to the MFPT definition for $\alpha = 2$ and the circles (analogue) and triangles (digital) for $\alpha = 1.2$. The dot-dashed line is the deterministic result for $\alpha = 2$. First, it should be noted that the simulation results are in good agreement with the MFPT theory presented in Section V.B. Second, the $\alpha = 2$ MFPT results lie above those results based on the moments definition and hence show an increased delay in the bifurcation point. It is interesting to note that the MFPT results can also lie above the line representing the deterministic case and hence, using this criterion, one is led to conclude that the effect of noise can lead to a stabilization of the system. This surprising result can be explained by considering Fig. 18b, where it can be seen that the FPT distribution is in fact asymmetrical with an extended wing at large t values. This effect biases $\langle t \rangle$ and hence $\langle z \rangle$ beyond the value at which the peak (this approximately represents the deterministic value) in the FPT appears. However, besides these general differences, the shapes of the curves that represent the MFPT and ZMVB theory are qualitatively similar, with both tending to a limiting value at large a and decreasing with increasing a.

The results for $\alpha = 1.03$ are again very different. In this case the MFPT results fall below those based on the moments approach, an effect that is due solely to a noise-activated phenomenon. Just after the circuit is released at time $t = 0$ there is a finite chance that a trajectory will cross before its initial relaxation simply because x_{th}^2 is only slightly greater than $x^2(0)$. In this case, although the MFPT from $x(0)$ to x_{th} is still well defined, the MFPT approach takes no account of the dynamics of the instability and consequently, in this sense, becomes a meaningless characterization of the bifurcation point. Therefore, the MFPT approach is meaningful only if the probability of a trajectory crossing x_{th}^2 due solely to noise activation is small.

b.   *Colored Noise.* Figure 23a shows the dependence of $\langle z \rangle$ on the log of the scaled noise color c. The results are for $a = -1.7$, $\alpha = 1.2$ and for two different values of b ($b = 0.05$ upper two curves, $b = 0.5$ lower two curves). The dashed lines correspond to the theory of ZMVB and the full lines are the MFPT theory calculated using Eq. (96); the crosses, squares, circles, and diamonds are data obtained from the analogue simulation, while the pluses and triangles correspond to the digital simulation.

In the limit of large noise color all curves tend to the deterministic value 1.76. This is easily understood since $\langle \xi^2 \rangle = D_\xi / \tau_\xi$ and hence the variance of the noise tends to zero as $\tau_\xi \to \infty$, which is essentially a noiseless limit. When c is increased one of the main qualitative differences between the MFPT approach and the moments approach is observed. The value of $\langle z \rangle$ initially increases with increasing c in the case of the MFPT approach, while it decreases using the moment approach. In fact, it was demonstrated by ZMVB that this decrease always occurs for the moments approach, regardless of the parameter values. Thus, a finite noise color can increase the delay of the bifurcation in the case of the MFPT characterization. When c is further increased the MFPT curves reach a maximum value and then decrease monotonically to their white noise ($\tau_\xi = 0$) values in a similar fashion to the ZMVB theory. Again it can be seen that the MFPT theory is in good agreement with experiment; the discrepancy between the MFPT theory and experiment, for $b = 0.5$, in the white noise limit, is again due solely to noise-induced crossings becoming important.

Results for $a = -2.0$, $b = 0.5$, $\alpha = 1.2$ are shown in Fig. 23b. The $a = -1.7$, $b = 0.5$ results are also displayed for comparison. The qualitative behavior of the results is similar to that of Fig. 23a. However, the

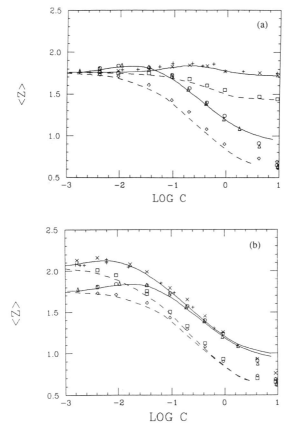

Figure 23. (a) Results for external colored noise, $a = -1.7$, $\alpha = 1.2$ and $b = 0.05$ and 0.5. The scaled bifurcation point $\langle z \rangle$ is plotted against log of the scaled color parameter c. The dashed lines correspond to the theory of ZMVB; the top curve for $b = 0.05$ and the bottom curve for $b = 0.5$; the squares and diamonds are the corresponding analogue simulation results. The full lines are the theory for the MFPT, top curve $b = 0.05$ and bottom $b = 0.5$, calculated using Eqs. (86) and (114)–(117); the pluses (digital) and the crosses (analogue) are the MFPT simulation results for $b = 0.05$ and the circles (analogue) and the triangles (digital) are MFPT simulation results for $b = 0.5$. (b) This shows the same plot as in (a) except for $b = 0.5$ and two values of a, $a = -1.7$ and -2.0. The $b = 0.5$, $a = -1.7$ results are the same as in (a) and have been shown for ease of comparison. The top solid line is the MFPT theory for $a = -2.0$ calculated using Eqs. (86) and (114)–(117) and the crosses (analogue) and the pluses (digital) are the associated simulation results. The top dashed line is the ZMVB theory for $a = -2.0$ and the squares are the corresponding analogue simulation results. (After Ref. 47.)

increase in $|a|$ can be seen to have shifted the maximum that occurs in the MFPT results to a larger value of noise color. Both criteria yield results that are effectively independent of a in the white noise limit. This is a consequence of the plateau that occurs at large enough values of $|a|$. The fact that the dependence on a becomes more prominent at larger noise colors implies that the plateau occurs at larger values of $|a|$ for finite noise color. Again it is observed that the MFPT results lie above the results based on the moments definition.

2. Internal Noise

a. White Noise. Figure 24 shows results for two values of the effective noise intensity β. The dashed curves, the top one for $\beta = 0.1$ and the lower one for $\beta = 0.35$, are the theory of ZMVB, while the full curve represents the MFPT theory. The crosses ($\beta = 0.1$, MFPT), squares ($\beta = 0.1$, moments), and circles ($\beta = 0.35$, MFPT) represent the analogue data, while the pluses ($\beta = 0.1$, MFPT) and the triangles ($\beta = 0.35$, MFPT) represent the results of the digital simulation. Considering the results for $\beta = 0.1$ it is obvious that qualitatively the results for the moments and the MFPT definition of a dynamical bifurcation point are

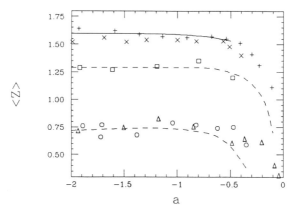

Figure 24. Results for internal white noise for $\beta = 0.1$ and 0.35. The scaled bifurcation point $\langle z \rangle$ is plotted against the scaled initial curvature a. The dashed lines are the theory of ZMVB, the top curve for $\beta = 0.1$ and the bottom curve for $\beta = 0.35$; the squares ($\beta = 0.1$) are the corresponding analogue simulation results. The full line is the MFPT theory for $\beta = 0.1$ calculated using Eqs. (85), (120), (122), and (124); the associated simulation results are the pluses (digital) and the crosses (analogue). The circles (analogue) and triangles (digital) are the MFPT simulation results for $\beta = 0.35$, where the MFPT theory is inapplicable; the fact that they fall on the lower ZMVB theoretical curve is a coincidence. (After Ref. 47.)

very similar, both curves increasing rapidly for decreasing $|a|$ and quickly reaching their limiting values. However, one notes how different their two limiting values are, for the moments definition $\lim_{a \to -\infty} \langle z \rangle = 1.29$ and in the case of the MFPT definition $\lim_{a \to -\infty} \langle z \rangle = 1.60$. The situation is quite different for $\beta = 0.35$, where it seems that the ZMVB theory coincides with the MFPT data; it is believed, however, that this agreement is purely coincidental. At this value of β solely noise-activated threshold crossings are becoming important, which has the effect of pulling down the MFPT results to the curve representing the ZMVB theory. For this same reason the theory developed in Section V.B is inapplicable in this regime.

b.   Colored Noise. The dependence of the mean bifurcation time on the effective noise color c is demonstrated in Fig. 25a for $b = 0.01$ and Fig. 25b for $b = 0.35$. The solid lines correspond to the MFPT theory and the dashed lines are the theory of ZMVB. The pluses and squares are the analogue data for the MFPT and the moments definitions, respectively, while the crosses are the digital simulation results. As in the white noise case the qualitative behavior of the curves based on the two definitions is similar. For both definitions the curves decrease monotonically for decreasing noise color. In the limit $c \to \infty$ one would expect $\langle z \rangle \to \infty$, which is in contrast to the external noise case where the deterministic value of $\langle z \rangle$ was obtained. This effect is easily understood since the initial distribution of starting points at $t = 0$ is determined by the noise, with a probability distribution given by

$$P(x(0)) = \frac{1}{\sqrt{2\pi(D/|\mu(0)|\tau)}} \exp\left(\frac{\mu(0)\tau x^2(0)}{2D}\right) \tag{131}$$

Consequently in the limit $\tau \to \infty$, $P(x,0) \to \delta(x)$. This is again a noiseless limit but with an initial starting condition of $x(0) = 0$. Thus, although the system is in an unstable condition, no infinitesimal perturbations are present to start the decay process and hence the bifurcation time diverges as this limit is approached.

The main difference between the two definitions is again an increase in $\langle z \rangle$ for the MFPT results with respect to the moments results. This quantitative difference between the two curves seems to show little dependence on noise color. Finally, it is interesting to note that the theory developed in Section V.B is now applicable for $\beta = 0.35$. This is due to the noise color reducing the probability of multiple crossings.

c.   Multiplicative Noise. Results for the dependence of $\langle z \rangle$ on c are shown in Fig. 26 for values of $\delta = 0.1$ and $\delta = 0.25$, $\alpha = 1.2$, and

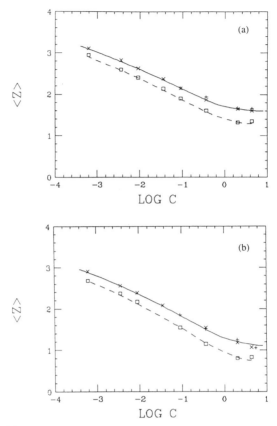

Figure 25. (a) Results for internal colored noise with $\beta = 0.1$. The scaled bifurcation point $\langle z \rangle$ is plotted against the log of the scaled color parameter c. The full line is the MFPT theory calculated using Eqs. (85) and (120)–(123); the crosses (analogue) and the pluses (digital) are the corresponding simulation results. The dashed line is the theory of ZMVB and the squares are the associated analogue simulation results. (b). This is the same as (a) except $\beta = 0.35$; all symbols have the same meaning. (After Ref. 47.)

$\gamma = -2.76$ where $\delta = D_\eta / v$ and $\gamma = \mu(0)/v$. The dashed lines again represent the theory of ZMVB, while the full lines represent the MFPT theory. The crosses and squares represent the MFPT analogue data for $\delta = 0.1$ and 0.25, respectively, and the pluses are the moments analogue data for $\delta = 0.1$. The MFPT results again lie above those given by the ZMVB theory. However, it is also apparent that dependence on noise color of the MFPT approach is much weaker. Only in a very narrow range of τ do the MFPT curves decrease to their white noise values. For $\tau \sim 10$

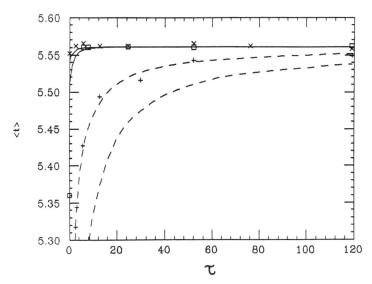

Figure 26. Results for colored multiplicative noise. Here the bifurcation time $\langle t^* \rangle$ is plotted directly against the noise correlation time τ for $\alpha = 1.2, \gamma = 2.76, \delta = 0.1$ and 0.25, where $\delta = D_\eta/\nu$ and $\gamma = \mu(0)/\nu$. The full lines (almost coincident) are the MFPT results calculated using Eqs. (85) and (125), (126); the top one for $\delta = 0.1$ and the bottom one for $\delta = 0.25$; the crosses and squares are the MFPT analogue simulation results. The dashed lines are the ZMVB theory, the upper one for $\delta = 0.1$ and the lower one for $\delta = 0.25$; the pluses are the analogue simulation results based on the moments definition for $\delta = 0.1$. (After Ref. 47.)

the MFPT results have reached their deterministic value, while the ZMVB results are still appreciably below this value. It is also apparent that the MFPT results are less sensitive to changes in δ.

E.   Conclusions

We have considered two different characterization schemes of a dynamical bifurcation for the three situations of external noise, internal noise, and multiplicative noise forcing. In all three cases it was found that, within a parameter range where an MFPT is a meaningful characterization of a dynamical bifurcation, the MFPT results always showed an increased delay in the bifurcation point with respect to the moments definition adopted by ZMVB. More precisely, for the case of external noise, it was demonstrated that the MFPT characterization could also yield results that showed an increase in the bifurcation point with respect to the deterministic (noise free) case. This effect was observed by a suitable selection of the effective initial curvature a of the potential or by an appropriate choice for the noise color τ. It was attributed to an asymmetry that exists in the FPT distribution, an asymmetry that biases the bifurcation time $\langle t \rangle$ to values beyond the maximum in the FPT distribution, which approximately coincides with the deterministic values. This is in contrast to the results based on a moments definition where it has been demonstrated by ZMVB that the effect of noise is always to decrease the bifurcation point and hence reduce the stability of the system. From the MFPT results, the conclusion is that the effect of noise can be to stabilize the system. Consequently, the question as to which statement is correct depends on what is actually asked. If one inquires about the average time at which an individual trajectory crosses the threshold, then one concludes that noise can stabilize the system. Alternatively, if one inquires as to the time at which an averaged trajectory $\langle x^2(t) \rangle$ crosses the threshold x_{th}^2 then one concludes that the noise always has the effect of destabilizing the system.

As already mentioned, there is in principle no reason to prefer one definition over the other, both averages being based on quite reasonable questions asked about the system. However, the MFPT characterization arguably provides a more complete description of the bifurcation process since one can, in principle, obtain information about higher-order moments of the FPT distribution. The MFPT description is also the one commonly used by experimentalists. The MFPT approach does, however, have certain disadvantages. In parameter regimes where multiple crossings of the threshold x_{th}^2 are common the MFPT approach can become meaningless. This was particularly noticeable with the external noise results where results based on the MFPT criteria fell well below those based on the moments approach when $x^2(0) \sim x_{\text{th}}^2$. In this situation the MFPT no longer reflects the dynamics of the system since crossings, at times only slightly greater than $t = 0$, occur that are solely noise activated. For this reason it does not seem unreasonable to drop the MFPT approach in favor of a mean passage time (MPT) approach, and hence incorporate multiple crossings into the definition of the bifurcation time. This is just the approach adopted in the theory developed in Section V.B, which has been shown to approximate accurately the MFPT results in the parameter range where a MFPT approach is meaningful. The advantage of this approach is that it gives a meaningful result over the whole parameter range.

VI.   STOCHASTIC RESONANCE

Stochastic resonance (SR) is a remarkable phenomenon in which a weak periodic signal, usually in a bistable system, can be enhanced by the addition of external noise. This highly counter intuitive effect was originally introduced in connection with the earth's ice-age cycle.[160, 161] Several years later, McNamara et al.[29] discovered that SR can also occur in a bistable ring laser. Their paper stimulated much new work in the area, which has led to the discovery of SR in passive optical systems[30] and a laser with saturable absorber,[162] as well as in an electron spin resonance spectrometer,[163] in a magnetoelastic strip,[164] and in periodically stimulated sensory neurons.[71] The effect arises because the periodic signal modulates the probabilities of fluctuational transitions between the coexisting stable states, and hence the populations of those states; consequently, there is a comparatively strong periodic modulation of the relevant variable, with an amplitude proportional to the distance between the stable positions.[23, 27] The phenomenon has been investigated extensively through simulation techniques, both analogue[43, 65–67, 71, 165–169] and digital.[170, 171] For a fuller bibliography of what is a rapidly growing field, the reader is referred to a recent review[172] and to a special issue[31] of the *Journal of Statistical Physics*.

The theory of SR has been developed, most commonly, for a discrete two-state model or, in the case of continuous systems, has been based on approximate or numerical solution of the Fokker-Planck equation[160, 161, 171, 172–176]; again, reference should be made to Refs. 31 or 172 for fuller lists of citations to some of the earlier work. An alternative, radically different, approach to the theory of SR is embodied in the suggestion[66] that it be treated by linear response theory (LRT) in terms of a susceptibility. LRT has the advantage not only of being simple, but also that, for thermally equilibrium (or quasiequilibrium) systems, the susceptibility can be related to the spectral density of the fluctuations (SDF) of the system under study. Although the utility of this approach was initially regarded with some suspicion,[176, 177] it was subsequently vindicated by the demonstration that it not only provides a good description of the signal/noise ratio R (Ref. 66), but can also describe correctly the dependence on noise intensity of the phase shift ϕ between the periodic input signal and the periodic response of the system.[67] We now describe the application of the LRT approach to SR in a model bistable optical system.

It should be borne in mind throughout the rest of this section that the discussion refers to the limit—commonly found in practice—where the amplitude of the periodic driving force is *small*.

A. Stochastic Resonance as a Linear Response Phenomenon

According to LRT, if a system with a coordinate q is driven by a weak force $A \cos \Omega t$ (the addition to the Hamiltonian function of the system being of the form of $-Aq \cos \Omega t$), there arises a small periodic term in the ensemble-averaged value of the coordinate, $\delta \langle q(t) \rangle$, oscillating at the same frequency Ω and with amplitude a proportional to that of the force[8]:

$$\delta\langle q(t)\rangle = a\cos(\Omega t + \phi) \qquad A \to 0 \tag{132}$$

$$a = A|\chi(\Omega)| \qquad \phi = -\arctan\frac{\operatorname{Im}\chi(\Omega)}{\operatorname{Re}\chi(\Omega)} \tag{133}$$

The quantity $\chi(\Omega)$ here is the *susceptibility* of the system. Equation (132) holds for dissipative and fluctuating systems that do not display persistent periodic oscillations in the absence of the force $A\cos\Omega t$; it is SR in bistable systems of this kind ("conventional" SR) that is considered below ("nonconventional" SR has been considered elsewhere[169]). The function $\chi(\Omega)$ contains, basically, all information on the response of the system to a weak driving force. It gives both the *amplitude* of the signal, a, and its *phase lag* with respect to the force ϕ. In turn, the value of $\frac{1}{4}a^2$ gives the *intensity* (i.e., the area) of the delta-shaped spike in the spectral density of fluctuations (SDF) $Q(\omega)$ of the system at the frequency Ω of the driving force,

$$Q(\omega) = \lim_{\tau\to\infty}(4\pi\tau)^{-1}\left|\int_{-\tau}^{\tau}dt\,q(t)\exp(i\,\omega t)\right|^2 \tag{134}$$

The onset of such a spike follows immediately from Eq. (132) with account taken of the principle of the decay of correlations:

$$\langle q(t)q(t')\rangle \to \langle q(t)\rangle\langle q(t')\rangle \qquad \text{for } |t - t'| \to \infty \tag{135}$$

Following Ref. 170, the response of the system, in the context of SR, is often characterized by the ratio R of the area of the above spike to the value $Q^{(0)}(\omega)$ of the SDF at the given frequency Ω in the absence of periodic driving, i.e., by the signal-to-noise ratio. It is evident from Eqs. (132)–(134) that R may be expressed in terms of a susceptibility $\chi(\Omega)$:

$$R = \frac{1}{4}A^2|\chi(\Omega)|^2/Q^{(0)}(\Omega) \qquad (A \to 0) \tag{136}$$

Therefore, the evolution of the susceptibility and of $Q^{(0)}(\Omega)$ with varying noise intensity D shows immediately whether SR (i.e., an increase and subsequent decrease in R with increasing D) is to be expected at a given frequency.

An important advantage of describing SR in terms of a susceptibility is that such a description relates SR to standard linear-response phenomena (conductivity, magnetic susceptibility, etc.) investigated in physical kinetics. Another advantage is that quite often the systems investigated are in thermal equilibrium or in quasiequilibrium. In this case the susceptibility can be expressed immediately in terms of the SDF $Q^{(0)}(\Omega)$ in the absence of periodic driving via the fluctuation–dissipation relations[8]:

$$\operatorname{Re} \chi(\omega) = \frac{2}{T} P \int_0^\infty d\omega_1 \, Q^{(0)}(\omega_1) \omega_1^2 \left(\omega_1^2 - \omega^2\right)^{-1}$$

$$\operatorname{Im} \chi(\omega) = \frac{\pi\omega}{T} Q^{(0)}(\omega)$$

(137)

where P implies the Cauchy principal value and T is temperature in energy units. It follows from Eqs. (136)–(137) that the onset of SR can be predicted from purely experimental data on the evolution of the SDF $Q^{(0)}(\omega)$ of a system with temperature without assuming anything at all about the equations that describe its dynamics, i.e., for a system treated simply as a "black box."

The relevance of this approach to SR is seen from Fig. 27, where some data from analogue experiments[66] for an electronic system simulating Brownian motion in a bistable potential are shown. The system simulated is quasithermal equilibrium, with noise intensity D standing for tempera-

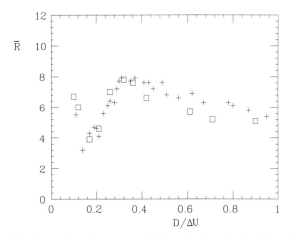

Figure 27. Signal-to-noise ratio vs. normalized noise intensity for Brownian motion in the double-well potential $U(q)$ of Eq. (138); $\ddot{q} + 2\Gamma\dot{q} + U'(q) = (2\Gamma)^{1/2}f(t) + A \cos \Omega t$. The values of $\bar{R} = 6.51 \times 10^{-4}R$ are given for $\Omega = 0.0695$, $A = 0.1$, $\Gamma = 0.125$. The squares are direct measurements; the pluses are data calculated from the *measured* $Q^{(0)}(\omega)$ via the fluctuation–dissipation relations. (After Ref. 66.)

ture T (as in Eq. (139) below). The "plus" data points are values of the scaled signal/noise ratio $\bar{R} = 6.51 \times 10^{-4}R$ calculated from Eq. (137) using measurements of the SDF $Q^{(0)}(\omega)$ in the absence of the periodic force. The "square" data points represent direct measurements of $\bar{R}$, with the periodic force applied to the system.

It is immediately evident from Fig. 27 that there is a range within which $\bar{R}$ increases with increasing noise intensity, and it is this remarkable effect that constitutes SR. The general behavior of the system may be interpreted as follows. At very small D, where transitions between the stable states hardly occur at all, an increase in the noise is bound to reduce the signal/noise ratio; when D becomes large enough to facilitate fluctuational transitions with a probability of the order of the frequency Ω of the driving field, the field effectively synchronizes the transitions through its periodic modulation of the barrier height, so that $\bar{R}$ rises corresponding to the onset of SR; for still larger D, when the transition probabilities substantially exceed Ω, the synchronization becomes ineffective, and, because the noise intensity is still increasing, $\bar{R}$ decreases smoothly. The parameter values for which the results of Fig. 27 were recorded were such[66] that the SR increase of $\bar{R}$ was relatively modest. With a larger damping constant (or for an overdamped system), and a smaller frequency Ω (see Ref. 170), considerably larger increases of $\bar{R}$ with D may[172] be obtained.

The agreement obtained, with no adjustable parameters, between the two very different forms of measurement in Fig. 27 demonstrates that SR in the signal-to-noise ratio is described *quantitatively* by Eq. (136) and by the fluctuation–dissipation relations Eq. (137), even in the range of large D where an explicit analytic calculation of the susceptibility of the system was impossible because there are no good approximations for $Q^{(0)}(\omega)$. An additional test of these ideas, which we discuss in the next subsection, is to check that the phase lag between the driving force and the response takes the form given by Eqs. (132), (133), and (137).

B.   Phase Shifts in Stochastic Resonance

The presence or absence of phase shifts in SR—a conundrum of many years' standing—was effectively and convincingly resolved through an analogue electronic experiment.[67] The question is interesting from the viewpoint of relating SR to standard resonance phenomena (cf. Ref. 173). It is well known in physics that when the frequency Ω of an external driving force is swept through the resonant frequency of a system the phase lag ϕ of the signal in the system decreases monotonically from nearly zero for small Ω to nearly $-180°$ for large Ω passing through $\approx -90°$ at the resonant value of Ω. SR in bistable systems arises because, with increasing noise intensity, the probabilities of fluctuational transitions between the stable states become of the same order of magnitude or larger than the frequency Ω of the driving force, thereby switching on the mechanism of strong response associated with the transitions. So, the physics is different from that in a standard resonance, and the dependence of the phase lag on the noise intensity would not necessarily be expected to be the same as $\phi(\Omega)$ in a conventional resonating system.

In this section we show in considerable detail, both experimentally and theoretically, that phase shifts do indeed accompany SR; however, in continuous systems, they take a form completely different from that predicted for two-state systems.[161] We treat the simplest nontrivial system: an overdamped Brownian particle moving in a symmetric bistable potential and, in addition, driven periodically,

$$\dot{q} + U'(q) = A \cos \Omega t + f(t) \qquad U(q) = -\tfrac{1}{2}q^2 + \tfrac{1}{4}q^4 \qquad (138)$$

where $f(t)$ is zero-mean Gaussian noise of intensity D,

$$\langle f(t)f(t') \rangle = 2D\delta(t - t') \qquad (139)$$

In the absence of periodic forcing, the system of Eqs. 138 and 139 is quasithermal, irrespective of the particular form of the (confining) potential $U(q)$. Its distribution over energy $U(q)$ (an overdamped system has potential energy only) is Gibbsian, with temperature

$$T = D$$

Therefore, the fluctuation–dissipation relations of Eq. (137) hold, and for weak periodic driving force, i.e., for small A in Eq. (138), the susceptibility $\chi(\Omega)$ can be expressed in terms of the spectral density of fluctuations $Q^{(0)}(\omega)$ for $A = 0$. Explicit expressions for $Q^{(0)}(\omega)$, $\chi(\omega)$ can be obtained analytically for small noise intensities (low temperatures) $D \ll \Delta U$, where for a general double-well potential ΔU is the depth of the (shallower) potential well and $\Delta U = \tfrac{1}{4}$ for the model Eq. (138). In this range $Q^{(0)}(\omega)$ and $\chi(\omega)$ are given[23, 27, 28] by the sums of partial contributions from fluctuations about the equilibrium positions $q_n, [U'(q_n) = 0, U''(q_n) > 0, n = 1, 2; q_n = (-1)^n$ for the potential Eq. (138)] and of the contribution from interwell transitions,

$$Q^{(0)}(\omega) = \sum_{n=1,2} w_n Q_n^{(0)}(\omega) + Q_{tr}^{(0)}(\omega)$$

$$\chi(\omega) = \sum_{n=1,2} w_n \chi_n(\omega) + \chi_{tr}(\omega) \qquad (140)$$

Here, w_n is the population of the nth stable state. The susceptibilities $\chi_n(\omega)$ and $\chi_{tr}(\omega)$ are expressed in terms of $Q_n^{(0)}(\omega), Q_{tr}^{(0)}(\omega)$ by Eq. (137), and therefore only the spectral densities of fluctuations will be written down here in explicit form. For the model Eq. (138), $w_1 = w_2 = \frac{1}{2}$, $Q_1^{(0)}(\omega) = Q_2^{(0)}(\omega)$, and $\chi_1(\omega) = \chi_2(\omega)$.

The SDF for the intrawell vibrations $Q_n^{(0)}(\omega)$ can be obtained by expanding $U(q)$ about the equilibrium position q_n. If we assume the nonlinear terms to be small, and allow for them by perturbation theory,[28] we obtain

$$Q_n^{(0)}(\omega) \simeq L_n(\omega) - \pi L_n^2(\omega)\left[U_n^{(IV)} - 9U_n'''^2 U_n''(4U_n''^2 + \omega^2)^{-1}\right]$$

$$L_n(\omega) = \frac{1}{\pi}D(U_n''^2 + \omega^2)^{-1}$$

(141)

where all derivatives $U_n^{(k)} \equiv U^{(k)}(q_n)$ are evaluated for $q = q_n$. The contribution from interwell transitions is

$$Q_{tr}^{(0)}(\omega) = \frac{1}{\pi}w_1 w_2 (\langle q \rangle_1^{(0)} - \langle q \rangle_2^{(0)})^2 W^{(0)}/(W^{(0)2} + \omega^2) \qquad \omega \ll U_{1,2}''$$

$$W^{(0)} \equiv W^{(0)}(D) = W_{12}^{(0)} + W_{21}^{(0)} \qquad \langle q \rangle_n^{(0)} = q_n - \frac{1}{2}DU_n'''(U_n'')^{-2}$$

(142)

Here, $\langle q \rangle_n^{(0)}$ is the average value of the coordinate in the nth well, neglecting interwell transitions, and $W_{nm}^{(0)}$ is the probability of the transition $n \to m$ in the absence of periodic forcing (corrections $\sim D/\Delta U$ to the Kramers expression for the transition probabilities[178] are required in Eq. (142)). In deriving the latter equation we have utilized the inequality $W^{(0)} \ll \Omega_r = \min(U_{1,2}'')$, implying that the transition probabilities are very much smaller than the relaxation rate of the system Ω_r (a condition that is necessary for the concept of a transition between well-defined metastable states to be meaningful).

To lowest order in $D/\Delta U$, to zeroth order in Ω/Ω_r, but for arbitrary $\Omega/W^{(0)}$, Eqs. (136) and (132) for the signal-to-noise ratio R and for the phase lag ϕ, respectively, resulting from Eqs. (137) and (140)–(142) for the model Eq. (138) become

$$R = \frac{\pi A^2}{4D^2}\frac{\Omega_r^2 W^{(0)2} + \Omega^2 D^2}{\Omega_r^2 W^{(0)} + \Omega^2 D} \qquad \Omega, D \ll \Omega_r \qquad W^{(0)} \ll D \quad (143)$$

$$\phi = -\arctan\left[\frac{(\Omega/\Omega_r)(\Omega_r^2 W^{(0)} + \Omega^2 D)}{\Omega_r W^{(0)2} + \Omega^2 D}\right]$$

(144)

where $\Omega_r = U_1'' = U_2'' = 2$. For very small D, where $W^{(0)} \ll (\Omega^2/\Omega_r^2)D$, it follows from Eq. (143) that $R \simeq \pi A^2/4D$, $\phi \simeq -\Omega/\Omega_r$. Thus, for a fixed forcing frequency Ω, R decreases with increasing D, whereas ϕ remains small and nearly independent of D.

For larger values of D, on the other hand,

$$R = \frac{\pi A^2}{4D^2}\left(W^{(0)} + \frac{\Omega^2 D^2}{\Omega_r^2 W^{(0)}}\right)$$

$$\phi = -\arctan\left(\frac{\Omega\Omega_r W^{(0)}}{\Omega_r W^{(0)2} + \Omega^2 D}\right) \tag{145}$$

$$D \gg W^{(0)} \gg \frac{\Omega^2}{\Omega_r^2}D$$

The behavior of R and ϕ as given by Eq. (145) depends on the ratio $\alpha = W^{(0)}\Omega_r/\Omega D$. When it is small, R is sharply decreasing while $|\phi|$ is sharply increasing with increasing D,

$$\phi \simeq -\alpha \equiv -\frac{W^{(0)}\Omega_r}{\Omega D} \qquad \alpha \ll 1 \tag{146}$$

For $\alpha \simeq 1$, R passes its minimum and then increases with D (i.e., with α) up to comparatively large $D \sim \Delta U$ where the weak-noise approximation of Eqs. (140)–(142) is inapplicable. It is this increase that is associated with conventional SR.

The central interest of the present subsection relates, however, to the behavior of ϕ. It follows from Eq. (145) that the sharp increase of $|\phi|$ with increasing D saturates in the range $\alpha \gg 1$, and, for $D \equiv D_{max} \ll \Delta U$, $|\phi|$ reaches its maximum:

$$(-\phi)_{max} = \arctan\left[\frac{1}{2}\left(\frac{\Omega_r}{D_{max}}\right)^{1/2}\right] \qquad W^{(0)}(D_{max}) = \Omega\left(\frac{D_{max}}{\Omega_r}\right)^{1/2} \tag{147}$$

(note that, in contrast to the behavior of R vs. D, the weak-noise theory Eqs. (140)–(142) holds in the vicinity of the *maximum* of $|\phi|$; we stress that it is the approximation Eqs. (140)–(142) that fails for strong noise, not LRT). The decrease of $-\phi$ for $D > D_{max}$ is seen from Eq. (145) to be much less steep than the increase described by Eq. (146). Overall, it follows from Eqs. (144) and (145) (see also Fig. 28 where ϕ vs. D as given

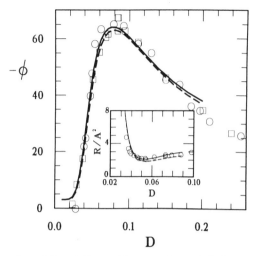

Figure 28. The phase shift $-\phi$ (degrees) between the periodic force of amplitude A and the averaged coordinate $\langle q(t) \rangle$ of the overdamped bistable system (138) measured as a function of noise intensity D in the electronic experiment for $\Omega = 0.1$ and $A = 0.04$ (circles), $A = 0.2$ (squares). The dashed curve represents the theoretical prediction based on LRT and the fluctuation dissipation theorem; the full curve takes account of nonlinear corrections for $A = 0.2$. The inset shows the normalized signal-to-noise ratio in the region of the minimum in R. (After Ref. 168.)

by Eq. (144) is plotted) that the phase shift displays a resonance-type (nonmonotonic) behavior as a function of the noise intensity D. This prediction is in contrast with the earlier theories[161, 170] for two-state systems displaying SR in the signal-to-noise ratio, but exhibiting a monotonic dependence of $|\phi|$ on D; the phase shift in these theories is described by Eq. 145 with Ω_r set equal to ∞ (if the intrawell relaxation was infinitely fast the intrawell motion would not come into play and the system would behave as a two-state one):

$$(\phi)_{\text{two-state}} = -\arctan\left(\frac{\Omega}{W^{(0)}}\right)$$

The LRT predictions of Eqs. (144) and (145) have been tested by means of an electronic experiment, using a circuit of conventional design to model Eq. (138). It is immediately evident from the measurements (Fig. 28), first, that contrary to Refs. 166, 167, 175, and 177, large phase shifts do indeed occur as D is varied and, second, that the LRT prediction describes the data remarkably well. For Eq. (138) with the parameters used in the experiment, a maximum value of $-\phi$ is predicted by LRT to be equal to 68° and to occur at $D_{max} = 0.08$, which is to be compared with the experimental observation for $A = 0.04$ of $(-\phi)_{max} = (66 \pm 2)°$ at $D = 0.08 \pm 0.01$. In accordance with the LRT prediction, the decrease of $|\phi|$ for $D > D_{max}$ is much more gradual than the rapid increase seen below D_{max}. The measured ϕ is relatively insensitive to A for the chosen frequency Ω/Ω_r; nonlinear effects under SR are discussed in Refs. 66, 168, 179, and 180.

The physical origin of the nonmonotonic behavior of the phase lag with increasing noise intensity D can be readily understood if one notices that, for very small $D \ll \Delta U$, the system is effectively confined to a single well and ϕ would be expected to be small because Ω is small compared with the reciprocal characteristic time of intrawell motion; in the opposite limit

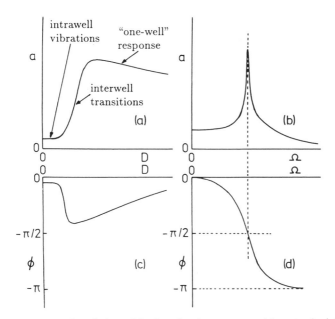

Figure 29. Comparison (schematic) of stochastic resonance with a standard (e.g., mechanical) resonance. (a) In SR, the response a of the system exhibits a maximum when plotted as a function of noise intensity D; (b) in a standard resonance, the response a shows a maximum when plotted as a function of the driving frequency Ω; (c) the phase shift ϕ in SR varies nonmonotonically with D, and the maximum lag (for small Ω, where SR is pronounced) is reached for a smaller value of D than that at which the maximum in $a(D)$ occurs; (d) in a standard resonance, the phase shift varies monotonically with Ω and is equal to $-\pi/2$ where $a(\Omega)$ passes through its resonance maximum. The different processes responsible for creating the SR maximum are indicated in (a). (After Ref. 168.)

of very large $D \gg \Delta U$ the double-well character of the potential becomes irrelevant and $\phi \simeq 0$, for the same reason; so, at the intermediate values of D where the interwell transitions play a substantial role and their probabilities are of the order of the frequency Ω, so that the field modulates the populations of the states effectively, a phase lag associated with this modulation must inevitably give rise to a maximum in $|\phi|$, just as observed. We note that this behavior and the *decrease* in the signal-to-noise ratio for very small D are both related to the continuity of the system, and they are not described by a two-state theory.

The observation of the phase lags for SR in bistable systems shows explicitly the difference between this "conventional" SR and standard resonance: Because Ω does not match any internal characteristic *vibrational* frequency of the system,[173] it should be only of the same order of magnitude or less than the Kramers hopping frequency. This is to be contrasted with SR in underdamped monostable systems,[70, 169] which is a true resonance phenomenon where external noise is used to tune the natural oscillation frequency of the system to that of the periodic force. The main differences between stochastic resonance and a standard (e.g., mechanical) resonance are summarized in Fig. 29.

C. Conclusions

Several conclusions may be drawn from the above results. First, the fact that the LRT approach provides a good description of the experimental results—both the signal/noise ratio and the phase shift—in the model electronic system means that it should be equally applicable to nonlinear optical (and other) systems describable in terms of a static bistable potential. To our knowledge, evidence for phase shifts has not yet been sought in, for example, a ring laser; but we would have no hesitation, on the basis of the present work, in predicting both their existence and the variation of their magnitude as a function of D.

Second, we would emphasize that LRT can readily be applied even to systems for which the SDF in the absence of the periodic force $Q^{(0)}(\omega)$ cannot be calculated. In such cases, $Q^{(0)}(\omega)$ can often be measured experimentally without too much difficulty, whereupon the susceptibility can immediately be found from Eq. (137).

Third, the perception of SR as a linear response phenomenon is very general. It is clear from inspection of Eq. (137) that SR phenomena are to be expected in *any* system displaying sharp peaks in its SDF $Q^{(0)}(\omega)$ that rise rapidly with increasing noise intensity. Provided Ω for the weak periodic force is in the immediate vicinity of one of these peaks, stochastic amplification (i.e., SR) is to be anticipated. Thus, the LRT approach provides a basis on which to search for SR in quite new types of physical systems that need not necessarily be bistable[70] and need not even be describable in terms of a static potential (see Section VII.B). It seems reasonable to predict that yet new forms of SR phenomena will be found on this basis in the near future, and that at least some of them will be demonstrable in nonlinear optical systems.

VII. HIGH-FREQUENCY CRITICAL PHENOMENA IN BISTABLE SYSTEMS

Many different physical quantities can exhibit optical bistability, especially in passive optical systems, for example, the transmission and reflection coefficients of a Fabry-Pérot cavity containing a medium with nonlinear refraction and/or absorption coefficient. In addition, the *oscillations* in the polarization of the medium are bistable. Their frequency is fixed at the light frequency, but their amplitude and phase can each assume either of two values. One would therefore expect to observe fluctuation phenomena and also critical phenomena at high (optical) frequencies. In particular, high-frequency critical phenomena related to fluctuational transitions between coexisting stable states could occur. As a result of a transition the amplitude and the phase of the vibrations change substantially, and therefore high-frequency fluctuations in the range of the kinetic phase transition (KPT), where the populations of the coexisting states are of the same order of magnitude, would be expected to be large. One would expect critical phenomena to be most easily observable in lumped-parameter systems where the transition probabilities are comparatively large.

The simplest model of a lumped-parameter system that displays bistability in a periodic field is the underdamped Duffing oscillator.[181] It has been traditionally used as a generic model of optical bistability and four wave mixing.[182, 183] (An exhaustive list of references appears in Ref. 12.) The equation of motion of a Duffing oscillator driven by a regular sinusoidal force plus noise is

$$\ddot{q} + 2\Gamma\dot{q} + \omega_0^2 q + \gamma q^3 = F \cos \omega_F t + f(t) \qquad (148)$$

where $f(t)$ is Gaussian white noise:

$$\langle f(t)f(t')\rangle = 4\Gamma T\delta(t - t') \qquad (149)$$

In Section VII.A we analyze the spectral density of fluctuations of the system Eq. (148) and discuss the supernarrow spectral peak that arises in the KPT range. We then go on, in Section VII.B, to show that the signal induced by an additional *weak* periodic force on the right side of Eq. (148) can sometimes be amplified by an increase in the noise term $f(t)$.

A. Supernarrow Spectral Peaks Near a Kinetic Phase Transition

We now embark on a calculation of the spectral density

$$Q(\omega) = \frac{1}{2\pi} \int_{-\infty}^{\infty} d\tau \exp(i\,\omega\tau) C(\tau)$$

$$C(\tau) = \lim_{t' \to \infty} \frac{1}{2t'} \int_{-t'}^{t'} dt [q(t+\tau) - \langle q(t+\tau)\rangle][q(t) - \langle q(t)\rangle] \tag{150}$$

This definition of spectral density, appropriate to a periodically oscillating coordinate q, differs slightly from the one used previously (Eq. (134)).

In the absence of noise and nonlinearity, the steady state solution of Eq. (148) is simply $q = \text{const} \times \cos(\omega_F t + \phi)$, suggesting that the full equation may be tackled by transforming to the rotating frame. In other words the position q and velocity $\dot{q}$ are replaced by two new variables u, u^*:

$$q = \left(\frac{2\omega_F(\omega_F - \omega_0)}{3\gamma}\right)^{1/2} [u \exp(i\,\omega_F t) + u^* \exp(-i\,\omega_F t)]$$

$$\dot{q} = i\,\omega_F \left(\frac{2\omega_F(\omega_F - \omega_0)}{3\gamma}\right)^{1/2} [u \exp(i\,\omega_F t) - u^* \exp(-i\,\omega_F t)] \tag{151}$$

(For definiteness we take $\gamma > 0$ and $\omega_F > \omega_0$.) Substitution in Eq. (148) shows that the complex envelopes u and u^* are determined by first-order differential equations, whose stationary solutions in the absence of noise satisfy

$$-\theta^{-1}u + i\,u(|u|^2 - 1) - i\,\beta^{1/2} = 0 \tag{152}$$

Here θ is the dimensionless frequency detuning and β the dimensionless field strength:

$$\theta = \frac{\omega_F - \omega_0}{\Gamma} \qquad \beta = \frac{3\gamma F^2}{32\omega_F^3(\omega_F - \omega_0)^3} \tag{153}$$

Equation (152) can be rewritten in terms of the squared amplitude of

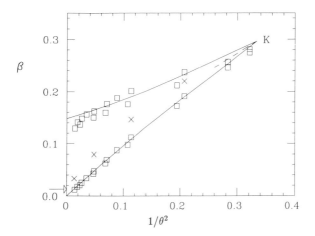

Figure 30. Phase diagram showing the region of bistability of the system Eq. (148) in terms of the reduced parameters of Eq. (153); the experimental data (squares) are compared with theory (solid lines). The theoretical phase transition line interpolates from the dashed curve to the arrowed point; the measured points at which the transition rates between attractors are equal are shown by crosses. (After Ref. 75.)

oscillation $x = |u|^2$:

$$x(x - 1)^2 + \theta^{-2}x - \beta = 0 \qquad (154)$$

This cubic equation has three real roots if the parameters β and θ lie within the region enclosed by solid lines in Fig. 30. Both the largest and smallest roots correspond to stable limit cycle solutions of Eq. (148) (labeled 1 and 2 in what follows) while the intermediate root represents an unstable limit cycle. Thus we have a bistable oscillatory system.

The effect of weak noise is twofold: It causes small fluctuations about each limit cycle, and occasional large outbursts of noise result in transitions from one limit cycle to the other. The spectral density of fluctuations Eq. (150) can be split into contributions $Q_i(\omega)$ formed by small fluctuations about the limit cycles $i = 1, 2$ and an additional term due to fluctuational transitions between the limit cycles:

$$Q(\omega) = \sum_i w_i Q_i(\omega) + Q_{tr}(\omega) \qquad (155)$$

where w_i is the population of limit cycle i. An expression for $Q_i(\omega)$ can easily be obtained by linearizing the equation of motion (148) about the stable state i:

$$Q_i = \frac{\Gamma T}{2\pi\omega_F^2} \frac{(\omega - \omega_F)^2 + 2(\omega - \omega_F)\Gamma\theta(2|u_i|^2 - 1) + \Gamma^2(\nu_i^2 + 2\theta^2|u_i|^4)}{\left[(\omega - \omega_F)^2 - \Gamma^2\nu_i^2\right]^2 + 4\Gamma^2(\omega - \omega_F)^2}$$

$$\nu_i^2 = 1 + \theta^2(3|u_i|^4 - 4|u_i|^2 + 1) \tag{156}$$

$Q_{tr}(\omega)$ can be calculated using a two-state approximation: The system occupies state 1 with probability w_1 or state 2 with probability $w_2 = 1 - w_1$. The populations w_1, w_2 are related to the transition rates W_{ij} from state i to state j by the master equation

$$\begin{aligned}
\dot{w}_1 &= -W_{12}w_1 + W_{21}w_2 \\
\dot{w}_2 &= W_{12}w_1 - W_{21}w_2
\end{aligned} \tag{157}$$

For an arbitrary initial state $q(t)$, the state of the system at a later time $t + \tau$ is given by the solution of Eq. (157). To ensemble average over $q(t)$ we also need the *stationary* solution of Eq. (157):

$$w_1 = \frac{W_{21}}{W_{12} + W_{21}} \qquad w_2 = \frac{W_{12}}{W_{12} + W_{21}} \tag{158}$$

The ensemble averaged correlation function $\langle[q(t + \tau) - \langle q(t + \tau)\rangle][q(t) - \langle q(t)\rangle]\rangle$ can now be evaluated. It can be converted into a time-averaged correlation function, as required by Eq. (148), by neglecting terms that are oscillatory functions of t. Fourier transformation then results in the spectral density

$$Q_{tr}(\omega) = \frac{2\omega_F(\omega_F - \omega_0)}{3\pi\gamma}|u_1 - u_2|^2 w_1 w_2 \frac{W_{12} + W_{21}}{(W_{12} + W_{21})^2 + (\omega - \omega_F)^2} \tag{159}$$

which has a peak centered on $\omega = \omega_F$. The transition rates W_{ij} are of activation type

$$W_{ij} = \text{const} \times \exp\left(-\frac{R_i}{\alpha}\right) \tag{160}$$

Here $\alpha = 3\gamma T/8\omega_F^3\Gamma$ is the effective noise strength in the rotating frame.

The width of the spectral peak is proportional to $W_{12} + W_{21}$ and is therefore exponentially small if the noise is weak. Such peaks were termed supernarrow in Ref. 75. The intensity of the peak is proportional to $w_1 w_2$, the product of the limit cycle populations. If we first examine their ratio

$$\frac{w_1}{w_2} = \frac{W_{21}}{W_{12}} \propto \exp\left(\frac{R_2 - R_1}{\alpha}\right) \qquad (161)$$

we see that one population is exponentially small and the supernarrow peak is of correspondingly low intensity, except in the region of the kinetic phase transition (KPT) defined by $R_1 = R_2$. Here the populations are roughly equal and a strong peak exists at $\omega = \omega_F$. Calculation of the exact form of the activation energies R_i is beyond the scope of this review. It is sufficient to note that they can be found analytically in the $\theta \to \infty$ limit and near the point K ($\beta = \frac{8}{27}, \theta = \sqrt{3}$) where the two stable limit cycles merge. This gives two ranges in parameter space at which $R_1 = R_2$, plotted in Fig. 30, and one can imagine the kinetic phase transition line $\beta = \beta_{\mathrm{KPT}}(\theta)$ being drawn between them.

The intensity increases exponentially as the KPT line is approached.

$$w_1 w_2 \simeq \exp\left[-\frac{|R'_1 - R'_2||\beta - \beta_{\mathrm{KPT}}(\theta)|}{\alpha}\right] \qquad (162)$$

where $R'_i = (\partial R_i / \partial \beta)|_{\beta = \beta_{\mathrm{KPT}}(\theta)}$. The first derivative of its logarithm with respect to distance from the phase transition line exhibits a cusp, which is a characteristic feature of a first-order phase transition.

These theoretical predictions were tested by electronic analogue simulation.[75] A circuit modeling the Duffing oscillator Eq. (148) was driven by a sinusoidal periodic force from an HP3325 frequency synthesizer. The fluctuating voltage in the circuit representing the coordinate $q(t)$ was digitized and the power spectral density $Q(\omega)$ of the fluctuations $q(t) - \langle q(t) \rangle$ was computed and averaged using a Nicolet 1080 data processor. The region in which the oscillator was found to be bistable is enclosed by the square data points in Fig. 30. Its boundaries agree well with the theoretical prediction (solid lines).

Pseudo-white Gaussian noise was then added to the circuit using a feedback shift-register noise generator, causing transitions between the stable states. The dependence of the reciprocal average residence time $\langle T_i \rangle^{-1} = W_{ij}$ in the ith attractor was confirmed to be of activation type. The experimentally determined KPT line along which $\langle T_1 \rangle = \langle T_2 \rangle$ is shown by the crosses in Fig. 27.

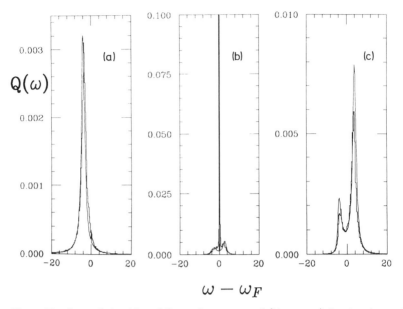

$$\omega - \omega_F$$

Figure 31. Spectral densities of fluctuations measured (histograms) for the electronic circuit model with $\theta = 4.574, \alpha = 8.69 \times 10^{-2}$ for (a) $\beta = 0.048$, (b) $\beta = 0.078$, and (c) $\beta = 0.150$. The solid lines represent theoretical predictions. (After Ref. 75.)

Some typical spectra showing the variation of $Q(\omega)$ with β are plotted (histograms) in Fig. 31 and compared with the theory (solid lines). The agreement is excellent; there are no adjustable parameters.

The most striking feature of the measured spectra in Fig. 31 is the supernarrow peak that appears in the KPT range. Its width (unresolvable by the data system) is much smaller than the width of the other peaks, the experimentally determined damping constant Γ, and the frequency detuning $\omega_F - \omega_0$. Figure 32 demonstrates the experimental dependence of the intensity of this peak on the distance from the phase transition line. The cusplike behavior is well described by Eq. (162) with experimental data used to determine the derivatives R'_1, R'_2. (At the time when Ref. 75 was published, R had not been calculated at this value of θ; such calculations have now been accomplished and will be published shortly[184].)

The fine structure of $Q(\omega)$ in Fig. 31 is also of interest. It contains between 1 and 5 peaks. The explanation is as follows: Each partial spectrum arises from the modulation of forced oscillations at frequency ω_F by relatively slow fluctuations (frequency $\sim \omega_F - \omega_0$), and hence contains two peaks. Superposition of the two partial spectra results in up to four peaks, depending on the values of β and θ.

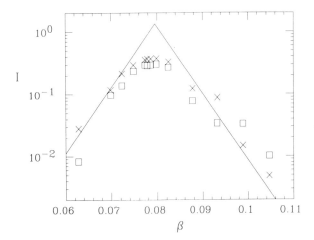

Figure 32. Variation of the intensity I of the supernarrow peak with distance from the phase-transition line for $\theta = 4.574$, $\alpha = 8.69 \times 10^{-2}$. The squares are direct measurements; the crosses are derived from Eq. (159), based on measured transition rates; the solid line also represents Eq. (159) but for $\ln(w_1 w_2)$ given by Eq. (162) with measured $R_{1,2}$. (After Ref. 69.)

B. High-Frequency Stochastic Resonance for Periodic Attractors

The possibility of stochastic resonance (see Section VI) in the Duffing oscillator can be explored[69, 169] by introducing an additional weak periodic force $A \cos \Omega t$ on the right side of (148). We work to first order in A and consider the case where Ω is close to ω_0 and hence also ω_F. Because of nonlinearity, the response of the system q will contain oscillatory components at the combination frequencies $|\Omega \pm \omega_F|$, $|\Omega \pm 2\omega_F| \ldots$ as well as the frequency Ω of the trial force. However, the dominant contributions occur at the near resonant frequencies Ω and $|\Omega - 2\omega_F|$. The change in the ensemble averaged response can therefore be written in terms of susceptibilities $\chi(\omega)$ and $X(\omega)$ as

$$\delta\langle q(t)\rangle = A\,\mathrm{Re}\big[\chi(\Omega)\exp(-\mathrm{i}\,\Omega t) + X(\Omega)\exp(\mathrm{i}(2\omega_F - \Omega)t)\big] \quad (163)$$

The calculation of the susceptibilities closely parallels that of the spectral density in the previous subsection. For small α (weak noise intensity) the trial force produces two distinct effects. First, it causes small periodic oscillations about the stable states. Second, by modulating the probability of fluctuational transitions between the stable states, it causes a periodic modulation of their populations. As a result the susceptibilities can be written in a form similar to Eq. (140):

$$
\chi(\Omega) = \sum_j w_j \chi_j(\Omega) + \chi_{tr}(\Omega)
$$
$$
X(\Omega) = \sum_j w_j X_j(\Omega) + X_{tr}(\Omega)
$$

(164)

Linearizing the equation describing the evolution of the complex amplitudes u, u^* about the solution in the absence of the trial force gives the following expressions for the partial susceptibilities:

$$
\chi_j(\Omega) = \frac{i}{2\omega_F} \frac{\Gamma - i(\Omega - \omega_F) - i(2|u_j|^2 - 1)(\omega_F - \omega_0)}{\Gamma^2 \nu_j^2 - 2i\Gamma(\Omega - \omega_F) - (\Omega - \omega_F)^2}
$$
$$
X_j(\Omega) = -\frac{1}{2\omega_F} \frac{u_j^2(\omega_F - \omega_0)}{\Gamma^2 \nu_j^2 - 2i\Gamma(\Omega - \omega_F) - (\Omega - \omega_F)^2}
$$

(165)

As far as the rotating frame is concerned, the trial force is of low frequency, equal to $|\Omega - \omega_F|$. It therefore induces slowly varying periodic contributions to the activation energies R_1 and R_2. In turn, these produce periodic perturbations in the transition rates W_{ij} and the populations w_1 and w_2. Solving the master equation (157) to first order in A determines the transitional components of the susceptibilities

$$
\chi_{tr}(\Omega) = \frac{w_1 w_2}{2\omega_F(\omega_F - \omega_0)} (u_1^* - u_2^*) \frac{\mu_1 - \mu_2}{\alpha} \left[1 - \frac{i(\Omega - \omega_F)}{W_{12} + W_{21}} \right]^{-1}
$$
$$
X_{tr}(\Omega) = \frac{u_1 - u_2}{u_1^* - u_2^*} \chi_{tr}(\Omega)
$$
$$
\mu_j = \sqrt{\beta} \frac{\partial R_i}{\partial \beta}
$$

(166)

The transition-induced contribution (166) will be significant only if the system is in the KPT range, because the $w_1 w_2$ factor will otherwise be exponentially small. In the vicinity of the KPT the contribution (166) will be substantial since it contains the large factor $|\mu_1 - \mu_2|/\alpha \gg 1$. Thus, noting the extremely rapid increase of W_{ij} (Eq. (160)) with noise intensity, Eq. (166) suggests that there will be a range of noise intensity where the strengths (integrated powers) of the periodic signals at Ω and the mirror-reflected frequency $(2\omega_F - \Omega)$

$$
S(\Omega) = \tfrac{1}{4} A^2 |\chi(\Omega)|^2
$$
$$
S(2\omega_F - \Omega) = \tfrac{1}{4} A^2 |X(\Omega)|^2
$$

(167)

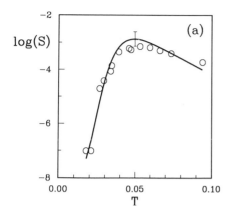

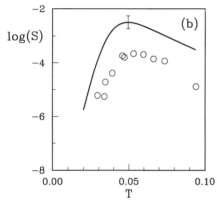

Figure 33. The logarithm of the responses S of the system given by Eq. (148) to a weak trial force at frequency Ω, as a function of noise intensity T: (a) at the trial frequency Ω; (b) at the "mirror-reflected" frequency $(2\omega_F - \Omega)$. The data points are experimental results from an electronic model. The curves represent the theory, incorporating measured values of the activation energies R_i. (After Ref. 69.)

will increase very rapidly with T. In other words, a form of stochastic resonance should occur, which we refer to as high-frequency stochastic resonance (HFSR), since it exists at frequencies close to ω_0, rather than the much lower frequency of interattractor hopping that characterizes conventional SR.

The analogue experiment of the previous section was extended to investigate HFSR.[69, 169] The weak trial force was supplied by a second HP3325 frequency synthesizer. The circuit parameters were $2\Gamma = 0.0397$, $\omega_0 = 1.00$, $\gamma = 0.1$, $\omega_F = 1.07200$, $\Omega = 1.07097$, $F = 0.068$, and $A = 0.006$. Because the three frequencies $\omega_0, \omega_F, \Omega$ are (necessarily) close, the data was analyzed using the superior frequency resolution offered by a Nicolet 1280 data processor.

The signal strengths were determined by measuring the heights of the delta function spikes in the power spectrum at frequencies Ω and $2\omega_F - \Omega$. They are plotted as functions of noise intensity T in Fig. 33.

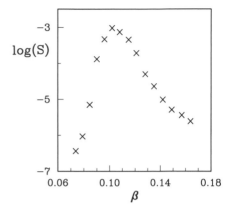

Figure 34. Variation of the response S measured at the trial force frequency Ω with distance from the kinetic phase transition, indicated by β, for fixed noise intensity $T = 0.05$. It takes the form of a cusp (note the log scale), rounded by noise. (After Ref. 69.)

HFSR is indeed observed; as T increases, the signal strength S displays a rapid rise followed by a slower fall. The form of the $S(T)$ curve is remarkably similar to standard stochastic resonance (Fig. 27). The solid lines in Fig. 33 represent the theoretical calculation, except that *measured* values of R_1 and R_2 have again been employed in the absence of their theoretical counterparts. The agreement between theory and experiment can be considered excellent, bearing in mind the difficulty of measuring β accurately—it contains $\omega_F - \omega_0$, a difference between two nearly equal quantities, raised to the third power. The signal/noise ratios have also been measured; each increases with T by a factor of about 25 between its minimum and maximum value.

The magnitude of the fluctuation-induced signal at Ω has been measured as a function of the distance from the KPT $\beta - \beta_{KPT}$. The result, displayed in Fig. 34 shows a rapid cusp-like decrease of S (note the log scale) as β moves away from its critical value, demonstrating that, like the onset of the supernarrow peak itself, HFSR for periodic attractors has the character of a critical phenomenon.

VIII.   A FABRY-PÉROT CAVITY WITH A MULTIVALUED POTENTIAL

An intriguing feature of the passive all-optical double-cavity membrane system (DCMS)[21] for which the onset of stochastic resonance (see Section VI) was demonstrated[30] is that the theory of fluctuations in the transmitted radiation due to internal fluctuations was developed in terms of a *multibranch* effective potential. Some remarkable consequences followed from this hypothesis. In particular, it was predicted that the probability density distribution $P(I_T)$ of the transmitted light intensity I_T should be confined within boundaries, that it should be singular at these boundaries, and that it might possess up to four separate maxima. These predictions could not be fully tested in the DCMS itself because of various smearing effects, in particular those due to fluctuations in the input light intensity. Partly for this reason, and partly because the theory of the DCMS involves two periodic functions of the phase gain ϕ of light in the cavity, $M(\phi)$, $N(\phi)$ (Ref. 21) whose exact form is unknown, it was impossible to make a direct comparison of experiment and theory.

An analogue electronic experiment[76] enabled us to model an idealized cavity in the absence of input intensity noise and with $M(\phi)$, $N(\phi)$ chosen in the (archetypal) forms corresponding to a Fabry-Pérot cavity. It was thus possible to investigate for the first time the interesting phenomena associated with a multibranch effective potential and to compare the results, both qualitatively and quantitatively, with theoretical predictions. As we demonstrate below, the experiment confirmed the unusual singular and multipeaked character of $P(I_T)$ arising from fluctuations of the intracavity phase gain ϕ and revealed features of the inner workings of the cavity system, such as the phase distribution, that are inaccessible through conventional optical measurements. Because phase noise is inherent to all optical cavities, the results obtained are of wide relevance.

The model considered is a nonlinear Fabry-Pérot resonator (cf. Fig. 2) displaying dispersive OB. In modeling the mechanism of OB we assume that the internal medium of the cavity has a refractive index that depends linearly on light intensity, and that relaxation of the intracavity phase gain of the radiation can be described by the Debye equation; see, e.g., Ref. 12 and references therein. (Note that this is actually a different OB mechanism to that of the air-spaced DCMS device.[21]) Our system is then described by

$$\dot{\phi} + \frac{1}{\tau}(\phi - \phi^{(0)}) = IM(\phi) + I_m(t)$$

$$I_T = IN(\phi)$$

$$M(\phi) = A_M\left[1 + \tfrac{1}{2}F(1 - \cos\phi)\right]^{-1} \qquad (168)$$

$$N(\phi) = A_N\left[1 + \tfrac{1}{2}F(1 - \cos\phi)\right]^{-1}$$

Here, I and I_T are the intensities of the incident and transmitted radiation, ϕ is the intracavity phase gain, which takes the value $\phi^{(0)}$ in the limit $I \to 0$, and F is the finesse of the cavity. The functions $M(\phi)$ and $N(\phi)$ relate the intensities of the transmitted light and of the intracavity field (which drives the optically nonlinear medium and causes the changes in ϕ) to that of the incident light; the expressions for $M(\phi)$, $N(\phi)$ are standard[12] for a Fabry-Pérot cavity; and we have included the nonlinearity parameter of the medium within A_M. The function $I_m(t)$ describes the noise driving the intracavity phase gain ϕ of the radiation. It is assumed white and Gaussian:

$$
\begin{aligned}
I_m(t) &= \bar{I}_m + \delta I_m(t) \qquad \langle \delta I_m(t) \rangle = 0 \\
\langle \delta I_m(t) \delta I_m(t') \rangle &= 2D\delta(t - t')
\end{aligned}
\tag{169}
$$

The noise can originate in various ways, e.g., from thermal fluctuations in the nonlinear medium, or through random vibrations of the mirrors resulting in variations of the intracavity optical length. The crucial point here is that the fluctuations of the transmitted light intensity I_T are due only to phase fluctuations, while the intensity of incident light I remains constant.

The probability distribution of the phase $p(\phi)$ has the form

$$
\begin{aligned}
p(\phi) &= Z^{-1} \exp(-U(\phi)/D) \\
Z &= \int d\phi \exp(-U(\phi)/D)
\end{aligned}
\tag{170}
$$

where $U(\phi)$ is the effective potential for the dynamics of the phase:

$$
U(\phi) = -I \int_0^\phi d\phi' \, M(\phi') + \frac{1}{2\tau}\phi^2 - \frac{1}{\tau}(\phi_0 + \bar{I}_m \tau)\phi \tag{171}
$$

In the range of optical bistability the potential $U(\phi)$ has two minima lying at $\phi = \phi_{1,2}$, where $\phi_{1,2}$ are the stable solutions of Eq. (168) with $I_m(t) = \bar{I}_m$.

The quantity of greatest physical interest is the probability distribution $P(I_T)$ of the transmitted light intensity, rather than $p(\phi)$, since it is $P(I_T)$ that can be measured in optical experiments. This quantity can be calculated immediately from Eqs. (168), (170), and (171) and has the form

$$
\begin{aligned}
P(I_T) &= K(I_T)\exp\left[-\frac{V(I_T)}{D}\right] \\
V(I_T) &= U(\phi(I_T)) \\
K(I_T) &= I_T^{-2}Z^{-1}\frac{2IA_N}{F|\sin\phi|}
\end{aligned}
\tag{172}
$$

It is evident from Eqs. (168)–(172) that there are dramatic differences between the phase distribution and that of the transmitted light intensity. This is due to the periodicity of the transmission coefficient $N(\phi)$ in Eq. (168): $N(\phi + 2\pi) = N(\phi)$. As a consequence, the whole ϕ axis is mapped by the relation $I_T = IN(\phi)$ onto the interval $(I_{T\min}, I_{T\max})$ of I_T

$$I_{T\min} = \frac{IA_N}{1 + F} \qquad I_{T\max} = IA_N \qquad (173)$$

and the distribution $P(I_T) = 0$ for all I_T lying outside this range. At the same time, the effective potential for the transmitted light $V(I_T)$ is a multibranch function, and the values of $I_{T\min}, I_{T\max}$ (i.e., those for $\phi = n\pi$) correspond to its branching points. A comparison of $U(\phi)$ and $V(I_T)$ is shown in Fig. 35.

It is the multibranch character of the effective potential $V(I_T)$ that accounts for some very unusual features seen in the distribution of the transmitted light. The distribution still has two maxima at $I_T = I_{T1,2} \equiv IN(\phi_{1,2})$ corresponding[185] to the minima of $V(I_T)$, but they can now lie on different branches of $V(I_T)$ (see Fig. 35b). The branching points give rise to the singularities of the distribution, because they correspond to an infinite "density of states" for the transmitted light intensity near boundaries (see prefactor in Eq. (172)). Correspondingly, $P(I_T)$ diverges for $I_T \to I_{T\min}, I_{T\max}$. It should thus be possible to obtain a *four*-peaked distribution for a double-well potential under OB conditions. While investigating this interesting theoretical prediction it is important to bear in mind that the distribution as determined from experiment $P_{ex}(I_T)$ is the coarse-grained "bare" distribution $P(I_T)$:

$$P_{ex}(I_T) = \frac{1}{2\Delta I_T} \int_{I_T - \Delta I_T}^{I_T + \Delta I_T} P(I_T') \, dI_T' \qquad (174)$$

Thus, in experiments, rather than singularities, we can expect to see very strong additional peaks near boundaries of the light distribution. We note that the onset of the four-peaked distribution depends on the system becoming optically bistable (for optically monostable systems, the distribution would be expected to have three peaks). Thus, in the model Eqs. (168) and (169) there is no special requirement on the finesse for the four-peaked distribution to arise; in real systems, however, which have losses, a minimum value of the finesse would need to be exceeded. All of these theoretical predictions, including the appearance of the four-peaked dis-

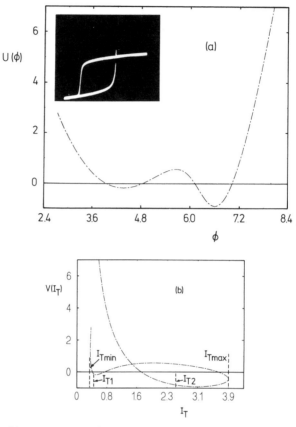

Figure 35. (a) The potential $U(\phi)$ for the distribution over the phase in the nonlinear Fabry-Pérot cavity described by Eq. (168) ($\tau IA_M = 3.925$, $\phi^{(0)} = 3.4$ rad, $F = 10.1$); (b) the corresponding effective multibranch potential $V(I_T)$ for the probability distribution of transmitted light intensity, which is bounded at $I_{T\min}$, $I_{T\max}$. The positions $I_{T1,2}$ of the inner maxima in $P(I_T)$ are close to, but do not coincide exactly[185] with, the corresponding minima in $V(I_T)$ shown in (b). The insert in (a) shows the experimental hysteresis loop measured for the electronic model using the same parameters. (After Ref. 76.)

tribution, have been tested[76] with the aid of an electronic analogue model of Eqs. (168) and (169).

The experimental hysteresis loop in the absence of noise is shown in the insert in Fig. 35a. It contains a characteristic[12] spike at the bifurcation value of the incident radiation intensity corresponding to the disappearance of the lower-transmission branch. The transient transmission exceeds that on the upper branch because, on its way from the lower-I_T stable state ($I_T = I_{T1}$ in Fig. 35b) to the higher-I_T one ($I_T = I_{T2}$ in Fig. 35b), along the input/output characteristic, it is obvious from Fig. 35 that the system must pass the value $I_{T\,max} > I_{T2}$. When noise was added to the system, transitions occurred between the stable states. Their reciprocal average lifetimes were measured and, in common with those from the earlier experiments on the DCMS,[21] were found to be of the activation type; a Lorentzian-shaped zero-frequency peak was observed in the measured spectral density. To this extent, the results obtained were in qualitative agreement with those from the DCMS. In the latter work it was not possible to compare the shape of the distribution of the transmitted light intensity quantitatively with theory because of the input intensity fluctuations and because, as mentioned above, the forms of $M(\phi), N(\phi)$ for the DCMS are unknown and presumably do not correspond to those assumed in the simple model of Eq. (168). Nor was it possible to observe the predicted 4-peaked distribution $P(I_T)$ for the DCMS.

The distribution $P(I_T)$ measured for the electronic model is shown by the jagged curve of Fig. 36a. Unlike the (smeared) results obtained in Ref. 21, the multipeaked structure of $P(I_T)$, expected to result from the branching of the effective potential $V(I_T)$ given by Eq. (172), is clearly resolved. Three maxima are immediately apparent; closer inspection reveals that the left hand one is actually a double peak, i.e., there are four maxima in total. It is also evident that the outer peaks are highly asymmetric. At their outer boundaries they are, in fact, singular within the resolution of the experiment, just as predicted. The full curve, representing the theory, is in good quantitative agreement with the measurements. The distribution of phase $p(\phi)$ has also been measured for the electronic model and is shown by the jagged curve in Fig. 36b. In striking contrast to $P(I_T)$, it exhibits a conventional double-peaked structure, consistent with the ordinary single-valued double-well effective potential Eq. (171) from which it is derived. Again, the measurements are in excellent quantitative agreement with the theoretical prediction (full curve).

We would point out that, although the investigations have related specifically to the model Eq. (168) of optical bistability in a Fabry-Pérot cavity, very similar phenomena to those discussed above—notably, the appearance of multipeaked distributions singular at their outer boundaries —may also be expected to arise in other types of passive optical resonator and also in lasers, as well as in nonoptical systems, whenever the fluctuating quantity (corresponding to the phase gain in the above analysis) is a multibranch function of the observed quantity (the transmitted light intensity).

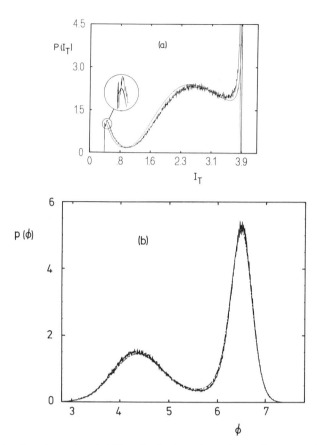

Figure 36. Experimental probability distributions (jagged curves), measured for the electronic model of Eq. (168) with $\tau IA_M = 3.4$, compared with the corresponding theoretical predictions (full curves): (a) the distribution of transmitted light intensity $P(I_T)$ displays four separate maxima confined between singular outer boundaries (note that the left peak is actually a double one, in both experiment and theory, as shown in the expanded region); (b) the variation of the phase takes the form of a standard double-peaked distribution $p(\phi)$. (After Ref. 76.)

IX. CONCLUSION

The two most general conclusions that might be drawn from the above results are that (1) in nonlinear optical systems with a few dynamical degrees of freedom there arises a whole *range* of phenomena that might be reasonably called critical, since they are associated with large fluctuations caused by relatively weak noise and come into play only within a narrow range of the parameters of a system; and (2) the role of simulation in investigating these phenomena has become extremely important. Simulations not only provide a tool for testing theoretical models, but they have also been shown to possess *predictive* power. The latter is most clearly illustrated, perhaps, by the results of Section III on the onset of the twin-peaked probability distribution near the critical point of a colored-noise-driven system; notwithstanding the theoretical discussion in Section III.B, this effect has not been fully described analytically so far. This means that in nonlinear optics, too, just as in condensed-matter physics and nuclear physics, for example, simulation becomes an equal partner in the chain *theory–simulation–experiment*. The results on critical phenomena in nonlinear optics presented above, and in particular those on the multibranch potential and stochastic resonance, show clearly how stimulating each link of this chain can be for the other two.

One advantageous feature of nonlinear optics, from the viewpoint of investigation through simulation, is that the systems of interest can often be described with the aid of only a few time-dependent dynamical variables, so that not only qualitative, but also quantitative, agreement of the theoretical, simulational and experimental results on fluctuations can be achieved, as has been clearly demonstrated above. The other significant feature of optics is that the underlying true quantities of physical interest, the electromagnetic fields, are fast-oscillating. Consequently, there can arise not only low-, but also high-frequency critical phenomena. In other words, optics adds an "extra dimension" to nonlinear dynamics by making it interesting to investigate phenomena that are induced by strong periodic fields and that occur at high frequencies. One of these is the onset of a signal at the "mirror" frequency analyzed in Section VII.B in the context of high-frequency stochastic resonance. It corresponds in optics to the idler wave in four-wave mixing. The onset of the peak (as a function of the noise intensity) in this signal corresponds to the onset of the peak in the corresponding four-wave scattering, which, of course, has no analogues in nonequilibrium systems with static attractors, and certainly none in thermal equilibrium systems. Another example of a critical phenomenon related to an attractor that is a limit cycle, rather than a steady state, is the transient frequency modulation in the switch-on of a single-mode laser,[186] and there are more such examples, of course.

Taken as a whole, the investigation of critical phenomena in optics is a fruitful and most promising area of scientific endeavor. We hope that the chapter has clearly demonstrated this, in addition to illustrating some modern approaches with examples of recent results obtained through the aid of simulation techniques.

Acknowledgments

In the course of the work described in this chapter we have benefitted from stimulating interactions with many colleagues and collaborators including, especially, Igor Fedchenia, Leone Fronzoni, Paolo Grigolini, Akito Igarashi, Colin Lambert, Luigi Lugiato, Frank Moss, Paolo Martano, Jose-Maria Sancho, and Slava Soskin. To all of them, and to many others, we are extremely grateful. The research, in its different aspects, has been supported by the Science and Engineering Research Council (UK), the Istituto Nazionale Fisica della Materia, the European Community, the Royal Society of London, NATO, the Ukrainian Academy of Sciences, and the British Council.

References

1. A. Einstein, *Ann. Phys.* **17**, 549 (1905).

2. M. von Smoluchowski, *Bull. Intern. Acad. Sci. Cracovie* (A) 418 (1913).

3. P. Langevin, *Comptes Rendus* **146**, 530 (1908).

4. W. H. Louisell, *Radiation and Noise in Quantum Electronics*, McGraw-Hill, New York, 1964.

5. M. O. Scully and W. E. Lamb, Jr., *Phys. Rev.* **159**, 208 (1967).

6. M. Lax, *Fluctuation and Coherence Phenomena in Classical and Quantum Physics*, Gordon & Breach, New York, 1968.

7. H. Haken, in S. Flugge (Ed.), *Encyclopaedia of Physics*, Vol. 25/2c, Springer, Berlin, 1970.

8. L. D. Landau and E. M. Lifshitz, *Statistical Physics*, 3d ed., part 1, Pergamon, New York, 1980.

9. R. Short, L. Mandel, and R. Roy, *Phys. Rev. Lett.* **49**, 647 (1982).

10. See, for example, N. B. Abraham, E. Garmire, and P. Mandel, (Eds.), *OSA Proceedings on Nonlinear Dynamics in Optical Systems*, Optical Society of America, Washington, DC, 1991, and subsequent volumes in this proceedings series.

11. L. A. Lugiato, in E. Wolf (Ed.), *Progress in Optics*, **21**, 69 (1984).

12. H. M. Gibbs, *Optical Bistability: Controlling Light with Light*, Academic, New York, 1985.

13. H. J. Carmichael, in R. W. Boyd, M. G. Raymer, and L. M. Narducci (Eds.), *Optical Instabilities*, Cambridge University Press, Cambridge, UK, 1986, p. 111.

14. L. A. Lugiato, P. Mandel, S. T. Dembinski, and A. Kossakowski, *Phys. Rev. A* **18**, 238 (1978).

15. R. Roy and L. Mandel, *Opt. Commun.* **34**, 133 (1980); P. Lett, W. Christian, S. Singh, and L. Mandel, *Phys. Rev. Lett.* **47**, 1892 (1981).

16. P. Lett, E. C. Gage, and T. H. Chyba, *Phys. Rev. A* **35**, 746 (1987).

17. S. L. McCall, H. M. Gibbs, F. A. Hopf, D. L. Kaplan, and S. Ovadia, *Appl. Phys. B* **28**, 99 (1982); S. L. McCall, S. Ovadia, H. M. Gibbs, F. A. Hopf, and D. L. Kaplan, *IEEE J. Quantum Electron.* **QE-21**, 1441 (1985).

18. W. Lange, F. Mitschke, R. Deserno, and J. Mlynek, *Phys. Rev. A* **32**, 1271 (1985); F. Mitschke, R. Deserno, W. Lange, and J. Mlynek, *Phys. Rev. A* **33**, 3219 (1986).

19. G. Broggi and L. A. Lugiato, *Phys. Rev. A* **29**, 2949 (1984); G. Broggi, L. A. Lugiato, and A. Colombo, *Phys. Rev. A* **32**, 2803 (1985); L. A. Lugiato, A. Colombo, G. Broggi, and R. J. Horowicz, in *Frontiers in Quantum Optics*, Adam Hilger, Bristol, 1986, p. 231.

20. E. Arimondo, D. Dangoisse, and L. Fronzoni, *Europhys. Lett.* **4**, 287 (1987).

21. M. I. Dykman, G. P. Golubev, D. G. Luchinsky, A. L. Velikovich, and S. V. Tsuprikov, *Phys. Rev. A* **44**, 2439 (1991).

22. G. P. Golubev, D. G. Luchinsky, A. L. Velikovich, and M. A. Liberman, *Opt. Commun.* **64**, 181 (1987); A. L. Velikovich, G. P. Golubev, V. P. Golubchenko, and D. G. Luchinsky, *Opt. Commun.* **80**, 444 (1991).

23. M. I. Dykman and M. A. Krivoglaz, *Sov. Phys. JETP* **50**, 30 (1979).

24. R. Bonifacio and L. A. Lugiato, *Phys. Rev. Lett.* **40**, 1023 (1978).

25. F. T. Arecchi and F. Lisi, *Phys. Rev. Lett.* **49**, 94 (1982); M. R. Beasley, D. D'Humieres, and B. A. Huberman, *Phys. Rev. Lett.* **50**, 1328 (1983); R. F. Voss, *Phys. Rev. Lett.* **50**, 1320 (1983); F. T. Arecchi, R. Badii, and A. Politi, *Phys. Rev. A* **32**, 402 (1985).

26. P. Hänggi and H. Thomas, *Phys. Rep. C* **88**, 207 (1982).

27. M. I. Dykman and M. A. Krivoglaz, in I. M. Khalatnikov (Ed.), *Soviet Physics Reviews*, Vol. 5, Harwood, New York, 1984, p. 265.

28. M. I. Dykman, M. A. Krivoglaz, and S. M. Soskin, in F. Moss and P. V. E. McClintock (Eds.), *Noise in Nonlinear Dynamical Systems*, Vol. 2, Cambridge University Press, Cambridge, UK, 1989, Chapter 13.

29. B. McNamara, K. Wiesenfeld, and R. Roy, *Phys. Rev. Lett.* **60**, 2626 (1988).

30. M. I. Dykman, A. L. Velikovich, G. P. Golubev, D. G. Luchinsky, and S. V. Tsuprikov, *Sov. Phys. JETP Lett.* **53**, 193 (1991).

31. A. Bulsara and F. Moss (Eds.), Proc. of Workshop on Stochastic Resonance and its Applications in Physics and Biology, San Diego, 1992, *J. Stat. Phys.* **70** (special issue), in press.

32. L. Fronzoni in F. Moss and P. V. E. McClintock (Eds.), *Noise in Nonlinear Dynamical Systems*, Vol. 3, Cambridge University Press, Cambridge, UK, 1989, Chapter 8.

33. P. V. E. McClintock and F. Moss in F. Moss and P. V. E. McClintock (Eds.), *Noise in Nonlinear Dynamical Systems*, Vol. 3, Cambridge University Press, Cambridge, UK, 1989, Chapter 9.

34. R. Mannella in F. Moss and P. V. E. McClintock (Eds.), *Noise in Nonlinear Dynamical Systems*, Vol. 3, Cambridge University Press, Cambridge, UK, 1989, Chapter 7.

35. R. Landauer, *J. Appl. Phys.* **33**, 2209 (1962).

36. R. L. Stratonovich, *Topics in the Theory of Random Noise*, Vol. 1, Gordon & Breach, New York, 1963.

37. R. L. Stratonovich, *Topics in the Theory of Random Noise*, Vol. 2, Gordon & Breach, New York, 1967.

38. J. B. Morton and S. Corrsin, *J. Math. Phys.* **10**, 361 (1969).

39. T. Kawakubo, S. Kabashima, and N. Nishimura, *J. Phys. Soc. Jpn.* **34**, 1149 (1973).

40. F. T. Arecchi, A. Politi, and L. Ulivi, *Nuovo Cimento* **71B**, 119 (1982).

41. J. M. Sancho, M. San Miguel, H. Yamazaki, and T. Kawakubo, *Physica* **116A**, 560 (1982).

42. W. Lange in F. Moss and P. V. E. McClintock (Eds.), *Noise in Nonlinear Dynamical Systems*, Vol. 3, Cambridge University Press, Cambridge, UK, 1989, Chapter 6.

43. S. Fauve and F. Heslot, *Phys. Lett.* **97A**, 5 (1983).

44. P. Horowitz and W. Hill, *The Art of Electronics*, Cambridge University Press, Cambridge, UK, 1980.

45. Analog Devices Inc., Box 9106, Norwood, MA 02062, USA.

46. Burr-Brown Inc., Box 11400, Tucson, AZ 85734, USA.

47. N. G. Stocks, *Experiments in Stochastic Nonlinear Dynamics*, Thesis, Lancaster University, Lancaster, UK, 1990.

48. S. Faetti, C. Festa, L. Fronzoni, P. Grigolini, and P. Martano, *Phys. Rev. A* **30**, 3252 (1984).

49. J. Casademunt, J. I. Jiménez-Aquino, J. M. Sancho, C. J. Lambert, R. Mannella, P. Martano, P. V. E. McClintock, and N. G. Stocks, *Phys. Rev. A* **40**, 5915 (1989).

50. M. I. Dykman, P. V. E. McClintock, N. D. Stein, and N. G. Stocks, *Phys. Rev. Lett.* **67**, 933 (1991); and M. I. Dykman, R. Mannella, P. V. E. McClintock, N. D. Stein, and N. G. Stocks, to be published.

51. A. Igarashi and T. Munakata, *J. Phys. Soc. Jpn.*, **57**, 2439 (1988).

52. W. Ebeling and L. Schimansky-Geier in F. Moss and P. V. E. McClintock (Eds.), *Noise in Nonlinear Dynamical Systems*, Vol. 1, Cambridge University Press, Cambridge, UK, 1989, Chapter 8; L. Schimansky-Geier and Ch. Zülicke, *Z. Phys. B* **79**, 451 (1990).

53. S. Kai in F. Moss and P. V. E. McClintock (Eds.), *Noise in Nonlinear Dynamical Systems*, Vol. 3, Cambridge University Press, Cambridge, UK, 1989, Chapter 2.

54. J. Smythe, F. Moss, and P. V. E. McClintock, *Phys. Rev. Lett.* **51**, 1062 (1983).

55. J. Smythe, F. Moss, P. V. E., McClintock, and D. Clarkson, *Phys. Lett.* **97A**, 95 (1983).

56. P. V. E. McClintock and F. Moss, *Phys. Lett.* **107A**, 367 (1985).

57. S. D. Robinson, F. Moss, and P. V. E. McClintock, *J. Phys. A: Math. Gen* **18**, L89 (1985).

58. F. Moss, P. Hänggi, R. Mannella, and P. V. E. McClintock, *Phys. Rev. A* **33**, 4459 (1986).

59. K. Vogel, H. Risken, W. Schleich, M. James, F. Moss, R. Mannella, and P. V. E. McClintock, *J. Appl. Phys.* **62**, 721 (1987).

60. P. Grigolini, L. A. Lugiato, R. Mannella, P. V. E. McClintock, M. Merri, and M. Pernigo, *Phys. Rev. A* **38** 1966 (1988).

61. D. K. Kondepudi, F. Moss, and P. V. E. McClintock, *Physica* **21D**, 296 (1986).

62. L. Fronzoni, R. Mannella, P. V. E. McClintock, and F. Moss, *Phys. Rev. A* **36**, 834 (1987).

63. N. G. Stocks, R. Mannella, and P. V. E. McClintock, *Phys. Rev. A* **40**, 5361 (1989).

64. N. G. Stocks, R. Mannella, and P. V. E. McClintock, *Phys. Rev. A* **42**, 3356 (1990).

65. L. Gammaitoni, F. Marchesoni, E. Menichella-Saetta, and S. Santucci, *Phys. Rev. Lett.* **62**, 349 (1989).

66. M. I. Dykman, R. Mannella, P. V. E. McClintock, and N. G. Stocks, *Phys. Rev. Lett.* **65** 2606 (1990); M. I. Dykman, P. V. E. McClintock, R. Mannella, and N. G. Stocks, *Sov. Phys. JETP Lett.* **52**, 141 (1990).

67. M. I. Dykman, R. Mannella, P. V. E. McClintock, and N. G. Stocks, *Phys. Rev. Lett.* **68**, 2985 (1992).

68. M. I. Dykman, D. G. Luchinsky, P. V. E. McClintock, N. D. Stein, and N. G. Stocks, *Phys. Rev. A* **46**, R1713 (1992).

69. M. I. Dykman, D. G. Luchinsky, R. Mannella, P. V. E. McClintock, N. D. Stein, and N. G. Stocks, to be published.

70. N. G. Stocks, N. D. Stein, and P. V. E. McClintock, *J. Phys. A* **26**, L385 (1993).

71. A. Longtin, A. Bulsara, and F. Moss, *Phys. Rev. Lett.* **67**, 656 (1991).

72. P. J. Jackson, C. J. Lambert, R. Mannella, P. Martano, P. V. E. McClintock, and N. G. Stocks, *Phys. Rev. A* **40**, 2875 (1989).

73. H. K. Leung, *Phys. Rev. A* **37**, 1341 (1988).

74. R. C. M. Dow, C. J. Lambert, R. Mannella, and P. V. E. McClintock, *Phys. Rev. Lett.* **59**, 6 (1987).

75. M. I. Dykman, R. Mannella, P. V. E. McClintock, and N. G. Stocks, *Phys. Rev. Lett.* **65**, 48 (1990).

76. M. I. Dykman, D. G. Luchinsky, P. V. E. McClintock, N. D. Stein, and N. G. Stocks, *Phys. Rev. A* **45**, R7678 (1992).

77. M. I. Dykman, P. V. E. McClintock, V. N. Smelyanski, N. D. Stein, and N. G. Stocks, *Phys. Rev. Lett.* **68**, 2718 (1992).

78. A. M. Smith and C. W. Gardiner, *Phys. Rev. A* **39**, 3511 (1989).

79. J. R. Klauder and W. P. Petersen, *SIAM (Soc. Ind. Appl. Math.) J. Numer. Anal.* **22**, 1153 (1985).

80. W. Rümelin, *SIAM (Soc. Ind. Appl. Math.) J. Numer. Anal.* **19**, 604 (1982).

81. M. Beccaria, G. Curci, and A. Vicere, *Phys. Rev. A*, in press.

82. A. S. Kronfeld, *Prog. Theor. Phys.*, Suppl No. 111, p. 293 (1993).

83. R. Mannella, Tesiperil Dottorato di Ricerca in Fisica, Università de Pisa (1990).

84. E. K. Blum, *Numerical Analysis and Computational Theory and Practice*, Wiley, New York, 1972.

85. R. Mannella and V. Palleschi, *Phys. Rev. A* **40**, 3381 (1989).

86. R. F. Fox, I. R. Gatland, R. Roy, and G. Vemuri, *Phys. Rev. A* **38**, 5938 (1988).

87. R. L. Honeycutt, *Phys. Rev. A* **45**, 600 (1992).

88. R. L. Honeycutt, *Phys. Rev. A* **45**, 604 (1992).

89. J. Müller and J. Schnakenberg, *Ann. Phys.*, **2**, 92 (1993).

90. K. Y. R. Billah and M. Shinozuka, *Phys. Rev. A* **42**, 7492 (1990).

91. R. Mannella and V. Palleschi, *Phys. Rev. A*, **46**, 8028 (1992).

92. K. Y. R. Billah and M. Shinozuka, *Phys. Rev. A*, **46**, 8031 (1992).

93. W. H. Press, B. P. Flannery, S. A. Teukolsky, and W. T. Vetterling, *Numerical Recipes: The Art of Scientific Computing*, Cambridge University Press, Cambridge, UK, 1986.

94. D. E. Knuth, *The Art of Computer Programming*, Addison-Wesley, New York, 1981.

95. G. Marsaglia and W. W. Tsang, *SIAM (Soc. Ind. Appl. Math.) J. Sci. and Stat. Comp.* **5**, 349 (1984).

96. E. Ott, C. Grebogi and J. A. Yorke, *Phys. Rev. Lett.* **64**, 1196 (1990).

97. W. Horsthemke and R. Lefever, *Noise-Induced Transitions: Theory and Applications in Physics, Chemistry and Biology*, Springer, Berlin, 1984.

98. H. Haken, *Synergetics: An Introduction*, Springer, Berlin, 1977.

99. K. Kitahara, W. Horsthemke, and R. Lefever, *Phys. Lett.* **70A**, 377 (1979).

100. K. Kitahara, W. Horsthemke, R. Lefever, and Y. Inaba, *Prog. Theor. Phys.* **64**, 1233 (1980).

101. J. M. Sancho, M. San Miguel, S. L. Katz, and J. D. Gunton, *Phys. Rev. A* **26**, 1589 (1982).

102. P. Jung and H. Risken, *Phys. Lett.* **103A**, 38 (1984).

103. H. M. Gibbs, S. L. McCall, and T. N. C. Venkatesan, *Phys. Rev. Lett.* **36**, 1135 (1976).

104. L. A. Lugiato and R. J. Horowicz, *J. Opt. Soc. Am.* **B2**, 971 (1985).

105. J. M. Sancho and M. San Miguel in F. Moss and P. V. E. McClintock (Eds.), *Noise in Nonlinear Dynamical Systems*, Vol. 1, Cambridge University Press, Cambridge, UK, 1990, Chapter 3.

106. F. Moss and P. V. E. McClintock (Eds.), *Noise in Nonlinear Dynamical Systems*, Vols. 1–3, Cambridge University Press, Cambridge, UK, 1989.

107. S. Faetti, L. Fronzoni, P. Grigolini, and R. Mannella, *J. Stat. Phys.* **52**, 951 (1988).

108. S. Faetti, L. Fronzoni, P. Grigolini, V. Palleschi, and G. Tropiano, *J. Stat. Phys.* **52**, 979 (1988).

109. K. Lindenberg and B. J. West, *Physica* **119A**, 485 (1983).

110. P. Grigolini, *Phys. Lett.* **119A**, 157 (1986).

111. R. F. Fox, *Phys. Rev. A* **33**, 467 (1986).

112. R. F. Fox, *Phys. Rev. A* **34**, 4525 (1986).

113. G. P. Tsironis and P. Grigolini, *Phys. Rev. Lett.* **61**, 7 (1988).

114. G. P. Tsironis and P. Grigolini, *Phys. Rev. A* **38**, 3749 (1988).

115. R. Mannella, V. Palleschi, and P. Grigolini, *Phys. Rev. A* **42**, 5946 (1990).

116. M. M. Kløsek-Dygas, B. J. Matkowsky, and Z. Schuss, *Phys. Rev. A* **38**, 2605 (1988).

117. M. M. Kløsek-Dygas, B. J. Matkowsky, and Z. Schuss, *J. Stat. Phys.* **54**, 1309 (1989).

118. P. Jung and P. Hänggi, *Phys. Rev. Lett.* **61**, 11 (1988).

119. S. Faetti and P. Grigolini, *Phys. Rev. A* **36**, 441 (1987).

120. P. Hänggi, F. Marchesoni, and P. Grigolini, *Z. Phys. B* **56**, 333 (1984).

121. The routine MINUIT is from the CERN Library.

122. F. T. Arecchi and V. Degiorgio, *Phys. Rev. A* **3**, 1108 (1971).

123. F. T. Arecchi, V. Degiorgio, and B. Querzola, *Phys. Rev. Lett.* **19**, 1168 (1967).

124. D. Meltzer and L. Mandel, *Phys. Rev. Lett.* **25**, 1151 (1970).

125. D. Meltzer and L. Mandel, *Phys. Rev. A* **3**, 1763 (1971).

126. R. Roy, A. W. Yu, and S. Zhu in F. Moss and P. V. E. McClintock (Eds.), *Noise in Nonlinear Dynamical Systems*, Vol. 3, Cambridge University Press, Cambridge, UK, 1989 chapter 4.

127. M. James, F. Moss, P. Hänggi, and C. Van den Broeck, *Phys. Rev. A* **38**, 4690 (1988).

128. J. M. Sancho and M. San Miguel, *Phys. Rev. A* **39**, 2722 (1989).

129. A. Hernández-Machado, M. San Miguel, and J. M. Sancho, *Phys. Rev. A* **29**, 3388 (1984).

130. J. Casademunt, R. Mannella, P. V. E. McClintock, F. E. Moss, and J. M. Sancho, *Phys. Rev. A* **35**, 5183 (1987).

131. C. R. Doering, P. S. Hagan, and C. D. Levermore, *Phys. Rev. Lett.* **59**, 2129 (1987); M. M. Kløsek-Dygas,, B. J. Matkowsky, and Z. Schuss, *SIAM J. Appl. Math.* **48**, 425 (1988); J. F. Luciani and A. D. Verga, *J. Stat. Phys.* **50**, 567 (1988); K. Lindenberg, B. J. West, and G. P. Tsironis, *Rev. Solid State Sci.* **3**, 143 (1990).

132. K. Binder, *Phys. Rev. B* **8**, 3423 (1973); Z. Rácz, *Phys. Rev. B* **13**, 263 (1976).

133. J. I. Jiménez-Aquino, J. Casademunt, and J. M. Sancho, *Phys. Lett. A* **133**, 364 (1988).

134. J. Casademunt, J. I. Jiménez-Aquino, and J. M. Sancho, *Physica A* **156**, 628 (1989).

135. J. Casademunt and J. M. Sancho, *Phys. Rev. A* **39**, 4915 (1989).

136. J. Casademunt, J. I. Jiménez-Aquino, and J. M. Sancho, *Phys. Rev. A* **40**, 5905 (1990).

137. P. Colet, M. San Miguel, J. Casademunt, and J. M. Sancho, *Phys. Rev. A* **39**, 149 (1989).

138. M. Suzuki, Y. Liu, and T. Tsuno, *Physica A* **138**, 433 (1986).

139. A. K. Dhara and S. V. G. Menon, *J. Stat. Phys.* **46**, 743 (1987).

140. P. Martano, thesis, Università di Pisa, 1985.

141. S. Faetti, C. Festa, L. Fronzoni, and P. Grigolini, in M. W. Evans, P. Grigolini, and G. Pastori-Paravicini (Eds.), *Memory Function Approaches to Stochastic Problems in Condensed Matter*, Wiley, New York, 1985, pp. 445–475.

142. O. Rice, *Bell Syst. Tech. J.* **23**, 282 (1944).

143. P. Mandel, in N. B. Abraham, F. T. Arecchi, and L. A. Lugiato (Eds.), *Instabilities and Chaos in Quantum Optics*, Vol. 2, Plenum, New York, 1988.

144. N. J. Halas, S.-N. Liu, and N. B. Abraham, *Phys. Rev. A* **28**, 2915, (1983).

145. E. Arimondo, F. Casagrande, L. A. Lugiato, and P. Glorieux, *App. Phys. B* **30**, 57 (1983).

146. T. Erneux and P. Mandel, *Phys. Rev. A* **30**, 1902 (1984).

147. P. Mandel and T. Erneux, *Phys. Rev. Lett.* **53**, 1818 (1984).

148. D. Dangoisse and P. Glorieux, *Phys. Lett.* **116A**, 311 (1986).

149. B. Morris and F. Moss, *Phys. Lett.* **118A**, 117 (1986).

150. C. Van Den Broeck and P. Mandel, *Phys. Lett.* **122A**, 36 (1987).

151. G. Broggi, A. Colombo, L. A. Lugiato, and P. Mandel, *Phys. Rev. A* **33**, 3635 (1986).

152. R. Mannella, P. V. E. McClintock, and F. Moss, *Phys. Lett.* **120A**, 11 (1987).

153. R. Mannella, F. Moss, and P. V. E. McClintock, *Phys. Rev. A* **35**, 2560 (1987).

154. H. Zeghlache, P. Mandel, and C. Van Den Broeck, *Phys. Rev. A* **40**, 286 (1989).

155. M. C. Torrent, F. Sagués, and M. San Miguel, *Phys. Rev. A* **40**, 6662 (1989).

156. M. C. Torrent and M. San Miguel, *Phys. Rev. A* **38**, 245 (1988).

157. F. T. Arecchi, W. Gadomski, R. Meucci, and J. A. Roversi, *Phys. Rev. A* **39**, 4004 (1989).

158. A. Mecozzi, S. Piazzolla, A. D'Ottavi, and P. Spano, *Phys. Rev. A* **38**, 3136 (1988).

159. M. Abramowitz and I. A. Stegun, *Handbook of Mathematical Functions*, Dover, New York, 1970.

160. R. Benzi, A. Sutera, and A. Vulpiani, *J. Phys. A* **14**, L453 (1981); R. Benzi, G. Parisi, A. Sutera, and A. Vulpiani, *Tellus* **34**, 10 (1982).

161. C. Nicolis, *Tellus* **34**, 1 (1982).

162. A. Fioretti, R. Mannella, L. Guidoni, L. Fronzoni, and E. Arimondo, *J. Stat. Phys.* **70**, 799 (1993).

163. L. Gammaitoni, M. Martinelli, L. Pardi, and S. Santucci, *Phys. Rev. Lett.* **67**, 1799 (1991); L. Gammaitoni, F. Marchesoni, M. Martinelli, L. Pardi, and S. Santucci, *Phys. Lett.* **158A**, 449 (1991).

164. W. L. Ditto et al., *J. Stat. Phys.*, in press.

165. G. Debnath, T. Zhou, and F. Moss, *Phys. Rev. A* **39**, 4323 (1989).

166. C. Presilla, F. Marchesoni, and L. Gammaitoni, *Phys. Rev. A* **40**, 2105 (1989).

167. L. Gammaitoni, E. Menichella-Saetta, S. Santucci, F. Marchesoni, and C. Presilla, *Phys. Rev. A* **40**, 2114 (1989).

168. M. I. Dykman, D. G. Luchinsky, R. Mannella, P. V. E. McClintock, N. D. Stein, and N. G. Stocks, *J. Stat. Phys.* **70**, 463 (1993).

169. M. I. Dykman, D. G. Luchinsky, R. Mannella, P. V. E. McClintock, N. D. Stein, and N. G. Stocks, *J. Stat. Phys.* **70**, 479 (1993).

170. B. McNamara and K. Wiesenfeld, *Phys. Rev. A* **39**, 4854 (1989).

171. G. Vemuri and R. Roy, *Phys. Rev. A* **39**, 4668 (1989).

172. F. Moss in G. H. Weiss (Ed.), *Some Problems in Statistical Physics*, SIAM, Philadelphia, 1992.

173. R. F. Fox, *Phys. Rev. A* **39**, 4148 (1989).

174. Hu Gang, G. Nicolis, and C. Nicolis, *Phys. Rev. A* **42**, 2030 (1990).

175. P. Jung and P. Hänggi, *Phys. Rev. A* **41**, 2977 (1990).

176. P. Jung and P. Hänggi, *Phys. Rev. A* **44**, 8032 (1991).

177. L. Gammaitoni, F. Marchesoni, E. Menichella-Saetta, and S. Santucci, *Phys. Rev. Lett.* **65**, 2607 (1990).

178. H. Risken, *The Fokker-Planck Equation*, 2d ed., Springer, Berlin, 1989.

179. T. Zhou, F. Moss, and P. Jung, *Phys. Rev. A* **42**, 3161 (1990).

180. M. I. Dykman, R. Mannella, P. V. E. McClintock, N. D. Stein, and N. G. Stocks, *Phys. Rev. E* **47**, 1629 (1993).

181. L. D. Landau and E. M. Lifshitz, *Mechanics*, 3d ed., Pergamon, Oxford, UK, 1978; see especially Section 29.

182. C. Flytzanis and C. L. Tang, *Phys. Rev. Lett.* **45**, 441 (1980).

183. J. A. Goldstone and E. Garmire, *Phys. Rev. Lett.* **53**, 910 (1984).

184. M. I. Dykman, R. Mannella, P. V. E. McClintock, N. D. Stein, and N. G. Stocks, in preparation.

185. Note that, because of the strong dependence on I_T of the prefactor $K(I_T)$ in (172), the positions $I_{T1,2}$ of the maxima in $P(I_T)$ do not coincide precisely with the positions of the minima in $V(I_T)$.

186. S. Balle, F. de Pasquale, N. B. Abraham, and M. San Miguel, *Phys. Rev. A* **45**, 1955 (1992).

QUANTUM THEORY OF NONLINEAR DISPERSIVE MEDIA

P. D. DRUMMOND

Physics Department, The University of Queensland, St. Lucia, Queensland, Australia

CONTENTS

I. INTRODUCTION

Over the last decade, a revolution in laser physics has taken place. It is now common for laser physics experiments to have sufficiently low noise to show completely nonclassical statistical properties in the radiation field.[1] This means that the shot noise limit of independent photon arrivals is now a limiting factor in many experiments. These quantum effects are often amplified to a macroscopic size, so that they are practically unavoidable. Quantum noise can also be reduced below the shot noise level through the use of correlated photons. Fields with reduced noise properties are termed "antibunched" or "squeezed," depending on the techniques used. Just as it has long been recognized that the atoms must be quantized, these

Modern Nonlinear Optics, Part 3, Edited by Myron Evans and Stanisław Kielich. Advances in Chemical Physics Series, Vol. LXXXV.
ISBN 0-471-30499-9 © 1994 John Wiley & Sons, Inc.

technological changes mean that it is also necessary to treat the radiation field as a quantum field.

Since the binding energies are low, the underlying quantum theory of a nonlinear dielectric is quantum electrodynamics, or QED.[2] However, while QED is well-understood for microscopic systems, it is nontrivial to derive the nonlinear quantum theory of electro-magnetic propagation in a real dielectric from first principles. Of course, this microscopic route is always possible in principle. The difficulty is that the detailed structural properties of a dielectric must then be included, even though this level of information may be irrelevant and varies widely from one physical system to another. Results can be also obtained with simplified models, but these often sacrifice essential features of real, nonlinear dispersive media.

Instead, it is useful to start from the viewpoint of macroscopic quantization, as a route to obtaining the simplest quantum theory compatible with known dielectric properties. In macroscopic quantization, the Dirac quantization procedure[3] is implemented at the level of the average electromagnetic fields and polarizations, rather than at the microscopic level of electrons and atoms. This method has the particular advantage that it allows the theory of a wide range of solids to be calculated in a unified, structure-independent way. The result has the same relation to QED as QED has to quantum chromodynamics, or QCD.[4] In other words, it is a low-energy approximation with a restricted but well-defined region of validity.

Surprisingly, the macroscopic quantum theory obtained this way has features that are often much more familiar to traditional quantum field theorists than to practitioners of quantum optics. The reason for this is that a large number of model quantum field theories have been investigated theoretically, outside of the familiar world of QED and QCD. Only a few are both realistic and soluble.[5] Most quantum field theories investigated prove to be either tractable but unphysical, or physical but intractable. The new theoretical and experimental techniques of quantum optics are changing this situation. We can now find experimentally testable physical examples of many known model field theories.

An excellent example of this is the discovery of quantum properties of fiber-optical solitons, whose quantization will be given in detail. In agreement with theoretical predictions,[6-10] recent experiments[11] at IBM Research Laboratories have led to the first evidence of quantum evolution of a soliton. That is, the experiments have led to results that can be explained only by quantizing the nonlinear Schrödinger equation. This appears to be the first direct evidence for quantum solitons. It is certainly true that various solid-state transport properties are soliton enhanced, but transport involves solitons only indirectly. The IBM discovery has ramifications in many areas of physics, since solitons have a universal character in nonlinear quantum field theories.

The interesting thing that stands out in the IBM experiments, and in even more recent experiments at NTT Laboratories,[12] is that the experiments involve solitons—nonlinear bound states of a quantum field—in the clearest possible way. Pulses are injected into the nonlinear medium. They then propagate as solitons for a period of time, are extracted, and a quantum measurement is performed. The initial state is a coherent state, corresponding to a linear superposition of different photon numbers. In the IBM experiments, quadrature squeezing was observed. This originates in quantum interference between the underlying fundamental quantum solitons. In the NTT experiments, quantum properties of soliton collisions[13] were measured.

More than 10^9 photons are bound together in each quantum soliton, making what can be thought of as a photonic macromolecule. An experimental input soliton is in a complex superposition of bound and continuum photons, with total photon number varying over a range of $\pm 10^5$. Thus, as well as testing quantum field theory, these experiments test how well quantum mechanics works at large particle number. Other topical research areas in quantum optics, such as photonic bandgap theory,[14] microcavity QED,[15] pulsed squeezing,[16] and quantum chaos,[17] involve similar nonlinearities. Quantum optical measurements, therefore, complement the more usual phase-insensitive measurements of small numbers of particles interacting at high energies, as performed in traditional particle scattering experiments. It seems likely that even more interesting behavior will be found as these methods become extended to include higher dimensions and larger numbers of interacting fields.

II. CLASSICAL NONLINEAR DIELECTRIC THEORY

To understand the quantum effects, we must first investigate the classical field equations. Canonical quantization then allows us to obtain a simple quantum theory without having to introduce a detailed lattice model,[18] which often entails a high degree of complexity. This is particularly true, for example, in optical fiber soliton experiments, which take place in silica fibers. Not only are the fibers inhomogeneous, but silica has an amorphous lattice structure. In this case, there is not even a periodic structure in the underlying microscopic lattice. Despite this, silica has a well-defined refractive index.

The essential point of Dirac canonical quantization is that the equations of motion in the classical limit are obtained from a Lagrangian. This must generate a canonical Hamiltonian corresponding to the system energy. Canonical commutation relations can then be imposed to give the quantum behavior, as in Jauch and Watson's[19] early approach to this problem. Both requirements—the correct equations of motion and the correct energy—are necessary steps in canonical quantization. A number of examples are known of a Hamiltonian with the right energy, but the wrong equations of motion,[20] or vice versa.[21] Surprisingly, these techniques are still commonly used.

The problem with an incorrect equation of motion is clear, since we must regain the classical equations in the large photon-number limit. It has been known for some time that incorrect classical equations result if one simply includes nonlinear terms in the Hamiltonian while retaining the usual expansion of the E field in mode operators. This procedure is still followed in some current papers in quantum optics. It is easy to check that it can lead to completely incorrect results. For example, it predicts that solitons ought to form with normal dispersion optical fibers. This contradicts the usual classical equations, as well as the experimental evidence.[22]

With correct equations of motion but an incorrect Hamiltonian, more subtle problems can occur. In this case there is no natural energy scale for the Hamiltonian. That is, there is no guarantee that single quanta will have the usual $E = \hbar\omega$ relationship. Despite this, the method works if the energy scale is guessed correctly. However, it is not a generally reliable quantization procedure. Accordingly, in this review I shall focus on canonical theories where the classical Hamiltonian equals the energy.

Because the equations of interest are the nonlinear, dispersive version of Maxwell's equations that hold for a real dielectric, the usual free-space Lagrangian is inapplicable. I neglect, for simplicity, any dielectric internal degrees of freedom apart from those already included in Maxwell's equations. In addition, the theory is one-dimensional, in terms of fields scaled by a nominal transverse scale length R. This method is easily adapted to treat inhomogeneous three-dimensional dielectrics of arbitrary structure.[23] A simplified model of a one-dimensional, uniform dielectric is discussed in detail here, to clarify the fundamental issues involved.

In one dimension, Maxwell's equations are

$$\partial_x E_z(t, x) = \mu \partial_t H_y(t, x)$$

$$\partial_x H_y(t, x) = \partial_t D_z(t, x) \tag{1}$$

where

$$E(t, x) = E[D]$$

$$= \sum_{n>0} \int_0^\infty \int_0^\infty \eta^{(n)}(\tau_1, \ldots, \tau_n) D(t - \tau_1, x) \cdots D(t - \tau_n, x) \, d^n\tau \tag{2}$$

Here, just a single polarization mode is treated, so that all electric fields are z-polarized and all magnetic fields are y-polarized. From now on, therefore, the polarization indices are omitted. The magnetic susceptibility is μ, and the nonlinearity is provided through the relationship of D to E given above. It is most useful to regard D as a canonical coordinate, with E as a local function of D. This is similar to the mechanical problem of a position dependent force, with D analogous to the position. Just as with position, it is possible to specify the displacement field (via the deposition of charges along the z coordinate) and measure the resulting force (via the induced potential difference).

It is operationally meaningful to define a causal dielectric response function as a functional Taylor series, which gives the electric field E in terms of the displacement field D at earlier times. This expansion is an alternative to the Bloembergen[24] expansion of nonlinear optics. Here, both are assumed strictly local, for simplicity. The relationship between this expansion and the earlier one is through a canonical transformation, which is necessary since Bloembergen's work started with the microscopic Hamiltonian in the $\mathbf{p} \cdot \mathbf{A}$ form. A simple variable change or canonical transformation, first introduced by Goeppert-Meyer[25] gives the dipole-coupled or multipolar-coupled[26] interaction Hamiltonian, which is widely used in modern calculations.[27] This canonical transformation leads to the above expansion.

The dipole coupling is often written—incorrectly—in terms of the *total* $\mathbf{E}$ field. This makes it seem logical to expand the polarization in terms of the $\mathbf{E}$ field. However, the form of the response function given above is better justified microscopically. The Hamiltonian dipole coupling of the polarization is really to the *displacement* field $\mathbf{D}$, not the total electric field. Therefore, microscopic calculations in the dipole-coupled form of the Hamiltonian yield response functions in the above form, which we will also use macroscopically. Accordingly, (1) and (2) completely define the equations of motion that are used here, for an arbitrary dispersive, locally nonlinear dielectric.

As one might expect, the electromagnetic energy is simply composed of magnetic plus electrostatic terms.[24] The nonlinear electrostatic term arises from integrating the force (E) with respect to the displacement (D). The total energy in length L is, therefore,

$$W = \int_0^L \left[\frac{1}{2}\mu H^2 + \int_{t_0}^t E(\tau)\dot{D}(\tau)\,d\tau \right] dx \qquad (3)$$

It is important to note that the energy depends on the time evolution of E and D. This means that the energy may not be a state function of D in a dispersive medium, which will necessitate careful treatment later.

III.   NONDISPERSIVE QUANTIZATION

The central problem of macroscopic quantization is to find a Lagrangian that generates the classical equations of motion (1) and (2), while defining a canonical Hamiltonian equal to the energy (3). It is simplest to start with the homogeneous, nondispersive case, where the response function $\eta^{(n)}$ are instantaneous. In this (rather unphysical) situation, the correct Lagrangian is known from an insightful paper of Hillery and Mlodinow.[28] Although their result is obtainable in more than one way, it is most straightforward in the case of a medium without free charges, in which case the Lagrangian can be written in terms of the dual potential. This potential is similar to the magnetic vector potential, except that the roles of magnetic and electric field are reversed. The crucial requirement, valid in a medium without free charges, is that $\mathbf{D}$ is divergenceless. This allows the definition of a dual potential $\mathbf{\Lambda}$. In one dimension, the defining equations for $\mathbf{\Lambda}$ are

$$\begin{aligned} \partial_t \Lambda &= H \\ \partial_x \Lambda &= D \end{aligned} \qquad (4)$$

A possible Lagrangian $\mathscr{L}$ that generates the classical equations (1) and (2) is

$$\mathscr{L} = \int_0^L \left[\frac{1}{2}\mu \dot{\Lambda}^2 - U(\partial_x \Lambda) \right] dx \qquad (5)$$

where

$$U(D) \equiv \int_0^D E[D]\,dD \qquad (6)$$

It is easy to verify that the Euler-Lagrange equations of $\mathscr{L}$ generate the original wave equations, and it is not even necessary to require U to be a homogeneous (spatially uniform) function. The theory is still correct if all the linear and nonlinear dielectric properties are heterogeneous.[23]

At this stage, the Lagrangian is nonunique, since multiplying it by a constant does not change the equations of motion. The scale factor is fixed by the requirement that the corresponding Hamiltonian in the nondispersive case is exactly equal to the total energy. With this requirement in mind, it is clear that the Hamiltonian already has the correct form, since

$$\mathscr{H} = \int_0^L \left[\frac{1}{2\mu} \Pi^2 + U(\partial_x \Lambda) \right] dx \qquad (7)$$

This agrees with (3) including the coupling of the electric field to the polarization term in the displacement field, so no further rescaling is necessary. On quantization, commutators are introduced instead of Poisson brackets. The classical fields Λ become quantum operators $\hat{\Lambda}$ in a Hilbert space. The canonical momentum and equal-time commutators are

$$\hat{\Pi} = \mu \dot{\hat{\Lambda}}$$
$$\left[\hat{\Lambda}(x), \hat{\Pi}(x') \right] = i\hbar\delta(x - x') \qquad (8)$$

Accordingly, these equations define a one-dimensional quantum field theory which could describe a nonlinear dielectric waveguide, like an optical fiber—if it were nondispersive!

This theory has some unusual features compared to the nonlinear field theories found in particle physics. The Lagrangian is nonlinear in its spatial derivatives, and so is not Lorenz-invariant. This is natural, and is due to the preferred rest-frame of the dielectric itself. It is also possible to formulate the theory using the vector potential, but the results have a more complicated character and are not presented here, since they are treated elsewhere.[28, 29]

IV. DISPERSION

There is an unexpected feature of the Hillery-Mlodinow theory that still needs to be explained: While the commutators of electric displacement and magnetic fields are the usual ones, the electric field commutation relation with the magnetic field is modified from its free field value.[30] An interaction does not alter equal-time commutation relations, unless it involves time derivatives. What has happened to the commutators? Surely, equal-time electromagnetic commutation relations must be invariant when interactions are introduced with material fields.

The answer to this commutator puzzle is that an instantaneous nonlinear field theory omits an important physical property of a real dielectric. This is the causal time delay between changes in D and corresponding changes in E. As a result, there is dispersion in the dielectric, due to the polarization being unable to respond instantaneously to changes in the displacement field. This physical fact was not included in the above Lagrangian. The neglect of dispersion also introduces some rather singular behavior in a continuum theory, due to the existence of phase matching over infinitely large bandwidths.

We now introduce a quantum theory that takes account of dispersion, although the treatment will be restricted to relatively weak dispersion, far away from absorption bands. Even then, the effects of dispersion are felt through the Hamiltonian. Since the dielectric can store energy too, its time-dependent response changes the integrated work that gives the total energy. Related effects occur in waveguides near cutoff, due to time-dependent variations in the transverse mode structure. These are not accounted for in the present one-dimensional theory, although they can be readily included in more general theories. For our purposes, waveguide dispersion is neglected.

Dispersion is often attributed to the delayed response of the dielectric polarization to the *electric* field, rather than to the displacement field. This is really an inappropriate description. The total electric field of a macroscopic system includes the field due to the polarization. It is therefore more natural to regard the displacement field as causing the dielectric response of the medium. We also know, from the original work by Goeppert-Meyer[25] and Power et al.[26] in nonlinear optics, that the usual minimal coupling Hamiltonian can be transformed to a form having a coupling of the polarization field to the displacement field. Other possible forms of the microscopic Hamiltonian include hybrid combinations,[31] which interpolate between these two. For simplicity, it is preferable to use the displacement field as the fundamental canonical variable even in the dispersive case.

At the microscopic level, the dielectric response function is obtained on specifying the displacement field, then using the Hamiltonian time evolution to compute the polarization, and hence the total electric field. The use of macroscopic fields requires local-field corrections in order to be related to calculated microscopic quantities. In this review I take the pragmatic approach of utilizing the known frequency-dependent refractive indices to evaluate the response functions. These implicitly include local-field corrections already. Hence, a frequency-dependent response function is defined as the Fourier transform of the causal response:

$$\eta^{(n)}(\omega; \omega_1, \ldots, \omega_n) = \delta_{-\omega, \omega_1 + \cdots + \omega_n} \int e^{i(\omega_1 t_1 + \cdots + \omega_n t_n)} \eta^{(n)}(t_1, \ldots, t_n) \, d^n t$$

(9)

In comparison to the usual dielectric permittivity, $\varepsilon(\omega)$, we notice that:

$$\eta^{(1)}(\omega; -\omega) = [\varepsilon(\omega)]^{-1}$$

(10)

The puzzle of commutation relations is now resolved, since the linear response at zero delay is just the vacuum response:

$$\eta^{(1)}[t = 0] = \varepsilon_0^{-1}$$

(11)

In addition, there can be no truly instantaneous nonlinear response, so that

$$\eta^{(n)}[t_1 = 0 \cdots t_n = 0] = 0[n > 1]$$

(12)

With these results, the electric field commutators at equal times must be related to the displacement field commutators in exactly the same way as they are for the free field case. It remains to quantize the dispersive nonlinear theory. To achieve this, a Lagrangian must be found for the dispersive equations. The simplest technique for doing this is to equate the real field Λ with the real part of a set of band-limited complex fields, and to approximate the Hamiltonian so that it is a state function of the new fields.

First, the equations of motion are expressed in terms of band-limited complex fields, just as in classical dispersive calculations. The dual potential Λ is therefore expanded in terms of a number of carrier frequencies ω as

$$\Lambda(t, x) = \sum_j [\Lambda^{\omega_j}(t, x) + (\Lambda^{\omega_j})^*(t, x)]$$

(13)

where

$$\langle \Lambda^{\omega_j}(t, x) \rangle \sim e^{-i\omega_j t}$$

(14)

Next, a Taylor series expansion in frequencies near ω_j is employed to simplify the wave equation. This technique is common in classical dispersion theory, and was first introduced into macroscopic field quantization by Kennedy and Wright.[21] The resulting wave equation in the rotating-wave approximation is, from Eq. (1),

$$
\mu\partial_t^2\Lambda^{\omega_j} = \partial_x\left[\sum_{n>1}\eta^{(n)}(-\omega_j;\omega_1,\ldots,\omega_n)(\partial_x\Lambda^{\omega_1})\cdots(\partial_x\Lambda^{\omega_n})\right.
$$
$$
\left.+\left(\eta_j + i\eta_j'\partial_t - \frac{1}{2}\eta_j''\partial_t^2\right)\partial_x\Lambda^{\omega_j} + \cdots\right]
$$
(15)

Here the terms η_j', η_j'' represent a quadratic Taylor series expansion, valid near ω_j. The coefficients are defined so that

$$
\eta^{(1)}(-\omega;\omega) \simeq \eta_j + \omega\eta_j' + \tfrac{1}{2}\eta_j''
$$
(16)

The careful reader will have recognized that the higher-order nonlinear terms have a similar expansion. These nonlinear dispersion terms are normally very small in dielectrics unless near an absorption band. In cases of high absorption, there is little hope of developing a theory of macroscopic quantization in the field alone. At the very least, it is necessary to include energy reservoirs, and the results will depend on the type of absorption. Nonlinear dispersion is therefore neglected here.

A straightforward modification of the earlier Lagrangian to allow for the new wave equation, still local in the fields, is

$$
\mathscr{L} = \int_0^L\left\{\sum_{\omega_j}\left[\mu|\dot\Lambda^{\omega_j}|^2 - \eta_j|\partial_x\Lambda^{\omega_j}|^2\right.\right.
$$
$$
\left.-\frac{i}{2}\eta_j'\left(\partial_x\dot\Lambda^{\omega_j}\cdot\partial_x\Lambda^{\omega_j*} - \partial_x\dot\Lambda^{\omega_j*}\partial_x\Lambda^{\omega_j}\right) - \frac{1}{2}\eta_j''|\partial_x\dot\Lambda^{\omega_j}|^2\right]
$$
$$
\left.-U^N(\partial_x\Lambda^{\omega_j},\partial_x\Lambda^{\omega_j*})\right\}dx
$$
(17)

where U^N represents the nonlinear part of the potential function, which is assumed nondispersive. As an example, in the simple case of just one carrier frequency ω_0, and a nonlinear refractive index,

$$
U^N(\partial_x\Lambda^{\omega_0},\partial_x\Lambda^{\omega_0*}) \equiv \tfrac{3}{2}\eta^{(3)}(\omega_0;\omega_0,\omega_0,-\omega_0)|\partial_x\Lambda^{\omega_0}|^4
$$
(18)

There are no terms in $\eta^{(2)}$. This is due to the choice of only one carrier frequency, together with the rotating-wave approximation. Such terms give rise to parametric amplification, a process that involves more carrier frequencies. Physically, parametric terms are absent in an inversion-symmetric dielectric.

To complete the canonical theory a Hamiltonian must be obtained. The canonical momentum fields are

$$\Pi^{\omega_j} = \frac{\delta \mathscr{L}}{\delta \dot{\Lambda}^{\omega_j}} = \mu \dot{\Lambda}^{\omega_j^*} - \frac{\mathrm{i}}{2} \eta_j' \partial_x^2 \Lambda^{\omega_j^*} - \frac{1}{2} \eta_j'' \partial_x^2 \dot{\Lambda}^{\omega_j^*} \tag{19}$$

From now on, the carrier frequency subscript will be omitted, and just the single carrier frequency case will be treated. The corresponding Hamiltonian is therefore

$$\mathscr{H} = \int_0^L \left\{ \mu |\dot{\Lambda}^{\omega_0}|^2 - \tfrac{1}{2} \eta'' |\partial_x \dot{\Lambda}^{\omega_0}|^2 + \eta |\partial_x \Lambda^{\omega_0}|^2 + U^{\mathrm{N}}(\partial_x \Lambda^{\omega_0}, \partial_x \Lambda^{\omega_0*}) \right\} \mathrm{d}x \tag{20}$$

The Hamiltonian is left in terms of the field derivatives, rather than being reexpressed as a function of the canonical momenta, as is customary. This allows a direct comparison with the classical energy of a dielectric, given a monochromatic excitation at frequency ω near ω_0.

In the case of a linear dielectric, the total energy—including the energy of the polarized medium—has a known expression. On averaging over a cycle, this is

$$\langle W \rangle_{\mathrm{cycle}} = \int_0^L \left[\mathscr{E}^*(x) \frac{\partial}{\partial \omega} [\omega \varepsilon(\omega)] \mathscr{E}(x) + \frac{1}{\mu} |\mathscr{B}(x)|^2 \right] \mathrm{d}x \tag{21}$$

Here $\mathscr{E}(x)$, $\mathscr{B}(x)$ are envelope functions, so that

$$\langle W \rangle_{\mathrm{cycle}} = \int_0^L \left[\mu |\dot{\Lambda}|^2 + \eta^{(1)}(\omega)^2 \frac{\partial}{\partial \omega} [\omega \varepsilon(\omega)] |\partial_x \Lambda|^2 \right] \mathrm{d}x \tag{22}$$

It is straightforward to verify that this expression is identical to that of our Hamiltonian in the linear case, and also in the nonlinear case if nonlinear dispersion is neglected, since

$$\eta^{(1)}(\omega)^2 \frac{\partial}{\partial \omega} [\omega \varepsilon(\omega)] \equiv \eta - \frac{1}{2} \omega^2 \eta'' \tag{23}$$

Clearly this Lagrangian generates both the correct equations and the correct energy.

V. DISPERSIVE QUANTIZATION

Having generated a Lagrangian that is valid inside a band of frequencies near ω, we now wish to quantize the theory. Since multiple carrier frequencies introduce no essentially new features, the following theory treats the case of one carrier frequency. While there are many ways to carry out the quantization, the most useful is to quantize the Fourier transform of $\Lambda^{\omega_0}(x)$. This method is simple to extend to three dimensions, where transversality is required; it also allows the spatial derivatives in the Lagrangian to be readily handled. Perhaps most important of all, the use of spatial modes permits the limitation of the range of frequencies that are excited, so that $\langle \hat{\Lambda}^{\omega_0}(x) \rangle \sim \mathrm{e}^{-\mathrm{i}\omega_0 t}$, as required.

The Lagrangian permits many possible excitation frequencies, only some of which are inside the required frequency band. Accordingly, the Hilbert space must be restricted in some way. For this approximate theory to be valid, modes that result in envelopes where $\langle \Lambda^{\omega_0}(t, x) \rangle \sim \mathrm{e}^{-\mathrm{i}\omega' t}$ with $\omega' \neq \omega_0$, must remain in the vacuum state. This restriction corresponds physically to the requirement that there is negligible inelastic scattering of photons into absorption bands. These modes can occur in the physical system, but will not be accurately treated in a theory that utilizes only refractive index type information. For this, a full lattice model of the actual dielectric is required.

The normal modes are the solutions that diagonalize the linear Hamiltonian. These have the structure (in a uniform medium) of

$$\hat{\Lambda}^{\omega_0}(t, x) = \sum_k \mathrm{e}^{\mathrm{i}kx}\left[\hat{a}_k \lambda_k + \hat{b}_k^\dagger \mu_k \right] \tag{24}$$

Here both $\hat{a}_k$ and $\hat{b}_k$ are boson annihilation operators, which vary as $\sim \mathrm{e}^{-\mathrm{i}\omega_0 t}$. Terms like $\hat{b}_k^\dagger$ must remain in the vacuum state, since they vary as $\sim \mathrm{e}^{\mathrm{i}\omega_0 t}$, and are therefore not in the required frequency band. This leaves a restricted set of modes of interest. In terms of these mode operators, the expansion is

$$\hat{\Lambda}^{\omega_0}(t, x) = \sum_k \mathrm{e}^{\mathrm{i}kx}\lambda_k \hat{a}_k \tag{25}$$

Calculation of λ_k shows that to obtain the usual form of

$$\hat{\mathscr{H}}_D = \sum \hbar \omega_k \hat{a}_k^\dagger \hat{a}_k \qquad (26)$$

it is necessary that

$$\lambda_k = \mathrm{i} \left[\frac{\hbar \partial \omega / \partial k}{2 L k \eta^{(1)}(-\omega_k, \omega_k)} \right]^{1/2} \qquad (27)$$

where $\hat{a}_k^\dagger$, $\hat{a}_{k'}$ have the standard commutators:

$$\left[\hat{a}_k, \hat{a}_{k'}^\dagger \right] = \delta_{k, k'} \qquad (28)$$

The final Hamiltonian is therefore written for modes with frequencies ω_k near ω_0 as

$$\hat{\mathscr{H}} = \sum_k \hbar \omega_k \hat{a}_k^\dagger \hat{a}_k + \int U^N \left(\partial_x \hat{\Lambda}^{\omega_0}, \partial_x \hat{\Lambda}^{\omega_0*} \right) \mathrm{d}x \qquad (29)$$

where

$$\omega_k = k \sqrt{\eta^{(1)}(-\omega_k, \omega_k)/\mu} \qquad (30)$$

This is straightforward to understand physically. The operators $\hat{a}_k^\dagger$, $\hat{a}_k$ generate the free-particle excitations of the coupled matter–field system for co-rotating frequencies near ω_0. Technically, these are polaritons propagating along the waveguide at a velocity equal to $\partial \omega / \partial k$, which is the group velocity corresponding to a wave vector k. Dispersion is effectively hidden in the mode-spacing, which depends on the wavelength; these modes are not equally spaced in frequency. Thus, the excitations that diagonalize the Hamiltonian of an interacting system have a mixed character—in this case, carrying both electromagnetic and polarization properties. In quantum optics, these are just called photons.

When there is a nonlinear refractive index, or $\eta^{(3)}$ term, the free particles interact via the Hamiltonian nonlinearity. It is this coupling that leads to soliton formation. The above Hamiltonian was originally used to obtain predictions of quantum phase diffusion and quadrature squeezing, which has now been verified in optical fiber soliton experiments. Although the details are deliberately omitted here, it is straightforward to generalize this procedure to treat coupling of different polarizations or different transverse modes.[23] It is also possible to include other types of nonlinearity, such as $\eta^{(2)}$ terms, that lead to second-harmonic and parametric interactions.[32]

In calculations that lead to practical applications, it is necessary to define photon density and flux amplitude fields, which have a direct interpretation in photodetection external to the dielectric. Clearly, if we wish to study the detailed theory of inhomogeneous waveguides with boundaries, it is desirable to find modal solutions to the corresponding inhomogeneous Maxwell equations. Thus, one should go through the detailed calculations of quantization in an inhomogeneous medium, leading to the asymptotic forms of mode operators. These details of dielectric boundaries are not treated here. We assume from energy conservation grounds that when the dielectric boundaries have anti-reflection coatings, polariton excitations propagate as ordinary photons external to the dielectric.

A polariton density field is simply defined as

$$\hat{\Psi}(t, x) = \sqrt{\frac{1}{L}} \sum_{k} e^{[i(k-k_0)x + i\omega_0 t]} \hat{a}_k \tag{31}$$

This has an equal-time commutator of

$$\left[\hat{\Psi}(x_1), \hat{\Psi}^\dagger(x_2)\right] = \tilde{\delta}(x_1 - x_2) \tag{32}$$

where $\tilde{\delta}$ is defined as a tempered version of the usual Dirac delta function:

$$\tilde{\delta}(x_1 - x_2) \equiv \frac{1}{L} \sum_{\Delta k} e^{i\Delta k(x_1 - x_2)} \tag{33}$$

The summation is over momenta $\Delta k = k - k_0$, where $k_0 = k(\omega_0)$ is the central wave number for the jth envelope field. The total polariton number operator is

$$\hat{N} = \int \hat{\Psi}^\dagger(t, x) \hat{\Psi}(t, x) \, dx \tag{34}$$

A polariton flux can also be approximately obtained as

$$\hat{\Phi}(t, x) = \sqrt{\frac{v}{L}} \sum_{k} e^{[i(k-k_0)x + i\omega_0 t]} \hat{a}_k \tag{35}$$

This has an equal-time commutator of

$$\left[\hat{\Phi}(x_1), \hat{\Phi}^\dagger(x_2)\right] = v\bar{\delta}(x_1 - x_2) \qquad (36)$$

Here, v is the central group velocity at the carrier frequency ω_0, so that $\langle\hat{\Phi}^\dagger(t, x)\Phi(t, x)\rangle$ is operationally the photon flux, or Poynting vector expectation value in units of photons/second. This definition is only approximate since we have assumed that the group velocity is nearly constant over each frequency band. It is possible to improve this situation by taking account of the effects of group-velocity dispersion in the definition of $\hat{\Phi}$. Because this complicates the equal-time commutators, with only minor changes in measurable quantities, this additional correction is not included here. However, there are treatments of dispersive media from a quasimicroscopic point of view that do include this effect. Since these authors[33] do not include any nonlinearities, their dispersive quantization results are not treated in detail, except to note that they are essentially equivalent to those presented here, apart from an improved treatment of the absorption bands.

Dimensionless variables are often used, especially in fiber optics applications. These are defined by an appropriate scaling of the dimensionless photon density or photon flux. A common choice is to define the dimensionless field $\hat{\phi}$ by the equivalence

$$\hat{\phi} = \hat{\Psi}\sqrt{\frac{vt_0}{\bar{n}}} \qquad (37)$$

Here $\bar{n}$ is a photon number scale, defined so that $\langle\hat{\phi}^\dagger\hat{\phi}\rangle$ is of order unity. Similarly t_0 is a time scale appropriate for the system. It is usually a typical pulse duration, since this is the most easily measured indication of the spatial extent of the pulse in the laboratory frame of reference.

This scaling transformation is accompanied by a change to a comoving coordinate frame, which simplifies the operator equations by removing group-velocity terms. This can be achieved in more than one way, depending on whether the space variable or the time variable is changed. The first choice, of an altered space variable, is the simplest in terms of normal Hamiltonian methods, giving

$$\tau = \frac{vt}{x_0}$$

$$\zeta_v = \frac{x/v - t}{t_0} \qquad (38)$$

Here a spatial length scale x_0 of typical pulse-shaping interaction distances is introduced to scale the interaction times. This moving frame transformation removes group-velocity terms exactly from the operator equations. A typical choice of x_0 is to scale relative to a dispersion length, so that

$$x_0 = t_0^2 / |k''| \tag{39}$$

where the group velocity dispersion parameter, k'' is defined as

$$k'' = \frac{\partial^2 k}{\partial \omega^2} = -\frac{1}{v^3} \frac{\partial^2 \omega}{\partial k^2} = -\frac{\omega''}{v^3} \tag{40}$$

An alternative moving frame transformation that is popular in laser applications is

$$\tau_v = \frac{t - x/v}{t_0}$$
$$\zeta = \frac{x}{x_0} \tag{41}$$

Later we see that this transforms the operator equations into an approximate form in which only first-order spatial derivatives appear, but cannot remove these terms exactly. This transformation is useful only when the characteristic pulse length scale $v t_0$ of interest in the field is short compared to the characteristic interaction scale length x_0. With either choice of variable, it is necessary to realize that the unsubscripted variable is a real coordinate in the laboratory, which indicates an interaction time or distance. The subscripted variable is a relative coordinate in the comoving frame, which indicates a pulse length or duration. As a rough guide, typical experimental numbers in silica fibers might be

$$t_0 = 1 \, \text{ps}$$
$$x_0 = 100 \, \text{m}$$
$$\frac{v t_0}{x_0} = 2 \times 10^{-6} \tag{42}$$

Clearly, with these numbers, a pulse is reshaped over distances much longer than the actual physical pulse width. As is shown in the next section, this permits some useful approximations to be used.

VI. OPERATOR EQUATIONS

I now wish to illustrate the quantization technique developed above by treating the case of a single-mode optical fiber. This has a nonlinear refractive index and anomalous dispersion at wavelengths longer than 1.5 μm, allowing solitons to form. While this is well understood classically,[22] the quantized behavior has only recently been understood for typical laser experiments with coherent pulse inputs.[6-10] In particular, there is an interplay of different effects in typical laser experiments, with a competition between quantum phase diffusion effects and thermally excited Raman processes. This means that it is often necessary to perform experiments at low temperatures, even though $kT \ll \hbar\omega$ at room temperature. Thus, a complete treatment involves both the electronic nonlinearity and phonon interactions.

The optical fiber treated will be a single-transverse mode fiber with dispersion and nonlinearity. Since boundary effects are usually negligible in experiments, it is useful to first take the infinite volume limit, which effectively replaces a summation over wave vectors with the corresponding integral. The effect of a transverse mode structure is also included, to show how the simplified theory is applicable in real three-dimensional fibers. The nonlinear Hamiltonian in this case is[23]

$$\hat{\mathscr{H}} = \int dk\, \hbar\omega(k)\hat{a}^\dagger(k)\hat{a}(k) + \tfrac{1}{4}\eta^{(3)}\int \hat{\mathbf{D}}^4(x)\,d^3\mathbf{x} \qquad (43)$$

Here $\omega(k)$ is the angular frequency of modes with wave vector k, describing the *linear* photon or polariton excitations in the fiber, including dispersion. Also, $\hat{a}(k)$ is a corresponding annihilation operator defined so that, at equal times,

$$\left[\hat{a}(k'), \hat{a}^\dagger(k)\right] = \delta(k - k') \qquad (44)$$

The coefficient $\eta^{(3)}$ is the nonlinear coefficient arising when the electronic polarization field is expanded as a function of the electric displacement. Compared to the commonly used Bloembergen[24] coefficient, $\eta^{(3)} = -\varepsilon_0\chi^{(3)}/\varepsilon^4$, although here the units are S.I. units, following current standard usage. In terms of modes of the waveguide, and neglecting modal dispersion, the electric displacement $\mathbf{D}(\mathbf{x})$ is expressed as

$$\hat{\mathbf{D}}(\mathbf{x}) = i\int dk\left(\frac{\hbar\varepsilon(k)kv(k)}{4\pi}\right)^{1/2}\hat{a}(k)\mathbf{u}(k,\mathbf{r})e^{ikx} + \text{h.c.}$$

where

$$\int |\mathbf{u}(k,\mathbf{r})|^2 \, d^2\mathbf{r} = 1 \qquad (45)$$

Here $v(k)$ is the group velocity, and $\varepsilon(k)$ is the dielectric permittivity. The mode function $\mathbf{u}(k,\mathbf{r})$ is included here in its usual three-dimensional form to show how the simplified one-dimensional quantum theory relates to vector mode theory.

In this infinite volume limit, the photon or polariton field is defined for a slowly varying envelope as

$$\hat{\Psi}(t,x) = \frac{1}{\sqrt{2\pi}} \int \hat{a}(k) e^{i(k-k_0)x + i\omega_0 t} \, dk \qquad (46)$$

This operator is a boson annihilation operator for the linear quasi-particle excitations of the fiber, corresponding to coupled excitations of the electromagnetic and polarization fields traveling at velocity $v(k)$. The quantum particle density is $\langle \hat{\Psi}^\dagger(t,x)\hat{\Psi}(t,x) \rangle$. Ignoring the implicit bandwidth limitations of the quantization method, the equal-time commutator is $[\hat{\Psi}(t,x), \hat{\Psi}^\dagger(t,x')] = \delta(x - x')$, as usual. From (43), the interaction Hamiltonian describing the evolution of $\hat{\Psi}$ in the slowly varying envelope and rotating-wave approximations is

$$\hat{\mathcal{H}}_I = \frac{\hbar}{2} \int \left\{ iv \left[\frac{\partial}{\partial x} \hat{\Psi}^\dagger \hat{\Psi} - \hat{\Psi}^\dagger \frac{\partial}{\partial x} \hat{\Psi} \right] \right.$$
$$\left. + \omega'' \frac{\partial}{\partial x} \hat{\Psi}^\dagger \frac{\partial}{\partial x} \hat{\Psi} - \chi_e \hat{\Psi}^{\dagger 2} \hat{\Psi}^2 \right\} dx$$

where

$$\chi_e \equiv \left[\frac{\hbar n_2 \omega_0^2 v^2}{\mathscr{A} c} \right] \equiv \left[\frac{3\hbar \chi^{(3)} w_0^2 v(k_0)^2}{4\varepsilon(k_0)c^2} \right] \left| \int |\mathbf{u}(\mathbf{r})|^4 \, d^2 r \right| \qquad (47)$$

Here $\mathscr{A}$ is the effective modal cross section, while n_2 is the refractive index change per unit field intensity. The free evolution part of the total Hamiltonian, which is removed here, just describes the carrier-frequency rotation at frequency ω_0. This is not needed, since it is already included in the definition (46). After taking this free evolution into account, we find the following Heisenberg equation of motion for the field operator propagating in the $+x$ direction:

$$\left[v \frac{\partial}{\partial x} + \frac{\partial}{\partial t} \right] \hat{\Psi}(t, x) = \left[\frac{iw''}{2} \frac{\partial^2}{\partial x^2} + i\chi_e \hat{\Psi}^\dagger \hat{\Psi} \right] \hat{\Psi}(t, x) \qquad (48)$$

where $v = v(k_0) = \partial w / \partial k|_{k=k_0}$, $w'' = \partial^2 w / \partial k^2|_{k=k_0}$, and $w(k)$ is expanded quadratically in a narrow band around $\omega_0 = \omega(k_0)$, which is the carrier frequency.

In a comoving reference frame defined by $x_v = x - vt$, this reduces to the usual quantum nonlinear Schrödinger equation:

$$i \frac{\partial}{\partial t} \hat{\Psi}(t, x_v) = \left[-\frac{\omega''}{2} \frac{\partial^2}{\partial x_v^2} - \chi_e \hat{\Psi}^\dagger \hat{\Psi} \right] \hat{\Psi}(t, x_v) \qquad (49)$$

This equation has a very simple physical meaning. In the moving frame, the particles of the theory have acquired a nonzero effective mass of typically about 10^{-34} kg:

$$m = \frac{\hbar}{\omega''} \qquad (50)$$

The nonlinear term χ_e describes an interaction potential that couples the particles together. This interaction potential is attractive when χ_e is positive, as it is in most Kerr media, and the value of the potential is

$$V(x_v, x_v') = -\chi_e \delta(x_v - x_v') \qquad (51)$$

It is known that this equation has bound states and is one of the simplest, exactly soluble known quantum field theories. The eigenstates, or quantum solitons, of the Hamiltonian are particle number states and were investigated by Yang,[34] who discussed the theory of bosons interacting with a delta-function potential. The method of solution involves Bethe's ansatz.[35]

In summary, the complications of dispersive quantization are now replaced by a remarkably simple physical picture of massive bosons attracted to each other with a local attractive potential. This is the only known physical system that is able to realize Yang's model in this straightforward way. In calculations it is preferable to scale the equations into a dimensionless form using (38), which results in

$$\frac{\partial}{\partial \tau} \hat{\phi}(\tau, \zeta_v) = \left[\frac{i}{2} \frac{\partial^2}{\partial \zeta_v^2} + i\hat{\phi}^\dagger \hat{\phi} \right] \hat{\phi}(\tau, \zeta_v) \tag{52}$$

Here I have obviously made the assumption that $\chi_e > 0$ and that $\omega'' > 0$, which is the case in the soliton-forming frequency region of anomalous dispersion, in dielectrics with a positive nonlinear index of refraction. The photon number scaling parameter $\bar{n}$ is defined to remove all nonlinear coefficients from the equation. This is obtained with the choice

$$\bar{n} = \frac{|k''|v^2}{\chi_e t_0} \tag{53}$$

There is another description of this physical system that has an approximate validity, and is more convenient for some purposes. The particle flux must be invariant at the dielectric boundaries because energy conservation demands it, even though the changing group velocity means that particle density must change. This leads to a description in terms of the flux operators, with an equation

$$\left(v \frac{\partial}{\partial x} + \frac{\partial}{\partial t} \right) \hat{\Phi}(t, x) = \left[\frac{i\omega''}{2} \frac{\partial^2}{\partial x^2} + \frac{i\chi_e}{v} \hat{\Phi}^\dagger \hat{\Phi} \right] \hat{\Phi}(t, x) \tag{54}$$

Because flux is most naturally regarded as evolving in space, it is also common in laser physics to make a somewhat different choice of coordinates. In this reference frame, the space variable is left unchanged, and the time variable is modified so that $t_v = t - x/v$. This reduces to an unusual form of the quantum nonlinear Schrödinger equation, as follows:

$$iv \frac{\partial}{\partial x} \hat{\Phi}(t_v, x) = \left[-\frac{\omega''}{2} \left(\frac{\partial^2}{\partial x^2} + \frac{\partial^2}{v^2 \partial t_v^2} - 2\frac{\partial^2}{v \partial t_v \partial x} \right) - \frac{\chi_e}{v} \hat{\Phi}^\dagger \hat{\Phi} \right] \hat{\Phi}(t_v, x) \tag{55}$$

Next, this equation is written in dimensionless form, using a scaled flux field and dimensionless coordinates. Provided that $x_0 \gg vt_0$, the derivative terms on the right side containing $\partial/\partial x$ have small coefficients, of order vt_0/x_0, and may be neglected. This leads to an approximate form of the quantum nonlinear Schrödinger equation, in which the time variable has changed places with the space variable:

$$\frac{\partial}{\partial\zeta}\hat{\phi}(\tau_v,\zeta) \simeq \left[\frac{i}{2}\frac{\partial^2}{\partial\tau_v^2} + i\hat{\phi}^\dagger\hat{\phi}\right]\hat{\phi}(\tau_v,\zeta) \qquad (56)$$

This version of the nonlinear Schrödinger equation (56) is often used as an alternative to the more precise (52). However, it requires more sophisticated techniques to handle the commutation relations in (56). Since the operators here have their standard meaning, they must have equal-*time* commutators. This would be inconsistent with an interpretation of this equation as corresponding to the quantum nonlinear Schrödinger equation with time and space interchanged. Obviously, if this were literally true, the operators would have equal-*space* commutators. However, normal quantum operators in nonlinear systems have well-defined commutators only at equal-time coordinates. This problem is discussed later, using operator representation theory.

To understand the physical meaning of the limit $x_0 \gg vt_0$, we now return to the original form of the Heisenberg equation in the time domain. In this limit, an input wave form is relatively undistorted on propagation over distances of the order of the physical wave-packet size. Thus, the time evolution of operators external to the dielectric is nearly identical to the time evolution a distance vt_0 into the dielectric. This is comparable to the physical pulse dimension. For this reason, an input coherent wave packet retains its shape on entering the dielectric, apart from a trivial compression by the factor v/c. Accordingly, the external pulse shape and statistics may be used as an appropriate initial condition for time evolution inside the dielectric. This also requires relatively weak dispersion, so that refractive-index matching is possible for all input frequency components.

VII. RAMAN SCATTERING

An important physical effect in propagation is Raman scattering off molecular excitations. In optical fibers this starts in the low-frequency region as an acoustic Brillouin effect, and extends up to a strong resonance at around 12–14 THz. For this reason, the nonlinear Schrödinger equation requires corrections due to refractive-index fluctuations for pulses longer than about 1 ps, and fails for pulse durations much shorter than this. In fact, it may not even be very useful for longer pulses, if high enough intensities are present. The concept of a soliton solution of the usual form then ceases to exist.

The reason is that part of the nonlinearity present in a fiber originates from Raman transitions. When the pulse is sufficiently long, the nonlinearity due to Raman contributions (around 19%) acts as though it was instantaneous. However, for short pulses the nonlinear term needs to be modified to take this time delay into account, as shown by the experiments of Mitschke and Mollenauer,[36] and described theoretically in a concise way by Gordon.[37] In this limit the high-frequency components of the pulse act as a Raman pump for the low-frequency components, transferring energy continuously down the spectrum. As the mean frequency of the pulse shifts accordingly, the effect has been termed self-frequency shifting and is observed to be in reasonable agreement with the classical theory developed by Gordon.[37]

The treatment of the quantum theory of propagation given so far was based on a Hamiltonian[6, 7] that successfully reproduced known results for the classical dynamics and energy of a nonlinear dispersive dielectric. This quantum theory is now extended to include the effects of thermal and quantum fluctuations due to Raman scattering. Although previous quantum treatments of Raman scattering have been given,[38, 39] it is necessary to modify these somewhat in the present situation. The Raman interaction energy[40, 41] of a fiber, in terms of atomic displacements from their mean lattice positions, is known to be

$$W_{R} = \sum_{j} \eta_{j}^{R} : \mathbf{D}(\bar{\mathbf{x}}^{j}) \mathbf{D}(\bar{\mathbf{x}}^{j}) \, \delta \mathbf{x}^{j} \tag{57}$$

Here $\mathbf{D}(\bar{\mathbf{x}}^{j})$ is the electric displacement at the jth mean atomic location $\bar{\mathbf{x}}^{j}$, $\delta \mathbf{x}^{j}$ is the atomic displacement operator, and $\mathbf{\eta}_{j}^{R}$ is a Raman coupling tensor.

To quantize this interaction with atomic positions using our macroscopic quantization method, we must now take into account the existence of a corresponding set of phonon operators. These diagonalize the atomic displacement Hamiltonian in each fiber segment, and have well-defined eigenfrequencies. In fact Bell and Dean[42] have actually calculated the frequency spectrum and normal modes of vibration for vitreous silica. They used physical models based on the random network theory of disordered systems, and their computed vibrational frequency spectrum appears remarkably similar to the observed Raman gain profile[43] for the same medium. In this review, the Raman gain $\alpha_{R}(\Omega)$ is normalized following Gordon.[37, 43]

The interaction does not involve any time derivatives, and hence does not change the canonical momenta. Thus, the Raman effect can be included macroscopically through a continuum Hamiltonian term coupling photons to phonons, of the form[44]

$$\hat{H}_R = \hbar \int_{-\infty}^{\infty} \int_0^{\infty} \hat{\Psi}^{\dagger}(z)\hat{\Psi}(z)r(z,w)\left[\hat{A}(z,w) + \hat{A}^{\dagger}(z,w)\right] dw\, dz$$

$$+ \hbar \int_{-\infty}^{\infty} \int_0^{\infty} w\left[\hat{A}^{\dagger}(z,w)\hat{A}(z,w)\right] dw\, dz$$

where

$$\left[\hat{A}(z,w), \hat{A}^{\dagger}(z',w')\right] = \delta(z-z')\delta(w-w') \tag{58}$$

Here, the Raman excitations are treated as an inhomogeneously broadened continuum of modes, localized at each longitudinal location (z). GAWBS (guided wave acoustic Brillouin scattering)[45] is a special case of this in the low-frequency limit. Since neither Raman nor Brillouin excitations are completely localized, this treatment implicitly requires a frequency and wave-number cutoff, so that the field operator $\hat{\Psi}$ is slowly varying on the phonon scattering distance scale. The macroscopic frequency dependent coupling $r(z,\omega)$ determines the Raman gain, which from now on is assumed to be uniform in space.

The corresponding coupled set of nonlinear operator equations are

$$\left[v\frac{\partial}{\partial z} + \frac{\partial}{\partial t}\right]\hat{\Psi}(t,z)$$

$$= \left[\frac{iw''}{2}\frac{\partial^2}{\partial z^2} + i\chi_e\hat{\Psi}^{\dagger}\hat{\Psi}\right]\hat{\Psi}(t,z) \tag{59}$$

$$- i\left[\int_0^{\infty} r(w)\left[\hat{A}(t,z,w) + \hat{A}^{\dagger}(t,z,w)\right] dw\right]\hat{\Psi}(t,z)$$

and

$$\frac{\partial}{\partial t}\hat{A}(t,z,w) = -iw\hat{A}(t,z,w) - ir(w)\hat{\Psi}^{\dagger}(t,z)\hat{\Psi}(t,z)$$

In summary, the original theory of nonlinear quantum field propagation is now extended to include both the nonlinear paths of propagation, i.e., the electronic and the Raman paths. The result is a modified Heisenberg equation with a delayed nonlinear response to the field due to the Raman coupling.

The relationship of macroscopic coupling $r(\omega)$ to measured Raman gain $\alpha_R(\Omega)$ is as follows:

$$r^2(\omega) = \frac{\chi\alpha_R(\omega t_0)}{2\pi} \tag{60}$$

where χ is the total effective nonlinear coefficient obtained from the low-frequency non-linear refractive index. The electronic or fast-responding nonlinear coefficient χ_e is now given by

$$\chi_e = \chi - 2\int_{-\infty}^{\infty}\int_0^{\infty} r^2(\omega)\sin(\omega t)\, d\omega\, dt \tag{61}$$

These equations also fully include the thermal noise introduced by the coupling, via the initial density matrix of the phonon modes. If these are assumed to be in a typical thermal Bose state, then the initial distribution gives rise to thermal occupation $n_{th}(\omega)$ of the phonon number states, where

$$n_{th}(\omega) = \left[\exp\left(\frac{\hbar\omega}{kT}\right) - 1\right]^{-1} \tag{62}$$

A result of this model is that the phonon operators do not have white noise behavior. In fact, this colored noise property is significant enough to invalidate the usual Markovian and rotating-wave approximations, which are therefore not used here. Of course, the photon modes may also be in a thermal state of some type. However, thermal effects are typically much more important at the low frequencies that characterize Raman and Brillouin scattering than they are at optical frequencies. In addition, if the input is a photon field generated by a laser, any departures from coherent statistics will be rather specific to the laser type, instead of having the generic properties of thermal fields.

Another effect has so far been neglected. This is the ultralow frequency tunneling due to lattice defects.[45] Because this is not strictly linear, it cannot be included accurately in our macroscopic Hamiltonian. Despite this, the effects of this $1/f$-type noise may be included approximately for any predetermined temperature. This can be achieved by modifying the coupling term $r(\omega)$ at low frequencies, so that it generates the known refractive index fluctuations. This approach requires caution, however: The present phonon model strictly requires a *linear* harmonic oscillator-type behavior in the phonon bath.

VIII.   PHASE-SPACE DISTRIBUTIONS

In practical terms, the known exact solutions[35, 34] of the quantum nonlinear Schrödinger equation have little utility at typical photon numbers of 10^9. This is because the initial field is usually a coherent state[46, 47] rather than a number state in experiment. Instead, it is often more useful to employ phase-space distributions or operator representations to calculate observable quantities such as quadrature variances and phase fluctuations. Several techniques are known, including the Wigner representation[48] and the Glauber-Sudarshan P-representation.[47] However, the first of these does not give positive distributions, and the second has a singular behavior for this Hamiltonian. Accordingly, a nondiagonal coherent state expansion of the density matrix is preferable. This results in a generalized P-representation Fokker-Planck equation, which is an extension of the diagonal Glauber-Sudarshan P-representation.

A number of different types of generalized P-representation are known.[49, 50] In this review the positive P-representation is used. This operator representation is able to treat all types of nonclassical radiation as a positive distribution on a nonclassical phase space, and is amenable to numerical simulation.[51, 52] Using this method, the operator equations are transformed to complex Ito stochastic equations, which involve only c-number (commuting) variables. Thus, while quantum effects require that the usual classical equation be reinterpreted as an operator equation, this in turn can be transformed to an equivalent pair of c-number stochastic equations.[6, 7]

Using these techniques, and assuming vanishing boundary terms, the equivalent Ito stochastic equations is

$$\frac{\partial}{\partial t}\Psi(t, x_v) = \left[\frac{i\omega''}{2}\frac{\partial^2}{\partial x_v^2} + i\chi_e\Psi^{\dagger}\Psi + \sqrt{i\chi_e}\,\Gamma(t, x_v)\right]\Psi(t, x_v) \quad (63)$$

where

$$\langle \Gamma(t, x_v)\Gamma(t', x_v')\rangle = \delta(t - t')\delta(x_v - x_v') \quad (64)$$

together with a similar equation for the complex c-number field $\Phi^{\dagger}$, representing the conjugate field $\hat{\Phi}^{\dagger}$, having a corresponding independent stochastic noise term $\Gamma^{\dagger}$.

It is often convenient to work in a frame in which the role of x and t variables are reversed. Just as with the operator equations, the resulting equations are only approximately valid, under conditions of slow spatial variation relative to the pulse length in space. These have the form

$$\frac{\partial}{\partial x}\Phi(t_v, x) \simeq \left[-i\frac{k''}{2}\frac{\partial^2}{\partial t_v^2} + i\frac{\chi_e}{v^2}\Phi^\dagger\Phi + \frac{\sqrt{i\chi_e}}{v}\Gamma(t_v, x) \right]\Phi(t_v, x) \quad (65)$$

together with a similar equation for $\Phi^\dagger$. With these stochastic equations, there is no difficulty with commutation relations, as there was in the earlier operator equations. These are now included in the properties of the noise sources, which have an exact symmetry on interchange of time and space coordinates. Thus, the equations are precisely equivalent to a suitably modified form of the quantum nonlinear Schrödinger equation, obtained from a "Hamiltonian" that causes translation in position,[53] not in time. The implication of this is that there are an equivalent set of new operators that create well-defined fields at temporally extended boundaries, rather than the usual spatially extended initial conditions. The corresponding commutation relations are therefore implicitly equal-space commutators.

Despite the intuitive attraction of this interpretation, it is the original equal-time version of the commutation relation that is fundamental, and the stochastic or operator equations that are first order in x are only approximately valid. For this reason, the equal-space commutator analogy should not be taken too far. There are causality problems involved when quantum theory is turned on its side in this way, and clearly the method is restricted to one-dimensional dielectrics. However, the x-dependent forms of the stochastic equations are often useful for computational purposes. In either form of the equation, soliton formation takes place only when $k'' < 0$ or $\omega'' > 0$, a condition known as anomalous dispersion.

For the remainder of this review, the coordinates used are those giving approximate first order in x equations. These are physically applicable to short pulse experiments. For this reason the v subscript on the τ variable will be understood implicitly. By now, it should be clear how to regain the standard equations in time, where necessary. Just as with the operator equations, it is useful to transform to a dimensionless form of the Ito stochastic equations, giving

$$\frac{\partial}{\partial\zeta}\phi(\tau, \zeta) \simeq \left[\frac{i}{2}\frac{\partial^2}{\partial\tau^2} + i\phi^\dagger\phi + \sqrt{\frac{i}{\bar{n}}}\,\Gamma(\tau, \zeta) \right]\phi(\tau, \zeta) \quad (66)$$

where

$$\langle\Gamma(\tau', \zeta')\Gamma(\tau, \zeta)\rangle = \delta(\tau - \tau')\delta(\zeta - \zeta') \quad (67)$$

together with a Hermitian conjugate equation obtained on replacing ϕ with $\phi^\dagger$, Γ with $\Gamma^\dagger$, and i with $-$i. The dimensionless field ϕ is related to $\hat{\Psi}$ by an equivalence of normally ordered operator products with stochastic moments, where

$$\phi \sim \hat{\Psi}\sqrt{\frac{vt_0}{n}} \tag{68}$$

For coherent inputs, the initial condition on these equations is a deterministic field $\phi_0(\tau)$, which can model laser inputs of arbitrary shape. At large photon number, the noise terms are relatively small. This means that either computer simulation techniques can be used, or methods utilizing the known exact classical inverse-scattering[54] solutions with linearized perturbations.[55] These equations successfully predicted phase diffusion and squeezing in propagating solitons,[6-10] which were recently observed experimentally.[11]

As pointed out in the last section, a strong practical limitation on experiment originates from the refractive-index fluctuations and Raman scattering in real media. These can be easily treated on extending the stochastic form of the equations to include Raman effects. By taking the equation for the density operator $\hat{\rho}$ describing the above Hamiltonian and introducing a coherent state expansion in the form of the positive P-representation,[49, 50] a Fokker-Planck equation can be readily derived. Thus, equivalent stochastic differential equations for the associated c-number fields are obtainable. Substituting the integrated phonon variables into the equations for the photon field gives the following equation for ϕ:

$$\frac{\partial}{\partial\zeta}\phi(\tau,\zeta) \simeq \left[if\phi^\dagger\phi \pm \frac{i}{2}\frac{\partial^2}{\partial\tau^2} \right.$$

$$+ i\int_{-\infty}^{\tau} d\tau' h(\tau - \tau')\phi^\dagger(\tau',\zeta)\phi(\tau',\zeta)$$

$$\left. + \sqrt{\frac{if}{n}}\,\Gamma(\tau,\zeta) + i\Gamma_R(\tau,\zeta) \right]\phi(\tau,\zeta) \tag{69}$$

There is a corresponding Hermitian conjugate equation for $\phi^\dagger$, obtained by making the substitutions $\phi \to \phi^\dagger$, $i \to -i$, $\Gamma \to \Gamma^\dagger$, and $\Gamma_R \to \Gamma_R^\dagger$. Here dimensionless variables are used following standard notation in fiber optics applications, with

$$h(\tau) = 2 \int_0^\infty r^2 \left(\frac{\nu}{t_0} \right) \sin(\nu\tau) \frac{d\nu}{\chi} \qquad \bar{n} = \frac{|k''|v^2}{\chi t_0}$$

$$\zeta = \frac{z}{z_0} \qquad\qquad f = \frac{\chi_e}{\chi} \qquad (70)$$

$$z_0 = \frac{t_0^2}{|k''|} \qquad\qquad k'' = -\frac{w''}{v^3}$$

Here, as previously, χ is the overall nonlinear coefficient. The fractional quantity $f = \chi_e/\chi$ describes the contribution to the nonlinearity due to direct electronic (rather than Raman) transitions. This is typically ≈ 0.81 in silica.

The integral containing $h(\tau)$ supplies the Raman contribution to the overall nonlinearity and is responsible for self-frequency shifting. This response function has the following properties:

$$\int_0^\infty d\tau\, h(\tau) = \frac{\chi_R}{\chi} \qquad \text{where } \chi_R + \chi_e = \chi$$

$$\sqrt{2\pi}\, h(\tau) = \int_{-\infty}^\infty d\nu\, \tilde{h}(\nu)\, e^{-i\nu\tau} \qquad (71)$$

$$\sqrt{2\pi}\, \tilde{h}(\nu) = \frac{i\,\mathrm{sgn}(\nu)}{2} \alpha_R(|\nu|) + \int_0^\infty d\Omega \frac{\alpha_R(\Omega)\Omega}{\pi(\Omega^2 - \nu^2)}$$

The expression for $\tilde{h}(\nu)$ has been attained by taking the limit as the width of the vibrational Raman modes (caused by any coupling to phonon reservoirs) goes to zero for all frequencies Ω. In this way $\alpha_R(\Omega)$ includes an inhomogeneous frequency distribution of the phonon states. The result for $\tilde{h}(\nu)$ is a form of the Kramers-Kronig relation.

The last terms appearing in the equation are stochastic functions representing noise sources with different origins. Γ represents the quantum noise of the field introduced by the electronic nonlinearity, and Γ_R is the thermal noise due to the phonon coupling. The correlation functions for Γ are, as before,

$$\langle \Gamma(\zeta',\tau')\Gamma(\zeta,\tau) \rangle = \langle \Gamma^\dagger(\zeta',\tau')\Gamma^\dagger(\zeta,\tau) \rangle$$
$$= \delta(\zeta - \zeta')\delta(\tau - \tau') \qquad (72)$$

The correlations for Γ_R, $\Gamma_R^\dagger$ are more easily interpreted in the Fourier domain where they take the form

$$\left\langle \tilde{\Gamma}_R(\nu,\zeta)\tilde{\Gamma}_R(\nu',\zeta') \right\rangle = \left\langle \tilde{\Gamma}_R^\dagger(\nu,\zeta)\tilde{\Gamma}_R^\dagger(\nu',\zeta') \right\rangle^*$$

$$= \frac{1}{n}\delta(\zeta - \zeta')\delta(\nu + \nu')$$

$$\times \left(\left[n_{th}\left(\frac{|\nu|}{t_0}\right) + \Theta(\nu) \right]\alpha_R(|\nu|) - i\sqrt{2\pi}\,\tilde{h}(-\nu) \right)$$

$$(73)$$

$$\left\langle \tilde{\Gamma}_R^\dagger(\nu,\zeta)\tilde{\Gamma}_R(\nu',\zeta') \right\rangle = \frac{1}{n}\delta(\zeta - \zeta')\delta(\nu + \nu')\left[n_{th}\left(\frac{|\nu|}{t_0}\right) + \Theta(\nu) \right]\alpha_R(|\nu|)$$

$$(74)$$

where

$$\sqrt{2\pi}\,\tilde{\Gamma}_R(\nu,\zeta) = \int_{-\infty}^{\infty} d\tau\, e^{i\nu\tau}\Gamma_R(\tau,\zeta) \tag{75}$$

Here $\Theta(\nu)$ is the step function with value unity for positive arguments and zero otherwise.

Equation (73) is an expected result,[56] since it states that when $\nu > 0$ the spectral intensity of noise due to the Stokes process, in which a photon is downshifted in frequency by an amount ν with the production of a phonon of the same frequency, is proportional to $n_{th} + 1$. However, the anti-Stokes process, in which a phonon is absorbed ($\nu < 0$), is proportional only to n_{th}. Here $n_{th}(|\nu|/t_0) = [\exp(\hbar|\nu|/kTt_0) - 1]^{-1}$ as usual, ν being dimensionless.

These stochastic equations can thus be used to investigate the propagation behavior of quantum fluctuations in an optical fiber, including Raman scattering. However, the use of an enlarged nonclassical phase space can increase computation times in practical applications. For this reason, the Wigner function defined on a classical phase space is useful when treating intense, nearly classical fields. The Wigner function, which is a complex Gaussian convolution of the positive P-representation distribution, does *not* have an exact stochastic equation. This is because there are third-order derivative terms in the Wigner function Fokker-Planck equation that have no stochastic equivalent. Despite this, in sufficiently intense fields the additional terms can be neglected, giving rise to a positive distribution with approximate Fokker-Planck and stochastic equations.

The detailed calculation of the Wigner equations can be carried out following standard procedures.[57] After truncation, the resulting Wigner theory is formally similar to the corresponding positive P-function theory, except for small frequency-shift corrections to the deterministic terms, which we can neglect here. In addition, all diffusion terms in the corresponding positive P-representation equations involving a thermal occupation number n are changed to a corresponding term that has $n + \frac{1}{2}$ in the Wigner equations. Every other diffusion term vanishes in this approximation. This rule is also applied to the photon and phonon initial conditions, which are given an effective occupation number of $n_{\text{th}} + \frac{1}{2}$ if initially in thermal equilibrium at temperature T, corresponding to the well-known thermal fluctuation level in symmetric ordering.

Applying this to the previous P-representation result[57] in reduced variables, the following approximate Wigner equations, suitable for calculating symmetrically ordered correlations are obtained:

$$
\frac{\partial}{\partial \zeta} \phi(\tau, \zeta)
$$

$$
\simeq \left[\mathrm{i} f \phi^* \phi \pm \frac{\mathrm{i}}{2} \frac{\partial^2}{\partial \tau^2} + \mathrm{i} \int_{-\infty}^{\tau} \mathrm{d}\tau' \, h(\tau - \tau') \phi^*(\tau', \zeta) \phi(\tau', \zeta) \right.
$$

$$
\left. + \mathrm{i} \Gamma_W(\tau, \zeta) \right] \phi(\tau, \zeta) \tag{76}
$$

It should be emphasized again that these equations have been transformed to an approximate form. This form is valid when the characteristic length scales of interest in the field are short compared to the characteristic pulse-shaping distance scales, and when photon numbers are large. The initial fields at $\zeta = 0$ are assumed to be in a coherent state, and have vacuum fluctuations correlated according to

$$
\langle \Delta\phi(\tau, 0) \Delta\phi(\tau', 0) \rangle = 0
$$

$$
\langle \Delta\phi(\tau, 0) \Delta\phi^*(\tau', 0) \rangle = \frac{1}{2\bar{n}} \delta(\tau - \tau') \tag{77}
$$

The photon flux (in photons/second) is represented by $\langle \bar{n} | \phi(\tau, \zeta) |^2 - 1/2\Delta t \rangle$, where Δt^{-1} is the fundamental frequency cutoff in the theory. This correlation factor is necessary because the Wigner function represents symmetrically ordered operators, which have a diverging vacuum noise term as the cutoff is taken to infinity. The initial thermal state of the

Raman phonons must be included also. This causes noise correlations given by

$$\left\langle \Gamma_{\mathrm{W}}(\nu,\zeta)\left[\Gamma_{\mathrm{W}}(-\nu',\zeta')\right]^{*}\right\rangle = \left\langle \Gamma_{\mathrm{W}}(\nu,\zeta)\Gamma_{\mathrm{W}}(\nu',\zeta')\right\rangle$$

$$= \frac{1}{\bar{n}}\delta(\zeta - \zeta')\delta(\nu + \nu')\left[n_{\mathrm{th}}\!\left(\frac{|\nu|}{t_0}\right) + \frac{1}{2}\right]$$

$$\times \alpha_{\mathrm{R}}(|\nu|)$$

where

$$\Gamma_{\mathrm{W}}(\nu,\zeta) = \frac{1}{\sqrt{2\pi}}\int\Gamma_{\mathrm{W}}(\tau,\zeta)\,\mathrm{e}^{i\nu\tau}\,\mathrm{d}\tau \tag{78}$$

Computer simulations[44] of these equations reveal that there is good agreement between this technique and the earlier technique using the positive P-representation at large photon number. The results demonstrate that for these propagation distances and in the large photon-number limit, the semiclassical or truncated Wigner method is a reliable simulation method for small quantum noise effects. The method is relatively efficient, at least for low-frequency measurements, where the sampling error from the vacuum fluctuations is small. The computer simulations done so far are all in the region of $\zeta \ll \bar{n}^{1/2}$, where quantum fluctuations are relatively small. For larger quantum fluctuations, the neglected third-order derivative terms in the truncated Wigner equations are expected to cause errors, as they are known to do elsewhere.[52] For this reason, this approximation should be checked in general against more reliable techniques, such as the positive P-representation, that require no truncation.

IX. CONCLUSION

A direct treatment of quantization of macroscopic equations must be able to generate the correct equations of motion, together with a Hamiltonian corresponding to the classical energy. This type of theory can then be tested experimentally. The result in the case of the fiber soliton is that the predictions are in accordance with experiment. There are a number of quantization proposals that do not have a Hamiltonian corresponding to the classical energy, leading to severe nonuniqueness problems. Similarly, macroscopic quantization that does not lead to classical equations in the appropriate limit is unlikely to agree with experiment.

For simplicity, these nonstandard procedures have not been reviewed here. They are available in the references. Again, to reduce complication, many technical details have not been covered. The emphasis of this review is on the unity of these results with earlier quantum field theories, rather than detailed derivations. In practice, it is often necessary to include other modes and couplings as well as those treated here. However, the same general principles hold.

Most importantly, it is obvious that quantum optical techniques can be used to test quantum field theory predictions. These include novel results, for example, on equivalence of some interacting boson theories with fermion theories. In many cases, these theories are not accessible in traditional scattering experiments. This led to Coleman's complaint that quantum soliton theory did not have experimental application.[5] We can now remove this "embarrassment." Obvious cases that are of most interest include topological and higher-dimensional solitons, together with quantum chaos and nonequilibrium phase transitions.

References

1. Squeezed states and nonclassical light have been topics of several special journal issues and reviews: D. F. Walls, *Nature* **306**, 141 (1983); H. J. Kimble, D. F. Walls (eds.), *J. Opt. Soc. Am. B* **4**, 10 (1987); R. Loudon and P. L. Knight (eds.), J. Mod. Optics **34**, 707 (1987); E. Giacobino and C. Fabre (eds.), *Appl. Phys. B*, **55**, 189 (1992).

2. P. A. M. Dirac, Proc. R. Soc. (London) Ser. A **114**, 243 (1927).

3. P. A. M. Dirac. *The Principles of Quantum Mechanics*, Clarendon, Oxford, UK, 1958; *Lectures on Quantum Mechanics*, New York, Belfer Graduate School of Science, 1964.

4. I. J. R. Aitchison and A. J. G. Hey, *Gauge Theories in Particle Physics*, Adam Hilger, Bristol, 1982.

5. S. Coleman, *Aspects of Symmetry*, Cambridge University Press, Cambridge, UK, 1985.

6. S. J. Carter, P. D. Drummond, M. D. Reid, and R. M. Shelby, *Phys. Rev. Lett.* **58**, 1841 (1987).

7. P. D. Drummond and S. J. Carter, *J. Opt. Soc. Am. B* **1**, 1656 (1987).

8. P. D. Drummond, S. J. Carter, and R. M. Shelby, *Opt. Lett.* **14**, 373 (1989).

9. R. M. Shelby, P. D. Drummond, and S. J. Carter, *Phys. Rev. A* **42**, 2966 (1990).

10. Y. Lai and H. A. Haus, *Phys. Rev. A* **40**, 844 (1989); H. A. Haus and Y. Lai, *J. Opt. Soc. Am. B* **7**, 386 (1990).

11. M. Rosenbluh and R. M. Shelby, *Phys. Rev. Lett.* **66**, 153 (1991).

12. S. R. Friberg, S. Machida, and Y. Yamamoto, *Phys. Rev. Lett.* **69**, 3165 (1992).

13. K. Watanabe, H. Nakano, A. Honold, and Y. Yamamoto, *Phys. Rev. Lett.* **62**, 2257 (1989); H. A. Haus, K. Watanabe, and Y. Yamamoto, *J. Opt. Soc. Am. B* **6**, 1138 (1989).

14. E. Yablonovitch and T. Gmitter, *Phys. Rev. Lett.* **58**, 2486 (1987).

15. E. A. Hinds, in D. Bates and B. Bederson (Eds.), *Advances in Atomic, Molecular and Optical Physics*, Vol. 28, Academic, New York, 1990, p. 237.

16. R. E. Slusher, P. Grangier, A. LaPorta, B. Yurke, and M. J. Potasek, *Phys. Rev. Lett.* **59**, 2566 (1987).

17. M. Toda, S. Adachi, and K. Ikeda, *Prog. Theor. Phys. Suppl.* **98**, 323 (1989).

18. J. J. Hopfield, *Phys. Rev.* **112**, 1555 (1958).

19. J. M. Jauch and K. M. Watson, *Phys. Rev.* **74**, 950 (1948).

20. Y. R. Shen, *Phys. Rev.* **155**, 921 (1967).

21. T. A. B. Kennedy and E. M. Wright, *Phys. Rev. A* **38**, 212 (1988).

22. A. Hasegawa and F. Tappert, *Appl. Phys. Lett.* **23**, 142 (1973); L. F. Mollenauer, *Philos. Trans. R. Soc. London Ser. A* **315**, 435 (1985).

23. P. D. Drummond, *Phys. Rev. A* **42**, 6845 (1990).

24. N. Bloembergen, *Nonlinear Optics*, Benjamin, New York, 1965.

25. M. Goeppert-Meyer, *Ann. Phys.* **9**, 273 (1931).

26. E. Power and S. Zienau, *Philos. Trans. R. Soc. Lodon Ser. A* **251**, 427 (1959); R. Loudon, *The Quantum Theory of Light*, Clarendon, Oxford, UK, 1983.

27. S. Geltmann, *Phys. Lett.* **4**, 168 (1963).

28. M. Hillery and L. D. Mlodinow, *Phys. Rev. A* **30**, 1860 (1984).

29. R. J. Glauber and M. Lewenstein, *Phys. Rev. A* **43**, 467 (1991); I. H. Deutsch and J. C. Garrison, *Phys. Rev. A* **43**, 2498 (1991).

30. S. T. Ho and P. Kumar, *J. Opt. Soc. Am. B.* (August, 1993).

31. P. D. Drummond, *Phys. Rev. A* **39**, 2718 (1989).

32. C. M. Caves and D. D. Crouch, *J. Opt. Soc. Am. B* **4**, 1535 (1987); M. G. Raymer, P. D. Drummond, and S. J. Carter, *Opt. Lett.* **16**, 1189 (1991).

33. R. Loudon, *J. Phys. A* **3**, 233 (1970); K. J. Blow, R. Loudon, S. J. D. Phoenix, and T. J. Shepherd, *Phys. Rev. A* **42**, 4102 (1990); B. Huttner, J. J. Baumberg, and S. M. Barnett, *Europhys. Lett.* **16**, 177 (1991).

34. C. N. Yang, *Phys. Rev.* **168**, 1920 (1967).

35. H. A. Bethe, *Z. Phys.* **71**, 205 (1931).

36. F. M. Mitschke and L. F. Mollenauer, *Opt. Lett.* **11**, 659 (1986).

37. J. P. Gordon, *Opt. Lett.* **11**, 662 (1986).

38. T. von Foerster and R. J. Glauber, *Phys. Rev. A* **3**, 1484 (1971).

39. I. A. Walmsley and M. G. Raymer, *Phys. Rev. Lett.* **50**, 962 (1983).

40. M. D. Levenson, *Introduction to Nonlinear Laser Spectroscopy*, Academic, New York, 1982.

41. S. J. Carter and P. D. Drummond, *Phys. Rev. Lett.* **67**, 3757 (1991).

42. R. J. Bell and P. Dean, *Disc. Faraday Soc.* **50**, 55 (1970); P. Dean, *Rev. Mod. Phys.* **44**, 127 (1972).

43. R. H. Stolen, C. Lee, and R. K. Jain, *J. Opt. Soc. Am. B* **1**, 652 (1984), R. H. Stolen and E. P. Ippen, *Appl. Phys. Lett.* **22**, 276 (1973).

44. P. D. Drummond and A. D. Hardman, *Europhys. Lett.*, **21**, 279 (1993).

45. R. M. Shelby, M. D. Levenson, and P. W. Bayer, *Phys. Rev. B* **31**, 5244 (1985), *Phys. Rev. Lett.* **54**, 939 (1985).

46. E. Schroedinger, *Naturwissenschaften* **14**, 644 (1927).

47. R. J. Glauber, *Phys. Rev.* **130**, 2529 (1963); E. C. G. Sudarshan, *Phys. Rev. Lett.* **10**, 277 (1963).

48. E. P. Wigner, *Phys. Rev.* **40**, 749 (1932).

49. P. D. Drummond and C. W. Gardiner, *J. Phys. A* **13**, 2353 (1980). Note that the correspondence to the positive *P*-representation phase-space equations is exact only if boundary terms vanish, which has been verified computationally within numerical accuracy.

50. C. W. Gardiner, *Quantum Noise*, Springer, Berlin, 1992.

51. P. D. Drummond and I. K. Mortimer, *J. Comput. Phys.* **93**, 144 (1991).

52. P. Kinsler and P. D. Drummond, *Phys. Rev. A* **43**, 6194 (1991).

53. B. Yurke and M. J. Potasek, *J. Opt. Soc. Am.* **B6**, 1227 (1989); B. Huttner, S. Serulnik, and Y. Ben-Aryeh, *Phys. Rev. A* **42**, 5594 (1990); I. Abram and E. Cohen, *Phys. Rev. A* **44**, 500 (1991); N. Imoto, J. R. Jeffers, and R. Loudon, P. Tombesi and D. F. Walls (Eds.), in *Quantum Measurements in Optics*, Plenum, New York, 1992, p. 295.

54. P. D. Lax, *Commun. Pure Appl. Math.* **21**, 467 (1968); V. Zakharov and A. Shabat, *Sov. Phys. JETP* **34**, 62 (1972).

55. D. J. Kaup, *Phys. Rev. A* **42**, 5689 (1990).

56. C. Kittel, *Introduction to Solid State Physics*, 5th ed., Wiley, New York, 1976.

57. R. Graham, in G. Hohler (Ed.), *Quantum Statistics in Optics and Solid-State Physics*, *Springer Tracts in Modern Physics*, Vol. 66, Springer, New York, 1973, p. 1.

FOUR-WAVE MIXING AND LIGHT SQUEEZING

M. D. REID

*Department of Physics, The University of Queensland,
Queensland, Australia*

and

D. F. WALLS

*Department of Physics, University of Auckland,
Auckland, New Zealand*

CONTENTS

I. INTRODUCTION

The first experimental realization of squeezed light was obtained by Slusher et al.[1] in 1985. In this experiment the squeezing was generated via four-wave mixing using an atomic beam of sodium interacting with light inside a single-ended optical cavity. Measurements performed on the light

Modern Nonlinear Optics, Part 3, Edited by Myron Evans and Stanisław Kielich. Advances in Chemical Physics Series, Vol. LXXXV.
ISBN 0-471-30499-9 © 1994 John Wiley & Sons, Inc.

transmitted from the cavity through the single output mirror revealed sub-shot noise levels ("squeezing").[2] While only a relatively small amount of squeezing was obtained initially, the result represented a significant scientific and technological advance. Squeezing is obtained when the noise associated with a particular quadrature phase amplitude is reduced below that expected for a coherent state. Experimentally this is observed as a noise level that goes below the shot noise level. Squeezed light cannot be predicted using standard classical radiation theory and is hence evidence for the quantum theory of light. It assumes a special importance when possible applications are considered. These range from use in gravity-wave detectors,[3] precision measurements,[4] and ultrasensitive spectroscopy[5] to potential use in optical communication systems. The potential applications as well as the fundamental importance were a motivating force behind the enormous interest that developed in squeezed states and their possible experimental generation. The first experimental attempts to squeeze light used four-wave mixing with sodium vapor[6] and four-wave mixing in an optical fiber.[7] Both these experiments later generated squeezing. Subsequently, substantial squeezing has also been obtained using parametric oscillation,[5, 8, 9] in the output light of laser diodes[10] and in pulsed light fields[11] produced using $\chi^{(2)}$ and $\chi^{(3)}$ media. Further atomic four-wave mixing experiments were performed by Hope et al.[12] Recent experimental interest has expanded to consider applications.[4, 5, 13]

In this chapter we present a review of the quantum theory which was originally developed by us[14] to predict squeezing of light using four-wave mixing with an atomic $\chi^{(3)}$ medium in an optical cavity. Alternative theories have also been developed[6, 15, 16, 17] and have given the same results. The theories assume at some point a linearization of quantum fluctuations. This is justified with the current experimental regimes, since photon numbers are large and quantum noise is a small perturbation about semiclassical solutions.

II. QUANTUM MODEL FOR FOUR-WAVE MIXING

Yuen and Shapiro[18] first suggested the possibility of generating squeezed light using four-wave mixing. The simplest possible interaction Hamiltonian that might describe four-wave mixing is written as follows:

$$H = \hbar\chi\left(E^2 a_2^\dagger a_3^\dagger + E^{*2} a_2 a_3\right) \tag{1}$$

Here χ is the nonlinear susceptibility of the medium, E is the amplitude for the pump, which here is taken to be classical and nondepleting and is at frequency ω_p, and a_2, a_3 are the boson operators for quantized signal and idler modes at frequencies $\omega_p \pm \varepsilon$, respectively. From this simplistic model we can predict the existence of quantum correlations between the signal and idler photon pairs emitted. The correlations are evident as reduced quantum noise in the combined two-mode quadrature phase amplitude

$$X_\theta = a_2 e^{-i\theta} + a_3^\dagger e^{i\theta} \tag{2}$$

The relevant two-mode noise quantity, which is measurable by homodyne detection measurements schemes,[17] is given by

$$V(X_\theta) = \tfrac{1}{2}(\langle X_\theta X_\theta^+ \rangle + \langle X_\theta X_\theta^+ \rangle) - \langle X_\theta \rangle \langle X_\theta^\dagger \rangle \tag{3}$$

A variance $V(X_\theta)$ dropping below that predicted for a coherent state (here we have for a coherent state $V(X_\theta) = 1$) is said to correspond to "squeezing." In this case we have two-mode squeezing, the formalism for which was studied by Caves and Schumaker.[19] Most of the original squeezing experiments detected two-mode squeezing. If we solve for the equations of motion for system (1), we find

$$V(X_\theta) = e^{-r}$$
$$V(X_{\theta + \pi/2}) = e^{r} \tag{4}$$

where $r = E^2 \chi t$. The fluctuations in one quadrature are reduced below that of a coherent state, while the fluctuations in the orthogonal quadrature are increased.

The Hamiltonian (1) is far too simplistic to accurately describe four-wave mixing in most experimental systems. If the pump frequency is far from any atomic resonance in the medium, spontaneous emission is not important and the medium can be characterized by a classical susceptibility χ. This is the case for four-wave mixing in a glass fiber and for many nonlinear crystals. Nevertheless, there are still linear and nonlinear dispersion effects and absorption that must be taken into account. These effects can change the amount of squeezing obtainable and the phase angle for optimal squeezing. If the pump is tuned close to an atomic resonance, then spontaneous emission begins to play an important role and must be considered. In this case one requires a quantum model for the atomic medium. In this chapter we present such a treatment.

We will model the medium as N two-level atoms, placed inside a cavity that is used to enhance the interaction time between the field and medium. Our model will describe nondegenerate four-wave mixing where the signal and idler fields are shifted in frequency by $\pm \varepsilon$ from the pump frequency. In degenerate four-wave mixing, in which both pump and signal/idler fields are at the same frequency, the amount of squeezing possible is severely limited for situations where the Q of the cavity is such that the cavity lifetime is much greater than the atomic relaxation rates. The limitation is due to dephasing effects arising from spontaneous emission, and will be calculated from our model. The implication is that for good squeezing to be obtained in degenerate four-wave mixing in a high-Q cavity, one needs to use relatively high intensities and cavity cooperativity values. To obtain good squeezing for low intensities and cooperativities, it is preferable to use either nondegenerate four-wave mixing or a situation where the cavity relaxation rate is of the order of or less than that of the atoms.

We indicate intuitively why the nondegenerate four-wave mixing scheme can help in reducing spontaneous emission while generating good squeezing. Consider the spontaneous-emission spectrum of a two-level atom pumped by a intense, detuned laser field. The spectrum has an elastic peak at the pump frequency and two low-intensity inelastic peaks symmetrically displaced from the pump frequency by an amount equal to the atomic detuning. At higher intensities, approaching saturation, the inelastic part of the spectrum becomes three peaked, with two side peaks at the generalized Rabi frequencies and a central inelastic peak at the pump frequency. The central peak becomes proportionately larger at increasing intensities. It consists of randomly phased radiation scattered back at the pump frequency, which degrades squeezing for degenerate four-wave mixing, where the squeezing is measured as a weak field at the pump frequency. Squeezing is degraded even if this spontaneous emission is not observed directly in the cavity mode, since random dephasing of the induced polarization leads to extra noise in the squeezing mode. The spectral width of the spontaneous-emission peak is of the order of atomic linewidth, $\gamma_\perp$. Intuitively, we would expect to be able to detune the weak squeezed fields from the central pump frequency by more than several atomic linewidths to avoid this extra spontaneous-emission noise and hence obtain better squeezing at higher pump intensities. In a cavity where the cavity relaxation rate κ is much shorter than the atomic rates $\gamma_\perp$ the emitted cavity spectral linewidth will fall well within the atomic envelope which is of the order of $\gamma_\perp$. We can in this case achieve the necessary nondegenerate four-wave mixing only by using separate cavity modes for the signal and idler. For the situations where κ is much greater than or of the order of $\gamma_\perp$, we can achieve the necessary frequency shift ε using a single-cavity mode.

Nondegenerate four-wave mixing in the two-level atomic medium was first studied classically by Fu and Sargent[20] and Boyd et al.[21] We describe a quantum theory of nondegenerate four-wave mixing that adapts the techniques used by Haken[22] for laser theory and Drummond and Walls[23] for optical bistability. An alternative quantum theory of nondegenerate four-wave mixing was presented by Sargent et al.[15] who adapted techniques used by Scully and Lamb[24] in laser theory. The predictions of Holm and Sargent[15] are in agreement with our results. More recently Ho et al.[6] and Brambilla et al.[16] have presented a theory of quantum four-wave mixing, and Courty et al.[17] have developed a theory treating three-level atomic media.

We consider the following model of nondegenerate four-wave mixing in an optical cavity. The medium is modeled as N two-level atoms with resonance frequency ω_0, which are described by the pseudo spin operators σ_i, $\sigma_i^\dagger$, and σ_i^z. We consider three cavity modes with frequencies ω_1, $\omega_2 = \omega_1 + \varepsilon$, and $\omega_3 = \omega_1 - \varepsilon$ that are described by operators a_1, a_2, and a_3. The cavity detuning $\omega_1 - \omega_p$ is much smaller than the separation in frequency ε between adjacent cavity modes so that only mode a_1 is effectively pumped.

The Hamiltonian may be written in the electric-dipole and rotating-wave approximations as

$$H = \sum_{\mu=0}^{4} H_\mu$$

$$H_0 = \sum_{j=1}^{3} \hbar\omega_j a_j^\dagger a_j + \sum_{i=1}^{N} \hbar\omega_0 \sigma_{zi}$$

$$H_1 = i\hbar g \sum_{i=1}^{N} \left\{ \sigma_i \left[a_1^\dagger \exp(-ik_1 r_i) + a_2^\dagger \exp(-ik_2 r_i) \right.\right.$$
$$\left.\left. + a_3^\dagger \exp(-ik_3 r_i) \right] + \text{h.c.} \right\} \tag{5}$$

$$H_2 = i\hbar \left[a_1^\dagger E \exp(-i\omega_p t) - a_1 E^* \exp(i\omega_p t) \right]$$

$$H_3 = \sum_{i=1}^{N} \left(\sigma_i^\dagger \Gamma_i + \sigma_i \Gamma_i^\dagger \right)$$

$$H_4 = \sum_{j=1}^{3} \left(a_j \Gamma_c^\dagger + a_j^\dagger \Gamma_c \right)$$

where g is the dipole coupling strength. The atomic reservoir Γ describes the energy loss from the atoms through spontaneous emission. The loss of energy of the field cavity modes because of dissipation through the cavity mirrors is described by the field reservoir Γ_c. We assume that the modes are independently coupled to the external environment (Γ_c) and take the cavity-decay rates to be equal to a value κ.

A master equation for the density operator of the coupled atom–field system may be derived by using standard techniques. This may be converted to a c-number equation of the Fokker-Planck form, using techniques developed by Haken for laser theory. A scaling argument in powers of N (the total number of atoms interacting with the cavity mode, assumed large) allows one to ignore all but the first- and second-order derivatives. The standard representation, as used in laser theory, does not in general provide a Fokker-Planck equation with a positive-definite diffusion matrix. Thus, it is necessary to use a generalized P-representation[25] for which the resulting Fokker-Planck equation has a positive semidefinite diffusion matrix, and one can apply Ito rules to write the equivalent stochastic differential equations. A full derivation of these results is given in Ref. 23. The resulting stochastic differential equations for the variables representing the field amplitude α, atomic polarization ν, and atomic inversion D are

$$\dot{\alpha} = E\exp(-i\omega_p t) - (\kappa + i\omega_1)\alpha + g\nu + \Gamma_\alpha$$
$$\dot{\nu} = -(\gamma_\perp + i\omega_0)\nu + g\alpha D + \Gamma_\nu \tag{6}$$
$$\dot{D} = -\gamma_\parallel(D + N) - 2g(\nu^\dagger \alpha + \nu \alpha^\dagger) + \Gamma_D$$

where α, $\alpha^\dagger$, ν and $\nu^\dagger$ are independent complex variables in the generalized P-representation. The equations for α^+, ν^+, etc. (the c.c. equations) are obtained from those of α, ν, etc. by replacing α, ν, Γ_α, and Γ_ν with $\alpha^\dagger$, $\nu^\dagger \Gamma_{\alpha+}$, and $\Gamma_{\nu\dagger}$ and taking the complex conjugate. The terms Γ are Gaussian noise functions with zero mean, reflecting the quantum noise present. The nonzero correlations of the quantum noise terms are

$$\langle \Gamma_\nu(t)\Gamma_\nu(t')\rangle = 2g\alpha\nu\delta(t - t')$$
$$\langle \Gamma_D(t)\Gamma_D(t')\rangle = [2\gamma_\parallel(D + N) - \kappa g(\nu^\dagger\alpha + \nu\alpha^\dagger)]\delta(t - t') \tag{7}$$

where $\gamma_\perp$ and $\gamma_\parallel$ are the transverse and longitudinal relaxation rates of the two-level atom, respectively, and we will assume radiative damping where $\gamma_\perp = \gamma_\parallel/2$. In the nondegenerate situation described by the Hamiltonian (Eq. (5)), the polarization ν oscillates at an infinite number of frequencies. The problem can be simplified somewhat if we recognize that we are interested in the gain of the weak-field modes a_2 and a_3 in the presence of the strong central pump mode a_1. In this limit the strong pump mode a_1 is treated correctly to all orders, describing the saturation of the medium, whereas the expressions for the weak fields a_2 and a_3 are kept to first order only. Then the polarization oscillates at three dominant frequencies ω_p, $\omega_p \pm \varepsilon$. Thus, we expand the field amplitude, polarization, and inversion into Fourier components and write

$$
\begin{aligned}
\alpha &= \alpha_1 \exp(-i\omega_p t) + \alpha_2 \exp\left[-i(\omega_p - \varepsilon)t\right] + \alpha_3 \exp\left[-i(\omega_p + \varepsilon)t\right] \\
\nu &= \nu_1 \exp(-i\omega_p t) + \nu_2 \exp\left[-i(\omega_p - \varepsilon)t\right] + \nu_3 \exp\left[-i(\omega_p + \varepsilon)t\right] \\
D &= D_1 + D_2 \exp(-i\varepsilon t) + D_2^\dagger \exp(i\varepsilon t)
\end{aligned}
\tag{8}
$$

The c-number stochastic differential equations for α_i, ν_i, and D_i are obtained by substituting Eqs. (8) into Eqs. (6) and equating terms of the same frequency, retaining, to first order only, terms in α_2, α_3, ν_2, ν_3, and D_2.

The final nondegenerate c-number equations in the above approximations are (in the rotating frame defined by (6))

$$
\begin{aligned}
\dot{\alpha}_1 &= E - \kappa(1 + i\phi)\alpha_1 + g\nu_1 \\
\dot{\alpha}_2 &= -\kappa(1 + i\phi)\alpha_2 + g\nu_2 \\
\dot{\alpha}_3 &= -\kappa(1 + i\phi)\alpha_3 + g\nu_3 \\
\dot{\nu}_1 &= -\gamma_1\nu_1 + g\alpha_1 D_1 + \Gamma_{\nu 1} \\
\dot{\nu}_2 &= -\gamma_2\nu_2 + g\alpha_2 D_1 + g\alpha_1 D_2^\dagger + \Gamma_{\nu_2} \\
\dot{\nu}_3 &= -\gamma_3\nu_3 + g\alpha_1 D_2 + g\alpha_3 D_1 + \Gamma_{\nu_3} \\
\dot{D}_1 &= -\gamma_\parallel(D_1 + N) - 2g\left(\alpha_1\nu_1^\dagger + \alpha_1^\dagger\nu_1\right) + \Gamma_{D_1} \\
\dot{D}_2 &= -\gamma D_2 - 2g\left(\alpha_1\nu_2^\dagger + \alpha_2^\dagger\nu_1 + \alpha_1^\dagger\nu_3 + \alpha_3\nu_1^\dagger\right)
\end{aligned}
\tag{9}
$$

and c.c. equations. We define

$$
\phi = \frac{\omega_1 - \omega_p}{\kappa} \qquad \gamma_j = \gamma_\perp(1 + i\Delta_j) \qquad \gamma = \gamma_\parallel(1 - i\delta)
$$

$$
\Delta_1 = \frac{\omega_0 - \omega_p}{\gamma_\perp} \qquad \Delta_2 = \Delta_1 + 2\delta f \qquad \Delta_3 = \Delta_1 - 2\delta f \tag{10}
$$

$$
\delta = \frac{\varepsilon}{\gamma_\parallel} \qquad f = \frac{\gamma_\parallel}{2\gamma_\perp}
$$

Here

$$
\left\langle \Gamma_{\nu_2}(t)\Gamma_{\nu_3}(t') \right\rangle = \left\langle \Gamma_{\nu_1}(t)\Gamma_{\nu_1}(t') \right\rangle = 2g\alpha_1\nu_1\delta(t-t')
$$

$$
\left\langle \Gamma_{\nu_2^\dagger}(t)\Gamma_{\nu_3^\dagger}(t') \right\rangle = \left\langle \Gamma_{\nu_1^\dagger}(t)\Gamma_{\nu_1^\dagger}(t') \right\rangle = 2g\alpha_1^\dagger\nu_1^\dagger\delta(t-t')
$$

$$
\left\langle \Gamma_{D_2}(t)\Gamma_{D_2}(t') \right\rangle = \left\langle \Gamma_{D_2^\dagger}(t)\Gamma_{D_2^\dagger}(t') \right\rangle = \left\langle \Gamma_{D_1}(t)\Gamma_{D_1}(t') \right\rangle
$$

$$
= \left[2\gamma_\parallel(D_1 + N) - 4g\left(\nu_1^\dagger\alpha_1 + \nu_1\alpha_1^\dagger\right) \right]\delta(t-t') \tag{11}
$$

At this point we simplify the equations by considering the two special cases discussed above. First, we consider nondegenerate four-wave mixing occurring inside a high-Q cavity where $\gamma_\perp, \gamma_\parallel \gg \kappa$, and the mixing occurs between the distinct cavity sideband modes represented by α_2 and α_3. Second, we consider single-mode operation (so that only the α_1, ν_1, D_1 equations are considered) but where the relative ratio of κ to $\gamma_\perp$ is kept arbitrary. To obtain the equations in the first limit, we set $\dot{\nu}_j = \dot{D}_1 = \dot{D}_2 = 0$ and solve for the atomic variables in terms of the field variables. The final equations are

$$
\dot{\alpha}_1 = E - \kappa(1 + i\phi)\alpha_1 - \frac{2C\kappa\alpha_1}{(1 + i\Delta_1)\Pi(0)} + F_{\alpha_1}(t)
$$

$$
\dot{\alpha}_2 = -\kappa\gamma(\delta)\alpha_2 + \kappa b(\delta)\alpha_3^\dagger e^{2i\theta_0} + F_{\alpha_2}(t) \tag{12}
$$

$$
\dot{\alpha}_3 = -\kappa\gamma(-\delta)\alpha_3 + \kappa b(-\delta)\alpha_2^\dagger e^{2i\theta_0} + F_{\alpha_3}(t)
$$

and the corresponding c.c. equations. We define $\gamma(\delta) = 1 + i\phi + \gamma_R(\delta) + i\gamma_I(\delta)$, $\chi(\delta) = \chi_R(\delta) + i\chi_I(\delta)$, $b(\delta) = b_R(\delta) + ib_I(\delta)$ and $\Pi(0) = 1 + I/(1 + \Delta_1^2)$. $F_i(t)$ are fluctuating noises with $\langle F_i(t) \rangle = 0$ and the nonzero noise correlations for the sidebands are

$$
\left\langle F_{\alpha_2}(t)F_{\alpha_3}(t') \right\rangle = \kappa d\, e^{2i\theta_0}\delta(t-t') \qquad d = d_R + id_I
$$

$$
\left\langle F_{\alpha_2^\dagger}(t)F_{\alpha_3^\dagger}(t') \right\rangle = \kappa d^*\, e^{-2i\theta_0}\delta(t-t') \tag{13}
$$

$$
\left\langle F_{\alpha_2}(t)F_{\alpha_2^\dagger}(t') \right\rangle = \left\langle F_{\alpha_3}(t)F_{\alpha_3^\dagger}(t') \right\rangle = \kappa\Lambda\delta(t-t')
$$

We have defined $C = g^2N/2\gamma_\perp\kappa$, the cooperativity parameter of the cavity; $I = |\alpha_1^{ss}|2/n_0$ the steady-state intracavity pump intensity in units of n_0 the resonant saturation intensity, where $n_0 = \gamma_\parallel\gamma_\perp/4g^2$; and $\alpha_1^{ss} = |\alpha_1^{ss}|e^{i\theta_0}$ is the steady-state semiclassical solution of the equation (12) for the cavity pump mode. The solutions are

$$\gamma_R(\delta) = \frac{2C\{1 + I[a - c + \Delta_2(d - b)] + I^2[bd - ac + \Delta_2(ad + cb)]\}}{(1 + \Delta_2^2)\Pi(0)|\Pi(\delta)|^2}$$

$$\gamma_I(\delta) = \frac{2C\{-\Delta_2 + I[d - b - \Delta_2(a - c)] + I^2[ad + bc - \Delta_2(bd - ac)]\}}{(1 + \Delta_2^2)\Pi(0)|\Pi(\delta)|^2}$$

$$b_R(\delta) = \frac{2CI\{e - \delta fq + I[ae + bq + \delta f(be - aq)]\}}{\Pi(0)|\Pi(\delta)|^2(1 + \delta^2)(1 + \Delta_1^2)(1 + \Delta_2^2)(1 + \Delta_3^2)}$$

$$b_I(\delta) = \frac{2CI\{q + \delta_{fe} + I[aq - be + \delta f(be - aq)]\}}{\Pi(0)|\Pi(\delta)|^2(1 + \delta^2)(1 + \Delta_1^2)(1 + \Delta_2^2)(1 + \Delta_3^2)}$$

$$\Lambda = \frac{2CI^2\{1 + \Delta_3^2 + f - f\Delta_1\Delta_3 + \delta(\Delta_1 + \Delta_3) + If/2\}}{\Pi(0)|\Pi(\delta)|^2(1 + \delta^2)(1 + \Delta_1^2)(1 + \Delta_2^2)(1 + \Delta_3^2)}$$

$$d_R = \frac{-2CI[f(1 + \delta^2)r + IA(r, s) + I^2fB(r, s)]}{\Pi(0)|\Pi(\delta)|^2(1 + \delta^2)(1 + \Delta_1^2)(1 + \Delta_2^2)(1 + \Delta_3^2)}$$

$$d_I = \frac{-2CI\{-f(1 + \delta^2)s + I[A(-s, r) + P_1] + I^2f[B(-s, r) + P_2]\}}{\Pi(0)|\Pi(\delta)|^2(1 + \delta^2)(1 + \Delta_1^2)(1 + \Delta_2^2)(1 + \Delta_3^2)}$$

$$P_1 = \Delta_2 + \Delta_3 - \Delta_2\Delta_3 + 1 \qquad P_2 = \frac{\Delta_1 - 1}{4}$$

$$a = \frac{2 + \Delta_3^2 + \Delta_2^2 - \delta\Delta_2(1 + \Delta_3^2) + \delta\Delta_3(1 + \Delta_2^2)}{2(1 + \delta^2)(1 + \Delta_2^2)(1 + \Delta_3^2)}$$

$$b = \frac{-\delta(2 + \Delta_3^2 + \Delta_2^2) - \Delta_2(1 + \Delta_3^2) + \Delta_3(1 + \Delta_2^2)}{2(1 + \delta^2)(1 + \Delta_2^2)(1 + \Delta_3^2)} \qquad (14)$$

$$c = \frac{(\Delta_3 - \Delta_1)(\Delta_3 + \Delta_1 - \delta f + \delta\Delta_3\Delta_1)}{2(1 + \delta^2)(1 + \Delta_3^2)(1 + \Delta_1^2)}$$

$$d = \frac{(\Delta_3 - \Delta_1)(1 - \Delta_3\Delta_1 + \delta(\Delta_3 + \Delta_1))}{2(1 + \delta^2)(1 + \Delta_3^2)(1 + \Delta_1^2)}$$

$$e = 1 + \Delta_3\Delta_1 + \Delta_2\Delta_3 - \Delta_2\Delta_1 + \delta(\Delta_3 - \Delta_1 - \Delta_2 - \Delta_1\Delta_2\Delta_3)$$

$$q = \Delta_3 - \Delta_1 - \Delta_2 - \Delta_1\Delta_2\Delta_3 - \delta(1 + \Delta_3\Delta_1 + \Delta_2\Delta_3 - \Delta_2\Delta_1)$$

$$\Pi(\delta) = \Pi_R + i\Pi_I \qquad \Pi_R = 1 + aI \qquad \Pi_I = bI$$

$$A(r, s) = f(1 + \delta^2)2ar - \frac{f(rg + ms)}{\left(1 + \Delta_2^2\right)\left(1 + \Delta_3^2\right)} - 1 + \Delta_2\Delta_3$$

$$B(r, s) = \frac{1}{4} + (1 + \delta^2)(a^2 + b^2)r$$

$$- \frac{rag + rbh + ams + bns - r(1 - \Delta_2\Delta_3)/4 + s(\Delta_2 + \Delta_3)/4}{\left(1 + \Delta_2^2\right)\left(1 + \Delta_3^2\right)}$$

$$r = 1 - \Delta_2\Delta_3 - \Delta_1\Delta_2 - \Delta_1\Delta_3 \qquad s = \Delta_2 + \Delta_3 + \Delta_1 - \Delta_1\Delta_2\Delta_3$$

$$2g = \left(1 + \Delta_3^2\right)(1 - \Delta_2\delta) + (1 + \Delta_3\delta)\left(1 + \Delta_2^2\right)$$

$$2h = -\left(1 + \Delta_3^2\right)(\Delta_2 + \delta) + \left(1 + \Delta_2^2\right)(\Delta_3 - \delta)$$

$$2m = -\left(1 + \Delta_3^2\right)(\Delta_2 + \delta) - \left(1 + \Delta_2^2\right)(\Delta_3 - \delta)$$

$$2n = -\left(1 + \Delta_3^2\right)(1 - \Delta_2\delta) + \left(1 + \Delta_2^2\right)(1 + \Delta_3\delta)$$

Solutions to these linear equations are discussed in the Sections III and IV.

The second case of interest to us is the single-mode cavity of arbitrary κ/γ ratio. The equations for α_1, ν_1, D_1 are given by Eqs. (9)–(12). We are interested in the limit where the quantum fluctuation terms are small compared to the semiclassical deterministic terms. In this limit of small fluctuations one may solve for the transmitted spectrum by linearization of the equations about a stable semiclassical steady-state solution. Ignoring quantum fluctuations altogether in the first instance, one may obtain the steady-state semiclassical deterministic solutions a_0, ν_0, D_0 ($\alpha_0^\dagger = \alpha_0$, $\nu_0^\dagger = \nu_0$)

$$\nu_0 = \frac{g\alpha_0 D_0}{\gamma_\perp(1 + i\Delta)} \qquad D_0 = \frac{-N}{1 + I/(1 + \Delta^2)}$$

$$Y = I\left[\left(1 + \frac{2C}{1 + \Delta^2 + I}\right)^2 + \left(\phi - \frac{2C\Delta}{1 + \Delta^2 + I}\right)^2\right]^{1/2} \tag{15}$$

Here $2C = g^2(N/\gamma_\perp k)$ is the cavity cooperativity parameter, $I = |\alpha_0|^2/n_0$, $Y = |E|^2/\kappa^2 n_0$, and $n_0 = \gamma_\parallel\gamma_\perp/4g^2$ is the saturation intensity on resonance. The steady-state solution for the field is the optical bistability state equation in the mean-field theory approximation and has been well studied in previous works.

Solutions are obtained in the limit of small fluctuations by linearizing Eq. (9) about the steady-state solutions α_0, ν_0, D_0 of Eq. (15). Writing

$$\alpha = \alpha_0 + \delta\alpha \qquad \nu = \nu_0 + \delta\nu \qquad D = D_0 + \delta D \qquad (16)$$

we obtain the following equation, written in convenient matrix form, describing to first order the fluctuation in the field and atomic variables:

$$\frac{d\delta\boldsymbol{\alpha}}{dt} = -\mathbf{A}\delta\boldsymbol{\alpha} + \mathbf{B}\varepsilon(t) \qquad (17)$$

where

$$\delta\boldsymbol{\alpha} = \begin{pmatrix} \delta\alpha \\ \delta\alpha^\dagger \\ \delta\nu \\ \delta\nu^\dagger \\ \delta D \end{pmatrix}$$

$$\mathbf{A} = \begin{pmatrix} \kappa(1+i\phi) & 0 & -g & 0 & 0 \\ 0 & \kappa(1-i\phi) & 0 & -g & 0 \\ -gD_0 & 0 & \gamma_\perp(1+i\Delta) & 0 & -g\alpha_0 \\ 0 & -gD_0 & 0 & \gamma_\perp(1-i\Delta) & -g\alpha_0^\dagger \\ 2g\nu_0 & 2g\nu_0 & 2g\alpha_0^\dagger & 2g\alpha_0 & \gamma_\parallel \end{pmatrix}$$

$$F(t) = \mathbf{B}\varepsilon(t) = \begin{pmatrix} \Gamma_\alpha(t) \\ \Gamma_{\alpha^\dagger}(t) \\ \Gamma_\nu(t) \\ \Gamma_{\nu^\dagger}(t) \\ \Gamma_D(t) \end{pmatrix}$$

Here $\varepsilon(t)$ is a δ-correlated noise vector with zero mean and such that

$$\langle \varepsilon(t)\varepsilon^T(t') \rangle = \mathbf{I}\delta(t-t') \qquad (18)$$

where $\mathbf{I}$ is the identity matrix. The noise correlations are written

$$\langle F(t)F^{\mathrm{T}}(t')\rangle = \mathbf{B}\langle \varepsilon(t)\varepsilon^{\mathrm{T}}(t')\rangle\mathbf{B}^{\mathrm{T}} = \mathbf{D}\delta(t-t')$$

$$\mathbf{D} = \begin{pmatrix} 0 & d_{\alpha\alpha^{\dagger}} & 0 & 0 & 0 \\ d_{\alpha\alpha^{\dagger}} & 0 & 0 & 0 & 0 \\ 0 & 0 & d_{\nu\nu} & 0 & 0 \\ 0 & 0 & 0 & d_{\nu\nu}^{*} & 0 \\ 0 & 0 & 0 & 0 & d_{DD} \end{pmatrix} = \mathbf{B}\mathbf{B}^{\mathrm{T}} \qquad (19)$$

$$d_{\nu\nu} = 2g\alpha_0\nu_0$$

$$d_{DD} = 2\gamma_{\parallel}(D_0 + N) - 4g(\nu_0^*\alpha_0 + \nu_0\alpha_0^*)$$

$$d_{\alpha\alpha^{\dagger}} = 2\kappa n_{\mathrm{th}}$$

The diffusion array $\mathbf{D}$ is simply that of (9), but with α, ν, and D assuming their steady-state values (15). The procedure of assuming small fluctuations about a steady-state deterministic solution is consistent only if the deterministic solution ν_0, α_0, D_0 is stable. This is the case only if the eigenvalues of the matrix $\mathbf{A}$ have positive real parts.

Explicit analytical solutions for the spectral components of the field transmitted external to the cavity are readily derivable from these equations. The solutions will be expressible in terms of the coefficients, γ, χ, d, and Λ defined above. These solutions are discussed in Section IV.

III. EXTERNAL SQUEEZING SPECTRUM: MULTIMODE NONDEGENERATE FOUR-WAVE MIXING

The experimentally measured quantity is the squeezing spectrum of the transmitted field. Unlike the cavity field which has discrete well-separated frequencies, the external field is continuous and multimode, comprised of traveling waves of different frequencies. Squeezing may be observed using a homodyne-detection scheme in which the output sidebands ($a_{2\,\mathrm{out}}$, $a_{3\,\mathrm{out}}$) say, at frequencies $\omega_{\mathrm{p}} \pm \varepsilon$ respectively, beat with a local oscillator $E_{\mathrm{LO}} = E\exp(i\phi)$, phase shifted by ϕ with respect to the external driving field at frequency ω_{p}. The sidebands and local oscillator beat on the surface of a photodetector, forming a photocurrent $i(\varepsilon)$. A spectrum analyzer permits measurement of the fluctuations in the current $i(\varepsilon)$ at frequency ε. The spectral fluctuations $\langle i^2(\varepsilon)\rangle - \langle i(\varepsilon)\rangle^2$ are proportional to the variance

$V(X_\phi, \delta)$ in the quadrature phase

$$X_\phi = a_{2\text{out}} \exp(-i\phi) + a_{3\text{out}}^\dagger \exp(i\phi) \tag{20}$$

and the variance is defined by Eq. (3). The above spectral variance of the fluctuations in the quadratures may be calculated as follows. The stochastic differential equations (Eqs. (12)) describing the cavity modes α_2 and α_3 may be written in the following matrix form:

$$\frac{d}{dt}\boldsymbol{\alpha} = -\mathbf{A}\boldsymbol{\alpha} + \mathbf{D}^{1/2}\mathbf{f}(t) \tag{21}$$

where $\boldsymbol{\alpha} = (\alpha_2, \alpha_2^\dagger \alpha_3, \alpha_3^\dagger)$, $\mathbf{f}(t)$ is a delta-correlated noise force, $\mathbf{A}$ is the drift matrix, and $\mathbf{D}$ is the diffusion matrix whose elements determine the noise that correlates as

$$\langle \Gamma_i(t)\Gamma_j(t') \rangle = D_{ij}\delta(t - t') \tag{22}$$

The steady-state deterministic solutions for the side modes are

$$\alpha_2^{ss} = \alpha_2^{\dagger ss} = \alpha_3^{ss} = \alpha_3^{\dagger ss} = 0 \tag{23}$$

This corresponds to nondegenerate four-wave mixing in a cavity below threshold. To treat the above-threshold regime, it would be necessary to include the depletion of the pump.

We consider a fixed sideband frequency $(\omega_p + \varepsilon)$. The linearized expression for the noise spectrum centered at $(\omega_p + \varepsilon)$ (i.e., a frequency shift ω from $\omega_p + \varepsilon$) may be written as

$$\begin{aligned} S_{ij}(\omega, \delta) &= \int_{-\infty}^{\infty} e^{i\omega t}\langle \alpha_i(t)\alpha_j(0) \rangle \, dt \\ &= \left[(A - i\omega I)^{-1} D (A + i\omega I)^{-1} \right]_{ij} \end{aligned} \tag{24}$$

Here we use the notation $\alpha_1(t) = \alpha_2$, $\alpha_2(t) = \alpha_2^\dagger$, $\alpha_3(t) = \alpha_3$, and $\alpha_4(t) = \alpha_3^\dagger$. In the high-$Q$ cavity $(\kappa \ll \gamma_\perp)$ considered here, the spectrum $S(\omega, \delta)$ as a function of ω generally has a Lorentzian profile, and its width is order κ. Here we will calculate the minimum fluctuations occurring at $\omega = 0$, corresponding previously to the sideband frequencies.

The statistics of the output sideband modes can then be deduced from the boundary conditions at the cavity mirrors. We will use the theory developed by Collett and Gardiner and Collett and Walls[26] to deduce the external squeezing spectrum in the generalized P-representation used here. The squeezing for the optimal phase angle ϕ in the output at the sideband modes is then determined by the spectral variance

$$V(X_\phi, \delta) = 1 + 2\kappa[S_{12}(0, \delta) + S_{34}(0, \delta) - 2|S_{13}(0, \delta)|] \quad (25)$$

where we have taken the optimal situation of a single-port cavity. This gives an expression for the minimum variance (occurring at $\omega = 0$) as one sweeps the side-mode detuning δ. Squeezing is obtained when $V(X_0, \delta) < 1$, and perfect squeezing corresponds to $V(X_\phi, \delta) = 0$. The solutions and results presented here are derived and discussed more thoroughly in the next section. This is because since we consider here only the situation of $\omega = 0$, the solutions take on a simple relationship with those of the external field of the low-Q single-mode cavity where $\kappa \gg \gamma_\perp \gamma_\parallel$. This is shown explicitly in the next section and explicit analytical solutions are given. The discussion of results is deferred therefore to the Section IV. The multimode cavity discussed here does have more structure than the low-Q single-mode cavity, since one can study the dependence on ω. This structure is not discussed in the present work.

IV.   EXTERNAL SQUEEZING SPECTRA: THE SINGLE-MODE CAVITY OF ARBITRARY RATIO $\kappa/\gamma_\perp$

In the single-mode cavity situation, the external field is also composed of traveling waves of different frequencies. The homodyne detection operates similarly to that described in Section III. The measured squeezing spectrum in this single-mode case is often defined as follows:

$$V(X_\theta, \omega) = \int_{-\infty}^{\infty} e^{i\omega t} \langle X_\theta(t + \tau), X_\theta(t) \rangle \, dt \quad (26)$$

where

$$X_\theta(t) = a_{\text{out}}(t) e^{-i\theta} + a_{\text{out}}^\dagger(t) e^{i\theta}$$

is the quadrature phase amplitude of the transmitted (output) field denoted by $a_{\text{out}}(t)$. We use the notation $\langle X, Y \rangle = \langle XY \rangle - \langle X \rangle \langle Y \rangle$.

For stationary or (stable) steady-state solutions the statistics will be independent of t, and we have $\langle a_{\text{out}}(t + \tau), a_{\text{out}}(t) \rangle = \langle a_{\text{out}}(\tau), a_{\text{out}} \rangle$. The relation between the output, input, and internal fields is given by the boundary condition at the cavity mirror, which acts as an input/output port. We take the optimal situation, discussed originally by Yurke,[27] of a single-ended cavity with transmission at only one mirror. The input field

a_{in} in this case is the vacuum, and we assume the commutation relation

$$\left[a_{in}(t), a_{in}^{\dagger}(t') \right] = \delta(t - t') \tag{27}$$

The theory developed by Collett and Gardiner[24] shows the following key results for the transmitted field:

$$\left[a_{out}(t), a_{out}^{\dagger}(t') \right] = \delta(t - t'), \qquad \left\langle a_{out}(\tau), a_{out} \right\rangle = 2\kappa \left\langle \delta\alpha(\tau), \delta\alpha \right\rangle$$

$$\left\langle a_{out}^{\dagger}(\tau), a_{out} \right\rangle = 2\kappa \left\langle \delta\alpha^{\dagger}(\tau), \delta\alpha \right\rangle \tag{28}$$

The output correlations relate directly to the c-number averages defined in the normally ordered P-representation.[25] Thus the squeezing spectrum is

$$V(X_\theta, \omega) = 1 + 2\kappa \int_{-\infty}^{\infty} e^{i\omega\tau} \left[\left\langle \delta\alpha^{\dagger}, \delta\alpha(\tau) \right\rangle + \left\langle \delta\alpha^{\dagger}(\tau), \delta\alpha \right\rangle \right.$$

$$\left. + e^{-2i\theta} \left\langle \delta\alpha(\tau), \delta\alpha \right\rangle + e^{2i\theta} \left\langle \delta\alpha^{\dagger}(\tau), \delta\alpha^{\dagger} \right\rangle \right] d\tau$$

$$= 1 + 2\kappa \left[S_{12}(\omega) + S_{21}(\omega) + e^{-2i\theta} S_{11}(\omega) + e^{2i\theta} S_{22}(\omega) \right] \tag{29}$$

Here we define the spectral matrix $\mathbf{S}(\omega) = [S_{ij}(\omega)]$ as the Fourier transform of the two-time correlation function $\left\langle \delta\boldsymbol{\alpha}(\tau), \delta\boldsymbol{\alpha}^T \right\rangle$, where here $\delta\boldsymbol{\alpha} = (\delta\alpha, \delta\alpha^{\dagger})$.

$$S_{ij}(\omega) = \int_{-\infty}^{\infty} e^{i\omega t} \left\langle \delta\alpha_i(\tau), \delta\alpha_j \right\rangle d\tau \tag{30}$$

The solution for $\mathbf{S}(\omega)$ is readily derived for a stationary solution of the linear process (17). The result is

$$\mathbf{S}(\omega) = (\mathbf{A} - i\omega\mathbf{I})^{-1} \mathbf{D}(\mathbf{A}^T + i\omega\mathbf{I})^{-1} \tag{31}$$

It is informative to discuss the result for the squeezing spectrum in terms of frequency components. The Fourier components are defined as the following Fourier transforms:

$$a_{out}(\omega) = \frac{1}{\sqrt{2\pi}} \int_{-\infty}^{\infty} a_{out}(t) e^{i\omega t} dt \tag{32}$$

We write the Fourier transform of $a_{out}^{\dagger}(t)$ as $a_{out}^{\dagger}(\omega)$, and note that $a_{out}^{\dagger}(\omega) = [a_{out}(-\omega)]^{\dagger}$. We define similarly the Fourier components $a(\omega)$

of the internal field mode $a(t)$ and the Fourier amplitudes $\delta\alpha(\omega)$ and $\delta\alpha^\dagger(\omega)$ of the c-numbers $\delta\alpha(t)$ and $\delta\alpha^\dagger(t)$. The quadrature phase X_θ may be written in terms of its Fourier components as follows:

$$X_\theta(t) = \int_{-\infty}^{\infty} X_\theta(\omega)\, e^{-i\omega t}\, d\omega \tag{33}$$

where $X_\theta(\omega) = a_{\text{out}}(\omega)e^{-i\theta} + a_{\text{out}}^\dagger(\omega)e^{i\theta}$ is the Fourier transform of $X_\theta(t)$. We note that $X_\theta(\omega)$ is not Hermitean. $X_\theta(t)$, however, is, and we may rewrite

$$X_\theta(t) = \int_0^\infty \left\{ X_\theta(\omega)\, e^{-i\omega t} + \left[X_\theta(-\omega) \right]^\dagger e^{i\omega t} \right\} d\omega \tag{34}$$

The correlations of the frequency components for a stationary field are readily deduced (we define $\tau = t - t'$):

$$
\begin{aligned}
\langle a_{\text{out}}(\omega), a_{\text{out}}(\omega') \rangle &= \frac{1}{2\pi} \int_{-\infty}^{\infty}\int_{-\infty}^{\infty} e^{i\omega t}\, e^{i\omega' t'} \langle a_{\text{out}}(t), a_{\text{out}}(t') \rangle \, dt\, dt' \\
&= \delta(\omega + \omega')2\kappa \int_{-\infty}^{\infty} e^{i\omega t}\langle \delta\alpha(\tau), \delta\alpha \rangle\, d\tau \\
&= \delta(\omega + \omega')2\kappa S_{11}(\omega) \\
\langle a_{\text{out}}^\dagger(\omega), a_{\text{out}}(\omega') \rangle &= \delta(\omega + \omega')2\kappa S_{21}(\omega) \\
\langle a_{\text{out}}^\dagger(\omega), a_{\text{out}}^\dagger(\omega') \rangle &= \delta(\omega + \omega')2\kappa S_{22}(\omega) \\
\langle a_{\text{out}}(\omega), a_{\text{out}}^\dagger(\omega') \rangle &= \delta(\omega + \omega') + \langle a_{\text{out}}^\dagger(\omega'), a_{\text{out}}(\omega) \rangle \\
&= \delta(\omega + \omega') + \delta(\omega + \omega')2\kappa S_{21}(\omega)
\end{aligned}
\tag{35}
$$

The commutation relation $[a_{\text{out}}(\omega), a_{\text{out}}^\dagger(\omega')] = \delta(\omega + \omega')$ is derivable from the output field commutation relation (28) (or vice versa). It is straightforward to demonstrate $S_{11}(\omega) = S_{11}(-\omega)$, $S_{22}(\omega) = S_{22}(-\omega)$, and $S_{21}(\omega) = S_{12}(-\omega)$. We note that $a_{\text{out}}(\omega)$ couples only with $a_{\text{out}}(-\omega)$, i.e., the field at $\omega_L - \omega$ couples with the field at $\omega_L + \omega$. The transmitted stationary intensity spectrum is given by

$$\int_{-\infty}^{\infty} e^{i\omega t}\langle a_{\text{out}}^\dagger(\tau)a_{\text{out}} \rangle = 2\kappa|\alpha_0|^2\delta(\omega) + 2\kappa S_{12}(\omega) \tag{36}$$

The element $2\kappa S_{12}$ is the (incoherent) intensity of the transmitted field at the frequency $\omega_L + \omega$, and $2\kappa S_{21}(\omega)$ is the intensity at $\omega_L - \omega$. We find from the results (35),

$$\langle X_\theta(\omega), X_\theta(\omega') \rangle = \langle X_\theta(\omega), X_\theta \rangle \delta(\omega + \omega') \tag{37}$$

where

$$\langle X_\theta(\omega), X_\theta \rangle = 1 + 2\kappa \left[S_{12}(\omega) + S_{21}(\omega) + e^{-2i\theta} S_{11}(\omega) + e^{2i\theta} S_{22}(\omega) \right]$$

It is straightforward to show that Eq. (26) for the squeezing spectrum may now be expressed

$$V(X_\theta, \omega) = \langle X_\theta(\omega), X_\theta \rangle \tag{38}$$

for a stationary state. The result is identical to (29). For a coherent state, $V(X_\theta, \omega) = 1$ and hence squeezing occurs for $V(X_\theta, \omega) < 1$. Perfect squeezing corresponds to $V(X_\theta, \omega) = 0$. One may optimize θ to give the minimum variance.

The solutions (31) and (29) allow us to calculate the external squeezing spectra $V(X_\theta, \omega)$ in the linear approximation. However, a convenient means of obtaining the linearized solution is to consider the equations in frequency space. We define

$$\delta\boldsymbol{\alpha}(\omega) = \frac{1}{\sqrt{2\pi}} \int_{-\infty}^{\infty} \delta\boldsymbol{\alpha}\, e^{i\omega t}\, dt$$

$$F(\omega) = \mathbf{B}\varepsilon(\omega) \tag{39}$$

$$\varepsilon(\omega) = \frac{1}{\sqrt{2\pi}} \int_{-\infty}^{\infty} \varepsilon(t)\, e^{i\omega t}\, dt$$

where

$$\delta\boldsymbol{\alpha}^{\mathrm{T}}(\omega) = \left(\delta\alpha(\omega), \delta\alpha^\dagger(\omega), \delta\nu(\omega), \delta\nu^\dagger(\omega), \delta D(\omega) \right)$$

and

$$\mathbf{F}^{\mathrm{T}}(\omega) = \left(\Gamma_\alpha(\omega), \Gamma_\alpha(\omega), \Gamma_\nu(\omega), \Gamma_{\nu^\dagger}(\omega), \Gamma_D(\omega) \right)$$

Equation (17) in Fourier space becomes, for a stable stationary state

$$0 = (-\mathbf{A} - i\omega\mathbf{I})\delta\boldsymbol{\alpha}(\omega) + \mathbf{F}(\omega) \tag{40}$$

Rewriting explicitly, we have

$$
0 = \begin{pmatrix}
-\kappa(1 + i\phi(\omega)) & 0 & g & 0 & 0 \\
0 & -\kappa(1 - i\phi(-\omega)) & 0 & g & 0 \\
gD_0 & 0 & -\gamma_\perp(1 + i\Delta(\varpi)) & 0 & g\alpha_0 \\
0 & gD_0 & 0 & -\gamma_\perp(1 - i\Delta(\varpi)) & g\alpha_0^* \\
-2g\nu_0^* & -2g\nu_0 & -2g\alpha_0^* & -2g\alpha_0 & -\gamma_\parallel\left(1 - i\dfrac{\omega}{\gamma_\parallel}\right)
\end{pmatrix}
$$

$$
\times \begin{pmatrix}
\delta\alpha(\omega) \\
\delta\alpha^\dagger(\omega) \\
\delta\nu(\omega) \\
\delta\nu^\dagger(\omega) \\
\delta D(\omega)
\end{pmatrix}
+ \begin{pmatrix}
\Gamma_\alpha(\omega) \\
\Gamma_{\alpha^\dagger}(\omega) \\
\Gamma_\nu(\omega) \\
\Gamma_{\nu^\dagger}(\omega) \\
\Gamma_D(\omega)
\end{pmatrix}
\tag{41}
$$

Here we define $\phi(\omega) = \phi - \omega/\kappa$, $\Delta(\varpi) = \Delta - \varpi$ and $\varpi = \omega/\gamma_\perp$. The noise correlations in frequency space are derived from the definitions (39) and using the result (19), in matrix notation

$$
\begin{aligned}
\langle \varepsilon(\omega)\varepsilon^T(\omega')\rangle &= \frac{1}{2\pi}\int_{-\infty}^{\infty}\int_{-\infty}^{\infty} e^{i\omega t} e^{i\omega' t'} \langle \varepsilon(t), \varepsilon^T(t')\rangle \, dt \, dt' \\
&= I\frac{1}{2\pi}\int_{-\infty}^{\infty} e^{i(\omega+\omega')t} \, dt = I\delta(\omega + \omega')
\end{aligned}
\tag{42}
$$

and

$$
\begin{aligned}
\langle F(\omega)F^T(\omega')\rangle &= B\langle \varepsilon(\omega)\varepsilon^T(\omega')\rangle B^T \\
&= BB^T\delta(\omega + \omega') = D\delta(\omega + \omega')
\end{aligned}
$$

Thus, explicitly the nonzero noise correlations in frequency space are

$$
\begin{aligned}
\langle \Gamma_\alpha(\omega)\Gamma_{\alpha^\dagger}(\omega')\rangle &= \delta(\omega + \omega')2\kappa n_{th} \\
\langle \Gamma_\nu(\omega)\Gamma_\nu(\omega')\rangle &= \delta(\omega + \omega')2g\alpha_0\nu_0 \\
\langle \Gamma_{\nu^\dagger}(\omega)\Gamma_{\nu^\dagger}(\omega')\rangle &= \delta(\omega + \omega')2g\alpha_0^*\nu_0^* \\
\langle \Gamma_D(\omega)\Gamma_D(\omega')\rangle &= \delta(\omega + \omega')\left[2\gamma_\parallel(D_0 + N) - 4g(\nu_0^*\alpha_0 + \nu_0\alpha_0^*)\right]
\end{aligned}
\tag{43}
$$

We now solve the nonadiabatically eliminated equations (41) analytically in stepwise fashion. Only the field spectral correlations $(S_{22}(\omega),$

$S_{12}(\omega)$, $S_{21}(\omega)$ given by (30)) are of final interest to us. We therefore consider the last three lines of the matrix equation (41) and eliminate $\delta D(\omega)$, to express the polarization $\delta\nu(\omega)$ in terms of the field alone. We point out that this is the same algebraic procedure as that usually taken where one adiabatically eliminates the atoms ($\gamma_\perp, \gamma_\parallel \gg \kappa$). The latter situation corresponds to $\varpi = 0$. We make this comparison carefully later. The result is

$$
\begin{aligned}
g\delta\nu(\omega) &= -\kappa\gamma(\delta)\delta\alpha(\omega) - \kappa b(\delta)\delta\alpha^\dagger(\omega) + F_\nu(\omega) \\
g\delta\nu^\dagger(\omega) &= -\kappa\gamma^*(-\delta)\delta\alpha^\dagger(\omega) - \kappa b^*(-\delta)\delta\alpha^\dagger(\omega) + F_{\nu^\dagger}(\omega)
\end{aligned}
\tag{44}
$$

where

$$
\begin{aligned}
\gamma(\delta) &= \gamma_R(\delta) + i\gamma_I(\delta) \\
\chi(\delta) &= \chi_R(\delta) + i\chi_I(\delta)
\end{aligned}
$$

according to the definitions given in Eq. (12), but where here we define $\delta = \omega/2\gamma_\perp f$, $\Delta_1 = \Delta = (\omega_0 - \omega_L)/\gamma_\perp$, $\Delta_2 = \Delta_1 + 2\delta f$, and $\Delta_3 = \Delta_1 - 2\delta f$. The stochastic term has nonzero correlations given by

$$
\begin{aligned}
\langle F_\nu(\omega) F_\nu(\omega') \rangle &= \kappa d(\delta)\delta(\omega + \omega') \\
\langle F_{\nu^\dagger}(\omega) F_{\nu^+}(\omega') \rangle &= \kappa d^*(\delta)\delta(\omega + \omega') \\
\langle F_\nu(\omega) F_{\nu^+}(\omega') \rangle &= \kappa \Lambda(\delta)\delta(\omega + \omega')
\end{aligned}
\tag{45}
$$

where $d(\delta) = d_R(\delta) + id_I(\delta)$ and $\Lambda(\delta)$ are defined by Eq. (12). In this paper, we assume pure radiative damping ($f = 1$). We have taken, for convenience, the steady-state solution α_0 to be real. The phase of α_0 relative to the external driving field E is determined by the optical bistability equation (15).

We substitute the solution (44) into the top two lines of (41) to derive the final equations for the field alone:

$$
0 = -\kappa \begin{pmatrix} a(\delta) - \dfrac{i\omega}{\kappa} & b(\delta) \\[2mm] b^*(-\delta) & a^*(-\delta) - \dfrac{i\omega}{\kappa} \end{pmatrix} \begin{pmatrix} \delta\alpha(\omega) \\[2mm] \delta\alpha^\dagger(\omega) \end{pmatrix}
$$

$$
+ \begin{pmatrix} F_\nu(\omega) + - \Gamma_\alpha(\omega) \\[2mm] F_{\nu^\dagger}(\omega) + \Gamma_{\alpha^\dagger}(\omega) \end{pmatrix}
\tag{46}
$$

$$
a(\delta) = 1 + i\phi + \gamma_R(\delta) + i\gamma_I(\delta)
$$

The equations are now of reduced dimension, being defined over a two-dimensional complex phase space. These equations for the field spectrum, without adiabatic elimination, may be rewritten in the following matrix notation:

$$0 = [-\mathbf{A}(\varpi) + i\omega\mathbf{I}]\delta\boldsymbol{\alpha}_R(\omega) + F_R(\omega) \tag{47}$$

where

$$\mathbf{A}(\varpi) = \kappa \begin{pmatrix} a(\delta) & b(\delta) \\ b^*(-\delta)a^*(-\delta) \end{pmatrix} \qquad \delta\mathbf{a}_R(\omega) = \begin{pmatrix} \delta\alpha(\omega) \\ \delta\alpha^\dagger(\omega) \end{pmatrix}$$

$$\mathbf{F}_R(\omega) = \begin{pmatrix} F_\nu(\varpi) + \Gamma_\alpha(\omega) \\ F_{\nu^\dagger}(\varpi) + \Gamma_{\alpha^\dagger}(\omega) \end{pmatrix} = \mathbf{B}(\varpi)\varepsilon(\omega) \tag{48}$$

In this paper we assume negligible thermal noise ($n_{th} = 0$). The noise correlations are determined by the result (45). On comparison with the original equation (40), the reduced equation contains drift and diffusion parts now dependent on frequency, and we have introduced formally the notation $\mathbf{A}(\varpi)$ and $\mathbf{B}(\varpi)$. A frequency-dependent diffusion matrix $\mathbf{D}(\varpi)$ is defined as follows:

$$\langle \mathbf{F}_R(\omega)\mathbf{F}_R^T(\omega') \rangle = \mathbf{B}(\varpi)\langle \varepsilon(\omega)\varepsilon^T(\omega') \rangle \mathbf{B}^T(\varpi')$$
$$= \mathbf{B}(\varpi)\mathbf{B}^T(\varpi')\delta(\omega + \omega') = \mathbf{D}(\varpi)\delta(\omega + \omega') \tag{49}$$

Thus, $\mathbf{D}(\varpi) = \mathbf{B}(\varpi)\mathbf{B}^T(-\varpi)$, and the result (45) gives us

$$\mathbf{D}(\varpi) = \kappa \begin{pmatrix} d(\delta) & \bar{\Gamma}(\delta) \\ \bar{\Gamma}(\delta) & d^*(\delta) \end{pmatrix} \qquad \bar{\Gamma}(\delta) = \Lambda(\delta) + 2\kappa n_{th} \tag{50}$$

The solution for $\delta\alpha_R(\omega)$ is readily expressed in matrix form

$$\delta\boldsymbol{\alpha}_R(\omega) = [\mathbf{A}(\varpi) - i\omega\mathbf{I}]^{-1}\mathbf{F}_R(\omega) \tag{51}$$

Thus, the frequency correlations for the field are

$$\langle \delta\boldsymbol{\alpha}_R(\omega), \delta\boldsymbol{\alpha}_R^T(\omega') \rangle = [\mathbf{A}(\varpi) - i\omega\mathbf{I}]^{-1}\langle \mathbf{F}_R(\omega)\mathbf{F}_R(\omega') \rangle [\mathbf{A}^T(\omega') - i\omega'\mathbf{I}]^{-1}$$
$$= [\mathbf{A}(\varpi) - i\omega\mathbf{I}]^{-1}\mathbf{D}(\varpi)[\mathbf{A}^T(\varpi') - i\omega'\mathbf{I}]^{-1}\delta(\omega - \omega') \tag{52}$$

The explicit matrix solution for the field spectrum defined in Eqs. (38) and (29) is now apparent. We define $\delta\boldsymbol{\alpha}_R^T(t) = (\delta\boldsymbol{\alpha}(t), \delta\boldsymbol{\alpha}^\dagger(t))$ and use the definition (39), the result (52), and the properties of the δ function to show

$$
\begin{aligned}
\mathbf{S}(\omega) &= \int_{-\infty}^{\infty} e^{i\omega\tau} \langle \delta\boldsymbol{\alpha}_R(\tau), \delta\boldsymbol{\alpha}_R^T \rangle \, d\tau \\
&= [\mathbf{A}(\varpi) - i\omega\mathbf{I}]^{-1} \mathbf{D}(\varpi) [\mathbf{A}^T(-\varpi) + i\omega\mathbf{I}]^{-1}
\end{aligned}
\tag{53}
$$

The steps involved in (53), relating the spectrum $\mathbf{S}(\omega)$ directly to the frequency correlations $\langle \delta\boldsymbol{\alpha}(\omega), \delta\boldsymbol{\alpha}^T(\omega') \rangle$, are well known and hold true in general for a stationary field.

The solution (53) for the field is now readily expressed analytically. The solution has the same form as that derived by Drummond and Walls[21] for a high-Q cavity, but with the drift and diffusion matrices $\mathbf{A}$ and $\mathbf{D}$ becoming frequency dependent. The solution has the normalized form

$$
\kappa\mathbf{S}(\omega) = \left(\frac{\mathbf{A}(\varpi)}{\kappa} - \frac{i\omega\mathbf{I}}{\kappa} \right)^{-1} \frac{\mathbf{D}(\varpi)}{\kappa} \left(\frac{\mathbf{A}^T(-\varpi)}{\kappa} + \frac{i\omega\mathbf{I}}{\kappa} \right)^{-1}
\tag{54}
$$

from which the adiabatic limits of high-Q ($\gamma_\perp \gg \kappa$) and low-Q ($\gamma_\perp \ll \kappa$) cavities are readily deduced. The high-Q solution has drift and diffusion matrices at zero frequency ($\varpi = \omega/\gamma_\perp = 0$). The low-$Q$ solution has $i\omega\mathbf{I}/\kappa$ going to zero. The explicit result for the full solution is ($S_{22}(\omega) = S_{11}^*(\omega)$ and $S_{12}(\omega) = S_{21}(-\omega)$)

$$
\kappa S_{11}(\omega) = \frac{1}{|P(-i\omega)|^2} \left\{ d(\delta) \left[a^*(\delta) a^*(-\delta) + - q^2\varpi^2 \right] \right.
\tag{55}
$$

$$
+ d^*(\delta) b(\delta) b(-\delta) - \bar{\Gamma}(\delta) \left[b(\delta) a^*(\delta) + b(-\delta) a^*(-\delta) \right]
$$

$$
\left. - iq\bar{\omega}\bar{\Gamma}(\delta) \left[b(\delta) - b(-\delta) \right] + iq\bar{\omega} d(\delta) \left[a^*(-\delta) - a^*(\delta) \right] \right\}
$$

$$
\kappa S_{11}(\omega) = \frac{1}{|P(-i\omega)|^2} \left\{ -d(\delta) b^*(\delta) a^*(-\delta) - d^*(\delta) b(\delta) a(-\delta) \right.
$$

$$
- iq\bar{\omega}\bar{\Gamma}(\delta) \left[a(-\delta) - a^*(-\delta) \right] + iq\bar{\omega} \left[d(\delta) b^*(\delta) - d^*(\delta) b(\delta) \right]
$$

$$
\left. + \bar{\Gamma}(\delta) \left[q^2\bar{\omega}^2 + |a(-\delta)|^2 + |b(\delta)|^2 \right] \right\}
$$

where

$$P(-i\omega) = [a(\delta) - iq\varpi][a^*(-\delta) - iq\varpi] - b(\delta)b^*(-\delta)$$

$$\delta = \frac{\varpi}{2}$$

$$\varpi = \frac{\omega}{\gamma_\perp} \quad \text{and} \quad q = \frac{\gamma_\perp}{\kappa}.$$

The transmitted intensity and squeezing spectra κS_{12} and $V(X_\theta, \omega)$, respectively, are now readily calculated.

The functions $\gamma_R(\delta), \ldots, \Lambda(\delta)$ describe the response of the atoms to the intracavity field. They have been derived and discussed previously by Fu and Sargent,[20] Boyd et al.,[21] Sargent et al.,[15] and Reid and Walls.[14] The function $\gamma_R(\delta)$ is the atomic absorption profile. One sees (Fig. 1a) a large absorption peak at the atomic resonance ω_0. There is a small gain peak at the frequency $2\omega_L - \omega_0$ due to the scattering process depicted in Fig. 1b. The function $\gamma_I(\delta)$ is the dispersion profile, while $b(\delta)$ is the coupling (four-wave mixing) coefficient between weak fields at frequencies $+\omega$ and $-\omega$. Figure 1c plots $|b(\delta)|$ for large intensities saturating the atoms. There are three resonances, a small central peak at $\omega = 0$ and two large side peaks at the Rabi frequencies $\omega = \pm\gamma_\perp(\Delta^2 + 2I)^{1/2}$ indicating the Stark splitting of the atomic energy levels. The reason for particularly large coupling at the Rabi frequencies is the scattering depicting in Fig. 1d. There is an enhancement of a four-wave mixing process tending to emit a photon pair at frequencies $\omega_L \pm \Omega$, where Ω is the Rabi frequency. We show later that this leads to significant squeezing at frequencies $\omega_L \pm \Omega$ in a single-mode low-Q cavity. The atomic noise function $F_\nu(\omega)$ is the source of quantum fluctuations for the amplitude $\delta a(\omega)$ at ω. The strength of correlation between fluctuation $F_\nu(\omega)$ and $F_\nu(-\omega)$ is given by $d(\delta)$. This phase-sensitive term is responsible for quantum phenomena such as squeezing and is noise generated by the processes that couple $\delta\alpha(\omega)$ and $\delta\alpha(-\omega)$, such as depicted in Fig. 1d. Even if one could "turn off" all processes that couple $\delta\alpha(\omega)$ and $\delta\alpha(-\omega)$, the transmitted incoherent spectrum $2\kappa S_{12}(\omega)$ is not zero. There are photons remitted due to a phase-insensitive fluorescence arising from spontaneous emissions. In a saturated resonant cavity ($\phi = \gamma_R(\delta) = \gamma_I(\delta) = 0$), the transmitted intensity in the hypothetical absence of coupling ($b(\delta) = d(\delta) = 0$) is $2\kappa S_{12}(\omega)$ $= \Lambda(\delta)/[1 + (\omega/\kappa)^2]$. The term $\Lambda(\delta)$ (which determines the correlation $\langle F_\nu(\omega)F_{\nu\dagger}(\omega)\rangle$) may thus be thought of as the phase-sensitive fluorescence source term. It describes the noise generated at ω in the absence of all coupling between $\delta\alpha(\omega)$ and $\delta\alpha(-\omega)$. Figure 1e shows $\Lambda(\delta)$ for a pump

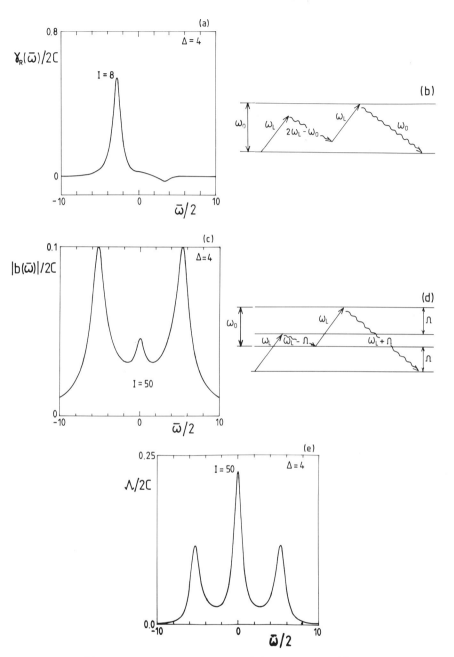

Figure 1. (a) The atomic absorption profile $\gamma_R(\varpi)$. $\Delta = 4$, $I = 8$. (b) Atomic scattering process for low intensities I. (c) The magnitude of the coupling coefficient $|b(\varpi)|$; $\Delta = 4$, $I = 50$. (d) Atomic scattering process at high intensities. (e) The atomic fluorescence term $\Lambda(\varpi)$ for a higher intensity; $\Delta = 4$, $I = 50$, $f = 1$. In (a), (c), and (e), we have plotted versus $\delta = \varpi/2$ and we have taken in this paper $f = 1$, pure radiative damping only.

intensity sufficient to saturate the atoms. The spectrum shows a clear central peak with Rabi side peaks at $\omega = \pm\gamma_\perp(2I + \Delta^2)^{1/2}$. This noise (which tends to detract from squeezing) is more significant at higher intensities and near the pump frequency ($\omega = 0$). The functions $\gamma_R(\delta), \ldots, \Lambda(\delta)$ describe the response of the atoms to the intracavity field and depend on the cavity variables (κ, ϕ, g) only in the sense of scaling I or C. The spectra $\gamma_R(\delta), \ldots$ show atomic resonances like those of usual resonance fluorescence and do not reflect cavity resonances.

The full atom-cavity behavior is given by the final solution (53)–(55). The linearized solution (55) holds only for a stable stationary state, and it is thus necessary to check the stability of the steady-state solution (15). The stability is determined by the eigenvalues of the drift matrix $\mathbf{A}$. The real parts of the eigenvalues must be positive to ensure stability. This criteria may be checked using the Hurwitz criteria or otherwise. The high-Q adiabatic limit ($\kappa \ll \gamma_\perp \gamma_\parallel$) displays instabilities only for regimes where the slope $\partial Y/\partial I$ of the state equation (15) is negative. It is known that regimes of positive-slope instabilities can occur in single-mode bistability for a more general choice of parameters. This is discussed by Lugiato et al.[28]

A. The High-Q Cavity ($\gamma_\perp/\kappa, \gamma_\parallel/\kappa \to \infty$)

There are two important limits of relative decay rates $\gamma_\perp$ and κ. Before discussing the full solution (55), we examine each of these limits. Perhaps the most commonly studied is the high-Q cavity, where the cavity relaxation rate κ of the cavity mode is much smaller than the relaxation rates $\gamma_\perp \gamma_\parallel$ for the atoms: $\kappa \ll \gamma_\perp \gamma_\parallel$. In this case the atomic variables may be adiabatically eliminated. This allows one to set $\delta\dot{\nu} = \delta\dot{\nu} = \delta\dot{D} = 0$ in Eq. (17) and then to eliminate δD to obtain an expression for $\delta\nu$ and $\delta\nu^\dagger$ and hence for the field. We notice that the equations obtained in this high-Q limit are simply the nonadiabatic elimination equations (44) with $\varpi = 0$ ($\delta = 0$).

The solution for the spectrum in this limit of elimination of atomic variables is thus simply (55), but with ω (and therefore δ) put equal to zero in the atomic functions $a(\delta), b(\delta), d(\delta),$ and $\Lambda(\delta)$. We have

$$\mathbf{S}(\omega) = [\mathbf{A}(0) - i\omega\mathbf{I}]^{-1}\mathbf{D}(0)[\mathbf{A}^T(0) + (\omega\mathbf{I})]^{-1} \qquad (56)$$

Thus, the explicit solution (55) has $\varpi = 0$ and $\delta = 0$ but is still a function of $\omega/\kappa = q\varpi$, which is finite. This solution was obtained previously by Drummond and Walls.[21] The result is perhaps not surprising. The photons transmitted from the cavity are detected and the photocurrent spectrum is analyzed. The zero-frequency component of the spectrum corresponds to a long detection counting time (relative to all other time scales of the system). In the adiabatic limit $\gamma_\perp \gamma_\parallel \gg \kappa$, the time a photon is stored in the cavity before being emitted is determined by $(2\kappa)^{-1}$. All times are long compared to the atomic relaxation times, and hence only the zero-frequency component of the atomic profile is seen.

The solution $S_{12}(\omega)$ for the high-Q incoherent intensity spectrum has been derived and discussed previously. Plots are shown of the intensity and squeezing spectrum in Fig. 2. A full discussion of the behavior is given by Reid.[14] To summarize, for intensities corresponding to the lower branch of the optical bistability curve (Fig. 2a), the intensity spectrum shows two symmetric sidepeaks resulting from a coherent inelastic scattering process (Fig. 2b) coupling frequencies ω'_c (the new cavity resonant frequency allowing for the refractive index) and $2\omega_L - \omega'_c$. The squeezing spectrum shows enhancement at these sidebands, the scattering providing a mechanism for a nondegenerate four-wave mixing process occurring between these sideband frequencies. As the intensity I is increased, the nonlinearity of the cavity increases and squeezing improves. Near the onset of the bistability threshold (Fig. 2c), the intensity spectrum $S_{12}(\omega)$ becomessingle-peaked and squeezing is maximum at $\omega = 0$. At higher intensities, the above threshold curve is reached (Fig. 2d), and moving along the upper branch the spectra again become double-peaked. This is the regime of atomic saturation, and the atomic function $\Lambda(0)$ becomes significant, representing an increase in phase-insensitive fluorescence that destroys squeezing. To achieve best squeezing then, one requires intensities I in the low-saturation regime. Yet one needs sufficient nonlinearity $b(0)$. Consequently, there is a window of optimal parameter range for I and $C(\Delta \ll I \ll \Delta^2, \Delta \ll 2C \ll \Delta^2)$. This has been discussed previously.[14]

Thus, the high-Q ($\kappa \ll \gamma_\perp$) limit is one where the effects of spontaneous emission may substantially limit the squeezing available. For the high-Q cavity the atomic profiles (absorption $\gamma_R(\delta)$, coupling $b(\delta), \ldots$) become broadband and show a flat spectrum compared to that of the cavity. The atomic profiles are seen at their zero-frequency ($\omega = 0$) components (since we are in a rotating frame, this corresponds to the pump frequency). As intensities increase toward saturation ($I \rightarrow \Delta^2$), there is enhancement of the phase-insensitive fluorescence $\Lambda(0)$ at zero frequency (corresponding to the center peak of the Stark triplet). While the Stark splitting provides a mechanism for significant enhancement of coupling $b(\delta)$ at the Rabi frequencies (Fig. 1d)), no similar enhancement occurs at zero frequency. Thus, in the high-Q cavity, $\Lambda(0)$ dominates at even moderate intensities and the squeezing reduces. The phase insensitive fluorescence Λ is seen by the cavity as a broadband noise source and

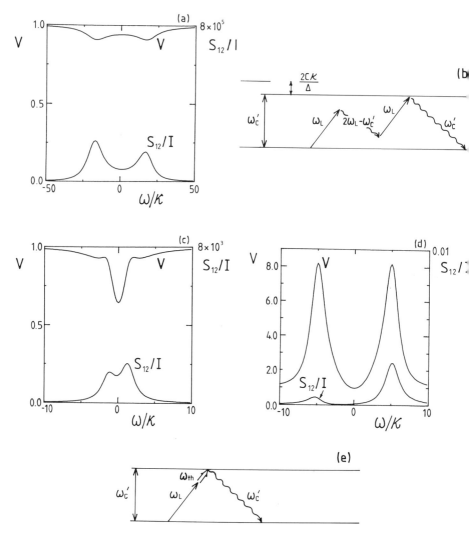

Figure 2. The high-Q cavity limit. (a) The squeezing $[V(X_\theta, \omega)]$ and incoherent normalized intensity $[S_{12}(\omega)/I]$ spectra; $\Delta = 100$, $C = 1100$, $\phi = 18$, $f = 1$, $n_{\mathrm{th}} = 0$, $I = 100$ (corresponding to the lower branch of the optical bistability curve). (b) Inelastic scattering process giving rise to symmetric sidebands at ω'_c and $2\omega_L - \omega'_c$. For high detunings Δ and low intensities, $\omega'_c = \omega_c - 2C\kappa/\Delta$. (c) $V(X_\theta, \omega)$ and $S_{12}(\omega)/I$; $\Delta = 100$, $C = 1100$, $\phi = 18$, $f = 1$, $n_{\mathrm{th}} = 0$, and $I = 2400$ (corresponding to nearer the turning point of the bistability curve). (d) $V(X_\theta, \omega)$ and $S_{12}(\omega)/I$; $\Delta = 100$, $C = 1100$, $\phi = 18$, $f = 1$, $n_{\mathrm{th}} = 0$, and $I = 4000$ (corresponding to the upper branch of the bistability curve). (e) Incoherent scattering process induced by broadband fluorescence (thermal-type) noise $\Lambda(0)$.

is analogous to a thermal noise term. Such noise tends to enhance (Fig. 2e) the transmitted incoherent intensity at the cavity resonance frequency, resulting in an asymmetric intensity spectrum with increased noise at the cavity frequency.

B. The Low-Q Cavity ($\gamma_\perp/\kappa$, $\gamma_\parallel/\kappa \to 0$)

The second limit of relative decay rates $\gamma_\perp$ and κ is the low-Q cavity for which $\gamma_\perp \gamma_\parallel \ll \kappa$. In this case the cavity-mode relaxation rate is much greater than the relaxation rate of the atoms. Hence, the field variables may be adiabatically eliminated. This allows one to set $\delta \dot{a} = \delta \dot{a}^\dagger = 0$ in Eq. (17).

The equations are rewritten in frequency space to obtain Eq. (41) with $\phi(\omega) = \phi(0)$. The solution for the $\delta\nu(\omega)$ in terms of the field $\delta\alpha(\omega)$ therefore takes the form (44) as before. Thus, the final equation for the field in frequency space is

$$
0 = -\kappa \begin{pmatrix} a(\delta) & -b(\delta) \\ b^*(-\delta) & a^*(-\delta) \end{pmatrix} \begin{pmatrix} \delta\alpha(\omega) \\ \delta\alpha^\dagger(\omega) \end{pmatrix}
$$
$$
+ \begin{pmatrix} F_\nu(\omega) + \Gamma_\alpha(\omega) \\ F_{\nu\dagger}(\omega) + \Gamma_{\alpha\dagger}(\omega) \end{pmatrix}
$$

(57)

where

$$
\langle F_\nu(\omega) F_\nu(\omega') \rangle = \kappa d(\delta) \delta(\omega + \omega')
$$
$$
\langle F_{\nu+}(\omega) F_{\nu+}(\omega') \rangle = \kappa d^*(\delta) \delta(\omega + \omega')
$$
$$
\langle F_{\nu+}(\omega) F_\nu(\omega') \rangle = \kappa \Lambda(\delta) \delta(\omega + \omega')
$$

The solution for the spectrum in the low-Q cavity is simply Eq. (55) but with ω/κ (i.e., $q\varpi$) put equal to zero and agrees with the limit $\kappa \to \infty$ of the full solution. The atomic functions $b(\delta)$, $d(\delta)$, and $\Lambda(\delta)$ retain the frequency dependence. We have

$$
\mathbf{S}(\omega) = [\mathbf{A}(\varpi)]^{-1} \mathbf{D}(\varpi) [\mathbf{A}^\mathrm{T}(\varpi)]^{-1}
$$

(58)

At this point we make the connection between the low-Q cavity results here and the zero-frequency results for the multimode nondegenerate four-wave mixing discussed in the last section. In the latter situation we have two cavity sideband modes separated in frequency from the pump by $\pm\varepsilon$. One may derive the transmitted spectrum (assuming stability) as discussed in Section III, and arrive at the result for $\mathbf{S}(0, \varepsilon/\gamma_\parallel)$ and

$V(X_\theta, \varepsilon/\gamma_\|)$, which is found to be identical to that given here for the low-Q cavity, but replacing ε with the ω defined above for the low-Q limit. Thus, $S(0, \varepsilon/\gamma_\|)$ is identical to the $S(\varepsilon)$ defined here and $V(X_\theta, \varepsilon/\gamma_\|)$ becomes $V(X_\theta, \varepsilon)$ defined by the solutions above. This is not surprising, since the intracavity side-mode field at ω has the same atomic profile as the single-mode frequency amplitude $\delta\alpha(\omega)$. The side modes, however, are independent modes (unlike the $\alpha(\omega)$ of the single-mode cavity) and build up in a different manner. In particular, the stability properties of the two cases differ. The cavity side modes generally become unstable at the cavity-mode frequencies ω, which correspond to the sidepeaks (and hence best squeezing) in the low-Q single-mode spectrum. This is because side modes at these frequencies see enhanced gain. Nevertheless, because the zero-frequency squeezing spectrum will be identical to the low-Q cavity result provided one has stability, we do not present a separate study of the results for this system.

The intensity spectrum $S_{12}(\omega)$ in the low-Q cavity limit has been derived and discussed previously. For small intensities $I \ll \Delta^2$, the spectrum is a broadened doublet. Both the linewidth and the splitting of the doublet are modified by the cavity detuning ϕ. Figure 3a plots the intensity and squeezing spectrum corresponding to $I = 100$, the lower branch of the optical bistability curve with $\Delta = 100$, $C = 1100$, and $\phi = 18$. The sidebands are explained by the inelastic scattering process of the type depicted in Fig. 1b, but replacing ω_0 with ω_0', the true atomic resonance incorporating the shift due to the cavity. The scattering depicted here (Fig. 3b) enhances both frequencies ω_0' and $2\omega_L - \omega_0'$, equally and the intensity spectrum is symmetric. The squeezing spectrum shows enhancement of squeezing at the sidebands due to the multiphoton process. As the intensity I increases, the peak situated at $\omega = 0$ appears and dominates as the turning point of the bistability equation is approached. The squeezing at the side bands improves with increasing intensity (Fig. 3c).

Sufficient I corresponds to the stable upper branch. Figure 3d plots the intensity and squeezing spectra corresponding to $I = 4000$, a point on the upper branch. Here there are three peaks. As the bistable region is approached, the sidepeaks move in closer to the central peak. Far enough above the bistable region on the upper branch, the atoms saturate, and the spectrum tends to the usual triplet of one-atom fluorescence. There is a significant central peak at $w = 0$ and two sidepeaks located at the Rabi frequencies

$$\frac{\omega}{\gamma_\perp} = \pm\frac{\Omega}{\gamma_\perp} = \pm\left(\Delta^2 + 2I\right)^{1/2}$$

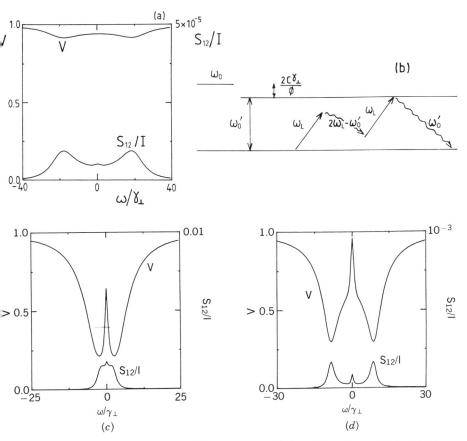

Figure 3. The low-Q cavity limit. (a) The squeezing [$V(X_\theta, \omega)$] and intensity [$S_{12}(\omega)/I$]
spectra; $\Delta = 100$, $C = 1100$, $\phi = 18$, $f = 1$, $n_{th} = 0$, and $I = 100$. (b) Scattering giving rise
to sidebands at ω_0' and $2\omega_L - \omega_0'$. For high detunings ϕ and low intensities, we have
approximately $\omega_0' = \omega_0 - 2C\gamma_\perp/\phi$. (c) $V(X_\theta, \omega)$] and $S_{12}(\omega)/I$; $\Delta = 100$, $C = 1100$, $\phi = 18$,
$f = 1$, $n_{th} = 0$, and $I = 2400$. (d) $V(X_\theta, \omega)$] and $S_{12}(\omega)/I$; $\Delta = 100$, $C = 1100$, $\phi = 18$,
$f = 1$, $n_{th} = 0$, and $I = 4000$. The phase angle θ for optimal squeezing at the sidepeaks is 0.
(e) $V(X_\theta, \omega)$] for $\Delta = 4$, $C = 5$, $\phi = 0$, $f = 1$, $n_{th} = 0$. Significant squeezing is obtainable in
the low-Q cavity limit even with very low atomic detunings Δ and C values. The mechanism
for the enhancement in squeezing is depicted in Fig. 1d. (f) $V(X_\theta, \omega)$] for $\Delta = 8$, $\phi = 0$,
$f = 1$, $n_{th} = 0$, and $I = 64$.

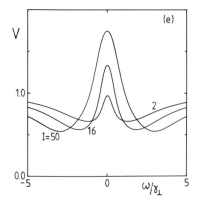
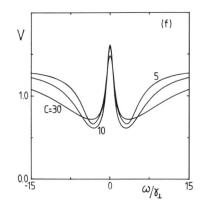

Figure 3. *(Continued)*

At such intensities I saturating the atoms, there is a Stark splitting of the (dressed) energy levels of the two level atom as depicted in Fig. 1d. The resonant inelastic scattering now involves absorption of two laser photons at frequency ω_L and emission of photons at the sideband frequencies $\omega_L \pm \Omega$. The squeezing is enhanced for these sidebands since they are generated via a two-photon process. The dressed atom also has energy levels separated by the laser frequency and thus the presence of the center peak at $\omega = 0$, but we note that this central peak corresponds to increased noise (reduced squeezing). Unlike the sidepeak photons, the photons scattered at the pump frequency ($\omega = 0$) are not scattered as part of a process resulting in the coherent emission of two photons.

The avoidance of the phase-insensitive fluorescence $\Lambda(0)$ in the low-Q cavity and the coupling possible at the sidepeaks for large intensities permits good squeezing to be obtained for a much wider range of cavity parameters, Δ, C, ϕ, and I than is possible in the high-Q case. Good squeezing is possible for low Δ, C, and I values (Figs. 3e and f). We notice from (57) that the particular amplitude $\delta\alpha(\omega)$ will have atomic dispersion $\gamma_I(\varpi)$, fluorescence $\Lambda(\varpi)$. The effective detuning of $\delta\alpha(\omega)$, from the empty cavity, however, is ϕ. The $\gamma_I(\varpi)$ is the change in the resonance frequency cavity due to the atoms, as seen at frequency ω in the absence of coupling $b(\omega)$. In general, the greater nonlinearity (hence squeezing) is achieved for $\phi + \gamma_I(\omega) \sim 0$. At low intensities, we have $\gamma_I \sim (-2C)/\Delta$ for low frequencies and $\phi \sim (2C)/\Delta$ is the optimum cavity detuning. At higher intensities γ_I saturates (more readily than the coupling b), and the better nonlinearity (squeezing) is obtained for smaller cavity detunings, $\phi \to 0$ (Fig. 3e). At the narrow resonance (Fig. 1c) corresponding to the

Rabi frequency, the coupling is large, and good squeezing is obtained at the Rabi side peaks for a range of ϕ. However, the smaller values of ϕ will correspond to broader sidepeaks and hence broader-band squeezing. Figure 3f demonstrates a variation in behavior with C. Lower-C cavities require greater intensities for sufficient coupling, which reduces squeezing at the center frequency, but considerable squeezing is obtainable at the sidepeaks. Higher-C cavities ($2C > \Delta^2$) have large absorption γ_R at low and moderate intensities $I \leq \Delta^2$ and hence reduced squeezing. However, at much higher intensities the absorption saturates more quickly than the coupling and good squeezing is possible at the sidepeaks.

C. The General Cavity

We now discuss the squeezing spectra possible for situations where κ and $\gamma_\perp$ are of the same order. An interesting feature at low intensities is vacuum-field Rabi splitting, which provides a signal-idler coupling mechanism that can enhance squeezing. This possibility was first discussed by Raizen et al.[29] The vacuum-field Rabi splitting that can occur with an atomic medium inside an optical cavity in the absorptive limit was studied by Carmichael.[30]

The dynamics of the general cavity is determined by five eigenvalues which are solved by considering the characteristic polynomial. In the zero-intensity limit, one may obtain some analytical results. Upon examining the absorptive limit for which $\Delta = \phi = 0$, it is apparent that for a range of intermediate q values the eigenvalues develop a nonzero imaginary component, the eigenvalues being

$$\lambda_1 = -\gamma_{11}$$

$$\lambda_{2,3} = \lambda_{4,5} = -\frac{(\gamma_\perp + \kappa)}{2} \pm \frac{i\kappa}{2}\left[8Cq - (1-q)^2\right]^{1/2} \qquad (59)$$

The spectrum exhibits sidepeaks, with positions dependent on C. This is in contrast to the high-Q and low-Q adiabatic elimination spectra which never show sidepeaks in the absorptive limit for intensities below saturation (of course, at saturation intensities the low-Q absorptive spectrum shows the three-peak Stark splitting with sidepeaks at $\omega = \pm\gamma_\perp\sqrt{2I}$). The sidepeaks in this zero-intensity limit arise because of a splitting in the degenerate first-excited energy level of the atom–field system. This vacuum-field Rabi splitting was first discussed by Sanchez-Mondragan et al.,[31] who considered a Rydberg atom interacting with a single mode of a

Figure 4. The dressed states for a single atom inter-
acting with a cavity mode ($\omega_L = \omega_0 = \omega_c$ and $\gamma_\perp = \gamma_\parallel = \kappa = 0$). The splitting of the first and second excited
states are shown.

lossless cavity. The system is modeled by the Hamiltonian

$$H = \hbar g(\sigma a^\dagger + \sigma^\dagger a) + \hbar\omega_0 a^\dagger a + \tfrac{1}{2}\hbar\omega_0 \sigma_z \qquad (60)$$

It is well known that the degenerate eigenstate with energy $\hbar\omega_0(n + 1)$ is
split by the atom–field interaction. The energy splitting is $\pm\hbar\Omega/2$, where
$\Omega = 2g\sqrt{n + 1}$ (Fig. 4). The first excited state is split by $\pm\hbar g$. The
spectrum calculated by Sanchez-Mondragan et al. shows side peaks at the
frequencies $\pm g$. Agarwal[30] showed the effect to be cooperative. With N
atoms the splitting increases to $\pm g\sqrt{N}$. The vacuum-field Rabi splitting is
not observable for the situation of a radiatively damped atom. However,
Carmichael[28] has pointed out that the splitting might be evidenced at
optical frequencies by placing the atoms in a cavity with a relaxation time
comparable to that of the atomic damping. He shows that the system in
the low-intensity limit may be modeled as a pair of damped coupled
harmonic oscillators representing the field and atomic polarization. The
normal modes of vibration in the nondamped situation oscillate at fre-
quencies $\omega_L \pm g\sqrt{N}$. This corresponds to a splitting (Fig. 5a) of $\pm g\sqrt{N}$ in
the degenerate first excited state of the atom–field system. This may be
recognized in Eq. (3) by noting that in the low-intensity limit $D = -N$
and the resulting coupling equations in α and v correspond to the
linearized equations (17) in the limit α_0 and $D_0 = -N$. Theseequations
have eigenvalues, which in the nondissipative limit ($\gamma_\perp \gamma_\parallel = \kappa = 0$) are
$\pm i g\sqrt{N}$. Carmichael then includes damping in his model and shows the
splitting effect to be optimized for $\gamma_\perp = \kappa$, the normal modes being
decoupled and independently damped and the frequencies of oscillation
still $\omega_L \pm g\sqrt{N}$. For the more general case the normal modes are coupled
and the new term $(1 - q)^2$ appearing in (59) will eventually destroy the
vacuum-field Rabi splitting as the ratio $\gamma_\perp = \kappa$ becomes too large or too
small. The vacuum-field splitting has now been experimentally evidenced
both in Rydberg atoms[33] and in the system of optical bistability.[34]

Figure 5 plots the intensity and squeezing spectra in the absorptive
situation, where $\Delta = \phi = 0$, for various q and C values. The vacuum-field
Rabi splitting is clearly apparent in Figs. 5b and c for low-intensities

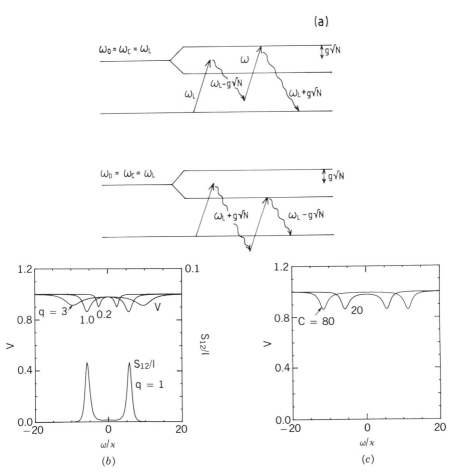

Figure 5. Vacuum-field splitting in the absorptive limit. (a) Scattering processes give rise to sidepeaks in the spectrum at frequencies $\omega \pm g\sqrt{N}$ (for $\gamma_\perp = \kappa$). (b) Vacuum-field Rabi splitting is evident in the squeezing $V(X_\theta, \omega)$ and incoherent intensity spectra $S_{12}(\omega)/I$, for intermediate q values; $\Delta = \phi = 0$, $C = 20$, $I = 2$, $f = 1$, $n_{th} = 0$. (Only the incoherent part $S_{12}(\omega)$ of the intensity spectrum is plotted; the δ-function component at $\omega = 0$ is not displayed.) (c) Vacuum-field Rabi splitting. Squeezing spectrum $V(X_\theta, \omega)$ plotted for $\Delta = \phi = 0$, $I = 2$, $q = 1$, $f = 1$, $n_{th} = 0$, and for various cavity cooperativity C.

regime and $q \sim 1$, in accordance with the eigenvalues. We note increased squeezing at these sidepeaks. The increase occurs because of the scattering process depicted in Fig. 5a where there is enhancement of coupling between frequencies $\omega_L + g\sqrt{N}$ and $\omega_L - g\sqrt{N}$. For larger q values the linewidth is greater than the splitting, the doublet is not resolved, and the vacuum-field splitting is not evident.

Of particular interest to us in this chapter is the behavior of the spectra including nonzero atomic and cavity detunings. The effect of vacuum-field splitting in the dispersive situation was first discussed and experimentally investigated by Raizen et al.[29] and Brecha et al.[34] and the reader is referred to these works. We study in the first instance the situation of equal cavity and atomic frequencies ($\omega_0 = \omega_c$, $\kappa\phi = \gamma_\perp \Delta$, or $q = \phi/\Delta$). For the situation where $8Cq \gg (1 - q)^2$ one has sufficient cavity cooperatively that vacuum-field Rabi splitting may be evidenced. There are now four sidepeaks in the spectrum, at approximately $\omega_L \pm (\omega/\kappa)$ where $(\omega/\kappa) = -\kappa\phi \pm \sqrt{2Cq}$. The relevant scattering processes are depicted in Fig. 6a, which illustrates the particular case where the pump frequency is almost resonant with the dressed energy state: $\omega_L \sim \omega_0 - g\sqrt{N}$ (i.e., $\kappa\phi \sim \kappa\sqrt{2Cq}$ or $\phi \sim 2C/\Delta$). For this choice of ϕ, we have sidepeak pairs well separated in frequency, at $|\omega| > 0$ and $|\omega| - 2\kappa\phi$. The splitting $\pm\kappa\sqrt{2Cq}$ about the positions $\pm\kappa\phi$ is vacuum-field Rabi splitting and is predicted to increase with C. The squeezing spectrum for a situation of this type is depicted in Fig. 6b, for $C = 1100$, $I = 300$, $\Delta = 100$, and $\phi = 18$. The value $q = 0.18$ corresponds to $\omega_0 = \omega_c$, and we see indeed four sidepeaks (vacuum Rabi splitting). The outer sidepeaks are a long way detuned from the pump frequency and show considerably less squeezing than the inner peaks. This might be expected since (Fig. 6a) the scattered photon at $\omega_L - \kappa\phi - g\sqrt{N}$ is a long way from resonance with any of the transitions between the dressed energy levels.

An interesting situation enabling a enhancement of the scattering processes (and hence strong squeezing) is where the cavity is detuned oppositely to the atoms ($\phi/\Delta < 0$, as in Fig. 7) and the laser is tuned to $(\omega_0 + \omega_c)/2$ (i.e., $q = -\phi/\Delta$). With sufficient cooperativity C that $g\sqrt{N} \gg |\gamma_\perp - \kappa|/2$ holds, we have

$$\text{Re } \lambda_{2,3} = \frac{-(\gamma_\perp + \kappa)}{2}$$

$$\text{Im } \lambda_{2,3} = \pm\left(g^2 N + \frac{(\omega_0 - \omega_c)^2}{4}\right) \qquad (61)$$

$$= \pm\kappa(2Cq + \phi^2)^{1/2}$$

The imaginary parts of two pairs of eigenvalues coincide and we see a single pair of sidepeaks at

$$\omega = \pm\left[g^2 N + \frac{(\omega_0 - \omega_c)^2}{4}\right]^{1/2} \qquad (62)$$

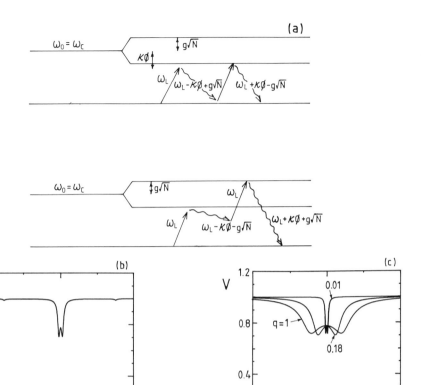

Figure 6. (a) Vacuum-field Rabi splitting for $\omega_0 = \omega_c \neq \omega_L$. We have taken the case $g^2N \gg (\kappa - \gamma_\perp)^2/4$. The two possible scattering processes are depicted, giving rise to two pairs of sidepeaks in the spectrum. The pump is tuned close to resonance with the dressed energy level. (b) Squeezing spectrum $V(X_\theta, \omega)$ for $\Delta = 100$, $C = 1100$, $\phi = 18$, $I = 300$, $f = 1$, $n_{th} = 0$, $q = 0.18$. Two pairs of sidepeaks are apparent, due to the vacuum-field Rabi splitting depicted in (a). (c) $V(X_\theta, \omega)$. $\Delta = 100$, $C = 1100$, $\phi = 18$, $I = 300$, $f = 1$, $n_{th} = 0$. Various q values.

(vacuum-field Rabi splitting). This corresponds to a single doubly resonant scattering as depicted in Fig. 7. The squeezing spectrum is plotted for $\Delta = 8$, $C = 30$, $I = 10$, $q = 0.125$, and $\phi = -1$ in Fig. 7b. For higher q values, we again see two pairs of sidepeaks, split from the position $(2Cq + \phi^2)^{1/2}$ by $\kappa(\phi + q\Delta)$ in agreement with predictions from the eigenvalues. There is reduced squeezing as the resonance is lost. Figure 7c

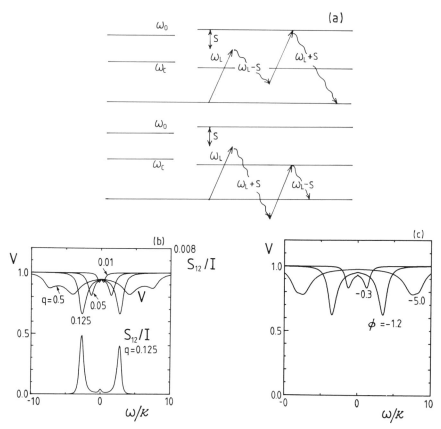

Figure 7(a). Vacuum-field splitting for $\omega_L = (\omega_0 + \omega_c)/2$. The processes depicted reinforce and give rise to spectral sidebands at frequencies $\omega_L \pm S$. (b) Squeezing $V(X_\theta, \omega)$ and incoherent intensity $S_{12}(\omega)$ for the situation depicted in Fig. 7a; $\Delta = 8$, $\phi = -1$, $I = 10$, $C = 30$, $f = 1$, $n_{th} = 0$, $q = 0.125$. Also plotted are spectra for $q = 0.01$, 0.05, 0.5. The optimal phase angle for $q = 0.125$ is $\theta = 127°$, at $\omega/\kappa = 2.7$. (c) $V(X_\theta, \omega)$ for the resonant situation $\omega_L = (\omega_0 + \omega_c)/2$ depicted in (a); $\Delta = 8$, $C = 30$, $I = 10$, $f = 1$, $n_{th} = 0$. Various ϕ are plotted $q = -\phi/\Delta$. (d) $V(X_\theta, \omega)$ for $\Delta = 8$, $C = 60$, $I = 10$, $f = 1$, $n_{th} = 0$, and $q = -\phi/\Delta$. (e) $V(X_\theta, \omega)$ for $\Delta = 8$, $I = 10$, $f = 1$, $n_{th} = 0$, $\phi = -1.2$, $q = -\phi/\Delta$.

shows squeezing spectra for $\Delta = 8$, $C = 30$, $I = 10$, and various cavity detunings ϕ, where one has in each case the situation depicted in Fig. 7a, $q = -\phi/\Delta$ for resonance with both transitions. Squeezing reduces as ϕ becomes comparable to $2Cq$ and as we approach the adiabatic elimination limits. Figure 7d plots the squeezing spectra for a higher value of $C(C = 60)$, showing good squeezing to be obtainable for greater cavity detuning

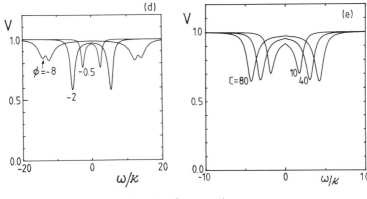

Figure 7. *(Continued)*

ϕ. Figure 7e shows squeezing spectra for a range of values. The squeezing in this case of $\phi/\Delta < 0$ is perhaps only somewhat better than that obtainable if one sets an optimal cavity detuning of the same sign as Δ. The enhancement, however, provides evidence for vacuum-field splitting. Enhancement of squeezing in this vacuum-field splitting regime has been experimentally demonstrated by Raizen et al.[29]

The spectra for higher values of intensity I are illustrated in Fig. 8. We point out that to get squeezing at all one requires a nonzero intensity, and possibilities of good squeezing often improve with intensity. As the intensity is increased, transitions between the higher excited states of the dressed atom-field system occur (cf. Figs. 4 and 1d). This is evidenced as an extra peak at $\omega = 0$ in the spectra. The appearance of a central fluorescence peak is evidenced in Fig. 8a. For such intensities the high-Q adiabatic solution exhibits increased noise (and no squeezing) at the sidepeaks corresponding to the cavity resonance ω_c' and $2\omega_L - \omega_c'$ (Fig. 2e). This noise is due to the thermal-type fluorescence ($\Lambda(0)$) (central Rabi peak of the atomic fluorescence spectrum $\Lambda(\delta)$ of Fig. 1e) and has been discussed previously. For relatively high $-q$ values the resonance frequency will fall within the center peak and the squeezing and intensity spectra will show the features of the high-Q adiabatic elimination spectra discussed in Section IV. For lower-Q values ($q \leq 1$), however, the resonance corresponds to frequencies well outside the atomic linewidth and squeezing is possible at resonances of the type depicted in Fig. 7a. We notice in Fig. 8a the sidepeaks for this moderate intensity have essentially the width and position indicated by the vacuum-field splitting. Figure 8b plots spectra for still higher values of intensity. The positions of the

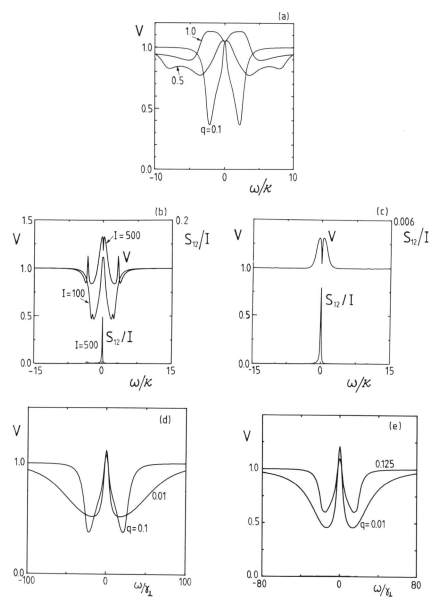

Figure 8. Spectra at higher intensities. (a) $V(X_\theta, \omega)$ for $\Delta = 8$, $C = 30$, $\psi f = -1$, $f = 1$, $n_{th} = 0$, $I = 50$. Various $q = \gamma_\perp/\kappa$. The noise due to scattering of photons at $\omega = 0$ is apparent. (b) $V(X_\theta, \omega)$ for various intensities; $\Delta = 8$, $C = 30$, $\phi = -1$, $f = 1$, $n_{th} = 0$, $q = 0.1$. (c) Squeezing and intensity spectra for $D = 8$, $C = 30$, $\phi = -1$, $f = 1$, $n_{th} = 0$, $q = 0.1$, $I = 5000$. (d) $V(X_\theta, \omega)$ for $\Delta = 8$, $I = 64$, $C = 30$, $f = 1$, $n_{th} = 0$. The curve $q = 0.1$ has $\phi = -1$. The curve $q = 0.01$ has $\phi = 0$ and approaches the low-Q limit. (e) $V(X_\theta, \omega)$ for $\Delta = 8$, $I = 64$, $C = 10$, $f = 1$, $n_{th} = 0$. The two curves correspond to parameters $(q = 0.125, \phi = -1)$ and $(q = 0.01, \phi = 0)$.

sidepeaks (barely visible) indicate Stark splitting as I becomes sufficiently large relative to C (Fig. 8c). Figures 8d and e make comparisons with spectra possible for lower-q values approaching the low-Q limit. Similar orders of squeezing are obtainable. The mechanism for squeezed side-peaks is a Rabi splitting increasing with both C and I. The lower-q values become more sensitive to the splitting in I and may become advantageous for lower C values (cf. Figs. 8d, 8e, and 3f).

V. CONCLUSION

We have presented a quantum theory enabling prediction of the squeezing obtainable from nondegenerate four-wave mixing with an atomic two-level medium. The squeezing is measured in the external field transmitted through a single cavity-end-mirror. Because in even a single-mode cavity, the external field is composed of traveling waves at different frequencies, nondegenerate four-wave mixing can occur in a single-mode cavity as well as a multimode cavity. We have thus examined both the single-mode cavity results and the multimode predictions. Both types of systems have been used experimentally to generate squeezing.[1, 27]

Because we include a quantized treatment of the medium, our theory predicts the effect of spontaneous emission on the squeezing possible. To summarize because spontaneous emission tends to dominate only at certain frequencies, it is usually possible to tune the parameter regime to avoid this, although this is more difficult in single-mode high-Q cavities. Squeezing tends to be enhanced at the sideband frequencies resonant with a coherent scattering process. An example of this is the squeezing possible at frequencies corresponding to either Rabi or vacuum-field Rabi splitting. This means it is almost always possible to adjust parameters to obtain squeezing in single-mode cavities, provided $\kappa \geq \gamma_\perp$. This prediction is supported by experimental results. No significant squeezing has been found in single-mode high-Q situations where $\kappa \ll \gamma_\perp$. Squeezing has been obtained by Slusher et al.[1] in multimode situations equivalent to the single-mode low-Q cavity, and by Raizen et al.[29] in single-mode cavities where $\kappa - \gamma_\perp$.

Recent experiments have detected a different type of squeezing to that discussed here, where there is only one quadrature phase measured. The experiments of Vallet et al.[35] use nondegenerate four-wave mixing with atomic vapor to generate a noise reduction that appears in the signal-idler intensity difference.

It is worth making a final remark about the applicability of the theory. The model assumes a large number of two-level atoms interacting with plane waves in a ring cavity. It does not take into account atomic motion or Doppler broadening. The original experiment of Slusher et al.[1] reduced this effect in an atomic beam experiment. The theory described here has been modified by Zwang and Walls[36] to include Doppler broadening, which is an important effect in experiments such as that of Vallet et al.[35] using atomic vapor. The theory also does not take into account standing waves or Gaussian mode functions,[37] though presumably modification to include these effects would not be too difficult.

References

1. R. E. Slusher, L. W. Hollberg, B. Yurke, J. C. Mertz, and J. F. Valley, *Phys. Rev. Lett.* **55**, 2409 (1985).

2. H. P. Yuen, *Phys. Rev. A* **13**, 2226 (1976); D. Stoler, *Phys. Rev. D* **1**, 3217 (1970); see Special issues on squeezed states of the electromagnetic field, *J. Opt. Soc. Am. B* **4** (1987); *J. Mod. Opt.* **34** (1987) and references therein.

3. C. M. Caves, *Phys. Rev. D* **230**, 1693 (1981).

4. M. Xiao, L. Wu, and H. J. Kimble, *Phys. Rev. Lett.* **59**, 278 (1987).

5. A. Heidmann, R. J. Horowicz, S. Reynaud, E. Giacobino, C. Fabre, and G. Camy, *Phys. Rev. Lett.* **59**, 2555 (1987).

6. S. Ho, P. Kumar, and J. H. Shapiro, *Phys. Rev. A* **35**, 3982 (1987); M. W. Maeda, P. Kumar, and J. H. Shapiro, *Opt. Lett.* **12**, 161 (1987).

7. R. M. Shelby, M. D. Levenson, S. H. Perlmutter, R. S. DeVoe, and D. F. Walls, *Phys. Rev. Lett.* **57**, 691 (1986).

8. L. A. Wu, H. J. Kimble, J. L. Hall, and H. Wu, *Phys. Rev. Lett.* **57**, 2520 (1986).

9. R. Movshovich, B. Yurke, P. G. Kaminksy, A. D. Smith, A. H. Silver, R. W. Simon, and M. V. Scheider, *Phys. Rev. Lett.* **65**, 1419 (1990).

10. W. Richardson, S. Machida, and Y. Yamamoto, *Phys. Rev. Lett.* **66**, 2867 (1991).

11 P. Grangier, R. E. Slusher, B. Yurke, and A. La Porta, *Phys. Rev. Lett.* **59**, 2153 (1987); M. Rosebluh and R. Shelby, *Phys. Rev. Lett.* **66**, 153 (1991).

12. D. M. Hope, H. A. Bachor, P. J. Manson, D. E. McClelland and P. T. H. Fisk, *Phys. Rev. A* **46**, 1181 (1992).

13. M. D. Levenson, R. M. Shelby, M. D. Reid, and D. F. Walls, *Phys. Rev. Lett.* **57**, 2473 (1986); Y. Yamamoto, N. Imoto and S. Machida, *Phys. Rev. A* **33**, 3242 (1986); A. La Porta, R. E. Slusher, and B. Yurke, *Phys. Rev. Lett.* **62**, 28 (1989).

14. M. D. Reid and D. F. Walls, *Phys. Rev. A* **32**, 396 (1985); M. D. Reid and D. F. Walls, *Phys. Rev. A* **34**, 4929 (1986); M. D. Reid, *Phys. Rev. A* **37**, 4792 (1988).

15. M. Sargent, D. A. Holm, and M. Zubairy, *Phys. Rev. A* **31**, 3112 (1985); S. Stenholm, D. A. Holm, and M. Sargent, *Phys. Rev. A* **31**, 3124 (1985); D. A. Holm and M. Sargent, *Phys. Rev. A* **35**, 2510 (1987).

16. M. Brambilla, F. Castelli, L. A. Lugiato, F. Prati and G. Strini, *Optics Comm.*, **83**, 367 (1991).

17. J. M. Courty, P. Grangier, L. Hilico, and S. Reynaud, to be published.

18. H. P. Yuen and J. H. Shapiro, *Opt. Lett.* **4**, 334 (1979).

19. C. M. Caves and B. L. Schumaker, *Phys. Rev. A* **31**, 3068 (1985); B. L. Shumaker and C. M. Caves, *Phys. Rev. A* **31**, 3093 (1985).

20. T. Fu and M. Sargent, *Opt. Lett.* **4**, 366 (1979).

21. R. W. Boyd, M. G. Raymer, P. Narum, and D. J. Marten, *Phys. Rev. A* **24**, 411 (1981).

22. H. Haken, *Handbook der Physik*, Vol. 25/2C, Springer, Berlin, 1970.

23. P. D. Drummond and D. F. Walls, *Phys. Rev. A* **23**, 2563 (1981).

24. M. O. Scully and W. Lamb, *Phys. Rev.* **159**, 208 (1967).

25. P. D. Drummond and C. W. Gardiner, *J. Phys. A* **13**, 2353 (1980).

26. M. J. Collett and C. W. Gardiner, *Phys. Rev. A* **30**, 1385 (1984); M. J. Collett and D. F. Walls, *Phys. Rev. A* **32**, 2887 (1985).

27. B. Yurke, *Phys. Rev. A* **32**, 300 (1985).

28. L. A. Lugiato, R. J. Horowicz, G. Strini, and L. M. Narducci, *Phys. Rev. A* **30**, 1366 (1984).

29. M. G. Raizen, L. Orozco, M. Xiao, T. L. Boyd, and H. J. Kimble, *Phys. Rev. Lett.* **59**, 198 (1987).

30. H. J. Carmichael, *Phys. Rev. A* **33**, 3262 (1986).

31. J. J. Sanchez-Mondragan, M. B. Marozhny, and J. H. Eberly, *Phys. Rev. Lett.* **51**, 551 (1983).

32. G. S. Agarwal, *Phys. Rev. Lett.* **53**, 1732 (1984).

33. Y. Kaluzny, P. Goy, M. Gross, J. M. Raimond and S. Haroche, *Phys. Rev. Lett.* **51**, 1175 (1983).

34. R. Brecha, L. A. Orozco, M. G. Raizen, M. Xiao, and H. J. Kimble, *J. Opt. Soc. Am. B* **3**, 238 (1986).

35. M. Vallet, M. Pinard, and G. Grynberg, *Europhys. Lett.* **11**, 739 (1990).

36. W. Zhang and D. F. Walls, *Phys. Rev. A* **41**, 6385 (1990).

37. A. T. Rosenberger, L. A. Orozco, H. J. Kimble, and P. D. Drummond, *Phys. Rev. A* **43**, 6284 (1991).

HYPER-RAYLEIGH SCATTERING IN SOLUTION

KOEN CLAYS
ANDRE PERSOONS

Laboratory of Chemical and Biological Dynamics, University of Leuven, Leuven, Belgium

and

LEO DE MAEYER

Max-Planck Institut für Biophysikalische Chemie, Göttingen, Germany

CONTENTS

Modern Nonlinear Optics, Part 3, Edited by Myron Evans and Stanisław Kielich. Advances in Chemical Physics Series, Vol. LXXXV.
ISBN 0-471-30499-9 © 1994 John Wiley & Sons, Inc.

I. INTRODUCTION

Hyper-Rayleigh scattering (HRS), i.e., elastic second-harmonic light scattering from macroscopically isotropic media was discovered soon after the introduction of megawatt peak power pulsed lasers.[1, 2] Theoretical aspects of the relevant mechanisms were discussed by several authors.[3-7] The phenomenon has regained attention as a method for determining first hyperpolarizabilities of dissolved organic molecules without needing recourse to superimposed dc fields or other means for producing preferential orientations.[8, 9] In the following a derivation of the observable intensity and depolarization of second-harmonic scattering as compared to linear anisotropic scattering is given for dissolved molecules of simple symmetry. The development of the derivation using a simple example is intended to show the similarities and differences with classical linear light scattering without getting too much diverted by the tensorial notations required for the general treatment. The influence of solvent-dependent factors is discussed in terms of the reaction field. Based on the results obtained on small solvent molecules with the internal reference method (IRM) we have extended the previously proposed equivalent internal field (EIF) model to small molecules with only σ-electron related polarizabilities. The potential of the HRS technique for the experimental determination of first hyperpolarizabilities is demonstrated with recently obtained results on molecules without dipole moment but with an octopolar charge distribution. Unique to the HRS technique is also the possibility of measuring ionic species. Absorption of the second-harmonic scattered light, often observed with ionic species, can be accounted for in the data analysis. Hyper-Rayleigh scattering results on synthetic polymers and natural proteins are reported for the first time. The experimental aspects of hyper-Rayleigh scattering in solution are also discussed, with emphasis on the specific requirements imposed on the pulsed laser system.

II. SCATTERING BY A SINGLE MOLECULE

We consider the simplest case of a single molecule in vacuum with a single axis of anharmonic anisotropy and with rotational symmetry $C_{\infty v}$ around this axis. A rectangular coordinate system ξ, η, ζ is associated with the molecule. As is shown in Fig. 1 (for a molecule with lower symmetry as a practical example), the ζ axis is chosen parallel to the anisotropy axis. The orientation of the molecular coordinate system with respect to the laboratory reference system x, y, z can be expressed by appropriate transformations. Because of the molecular shape (for $C_{\infty v}$ symmetry), the directions of ξ and η are not imposed by any part of the molecule. They may thus be

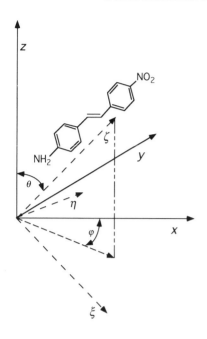

Figure 1. The relationship between the molecular coordinate system ξ, η, ζ and laboratory coordinate system x, y, z for a typical molecule with second-order nonlinear optical properties.

chosen to be in, resp. orthogonal, to the z, ζ plane. The ζ axis makes an angle $0 \leq \theta \leq \pi$ with the z axis; the x axis makes an angle φ with the projection of the ζ axis on the x, y plane. The transformation matrices to be used in the general case are given in Appendix A.

A polarized light beam of intensity I_z, with its electric vector polarized along the z axis and progressing along the y axis, is incident on the molecule. Buckingham and Pople originally introduced the expansion of the total dipole moment of a molecule in a uniform field in the tensor form[10]

$$m_\alpha = \mu_\alpha + \alpha_{\alpha\beta}E_\beta + \tfrac{1}{2}\beta_{\alpha\beta\gamma}E_\beta E_\gamma + \tfrac{1}{6}\gamma_{\alpha\beta\gamma\delta}E_\beta E_\gamma E_\delta + \cdots$$

In this chapter we include the numerical factors $\tfrac{1}{2}, \tfrac{1}{6}$, etc. in the tensor coefficients β, γ, etc., since this has become customary in recent literature.[11, 12] Of course, one has to be aware of the definition used in each case when comparing published values of β. For our simple molecule with a single axis of anharmonic anisotropy, the induced dipole moment is given by

$$\mu = (\alpha + \beta E_z)E_z$$

Because of the rotational symmetry only the diagonal elements of the first-order polarizability tensor α have nonzero values. For an ellipsoid of revolution we will have

$$\alpha_{11} = \alpha_{22} \neq \alpha_{33}$$

In the simplest case we may suppose that the field dependence of the polarizability is much larger along the ζ axis than along the ξ and η axes. We may then neglect all but one component of the second-order polarizability anharmonicity tensor β and write

$$\mu_\xi = \alpha_{11} E_\xi = -\alpha_{11} E_z \sin \theta$$

$$\mu_\eta = \alpha_{22} E_\eta = 0$$

$$\mu_\zeta = \alpha_{33} E_\zeta + \beta_{333} E_\zeta^2 = \alpha_{33} E_z \cos \theta + \beta_{333} E_z^2 \cos^2 \theta$$

The components of polarization in the laboratory frame x, y, z are

$$\mu_x = (\alpha_{33} - \alpha_{11}) E_z \cos \theta \sin \theta \cos \varphi + \beta_{333} E_z^2 \cos^2 \theta \sin \theta \cos \varphi$$

$$\mu_y = (\alpha_{33} - \alpha_{11}) E_z \cos \theta \sin \theta \sin \varphi + \beta_{333} E_z^2 \cos^2 \theta \sin \theta \sin \varphi$$

$$\mu_z = (\alpha_{11} \sin^2 \theta + \alpha_{33} \cos^2 \theta) E_z + \beta_{333} E_z^2 \cos^3 \theta$$

With the substitutions

$$\alpha_0 = \frac{\alpha_{11} + \alpha_{22} + \alpha_{33}}{3}$$

$$\delta\alpha = \frac{\alpha_{33} - \alpha_{11}}{3}$$

the polarization components of a single molecule can be split in its isotropic, anisotropic, and anharmonic parts:

$$\mu_x^{\text{isotrop.}} = 0$$

$$\mu_x^{\text{anisotr.}} = 3\delta\alpha E_z (\cos \theta \sin \theta \cos \varphi)$$

$$\mu_x^{\text{anharm.}} = \beta_{333} E_z^2 (\cos^2 \theta \sin \theta \cos \varphi)$$

$$\mu_y^{\text{isotrop.}} = 0$$

$$\mu_y^{\text{anisotr.}} = 3\delta\alpha E_z (\cos \theta \sin \theta \sin \varphi)$$

$$\mu_y^{\text{anharm.}} = \beta_{333} E_z^2 (\cos^2 \theta \sin \theta \sin \varphi)$$

$$\mu_z^{\text{isotrop.}} = \alpha_0 E_z$$

$$\mu_z^{\text{anisotr.}} = \delta\alpha E_z (2 \cos^2 \theta - \sin^2 \theta)$$

$$\mu_z^{\text{anharm.}} = \beta_{333} E_z^2 (\cos^3 \theta)$$

The probability that a randomly oriented molecule makes an angle between θ and $\theta + d\theta$ with the polarization plane of the incident optical field is

$$p(\theta)\, d\theta = \tfrac{1}{2} \sin\theta\, d\theta \qquad 0 \le \theta \le \pi$$

the probability to make an angle between φ and $\phi + d\phi$ is

$$p(\varphi)\, d\varphi = \frac{d\varphi}{2\pi} \qquad 0 \le \varphi \le 2\pi$$

The expectation value of the intensity of light scattered by the molecule and polarized along the x, y, or z directions is proportional to the expectation values $\langle \mu_x^2 \rangle$, $\langle \mu_y^2 \rangle$, and $\langle \mu_z^2 \rangle$. These values are found by multiplying μ_x^2, μ_y^2 resp. μ_z^2 with $p(\varphi)p(\theta)$ and integrating:

$$\langle \mu_x^2 \rangle = \frac{E_z^2}{4\pi} \int_0^{2\pi} \cos^2\varphi\, d\varphi \int_0^\pi 9\delta\alpha^2 \cos^2\theta \sin^3\theta\, d\theta$$

$$+ E_z^3 \int_0^{2\pi} \cos^2\varphi\, d\varphi \int_0^\pi 3\delta\alpha\beta_{333} \cos^3\theta \sin^3\theta\, d\theta$$

$$+ \frac{E_z^4}{4\pi} \int_0^{2\pi} \beta_{333}^2 \cos^2\varphi\, d\varphi \int_0^\pi \cos^4\theta \sin^3\theta\, d\theta$$

$$\langle \mu_y^2 \rangle = \frac{E_z^2}{4\pi} \int_0^{2\pi} \sin^2\varphi\, d\varphi \int_0^\pi 9\delta\alpha^2 \cos^2\theta \sin^3\theta\, d\theta$$

$$+ E_z^3 \int_0^{2\pi} \sin^2\varphi\, d\varphi \int_0^\pi 3\delta\alpha\beta_{333} \cos^3\theta \sin^3\theta\, d\theta$$

$$+ \frac{E_z^4}{4\pi} \int_0^{2\pi} \beta_{333}^2 \sin^2\varphi\, d\varphi \int_0^\pi \cos^4\theta \sin^3\theta\, d\theta$$

$$\langle \mu_z^2 \rangle = \frac{E_z^2}{2} \int_0^\pi \left[\alpha_0 + \delta\alpha(2\cos^2\theta - \sin^2\theta) \right]^2 \sin\theta\, d\theta$$

$$+ \frac{E_z^3}{2} \int_0^\pi \left[\alpha_0\beta_{333} \cos^3\theta + 2\beta_{333}\delta\alpha(2\cos^2\theta - \sin^2\theta)\cos^3\theta \right]$$

$$\times \sin\theta\, d\theta + \frac{E_z^4}{2} \int_0^\pi \beta_{333}^2 \cos^6\theta \sin\theta\, d\theta$$

The terms in E_z^3 vanish, and the result is

$$\langle \mu_x^2 \rangle = \langle \mu_y^2 \rangle = \tfrac{3}{5}\delta\alpha^2 E_z^2 + \tfrac{1}{35}\beta_{333}^2 E_z^4$$

$$\langle \mu_z^2 \rangle = \alpha_0^2 E_z^2 + \tfrac{4}{5}\delta\alpha^2 E_z^2 + \tfrac{1}{7}\beta_{333}^2 E_z^4$$

The average intensity (over all possible orientations of the molecule) of the second-harmonic light scattered in the x direction will be

$$\langle I_{2\omega} \rangle_x = \tfrac{6}{35}\beta_{333}^2 G(r) I_z^2$$

with a depolarization factor of $\tfrac{1}{5}$ (The induced moment μ_x does not contribute to scattering in the x direction). In the z direction the average intensity is

$$\langle I_{2\omega} \rangle_z = \tfrac{2}{35}\beta_{333}^2 G(r) I_z^2$$

In this direction there is complete depolarization. The factor $G(r)$ may be obtained with the equations for the energy flux radiated by an oscillating dipole, resulting in

$$G(r) = \frac{32\pi^2}{\varepsilon_0^3 c \lambda_\omega^4 r^2}$$

where r is the distance from the dipole, c the velocity of light, and λ_ω the wavelength of the exciting light.

The factor $G(r)$ is not dimensionless. It is given here in mks units $(V^3 \cdot A^{-3} \cdot s^{-2} \cdot m^{-4})$ if mks units for ε_0, c, λ_ω, and r are used. $I_{2\omega}$ and I_ω must then also be given in mks units $(W \cdot m^{-2})$ and β is obtained in $A \cdot s \cdot m^3 \cdot V^{-2}$. To convert $G(r)$ to unrationalized cgs units, replace ε_0 by $(4\pi)^{-1}\lambda_\omega$ and r must then be given in cm, c in cm $\cdot s^{-1}$, and I in erg $\cdot s^{-1} \cdot cm^{-2}$.

For equal frequencies of the exciting fields and no absorption in the frequency range ω_0 to $2\omega_0$ and retaining all nonzero tensor coefficients for molecules with $C_{\infty v}$ symmetry (β_{333}; $\beta_{113} = \beta_{131} = \beta_{311} = \beta_{223} = \beta_{232} = \beta_{322}$), we find for the full anharmonicity contributions:

$$\mu_\xi^{\text{anharm.}} = (\beta_{113} + \beta_{131})E_\xi E_\zeta = -2\beta_{113}E_z^2 \sin\theta\cos\theta$$

$$\mu_\eta^{\text{anharm.}} = (\beta_{223} + \beta_{232})E_\eta E_\zeta = 0$$

$$\mu_\zeta^{\text{anharm.}} = \beta_{311}E_\xi^2 + \beta_{322}E_\eta^2 + \beta_{333}E_\zeta^2 = \beta_{113}E_z^2 \sin^2\theta + \beta_{333}E_z^2 \cos^2\theta$$

giving

$$\mu_x^{\text{anharm.}} = (\beta_{333} - 2\beta_{113}) E_z^2 \sin\theta \cos^2\theta \cos\varphi + \beta_{113} E_z^2 \sin^3\theta \cos\varphi$$

$$\mu_y^{\text{anharm.}} = (\beta_{333} - 2\beta_{113}) E_z^2 \sin\theta \cos^2\theta \sin\varphi + \beta_{113} E_z^2 \sin^3\theta \sin\varphi$$

$$\mu_z^{\text{anharm.}} = \beta_{333} E_z^2 \cos^3\theta + 3\beta_{113} E_z^2 \sin^2\theta \cos\theta$$

and

$$\langle \mu_x^2 \rangle = \langle \mu_y^2 \rangle = \tfrac{1}{105}\left(3\beta_{333}^2 - 4\beta_{113}\beta_{333} + 20\beta_{113}^2\right) E_z^4$$

$$\langle \mu_z^2 \rangle = \tfrac{1}{35}\left(5\beta_{333}^2 + 12\beta_{113}\beta_{333} + 24\beta_{113}^2\right) E_z^4$$

The depolarization factor ρ in the x direction can be expressed as a function of $k = \beta_{113}/\beta_{333}$:

$$\rho_x = \frac{3 - 4k + 20k^2}{15 + 36k + 72k^2}$$

The total second-harmonic intensities in the x resp. z directions are

$$\langle I_{2\omega} \rangle_x = \tfrac{1}{105}\left(6\beta_{333}^2 - 8\beta_{113}\beta_{333} + 40\beta_{113}^2\right) G(r) I_z^2$$

$$\langle I_{2\omega} \rangle_z = \tfrac{1}{105}\left(18\beta_{333}^2 + 28\beta_{113}\beta_{333} + 104\beta_{113}^2\right) G(r) I_z^2$$

III. SCATTERING BY A MACROSCOPIC COLLECTION OF MOLECULES

Let us consider a collection of many molecules of the kind described above, with relatively large distances between them as in a gas or in solution in a liquid medium. This system is illuminated by an incident light beam with intensity I_z polarized in the z direction. Since the molecules are distributed in space they will be polarized by the incident field with different phases, depending on their position. The second-harmonic fields emitted in a given direction must be added for each polarization, taking into account the appropriate retardations.

For simplicity, we neglect the momentary translational and rotational motions of the molecules. Actually, these motions produce Doppler effects, spreading the frequency of the emitted radiation over a small range and resulting in temporal modulations of the scattered intensity.

From a volume element whose linear dimensions are small compared to the wavelength, but still containing a large number of molecules, retardations are such that all fields add coherently. The phase of the 2ω field of molecules oriented with $0 < \theta < \pi/2$ is the opposite of those with $\pi/2 < \theta < \pi$. Consequently, their emitted fields interfere destructively, unless, as a consequence of fluctuations in orientation, there are at some instant more molecules pointing with their positive ζ axes in one of the half-spheres over the x, y plane than in the other. Second-harmonic scattering from a macroscopically isotropic collection of anharmonic molecules depends on orientation fluctuations. The Fourier amplitude of the induced dipole moment per unit volume at 2ω is

$$P_i(2\omega) = B_{ikl}(-2\omega; \omega, \omega) E_k(\omega) E_l(\omega)$$

where B_{ikl} is the ikl component of the macroscopic second-order susceptibility. In isotropic media, only the average value of B_{ikl} is zero, due to these orientation fluctuations.[1] As in the case of quasi-elastic scattering, due to density fluctuations, or depolarized scattering by anisotropy fluctuations, second-harmonic scattering in a particular direction requires a corresponding nonvanishing component of the instantaneous Fourier expansion of the fluctuating spatial distribution of the molecular orientations.

We are dealing with a fluctuating pattern of orientations, described by its spatial Fourier spectrum. The individual spectral terms resemble wave-like patterns, although they do not propagate like waves, since there is no coupling of neighboring orientations by momentum transfer between neighboring volume elements, as in sound waves. Electromagnetic couplings, as described by Maxwell's field equations, can also be neglected, since the rotational motions of dissolved dipolar molecules are too slow to respond to fluctuating electromagnetic waves of the required wavelength.

Besides the fluctuation of orientations, there will be a fluctuation of the number of molecules in the volume element considered. This affects the isotropic part of the polarization, which, for a collection of molecules, represents the main interaction with the forward propagating field. It leads to inhomogeneity scattering, due to local variations in the number of scatterers. This contribution is usually derived from thermodynamic expressions for density fluctuations. When the average number of molecules in a volume element in the equilibrium state is $\langle N \rangle$, increasing that number by an amount δN requires an amount of work equal to $\frac{1}{2}\delta\gamma\delta N = \frac{1}{2}(\delta\gamma/\delta N)_{T,V}\delta N^2$, where γ is the chemical potential (free energy per particle). We will consider the isotropic term later and focus our attention here on the orientation-dependent terms only. At the moment, we assume that the molecules do not have orientation-dependent interactions, allowing us to apply the same angular probability distributions as before to each molecule in the collection.

For a physical volume element v, whose linear dimensions are much smaller than the wavelength, containing N independent randomly oriented molecules, the expressions for x, y, and z components of the induced polarization are

$$P_x = A_x E_z + B_x E_z^2$$

$$P_y = A_y E_z + B_y E_z^2$$

$$P_z = \frac{N\alpha_0}{v} E_z + A_z E_z + B_z E_z^2$$

with

$$A_x = \frac{3\delta\alpha}{v} \sum_{i=1}^{N} \cos\theta_i \sin\theta_i \cos\varphi_i$$

$$A_y = \frac{3\delta\alpha}{v} \sum_{i=1}^{N} \cos\theta_i \sin\theta_i \sin\varphi_i$$

$$A_z = \frac{\delta\alpha}{v} \sum_{i=1}^{N} \left(2\cos^2\theta_i - \sin^2\theta_i\right)$$

$$B_x = \frac{\beta_{333}}{v} \sum_{i=1}^{N} \cos^2\theta_i \sin\theta_i \cos\varphi_i$$

$$B_y = \frac{\beta_{333}}{v} \sum_{i=1}^{N} \cos^2\theta_i \sin\theta_i \sin\varphi_i$$

$$B_z = \frac{\beta_{333}}{v} \sum_{i=1}^{N} \cos^3\theta_i$$

The averages and variances of the polarization components are defined in Appendix B. When the integrations are performed with the probability densities for random orientation, the results are

$$\langle P_x^2 \rangle = \langle P_y^2 \rangle = \frac{N}{v^2}\left(\tfrac{3}{5}\delta\alpha^2 E_z^2 + \tfrac{1}{35}\beta_{333}^2 E_z^4\right)$$

$$\langle P_z^2 \rangle = \frac{N}{v^2}\left(N\alpha_0^2 E_z^2 + \tfrac{4}{5}\delta\alpha^2 E_z^2 + \tfrac{1}{7}\beta_{333}^2 E_z^4\right)$$

This leads us to the important conclusion that anisotropic and second-harmonic light scattering, caused by the fluctuations of the average orientation of a collection of noninteracting, independent randomly oriented molecules, appears identical to the sum of N individual scatterers. Therefore, the intensity of the second-harmonic scattered light, polarized along the J direction ($J = X$ or Z) and collected over a certain solid angle, $I_J(2\omega)$, can be written as

$$I_J(2\omega) = g B_{JZZ}^2 I_Z^2$$

The factor g depends on the scattering geometry (scattering angle and solid angle of photon collection). For dipolar nonlinear optical molecules (in contrast with the more recently studied apolar octopolar molecules, see Section VII) that derive their nonlinearity from the difference in dipole moment between the ground and excited state, the component of the first hyperpolarizability β along the molecular charge transfer axis ξ, β_{333}, is much larger than any other tensor element. The hyper-Rayleigh scattering signal $S_J(2\omega)$ can then be written as

$$S_J(2\omega) = G_J B^2 I_Z^2 = G_J \sum_s N_s \beta_{333,s}^2 I_Z^2$$

N_s is the number density of species s (with $C_{\infty v}$ symmetry) with first hyperpolarizability $\beta_{333,s}$. The factor G_J includes the previous factor g, all instrumental factors, and the appropriate averages of the products of the direction cosines ($\frac{1}{7}$ for $J = Z$ and $\frac{1}{35}$ for $J = X$). A quadratic dependence of the HRS signal $S_J(2\omega)$ on the incident fundamental light I_Z, as shown in Fig. 2 for different concentrations of *para*-nitroaniline in methanol, is observed.

For a two-component solvent–solute system (e.g., *para*-nitroaniline in methanol), with N_S and $\beta_{333,S}$ the number density, resp., the first hyperpolarizability of the solvent, and N_s and $\beta_{333,s}$ the number density, resp., the first hyperpolarizability of the solute $S_J(2\omega)$ is written as

$$S_J(2\omega) = G_J B^2 I_Z^2 = G_J \left(N_S \beta_{333,S}^2 + N_s \beta_{333,s}^2 \right) I_Z^2$$

For the low concentrations of solute used, the number density of the solvent N_S is approximately constant. Measurements of the second-harmonic scattered light intensity $S_J(2\omega)$ as a function of incident light intensity I_Z at different number densities of the solute N_s then show a linear dependence of the quadratic coefficient $G_J B^2$ on N_s. Figure 3

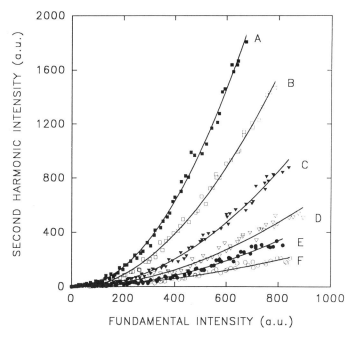

Figure 2. Hyper-Rayleigh scattering signal $S(2\omega)$ for *para*-nitroaniline in methanol at 293 K at different number densities in units of 10^{18} cm^{-3}: (A) 92; (B) 46; (C) 23; (D) 9.2; (E) 4.6; (F) 1.8; the solid lines are fitted quadratic curves.

shows this linear dependence for the quadratic coefficients obtained from the data shown in Fig. 2. From the intercept $G_J N_S \beta_S^2$ and the slope $G_J \beta_s^2$, β_s is calculated when β_S is known, or vice versa. Since no electric field directing the dipoles has to be applied, the local-field correction factor at zero frequency is eliminated.

However, at optical frequencies, an important difference must still be considered. For an individual molecule in vacuo, the excitation field E_z is the vacuum field, but the field in a medium whose volume elements contain many molecules is different. In a continuous medium the vacuum field must be replaced by the local optical field that acts on a molecule and contains contributions from the polarized neighboring molecules. These molecules may have orientation-dependent polarizabilities. The effective field E_z is obtained by averaging over the distribution of orientations of the neighboring molecules. This averaging takes place since each of the N molecules of the collection considered here is surrounded by a different configuration of neighboring molecules.

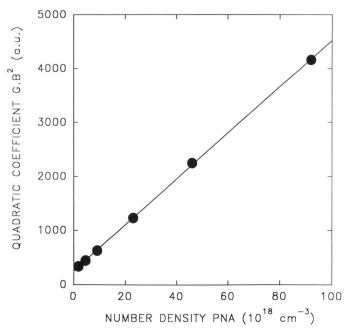

Figure 3. Quadratic coefficient GB^2, obtained from the curves of Fig. 2, as a function of number density of *para*-nitroaniline N_s; the solid line is a linear fit.

IV.   THE INTERNAL EXCITATION FIELD

The field E to which a molecule in a dense medium is subjected, is not the vacuum field of the incident radiation. It is the macroscopic field E_0 augmented by the effect X of the macroscopic polarization of the environment on the local position of the molecule: $E = E_0 + X$. In calculating E and X at the position of the molecule, one must consider the fluctuating collection of which it is part, and not only the isotropic medium, making it quite difficult to obtain the appropriate expressions. Considerable effort has been invested in solving this problem, which has been controversial for a long time.

The field E_0 is the field generated by external sources. Under certain conditions it may be identical with the field that the external source (in the present case; the illumination source) would produce at the same position if the specimen were absent. In general, however, this is not the case. The macroscopic field inside the specimen depends strongly on the geometric form of its boundary. In different specimens of the same form, but with different refractive index, E_0 will change due to different beam divergence in focusing. In principle, one can correct for this in the experimental setup and compare experiments in which E_0 is the same.

In electrostatics, the field X is identified with the field produced at a specific location by all dipoles in the specimen, averaged over their distribution in the field E_0. If one neglects local correlations between the positions and orientations of nearest-neighbor molecules, then this field is the same as that produced by the average polarization of a dielectric continuum:

$$X = \frac{P}{3\varepsilon_0}$$

With $P = \varepsilon_0(\varepsilon - 1)E_0$, this leads to the classical Lorentz expression

$$E = \left(\frac{(\varepsilon + 2)}{3}\right)E_0.$$

But since the scattering process depends on fluctuations, we must retain in P not only the average part but also the local variation. We consider a volume element v which is small compared to the wavelength, but contains a large number of molecules. The fluctuating part in P contributes a fluctuating part in E. This part, δE, will be the local contribution to the scattered field and has components not only in the z direction, but also in the x and y directions. Defining $E_z = \overline{E}_z + \delta E_z$ and introducing this value in P_z leads to

$$\overline{E}_z + \delta E_z = E_0 + \frac{\Sigma N_k \alpha_k^0}{3\varepsilon_0 v}\overline{E}_z$$

$$+ \frac{1}{3\varepsilon_0}\left[\frac{\Sigma N_k \alpha_k^0}{v}\delta E_z + A_z(\overline{E}_z + \delta E_z) + B_z(\overline{E}_z + \delta E_z)^2\right.$$

$$\left. + \frac{\Sigma \delta N_k \alpha_k^0}{v}(\overline{E}_z + \delta E_z)\right]$$

The terms containing $\Sigma N_k \alpha_k^0$ require a summation over all molecular species, including the solvent, in the volume v. All molecules contribute to isotropic polarization. If there is more than one kind of anisotropic or anharmonic molecule, the terms in A_z and B_z also represent sums over the individual species with their respective $\delta\alpha$ and β_{333}. The last term in this equation represents inhomogeneity scattering, included here for completeness.

The first two terms on the right side of this equation represent the nonfluctuating part of E_z:

$$\bar{E}_z = E_0 + \frac{\Sigma N_k \alpha_k^0}{3\varepsilon_0 v} \bar{E}_z$$

giving

$$\bar{E}_z = FE_0 \quad \text{with } F = \left(1 - \frac{\Sigma N_k \alpha_k^0}{3\varepsilon_0 v}\right)^{-1}$$

All other terms depend on local variations in orientation or particle density.

Leaving out the terms that depend only in second order on fluctuations, we have

$$\delta E_z = \frac{\Sigma N_k \alpha_k^0}{3\varepsilon_0 v} \delta E_z + \frac{A_z}{3\varepsilon_0} \bar{E}_z + \frac{B_z}{3\varepsilon_0} \bar{E}_z^2 + \frac{\Sigma \alpha_k^0 \delta N_k}{3\varepsilon_0 v} \bar{E}z$$

or

$$\delta E_z = \frac{A_z}{3\varepsilon_0} F^2 E_0 + \frac{B_z}{3\varepsilon_0} F^3 E_0^2 + \frac{\Sigma \alpha_k^0 \delta N_k}{3\varepsilon_0 v} F^2 E_0$$

It would be more accurate to consider the frequency dependence of α_0 and to split δE_z in two parts:

$$\delta E_z = \delta E_{z,\omega} + \delta E_{z,2\omega}$$

giving correction factors

$$F_\omega = \left(1 - \frac{\Sigma N_k \alpha_k^{0,\omega}}{3\varepsilon_0 v}\right)^{-1} \qquad F_{2\omega} = \left(1 - \frac{\Sigma N_k \alpha_k^{0,2\omega}}{3\varepsilon_0 v}\right)^{-1}$$

F_ω is used in $\bar{E}_z = F_\omega E_0$ and in $\delta E_{z,\omega}$ for anisotropy and inhomogeneity scattering. $F_{2\omega}$ is used in $\delta E_{z,2\omega}$ for anharmonic scattering.

The second moments of the quantities A_z and B_z that depend on the fluctuating θ_i may be calculated as before, giving

$$\langle \delta E_z^2 \rangle = \frac{\langle A_z^2 \rangle}{9\varepsilon_0^2} F_\omega^4 E_0^2 + \frac{\langle B_z^2 \rangle}{9\varepsilon_0^2} F_\omega^4 F_{2\omega}^2 E_0^4 + \frac{\alpha_0^2}{9\varepsilon_0^2} \left\langle \left(\frac{\delta N}{v}\right)^2 \right\rangle F_\omega^4 E_0^2$$

The same approach is used for calculating the fluctuating components δE_x and δE_y. $\overline{E}_x$ and $\overline{E}_y$ are zero for a z-polarized source field E_0. One has, with $I = x$ or y

$$E_I = \delta E_I = \frac{P_I}{3\varepsilon_0} = \frac{\sum N_k \alpha_k^0}{3\varepsilon_0 v} \delta E_I + \frac{A_I}{3\varepsilon_0} \overline{E}_z + \frac{B_I}{3\varepsilon_0} \overline{E}_z^2$$

which may be separated into

$$\delta E_{I,\omega} = \frac{\sum N_k \alpha_k^0}{3\varepsilon_0 v} \delta E_{I,\omega} + \frac{A_I}{3\varepsilon_0} \overline{E}_z \qquad \delta E_{I,2\omega} = \frac{\sum N_k \alpha_k^0}{3\varepsilon_0 v} \delta E_{I,2\omega} + \frac{B_I}{3\varepsilon_0} \overline{E}_z$$

Using the correction factors F_ω and $F_{2\omega}$, these expressions become

$$\delta E_{I,\omega} = \frac{A_I}{3\varepsilon_0} F_\omega^2 E_0 \qquad \delta E_{I,2\omega} = \frac{B_I}{3\varepsilon_0} F_\omega^2 F_{2\omega} E_0^2$$

$$\langle \delta E_{I,\omega}^2 \rangle = \frac{\langle A_I^2 \rangle}{9\varepsilon_0^2} F_\omega^4 E_0^2 \qquad \langle \delta E_{I,2\omega}^2 \rangle = \frac{\langle B_I^2 \rangle}{9\varepsilon_0^2} F_\omega^4 F_{2\omega}^2 E_0^4$$

The derivations for $\langle A_I^2 \rangle$ are as before:

$$\langle A_x^2 \rangle = \langle A_y^2 \rangle = \frac{3}{5}\left(\frac{N}{v^2}\right)\delta\alpha^2 \qquad \langle A_z^2 \rangle = \frac{4}{5}\left(\frac{N}{v^2}\right)\delta\alpha^2$$

$$\langle B_x^2 \rangle = \langle B_y^2 \rangle = \frac{1}{35}\left(\frac{N}{v^2}\right)\beta_{333}^2 \qquad \langle B_z^2 \rangle = \frac{1}{7}\left(\frac{N}{v^2}\right)\beta_{333}^2$$

The correction factors F_ω and $F_{2\omega}$ are related to the refractive index

$$F_\omega = \frac{n_\omega^2 + 2}{3} \qquad F_{2\omega} = \frac{n_{2\omega}^2 + 2}{3}$$

The theoretical arguments for including the factor F_ω^4 in the Rayleigh ratio of depolarized scattering due to anisotropy have been discussed at length in the literature, especially since the experimental data do not strongly support this dependence. There is, of course, the problem that the anisotropy $\delta\alpha$ of a molecule is not necessarily the same in a dense medium as in the gas phase. It follows from experimental observations that the measured effective anisotropy seems to decrease with increasing optical polarizability of the solvent. A theory for this consistent effect has been proposed,[13, 14] which is based on the assumption of pair correlations between an anisotropic molecule of nonspherical shape and the possible positions of its neighbors. As a consequence, the local field generated by these neighbors is affected. The orientational fluctuation changes the local symmetry determining the internal field.

The additional correction depends on shape factors and is therefore difficult to predict accurately. For pure packing interactions of nonspherical molecules the effective anisotropy and effective anharmonicity will not necessarily be correlated.

V. MOLECULES WITH PERMANENT DIPOLE MOMENT

In almost all cases molecules with uniaxial anharmonic polarizability as considered before will also possess a finite dipole moment. This introduces interactions that, already in the gaseous phase, contribute for a considerable part to the cohesion forces. It must be verified if, or under what conditions, the assumption of independent random orientation of a solute can be maintained. Besides the possible influence on orientational fluctuations, the effect of the reaction field E_{rf} of a polarizable medium on the hyperpolarizability of a solute dipolar molecule must be considered.

A. The Dipole Reaction Field

In a dense medium, the polarization induced by a dipole in its environment creates a reaction field, which enhances the dipole moment of the molecule. The reaction field does not exert orienting forces on the inducing dipole, since it always has the same direction as the dipole moment. The enhancement is proportional to the polarizability of the inducing molecule. When the molecule is anisotropic, the relative directions of the dipole axis and the anisotropy axis must be taken into account.

The reaction field E_{rf} changes the charge distribution in the central molecule. This affects not only its effective dipole moment, but also its polarizability and hyperpolarizability. Formally, the effect of a polarizable or polar solvent on a dissolved dipolar molecule may be expressed by combining the reaction field E_{rf} with the local internal field E in the original expression of the field-perturbed molecular moment. As before, the local field E is the result of external sources, and includes the field created at the position of the molecule by the polarized medium. The reaction field E_{rf} and the field $E = E_0 + X$ will, in general, have different directions. E_{rf}, as well as μ_0, is in the ζ direction of the molecule E and has the same direction as E_0. We then have

$$
\begin{aligned}
\mu &= \mu_0 + \alpha(E_{rf} + E) + \beta(E_{rf} + E)(E_{rf} + E) \\
&\quad + \gamma(E_{rf} + E)(E_{rf} + E)(E_{rf} + E) \\
&= \mu_0 + \alpha E_{rf} + \beta E_{rf}E_{rf} + \gamma E_{rf}E_{rf}E_{rf} + (\alpha + 2\beta E_{rf} + 3\gamma E_{rf}E_{rf})E \\
&\quad + (\beta + 3\gamma E_{rf})EE
\end{aligned}
$$

The effective, solvent-dependent, anharmonicity of the uniaxial molecule considered here becomes

$$
\beta_{333}^{eff} = \beta_{333} + 3\gamma_{eff}E_{rf}
$$

The reaction field of a polarizable dipole in the order of approximation needed here is

$$
E_{rf} = f(\mu_0 + \alpha_0 E_{rf})
$$

where the factor f, as derived by Onsager for a spherical molecule with radius a, is given by

$$
f = \left(\frac{1}{a^3}\right)\left(\frac{2(\varepsilon - 1)}{2\varepsilon + 1}\right)
$$

Böttcher[15] has given an expression for E_{rf} (in esu) in the desired approximation, considering that the molecular radius can be obtained from the molecular weight M and the density ρ of a dense liquid or solid phase of the dipolar solute molecules. The average polarizability α_0 is also related to the refractive index of such a phase, giving

$$
\beta_{eff} = \beta_{333} + 3\mu_0\gamma_{eff}\left[\left(\frac{4\pi}{3}\right)\left(\frac{\rho N_{Avo}}{M}\right)\left(\frac{2(\varepsilon - 1)}{2\varepsilon + n^2}\right)\left(\frac{n^2 + 2}{3}\right)\right]
$$

To be correct, the refractive index n should be taken here at a frequency where the atomic polarizability is included, since the reaction field has a fixed orientation with respect to the molecule and has to adapt only to the rotational diffusion rate of the molecule in its environment. For the same reason ε in this expression is the low-frequency dielectric constant of the solution.

TABLE I

First Hyperpolarizability β for *para*-Nitroaniline (PNA), 4-Methoxy-4'-nitrostilbene (MONS), and 4-Hydroxy-4'-nitrostilbene (HONS) in Different Solvents

	$\beta(10^{-30}$ esu$)$		
Solvent	PNA	MONS	HONS
para-Dioxane	16.9 ± 0.4[a]	81 ± 8[b]	
Chloroform	23 ± 3	105 ± 35	93 ± 30
1,2-Dichloroethane		119 ± 17[c]	
Methanol	34.5 ± 4[d]		

[a] From Ref. 44.
[b] From Ref. 45.
[c] From Ref. 46.
[d] From Ref. 22.

We may conclude that the rather strong dependence on solvent polarity, which has been reported for β values measured by second-harmonic light-scattering (Table I),[8] is caused by this reaction-field effect, rather than by orientational correlations between the solute molecules. Orientational correlations with neighboring solvent, induced by shape and dipole moment of a solute molecule, may require corrections to the last equation, where the bulk value of ε does not necessarily apply to the local environment of the solute.

The experimental hyperpolarizability data in different solvents have been analyzed according to the previous equation. The solvent data (density ρ, molecular weight M, dielectric constant ε, and refractive index n) are tabulated in Table II. In Fig. 4, the effective first hyperpolarizability β_{eff} data are plotted as a function of

$$\frac{24\pi\rho N_{\text{Avo}}(\varepsilon - 1)(n^2 + 2)}{9M(2\varepsilon + n^2)}$$

TABLE II

Values for the Density ρ, Molecular Weight M, Dielectric Constant ε, and Refractive Index n of the Solvents from Table I

Solvent	ρ	M	ε	n
para-Dioxane	1.033	88.12	2.2	1.422
Chloroform	1.483	119.38	4.8	1.446
1,2-Dichloroethane	1.235	98.96	10.6	1.445
Methanol	0.791	32.04	32.6	1.329

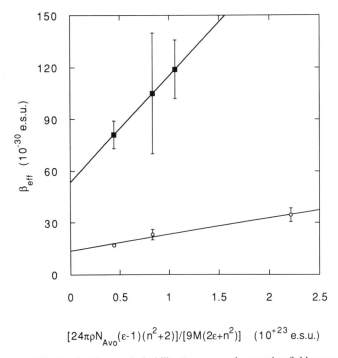

Figure 4. Effective first hyperpolarizability β_{eff} versus the reaction field parameter

$$\frac{24\pi\rho N_{\text{Avo}}(\varepsilon - 1)(n^2 + 2)}{9M(2\varepsilon + n^2)}$$

for 4-methoxy-4'-nitrostilbene (∇) and *para*-nitroaniline ($\bullet$). The reaction field parameters are calculated with the density ρ, the molecular weight M, the dielectric constant ε, the refractive index n, given in Table II. The solid line is a linear fit to the data. The second hyperpolarizability calculated from the slope of this fit and the dipole moment, γ_{calc}, is given in Table III.

such that the slope would be determined by the product $\mu_0\gamma_{\text{eff}}$. The second hyperpolarizability γ_{eff} calculated with literature values for the dipole moment μ_0 can then be compared with reported values for γ_{eff}. From the observed linear dependence of β_{eff} on $3E_{\text{rf}}/\mu_0$, we can already conclude that the reaction-field model holds qualitatively. The value for γ_{eff}, as calculated from the observed slope and the dipole moment μ_0, is $(14 \pm 2) \times 10^{-36}$ esu for *para*-nitroaniline and $(120 \pm 20) \times 10^{-36}$ esu for 4-methoxy-4'-nitrostilbene. This is in good agreement with literature values for the second hyperpolarizability (see Table III), especially taking

TABLE III

Values for the Dipole Moment μ, Second Hyperpolarizability Calculated from the slope of Fig. 4, γ_{calc}, and Comparison with Literature Values for Second Hyperpolarizability, γ_{ref}

	μ $(10^{-18}$ esu)	γ_{calc} 10^{-36} esu)	γ_{ref} $(10^{-36}$ esu)
PNA	6.2^a	14 ± 2	15^a
	6.2^b		48^b
MONS	4.5^c	120 ± 20	79^c
	5.7^d		90 ± 4^e

[a] Determined by THG at 1.91 μm in acetone.[47]

[b] Determined by EFISHG at 1.064 μm in methanol.[22]

[c] Determined by THG at 1.91 μm in ρ-dioxane.[47]

[d] Determined by EFISHG at 1064 nm in ρ-dioxane.[45]

[e] Determined by degenerate four-wave mixing in nonlinear interferometer at 1.064 μm in chloroform (to be published).

into consideration the different measurement techniques (third-harmonic generation, THG; electric-field-induced second-harmonic generation, EFISHG) and hence the different reference materials and the different dispersion in γ_{eff}. We have determined $\gamma(-2\omega_{opt} \pm \omega_{rf}; \omega_{opt}, \omega_{opt}, \pm \omega_{rf})$, while in THG and EFISHG, $\gamma(-3\omega_{opt}; \omega_{opt}, \omega_{opt}, \omega_{opt})$ and $\gamma(-2\omega_{opt}; \omega_{opt}, \omega_{opt}, 0)$ are determined, respectively. The second hyperpolarizability γ_{eff}, corrected for cascading effects,[16] is, as a scalar quantity, not expected to show such a significant solvent dependence as the first hyperpolarizability β with its vectorial part.

We conclude that from the solvent dependence of the first hyperpolarizability β, as described by the effect of the reaction field E_{rf} of the dissolved dipole with dipole moment μ_0, it is possible to determine the second hyperpolarizability γ of a noncentrosymmetric molecule. On the other hand, it should be possible to use a centrosymmetric molecule with a known second hyperpolarizability γ as a probe for the field strength E_{rf} and the dynamics of the reaction field.[8]

In contrast to the simplified uniaxial case considered here, molecules with dipole moments in directions other than the main anharmonicity axis may exhibit different depolarization ratios of the scattered second-harmonic light in solvents of different polarity.

When a molecule is transferred from vacuum to a polar medium, its internal charge distribution is modified. This leads to solvatochromic shifts in the absorption spectrum, depending on whether the ground state or the first excited state is the more polar one.[17] This effect has been used to characterize the polarity of solvents. The term polarity is used in this context as a description for the ability of a solvent to solvate a polar (ionic or dipolar) solute. Dimroth's parameter E_T is the transition energy (in kcal mol^{-1}) of the longest-wavelength solvatochromic absorption of a dissolved betaine dye (pyridinium-N-phenoxide betaine). Kosower's parameter Z is the transition energy for the charge-transfer band of 1-ethyl-1-methoxycarbonyl pyridinium iodide. It is not feasible to express these parameters in quantitative terms as a function of the dielectric properties of each solvent, although there seems to be a good correlation with the product of density and dipole moment $q_{solv} = \mu_{solv} p_{solv}$ within different classes of solvents (hydrogen-bonded or non-hydrogen-bonded). The quantity q_{solv}/ε_0 is dimensionally related to the average electric field at the surface of a solvent molecule.

B. The Dipolar Interactions

The interactions of neighboring dipoles have been studied for a long time. The simplest model, representing a molecule by a nonpolarizable sphere, carrying a fixed dipole in its center, was used by Keesom for investigating the possible source of van der Waals forces.[18] A similar model with polarizable spheres was used by Falkenhagen.[19] These calculations were aimed at obtaining the contributions to the virial coefficients for gaseous systems. The calculations show that the lowest energy of a pair of molecules in contact is the parallel configuration, with the dipoles pointing along the line connecting the centers. The interaction energy of the antiparallel configuration, with the dipole axes perpendicular to the connecting line, is only half as large. This result, however, applies only to spherical molecules, where the distance of nearest approach of the two configurations is necessarily the same. The orientational correlation of a pair of neighboring particles in contact is shape dependent.[20, 21] For rotationally symmetric ellipsoidal particles with major and minor axes a resp. b, and with their dipole moment pointing in the direction of the major axis, the parallel and the antiparallel configurations are equally probable for $a/b = 2^{1/3}$. The antiparallel orientation becomes the most stable one when $a/b < 2^{1/3}$. The theory of short-range and long-range orientational correlations in liquids is rather involved and not complete. In the case of solutions of anharmonic molecules, the effects of mutual interactions between dissolved molecules are concentration dependent. In solution, the presence of a polarizable solvent weakens the interactions between the solute molecules, but, at the same time, it may introduce correlations between the orientation of a solute molecule and its surrounding solvent neighbors. In dilute solutions mutual orientational correlations induced by encounters between solute molecules should disappear.

We will verify the assumption of independent orientation by deriving the amplitude of fluctuations in a different way. For anisotropic scattering it does not matter whether the ζ axis of a molecule is parallel or antiparallel to the z axis. For anharmonic scattering, the phase of the second-harmonic field is opposite in these two cases. To lead to second-harmonic scattering, a fluctuation in orientation must necessarily involve a local excess of molecules with parallel orientations, or, in the case of dipolar molecules, the creation of a spontaneous local electric moment.

Compared with the induced polarization at optical frequencies by the incident light wave, spontaneous dipolar polarization forms and relaxes with the much lower rate of dipolar reorientation. Therefore, the energy for its formation will be related to the static dielectric constant of the medium, rather than to its refractive index.

The probability of a fluctuation may be derived from Boltzmann's equation. If p_0 is the probability of the isotropic state of a volume element, in which all dipolar orientations balance each other, so that the macroscopic electric moment of the volume element is zero, then the probability of a state with macroscopic moment m is given by

$$p(m) = p_0 \exp\left\{ -\frac{W_m - W_0}{kT} \right\}$$

At constant volume and temperature, $W_m - W_0$ is the difference in free energy of the volume element in the two states. From electrostatics we know that the work required to produce an electric moment m in a dielectric with polarizability $\bar{\alpha}$ is

$$W_m - W_0 = \frac{m^2}{2\bar{\alpha}}$$

The entropic part of the free energy change is included in this expression. It is contained in $\bar{\alpha}$, which depends on the degrees of freedom in the system that allow a change in the charge distribution. We therefore have for the *probability*

$$p(m) = C \exp\left\{ -\frac{m^2}{2\bar{\alpha}kT} \right\}$$

where C is the normalization constant. The normalized *probability* can then be written as a Gaussian distribution with variance $\sigma^2 = \bar{\alpha}kT$

$$p(m) = \frac{1}{\sigma\sqrt{2\pi}} \exp\left\{ -\frac{m^2}{2\sigma^2} \right\}$$

Let us compare this probability density with one derived on the assumption of independent orientation. We associate with each molecule i in the specimen a unit vector $\mathbf{e}_i$ along its positive ζ axis, making an angle θ_i ($0 \leq \theta_i \leq \pi$) with the z axis. For uniform distribution of orientations in space angle 4π, the probability density $p(\theta_i)$ is

$$p(\theta_i) = \tfrac{1}{2} \sin \theta \, d\theta$$

The probability density $p(\varphi_i)$ that the projection of the vector $\mathbf{e}_i$ on the x, y plane makes an angle φ_i ($0 \leq \varphi_i \leq 2\pi$) with the x axis is $1/2\pi$. The sum $L_i(t)$ of the projections of N molecules on the i axis ($i = x, y$, or z) at any instant t is

$$L_x(t) = \sum_{i=1}^{N} \sin \theta_i(t) \cos \varphi_i(t)$$

$$L_y(t) = \sum_{i=1}^{N} \sin \theta_i(t) \sin \varphi_i(t)$$

$$L_z(t) = \sum_{i=1}^{N} \cos \theta_i(t)$$

$\theta_i(t), \varphi_i(t)$ and $\theta j(t), \varphi j(t)$ are statistically independent. The expectation value for the sum is then

$$\langle L_x \rangle = \sum_{i=1}^{N} \langle \sin \theta_i \cos \varphi_i \rangle = \sum_{i=1}^{N} \left(\frac{1}{2} \int_0^{\pi} \sin^2 \theta \, d\theta \int_0^{2\pi} \frac{\cos \varphi}{2\pi} \, d\varphi \right) = 0$$

$$\langle L_y \rangle = \sum_{i=1}^{N} \langle \sin \theta_i \sin \varphi_i \rangle = \sum_{i=1}^{N} \left(\frac{1}{2} \int_0^{\pi} \sin^2 \theta \, d\theta \int_0^{2\pi} \frac{\sin \varphi}{2\pi} \, d\varphi \right) = 0$$

$$\langle L_z \rangle = \sum_{i=1}^{N} \langle \cos \theta_i \rangle = \frac{N}{2} \int_0^{\pi} \cos \theta \sin \theta \, d\theta = 0$$

Since the average values are zero, the variances of the projections are equal to the second moments:

$$\langle L_x^2 \rangle = \left\langle \left(\sum_{i=1}^{N} \sin \theta_i \cos \varphi_i \right)^2 \right\rangle$$

$$= \left\langle \sum_{i=1}^{N} \sin^2 \theta_i \cos^2 \varphi_i + \sum_{i=1}^{N} \sum_{j \neq i}^{N-1} \sin \theta_i \sin \theta_j \cos \varphi_i \cos \varphi_j \right\rangle$$

$$= \left\langle \sum_{i=1}^{N} \sin^2 \theta_i \cos^2 \varphi_i \right\rangle = N \langle \sin^2 \theta_i \cos^2 \varphi_i \rangle$$

$$= N \frac{1}{2} \int_0^{\pi} \sin^3 \theta \, d\theta \int_0^{2\pi} \frac{\cos^2 \varphi}{2\pi} \, d\varphi = \frac{N}{3}$$

$$\langle L_y^2 \rangle = \left\langle \left(\sum_{i=1}^{N} \sin \theta_i \sin \varphi_i \right)^2 \right\rangle$$

$$= N \frac{1}{2} \int_0^{\pi} \sin^3 \theta \, d\theta \int_0^{2\pi} \frac{\sin^2 \varphi}{2\pi} \, d\varphi = \frac{N}{3}$$

$$\langle L_z^2 \rangle = \left\langle \left(\sum_{i=1}^{N} \cos \theta_i \right)^2 \right\rangle$$

$$= \left\langle \left(\sum_{i=1}^{N} \cos^2 \theta_i + \sum_{i=1}^{N} \sum_{j \neq i}^{N-1} \cos \theta_i \cos \theta_j \right) \right\rangle$$

$$= \left\langle \sum_{i=1}^{N} \cos^2 \theta_i \right\rangle = \frac{N}{2} \int_0^{\pi} \cos^2 \theta \sin \theta \, d\theta = \frac{N}{3}$$

The probability distributions for projections L_x, L_y, and L_z are equal and Gaussian with variance $\langle L_i^2 \rangle = N/3$.

In view of the central limit theorem, we may approximate the probability distribution $p(L)$ by a Gaussian distribution ($0 \leq L \leq N$; L cannot go to infinity unless the specimen is infinite):

$$p(L) = \frac{1}{\sqrt{2\pi \langle L^2 \rangle}} \exp\left\{ -\frac{L^2}{2 \langle L^2 \rangle} \right\} = \frac{1}{\sqrt{2\pi (N/3)}} \exp\left\{ -\frac{3L^2}{2N} \right\}$$

Let the unit vector $\mathbf{e}_i$ associated with each molecule now indicate the direction of the dipole moment μ of the molecule. The variance of the fluctuating moment produced by N independently oriented dipole molecules is then $\mu^2(N/3)$, and the probability density for the instantaneous moment measurable in an arbitrarily selected direction is

$$p(m) = \frac{1}{\sigma'\sqrt{2\pi}} \exp\left\{-\frac{m^2}{2\sigma'^2}\right\}$$

where we now have $\sigma'^2 = \mu^2(N/3)$. Compared with the previous equation, we find

$$\frac{\sigma'^2}{\sigma^2} = \frac{\mu^2 N}{3\bar{\alpha}kT}$$

If we substitute for $\bar{\alpha}$ the value derived from the linearized Langevin function for the electric polarizability of a collection of dipoles in a small directing field E_r

$$\langle m \rangle = \bar{\alpha}E_r = \frac{\mu^2 N}{3kT}E_r$$

we find that the two derivations lead to the same result. This is, of course, due to the fact that Langevin's function does not consider couplings between the dipole orientations other than those mediated by the direction field E_r and those included in the "effective" dipole moment μ of a molecule. For a solution in a polar medium, $\bar{\alpha}$ and $\langle m^2 \rangle$ contain contributions from the permanent dipole moments and the molecular polarizabilities of the solvent and the dissolved molecules. Although we may neglect direct couplings between the dissolved molecules in a dilute solution, these molecules remain strongly coupled to the solvent. As compared to a system of identical orientable molecules, there are now additional degrees of freedom for fluctuations, e.g., those in which a local excess moment generated by the solvent molecules produces a correlated orientation of dissolved molecules that may add to or partly compensate the other fluctuations in the electric moment:

$$\langle m^2 \rangle = \langle (m_S + m_s)^2 \rangle = \langle m_S^2 + m_s^2 + 2m_S m_s \rangle$$

In such a system, can the variance of the deviation from the isotropic distribution of the dissolved molecules become larger than that of the same molecules at the same density in a gas? In a gas of uncoupled molecules, the variance expresses a pure entropy term. This system's entropy change is of the order of Boltzmann's constant k for each mode of fluctuation. As long as different orientations are degenerate (i.e., do not differ in energy), the fluctuation variance is temperature independent and limited by the loss of configurational entropy. In the coupled system, the free energy is lowered due to the interactions. Fluctuations of orientation will become larger only if, one molecule having a given orientation, another molecule assumes the same orientation with greater probability. Correlations between the orientations of the dissolved anharmonic molecules could be mediated by the solvent, but we would expect this effect to be concentration dependent and to disappear at infinite dilution. In concentrated solutions and in pure liquids of dipolar molecules with anharmonic polarizability, the sign and extent of orientational correlation could be studied by comparing the relative amounts of anisotropic and anharmonic scattering.

VI. THE EQUIVALENT INTERNAL FIELD MODEL EXTENDED

Based on the internal reference method (IRM), it is possible to accurately determine the first hyperpolarizability of small solvent molecules. A solute molecule with a known β_s value is then used as the internal reference for the experimental determination of the first hyperpolarizability of the solvent β_S. Since reference and unknown values are measured in the same local field, no local field corrections are necessary. From the literature value of 34.5×10^{-30} esu for *para*-nitroaniline in methanol,[22] an absolute value of $(0.69 \pm 0.07) \times 10^{-30}$ esu for the first hyperpolarizability of methanol was obtained.

Without internal reference, it is still possible to calibrate against a solution with known B^2. Since the local field may be different then, local field corrections at optical frequency have to be applied. Note that with HRS, local field correction factors at zero frequency are never needed. In this way, the first hyperpolarizability of nitromethane was determined to be $(1.82 \pm 0.33) \times 10^{-30}$ esu. These values are in good agreement with reported values for other small, saturated molecules.

A linear relationship between the first hyperpolarizability β and the dipole moment μ for monosubstituted benzene derivatives has been derived on the basis of the equivalent internal field (EIF) model.[23] This model has been shown to hold also qualitatively for monosubstituted stilbene derivatives.[24] The results obtained for the small saturated solvent molecules, tabulated in Table IV, suggest the same relationship. The case of methane derivatives has not been considered in the EIF model, where the first hyperpolarizability β is shown to be equal to $3(\gamma_\pi/\alpha_\pi)\,\Delta\mu_x$, with

TABLE IV
Values for the Dipole Moment μ and Absolute Value of First Hyperpolarizability
$|\beta|$ for Small Saturated Molecules

Solvent	μ (10^{-18} esu)	$|\beta|$ (10^{-30} esu)
Methane	0.0	0.0^a
Chloroform	1.01	0.49 ± 0.05^b
Methanol	1.70	0.69 ± 0.07
Nitromethane	3.46	1.82 ± 0.33

$^a\beta_{xyz} = 0.01 \times 10^{-30}$ esu.[7]
^{b}From Ref. 48.

$\Delta\mu_x = \alpha_\pi E_0$ the *mesomeric* moment caused by the substitution on the polarizable π-electron system, resulting in an equivalent field E_0. The same concept can also be applied to a saturated molecule where only σ-electron-related polarizability is present. The induced dipole moment in the case of methane is $\mu_{ind} = \alpha_\sigma E_{tot} + \gamma_\sigma E_{tot}^3$. This results in an analogous expression for the first hyperpolarizability β in terms of α_σ and γ_σ,

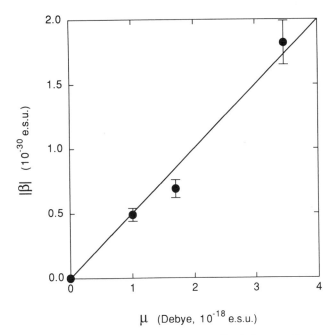

Figure 5. Absolute value of the first hyperpolarizability $|\beta|$ versus dipole moment for CH_4, $CHCl_3$, CH_3OH, and CH_3NO_2. $\cdot$, Experimental points; —, fitted curve with slope 0.50×10^{-12} esu.

$\beta = 3(\gamma_\sigma/\alpha_\sigma)\Delta\mu_x$. An important difference is that $\Delta\mu_x$ is now the *electromeric* moment, originating solely from localized, but polarizable σ-electrons. The assumptions can then be made that $\Delta\mu_x$ is equal to the static dipole moment μ of the saturated molecule.[25] A plot of the first hyperpolarizability β as a function of the dipole moment μ (see Fig. 5) then should be linear with a slope equal to $3(\gamma_\sigma/\alpha_\sigma)$. The reported values for α_σ, the polarizability of methane, 2.6×10^{-24} esu,[26] and for γ_σ, the second hyperpolarizability of methane, 0.42×10^{-36} esu,[27] result in a theoretical slope of 0.48×10^{-12} esu. The experimentally determined slope is $(0.50 \pm 0.02) \times 10^{-12}$ esu. Although the approximation that the first hyperpolarizability tensor for monosubstituted methane derivatives is dominated by only one component is a fairly crude one, the excellent agreement between experimental and theoretical slope seems to justify this approximation and suggests the extension of the equivalent internal field model from conjugated π-electron systems to fully saturated molecules with only σ-electron-related polarizabilities.

VII. MOLECULES WITHOUT DIPOLE MOMENT: OCTOPOLES

The third-rank tensor β_{ijk} must conform to the symmetry of the molecule. As a consequence, for many symmetries, some of the elements of β_{ijk} will vanish and others will be required to have identical values. For equal frequencies of the exciting fields E_j and E_k, β_{ijk} must be symmetric in the second and third index j and k. If the imaginary parts of β_{ijk} vanish (no absorption in the frequency range ω_0 to $2\omega_0$), then β_{ijk} must be symmetric with respect to a permutation of all three indices (Kleinmann symmetry).[28] Under these conditions, β_{ijk} is a totally symmetric tensor, even for molecules lacking all symmetry operations except the identity operator. There are then at most 10 independent components. They consist of at most three dipolar (β_{111}, β_{222}, and β_{333}) and at most seven octpolar (β_{121}, β_{122}, β_{131}, β_{132}, β_{133}, β_{232}, and β_{233}) contributions. As a result, molecular symmetries excluding the dipolar but allowing octopolar contributions will still give rise to second-harmonic scattering.[29]

The classical molecular requirements for second-order NLO applications have led to the design and optimization of highly polarizable, asymmetric organic molecules, typically asymmetrically *para*-substituted polar benzene, azobenzene, stilbene derivatives and polyenes.[11] Octopolar molecules represent a new class of molecules that are potentially useful for NLO applications, since these nonpolar molecules combine favorable NLO properties with a strict cancellation of all vectorlike observables, including the ground- and excited-state dipole moment. The advantages of using nonpolar species for NLO applications include easier (noncentro-symmetric) crystallization, no dipolar interaction toward (centrosymmetric) aggregate formation, better ratio of off-diagonal versus diagonal tensor components, and an improved efficiency-transparency trade-off.[29] Based on these advantages, octopolar molecules are expected to become basic components of novel NLO materials.

Molecules with tetrahedral (T) or with D_{3h} symmetry belong to this class. Since they do not have a permanent dipole moment, they are not oriented by an electric field. It is therefore not possible to measure their hyperpolarizability by the usual EFISHG technique. The HRS technique is the only technique that makes the first hyperpolarizability tensor components of octopoles experimentally accessible. For a tetrahedral molecule, there is all but one nonzero tensor component ($\beta = \beta_{123}$). Using the same experimental conditions as before (Section II), we obtain

$$\langle \mu_z^2 \rangle = \tfrac{12}{35}\beta^2 E_z^4$$

$$\langle \mu_x^2 \rangle = \langle \mu_y^2 \rangle = \tfrac{8}{35}\beta^2 E_z^4$$

with a depolarization ratio of $\tfrac{2}{3}$.

In the case of D_{3h} symmetry, all but four components of $\beta_{(D_{3h})}$ vanish, and they all have the same absolute value. Depending on the definition of the axes ξ and η of the molecular coordinate system (the ζ axis coinciding with the threefold rotational symmetry axis), several representations of $\beta_{(D_{3h})}$ are possible. One of these, with the y axis coinciding with one of the dihedral axes, leads to $\beta_{112} = \beta_{121} = \beta_{211} = -\beta_{222} = \beta$. The usual experimental conditions yield

$$\langle \mu_z^2 \rangle = \tfrac{8}{35}\beta^2 E_z^4$$

$$\langle \mu_x^2 \rangle = \langle \mu_y^2 \rangle = \tfrac{16}{105}\beta^2 E_z^4$$

with a depolarization ratio of $\tfrac{2}{3}$.

Although molecules with D_{3h} or tetrahedral symmetry do not have a dipole moment, their measured hyperpolarizability from dilute solution scattering may still be influenced by the fields and hyperpolarizabilities of neighboring polar solvent molecules. However, based on the absence of a dipole moment, this solvent dependence should be reduced with respect to dipolar molecules. A number of β measurements on trinitrophloroglucinol (2,4,6-trinitro-1,3,5-trihydroxybenzene) in different solvents indicate that, within experiment error, the obtained values for β are independent of the

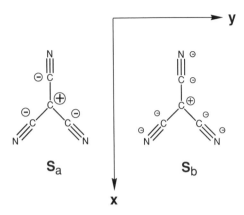

Figure 6. Molecular structure and definition of molecular coordinates for the octopolar tricyanomethanide anion $[C(CN_3)]^-$, together with a pictorial representation of the charge transfers in the two excited states that provide the most significant contributions to the first hyperpolarizability β. The plus (minus) signs represent a decrease (increase) in π-electron charge.

solvent.[30] This is an additional indication of the validity of the theory developed in Section V for the solvent dependence of the β values in terms of the reaction field.

The first hyperpolarizability of an octopolar ion was recently reported. Hyper-Rayleigh scattering from the planar, nonpolar, but highly polarizable tricyanomethanide $[C(CN)_3]^-$ ion was observed in organic and aqueous solution.[31] The molecular structure and the definition of the molecular coordinates are shown in Fig. 6, together with a pictorial representation of the charge transfers in the two excited states that contribute most significantly to the first hyperpolarizability. The INDO/SCI results indicate that the first state (S_a) is located 4.6 eV above the ground state and involves mostly a polarization of the 3 C–C bonds, with the central carbon becoming more positive. The second state (S_b), located 7.0 eV above the ground state, corresponds to a polarization along all the bonds, with the central carbon also becoming more positive, but to a lesser extent. The experimental values of the first hyperpolarizability β ($(7 \pm 1.5) \times 10^{-30}$ esu, independent of the solvent) are in good agreement with the results from quantum chemical calculations.[31] The observed depolarization ratio of 0.7 ± 0.1, in full agreement with $\frac{2}{3}$, as expected on the basis of the symmetry of the scatterers, is an indication that there are no molecular correlations.[5]

VIII.   IONIC AND ABSORBING SPECIES

An important advantage of the HRS technique over the classical EFISHG technique is the possibility of β measurements in conducting solutions and, more specifically, on ionic NLO chromophores. Such ionic species may be of importance for NLO applications since denser molecular packing can result from ionic interactions between chromophores. Ionic species often have an absorption spectrum that is red-shifted with respect to the spectrum of the neutral species. While this has a positive effect on the value of the first hyperpolarizability β through resonance enhancement, the proximity of an electronic resonance often results in absorption of the second-harmonic scattered light. The expression for the HRS signal, i.e., the detected second-harmonic light intensity, has to be slightly modified to account for absorption at the second-harmonic wavelength:

$$S_J(2\omega) = G_J\left(N_S\langle\beta_{zzz}^2\rangle_S + N_s\langle\beta_{zzz}^2\rangle_s\right)e^{-\sigma(2\omega)lN_s}I_Z^2$$

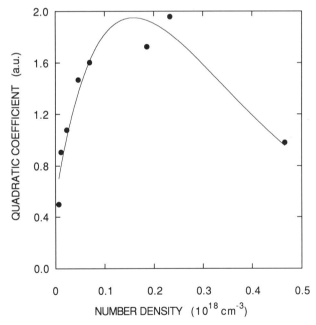

Figure 7.   Quadratic coefficient GB^2 as a function of number density N_s of the 4-oxy-4'-nitrostilbene anion in methanol at 293 K.

TABLE V
First Hyperpolarizablity β of 4-Hydroxy-4'-nitrostilbene and Its Anion at 1064 nm
and Extinction Coefficient ε at 532 nm Determined by Spectrophotometry
and Obtained from HRS Measurements in Chloroform

β_{HONS} $(10^{-30}$ esu)	287 ± 80
β_{ONS^-} $(10^{-30}$ esu)	847 ± 250
$\varepsilon_{ONS^-, HRS}$ $(M^{-1}$ cm$^{-1})^a$	2900 ± 300
$\varepsilon_{ONS^-, SPEC}$ $(M^{-1}$ cm$^{-1})$	3290

[a]Assuming an optical path length in HRS cell of 0.5 ± 0.1 cm.

In this formula, $\sigma_{2\omega}$ is the absorption cross section in cm^2 of the solute molecule and l is the effective pathlength.

From the appearance of the last equation, it is clear the HRS signal as a function of number density of the solute will display a maximum. This was experimentally confirmed for the 4-tetrabutylammonium-oxy-4'-nitrostilbene, as shown in Fig. 7. This NLO chromophore dissociates in solution to form the 4-oxy-4'-nitrostilbene anion. The data analysis results in the experimental value of the first hyperpolarizability β and the absorption cross section of the solute $\sigma_{2\omega}$. This value can be compared with the spectrophotometrically determined extinction coefficient for the solution. These values agree well, within the experimental uncertainty of the path length from focus to cell window. The experimental results (β values and extinction coefficients) are given in Table V.

IX.   SYNTHETIC POLYMERS AND PROTEINS

The potential of the HRS technique is also demonstrated by the possibility of measuring first hyperpolarizabilities of NLO molecules covalently linked to synthetic polymers and of biopolymers with an NLO moiety. The limited applicability of second-harmonic scattering for the determination of the shape of macromolecules was realized early on.[32] Since it proved to be difficult to measure β values with the first-generation pulsed power lasers, it was suggested that only the angular dependence of the second-harmonic scattered light be measured to determine the correlation length of the orientation distribution function of the polymer segments.[5] Now, with the possibility of measuring actual values for β of NLO chromophores with HRS, also in polymers, the problem of the orientation correlation can be addressed by direct determination of an effective hyperpolarizability β_{eff}. The β values for a monomeric chromophore and the β_{eff} value for this chromophore covalently attached to a polymeric backbone at different loading levels are given in Table VI. The structure

TABLE VI
First Hyperpolarizability β for Monomeric Chromophore (4-dimethylamino-4'-cyanostilbene, DMACS) and Effective First Hyperpolarizability β_{eff} for the Same Chromophore Covalently Attached to Poly(methyl methacrylate) (PMMA) Backbone at Different Loading Fractions x

Material	β $(10^{-30}$ esu)
DMACS	240 ± 20[a]
OMACS[b]	250 ± 50
PMMA-MACS, $x = 0.103$	220 ± 30
PMMA-MACS, $x = 0.164$	230 ± 40
PMMA-MACS, $x = 0.225$	230 ± 40

[a]Determined by EFISHG.
[b]4-Octadecylmethylamino-4'-cyanostilbene, same chromophore with aliphatic chain for incorporation in Langmuir-Blodgett film.

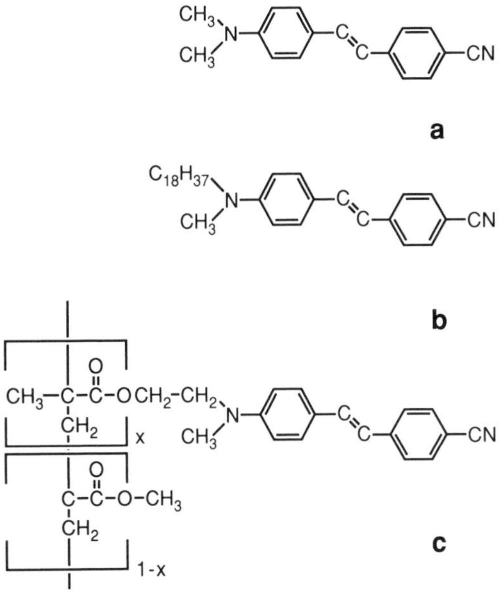

Figure 8. Molecular structure of (a) the NLO chromophore 4-dimethylamino-4'-cyanostilbene, DMACS; (b) the same NLO chromophore with an aliphatic $-C_{18}H_{37}$ chain covalently linked, 4-octadecylmethylamino-4'-cyanostilbene, OMACS; and (c) the poly(methyl methacrylate) polymer backbone, covalently functionalized with the same NLO chromophore as pendant side chains PMMA-MACS, with varying loading fraction x.

of the NLO chromophore, 4-dimethylamino-4′-cyanostilbene (DMACS), and of the side chain methyl methacrylate polymer are shown in Fig. 8. The agreement between the β value for the chromophore, as measured both by HRS and EFISHG, and the β_{eff} value for the chromophore in the polymer, as measured for the first time by HRS, seems to indicate that even for a relatively high degree of covalent functionalization of the repeat units with NLO chromophores, there is little correlation in the side-chain orientation fluctuation.

The experimental determination of the first hyperpolarizability of proteins is another possibility offered by the HRS technique. Proteins often carry a net charge. The application of an external dc field to orient the proteins then induces a charge migration rather than a preferred dipolar orientation. Even at the isoelectric point of the protein (for that pH of the solution at which there is no net charge, depending on the amino acid content of the protein), the external field causes orientation of the protein determined by the location of the charged amino acids, rather than by the static dipole moment of the NLO chromophore embedded in the protein. An indirect method for the determination of the first hyperpolarizability β of a charged protein has been devised, based on the relation between β and the two-photon absorptivity under near resonance condition.[33, 34] We have demonstrated the applicability of the HRS technique for the direct and, hence, much simpler and potentially more accurate determination of β of proteins. The value for the first hyperpolarizability obtained through HRS measurements for bacteriorhodopsin, suspended in Triton X 100 in acetate buffer, is $(2050 \pm 100) \times 10^{-30}$ esu. This value is in agreement with results from SHG experiments on purple membrane poly(vinyl alcohol) films,[35] and confirms the potential of the protein for NLO applications such as second-harmonic generation and electric field-induced spatial light modulation. Comparing this β value for bacteriorhodopsin with the effective first hyperpolarizability β of the NLO moiety, *all-trans* retinal, in different solvents, with different values for the reaction field E_{rf} will allow the assessment of the internal field in the protein. Different orientational fluctuations for the retinal in solvents and in the protein will have to be considered, although the agreement of the β values obtained from SHG in thin films[35] and from HRS in solution do seem to indicate that the HRS measurements are performed in a regime where the amount of orientation fluctuation has no effect on the HRS signal intensity.

X. EXPERIMENTAL ASPECTS OF HYPER-RAYLEIGH SCATTERING IN SOLUTION

Since second-order nonlinear effects are forbidden in the electric dipole approximation,[3] the efficiency for hyper-Rayleigh scattering in solution is fairly low. This is why the first reports of hyper-Rayleigh scattering had to await the advent of megawatt peak power pulsed lasers.[1] However, the mere observation of second-harmonic scattered light does not enable the experimental determination of the first hyperpolarizability β of an NLO molecule. The early reports of nonlinear light scattering often only presented depolarization ratios for the nonlinearly scattered light. This depolarization ratio is, in general, a function of a linear combination of all components β_{ijk} of the hyperpolarizability tensor β.[5] In the case of tetrahedral T_d symmetry, there is only one component, β_{123}, of the hyperpolarizability tensor β. This reduces the depolarization ratio for this symmetry class to the constant $\frac{2}{3}$. In the early reports of nonlinear light scattering, a number of depolarization ratios have been reported for the tetrahedral symmetry class, and the deviation from the theoretical value of $\frac{2}{3}$ has been interpreted as indicative of molecular interactions.[1, 5] Quantitative data of hyperpolarizabilities were difficult to obtain with the pulse profile of the mechano-optically Q-switched ruby lasers used at that time.[5, 36] Some initial effort was already put into shaping the temporal and spatial uniformity of the output pulse, but it still proved impossible to do accurate intensity measurements, because useful data could be obtained only when operating the laser at levels above the threshold for stimulated Brillouin scattering (SBS).[37] This previous work thus concentrated on relaxation times, on linewidths, and again on depolarization ratios for small molecules with relatively high symmetry or for other materials chosen for their high threshold for dielectric breakdown.

An interesting technique for obtaining a value for the first hyperpolarizability β was introduced by Mayer,[38] based on the observation of optical harmonic generation in calcite with a dc electric field applied over the crystal.[36] This was the first report of actual values for the (second) hyperpolarizability γ of a series of molecules. The possibility of calculating values for the first hyperpolarizability β was proposed on the basis of the partial alignment of the dipoles of polar molecules. Experimental difficulties are related to the high fields necessary to observe a useful signal: The amplitude of the electrical field to orient the dipoles of the polar molecules has to be close to electrical discharge threshold between the electrodes, and the optical fields are close to the optical breakdown in the solution.

With the increasing interest in possible applications of NLO techniques for optical information processing, the dc electrical-field-induced second-harmonic generation (EFISHG) technique became the standard technique for the experimental determination of the first hyperpolarizability β. Commercially available Nd^{3+}-YAG lasers with active electronic Q-switching[39] and extrapolation procedures[40] for data evaluation were instrumental for more reliable and accurate values of the first hyperpolarizability of series of NLO molecules. An enormous wealth of hyperpolarizability data on different classes of polar molecules resulted in the deduction of important structure–property relationships for large optical nonlinearities.[27] On the basis of the observed trends in the hyperpolarizability, theoretical models were proposed.[22–24] These models led to the design and synthesis of new candidate NLO molecules. The EFISHG technique was then used again for the characterization of these optimized molecules. The choice of this technique was based on necessity rather than convenience, since there are disadvantages associated with the EFISHG technique. This technique is fundamentally limited to dipolar and nonionic species, since the symmetry of the isotropic solution is reduced by the interaction of a dc electrical field with the dipoles of the molecules. The theoretical analysis for the determination of the first hyperpolarizability β then also includes the value of the dipole moment μ and the second hyperpolarizability γ. In most cases, this latter value can be assumed to be negligible. The experimental determination of the dipole moment is often tedious, and the estimated uncertainty on the value rather large, partly due to the increase of the dipole moment, caused by its own reaction field.[15] The presence of a dc electrical field requires an additional local field correction factor at zero frequency. Furthermore, the only value that is really obtained in EFISHG is the scalar product of the dipole moment vector μ with the vectorial part of the hyperpolarizability tensor β. To calculate an actual value for β, an assumption has to be made about the relative orientation of the dipole moment vector μ and the third-rank tensor β (odd-rank tensor with vector properties). For the NLO molecules, designed on the basis of the derived structure–property relations, collinearity of the two vectors could be readily assumed. But, in general, an arbitrary angle between the dipole moment vector and the vector part of the first hyperpolarizability precludes the resolution of the $\mu\beta$ scalar product.

To eliminate these experimental complications and fundamental limitations associated with the presence of the dc electrical field, we recurred to the original nonlinear light scattering experiments in solution. The technique of injection-seeding of Q-switched Nd^{3+}-YAG lasers has dramatically improved the temporal profile of the fundamental laser pulse. Injection seeding is a convenient method to obtain single longitudinal mode operation of a laser.[41, 42] Low-power light from a stable single-frequency master oscillator is injected in a high power Q-switched slave oscillator cavity during the pulse buildup period. The master oscillator is of the monolithic isolated single-mode end-pumped ring (MISER) design.[43] This design, incorporating an effective half-wave plate polarization rotator, Faraday rotator, and polarizer, ensures unidirectional single-mode operation. Spatial hole burning is eliminated in this traveling wave resonator, allowing for higher output power in a single longitudinal mode. The monolithic design with reflective coating applied directly on the surfaces of the Nd:YAG and the use of a GaAlAs diode laser as the pump source provide the necessary frequency stability of the output. A permanent-magnet Faraday isolator between the master oscillator and the slave oscillator protects the master laser from any backward propagating radiation from the slave laser and prevents destabilization feedback into the master oscillator. Finally, temperature control of the diode laser pump, the ring laser, the Faraday isolator, and the internal cooling water ensures a frequency detuning between the master and slave oscillator within the range of the feedback loop of the active stabilization technique based on minimizing the Q-switched pulse buildup time.

The specific advantage of injection seeding for the HRS experiments is not the single-frequency operation as such, but as a consequence of the absence of longitudinal mode beating, the improved temporal pulse profile. The smoother pulse allows the use of higher effective fundamental wavelength pulse amplitudes without the peaks associated with mode beating reaching the threshold of other nonlinear effects or dielectric breakdown. This gives us a wider range of fundamental light intensities, enabling a more accurate determination of the quadratic coefficient in a HRS experiment. It also ensures that all experiments can be performed well below the threshold for stimulated (forward and backward) Raman scattering, stimulated (backward) Brillouin scattering, self-focusing or self-defocusing, and, eventually, breakdown.

The fundamental wavelength of 1064 nm for a Nd^{3+}-YAG laser can be converted to other wavelengths to study the dispersion of the hyperpolarizability. However, the techniques for shifting the Nd^{3+}-YAG laser wavelength to lower (anti-Stokes Raman shifting, frequency doubling, dye laser pumping with the second harmonic, and Raman shifting the dye laser output) or higher wavelength (Stokes Raman shifting) all employ (multiple) nonlinear techniques. Without injection seeding, the amplitude modulation within a single pulse can amount to 100%. We therefore recommend the use of the injection-seeding technique for all types of experiments that use these frequency conversion techniques, and especially NLO experiments, involving a nonlinear interaction of second or third order, since the high-frequency intensity fluctuation due to mode beating would show up quadratically, resp. cubically, enhanced in the signals of these experiments.

It is customary practice to record the HRS signal (intensity of the second-harmonic scattered light) as a function of fundamental light intensity for a number of different intensities, although we also often mea-

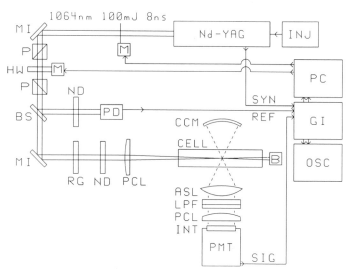

Figure 9. Experimental setup for hyper-Rayleigh scattering in solution. ASL, aspheric lens condenser; B, beamstop; BS, beamsampler; CCM, concave mirror; GI, gated integrator; HW, half-wave plate; INJ, injection seeder; INT, interference filter; LPF, low-pass filter; M, step motor; MI, mirror; ND, neutral density filter; OSC, fast oscilloscope; P, polarizer; PC, personal computer with ADC board and step motor control; PCL, planoconvex lens; PD, fast photodiode; PMT, photomultiplier tube; REF, reference signal; RG, high-pass filter; SIG, hyper-Rayleigh scattering signal; SYN, synchronization signal.

sure the HRS signal at a fixed fundamental intensity. Both procedures allow for the determination of an experimental error, but the former method has the advantage that the quadratic dependence can be checked. With an injection-seeded laser, it is no longer possible to vary the output power by simply varying the flash lamp current. The laser should always be run at a specified lamp current, resulting in a constant average output power. A convenient way to change the fundamental light intensity is then by rotation of a half-wave plate in front of a polarizer (See Fig. 9). The output of the Q-switched laser is not completely polarized. To obtain complete extinction, an additional polarizer in front of the half-wave plate is then necessary. Automatic variation of the fundamental light intensity is possible by (computer-controlled) rotation of the half-wave plate. A small fraction of the fundamental intensity is sampled and directed onto a fast photodiode for measuring the fundamental light intensity and monitoring the injection-seeding process. Stable seeding results in a temporally smooth pulse that is developed in a shorter buildup time.[41]

The low efficiency of the incoherent scattering of second-harmonic photons suggests the use of a condenser system for the efficient collection of useful photons over a wide solid scattering angle. A condenser consists of a concave mirror to retroreflect any light scattered away from the detector, an aspheric lens to condense the light scattered over a wide solid angle, and a planoconvex lens to focus the light on the detector. Since hyper-Rayleigh scattering is not a coherent process, a long interaction length is not required. It is preferable, however, not to use a short focal length lens (with smaller beam waist) to achieve a higher power density. The position of the focus is a function of the refractive index of the solution, but this effect can be minimized with the use of a long focal length lens. This lens is mounted on a translation stage, as are the retroreflecting mirror and aspheric lens of the condenser. This allows for optimal spatial overlap of the foci of mirror and lenses. Spectral discrimination is achieved through the use of appropriate optical filters to eliminate linearly scattered infrared photons. To ensure that only the second-harmonic photons of the right wavelength can be detected, a 3-nm-bandwidth interference filter is screwed on the μ-metal shield of the photomultiplier tube (PMT). This whole setup is covered in a light-tight box with one single optical window with infrared bandpass filter. The choice of the PMT was dictated by low dark current and high linear range.

The actual (relative) values of the amplitudes of the pulsed signals is measured with gated integrators. The timing reference for the gate window is obtained from the sync output of the laserhead. It did not prove advantageous to use optical triggering derived from a reference photodiode. The measurement procedure is completely automated, including laser beam blanking to adjust offset levels, variation of the fundamental light intensity by rotation of the half-wave plate, and analog-to-digital conversion of the fundamental and second-harmonic intensities. A detailed discussion of the HRS instrument, with description of the optical components, has been given previously.[9]

XI. CONCLUSION

With its experimental simplicity and straightforward data analysis, the technique of hyper-Rayleigh scattering in solution offers an attractive alternative over the EFISHG technique for the experimental determination of the first hyperpolarizability β of classical, dipolar, NLO molecules. The HRS technique has been used to show the applicability of the equivalent internal field model on small saturated molecules and to explain the solvent dependence of the first hyperpolarizability β in terms of the reaction field. Hyper-Rayleigh scattering is the only technique that allows the experimental determination of the first hyperpolarizability β of molecules without a dipole moment or of ionic molecules. This allows the experimental study of new classes of promising materials for NLO applications. Hyper-Rayleigh scattering can also be used for the characterization of synthetic polymers functionalized with NLO chromophores and of proteins with an NLO moiety. The extremely broad applicability of the HRS technique, with new possibilities emerging every day, makes it a very promising technique in the field of NLO characterization.

APPENDIX A

The angles ϕ and θ defined here are chosen to correspond to the first and second Eulerian angles for transforming the right-handed coordinate system x, y, z into ξ, η, ζ. The third Eulerian angle χ remains zero. The transformation matrices T resp. $T^{-1} = T^{\text{transpose}}$ are then

$$\begin{pmatrix} E_\xi \\ E_\eta \\ E_\zeta \end{pmatrix} = \begin{pmatrix} \cos\varphi\cos\theta & \sin\varphi\cos\theta & -\sin\theta \\ -\sin\varphi & \cos\varphi & 0 \\ \cos\varphi\sin\theta & \sin\varphi\sin\theta & \cos\theta \end{pmatrix} \begin{pmatrix} E_x \\ E_y \\ E_z \end{pmatrix}$$

$$\begin{pmatrix} \mu_x \\ \mu_y \\ \mu_z \end{pmatrix} = \begin{pmatrix} \cos\varphi\cos\theta & -\sin\varphi & \cos\varphi\sin\theta \\ \sin\varphi\cos\theta & \cos\varphi & \sin\varphi\sin\theta \\ -\sin\theta & 0 & \cos\theta \end{pmatrix} \begin{pmatrix} \mu_\xi \\ \mu_\eta \\ \mu_\zeta \end{pmatrix}$$

For molecules with lower symmetry, the orientation of a molecule with respect to the x, y, z axes must be specified with the full set of Eulerian angles ϕ, θ, χ with

$$T = \begin{pmatrix} \cos\varphi\cos\theta\cos\chi - \sin\varphi\sin\chi & \sin\varphi\cos\theta\cos\chi + \cos\varphi\sin\chi & -\sin\theta\cos\chi \\ -\cos\varphi\cos\theta\sin\chi - \sin\varphi\cos\chi & -\sin\varphi\cos\theta\sin\chi + \cos\varphi\cos\chi & \sin\theta\sin\chi \\ \cos\varphi\sin\theta & \sin\varphi\sin\theta & \cos\theta \end{pmatrix}$$

and the transposed matrix $T^{\text{transpose}}$, giving

$$E_{(\xi, \eta, \zeta)} = TE_{(x, y, z)}$$

and

$$\mu_{(x, y, z)} = T^{\text{transpose}}\mu_{(\xi, \eta, \zeta)}$$

$$T^{-1} = T^{\text{transpose}}$$

Writing the transformation matrix $T = t_{iK}$ as a mixed second-order tensor

t_K^i, and using the summing convention over repeated upper and lower indices, the required transformations may be written as

$$\mu_K = \beta_{ijk} t_{KLM}^{ijk} E^L E^M = \beta_{KLM} E^L E^M$$

where t_{KLM}^{ijk} is the product $t_K^i t_L^j t_M^k$. The lowercase indices i, j, k refer to molecule-fixed coordinates; the uppercase indices K, L, M refer to laboratory-fixed coordinates.

APPENDIX B

$$\langle A_x \rangle = \frac{3\,\delta\alpha}{v} \left\langle \sum_{i=1}^{N} \cos\theta_i \sin\theta_i \cos\varphi_i \right\rangle$$

$$= \frac{3\,\delta\alpha}{v} \frac{N}{4\pi} \int_0^{2\pi} \cos\varphi\,\mathrm{d}\varphi \int_0^{\pi} \cos\theta \sin^2\theta\,\mathrm{d}\theta = 0$$

$$\langle A_y \rangle = \frac{3\,\delta\alpha}{v} \left\langle \sum_{i=1}^{N} \cos\theta_i \sin\theta_i \sin\varphi_i \right\rangle$$

$$= \frac{3\,\delta\alpha}{v} \frac{N}{4\pi} \int_0^{2\pi} \sin\varphi\,\mathrm{d}\varphi \int_0^{\pi} \cos\theta \sin^2\theta\,\mathrm{d}\theta = 0$$

$$\langle A_z \rangle = \frac{\delta\alpha}{v} \left\langle \sum_{i=1}^{N} \left(2\cos^2\theta_i - \sin^2\theta_i\right) \right\rangle$$

$$= \frac{\delta\alpha}{v} \left(N \int_0^{\pi} \cos^2\theta \sin\theta\,\mathrm{d}\theta - \frac{N}{2} \int_0^{\pi} \sin^3\theta\,\mathrm{d}\theta \right) = 0$$

$$\langle B_x \rangle = \frac{\beta_{333}}{v} \left\langle \sum_{i=1}^{N} \cos^2\theta_i \sin\theta_i \cos\varphi_i \right\rangle$$

$$= \frac{\beta_{333}}{v} \frac{N}{4\pi} \int_0^{2\pi} \cos\varphi\,\mathrm{d}\varphi \int_0^{\pi} \cos^2\theta \sin^2\theta\,\mathrm{d}\theta = 0$$

$$\langle B_y \rangle = \frac{\beta_{333}}{v} \left\langle \sum_{i=1}^{N} \cos^2\theta_i \sin\theta_i \sin\varphi_i \right\rangle$$

$$= \frac{\beta_{333}}{v} \frac{N}{4\pi} \int_0^{2\pi} \sin\varphi\,\mathrm{d}\varphi \int_0^{\pi} \cos^2\theta \sin^2\theta\,\mathrm{d}\theta = 0$$

$$\langle B_z \rangle = \frac{\beta_{333}}{v} \left\langle \sum_{i=1}^{N} \cos^3\theta_i \right\rangle$$

$$= \frac{\beta_{333}}{v} \frac{N}{2} \int_0^{\pi} \cos^3\theta \sin\theta\,\mathrm{d}\theta = 0$$

The average polarization of all components is zero. The variances are then equal to the second moments:

$$
\langle A_x^2 \rangle = \frac{9\,\delta\alpha^2}{v^2} \left\langle \left(\sum_{i=1}^{N} \cos\theta_i \sin\theta_i \cos\varphi_i \right)^2 \right\rangle
$$

$$
= \frac{9\,\delta\alpha^2}{v^2} \left\langle \sum_{i=1}^{N} \cos^2\theta_i \sin^2\theta_i \cos^2\varphi_i \right.
$$

$$
\left. + \sum_{i=1}^{N} \sum_{j\neq i}^{N-1} \left(\cos\theta_i \sin\theta_i \cos\varphi_i\right)\left(\cos\theta_j \sin\theta_j \cos\varphi_j\right) \right\rangle
$$

$$
= \frac{9\,\delta\alpha^2}{v^2} \left(\frac{N}{4\pi} \int_0^{2\pi} \cos^2\varphi \, \mathrm{d}\varphi \int_0^{\pi} \cos^2\theta \sin^3\theta \, \mathrm{d}\theta \right.
$$

$$
\left. + \frac{N(N-1)}{16\pi^2} \left\{ \int_0^{2} \cos\varphi \, \mathrm{d}\varphi \int_0^{\pi} \cos\theta \sin^2\theta \, \mathrm{d}\theta \right\}^2 \right)
$$

$$
= \frac{3}{5} \frac{N\,\delta\alpha^2}{v^2}
$$

$$
\langle A_y^2 \rangle = \frac{9\,\delta\alpha^2}{v^2} \left\langle \left(\sum_{i=1}^{N} \cos\theta_i \sin\theta_i \sin\varphi_i \right)^2 \right\rangle
$$

$$
= \frac{3}{5} \frac{N\,\delta\alpha^2}{v^2}
$$

$$
\langle A_z^2 \rangle = \frac{\delta\alpha^2}{v^2} \left\langle \left(\sum_{i=1}^{N} \left(2\cos^2\theta_i - \sin\theta_i\right) \right)^2 \right\rangle
$$

$$
= \frac{4}{5} \frac{N\,\delta\alpha^2}{v^2}
$$

$$
\langle B_x^2 \rangle = \frac{\beta_{333}^2}{v^2} \left\langle \left(\sum_{i=1}^{N} \cos^2\theta_i \sin\theta_i \cos\varphi_i \right)^2 \right\rangle
$$

$$
= \frac{1}{35} \frac{N\beta_{333}^2}{v^2}
$$

$$\langle B_y^2 \rangle = \frac{\beta_{333}^2}{v^2} \left\langle \left(\sum_{i=1}^{N} \cos^2 \theta_i \sin \theta_i \sin \varphi_i \right)^2 \right\rangle$$

$$= \frac{1}{35} \frac{N\beta_{333}^2}{v^2}$$

$$\langle B_z^2 \rangle = \frac{\beta_{333}^2}{v^2} \left\langle \left(\sum_{i=1}^{N} \cos^3 \theta_i \right)^2 \right\rangle$$

$$= \frac{1}{7} \frac{N\beta_{333}^2}{v^2}$$

Acknowledgments

This research was supported by research grants from the Belgian government (GOA 87/91-109) and from the Belgian National Science Foundation (FKFO 2.9003.90). K. Clays is a research associate of the Belgian National Fund for Scientific Research.

References

1. R. W. Terhune, P. D. Maker, and C. M. Savage, *Phys. Rev. Lett.* **14**, 681 (1965).

2. D. L. Weinberg, *J. Chem. Phys.* **47**, 1307 (1967).

3. J. A. Giordmaine, *Phys. Rev.* **138**, 1599 (1965).

4. S. J. Cyvin, J. E. Rauch, and J. C. Decius, *J. Chem. Phys.* **43**, 4083 (1965).

5. R. Bersohn, Y.-H. Pao, and H. L. Frisch, *J. Chem. Phys.* **45**, 3184 (1966).

6. S. Kielich, *Chem. Phys. Lett.* **1**, 441 (1967).

7. A. D. Buckingham and B. J. Orr, *Q. Rev. Chem. Soc.* **21**, 195 (1967).

8. K. Clays and A. Persoons, *Phys. Rev. Lett.* **66**, 2980 (1991).

9. K. Clays and A. Persoons, *Rev. Sci. Instrum.* **63**, 3285 (1992).

10. A. D. Buckingham and J. A. Pople, *Proc. Phys. Soc. London Ser. A* **68**, 905 (1955).

11. D. J. Williams, *Angew. Chem. Int. Ed. Engl.* **23**, 690 (1984).

12. D. J. Williams, *Introduction to Nonlinear Optical Effects in Molecules and Polymers*, Wiley, New York, 1991.

13. B. M. Landanyi and T. Keyes, *Mol. Phys.* **33**, 1063 (1977).

14. T. Keyes and B. M. Landanyi, in I. Prigogine and S. A. Rice (Eds.), *Advances in Chemical Physics*, Vol. 56, Wiley, New York, 1984, p. 411.

15. C. J. P. Böttcher, in *Theory of Electric Polarization*, Elsevier, Amsterdam, 1973, p. 129.

16. G. R. Meredith and B. Buchalter, *J. Chem. Phys.* **78**, 1938 (1983).

17. E. Lippert, *Z. Elecktrochem.* **61**, 962 (1957).

18. W. H. Keesom, *Phys. Z.* **22**, 129 (1921).

19. H. Falkenhagen, *Phys. Z.* **23**, 87 (1922).

20. R. M. Fuoss, *J. Am. Chem. Soc.* **56**, 1027 (1934).

21. R. M. Fuoss, *J. Am. Chem. Soc.* **56**, 1031 (1934).

22. J. L. Oudar and D. S. Chemla, *J. Chem. Phys.* **66**, 2664 (1977).

23. J. L. Oudar and D. S. Chemla, *Opt. Commun.* **13**, 164 (1975).

24. J. L. Oudar, *J. Chem. Phys.* **67**, 446 (1977).

25. B. F. Levine, *J. Chem. Phys.* **63**, 115 (1975).

26. A. T. Amos and R. J. Crispin, *J. Chem. Phys.* **63**, 1890 (1975).

27. B. F. Levine and C. G. Bethea, *J. Chem. Phys.* **63**, 2666 (1975).

28. D. A. Kleinman, *Phys. Rev.* **126**, 1977 (1962).

29. J. Zyss, *Nonlinear Opt.* **1**, 3 (1991).

30. A. Persoons, K. Clays, T. Verbiest, J. J. Wolff, F. Gredel, H. Irngartinger, T. Dreier, and D. W. Robinson, submitted to Angew. Chemie.

31. T. Verbiest, K. Clays, A. Persoons, F. Meyers, and J.-L. Brédas, *Opt. Lett.*, **18**, 525 (1993).

32. R. Bersohn, *J. Am. Chem. Soc.* **86**, 3505 (1964).

33. R. R. Birge, P. A. Fleitz, A. F. Lawrence, M. A. Masthay, and C. F. Zhang, *Mol. Cryst. Liq. Cryst.* **189**, 107 (1990).

34. R. R. Birge and C.-F. Zhang, *J. Chem. Phys.* **92**, 7178 (1990).

35. J. Y. Huang, Z. Chen, and A. Lewis, *J. Phys. Chem.* **93**, 3314 (1989).

36. R. W. Terhune, P. D. Maker, and C. M. Savage, *Phys. Rev. Lett.* **8**, 404 (1962).

37. P. D. Maker, *Phys. Rev. A* **1**, 923 (1970).

38. G. Mayer, *C. R. Acad. Sci. Paris* **267B**, 5457 (1968).

39. G. R. Meredith, *Rev. Sci. Instrum.* **53**, 48 (1982).

40. K. D. Singer and A. F. Garito, *J. Chem. Phys.* **75**, 3572 (1981).

41. R. L. Schmitt and L. A. Rahn, *Appl. Opt.* **25**, 629 (1986).

42. B. Zhou, T. J. Kane, G. J. Dixon, and R. L. Byer, *Opt. Lett.* **10**, 62 (1985).

43. T. J. Kane and R. L. Byer, *Opt. Lett.* **10**, 65 (1985).

44. C. C. Teng and A. F. Garito, *Phys. Rev. B* **28**, 6766 (1983).

45. R. A. Huijts and G. L. J. Hesselink, *Chem. Phys. Lett.* **126**, 209 (1989).

46. R. A. Huijts, unpublished result.

47. L.-T. Cheng, W. Tam, G. R. Meredith, G. L. J. A. Rikken, and E. W. Meijer, *Proc. Soc. Photo-Opt. Instrum. Eng.* **1147**, 68 (1990).

48. F. Kazjar, I. Ledoux, and J. Zyss, *Phys. Rev. A* **36**, 2210 (1987).

CORRELATED AND SQUEEZED COHERENT STATES OF TIME-DEPENDENT QUANTUM SYSTEMS

V. V. DODONOV

and

V. I. MAN'KO

Lebedev Physics Institute, Moscow, Russia

CONTENTS

I. DEFINITION AND MAIN PROPERTIES OF CORRELATED COHERENT STATES

Correlated coherent states (CCS) were introduced in Ref. 1 (see also Ref. 2) as states minimizing the left side of the precise Schrödinger-Robertson uncertainty relation[3-5]

$$\sigma_p \sigma_q - \sigma_{pq}^2 \equiv \sigma_p \sigma_q (1 - r^2) \geq \frac{h^2}{4} \qquad (1)$$

Modern Nonlinear Optics, Part 3, Edited by Myron Evans and Stanisław Kielich. Advances in Chemical Physics Series, Vol. LXXXV.
ISBN 0-471-30499-9 © 1994 John Wiley & Sons, Inc.

for the fixed correlation coefficient

$$r = \frac{\sigma_{pq}}{(\sigma_p \sigma_q)^{1/2}} \tag{2}$$

where q and p are the generalized canonical coordinate and momentum, and σ_q, σ_p, σ_{pq} are their variances and covariance, respectively. We assume hereafter that coordinates and momenta have been made dimensionless in a proper manner, so that Planck's constant formally equals unity, whereas operators $\hat{q}$ and $\hat{p}$ form the operators of annihilation and creation of physical quanta (which are registered in experiments) according to formulas

$$\hat{a} = 2^{-1/2}(q + ip) \qquad \hat{a}^+ = 2^{-1/2}(q - ip) \tag{3}$$

The states for which relation (1) becomes the equality turn out to be eigenstates of the operator[1, 2]

$$\hat{b}(r, \lambda) = \frac{1}{2\lambda}\left[1 - \frac{ir}{\sqrt{1 - r^2}}\right]\hat{q} + i\lambda\hat{p} \tag{4}$$

where

$$\sigma_q = \lambda^2 \qquad \sigma_p = \frac{1}{4\lambda^2(1 - r^2)} \qquad \sigma_{pq} = \frac{r}{2\sqrt{1 - r^2}} \tag{5}$$

so that r and λ are real parameters: $0 < \lambda < \infty$; $-1 < r < 1$.

The explicit expression for a correlated coherent state in the coordinate representation is

$$\psi_\beta(x) = (2\pi\lambda^2)^{-1/4}\exp\left[-\frac{x^2}{4\lambda^2}\left(1 - \frac{ir}{\sqrt{1 - r^2}}\right) + \frac{\beta x}{\lambda}\right.$$
$$\left. - \frac{1}{2}(\beta^2 + |\beta|^2)\right] \tag{6}$$

$$\hat{b}\psi_\beta = \beta\psi_\beta$$

Operators $\hat{a}$ and $\hat{a}^+$ are connected with operators $\hat{b}$ and $\hat{b}^+$ by means of the Bogoliubov canonical transformation

$$\hat{b} = u\hat{a} + v\hat{a}^+ \qquad \hat{b}^+ = u^*\hat{a}^+ + v^*\hat{a} \tag{7}$$

with coefficients

$$u = \frac{1}{2^{3/2}\lambda}\left(1 - \frac{ir}{\sqrt{1-r^2}}\right) + \frac{\lambda}{2^{1/2}}$$

$$v = \frac{1}{2^{3/2}\lambda}\left(1 - \frac{ir}{\sqrt{1-r^2}}\right) - \frac{\lambda}{2^{1/2}} \tag{8}$$

As a consequence we have the identity

$$|u|^2 - |v|^2 = 1 \tag{9}$$

ensuring the validity of the commutation relations

$$[\hat{a}, \hat{a}^+] = [\hat{b}, \hat{b}^+] = 1$$

Average values of the coordinate and momentum in state (6) are

$$\langle q \rangle = \sqrt{2}\,\mathrm{Re}[(u-v)\beta^*] = 2\lambda\,\mathrm{Re}\,\beta$$

$$\langle p \rangle = \sqrt{2}\,\mathrm{Im}[(u^*+v^*)\beta] = \lambda^{-1}\left[\mathrm{Im}\,\beta + \frac{r}{\sqrt{1-r^2}}\,\mathrm{Re}\,\beta\right] \tag{10}$$

Variances can be expressed in terms of the Bogoliubov transformation coefficients u, v according to

$$\sigma_q = \tfrac{1}{2} + |v|^2 - \mathrm{Re}(uv^*)$$

$$\sigma_p = \tfrac{1}{2} + |v|^2 + \mathrm{Re}(uv^*) \tag{11}$$

$$\sigma_{pq} = \mathrm{Im}(uv^*)$$

Now let us discuss the relations given above. The linear canonical transformation (7) is defined by two complex parameters satisfying constraint (9). Consequently, there are only three independent real parameters. However, one of them is trivial, since it corresponds to a multiplication of operator $\hat{b}$ by a factor $\exp(i\phi)$, ϕ being an arbitrary phase. Therefore, only two parameters are significant. This means that correlated coherent states can be considered as the most general pure quantum states described by Gaussian wave functions. In other words, any Gaussian pure state minimizes the generalized uncertainty relation (1), i.e., it is a correlated coherent state.

During the last quarter of this century various generalizations of Glauber's coherent states[6] were subjected to intensive investigations both by theoreticians and experimenters: see, e.g., original papers,[7-19] reviews,[20-24] and papers in Ref. 25. It turns out, however, that eigenstates of operators like (7) can be found in the literature under different names. Each name emphasizes some characteristic feature of these states. For example, the following terms are known: minimal uncertainty states,[13] two-photon states,[15] and squeezed states.[17, 18] The infinite-dimensional generalizations were named Gaussons.[19] Therefore, it is expedient to show the place of CCS in the set of numerous generalizations of Glauber's coherent states. The eigenstates of operators like (7) with $v \neq 0$ are now usually called "squeezed" states. This name is explained by the fact that if $r = 0$ and $\sigma_p \sigma_q = \frac{1}{4}$, then either σ_p or σ_q inevitably occurs less than $\frac{1}{2}$ for any non-Glauber's state. In other words, fluctuations in one of two quadrature components are "squeezed" by amplifying fluctuations of another component. Therefore, the term "squeezed states," in our opinion, well describes the essence of the phenomenon in the case of zero correlation coefficient, when parameters u, v can be assumed to be real (more precisely, their phases coincide). In this case a consequence of (9) is the natural parametrization

$$u = \cosh \tau \qquad v = \sinh \tau \qquad (12)$$

resulting in the following variant of Eqs. (11):

$$\sigma_q = \tfrac{1}{2} e^{-2\tau} \qquad \sigma_p = \tfrac{1}{2} e^{-2\tau} \qquad \sigma_{pq} = 0 \qquad (13)$$

A more general parametrization is chosen frequently in the form

$$u = \cosh \tau \qquad v = \sinh \tau e^{-i\vartheta} \qquad (14)$$

and corresponding states are still called "squeezed," but for $\vartheta \neq 0$ this name does not correspond to the point properly, since instead of (13) the following formulas hold:

$$\sigma_q = \tfrac{1}{2}(\cosh 2\tau - \sinh 2\tau \cos \vartheta)$$
$$\sigma_p = \tfrac{1}{2}(\cosh 2\tau + \sinh 2\tau \cos \vartheta) \qquad (15)$$

so that for values of phase ϑ close to $\pi/2$ both variances can be much greater than fluctuations in Glauber's coherent states. For $\vartheta \neq 0$ the correlation coefficient differs from zero:

$$\sigma_{pq} = \tfrac{1}{2} \sinh 2\tau \sin \vartheta$$

$$r = \frac{\sinh 2\tau \sin \vartheta}{\left[1 + \left(\sinh 2\tau \sin \vartheta\right)^2\right]^{1/2}} \tag{16}$$

For this reason, in our opinion, the term "correlated state," emphasizing the statistical dependence of quadrature components, i.e., their nonzero correlation coefficient, is the most adequate in the case of complex parameters u and v (with different phases). Different parametrizations of generalized coherent states and their unitary equivalence were considered in detail in Refs. 2 and 21.

Instead of parameter λ characterizing the variance of coordinate, i.e., the "absolute" squeezing of its fluctuations, a new parameter

$$k = \frac{\sigma_q}{\sigma_p} \tag{17}$$

characterizing the "relative" squeezing of quadrature components can be introduced. Then

$$\lambda = 2^{-1/2} \left(\frac{k}{1 - r^2}\right)^{1/4}$$

$$u = \frac{1}{2} \left(\frac{1 - r^2}{k}\right)^{1/4} \left(1 - \frac{ir}{\sqrt{1 - r^2}}\right) + \frac{1}{2} \left(\frac{k}{1 - r^2}\right)^{1/4} \tag{18}$$

$$v = \frac{1}{2} \left(\frac{1 - r^2}{k}\right)^{1/4} \left(1 - \frac{ir}{\sqrt{1 - r^2}}\right) - \frac{1}{2} \left(\frac{k}{1 - r^2}\right)^{1/4}$$

Higher moments of the coordinate in CCS can be calculated with the aid of[26]

$$\langle q^n \rangle = (-2)^{-n/2} \lambda^n \mathscr{H}_n\left((-2)^{1/2} \frac{\langle q \rangle}{2\lambda}\right)$$

$$= \left(\frac{i}{2}\right)^n \left(\frac{k}{1 - r^2}\right)^{n/4} \mathscr{H}_n\left(-i\sqrt{2}\, \mathrm{Re}\, \beta\right) \tag{19}$$

where $\mathscr{H}_n(2)$ is the Hermite polynomial.

II. GENERATING CORRELATED COHERENT STATES

Let us discuss in principle the problem of generating correlated coherent states. For this purpose consider a general problem of solving a nonstationary Schrödinger equation for an oscillator with time-dependent frequency $\Omega(t)$ (remember that in the stationary case the frequency is assumed to be equal to unity). The solution to this problem is well known. It was given in great detail in Ref. 26, where it was shown that there exists a complete normalized system of solutions (more precisely, an overcomplete system in the same sense as the usual system of Glauber's coherent states) in the form of Gaussian exponentials

$$\psi_\alpha(x,t) = \pi^{-1/4}\eta^{-1/2}\exp\left(\frac{i\dot{\eta}}{2\eta}x^2 + \frac{\sqrt{2}}{\eta}\alpha x - \frac{\alpha^2\eta^*}{2\eta} - \frac{|\alpha|^2}{2}\right) \quad (20)$$

determined by an arbitrary complex number α and an arbitrary complex solution of the classical equation of motion for a parametrically excited oscillator $\eta(t)$:

$$\frac{d^2\eta}{dt^2} + \Omega^2(t)\eta = 0 \quad (21)$$

Since the complex conjugated function $\eta^*(t)$ also satisfies the same equation, and Wronskian does not depend on time, it is convenient to impose on the solutions a normalization constraint preserved in time:

$$\dot{\eta}\eta^* - \dot{\eta}^*\eta = 2i \quad (22)$$

The choice of constant on the right side corresponds in the case of $\Omega = 1$ to the solution $\eta(t) = e^{it}$. Then function (20) is the Glauber coherent state. Any other solution $\eta(t)$ yields, in general, a correlated state due to[26]

$$\sigma_q = \frac{|\eta|^2}{2} \qquad \sigma_p = \frac{|\dot{\eta}|^2}{2} \qquad \sigma_{pq} = \frac{1}{2}\,\mathrm{Re}(\dot{\eta}\eta^*)$$

$$k = \left|\frac{\eta}{\dot{\eta}}\right|^2 \qquad r = \frac{\mathrm{Re}(\dot{\eta}\eta^*)}{|\eta\dot{\eta}|} \qquad r^2 = 1 - |\eta\dot{\eta}|^{-2}$$

$$(23)$$

which can be obtained if one compares Eqs. (6) and (20) (in formula for r^2

identity (22) is taken into account). Thus, the most natural way to create a correlated coherent state is to subject the oscillator to the parametric swinging. The possibility of obtaining states with variances less than in the ground state by means of parametric swinging was first investigated in detail in Refs. 7, 27, and 28, although the problem of parametric excitation of a quantum oscillator was investigated (from a different point of view) even earlier[29] (see also Refs. 26 and 30, where large references lists are given). In fact, all experimental methods of generating squeezed states correspond to one or the other way of parametric excitation (see, e.g., Refs. 22–24 and 31–34, and references therein).

In general, both the correlation coefficient and the squeezing coefficient vary in time. Let us find such dependences of frequency on time which allow us to construct states with time-independent parameters r and k. Begin with the correlation coefficient. Due to (22) and (23) condition $r = $ const is equivalent to $\mathrm{Re}(\dot{\eta}\eta^*) = $ const. Let us single out the modulus and phase of complex function $\eta(t)$:

$$\eta(t) = \rho(t)\exp[i\varphi(t)] \tag{24}$$

Then $\mathrm{Re}(\dot{\eta}\eta^*) = \rho\dot{\rho} = \frac{1}{2}\,d\rho^2/dt$. Consequently,

$$\rho^2(t) = 2bt + c \qquad b, c = \text{const.} \tag{25}$$

The equation for $\rho(t)$ is obtained from Eq. (21), if one takes into account the identity

$$\dot{\varphi} = \rho^{-2} \tag{26}$$

following from (22). As a result we have the equation

$$\frac{d^2\rho}{dt^2} + \Omega^2(t)\rho = \rho^{-3} \tag{27}$$

allowing us to find the dependence $\Omega(t)$ for the given function $\rho(t)$. For function (25) we get

$$\Omega(t) = \frac{\sqrt{1 + b^2}}{2bt + c}$$

Then correlation and squeezing coefficients can be expressed as

$$k = \frac{\rho^4}{\rho^2 \dot{\rho}^2 + 1} = \frac{(2bt + c)^2}{1 + b^2} = \frac{1}{\Omega^2(t)}$$

$$r = \frac{\rho\dot{\rho}}{\sqrt{\rho^2\dot{\rho}^2 + 1}} = \frac{b}{\sqrt{1 + b^2}}$$

(28)

Consequently, to preserve the correlation coefficient the frequency has to vary according to the law

$$\Omega(t) = \frac{\Omega_0}{2r\Omega_0 t + 1}$$

(29)

It should be mentioned that the arising CCS described by formulas (20) and (24)–(26) is in fact *unsqueezed*, in spite of relation (28). The point is that the squeezing coefficient in (28) was defined with respect to vacuum fluctuations of the oscillator with unit frequency. For the oscillator with frequency ω the vacuum state fluctuations $\sigma_q^{(0)}$ and $\sigma_p^{(0)}$ are related by formula $\sigma_p^{(0)} = \omega^2\sigma_q^{(0)}$ (for unit mass). Therefore, it is more natural to define the relative squeezing coefficient for the oscillator with frequency ω as

$$k_\omega = \frac{\omega^2\sigma_q}{\sigma_p} = \omega^2 k$$

(30)

In the example considered above $k_\omega \equiv 1$ for all instants of time. Moreover, the absolute squeezing is absent at all times, since

$$\sigma_q = \frac{|\eta|^2}{2} = \frac{\rho^2}{2} = (2\Omega)^{-1}[1 - r^2]^{-1/2}$$

$$\sigma_p = \frac{\Omega}{2}[1 - r^2]^{-1/2}$$

and both these values are $(1 - r^2)^{-1/2}$ times higher than zero fluctuations of coordinate and momentum for the oscillator with frequency Ω. Thus, this example demonstrates distinctly the difference between squeezed and correlated states.

Due to the first equality from (28), the condition of independence on the time of coefficient k (i.e., the relative squeezing normalized by the initial unit frequency) leads to

$$\rho\dot\rho = \sqrt{k^{-1}\rho^4 - 1} \qquad (31)$$

Its trivial solution $\dot\rho \equiv 0$, $\rho^4 = k$ corresponds to the constant frequency $\omega = k^{-1/2}$, i.e., to unsqueezed ($k_\omega = 1$) and noncorrelated states. If $\dot\rho \neq 0$, then the integration of (31) results in

$$\rho^2 = k^{1/2}\cosh(2tk^{-1/2})$$

(the initial instant of time is assumed to be zero). From (27) we get $\Omega^2 = -k^{-1}$, which corresponds to the motion in the centrifugal potential $U(x) = -x^2/2k$. In this case absolute values of both variances and the correlation coefficient grow in time monotonically:

$$\sigma_q = \tfrac{1}{2}k^{1/2}\cosh(2tk^{-1/2}) \qquad \sigma_p = \tfrac{1}{2}k^{-1/2}\cosh(2tk^{-1/2})$$

$$r = \tanh(2tk^{-1/2})$$

It is not difficult to derive a general dependence $\Omega(t)$ resulting in any given function $r(t)$. Indeed, due to Eqs. (5), (22), and (23), function $g = \rho^2$ satisfies the equation

$$\dot g = 2\rho\dot\rho = 2r(1 - r^2)^{-1/2} = 4\sigma_{pq}(t) \qquad (32)$$

Therefore, with the given dependence $r(t)$ or $\sigma_{pq}(t)$, function $g(t)$ can be obtained by means of a quadrature. After this function, $\Omega(t)$ can be found from Eq. (27), if one takes into account the relation

$$\ddot g = 4\dot\sigma_{pq} = 2\rho\ddot\rho + 2\dot\rho^2$$

That is,

$$\Omega^2(t) = \frac{1 + 4\sigma_{pq}^2 - 2g\dot\sigma_{pq}}{g^2} = \frac{1 - 4g\dot r(1 - r^2)^{-1/2}}{g^2(1 - r^2)} \qquad (33)$$

III. LIMITATIONS ON POSSIBLE SQUEEZING AND CORRELATION COEFFICIENTS

Certain inequalities for the squeezing and correlation coefficients can be found for arbitrary time dependence of the frequency in Eq. (21) if one takes into account that this equation turns into the Helmholtz equation describing the one-dimensional wave propagation through a nonhomogenious medium after the replacement $t \to x$, $\Omega(t) \to \kappa(x)$ (κ being the wave number). Suppose that function $\Omega(t)$ assumes constant values Ω_i in the remote past and Ω_0 in the future and the initial value of $\eta(t)$ is $\Omega_i^{-1/2}\exp(i\Omega_i t)$ when $t \to -\infty$. Then for $t \to \infty$ we have

$$\eta(t) = \Omega_0^{-1/2}\left[\mu \exp(i\Omega_0 t) + \nu \exp(-i\Omega_0 t)\right] \tag{34}$$

with time-independent coefficients satisfying the relation

$$|\mu|^2 - |\nu|^2 = 1 \tag{35}$$

resulting from (22). Then the ratio ν^*/μ^* can be treated as the amplitude reflection coefficient from the effective "potential barrier" represented by function $\Omega^2(t)$ for the wave going from right to left. On the other hand, Eq. (34) leads to the relation

$$\frac{\nu}{\mu} = \exp(2i\Omega_0)\frac{i\Omega_0\eta - \dot{\eta}}{i\Omega_0\eta + \dot{\eta}}$$

Consequently, the energy reflection coefficient $R = |\nu/\mu|^2$ (which does not depend on the direction of the wave) equals

$$R = \left|\frac{\nu}{\mu}\right|^2 = \frac{\Omega_0^2|\eta|^2 + |d\eta/dt|^2 - 2\Omega_0}{\Omega_0^2|\eta|^2 + |d\eta/dt|^2 + 2\Omega_0} \tag{36}$$

where relation (22) was taken into account. From (22) and (23) we get

$$|\eta|^2 = \left(\frac{k}{1-r^2}\right)^{1/2} \qquad \left|\frac{d\eta}{dt}\right|^2 = \left(k(1-r^2)\right)^{-1/2} \tag{37}$$

Then we can consider (36) as the equation determining k with the given values of R and r. Solving it, one gets for the reduced coefficient (see (30))

$k_* = \Omega_0^2 k$ the following relation:

$$k_* = \left\{ \frac{(1 + R)(1 - r^2)^{1/2} \pm \left[4R - r^2(1 + R)^2\right]^{1/2}}{1 - R} \right\}^2 \qquad (38)$$

The requirement that k_* must be a real number leads to the restriction on possible values of the correlation coefficient with the given value of the energy reflection coefficient from the barrier

$$|r| \le \frac{2R^{1/2}}{1 + R} \qquad (39)$$

The state with the maximum correlation coefficient turns out to be unsqueezed: $k_* = 1$. Analyzing expression (38) as the function of parameter r, one can see that with the given value of R the squeezing coefficient is confined within the limits[35]

$$\frac{1 - R^{1/2}}{1 + R^{1/2}} \le k_*^{1/2} \le \frac{1 + R^{1/2}}{1 - R^{1/2}} \qquad (40)$$

The extremal values are reached for noncorrelated states. Thus, we see that neither squeezing nor correlation can be obtained for reflectionless barriers.

Suppose the oscillator was initially in the vacuum state. Due to (23) its energy for $t \to \infty$ is

$$E = \tfrac{1}{2}\left(\sigma_p + \Omega_0^2 \sigma_q\right) = \tfrac{1}{4}\left[\Omega_0^2 |\eta|^2 + \left|\frac{d\eta}{dt}\right|^2\right] \qquad (41)$$

On the other hand,

$$E = \hbar\Omega_0\left(N + \tfrac{1}{2}\right) \qquad (42)$$

N being the number of quanta in the final state. Comparing (41) and (42) with (36) we see that the number of generated quanta is uniquely related to the energy reflection coefficient from the effective barrier:

$$N = |\nu|^2 = \frac{R}{1 - R} \qquad (43)$$

Then relations (39) and (40) can be rewritten as

$$|r_{max}| = \frac{2\sqrt{N(N+1)}}{1+2N} \qquad k_*^{max} = \left[\sqrt{N} + \sqrt{1+N}\right]^4 \qquad (44)$$

so that

$$1 - 4N^{1/2} \le k_* \le 1 + 4N^{1/2} \qquad \text{for } N \ll 1 \qquad (45)$$

$$(4N)^{-2} \le k_* \le (4N)^2 \qquad \text{for } N \gg 1 \qquad (46)$$

IV. EXPLICIT RESULTS FOR SPECIAL TIME DEPENDENCES OF FREQUENCY

If the frequency varies with time more or less monotonously and sufficiently slow, so that

$$\frac{d\Omega}{dt} \ll \Omega^2 \qquad (47)$$

then Eq. (21) can be solved in adiabatic approximation corresponding to the approximation of geometrical optics for the Helmholtz equation. The solutions of the zeroth order have the form

$$\eta(t) = \Omega^{-1/2}(t)\exp\left[\pm i\int^t \Omega(\tau)\, d\tau\right] \qquad (48)$$

The first-order corrections to these solutions yields the reflection coefficient, which can be expressed as[36]

$$R = \left|\frac{1}{4}\int_{-\infty}^{\infty}\frac{d\tau}{\Omega^{1/2}(\tau)}\frac{d}{d\tau}\left(\frac{d\Omega/d\tau}{\Omega^{3/2}}\right)\exp\left[-2i\int^\tau \Omega(x)\, dx\right]\right|^2 \qquad (49)$$

This formula holds provided $R \ll 1$. Then the average number of created quanta is also given by (37), as well as the probability to register a quantum.

As an example let us consider the case when the frequency changes in time according to an exponential law in the interval $0 \le t \le t_0$:

$$\Omega^2(t) = \begin{cases} 1 & t < 0 \\ \exp(\kappa t) & 0 \le t \le t_0 \\ \exp(\kappa t_0) & t > t_0 \end{cases} \tag{50}$$

In this case the main contribution to the integral (49) is given by two delta functions $\delta(t)$ and $\delta(t - t_0)$ arising due to the discontinuity of $d\Omega/dt$ at points $t = 0$ and $t = t_0$. The number of created quanta equals

$$N(t_0) = \frac{\kappa^2}{64}\left[1 + \exp(-\kappa t_0) - 2\exp\left(-\frac{\kappa t_0}{2}\right)\cos(2\varphi)\right]$$

$$\varphi = \frac{2}{\kappa}\left[\exp\left(\frac{\kappa t_0}{2}\right) - 1\right] \tag{51}$$

This formula is valid provided $\kappa \ll 1$.

The exact analytical expressions for the reflection coefficients are known, e.g., for the symmetric Epstein profile[36, 37]:

$$\Omega^2(t) = 1 - M\left[\cosh\left(\frac{\gamma t}{2}\right)\right]^{-2} \tag{52}$$

If $\gamma^2/16 < M < 1$, then the energy reflection coefficient is given by the expression[36, 37]

$$R = \frac{\cosh^2(\pi d_1)}{\cosh[\pi(d_1 + s)]\cosh[\pi(d_1 - s)]} \tag{53}$$

where

$$s = \frac{2}{\gamma} \qquad d_1 = \left[M\left(\frac{2}{\gamma}\right)^2 - \frac{1}{4}\right]^{1/2} \tag{54}$$

The number of created quanta due to (43) equals

$$N = \left(\frac{\cosh(\pi d_1)}{\sinh(\pi s)}\right)^2 \tag{55}$$

and in the adiabatic limit we get for $M < 1$

$$N = \exp\left[-(1 - M^{1/2})4\pi/\gamma\right] \qquad 4\pi/\gamma \gg 1 \qquad (56)$$

In the case of $M < 0$ the following formula is valid:

$$R = \frac{\cos^2(\pi d_2)}{\cos^2(\pi d_2)\cosh^2(\pi s) + \sin^2(\pi d_2)\sinh^2(\pi s)} \qquad (57)$$

$$d_2 = \left[\frac{1}{4} - M\left(\frac{2}{\gamma}\right)^2\right]^{1/2} \qquad (58)$$

It leads to the strongly oscillating number of quanta

$$N = \left[\frac{\cos(\pi d_z)}{\sinh(\pi s)}\right]^2 \qquad (59)$$

In the adiabatic limit $2\pi|M|^{1/2}/\gamma \gg 1$ one has

$$N = 4\cos^2\left(\frac{2\pi|M|^{1/2}}{\gamma}\right)\exp\left(-\frac{4\pi}{\gamma}\right) \qquad (60)$$

Oscillations disappear when $\gamma^2/16 \gg |M|$:

$$N = \left(\frac{2M}{\gamma}\right)^2 \ll 1 \qquad (61)$$

Another exactly solvable case corresponds to the transitional Epstein profile

$$\Omega^2(t) = 1 - p\frac{\exp(\gamma t)}{1 + \exp(\gamma t)} \qquad (62)$$

which can be considered as a smoothed variant of the exponential dependence (50). The reflection coefficient is given by[36, 37]

$$R = \left\{\frac{\sinh\left[\frac{1}{2}\pi s\left(1 - (1-p)^{1/2}\right)\right]}{\sinh\left[\frac{1}{2}\pi s\left(1 + (1-p)^{1/2}\right)\right]}\right\}^2 \qquad s = \frac{2}{\gamma} \qquad (63)$$

Consequently,

$$N = \frac{\left\{\sinh\left[\frac{1}{2}\pi s\left(1 - (1 - p)^{1/2}\right)\right]\right\}^2}{\sinh(\pi s)\sinh\left[\pi s(1 - p)^{1/2}\right]} \tag{64}$$

In the adiabatic limit

$$N = \exp\left[-\frac{4\pi(1 - p)^{1/2}}{\gamma}\right] \qquad \frac{4\pi}{\gamma} \gg 1 \tag{65}$$

The opposite limit $\gamma \to \infty$ transforms (62) into the sharp step function barrier. Such a frequency-jump case was considered in connection with the problem of squeezed states generation in Refs. 38–40. If the ratio of the final frequency to the initial one is $\Omega_f/\Omega_i = \xi$, then the reflection coefficient is given by the usual Fresnel formula, and the number of created quanta equals

$$N = \frac{(\xi - 1)^2}{4\xi} \tag{66}$$

An interesting effect arises if in some time τ the frequency restores (also instantly) its initial value. This situation is described by the usual formulas for the ideal Fabry-Perot resonator, and after simple algebra we get

$$N = \frac{(\xi^2 - 1)^2}{4\xi^2} \sin^2(\xi\tau) \tag{67}$$

We see that due to a kind of temporal interference for certain durations of frequently disturbances quanta are not created at all, as well as in the case of a smooth barrier (59).

It is worth noting that for a quite arbitrary dependence $\Omega(t)$, the combination

$$I = \sigma_p\sigma_q - \sigma_{pq}^2 \tag{68}$$

does not depend on time due to (21)–(23). This combination is in fact the simplest example of so-called universal quantum invariants, i.e., certain functions of variances that are conserved time independently of the concrete parameters of the Hamiltonian. For general multidimensional quadratic Hamiltonians (and even for more general Hamiltonians that are linear forms with respect to generators of arbitrary semisimple Lie algebras or superalgebras) with time-dependent coefficients such invariants were discovered and studied in Ref. 41. For vacuum or coherent initial states $I = \frac{1}{4}$.

V. SQUEEZING AND GENERATION OF QUANTA DUE TO PERIODIC VARIATION OF FREQUENCY

One of the most effective and practically achievable ways to generate quanta from the vacuum state is to subject the oscillator eigenfrequency to some periodic external action. Let us first consider the case when the eigenfrequency harmonically oscillates with twice the frequency:

$$\Omega^2(t) = \Omega_0^2[1 + \kappa \cos(2\Omega_0 t)] \qquad |\kappa| \ll 1 \tag{69}$$

We look for the solution of Eq. (21) in the form

$$\eta(t) = \Omega_0^{-1/2}[u(t)\exp(i\Omega_0 t) + v(t)\exp(-i\Omega_0 t)] \tag{70}$$

with slowly varying time-dependent amplitudes. Substituting (69) and (70) into (21) and performing averaging over fast oscillations (neglecting the second-order derivatives of slowly varying amplitudes), we arrive at

$$\frac{du}{dt} = -\frac{i\Omega_0 \kappa v}{4} \qquad \frac{dv}{dt} = \frac{i\Omega_0 \kappa u}{4} \tag{71}$$

whose solutions are

$$u(t) = \cosh\left(\frac{\Omega_0 \kappa t}{4}\right) \qquad v(t) = i \sinh\left(\frac{\Omega_0 t}{4}\right) \tag{72}$$

The variances σ_p and σ_q oscillate with twice the resonance frequency, but their normalized ratio k_* (31) is confined at every instant between the values

$$\exp(-\Omega_0 \kappa t) \le \frac{\Omega_0^2 \sigma_q}{\sigma_p} \le \exp(\Omega_0 \kappa t) \tag{73}$$

The number of created quanta according to (43) equals

$$N(t) = \left[\sinh\left(\frac{\Omega_0 \kappa t}{4}\right)\right]^2 \qquad (74)$$

For large values of parameter $\Omega_0 \kappa t$ this number increases exponentially with time.

These results are valid provided the depth of frequency modulation κ is sufficiently small. It is possible to obtain exact and explicit formulae in the case of strong modulation as well, if one considers the temporal version of the well-known Kronig-Penney model[42-44]:

$$\Omega^2(t) = \Omega_0^2 - 2\kappa \sum_{k=0}^{N-1} \delta(t - k\tau) \qquad (75)$$

where Ω_0 is a constant frequency and $\delta(t)$ is the Dirac delta function. Coupling constant κ now has the dimension of frequency.

For every interval of time $(k - 1)\tau < t < k\tau$ the solution for Eq. (21) is given by

$$\eta_k(t) = A_k e^{i\Omega_0 t} + B_k e^{-i\Omega_0 t} \qquad k = \overline{0, N}$$

Due to the continuity conditions

$$\eta_{k-1}(k\tau) = \eta_k(k\tau)$$

$$-\left[\dot{\eta}_k(k\tau) - \dot{\eta}_{k-1}(k\tau)\right] + 2\kappa \eta_{k-1}(k\tau) = 0$$

(obtained by integrating Eq. (21) over the infinitely small interval $n\tau - 0 < t < n\tau + 0$) coefficients A_k and B_k must satisfy some relations, which can be represented in matrix form as[45]

$$\begin{pmatrix} A_k e^{i\Omega_0 t_k} \\ B_k e^{-i\Omega_0 t_k} \end{pmatrix} = \begin{pmatrix} 1 - \dfrac{i\kappa}{\Omega_0} & -\dfrac{i\kappa}{\Omega_0} \\ \dfrac{i\kappa}{\Omega_0} & 1 + \dfrac{i\kappa}{\Omega_0} \end{pmatrix} \begin{pmatrix} A_{k-1} e^{i\Omega_0 t_k} \\ B_{k-1} e^{-i\Omega_0 t_k} \end{pmatrix}$$

After the sequence of δ kicks, coefficients A_n, B_n are connected with the initial coefficients A_0, B_0 through the equation

$$\begin{pmatrix} A_n \\ B_n \end{pmatrix} = \mathbf{S}^{(n)} \begin{pmatrix} A_0 \\ B_0 \end{pmatrix} = \mathbf{T}^{-(n-1)}(\mathbf{MT})^n \begin{pmatrix} A_0 \\ B_0 \end{pmatrix}$$

with matrices $\mathbf{T}$ and $\mathbf{M}$ are given by

$$\mathbf{M} = \begin{pmatrix} 1 - \dfrac{i\kappa}{\Omega_0} & -\dfrac{i\kappa}{\Omega_0} \\[2ex] \dfrac{i\kappa}{\Omega_0} & 1 + \dfrac{i\kappa}{\Omega_0} \end{pmatrix} \qquad \mathbf{T} = \begin{pmatrix} e^{i\Omega_0\tau} & 0 \\ 0 & e^{-i\Omega_0\tau} \end{pmatrix}$$

Matrix

$$\mathbf{S}^{(1)} \equiv \mathbf{S} = \mathbf{MT} = \begin{pmatrix} \left(1 - \dfrac{i\kappa}{\Omega_0}\right)e^{i\Omega_0\tau} & \left(-\dfrac{i\kappa}{\Omega_0}\right)e^{-i\Omega_0\tau} \\[3ex] \left(\dfrac{i\kappa}{\Omega_0}\right)e^{-i\Omega_0\tau} & \left(1 + \dfrac{i\kappa}{\Omega_0}\right)e^{-i\Omega_0\tau} \end{pmatrix}$$

is unimodular: $\det \mathbf{S} = 1$. It can be shown (see, e.g., Refs. 46 and 47) that all powers of two-dimensional unimodular matrices can be expressed as linear combinations of the matrices themselves and the unity matrix, the coefficients being Chebyshev's polynomials of the second kind, whose arguments are expressed in terms of traces of the initial matrices (here $\mathbf{E}$ means the unity matrix):

$$\mathbf{S}^n = U_{n-1}(\cos\phi)\mathbf{S} - U_{n-2}(\cos\phi)\mathbf{E} \qquad \cos\phi = \frac{1}{2}\operatorname{Tr}\mathbf{S} = \frac{\chi}{2}$$

We use the following definitions of the Chebyshev polynomials[48, 49]:

$$U_n(\cos\phi) = \frac{\sin[(n+1)\phi]}{\sin\phi}$$

$$U_n(\chi) = \frac{\left[\chi + (\chi^2 - 4)^{1/2}\right]^{n+1} - \left[\chi - (\chi^2 - 4)^{1/2}\right]^{n+1}}{2^{n+1}(\chi^2 - 4)^{1/2}}$$

In the case under study

$$\frac{\chi}{2} = \cos\phi = \cos\Omega_0\tau + \frac{\kappa}{\Omega_0}\sin\Omega_0\tau \tag{76}$$

Thus, matrix elements of matrix $\mathbf{S}^{(n)}$ can be written as

$$S_{11}^{(n)} = \left(1 - \frac{i\kappa}{\Omega_0}\right)U_{n-1}\left(\frac{\chi}{2}\right)e^{-i\Omega_0(n-2)\tau} - U_{n-2}\left(\frac{\chi}{2}\right)e^{-i\Omega_0(n-1)\tau}$$

$$S_{12}^{(n)} = -\frac{i\kappa}{\Omega_0}U_{n-1}\left(\frac{\chi}{2}\right)e^{-i\Omega_0 n\tau}$$

$$S_{21}^{(n)} = \frac{i\kappa}{\Omega_0}U_{n-1}\left(\frac{\chi}{2}\right)e^{i\Omega_0 n\tau}$$

$$S_{22}^{(n)} = \left(1 + \frac{i\kappa}{\Omega_0}\right)U_{n-1}\left(\frac{\chi}{2}\right)e^{i\Omega_0(n-2)\tau} - U_{n-2}\left(\frac{\chi}{2}\right)e^{i\Omega_0(n-1)\tau}$$

If $\eta(-0) = 1$ and $\dot{\eta}(-0) = i\Omega_0$ at the initial instant, then $A_0 = 1$, $B_0 = 0$, so $A_n = S_{11}^{(n)}$, $B_n = S_{21}^{(n)}$. In such a case Eqs. (23) result in the following expressions for variances of the coordinate after the sequence of n δ kicks (the arguments of Chebyschev's polynomials are omitted for brevity):

$$\sigma_q(t) = (2\Omega_0)^{-1}\left(U_{n-1}^2 + U_{n-2}^2 + \left(\frac{2\kappa}{\Omega_0}\right)U_{n-1}^2 \sin[2\Omega_0(t - (n-1)\tau)]\right.$$

$$+ \left(\frac{4\kappa^2}{\Omega_0^2}\right)U_{n-1}^2\{\sin[\Omega_0(t - (n-1)\tau)]\}^2 - \chi U_{n-1}U_{n-2}$$

$$\left. - \left(\frac{2\kappa}{\Omega_0}\right)U_{n-1}U_{n-2} \sin[2\Omega_0(t - (n-1/2)\tau)]\right)$$

The covariance between coordinate and momentum equals

$$\sigma_{qp} = \frac{1}{2}\left\{\left[\left(1 + \frac{2\kappa^2}{\Omega_0^2}\right)U_{n-1}^2 + U_{n-2}^2 - \chi U_{n-1}U_{n-2}\right]^2\right.$$

$$- \left[\left(\frac{2\kappa}{\Omega_0}\right)U_{n-1}^2 \sin[2\Omega_0(t - (n-1)\tau)]\right]$$

$$- \left(\frac{2\kappa^2}{\Omega_0^2}\right)U_{n-1}^2 \cos[2\Omega_0(t - (n-1)\tau)]$$

$$\left. - \left(\frac{2\kappa}{\Omega_0}\right)U_{n-1}U_{n-2} \sin[2\Omega_0(t - (n-1/2)\tau)]\right]^2 - 1\right\}^{1/2}$$

The total energy (41) after the nth kick equals ($\hbar = 1$)

$$
\begin{aligned}
E = \frac{\Omega_0}{2} \Bigg\{ & U_{n-1}^2\left(\frac{\chi}{2}\right) + U_{n-2}^2\left(\frac{\chi}{2}\right) \\
& + \left(\frac{4\kappa^2}{\Omega_0^2}\right) U_{n-1}^2\left(\frac{\chi}{2}\right) - \chi U_{n-1}\left(\frac{\chi}{2}\right) U_{n-2}\left(\frac{\chi}{2}\right) \Bigg\}
\end{aligned}
\tag{77}
$$

These formulae hold provided $n \geq 2$. The case of a small number of kicks (including nonperiodic situation) was investigated in Ref. 45.

If parameter $\chi/2$ does not belong to interval $[-1.1]$, the asymptotic formula for the Chebyshev polynomials [49]

$$
\begin{aligned}
U_n(z) \approx {} & \frac{n^{1/2}}{2(n-1)^{1/2}} (z^2 - 1)^{-1/2} \{(z+1)^{1/2} + (z-1)^{1/2}\} \\
& \times \left\{ z + (z^2 - 1)^{1/2} \right\}^{n-1/2} \qquad n \gg 1
\end{aligned}
$$

gives rise to the asymptotic expression for the total energy of fluctuations ($n \to \infty$)

$$
E \approx \frac{2\kappa^2}{\Omega_0} \frac{\left[z + (z^2 - 1)^{1/2} \right]^{2n}}{(z^2 - 1)} \qquad z = \frac{\chi}{2}
\tag{78}
$$

If we choose the simplest case, when the interval between the kicks satisfies the condition $\Omega_0 \tau = \pi/2 + 2k\pi$, $k = 0, 1, \ldots$, the arguments of Chebyshev's polynomials are equal to κ/Ω_0. When the strength of δ kicks κ is larger than the constant frequency Ω_0, one has asymptotic formula for total energy

$$
E \approx \frac{2\kappa^2}{\Omega_0} \frac{\left[\kappa/\Omega_0 + \left(\kappa^2/\Omega_0^2 - 1\right)^{1/2} \right]^{2n}}{\left(\kappa^2/\Omega_0^2 - 1\right)}
\tag{79}
$$

In the case when $\kappa/\Omega_0 \gg 1$, the total energy increases exponentially with the number of kicks:

$$
E = 2\Omega_0 \left(\frac{2\kappa}{\Omega_0}\right)^{2n} \qquad n \gg 1
\tag{80}
$$

If δ kicks take place at moments of time given by $\Omega_0 \tau = 2\pi m$, $m = 0, 1, \ldots$, then the argument of Chebyshev's polynomials is equal to unity independently on the strength of δ kicks. Then the energy increases much more slowly:

$$E = \frac{\Omega_0}{2}\left(1 + \frac{4\kappa^2}{\Omega_0^2}\right)n^2 \qquad (81)$$

In the case when the strength of δ kicks is small ($\kappa/\Omega_0 \ll 1$), and interval between kicks is given by $\Omega_0 \tau = \pi/2 + 2\pi k$, $k = 0, 1, \ldots$, the following inequality for maximum total energy can be obtained:

$$E_{\max} < \frac{\Omega_0}{2}\left(1 + \frac{4\kappa^2}{\Omega_0^2}\right)n^2 \qquad (82)$$

VI. GENERATION OF SQUEEZED STATES OF ELECTROMAGNETIC FIELD IN DIELECTRIC MEDIUM

In this section we consider squeezed states of electromagnetic field in dielectric media. As was shown above, to produce squeezed states one has to vary in time parameters of the quantum system under study. Therefore, first we briefly discuss the specific features relating to the problem of field quantization in time-dependent media.

Usually the problem of electromagnetic field quantization is considered in textbooks under the assumption that the field occupies some empty box. The case when the box is filled with a uniform dielectric medium was considered in Refs. 50 and 51. The quantization of the field in the medium consisting of two uniform dielectrics with different permeabilities was studied in Refs. 52–54. The case of an arbitrary inhomogenious (but time-independent) dielectric medium was investigated in Refs. 55 and 56 and especially in Refs. 57 and 58. The most general case of nonuniform and time-dependent media was investigated in Ref. 59. We confine ourselves here to the special case when spatial and temporal dependences of dielectric permeability can be separated (or factorized). In this case, explicit results can be obtained, in principle, without limitations on the magnitude of variations of dielectric permeability (in Ref. 59 only approximate solutions valid for small polarization of the medium were found).

The basis of the subsequent consideration is the system of Maxwell's equations in linear passive scalar nondispersive time-dependent dielectric and magnetic medium with external current (the field quantization in nonlinear stationary dielectric medium was investigated in Refs. 55, 57, and 60, and the most general approach suitable for nonstationary nonlinear medium was proposed recently in Ref. 61):

$$\operatorname{rot} \mathbf{E} = -\frac{1}{c}\frac{\partial \mathbf{B}}{\partial t} \qquad \operatorname{rot} \mathbf{H} = \frac{1}{c}\frac{\partial \mathbf{D}}{\partial t} + \frac{4\pi}{c}\mathbf{j}$$
$$\operatorname{div} \mathbf{D} = 0 \qquad \operatorname{div} \mathbf{B} = 0, \tag{83}$$
$$\mathbf{D} = \varepsilon(r,t)\mathbf{E} \qquad \mathbf{B} = \mu(r,t)\mathbf{H}$$

Introducing the vector potential according to the relations

$$\mathbf{B} = \operatorname{rot} \mathbf{A} \qquad \mathbf{E} = -\frac{1}{c}\frac{\partial \mathbf{A}}{\partial t} \tag{84}$$

and imposing gauge conditions

$$\operatorname{div}\left(\varepsilon\frac{\partial \mathbf{A}}{\partial t}\right) = 0 \qquad \phi = 0 \tag{85}$$

we can replace the system of the first-order equations (83) with the single second-order one:

$$\operatorname{rot}\left(\frac{1}{\mu}\operatorname{rot} \mathbf{A}\right) + \frac{1}{c^2}\frac{\partial}{\partial t}\left(\varepsilon\frac{\partial \mathbf{A}}{\partial t}\right) = \frac{4\pi}{c}\mathbf{j} \tag{86}$$

We can check that the vector equation (86) coincides with the set of Euler's equations

$$\frac{\partial}{\partial t}\frac{\partial L}{\partial(\partial_t A_\beta)} + \frac{\partial}{\partial x_\alpha}\frac{\partial L}{\partial(\partial_\alpha A_\beta)} - \frac{\partial L}{\partial A_\beta} = 0 \tag{87}$$

for the Lagrangian density

$$L = \frac{1}{8\pi}\left\{\frac{\varepsilon(r,t)}{c^2}\left(\frac{\partial \mathbf{A}}{\partial t}\right)^2 - \frac{(\operatorname{rot} \mathbf{A})^2}{\mu(r,t)}\right\} + \frac{1}{c}\mathbf{j}\mathbf{A} \tag{88}$$

in the case of quite arbitrary time and space dependences of the dielectric and magnetic permeabilities. Then introducing the canonically conjugated variable

$$\mathbf{P} = \frac{\partial L}{\partial(\partial_t \mathbf{A})} = \frac{\varepsilon(r,t)}{4\pi c^2}\frac{\partial \mathbf{A}}{\partial t} = -\frac{\mathbf{D}}{4\pi c} \tag{89}$$

one can construct the Hamiltonian density

$$\mathscr{H} = \mathbf{P}\frac{\partial \mathbf{A}}{\partial t} - L = \frac{1}{8\pi}\left\{\frac{\mathbf{D}^2}{\varepsilon(r,t)} + \frac{\mathbf{B}^2}{\mu(r,t)}\right\} - \frac{1}{c}\mathbf{jA} \tag{90}$$

which leads again to Eq. (86).

Now suppose that spatial and temporal dependences of functions ε and μ are factorized:

$$\varepsilon(\mathbf{r},t) = \tilde{\varepsilon}(\mathbf{r})\chi(t) \qquad \mu(\mathbf{r},t) = \tilde{\mu}(\mathbf{r})\nu(t) \tag{91}$$

Let us introduce the equation

$$\mathrm{rot}\left[\tilde{\mu}^{-1}(\mathbf{r})\mathrm{rot}\,\mathbf{g}\right] - k^2\tilde{\varepsilon}(\mathbf{r})\mathbf{g} = 0 \qquad k = \mathrm{const} \tag{92}$$

playing the same role for the generalized wave equation (86) as the Helmholtz equation in the case of the usual wave equation. It can be proved that solutions of Eq. (92) form a complete normalized set with scalar product given by

$$\int \tilde{\varepsilon}(\mathbf{r})\mathbf{g_k}(\mathbf{r})\mathbf{g_l}(\mathbf{r})\,\mathrm{d}^3\mathbf{r} = 4\pi\delta_{\mathbf{kl}} \tag{93}$$

Then it is natural to look for solutions of Maxwell's equations in the form of superpositions of basic solutions

$$\mathbf{D}(\mathbf{r},t) = c\tilde{\varepsilon}(\mathbf{r})\sum_{\mathbf{k}}k\mathbf{g_k}(\mathbf{r})q_{\mathbf{k}}(t) \qquad k = |\mathbf{k}| \tag{94}$$

$$\mathbf{A}(\mathbf{r},t) = \sum_{\mathbf{k}}k^{-1}\mathbf{g_k}(\mathbf{r})p_{\mathbf{k}}(t) \tag{95}$$

Putting these developments into (83) and taking into account (92), we obtain the following set of ordinary differential equations for time-dependent coefficients

$$\dot{q}_{\mathbf{k}} = \frac{p_{\mathbf{k}}}{\nu(t)} - f_{\mathbf{k}}(t) \tag{96}$$

$$\dot{p}_{\mathbf{k}} = -\frac{q_{\mathbf{k}}(ck)^2}{\chi(t)} \tag{97}$$

where $f_{\mathbf{k}}(t)$ are the coefficients in the generalized Fourier decomposition of current:

$$\mathbf{j}(\mathbf{r}, t) = \left(\frac{c}{4\pi}\right)\tilde{\varepsilon}(\mathbf{r}) \sum_{\mathbf{k}} k\mathbf{g}_{\mathbf{k}}(\mathbf{r})f_{\mathbf{k}}(t)$$

$$f_{\mathbf{k}}(t) = \int (ck)^{-1}\mathbf{g}_{\mathbf{k}}(\mathbf{r})\mathbf{j}(\mathbf{r})\, d^3\mathbf{r} \tag{98}$$

Now let us put the expansions (94), (95), and (98) into the Hamiltonian density (90) and integrate over spatial variables. We arrive at the Hamiltonian

$$\mathcal{H}(t) = \int \mathcal{W}(\mathbf{r}, t)\, d^3\mathbf{r}$$

$$= \sum_{\mathbf{k}} \left\{ \frac{1}{2}\left[\frac{p_{\mathbf{k}}^2}{\nu(t)} + \frac{(ck)^2}{\chi(t)}q_{\mathbf{k}}^2 \right] - f_{\mathbf{k}}(t)p_{\mathbf{k}} \right\} \tag{99}$$

leading to the same equations of motion (96) and (97). After this we may proclaim $q_{\mathbf{k}}$ and $p_{\mathbf{k}}$ *operators* satisfying canonical commutation relations, and the field will be *quantized*. We used this procedure in the special case when $\tilde{\varepsilon} = \tilde{\mu} = 1$ in Refs. 62 (without external current) and 63 (in the presence of current).

It should be emphasized that introduction of the Hamiltonian (99) appears possible only for factorizable media. In the case of quite arbitrary dependence $\varepsilon(\mathbf{r}, t)$ decompositions like (94) with time-dependent coefficients satisfying some dynamical equations do not hold, so that no quantum Hamiltonian exists. In this case only the quantum Heisenberg picture can be used[59] (see also discussion on this subject in Ref. 64, where the problem of field quantization in systems with moving boundaries was investigated for the first time).

Let us draw an attention to the following interesting features of Hamiltonian (99). First, it seems natural to choose the coefficients q_k of the *electric displacement vector* **D** decomposition (94) as generalized coordinates, whereas the generalized momenta p_k are naturally related to the vector potential **A**. (The role of vector **D** for the field quantization in dielectric media was emphasized in many papers; see, e.g., Refs. 50–61.) In such an interpretation the time-dependent factor of the *magnetic permeability* $\nu(t)$ plays the role of effective mass, and equation of motion for the generalized coordinate

$$\ddot{q}_k + \gamma(t)\dot{q}_k + \Omega_k^2(t)q_k = 0 \tag{100}$$

$$\Omega_k^2(t) = \frac{(ck)^2}{\nu(t)\chi(t)} \qquad \gamma(t) = \frac{\dot{\nu}}{\nu} \tag{101}$$

resembles the equation for a damped harmonic oscillator. Really, of course, we have, not "damping," but the effect of time-varying mass. Since the medium is supposed to be linear, magnetic effects are usually negligible, so in the most cases one may assume $\mu = 1$. However, the situation when $\mu \approx 1$, but γ is not extremely small, is also possible if $\dot{\nu}$ is sufficiently large.

Another interesting feature of Hamiltonian (99) is the form of linear coupling to the external generalized force (representing the external current): not through coordinate, but through *generalized momentum*. Exact solutions of the Schrödinger equation with time-dependent Hamiltonians like (99) were obtained and investigated in detail, e.g., in Refs. 26, 30, 63, and 65. Concrete results relating to squeezing and number of generated photons in the modes of electromagnetic field are given in the previous sections. Here we want to precise them, taking into account the specific values of parameters in formulas written above.

First let us look at formulas (52)–(65) giving the numbers of created quanta during the processes described by smooth Epstein's profiles of time dependences of the eigenmode frequencies. All results can be reproduced provided one multiplies in accordance with (101) the right-hand sides of (52) and (62) by $(kc)^2$ and replaces parameter γ by $\gamma/(kc)$ in expressions for the reflection coefficient and number of quanta. Particularly, in the adiabatic regime Eqs. (56) and (65) show Wien's spectrum of created photons

$$N(k) = \exp\left(-\frac{kc}{T_{\text{eff}}}\right) \qquad T_{\text{eff}} = \text{const}\,\gamma \tag{102}$$

in the low-temperature limit. This result is interesting in connection with a paper by Yablonovitch,[66] who proposed using a medium with a decreasing time refractive index (the so-called "plasma window") to simulate the Unruh effect, i.e., creation of quanta in accelerated frame of reference. Using some heuristic reasoning, he claimed that the spectrum of photons created in the plasma window would resemble Planck's spectrum with effective temperature proportional to $|(1/\chi)(d\chi/dt)|$, i.e., parameter κ in the exponential profile model (50) or parameter γ in Epstein's example. Equation (51) shows that the real spectrum of photons created in exponential model of the plasma window with $(1/\chi)(d\chi/dt) = \text{const}$ has nothing in common with Planck's spectrum even in the asymptotical limit of infinitely long time $t_0 \to \infty$. The exponentially small reflection coefficient (resulting not in Planck's, but in Wien's spectrum), as is known from the theory of adiabatic invariants, is possible only for those slowly varying functions $\chi(t)$ that have continuous derivatives of all orders[67] (as for Epstein's profiles, for example).

In the case of rapidly varying dielectric permeability the situation is much more intricate. Planck's spectrum does not arise in any model. Moreover, the case $M > 0$ in formula (52) for eigenfrequency (when Wien's spectrum can be obtained) corresponds not to a plasma window, but to a "dielectric window," since in this case $\varepsilon(t) > \varepsilon_{i_n}$. In the plasma case of $M < 0$, as demonstrated by Eqs. (59) and (60), the spectrum of generated photons is an oscillating function of the wave number, so that in certain modes photons are not generated at all, due to a peculiar interference in time.

Oscillations disappear in the case of rapid change of dielectric permeability. However, if the full variation of ε is finite, the number of created photons appears very small; see, e.g., Eq. (61) yielding the spectrum

$$N(k) = \left(\frac{2kcM}{\gamma}\right)^2 \ll 1 \qquad \frac{\gamma^2}{(4kc)^2} \gg |M| \qquad (103)$$

which resembles the Rayleigh-Jeans spectrum, but with effective temperature inversely proportional to the square of parameter γ characterizing the rate of change of dielectric permeability.

In the case of step change of dielectric function the number of created quanta due to (66) and (101) equals

$$N = \frac{(\varepsilon^{1/2} - 1)^2}{4\varepsilon^{1/2}} \qquad \varepsilon \equiv \frac{\varepsilon_{\text{fin}}}{\varepsilon_{\text{in}}} \qquad (104)$$

without any dependence on the wave number. For the temporal Fabry-Perot resonator we get, due to (67),

$$N(k) = \frac{(\varepsilon - 1)^2}{4\varepsilon} \sin^2\left(\frac{kc^\tau}{\varepsilon^{1/2}}\right) \qquad (105)$$

(one should remember that the effective refractive index in the time-dependent case is not $\varepsilon^{1/2}$ but $\varepsilon^{-1/2}$). We see again that, due to a kind of temporal interference, photons in certain modes are not created at all.

In real experiments time dependence of the dielectric permeability arises as a result of action on the nonlinear medium by an external pumping field. Thus, the situation considered in the paper, when the dielectric permeability is some prescribed beforehand function of time, is in fact a model of real experimental situations. However, this model presents a correct qualitative description of the processes of squeezing and photon creation. It is seen in the example of the parametric resonance at the twice resonator eigenfrequency considered in the beginning of Section V. In this case our results relating to the rate of photons generation are in qualitative agreement with the results of Ref. 68, where the pumping field was taken into account explicitly (the case of classical pumping was investigated in detail, e.g., in Refs. 27 and 28).

The effect of photon creation in time-dependent media may be interesting for the problem of propagating ultrashot strong light impulses in dielectrics, when the dielectric permeability can change in time under the influence of the impulse electric field. Then the energy of generated noise photons will be derived after all from the initial energy of the impulse, resulting in its extra attenuation.

VII. IMPULSE PROPAGATION IN TIME-DEPENDENT MEDIA

Here we want to discuss some interesting features relating to the electromagnetic wave propagation in spatially uniform but time-dependent media. The electric displacement for traveling waves can be expressed as

$$\mathbf{D}_k(x, t) = e^{ikx}\eta_k(t) \qquad (106)$$

where the time-dependent factor satisfies the equation

$$\frac{d^2\eta_k}{dt^2} + \frac{k^2c^2\eta_k}{\varepsilon(t)} = 0 \qquad (107)$$

Suppose $\varepsilon(t) = 1$ for $t < 0$. Then the traveling wave solution is

$$\mathbf{D}_k^{(\mathrm{i})}(x,t) = \exp[ik(x - ct)] \qquad t < 0 \qquad (108)$$

If the dielectric permeability changes in the interval $0 < t < T$ but assumes some constant value ε_0 for $t > T$, then function (108) will be transformed for $t > T$ into the superposition of two waves traveling in opposite directions:

$$\mathbf{D}_k^{(\mathrm{f})}(x,t) = \alpha \exp[ik(x - ct\varepsilon_0^{-1/2})] + \beta \exp[ik(x + ct\varepsilon_0^{-1/2})] \quad (109)$$

We see that the ratio $|\beta/\alpha|^2$, which was treated in Section III as the reflection coefficient from some conventional potential barrier, appears in the case under study as a genuine reflection coefficient, since the wave going in the opposite direction really exists. This is seen distinctly in the simplest example of an instant change of the dielectric permeability (which can be caused by an instant change of the medium density, temperature, or other parameter due to some external action; see, e.g., Ref. 69). Then an arbitrary initial wave packet $\mathbf{D}^{(\mathrm{i})}(x - ct)$, $t < 0$, will be transformed due to the nonstationary wave equation

$$\frac{\partial^2 \mathbf{D}}{\partial x^2} = \frac{\varepsilon(t)}{c^2} \frac{\partial^2 \mathbf{D}}{\partial t^2} \qquad (110)$$

and the continuity conditions into the function $(t > 0)$

$$\mathbf{D}(x,t) = \tfrac{1}{2}\Big[\big(1 + \varepsilon_0^{1/2}\big)\mathbf{D}^{(\mathrm{i})}\big(x - ct\varepsilon_0^{1/2}\big) \\ + \big(1 - \varepsilon_0^{1/2}\big)\mathbf{D}^{(\mathrm{i})}\big(x + ct\varepsilon_0^{1/2}\big)\Big] \qquad (111)$$

Since we consider nondispersive media, this solution is physically acceptable provided $\varepsilon_0 > 1$. Note that the transmitted wave packet (the first term on the right side of (111)) is amplified in comparison with the initial one. Moreover, for $\varepsilon_0 > 9$ the reflected wave packet is amplified too. The forms of both transmitted and reflected impulses are the same as the form of the initial packet, since the reflection coefficient does not depend on the wave number.

Now suppose that in time T function $\varepsilon(t)$ restores its initial unit value. Then coefficients α and β in (109) become dependent on the wave number:

$$\alpha = \tau_+ \exp(ik\delta_-) - \tau_- \exp(ik\delta_+)$$
$$\beta = \rho[\exp(-ik\delta_-) - \exp(-ik\delta_+)]$$

(112)

· where

$$\tau_\pm = \frac{\left(\varepsilon_0^{1/2} \pm 1\right)^2}{4\varepsilon_0^{1/2}} \qquad \rho = \frac{\varepsilon_0 - 1}{4\varepsilon_0^{1/2}} \qquad \delta_\pm = cT\left(1 \pm \varepsilon_0^{-1/2}\right)$$

(113)

The reflected wave disappears provided the condition

$$k(\delta_+ - \delta_-) = 2\pi m \qquad m = 1, 2, \ldots$$

(114)

is fulfilled. However, this is true only for a monochromatic initial wave (with an infinite extent in space). For bounded in space packets the situation can be elucidated in the frame of an exactly solvable example of a Gaussian initial packet:

$$D^{(i)}(x, t) = \exp\left[-\frac{(x - ct)^2}{\sigma^2} + ik_0(x - ct)\right]$$

(115)

Calculating the Fourier transform of (115) (which is a Gaussian exponential again) and replacing each Fourier component by expression (109) with coefficients given in (112) and (113), we can easily obtain the following explicit expressions for the transmitted D_t and reflected D_r waves,

$$D_t(x, t) = \tau_+ \exp\left[-\frac{(x - ct + \delta_-)^2}{\sigma^2} + ik_0(x - ct + \delta_-)\right]$$
$$- \tau_- \exp\left[-\frac{(x - ct + \delta_+)^2}{\sigma^2} + ik_0(x - ct + \delta_+)\right]$$

(116)

$$D_r(x, t) = \rho\left\{\exp\left[-\frac{(x + ct - \delta_+)^2}{\sigma^2} + ik_0(x + ct - \delta_+)\right]\right.$$
$$\left. - \exp\left[-\frac{(x + ct - \delta_-)^2}{\sigma^2} + ik_0(x + ct - \delta_-)\right]\right\}$$

(117)

We see that the initial impulse is split into four packets with the same shapes: two transmitted and two reflected. This phenomenon manifests itself in the most distinct form for narrow packets satisfying the condition

$$\sigma \ll \delta_+ - \delta_- = 2cT\varepsilon_0^{-1/2} \tag{118}$$

In this case no disappearance of the reflected wave is observed.

References

1. V. V. Dodonov, E. V. Kurmyshev, and V. I. Man'ko, *Phys. Lett. A* **79**, 150 (1980).

2. V. V. Dodonov, E. V. Kurmyshev, and V. I. Man'ko, in *Proc. Lebedev Phys. Inst.*, Vol. Moscow, Nauka, 1986, pp. 128–150. [Translation: Nova Science, Commack, NY, 1988, pp. 169–199.]

3. E. Schrödinger, in *Ber. Kgl. Acad. Wiss.*, Berlin, 1930 S.296–303.

4. H. P. Robertson, *Phys. Rev. A* **35**, 667 (1930).

5. V. V. Dodonov and V. I. Man'ko, in *Proceedings of Lebedev Physics Institute, 1987*, Vol. 183, pp. 5–70. [Translated by Nova Science, Commack, NY, 1989, pp. 3–101.]

6. R. J. Glauber, *Phys. Rev. A* **131**, 2766 (1963).

7. H. Takahasi, *Adv. Commun. Syst.* **1**, 227 (1965).

8. M. M. Miller and E. A. Mishkin, *Phys. Rev.* **152**, 1110 (1966).

9. Z. Bialynicka-Birula, *Phys. Rev.* **173**, 1207 (1968).

10. D. Stoler, *Phys. Rev. D* **1**, 3217 (1970).

11. P. P. Bertrand, K. Moy, and E. A. Mishkin, *Phys. Rev. D* **4**, 1909 (1971).

12. E. Y. C. Lu, *Lett. Nuovo Cim.* **2**, 1241 (1971).

13. D. Stoler, *Phys. Rev. D* **4**, 1309 (1971).

14. D. A. Trifonov, *Phys. Lett. A* **48**, 165 (1974).

15. H. P. Yuen, *Phys. Rev. A* **13**, 2226 (1976).

16. V. Canivell and P. Seglar, *Phys. Rev. D* **15**, 1050 (1977).

17. J. N. Hollenhorst, *Phys. Rev. D* **19**, 1669 (1979).

18. C. M. Caves, *Phys. Rev. D* **23**, 1693 (1981).

19. I. Bialynicki-Birula, in F. Haake, L. M. Narducci, and D. Walls (Eds.), *Coherence, Cooperation and Fluctuations*, Cambridge University Press, Cambridge, UK, 1986, pp. 159–170.

20. D. F. Walls, *Nature* **306**, 141 (1983).

21. B. L. Schumaker, *Phys. Rep.* **135**, 317 (1986).

22. R. Loudon and P. L. Knight, *J. Mod. Opt.* **34**, 709 (1987).

23. *Squeezed States of Electromagnetic field*, special issue of *JOSA B* **4**(10) (1987).

24. M. C. Teich and B. E. A. Saleh, *Quantum Opt.* **1**, 153 (1989).

25. *Workshop on Squeezed States and Uncertainty Relations*, D. Han, Y. S. Kim, and W. W. Zachary (Eds.), Proceedings of a workshop held at the University of Maryland (College Park), 28–30 March 1991, *NASA Conference Publication 3135* (1992).

26. V. V. Dodonov and V. I. Man'ko, in *Proc. Lebedev Phys. Inst.*, Vol. 183, Moscow, Nauka, 1987, pp. 71–181. [Translation: Nova Science, Commack, NY, 1989, pp. 103–261.]

27. M. T. Raiford, *Phys. Rev. A* **2**, 1541 (1970).

28. M. T. Raiford, *Phys. Rev. A* **9**, 2060 (1974).

29. K. Husimi, *Progr. Theor. Phys.* **9**, 381 (1953).

30. I. A. Malkin and V. I. Man'ko. *Dinamicheskie simmetrii i kogerentnye sostoyaniya kvantovykh sistem* [Dynamical symmetries and coherent states of quantum systems], Moscow, Nauka, 1979.

31. H. Paul, *Rev. Mod. Phys.* **58**, 209 (1986).

32. S. Kielich, R. Tanas, and R. Zawodny, *JOSA B* **4**, 1627 (1987).

33. D. F. Smirnov and A. C. Troshin, *Usp. Fiz. Nauk* **153**, 233 (1987).

34. H.-W. Lee, *Phys. Lett. A* **153**, 219 (1991).

35. V. V. Dodonov, A. B. Klimov, and V. I. Man'ko. *Phys. Lett. A* **134**, 211 (1989).

36. V. L. Ginzburg, *Propagation of Electromagnetic Waves in Plasma*, Nauka, Moscow, 1967, Chapter 4.

37. P. Epstein, *Proc. Nat. Acad. Sci. USA* **16**, 627 (1930).

38. J. Janszky and Y. Y. Yushin, *Opt. Commun.* **59**, 151 (1986).

39. R. Graham, *J. Mod. Opt.* **34**, 873 (1987).

40. X. Ma and W. Rhodes, *Phys. Rev. A* **39**, 1941 (1989).

41. V. V. Dodonov and V. I. Man'ko, in *Group Theoretical Methods in Physics*, Proceedings of the Second International Seminar, Vol. 1, Harwood Academic London, 1985, p. 591; *Proc. Lebedev Phys. Inst.* **167**, 7 (1987); **183**, 263 (1989) (Nova Science, Commack, NY).

42. R. Kronig and W. Penney, *Proc. R. Soc.* **130**, 499 (1931).

43. P. Schnupp, *Solid-State Electron.* **10**, 785 (1967).

44. S. Albeverio, F. Gesztesy, R. Hoegh-Krohn, and H. Holden, *Solvable Models in Quantum Mechanics*, Springer, Berlin, 1988.

45. V. V. Dodonov, O. V. Man'ko, and V. I. Man'ko, *J. Sov. Laser Res.* **13**, 196 (1992).

46. D. Sengupta and P. K. Ghosh, *Phys. Lett. A* **68**, 107 (1978).

47. A. Maitland and M. H. Dunn, *Laser Physics*, North Holland, Amsterdam, 1969, Appendix E.

48. A. Erde'lyi (Ed.), *Higher Transcendental Functions*, McGraw Hill, New York, Vol. 2, 1953, Vol. 3, 1955.

49. G. Szego, *Orthogonal Polynomials*, Vol. 23, American Mathematical Society, Colloquium Publ., New York, 1959.

50. J. M. Jauch and K. M. Watson, *Phys. Rev.* **74**, 950 (1948).

51. V. L. Ginzburg, *Theoretical Physics and Astrophysics*, Nauka, Moscow, 1975.

52. C. K. Carniglia and L. Mandel, *Phys. Rev. D* **3**, 280 (1971).

53. K. Ujihaa, *Phys. Rev. A* **12**, 148 (1975).

54. I. Abram, *Phys. Rev. A* **35**, 4661 (1987).

55. Y. R. Shen, *Phys. Rev.* **155**, 921 (1967).

56. L. Knoll, W. Vogel, and D.-G. Welsh, *Phys. Rev. A* **36**, 3803 (1987).

57. P. D. Drummond, *Phys. Rev. A* **42**, 6845 (1990).

58. R. J. Glauber and M. Lewenstein, *Phys. Rev. A* **43**, 467 (1991).

59. Z. Bialynicka-Birula and I. Bialynicki-Birula, *JOSA B* **4**, 621 (1987).

60. M. Hillery and L. D. Mlodinow, *Phys. Rev. A* **30**, 1860 (1984).

61. A. A. Lobashov and V. M. Mostepanenko, *Teoret. Mat. Fiz.* **86**, 438 (1991).

62. V. V. Dodonov, A. B. Klimov, and V. I. Man'ko, *J. Sov. Laser Res.* **12**, 439 (1991).

63. V. V. Dodonov, T. F. George, O. V. Man'ko, C. I. Um, and K. H. Yeon, *J. Sov. Laser Res.* **13**, 219 (1992).

64. G. T. Moore, *J. Math. Phys.* **11**, 2679 (1970).

65. V. V. Dodonov and V. I. Man'ko, *Phys. Rev. A* **20**, 550 (1979).

66. E. Yablonovitch, *Phys. Rev. Lett.* **62**, 1742 (1989).

67. R. M. Kulsrud, *Phys. Rev.* **106**, 205 (1957).

68. A. A. Lobashov and V. M. Mostepanenko, *Teoret. Mat. Fiz.* **88**, 340 (1991).

69. S. C. Wilks, J. M. Dawson, and W. B. Mori, *Phys. Rev. Lett.* **61**, 337 (1988).

QUANTUM-STATISTICAL THEORY OF RAMAN SCATTERING PROCESSES

A. MIRANOWICZ

and

S. KIELICH

Nonlinear Optics Division, Institute of Physics, Adam Mickiewicz University, Poznań, Poland

CONTENTS

I. INTRODUCTION: HISTORICAL DEVELOPMENTS

Almost simultaneously in 1928 Raman and Krishnan[1, 2] and Landsberg and Mandel'stamm[3] observed a new kind of scattering, now referred to as

Modern Nonlinear Optics, Part 3, Edited by Myron Evans and Stanisław Kielich. Advances in Chemical Physics Series, Vol. LXXXV.
ISBN 0-471-30499-9 © 1994 John Wiley & Sons, Inc.

(spontaneous) Raman scattering. For the last 65 years Raman scattering has unceasingly been in the forefront of both scientific and experimental investigations, particularly after the first observation of stimulated Raman scattering by Woodbury and Ng[4] (see also Ref. 5). Without exaggeration one can say that Raman scattering and spectroscopy constitute a completely autonomous discipline.

The literature on Raman scattering is quite prodigious. The theoretical principles and milestone experiments describing the Raman effect are summarized in a number of excellent monographs and reviews, for instance, by Bloembergen,[6] Kaiser and Maier,[7] Koningstein,[8] Grasyuk,[9, 10] Wang,[11, 12] Cardona,[13] Long,[14] Hayes and Loudon,[15] Penzkofer et al.,[16] Kielich,[17–19] Shen,[20] D'yakov and Nikitin,[21] and the most recent reviews by Raymer and Walmsley,[22] Peřina,[23] and Mostowski and Raymer.[24] We also refer the reader to the special issue of the *Journal of the Optical Society of America B*,[25] which is devoted entirely to Raman scattering. Although an extensive literature has accumulated dealing with Raman scattering, it should be emphasized that the understanding of the fundamental principles that govern the process is still incomplete.

There are several major groups of theories treating the Raman effect in the semiclassical and quantum approaches, and theories for standing waves and spatially propagating waves. Here, we discuss in detail the quantum theory of Raman scattering for several radiation modes only; this implies that the theory is the best suited for scattering in a tuned cavity. Nevertheless, some predictions from the standing wave model also can be applied for traveling wave models.[26–29]

Various methods have been applied to the Raman effect in each of the above theories. Taking into account the equation of motion as the basis for classification, we can distinguish the following approaches, based on the photon rate equation, the Schrödinger equation, the Heisenberg equation (Heisenberg-Langevin equation), the master equation (generalized Fokker-Planck equation), and the Maxwell-Heisenberg equation (Maxwell-Block equation); we refer to Refs. 22, 23, 30, 31. The above classification is obviously oversimplified. Firstly, there are many relations bridging these approaches. For instance, we shall apply the master equation approach from which we shall derive the Fokker-Planck equation and the photon rate equation. Secondly, there exist other alternative methods, which do not fit into our classification. Let us mention, for example, those developed by Mavroyannis[32–34] and Freedhoff.[35] Thirdly, one can classify the Raman effect theories in many other ways (see, e.g., Ref. 22).

We shall be considering the incident laser photons to be scattered by chaotic phonons or quantized chaotic vibrations in a crystal. The process leads to Stokes and anti-Stokes photons. To the description of Raman scattering, we use two trilinear Hamiltonians coupled via an infinite number of phonon modes; one Hamiltonian describes Stokes radiation, and the other describes anti-Stokes radiation. The problem of coupled Stokes and anti-Stokes modes were studied previously by Bloembergen and Shen,[36-38] who applied the coupled wave theory of nonlinear optics formulated by Armstrong et al.[39] Later, Mishkin and Walls[40] quantized the Stokes and anti-Stokes modes, but dealt with the laser mode as a constant amplitude (so-called the parametric approximation). In fact, they considered two bilinear Hamiltonians, coupled by way of a phonon mode. Stokes scattering was treated as a parametric amplifier, whereas anti-Stokes scattering was treated as a parametric frequency converter. A detailed study of quantum statistics of the bilinear Hamiltonians, proposed by Louisell et al.,[41] has been extensively carried out (e.g., Refs. 42–50) and applied to Raman scattering, in particular by Mishkin and Walls,[40] Walls,[51, 52] Peřina,[53-55] Kárská and Peřina,[56] and others. Walls[57] (see also Ref. 44) has extended the bilinear Hamiltonian to a trilinear form to describe Raman scattering. The dynamics of Raman processes with trilinear Hamiltonian has been studied by Szlachetka et al.,[58-60] Szlachetka and Kielich,[61] Szlachetka,[27] Trung and Schütte,[62] Tänzler and Schütte,[63] Reis and Sharma,[64] Peřina et al.,[65, 66] Peřina and Křepelka,[67, 68] Levenson et al.,[69] and others (for general analyses see also Refs. 23, 70, and 71).

We shall describe Raman scattering from phonons as collective phenomena involving the interaction of many molecules. Much attention has also been drawn to a microscopic picture of the Raman effect by considering the interaction with individual molecules. Shen,[26, 72] in his quantum-statistical theory of nonlinear phenomena, proposed the general $m + n$ photon Hamiltonian, describing m emissions and n absorptions, and atomic transitions of an ensemble of N f-level atoms. This microscopically correct Hamiltonian contains Bose operators of a field and Fermi operators for optically active electrons and therefore describes a variety of nonlinear phenomena, in particular Raman scattering (for two- or three-level atoms). The same general Hamiltonian has been used by Walls[52] and McNeil and Walls.[73] Raman scattering from a two-level molecular (atomic) system[74-84] and in a three-level molecular system[74, 85-92] has been extensively studied by various authors. Walls[74] has shown that a description of Raman scattering from two-level molecules with a large cooperation number (coherent molecular coupling) is markedly similar to the results for Raman scattering from phonons. This is because for coherent molecular coupling, sums of the Fermi operators for the individual molecules can be replaced by the collective operators approximately satisfying boson commutation relations.

Nonclassical properties of radiation, such as squeezing, sub-Poissonian photon statistics, and photon antibunching, remain central topics in quantum optics. The literature in this area is truly prodigious. The reader is referred to the articles published in Vol. 85 of this series and references therein, for instance, Refs. 93–95, as well as the reviews by Kielich et al.,[96] Leuchs,[97] Loudon and Knight,[98] Teich and Saleh,[99, 100] Zaheer and Zubairy,[101] and the topical issues of the *Journal of the Optical Society of America B*[102] and the *Journal of Modern Optics*.[103]

Squeezing properties of Raman scattering have been studied by Peřinová et al.,[65, 66] Peřina,[55] Kárská and Peřina,[56] Levenson et al.,[69] and Peřina and Křepelka.[67, 68] Sub-Poissonian photon-counting statistics and/ or photon anticorrelations (in particular antibunching) have been investigated within various approaches to the standing-wave Raman effect by Loudon,[30] Simaan,[75] Agarwal and Jha,[87] Trung and Schütte,[62] Szlachetka and Kielich,[61] Szlachetka et al.,[58–60] Gupta and Mohanty,[78, 79] Peřina,[53, 54] Tänzler and Schütte,[63] Germey et al.,[104] Mohanty et al.,[80] Král,[105] Gupta and Dash,[90, 106] Ritsch et al.,[92] and in papers already mentioned[55, 56, 65–68] (see also Ref. 23). We note that photons scattered in the hyper-Raman effect can also exhibit nonclassical photon-counting correlations[27, 60, 65, 106–116] or squeezing.[65, 113, 116]

We shall analyze, in particular, cross-fluctuations (cross-correlations) in quadratures and in photocount statistics between different radiation modes. The theory of coherent light scattering within the consistent multipole tensor formalism, developed by Kielich[117] (see also Refs. 27, 58, 60) was successfully applied to disclose a novel cross-fluctuation mechanism. Here, an analysis of cross-correlations is presented along the lines of Szlachetka et al.[58, 59] (see also Ref. 23), as well as Loudon.[30]

We shall be studying sub- or super-Poissonian photon-counting statistics. We shall not analyze photon antibunching or bunching. The inclusion, in our Raman scattering model, of standard (i.e., temporal) photon antibunching would pose no problem. Let us mention the difference between sub-Poissonian statistics and anti-bunching pointed out in Refs. 118 and 119, which enables us to claim that these are distinct phenomena, and definitions should not be confused. The Raman scattering model is not suitable for investigations of spatial antibunching as defined by Le Berre-Rousseau et al.[120] and Białynicka-Birula et al.[121] in terms of negative angular correlations of photons.

As stated above, we shall be considering the quantum statistics of Raman scattering from phonons. We shall concentrate on a statistical analysis within the master equation approach to the Raman effect proposed by Shen,[26] Walls,[122] and McNeil and Walls.[73] This approach has been studied by various authors, e.g., Simaan,[75] Schenzle and Brand,[31] Peřina,[23, 53, 54] Germey et al.,[104] Gupta and Dash,[106] Bogolubov et al.,[88] Grygiel,[83] Miranowicz,[84] and Kárská and Peřina.[56] Usually a master equation is converted to a classical differential equation. Here, we shall apply a transformation to a Fokker-Planck equation (FPE) for s-parametrized quasidistributions using the coherent-state technique and an alternative method of a master equation in terms of Fock states (or a rate equation for the conditional photon-number probabilities).

Walls[122] was the first to apply the FPE technique to Raman scattering. This approach was extensively developed by Peřina and coworkers (Ref. 23 and references therein). Unfortunately, a FPE for the Raman effect has been solved exactly under parametric approximation only; i.e., a pump depletion was not included. It means that Raman scattering is described as a competing process of parametric amplification (Stokes scattering) and parametric frequency conversion of light (anti-Stokes scattering) in a nonlinear crystal. This approximation seems to be a real shortcoming of the FPE approach. A problem of the existence of a solution of the FPE also arises. A diffusion matrix of the FPE for the s-parametrized quasidistributions (with $s \approx 1$) in many cases is not positive or positive semidefinite. Therefore, such a FPE cannot be interpreted as an equation of motion describing the Brownian motion under the influence of a suitable force.[123] For this reason the term pseudo- or generalized-FPE is used in the literature. It is sometimes argued that equations of this type are unphysical. However by doubling the phase space, it is possible to introduce a generalized P-representation (the positive P-representation).[124, 125] The equation of motion for this generalized P-representation is a FPE with a positive or positive semidefinite diffusion matrix. The nonpositive definite diffusion matrix plays an essential role in the production of nonclassical fields.[126]

The second method of an equation of motion in Fock representation has been applied to various multi-photon Raman processes.[30, 73, 75, 78–80, 90, 106, 107, 112, 114–116] The master equation (in terms of Fock states) for first-order Stokes scattering can be solved by applying the Laplace transform method. Solutions obtained by McNeil and Walls,[73] Simaan,[75] and others apply only to the diagonal elements of the density matrix $\hat{\rho}$, which is a serious drawback of these formulations.[23] The photocount statistics (sub-Poissonian photon statistics, antibunching, or anticorrelation) can be fully analyzed using the diagonal, in Fock representation, matrix elements of $\hat{\rho}$ only. However, the phase properties of the fields,[129–131] or squeezing properties[98, 132] (which are sensitive to the phase of the field) require the availability of the non-diagonal terms of the density matrix. We shall derive, for Raman scattering including depletion of the pump field, an exact solution of the master equation for the complete density matrix in Fock representation $\langle n, m | \rho | n', m' \rangle$ with arbitrary n, m, n', m'.

The classical description of Raman scattering into both the Stokes and anti-Stokes fields seems to be well understood,[6, 20] contrary to quantum description, which is hampered by the complexity of the underlying Hamiltonians and hence the complex structure of the equations of motion. One of the simplest nontrivial models describing the coupling of the Stokes and anti-Stokes scattering was proposed by Knight.[133] This Raman-coupled model, in which a single atom is coupled to a single-cavity mode by Raman type transitions, has attracted some attention and has been generalized to much more realistic experimental conditions in the subsequent papers by Phoenix and Knight,[134] Schoendorff and Risken,[135] Agarwal and Puri,[136, 137] as well as Gerry and Eberly,[138] Gerry,[139] Gerry and Huang,[140] and Gangopadhyay and Ray.[141] It is quite remarkable that there exists a strict *operator* solution[136, 137] of the master equation describing the evolution of the generalized Knight model, which describes the system of an atom undergoing Raman transitions between two degenerate levels on interaction with a quantized field in a lossy cavity driven by an external field including the effects of atomic dephasing collisions. An extension of the propagation theory of Raman effect[142, 143] to include anti-Stokes scattering has been developed by Kilin[144] and independently by Li et al.[145] As mentioned, we shall analyze another model of the Stokes-anti-Stokes coupling within the framework of the temporal theory of Raman effect proposed by Walls[122] and extensively studied by Peřina and coworkers (see Refs. 23 and 68).

We shall discuss only temporal variations of fields instead of full temporal and spatial analysis. The assumption of monochromatic pump, Stokes, and anti-Stokes fields restricts the validity of our theory to a cavity problem. However, a temporal evolution in a cavity problem can usually be converted to a corresponding steady-state propagation in a dispersionless medium by simply replacing the time variable t by $-z/c$, a "normalized" space variable z. This procedure permits us to address nonlinear optical phenomena, in particular Raman scattering, in a manner analogous to their classical treatment.[20, 26, 28, 29] Formal space–time analogies have also been pointed out in the differential equations for the propagation of short light pulses.[146] Obviously, a full quantum space–time description is considerably more complex and resides in solving equations of motion for an infinite number of creation (annihilation) operators of the single-mode radiation fields. The total spatially dependent field is a sum of the single-mode solutions.

Here we mention only some spatial propagation theories of Raman scattering. For a detailed analysis we refer the reader to the review by Raymer and Walmsley[22] and references therein. The temporal and spatial evolution of the radiation fields (laser, Stokes, and anti-Stokes fields) in Raman scattering was successfully described within the framework of the classical coupled wave theory developed by Bloembergen and Shen[36–38] (see also Ref. 20). The first quantum theories of Raman scattering including spatial propagation were proposed independently by von Foerster and Glauber[147] and Akhmanov et al.[148] using the analogy of Raman effect and optical parametric amplification processes. Another method was proposed by Emel'yanov and Seminogov[149] and Mostowski and Raymer[142, 143] using the analogy between Raman scattering and superfluorescent processes.[150] Spectacular predictions of the latter theory have been, in particular, macroscopic pulse-energy fluctuations of the emitted radiation in a manner reflecting the underlying spontaneous initiation.[151-153] The negative-exponential probability distribution (NEPD), derived by Raymer et al.,[151] describes the macroscopic fluctuations of the scattered radiation. Kárská and Peřina[56] pointed out that the NEPD corresponds to the generating function of the integrated intensity extensively used in this paper (see Section V.B). The standing-wave theory of Raman scattering properly describes the macroscopic fluctuations in the low-gain and high-gain regime (see Refs. 22, 23, and 145 and references therein). In the latter limit the quantum fluctuations of the generated fields can be thought of as arising from a classical noise process, contrary to the low-gain limit, where certain nonclassical effects occur.

Finally, we should mention certain crucial experiments revealing some manifestations of Raman scattering. For more details, see the review article by Raymer and Walmsley.[22] Experiments on the detection of fluctuations of Stokes pulse energies were carried out by Walmsley and Raymer[154, 155] and Fabricius et al.[156] The temporal and spatial fluctuations of the Stokes beam profile, the spectrum, and delay have been investigated in a number of experiments both for depleted and undepleted pump pulse (for references see Ref. 22). We mention these experiments because the theory of Raman scattering for cavity modes, to be presented here, correctly predicts the existence of macroscopic quantum fluctuations of the Stokes pulses.

Generation of Raman solitons in the heavily depleted pump pulse has recently been observed by MacPherson et al.[157] and Swanson et al.,[158] as predicted by Englund and Bowden.[159, 160] Cooperative effects in Raman scattering, referred to as cooperative Raman scattering, which is analogous to two-level superfluorescence,[161] occurs for a laser pump not significantly depleted. The effect was first observed by Kirin et al.[162] and then re-examined under fully convincing experimental conditions by Pivtsov et al.[163] In our analysis we clearly distinguish the two cases of the depleted and undepleted pump field, and therefore have listed some effects and experiments in which this condition for the intensity pump is crucial.

Hyper-Raman scattering, i.e., the three-photon analog of Raman scattering, was discovered in 1965 by Terhune et al.[164] (see also Ref. 165). This effect was predicted theoretically by Neugebauer,[166] Kielich,[167, 168] and Li[169] prior to its experimental detection. Since the discovery of hyper-Raman scattering, numerous papers have appeared reporting theoretical investigations and observations of the process in a variety of solids, liquids and vapours. Here, we shall not discuss higher-order Raman scattering processes. We refer to the reviews of Refs. 17, 23, 27, and 172 for details and literature.

This paper is organized as follows. In Section II, the standing-wave model of Raman scattering is constructed and the basic equation of motion (master equation) is derived. In Section III, we give a short account of multimode s-parametrized quasidistributions and s-parametrized characteristic functions. In Section IV, we introduce definitions of nonclassical properties of radiation such as quadrature ("usual" and principal) squeezing and sub-Poissonian photon statistics. In Section V, we present the s-parametrized quasidistribution formalism of Raman scattering either including (in Section V.A) or neglecting (in Section V.B) depletion of the pump laser beam in the process of scattering. In Section VI we develop the density matrix formalism of Raman scattering. We derive exact solutions of the master equation in Fock representation in Section VI.A.2. We also give short-time (in Section VI.A.1) and long-time (in Section VI.A.3) solutions of the master equation. In Section VI.B we present approximate solutions valid under parametric approximation, i.e., when pump depletion is neglected.

II. MODEL AND MASTER EQUATION

Let us analyze Raman scattering starting from a completely quantum Hamiltonian but describing phenomenologically only the net effect, i.e., ignoring the details of the scattering mechanism. We describe the interaction of three single-mode radiation fields: an incident laser beam at the frequency ω_L, a Stokes field at the frequency ω_S, and an anti-Stokes field at the frequency ω_A through an infinite phonon system at frequencies ω_{V_j}, after Walls[122] (see also Refs. 53–55), by the effective Hamiltonians:

$$\hat{H}_0 = \hbar\omega_L \hat{a}_L^+ \hat{a}_L + \hbar\omega_S \hat{a}_S^+ \hat{a}_S + \hbar\omega_A \hat{a}_A^+ \hat{a}_A + \hbar\sum_j \omega_{V_j} \hat{a}_{V_j}^+ \hat{a}_{V_j}$$

$$\hat{H}_S = \hbar\sum_j \lambda_{S_j} \hat{a}_L \hat{a}_S^+ \hat{a}_{V_j}^+ + \text{h.c.}$$

$$(1)$$

$$\hat{H}_A = \hbar\sum_j \lambda_{A_j}^* \hat{a}_L \hat{a}_A^+ \hat{a}_{V_j} + \text{h.c.} \tag{2}$$

$$\hat{H}_T = \hat{H}_0 - \hat{H}_S - \hat{H}_A \tag{3}$$

where $\hat{H}_S$ ($\hat{H}_A$) is the trilinear interaction Hamiltonian for Stokes (anti-Stokes) scattering and H_0 is the unperturbed Hamiltonian. For simplicity, we have dropped the zero-point contributions. The annihilation operators for the laser, Stokes, anti-Stokes, and phonon fields are denoted by $\hat{a}_L$, $\hat{a}_S$, $\hat{a}_A$, and $\hat{a}_{Vj}$, respectively (we label all Hilbert space operators with caret). The coupling coefficient $\lambda_{Sj}(\lambda_{Aj})$ denotes the strength of the coupling between the Stokes (anti-Stokes) mode and the optical phonon at the frequency ω_{Vj}. These coefficients depend on the actual interaction mechanism. In the Hamiltonians (2) we neglect terms describing higher-order Stokes scattering,[57, 64, 74] as well as terms describing hyper-Raman scattering.[27, 65, 106] In Section VI.A in the analysis of the Raman effect without parametric approximation we also neglect anti-Stokes production.

In our model we take into account only the radiation modes appropriate for a cavity. It should be kept in mind that the several radiation mode description is applied to the waves involved in the whole course of the interaction, not only at the beginning of the interaction process. This approximation is a shortcoming from the experimental point of view, since it is not very suitable for describing the most common experimental arrangements used when measuring stimulated Raman scattering.[20–22]

We apply the rotating wave approximation since in the interaction Hamiltonians (2) we have omitted terms of the form $\hat{a}_{Vj}\hat{a}_S^+\hat{a}_L$ + h.c. and $\hat{a}_{Vj}^+\hat{a}_A^+\hat{a}_L$ + h.c.. For weak coupling these terms are negligible because they vary rapidly as $\exp[\pm i(\omega_{Vj} + |\omega_L - \omega_{S,A}|)t]$, which implies that their average is approximately zero for times of evolution much greater than $|\omega_L - \omega_{S,A}|^{-1}$, contrary to the interaction Hamiltonians H_S and H_A (2), which vary as $\exp[\pm i(\omega_{Vj} - |\omega_L - \omega_{S,A}|)t]$ giving unity for $\omega_{Vj} \approx |\omega_L - \omega_{S,A}|$. We have also neglected terms of the form $\hat{a}_{Vj}^{\pm}\hat{a}_S\hat{a}_L$ + h.c. and $\hat{a}_{Vj}^{\pm}\hat{a}_A\hat{a}_L$ + h.c.. These terms, if included, would describe a process in which both the Stokes (anti-Stokes) and laser photons are annihilated and created in the scattering act.

The Hamiltonians (2) describe Raman scattering under the long wavelength approximation, which has several important implications.[147, 173, 174] Firstly, we can neglect the intermolecular interactions. Each optical vibrational mode of the medium is equivalent to a simple harmonic oscillator. Secondly, the optical phonon dispersion is negligible. A typical dispersion curve for optical phonons, $\omega_V(k_V)$, is almost flat for wave vectors k_V from the interval $(-1/\lambda, 1/\lambda)$, where λ is an optical wavelength. In other words, optical wave vectors occupy only a very small volume about the origin of the reciprocal lattice. Thirdly, a crystal can be treated as a continuum; thus, from the mathematical point of view, sums over lattice sites can be replaced by integrals over a volume of the crystal. This long wave approximation is quite realistic for optical processes, in particular Raman scattering.

A detailed derivation of the Hamiltonian from first principles has been given by von Foerster and Glauber[147] in their quantum propagation theory of Raman scattering from phonons. Although we deal with modes in a cavity, many aspects of their theory recur in our approach.

In the case of an unbounded medium the momentum is conserved in the interaction, i.e., the sum (difference) of the wave vectors $\mathbf{k}_S$ ($\mathbf{k}_A$) of the Stokes (anti-Stokes) photon and $\mathbf{k}_V$ of the photon involved in the scattering act is exactly equal to the laser light wave vector $\mathbf{k}_L$,

$$\mathbf{k}_L = \mathbf{k}_S + \mathbf{k}_V \qquad \mathbf{k}_L = \mathbf{k}_A - \mathbf{k}_V \tag{4}$$

This means that each laser mode interacts strongly only with phonons having a single wave vector (one and only one vibrational mode). This is the requirement of translational invariance. Momentum is no longer strictly conserved for interactions in a finite medium, since the introduction of boundaries destroys the translational invariance of the medium. The strongest interaction is still for those modes which conserve momentum (4) and energy ($\omega_{Vj} \approx |\omega_L - \omega_{S,A}|$); nevertheless, in this case the radiation modes are coupled to a certain range of optical phonons whose wave vectors may not satisfy the adequate conditions (4) by amounts of the order of the reciprocal of the dimensions of the medium.[147] The coupling constants λ_{Sj}, and λ_{Aj} contain these momentum mismatches via phase integrals[27, 122]:

$$\lambda_{Sj} \sim \int_V \exp\left[-i(\mathbf{k}_L - \mathbf{k}_S - \mathbf{k}_{Vj}) \cdot \mathbf{r}\right] d^3r$$

$$\lambda_{Aj} \sim \int_V \exp\left[-i(\mathbf{k}_L - \mathbf{k}_A + \mathbf{k}_{Vj}) \cdot \mathbf{r}\right] d^3r \tag{5}$$

Hence, the interaction Hamiltonians (2) are represented by sums over all optical vibrational modes that may scatter into or out of the desired mode. This means that the coupling of the radiation fields (in particular Stokes and anti-Stokes) through a large number of optical phonons is treated stochastically.

In the Hamiltonians (1)–(3) we have assumed all the radiation fields and phonons to be polarized linearly in the same direction. We have not included explicitly the polarization states of those photons, which might affect the photon-counting statistics,[175, 176] squeezing,[177, 178] and other properties (Ref. 179 and references therein). Obviously, this would require the discussion of correlation tensors, in place of correlation functions, involving the photon polarization states.[19, 180, 181]

The model under discussion is restricted to the approximation of electric-dipole transitions. In previous papers,[170, 171] Kielich has proposed and extensively developed the formal quantum theory of first-, second-, and higher-order processes (in particular Raman scattering) taking into account multipolar electric and magnetic quantum transitions.

A lot of attention has been devoted to a simpler completely boson Hamiltonian applied to the description of the statistical properties of Raman scattering by phonons treated as a single monochromatic mode (Refs. 23, 44, 57, 58, and 69 and references therein). It is clear that the use of a large number of phonon modes (a phonon bath) in the model Hamiltonians (2) provides a fuller picture of the scattering processes. In particular, the model describes the stochastic coupling of the Stokes and anti-Stokes modes through a phonon bath. The assumption of a single phonon mode implies that the Stokes and anti-Stokes fields are coupled in a deterministic manner, which seems to be a rather serious drawback.[122]

As a digression, let us mention that the same phenomenological Hamiltonians (2) have been used in the description of Brillouin scattering (see, for example, Refs. 23, 104, and 182 and references therein). The main difference between Brillouin and Raman scattering lies in different kinds of the scatterers responsible for these effects: acoustic phonons in the Brillouin effect, and optical phonons in the Raman effect. This difference is included in the frequencies, the coupling constants $\lambda_{Sj}, \lambda_{Aj}$, and the reservoir spectrum. More important, acoustic phonons exhibit much greater dispersion than optical phonons. In our approach to Raman scattering we neglect dispersion. This assumption applied to Brillouin scattering has considerably less validity.

We are interested only in the statistical properties of the radiation fields (the pump and scattered beams) considered as a system. We therefore remove the unnecessary information about the infinite system of optical phonons, treated as a reservoir (heat bath). The procedure leading to the master equation is widely used in quantum optics. For a general review of the master equation methods and the extensive bibliography see Refs. 23, 51, 183, 184. We rewrite the interaction Hamiltonians H_S and H_A in the interaction picture as

$$\hat{H}_S + \hat{H}_A = \hbar \sum_{k=1}^{4} \hat{F}_k \hat{Q}_k \qquad (6)$$

where

$$\hat{F}_1 = \hat{F}_2^+ = \sum_j \lambda_{Sj} \hat{a}_{Vj}^+ \exp\left[i\omega_{Vj}(t - t_0)\right]$$

$$\hat{F}_3 = \hat{F}_4^+ = \sum_j \lambda_{Aj}^* \hat{a}_{Vj} \exp\left[-i\omega_{Vj}(t - t_0)\right]$$

$$\hat{Q}_1 = \hat{Q}_2^+ = \hat{a}_L \hat{a}_S^+ \exp\left[-i\Omega_S(t - t_0)\right] \tag{7}$$

$$\hat{Q}_3 = \hat{Q}_4^+ = \hat{a}_L \hat{a}_A^+ \exp\left[i\Omega_A(t - t_0)\right]$$

The $\hat{Q}_i$ ($\hat{F}_i$) are respectively functions of the system (reservoir) operators only. The "cavity" frequencies Ω_S, Ω_A are equal to

$$\Omega_{S,A} = |\omega_L - \omega_{S,A}| \tag{8}$$

Since the system and the reservoir variables are mutually independent, as it follows from

$$\left[\hat{a}_i, \hat{a}_j^+\right] = \delta_{ij} \qquad \text{for } i, j = L, S, A, V_1, V_2, \ldots \tag{9}$$

we may trace, in standard manner, the complete density matrix over the reservoir leading to the reduced density matrix $\hat{\rho}(t)$. Obviously, we cannot obtain any reservoir averages from $\hat{\rho}(t)$. There are some Raman scattering models (e.g., Refs. 23, 104, and 147), where optical phonons are included in the system, whereas other crystal excitation modes, such as acoustical phonons, electric excitations, and other species of molecular vibrations, serve as a thermal reservoir.

The radiation fields are weakly coupled to the thermal reservoir. The anti-Stokes mode loses energy to the reservoir. The fluctuations in the reservoir also couple back into the system introducing noise from the reservoir. However, we apply the Markov approximation, a condition sufficient to ensure that energy that goes into the reservoir will not return to the radiation fields. This conclusion follows from the definition of the Markovian system as one that cannot develop memory—the future of the system is determined by the present and not its past.[183, 184] The importance of this assumption is sometimes stressed in the concept of a Schrödinger-Markov (or Heisenberg-Markov) picture, meaning the standard pictures under Markov approximation.[183] The importance of non-Markovian effects in Raman scattering has been recently studied by, e.g., Sugawara et al.[190] and Villaeys et al.[91] Obviously, the system operators $\hat{Q}_i$

obey the same commutation relations under this approximation as they did originally.

To obtain the equation of motion for the reduced density matrix $\hat{\rho}(t)$, one has to compute the reservoir spectral densities

$$\omega_{ij}^{+} = \int_0^{\infty} \dot{e}^{i\omega_i \tau} \langle \hat{F}_i(\tau) \hat{F}_j \rangle_R \, d\tau$$

$$\omega_{ji}^{-} = \int_0^{\infty} e^{i\omega_i \tau} \langle \hat{F}_j \hat{F}_i(\tau) \rangle_R \, d\tau$$

(10)

where $\langle \dots \rangle_R$ is the average over all reservoir operators; ω_i takes the values $\pm \Omega_{S,A}$. The infinite system of optical phonons is assumed to be densely spaced with the number of modes between ω_i and $\omega_i + d\omega_i$ equal to $g(\omega_i) d\omega_i$, so we may replace the sums over the optical vibrational modes by integrals

$$\sum_j (\dots) \approx \int_0^{\infty} d\omega_j \, g(\omega_j)(\dots)$$

(11)

Let us introduce two quantities: $\Delta\Omega$–the frequency mismatch and Ω–the medium "cavity" frequency, defined by

$$\Delta\Omega = \frac{\Omega_S - \Omega_A}{2}$$

$$\Omega = \frac{\Omega_S + \Omega_A}{2}$$

(12)

The frequency mismatch $\Delta\Omega$, in general, is not equal to zero. It is quite realistic for optical phonons to assume that the coupling constants $\lambda_{S,A}(\omega_j)$ and the phonon density of $g(\omega_j)$ are flat in the vicinity of Ω, so that we can write

$$g(\Omega \pm \Delta\Omega) \approx g(\Omega)$$

$$\lambda_k(\Omega \pm \Delta\Omega) \approx \lambda_k(\Omega) \qquad k = S, A$$

(13)

The reservoir is supposed to be at thermal equilibrium. The phonons are unaffected by interaction with the radiation fields. In the classical sense this means that the phonons are so quickly damped that they remain in their steady state.[20, 37, 38] The mean number of phonons in the reservoir mode at thermal equilibrium is defined by the Bose-Einstein distribution

$$\langle \hat{n}(\omega_{Vj}) \rangle = \left[\exp\left(\frac{\hbar \omega_{Vj}}{k_B T} \right) - 1 \right]^{-1} \tag{14}$$

where k_B is the Boltzmann constant and T is the temperature of the reservoir. Obviously, as the reservoir temperature approaches absolute zero, the mean number of phonons $\langle \hat{n}(\omega_{Vj}) \rangle$ tends to zero as well. In Section V.B, we analyze Raman scattering in a parametric approximation for a "noisy" reservoir ($\langle \hat{n}(\omega_{Vj}) \rangle \neq 0$), whereas we study Raman scattering including the pump depletion for "quiet" reservoir ($\langle \hat{n}(\omega_{Vj}) \rangle \approx 0$). After some algebra one obtains from (10),

$$\omega_{21}^+ = \left(\frac{\gamma_S}{2} + i\Delta\omega_S \right)(\langle \hat{n}_V \rangle + 1)$$

$$\omega_{12}^+ = \left(\frac{\gamma_S}{2} - i\Delta\omega_S \right)\langle \hat{n}_V \rangle$$

$$\omega_{43}^+ = \left(\frac{\gamma_A}{2} - i\Delta\omega_A \right)\langle \hat{n}_V \rangle$$

$$\omega_{34}^+ = \left(\frac{\gamma_A}{2} + i\Delta\omega_A \right)(\langle \hat{n}_V \rangle + 1)$$

$$\omega_{31}^+ = \left(\frac{\gamma_{AS}}{2} + i\Delta\omega_{AS} \right)(\langle \hat{n}_V \rangle + 1) \tag{15}$$

$$\omega_{13}^+ = \left(\frac{\gamma_{SA}}{2} - i\Delta\omega_{SA} \right)\langle \hat{n}_V \rangle$$

$$\omega_{42}^+ = \left(\frac{\gamma_{AS}}{2} - i\Delta\omega_{AS} \right)\langle \hat{n}_V \rangle$$

$$\omega_{24}^+ = \left(\frac{\gamma_{SA}}{2} + i\Delta\omega_{SA} \right)(\langle \hat{n}_V \rangle + 1)$$

$$\omega_{ij}^- = \left(\omega_{ij}^+ \right)^*$$

All other reservoir spectral densities, in particular the diagonal densities $\omega_{ii}^\pm$ (for $i = 1, \ldots, 4$), vanish. For simplicity we have denoted the mean number of phonons at frequency Ω by $\langle \hat{n}_V \rangle = \langle \hat{n}_V(\Omega) \rangle$. The gain constant for the Stokes mode γ_S, the damping constant for the anti-Stokes model γ_A, and the mutual damping constants for both scattered fields

γ_{SA}, γ_{AS} are

$$\gamma_k = 2\pi g(\Omega)|\lambda_k(\Omega)|^2 \qquad (k = S, A)$$
$$\gamma_{SA} = \gamma_{AS}^* = 2\pi g(\Omega)\lambda_S(\Omega)\lambda_A^*(\Omega) \tag{16}$$

where $g(\Omega)$, as earlier, denotes the density of the optical phonon modes (the reservoir spectrum) at frequency Ω. It is seen that the following simple relation between the single and mutual damping constants holds: $|\gamma_{SA}|^2 = |\gamma_{AS}|^2 = \gamma_A \gamma_S$. The frequency shifts, representing the Lamb shift in the frequency $\Omega \approx \Omega_j$, are expressed by the Cauchy principle value, $\mathscr{P}$ of the integrals:

$$\Delta\omega_k = -\mathscr{P}\int_0^\infty \frac{g(\omega_j)|\lambda_k(\omega_j)|^2}{\omega_j - \Omega_k}\, d\omega_j \qquad (k = S, A)$$

$$\Delta\omega_{SA} = (\Delta\omega_{AS})^* = -\mathscr{P}\int_0^\infty \frac{g(\omega_j)\lambda_S(\omega_j)\lambda_A^*(\omega_j)}{\omega_j - \Omega}\, d\omega_j \tag{17}$$

The only effect of the $\Delta\omega_i$ is to change slightly the frequency Ω, so we neglect them. Having calculated the reservoir spectral densities we can write the master equation for the reduced density matrix $\hat{\rho} = \hat{\rho}(\hat{a}_L, \hat{a}_S, \hat{a}_A, t)$ as

$$
\begin{aligned}
\frac{\partial}{\partial t}\hat{\rho} &= \tfrac{1}{2}\gamma_S\left(\left[\hat{a}_L\hat{a}_S^+, \hat{\rho}\hat{a}_L^+\hat{a}_S\right] + \text{h.c.}\right) \\
&+ \tfrac{1}{2}\gamma_A\left(\left[\hat{a}_L^+\hat{a}_A, \hat{\rho}\hat{a}_L\hat{a}_A^+\right] + \text{h.c.}\right) \\
&+ \tfrac{1}{2}\gamma_{SA}\,e^{-2i\Delta\Omega\,\Delta t}\left(\left[\hat{a}_L\hat{a}_S^+, \hat{\rho}\hat{a}_L\hat{a}_A^+\right] + \left[\hat{a}_L\hat{a}_S^+\hat{\rho}, \hat{a}_L\hat{a}_A^+\right]\right) \\
&+ \tfrac{1}{2}\gamma_{AS}\,e^{2i\Delta\Omega\,\Delta t}\left(\left[\hat{a}_L^+\hat{a}_A, \hat{\rho}\hat{a}_L^+\hat{a}_S\right] + \left[\hat{a}_L^+\hat{a}_A\hat{\rho}, \hat{a}_L^+\hat{a}_S\right]\right) \\
&- \langle\hat{n}_V\rangle\Big\{\tfrac{1}{2}\gamma_S\left(\left[\hat{a}_L^+\hat{a}_S, \left[\hat{a}_L\hat{a}_S^+, \hat{\rho}\right]\right] + \text{h.c.}\right) \\
&\qquad + \tfrac{1}{2}\gamma_A\left(\left[\hat{a}_L\hat{a}_A^+, \left[\hat{a}_L^+\hat{a}_A, \hat{\rho}\right]\right] + \text{h.c.}\right) \\
&\qquad + \gamma_{SA}\,e^{-2i\Delta\Omega\,\Delta t}\left[\hat{a}_L^+\hat{a}_S, \left[\hat{a}_L^+\hat{a}_A, \hat{\rho}\right]\right] \\
&\qquad + \gamma_{AS}\,e^{2i\Delta\Omega\,\Delta t}\left[\hat{a}_L\hat{a}_A^+, \left[\hat{a}_L\hat{a}_S^+, \hat{\rho}\right]\right]\Big\}
\end{aligned}
\tag{18}
$$

The term in γ_S represents the amplification of the Stokes mode; the term in γ_A describes the loss of energy from the anti-Stokes mode into the reservoir; the γ_{AS} and γ_{SA} terms represent the stochastic coupling between the Stokes and anti-Stokes modes through the reservoir; the remaining terms in $\langle \hat{n}_V \rangle \gamma_i$ represent the diffusion of fluctuations of the reservoir into the system modes. Eq. (18) describes, moreover, the evolution of the laser beam, i.e., the depletion of the laser field, the coupling of the field with the Stokes and anti-Stokes fields, as well as the diffusion of the reservoir fluctuations into the laser field. The interpretation of the γ_S (γ_A) terms as the amplification (attenuation) of the radiation fields is as yet intuitive, but will gain in precision on solution of the generalized Fokker-Planck equation. The master equation (18) could have been written in more compact form; albeit for purposes of interpretation the above form is more convenient.

The master equation (18), in the particular case of parametric approximation, reduces to the equation obtained by Walls[122] and Peřina,[53] and reduces to that of McNeil and Walls for Stokes scattering alone but with no need for the parametric approximation.[73] Our master equation (18) differs but slightly in the diffusion terms $\langle \hat{n}_V \rangle \gamma_i$ only from the special case of the master equation given by Agarwal[185] (see also Ref. 73).

The master equation may be solved by various techniques presented in standard textbooks.[23, 183, 184, 186–188] Here, we apply two methods. We convert the master equation to an associated classical equation. On the one hand, expressing the quantum equation in s-ordered form one obtains the generalized Fokker-Planck equation for the s-parametrized quasi-probability distribution, which can be exactly solved for a class of Ornstein-Uhlenbeck processes.[188, 191] On the other hand, one can express the master equation in Fock representation, which can be solved, for instance, by the Laplace transform method.[30, 73, 75]

In the following sections we analyze three cases. Firstly, we briefly describe coupling of the three quantum radiation fields: the laser, Stokes, and anti-Stokes beams. The problem simplifies considerably if one assumes narrow quasi-probability distributions. Secondly, we apply the parametric approximation, which means that the pump field is treated classically. We include the coupling of the Stokes and anti-Stokes field through the phonon bath. Thirdly, we separately describe either the laser and Stokes mode or the laser and anti-Stokes mode, but include the depletion of the pump laser light. In this case we assume the heat bath to be "quiet."

III. MULTIMODE s-PARAMETRIZED QUASIDISTRIBUTIONS

A description of the multimode fields via quasiprobability distributions (quasidistributions, QPDs) or equivalently via characteristic functions was first proposed by Glauber,[180, 192, 193] Cahill,[194] and Klauder et al.[195] General ordering theorems have been given by Agarwal and Wolf.[196] The s-parametrized single-mode quasidistributions and characteristic functions were introduced by Cahill and Glauber,[197] who extensively studied various ways of defining correspondences between the operators and functions. For a recent review of the multimode s-parametrized functional formalism we refer the reader to Ref. 23. Here, we list the basic definitions and properties of the s-parametrized multimode quasidistributions and characteristic functions useful for our further investigations.

To solve the master equation (18), i.e., the operator equation, we use the c-number representations $\mathscr{W}^{(s)}(\{\alpha_k\})$ and $\mathscr{C}^{(s)}(\{\beta_k\})$ of the density operator introduced by Cahill and Glauber.[197] These representations not only are useful as a calculation tool, but also provide insight into the interrelations between classical and quantum mechanics. By virtue of the multimode s-parametrized displacement operator

$$\hat{D}^{(s)}(\{\beta_k\}) = \Pi_k \hat{D}^{(s)}(\beta_k) = \prod_k \exp\left(\beta_k \hat{a}_k^+ - \beta_k^* \hat{a}_k + \frac{s}{2}|\beta_k|^2\right) \quad (19)$$

where the continuous parameter s belongs to the interval $\langle -1, 1\rangle$, one can define the s-parametrized multimode characteristic function as the mean value of $\hat{D}^{(s)}(\{\beta_k\})$,

$$\mathscr{C}^{(s)}(\{\beta_k\}) = \mathrm{Tr}\left[\hat{\rho}\hat{D}^{(s)}(\{\beta_k\})\right] \quad (20)$$

In our situation involving the three radiation modes laser ($k = L$), Stokes (S) and anti-Stokes (A), the simplified notation in Eqs. (19) and (20) stands for $(\{\beta_k\}) = (\beta_L, \beta_S, \beta_A)$. The Fourier transform of the characteristic function $\mathscr{C}^{(s)}(\{\beta_k\})$ (20) readily gives the s-parametrized multimode quasidistribution $\mathscr{W}^{(s)}(\{\alpha_k\})$,

$$\mathscr{W}^{(s)}(\{\alpha_k\}) = \int \mathscr{C}^{(s)}(\{\beta_k\}) \exp\left[\sum_k (\alpha_k \beta_k^* - \alpha_k^* \beta_k)\right] d^2\left\{\frac{\beta_k}{\pi}\right\} \quad (21)$$

For completeness we write the inverse Fourier transform, which enables us to determine $\mathscr{C}^{(s)}$ from $\mathscr{W}^{(s)}$, namely,

$$\mathscr{C}^{(s)}(\{\beta_k\}) = \int \mathscr{W}^{(s)}(\{\alpha_k\}) \exp\left[\sum_k (\alpha_k^* \beta_k - \alpha_k \beta_k^*)\right] d^2\left\{\frac{\alpha_k}{\pi}\right\} \quad (22)$$

where integration extends over α_k in the following sense:

$$d^2\left\{\frac{\alpha_k}{\pi}\right\} = \prod_{k=L,S,A} \pi^{-1} d^2\alpha_k = \pi^{-3} \prod_{k=L,S,A} d(\text{Re } \alpha_k) \, d(\text{Im } \alpha_k)$$

or over β_k similarly. The normalization is chosen to satisfy

$$\int \mathcal{W}^{(s)}(\{\alpha_k\}) \, d^2\left\{\frac{\alpha_k}{\pi}\right\} = \mathcal{C}^{(s)}(0) = 1 \tag{23}$$

In the three special cases of $s = -1, 0, 1$ one recognizes the well-known QPDs,[23, 197, 201] namely the Q function, the Wigner function, and the Glauber-Sudarshan P-function, respectively:

$$Q(\{\alpha_k\}) = \langle\{\alpha_k\}|\hat{\rho}|\{\alpha_k\}\rangle = \mathcal{W}^{(-1)}(\{\alpha_k\})$$
$$W(\{\alpha_k\}) = \mathcal{W}^{(0)}(\{\alpha_k\}) \tag{24}$$
$$P(\{\alpha_k\}) = \pi^{-M}\mathcal{W}^{(1)}(\{\alpha_k\})$$

with M denoting the number of modes (in our analysis M will be equal to 3, 2, or 1). One can say that the s-parametrized quasidistribution $\mathcal{W}^{(s)}$ (with s from the interval $\langle -1, 1\rangle$) is a continuous interpolation between the P- and Q-functions. The Q-function directly determines antinormally ordered expectation values, the P-function determines normally ordered averages, and the Wigner function can be used directly to calculate the averages of symmetrically ordered operators. The following relations hold for any parameter s:

$$\left\langle \prod_k (\hat{a}_k^+)^{m_k}(\hat{a}_k)^{n_k}\right\rangle_{(s)} = \text{Tr}\left[\hat{\rho}\left\{\prod_k (\hat{a}_k^+)^{m_k}(\hat{a}_k)^{n_k}\right\}_{(s)}\right]$$

$$= \int \mathcal{W}^{(s)}(\{\alpha_k\})\prod_k (\alpha_k^*)^{m_k}(\alpha_k)^{n_k} \, d^2\{\alpha_k/\pi\} \tag{25}$$

$$= \prod_k \frac{\partial^{m_k}}{\partial\beta_k^{m_k}} \frac{\partial^{n_k}}{\partial(-\beta_k^*)^{n_k}}\mathcal{C}^{(s)}(\{\beta_k\})\Bigg|_{\{\beta_k\}=0}$$

where $\{\beta_k\} = 0$, in the three-mode case, means that $\beta_L = \beta_S = \beta_A = 0$. The generally accepted criterion for the definition of a nonclassical field resides in the existence of a positive P-function, i.e., a classical state is one whose P-function is no more singular than a δ-function and is nonnegative definite (e.g., Refs. 23, 98, 202, and 203). This means that the quantum statistical properties of the nonclassical field cannot be described completely within the framework of a classical probability theory. A detailed discussion of the existence of quasidistributions $\mathcal{W}^{(s)}(\{\alpha_k\})$ for the Raman scattering model under consideration is presented in Section V.B. The Wigner function always exists as a nonsingular function, but may assume negative values, and in this sense is not a classical probability distribution (nevertheless, as was shown by Stenholm,[204] experiments always give a positive Wigner function). The Q-function has the properties of a well-behaved (bounded, nonnegative and infinitely differentiable) classical probability distribution.

Let us write down the relation between two s_1- and s_2-parametrized quasidistributions:

$$\mathcal{W}^{(s_2)}(\{\alpha_k\}, t) = \left(\frac{2}{s_1 - s_2}\right)^M \int \exp\left(-\frac{2}{s_1 - s_2}\sum_k |\alpha_k - \beta_k|^2\right)$$
$$\times \mathcal{W}^{(s_1)}(\{\beta_k\}, t)\, d^2\left\{\frac{\beta_k}{\pi}\right\} \tag{26}$$

where $s_2 < s_1$. It is seen that the quasidistribution $\mathcal{W}^{(s_2)}$ is given by the convolution of $\mathcal{W}^{(s_1)}$ with the multidimensional Gaussian distribution. The analogous relation for characteristic functions (20) is simpler and valid for any s_1 and s_2,

$$\mathcal{C}^{(s_2)}(\{\beta_k\}, t) = \mathcal{C}^{(s_1)}(\{\beta_k\}, t)\exp\left(\frac{s_2 - s_1}{2}\sum_k |\beta_k|^2\right) \tag{27}$$

Even in the case when s_1-parametrized QPDs do not exist, the calculation of the expectation values $\langle \hat{a}^{+m}\hat{a}^n \rangle_{(s_1)}$ in s_1 order poses no problem. They can be obtained from the corresponding s_1-parametrized characteristic function $\mathcal{C}^{(s_1)}$ in view of Eq. (25) or, equivalently, from an s_2-parametrized quasidistribution $\mathcal{W}^{(s_2)}$, which does exist, by means of the relation

$$\left\langle \prod_k (\hat{a}_k^+)^{m_k}(\hat{a}_k)^{n_k}\right\rangle_{(s_1)} = \int \sum_k m_k!\left(\frac{s_2 - s_1}{2}\right)^{m_k}$$
$$\times \alpha_k^{n_k - m_k}L_{m_k}^{n_k - m_k}\left(\frac{2|\alpha_k|^2}{s_1 - s_2}\right)\mathcal{W}^{(s_2)}(\{\alpha_k\})\, d^2\left\{\frac{\alpha_k}{\pi}\right\} \tag{28}$$

where $L_m^n(x)$ is the generalized Laguerre polynomial. Alternatively, to obtain the s_1-ordered moments $\langle \hat{a}^{+m}\hat{a}^n \rangle_{(s_1)}$ one can use the generalized P-representation (positive P-representation).[124-128]

IV. PHOTON-COUNTING STATISTICS AND SQUEEZING: DEFINITIONS

To investigate nonclassical phenomena such as sub-Poissonian photon-counting statistics or photon antibunching, one needs to know the diagonal matrix elements in Fock representation of the density matrix $\hat{\rho}(\{\hat{\alpha}_k\})$ only. We start from the probability distribution $p(n)$ of the photon-number n in the k-mode field within a given volume V of space at the time t, defined by

$$p(n) = \sum_{\{n_k\}} \langle \{n_k\} | \hat{\rho} | \{n_k\} \rangle \delta_{n, \Sigma n_k} \tag{29}$$

where $n_k = |\alpha_k|^2$. The s-parametrized quasidistribution $\mathscr{W}^{(s)}(\{\alpha_k\}, t)$ (21) can be readily transformed to the following s-parametrized integrated quasidistribution (intensity distribution) $\mathscr{W}^{(s)}(W, t)$ by means of the δ-function,

$$\mathscr{W}^{(s)}(W, t) = \int \mathscr{W}^{(s)}(\{\alpha_k\}, t) \delta\left(\sum_k |\alpha_k|^2 - W \right) d^2\left\{ \frac{\alpha_k}{\pi} \right\} \tag{30}$$

where the variable W can be interpreted as the integrated intensity. The photodetection equation gives a connection between the continuous integrated quasidistribution $\mathscr{W}^{(1)}(W, t)$ and the discrete photon-number distribution first derived by Mandel.[205, 206] This photodetection equation states that the photocount distribution $p(n)$ is the Poisson transform of the integrated quasidistribution $\mathscr{W}^{(1)}(W, t)$. A generalized photodetection equation for $\mathscr{W}^{(s)}(\{\alpha_k\}, t)$ or for $\mathscr{W}^{(s)}(W, t)$ can be written as[23, 207]

$$\begin{aligned}
p(n) &= \left(\frac{2}{1+s} \right)^M \left(\frac{s-1}{1+s} \right)^n \int \mathscr{W}^{(s)}(\{\alpha_k\}) \\
&\quad \times \exp\left(-\frac{2}{1+s} \sum_k |\alpha_k|^2 \right) L_n^{M-1}\left(\frac{4}{1-s^2} \sum_k |\alpha_k|^2 \right) d^2\left\{ \frac{\alpha_k}{\pi} \right\} \\
&= \left(\frac{2}{1+s} \right)^M \left(\frac{s-1}{1+s} \right)^n \int \mathscr{W}^{(s)}(W) \\
&\quad \times \exp\left(-\frac{2W}{1+s} \right) L_n^{M-1}\left(\frac{4W}{1-s^2} \right) d^2 W
\end{aligned} \tag{31}$$

with $L_n^{M-1}(x)$ denoting the generalized Laguerre polynomial. We formally identify the photon-number distribution (29) with the photocount distribution (30). There is some slight difference in their physical interpretation, since the former distribution describes the probability of having n photons in the mode volume V, whereas the latter distribution describes the probability of detecting n photons in the detector volume V_{det}, defined by its parameters (sensitive area, response time, quantum efficiency, etc.). It can be argued, however (e.g., Refs. 208–210), that there is perfect physical equivalence between the photon-number moments obtained from (29) and the photocount-number moments calculated from (31), under the assumption of ideal detectors.

The s-parametrized time-dependent generating function $\langle \exp(-\lambda W) \rangle_{(s)}$, defined by the Fourier transform of the s-parametrized quasidistribution $\mathscr{W}^{(s)}(\{\alpha_k\}, t)$ or characteristic function $\mathscr{C}^{(s)}(\{\beta_k\}, t)$:

$$
\begin{aligned}
\langle \exp(-\lambda W(t)) \rangle_{(s)} &= \int \mathscr{W}^{(s)}(\{\alpha_k\}, t) \exp\left(-\lambda \sum_k |\alpha_k|^2\right) d^2\left\{\frac{\alpha_k}{\pi}\right\} \\
&= \lambda^{-M} \int \mathscr{C}^{(s)}(\{\beta_k\}, t) \exp\left(-\frac{1}{\lambda} \sum_k |\beta_k|^2\right) d^2\left\{\frac{\alpha_k}{\pi}\right\}
\end{aligned}
$$

(32)

enables us to calculate the photon-number distribution $p(n, t)$ and the s-ordered photon-number moments $\langle \hat{n}^k \rangle_{(s)}$ in a particularly simple manner:

$$
p(n) = \frac{(-1)^n}{n!} \frac{d^n}{d\lambda^n} \left(1 + \frac{s-1}{2}\lambda\right)^{-M} \left\langle \exp\left(-\frac{\lambda}{1 + \frac{s-1}{2}\lambda} W\right)\right\rangle_{(s)}\Bigg|_{\lambda=1}
$$

(33)

$$
\langle W^k \rangle_{(s)} = (-1)^k \frac{d^k}{d\lambda^k} \langle \exp(-\lambda W) \rangle_{(s)}\Bigg|_{\lambda=0}
$$

(34)

Eq. (33) takes the simplest form for $s = 1$. Several parameters are widely used in the literature to describe the photon-number statistics, e.g.; the Mandel Q parameter, the Fano factor, or the normalized second-order correlation function. In our analysis we employ the normalized second-order factorial moment of the photon-number operators (or integrated intensity) (Refs. 23 and 210 and references therein)

$$\gamma_k^{(2)} = \frac{\langle (\Delta \hat{n}_k)^2 \rangle_{(1)}}{\langle \hat{n}_k \rangle^2} = \frac{\langle \hat{n}_k^2 \rangle_{(1)}}{\langle \hat{n}_k \rangle^2} - 1 = \frac{\langle \hat{n}_k (\hat{n}_k - 1) \rangle}{\langle \hat{n}_k \rangle^2} - 1 \quad (35)$$

and its generalization, the normalized pth order factorial moment of the kth and lth mode (the normalized two-mode cross-correlation function of pth order)

$$\gamma_{kl}^{(p)} = \frac{\langle \hat{n}_{kl}^p \rangle_{(1)}}{\langle \hat{n}_{kl} \rangle^p} - 1 = \frac{\langle \hat{n}_{kl} (\hat{n}_{kl} - 1) \ldots (\hat{n}_{kl} - p + 1) \rangle}{\langle \hat{n}_{kl} \rangle^p} - 1 \quad (36)$$

where $\hat{n}_{kl} = \hat{n}_k + \hat{n}_l$. The higher-order factorial moments (36) by comparison with the second-order moments (35) provide us with further information concerning the photon-number distributions. In view of the fact that $\hat{n}_{kl}$ is the sum of the single-mode photon-number operators, the factorial moment $\gamma_{kl}^{(2)}$ can be written as

$$\gamma_{kl}^{(2)} = \frac{\langle (\Delta \hat{n}_{kl})^2 \rangle_{(1)}}{\langle \hat{n}_{kl} \rangle^2} = \frac{\langle (\Delta \hat{n}_k)^2 \rangle_{(1)} + \langle (\Delta \hat{n}_l)^2 \rangle_{(1)} + 2 \langle \Delta \hat{n}_k \Delta \hat{n}_l \rangle}{\langle \hat{n}_k \rangle^2 + \langle \hat{n}_l \rangle^2 + 2 \langle \hat{n}_k \rangle \langle \hat{n}_l \rangle} \quad (37)$$

The Mandel Q parameter for the mode k is equal to $\gamma_k^{(2)} \langle \hat{n}_k \rangle$, whereas the Fano factor F is $(\gamma_k^{(2)} \langle \hat{n}_k \rangle + 1)$ (the photoefficiency η of the photodetector is assumed to be $\eta = 1$).

Light with photon-number fluctuations smaller than those of the Poisson distribution is called sub-Poissonian (or photon-number squeezed) and is described by a negative value of $\gamma^{(2)}$, both for $\gamma_k^{(2)}$ in the single-mode case and for $\gamma_{kl}^{(2)}$ in the two-mode case. In Section VI we analyze the two-mode model of the Raman effect that comprises the laser (L) and the Stokes mode (S). We show that the sum of photon-number operators in both modes is a constant of motion, which implies that the factorial moments $\gamma_{LS}^{(p)}$ are constant as well. Henceforth we shall be applying another definition to investigate two-mode cross-correlation, referred to as the interbeam degree of second-order coherence, given by (Ref. 30 and references therein)

$$g_{kl}^{(2)} = \frac{\langle \Delta \hat{n}_k \Delta \hat{n}_l \rangle}{\langle \hat{n}_k \rangle \langle \hat{n}_l \rangle} = \frac{\langle \hat{n}_k \hat{n}_l \rangle}{\langle \hat{n}_k \rangle \langle \hat{n}_l \rangle} - 1 \quad (38)$$

(Our definition deviates from those of Ref. 30 by the extra term -1.)

To investigate squeezing properties of light we introduce the Hermitian single- and two-mode operators:

$$\hat{X}_k(\theta) = \hat{a}_k \, e^{-i\theta} + \hat{a}_k^+ \, e^{i\theta} \tag{39}$$

$$\hat{X}_{jk}(\theta) = \hat{a}_{jk} \, e^{-i\theta} + \hat{a}_{jk}^+ \, e^{i\theta} = \hat{X}_j(\theta) + \hat{X}_k(\theta) \tag{40}$$

where $\hat{a}_{kl} = \hat{a}_k + \hat{a}_l$. The operator $\hat{X}_k(\hat{X}_{kl})$ for $\theta = 0$ corresponds to the in-phase quadrature component of the kth (kth and lth) mode (modes) of the field, whereas for $\theta = \pi/2$ it corresponds to the out-of-phase component. For brevity, we use the notation $\hat{X}_{k1} = \hat{X}_k(0)$, $\hat{X}_{k2} = \hat{X}_k(\pi/2)$, as well as $\hat{X}_{kl1} = \hat{X}_{kl}(0)$ and $\hat{X}_{kl2} = \hat{X}_{kl}(\pi/2)$. The following commutation rules hold:

$$\left[\hat{X}_{k1}, \hat{X}_{k2} \right] = 2i \tag{41}$$

$$\left[\hat{X}_{kl1}, \hat{X}_{kl2} \right] = 4i \tag{42}$$

Firstly, we shall discuss in brief the single-mode case. The variances of the θ-dependent quadrature (39) are

$$\left\langle \left(\Delta \hat{X}_k(\theta) \right)^2 \right\rangle = 2 \, \mathrm{Re} \left[e^{-2i\theta} \langle (\Delta \hat{a}_k)^2 \rangle \right] + \left\langle \{ \Delta \hat{a}_k^+, \Delta \hat{a}_k \} \right\rangle \tag{43}$$

which obviously give $\hat{X}_{k1}$ and $\hat{X}_{k2}$ in special cases. The Heisenberg uncertainty relation for quadratures,

$$\left\langle \left(\Delta \hat{X}_{k1} \right)^2 \right\rangle \left\langle \left(\Delta \hat{X}_{k2} \right)^2 \right\rangle \geq 1 \tag{44}$$

lays the basis for the definition of "usual" ("standard") squeezing. The state of the field is said to be squeezed if the variance of $\hat{X}_{k1}$ or $\hat{X}_{k2}$ becomes smaller than unity (in general, smaller than the square root of the right side of the uncertainty relation for the quadratures). Equivalently, light whose quantum fluctuations in the one quadrature are smaller than those associated with coherent light (minimizing the uncertainty relation) is called squeezed (in the usual meaning). Since, for a given quantum state, the variance (43) is still dependent on θ, the angle θ can be chosen in a way to minimize (or maximize) the variance. Differentiation with respect to θ leads to the angles θ_+ and θ_- for the maximal and minimal variances, respectively, given by the relation[211, 212]

$$\exp(2i\theta_\pm) = \pm\left(\frac{\langle(\Delta\hat{a}_k)^2\rangle}{\langle(\Delta\hat{a}_k^+)^2\rangle}\right)^{1/2} \tag{45}$$

where the difference between the angles θ_+ and θ_- is $\pi/2$. On inserting (45) into (43) one obtains the extremal variances

$$\langle(\Delta\hat{X}_{k\pm})^2\rangle \equiv \langle(\Delta\hat{X}_k(\theta_\pm))^2\rangle$$
$$= \pm 2|\langle(\Delta\hat{a}_k)^2\rangle| + \langle\{\Delta\hat{a}_k^+, \Delta\hat{a}_k\}\rangle \tag{46}$$

It is noteworthy that the θ-dependent variance (43) can be expressed in terms of the extremal variances

$$\langle(\Delta\hat{X}_k(\theta))^2\rangle = \langle(\Delta\hat{X}_{k-})^2\rangle\cos^2(\theta - \theta_-)$$
$$+ \langle(\Delta\hat{X}_{k+})^2\rangle\sin^2(\theta - \theta_-) \tag{47}$$

which is the equation for Booth's elliptical lemniscate in polar coordinates.[212] The principal squeezing, introduced by Lukš et al.,[211, 213] occurs if the minimum variance is less than unity:

$$\langle(\Delta\hat{X}_{k-})^2\rangle \le 1 \tag{48}$$

From (46) it follows that the principal squeezing requires the fulfillment of the condition

$$\langle\Delta\hat{a}_k^+\Delta\hat{a}_k\rangle < |\langle(\Delta\hat{a}_k)^2\rangle| \tag{49}$$

whereas the condition for standard squeezing, in view of (43), is

$$\min\left\{\langle\Delta\hat{a}_k^+\Delta\hat{a}_k\rangle \pm \text{Re}[\langle(\Delta\hat{a}_k)^2\rangle]\right\} < 0 \tag{50}$$

The mathematically elegant formalism of principal squeezing (in particular other equivalent conditions for principal squeezing) can be formulated using the generalized Heisenberg uncertainty relation (the Schrödinger uncertainty relation)[211, 214, 215]:

$$\left\langle \left(\Delta \hat{X}_{k1}\right)^2\right\rangle \left\langle \left(\Delta \hat{X}_{k2}\right)^2\right\rangle \geq \tfrac{1}{4}\left\langle \left\{\Delta \hat{X}_{k1}, \Delta \hat{X}_{k2}\right\}\right\rangle^2 + 1 \tag{51}$$

which includes the Wigner covariance (cross-correlation) of the quadratures $\hat{X}_{k1}$ and $\hat{X}_{k2}$ equal to

$$\left\langle \left\{\Delta \hat{X}_{k1}, \Delta \hat{X}_{k2}\right\}\right\rangle = 4\,\mathrm{Im}\left[\left\langle (\Delta \hat{a}_k)^2\right\rangle\right] \tag{52}$$

For extremal variances $\langle (\Delta \hat{X}_{k\pm})^2\rangle$ the generalized Heisenberg relation reduces to the standard uncertainty relation.

The generalization of the above definitions for the two-mode case is straightforward. By virtue of the commutator (42), twice as great as for the single-mode case (41), the standard and principal squeezing can be defined, respectively, as

$$\min\left\{\left\langle \left(\Delta \hat{X}_{kl1}\right)^2\right\rangle, \left\langle \left(\Delta \hat{X}_{kl2}\right)^2\right\rangle\right\} \leq 2 \tag{53}$$

$$\left\langle \left(\Delta \hat{X}_{kl-}\right)^2\right\rangle \leq 2 \tag{54}$$

We express the two-mode variances and the Wigner covariances in terms of the single-mode moments:

$$\left\langle \left(\Delta \hat{X}_{kl}\right)^2\right\rangle = \left\langle \left(\Delta \hat{X}_k\right)^2\right\rangle + \left\langle \left(\Delta \hat{X}_l\right)^2\right\rangle + 2\left\langle \Delta \hat{X}_k \Delta \hat{X}_l\right\rangle \tag{55}$$

$$\begin{aligned}\left\langle \left\{\Delta \hat{X}_{kl1}, \Delta \hat{X}_{kl2}\right\}\right\rangle &= \left\langle \left\{\Delta \hat{X}_{k1}, \Delta \hat{X}_{k2}\right\}\right\rangle + \left\langle \left\{\Delta \hat{X}_{l1}, \Delta \hat{X}_{l2}\right\}\right\rangle \\ &\quad + 2\left\langle \Delta \hat{X}_{k1} \Delta \hat{X}_{l2}\right\rangle + 2\left\langle \Delta \hat{X}_{k2} \Delta \hat{X}_{l1}\right\rangle\end{aligned} \tag{56}$$

where $\hat{X}_{kl}$ stands for $\hat{X}_{kl}(\theta)$ (in particular the quadratures). Relations such as (55) and (56) for the quadratures hold for the two-mode creation and annihilation operators $\hat{a}_{kl}^{\pm}$. The moments $\langle (\hat{X}_k)^2\rangle$ and $\langle \{\Delta \hat{X}_{k1}, \Delta \hat{X}_{k2}\}\rangle$ are given by (43) and (52). The remaining cross-correlations have the following form in terms of the annihilation and creation operators:

$$\begin{aligned}\langle \Delta \hat{X}_{k1} \Delta \hat{X}_{l1}\rangle &= 2\,\mathrm{Re}\left[\langle \Delta \hat{a}_k \Delta \hat{a}_l\rangle + \langle \Delta \hat{a}_k^+ \Delta \hat{a}_l\rangle\right] \\ \langle \Delta \hat{X}_{k2} \Delta \hat{X}_{l2}\rangle &= 2\,\mathrm{Re}\left[-\langle \Delta \hat{a}_k \Delta \hat{a}_l\rangle + \langle \Delta \hat{a}_k^+ \Delta \hat{a}_l\rangle\right] \\ \langle \Delta \hat{X}_{k1} \Delta \hat{X}_{l2}\rangle &= 2\,\mathrm{Im}\left[\langle \Delta \hat{a}_k \Delta \hat{a}_l\rangle + \langle \Delta \hat{a}_k^+ \Delta \hat{a}_l\rangle\right] \\ \langle \Delta \hat{X}_{k2} \Delta \hat{X}_{l1}\rangle &= 2\,\mathrm{Im}\left[\langle \Delta \hat{a}_k \Delta \hat{a}_l\rangle - \langle \Delta \hat{a}_k^+ \Delta \hat{a}_l\rangle\right]\end{aligned} \tag{57}$$

Substituting Eqs. (43), (52), and (57) into (55) and (56) one obtains explicit dependencies of the two-mode quadrature moments on the annihilation operators.[213]

Alternatively, the single-mode moments (43), (46), and (52) and the conditions (49) and (50) for single-mode squeezing can be generalized to a two-mode case by simple replacement of $\hat{a}_k$, $\hat{X}_k(\theta)$ by $\hat{a}_{kl}$, $\hat{X}_{kl}(\theta)$, showing complete analogy between the single- and two-mode descriptions. In particular, the two-mode extremal variances are

$$\left\langle \left(\Delta \hat{X}_{kl\pm} \right)^2 \right\rangle = \pm 2 \left| \left\langle \left(\Delta \hat{a}_{kl} \right)^2 \right\rangle \right| + \left\langle \left\{ \Delta \hat{a}_{kl}^+, \Delta \hat{a}_{kl} \right\} \right\rangle \tag{58}$$

by analogy to (46).

V. FOKKER-PLANCK EQUATION

A. Raman Scattering Including Pump Depletion

The master equation (ME) is the quantum equation of motion for operators and hence it is possible to solve it directly only for a small class of models. As an example we cite the Raman-coupled model of Knight[133] and its generalizations (Ref. 137 and references therein). Usually the quantum master equation is converted to a classical differential equation. Then, standard methods of mathematical analysis can be applied. In this section we present one of the most popular methods: transformation to a generalized Fokker-Planck equation (FPE) or equivalently to an equation of motion for characteristic functions. This method is extensively studied in a number of textbooks[23, 183, 184, 188] and consists of performing s-ordering of the field operators in the ME (18) and then applying the quantum–classical number correspondence of coherent-state technique. The rules for the transformation of the ME into Fokker-Planck equations for the s-parametrized quasidistribution $\mathscr{W}^{(s)}(\alpha, \alpha^*, \overline{A})$ are the following (e.g., Refs. 216 and 217):

$$\begin{aligned}
\left. \begin{matrix} \hat{A}\hat{a} \\ \hat{a}\hat{A} \end{matrix} \right\} &\mapsto \left(\alpha - \frac{s \pm 1}{2} \frac{\partial}{\partial \alpha^*} \right) \mathscr{W}^{(s)}(\alpha, \alpha^*, \overline{A}) \\[2mm]
\left. \begin{matrix} \hat{a}^+\hat{A} \\ \hat{A}\hat{a}^+ \end{matrix} \right\} &\mapsto \left(\alpha^* - \frac{s \pm 1}{2} \frac{\partial}{\partial \alpha} \right) \mathscr{W}^{(s)}(\alpha, \alpha^*, \overline{A})
\end{aligned} \tag{59}$$

where $\hat{A}$ is an arbitrary operator; in particular, $\hat{A}$ can be the density matrix $\hat{\rho}$; $\overline{A}$ is the classical function associated with the operator $\hat{A}$; the parameter s takes arbitrary values in the range $\langle -1,1\rangle$. If necessary, these rules can be applied repeatedly. Similarly, we list the rules of transformation of the master equation (18) to the equation of motion for the s-parametrized characteristic function $\mathscr{C}^{(s)}(\beta,\beta^*,\overline{A})$ (e.g., Ref. 216):

$$
\begin{Bmatrix} \hat{A}\hat{a} \\ \hat{a}\hat{A} \end{Bmatrix} \mapsto \left(-\frac{\partial}{\partial\beta^*} + \frac{s\pm 1}{2}\beta \right)\mathscr{C}^{(s)}(\beta,\beta^*,\overline{A})
$$

$$
\begin{Bmatrix} \hat{a}^+\hat{A} \\ \hat{A}\hat{a}^+ \end{Bmatrix} \mapsto \left(\frac{\partial}{\partial\beta} - \frac{s\pm 1}{2}\beta^* \right)\mathscr{C}^{(s)}(\beta,\beta^*,\overline{A})
$$

(60)

Applying repeatedly the rule (59) of transformation to the master equation (18) and after some lengthy algebra we finally arrive at the generalized Fokker-Planck equation for the s-parametrized quasidistribution $\mathscr{W}^{(s)} \equiv \mathscr{W}^{(s)}(\alpha_L,\alpha_S,\alpha_A,t)$:

$$
\begin{aligned}
\frac{\partial}{\partial t}\mathscr{W}^{(s)} = \frac{1}{2}\gamma_S &\left\{ \left[-\frac{\partial}{\partial\alpha_L}\alpha_L\frac{\partial}{\partial\alpha_S}\alpha_S \right.\right. \\
&+ \left(|\alpha_S|^2 + \frac{1+s}{2} \right)\frac{\partial}{\partial\alpha_L}\alpha_L - \left(|\alpha_L|^2 - \frac{1-s}{2} \right)\frac{\partial}{\partial\alpha_S}\alpha_S \\
&+ \frac{1-s^2}{4}\left(\frac{\partial}{\partial\alpha_L}\alpha_L\frac{\partial}{\partial\alpha_S}\frac{\partial}{\partial\alpha_S^*} - \frac{\partial}{\partial\alpha_L}\frac{\partial}{\partial\alpha_L^*}\frac{\partial}{\partial\alpha_S}\alpha_S \right) + \text{c.c.} \right] \\
&+ \left[(1-s)|\alpha_S|^2 + \frac{1-s^2}{2} \right]\frac{\partial}{\partial\alpha_L}\frac{\partial}{\partial\alpha_L^*} \\
&+ \left.\left[(1+s)|\alpha_L|^2 - \frac{1-s^2}{2} \right]\frac{\partial}{\partial\alpha_S}\frac{\partial}{\partial\alpha_S^*} \right\}\mathscr{W}^{(s)} \\
+ \frac{1}{2}\gamma_A &\left\{ \left[-\frac{\partial}{\partial\alpha_L}\alpha_L\frac{\partial}{\partial\alpha_A}\alpha_A \right.\right. \\
&- \left(|\alpha_A|^2 - \frac{1-s}{2} \right)\frac{\partial}{\partial\alpha_L}\alpha_L + \left(|\alpha_L|^2 + \frac{1+s}{2} \right)\frac{\partial}{\partial\alpha_A}\alpha_A \\
&- \frac{1-s^2}{4}\left(\frac{\partial}{\partial\alpha_L}\alpha_L\frac{\partial}{\partial\alpha_A}\frac{\partial}{\partial\alpha_A^*} - \frac{\partial}{\partial\alpha_L}\frac{\partial}{\partial\alpha_L^*}\frac{\partial}{\partial\alpha_A}\alpha_A \right) + \text{c.c.} \right]
\end{aligned}
$$

$$+ \left[(1+s)|\alpha_A|^2 - \frac{1-s^2}{2} \right] \frac{\partial}{\partial \alpha_L} \frac{\partial}{\partial \alpha_L^*}$$

$$+ \left[(1-s)|\alpha_L|^2 + \frac{1-s^2}{2} \right] \frac{\partial}{\partial \alpha_A} \frac{\partial}{\partial \alpha_A^*} \Bigg\} \mathscr{W}^{(s)}$$

$$+ \left\{ \frac{1}{2} \gamma_{SA} \exp(-2i\Delta\Omega\,\Delta t) \left[-\alpha_L^2 \left(\alpha_A^* \frac{\partial}{\partial \alpha_S} + \frac{\partial}{\partial \alpha_S} \frac{\partial}{\partial \alpha_A} - \alpha_S^* \frac{\partial}{\partial \alpha_A} \right) \right. \right.$$

$$+ \alpha_L \left((1+s)\alpha_A^* \frac{\partial}{\partial \alpha_S} + (1-s)\alpha_S^* \frac{\partial}{\partial \alpha_A} \right) \frac{\partial}{\partial \alpha_L^*}$$

$$+ \left(\frac{1-s^2}{4} \alpha_A^* \frac{\partial}{\partial \alpha_S} - \alpha_S^* \alpha_A^* - \frac{1-s^2}{4} \alpha_S^* \frac{\partial}{\partial \alpha_A} \right) \frac{\partial^2}{\partial \alpha_L^{*2}} \Bigg] + \text{c.c.} \Bigg\} \mathscr{W}^{(s)}$$

$$+ \gamma_S \langle \hat{n}_V \rangle \left\{ \left(\frac{1}{2} \frac{\partial}{\partial \alpha_L} \alpha_L - \frac{\partial}{\partial \alpha_L} \alpha_L \frac{\partial}{\partial \alpha_S} \alpha_S + \frac{1}{2} \frac{\partial}{\partial \alpha_S} \alpha_S + \text{c.c.} \right) \right.$$

$$+ |\alpha_S|^2 \frac{\partial}{\partial \alpha_L} \frac{\partial}{\partial \alpha_L^*} + |\alpha_L|^2 \frac{\partial}{\partial \alpha_S} \frac{\partial}{\partial \alpha_S^*} \Bigg\} \mathscr{W}^{(s)}$$

$$+ \gamma_A \langle \hat{n}_V \rangle \left\{ \left(\frac{1}{2} \frac{\partial}{\partial \alpha_L} \alpha_L - \frac{\partial}{\partial \alpha_L} \alpha_L \frac{\partial}{\partial \alpha_A} \alpha_A + \frac{1}{2} \frac{\partial}{\partial \alpha_A} \alpha_A + \text{c.c.} \right) \right.$$

$$+ |\alpha_A|^2 \frac{\partial}{\partial \alpha_L} \frac{\partial}{\partial \alpha_L^*} + |\alpha_L|^2 \frac{\partial}{\partial \alpha_A} \frac{\partial}{\partial \alpha_A^*} \Bigg\} \mathscr{W}^{(s)}$$

$$- \left\{ \gamma_{AS} \langle \hat{n}_V \rangle \exp(2i\Delta\Omega\,\Delta t) \left[\alpha_S^* \alpha_A^* \frac{\partial^2}{\partial \alpha_L^{*2}} + \alpha_L^2 \frac{\partial}{\partial \alpha_S} \frac{\partial}{\partial \alpha_A} \right. \right.$$

$$- \alpha_L \left(\alpha_A^* \frac{\partial^2}{\partial \alpha_S} + \alpha_S^* \frac{\partial^2}{\partial \alpha_A} \right) \frac{\partial}{\partial \alpha_L^*} \Bigg] + \text{c.c.} \Bigg\} \mathscr{W}^{(s)}$$

$$\tag{61}$$

which is a generalization of our former relation for $\gamma_A, \gamma_{SA} \neq 0$ and arbitrary parameter s (Ref. 218). For brevity, we refer to the generalized

Fokker-Planck equation[23] simply as the Fokker-Planck equation (FPE). The physical interpretation of (61) can be given in the same manner as the interpretation of the appropriate terms in the ME (18) given in Section II. The FPE (61) exhibits a highly complicated structure. Nonetheless, the equations of motion for the mean values $\langle \alpha_k \rangle, \langle \alpha_k \alpha_l \rangle, \langle \alpha_k^* \alpha_l \rangle$, (with $k, l = L, S, A$) can be calculated. In particular, we obtain

$$\frac{d}{dt}\left(\langle \hat{n}_L(t) \rangle + \langle \hat{n}_S(t) \rangle + \langle \hat{n}_A(t) \rangle\right) = 0 \qquad (62)$$

with $\langle \hat{n}_k(t) \rangle = \langle \alpha_k^* \alpha_k \rangle$. Eq. (62) states that the total mean number of photons (in all radiation modes) is a constant of motion.

The FPE (61) contains terms of the form

$$\frac{\partial}{\partial \alpha_i} \alpha_j \alpha_k \alpha_l \mathscr{W}^{(s)} \qquad \frac{\partial}{\partial \alpha_i}\frac{\partial}{\partial \alpha_j} \alpha_k \alpha_l \mathscr{W}^{(s)} \qquad \frac{\partial}{\partial \alpha_i}\frac{\partial}{\partial \alpha_j}\frac{\partial}{\partial \alpha_k} \alpha_l \mathscr{W}^{(s)}$$

where $\alpha_i, \alpha_j, \alpha_k, \alpha_l = \alpha_L, \alpha_L^*, \alpha_S, \alpha_S^*, \alpha_A, \alpha_A^*$. It is seen that most components of the drift vector are nonlinear to the third order in α, and most components of the diffusion matrix are nonlinear up to the second order. It is particularly difficult to solve a differential equation with such nonlinear diffusion and drift coefficients. Besides, the FPE (61) for $\mathscr{W}^{(s)} \equiv \mathscr{W}^{(s)}(\alpha_L, \alpha_S, \alpha_A, t)$ with the parameter $s \neq \pm 1$ contains third-order derivatives in the terms

$$\frac{\partial}{\partial \alpha_i}\frac{\partial}{\partial \alpha_j}\frac{\partial}{\partial \alpha_k} \alpha_l \mathscr{W}^{(s)}$$

This could be expected since in many models,[23, 19] for instance in the anharmonic oscillator model (for references see Ref. 94), there occur third-order derivatives in the FPEs for the Wigner function ($s = 0$).

The corresponding equation of motion for the s-parametrized characteristic function $\mathscr{C}^{(s)}(\beta_L, \beta_S, \beta_A, t)$ can be obtained either from the ME (18) by performing the transformation (60), or from the FPE (61) by means of the Fourier transformation (22) with respect to the variables $\alpha_L, \alpha_S, \alpha_A$.

Finally, we arrive at the equation of motion for $\mathscr{C}^{(s)} \equiv \mathscr{C}^{(s)}(\beta_L, \beta_S, \beta_A, t)$:

$$
\begin{aligned}
\frac{\partial}{\partial t}\mathscr{C}^{(s)} = \frac{1}{2}\gamma_S\Bigg\{&\Bigg[-\beta_L\beta_S\frac{\partial}{\partial\beta_L}\frac{\partial}{\partial\beta_S} + \beta_L\left(-\frac{1+s}{2} + \frac{\partial}{\partial\beta_S}\frac{\partial}{\partial\beta_S^*}\right)\frac{\partial}{\partial\beta_L}\\
&-\beta_S\left(\frac{1-s}{2} + \frac{\partial}{\partial\beta_L}\frac{\partial}{\partial\beta_L^*}\right)\frac{\partial}{\partial\beta_S}\\
&+\frac{1-s^2}{4}\left(|\beta_S|^2\beta_L\frac{\partial}{\partial\beta_L} - |\beta_L|^2\beta_S\frac{\partial}{\partial\beta_S}\right) + \text{c.c.}\Bigg]\\
&+|\beta_L|^2\left[-\frac{1-s^2}{2} + (1-s)\frac{\partial}{\partial\beta_S}\frac{\partial}{\partial\beta_S^*}\right]\\
&+|\beta_S|^2\left[\frac{1-s^2}{2} + (1+s)\frac{\partial}{\partial\beta_L}\frac{\partial}{\partial\beta_L^*}\right]\Bigg\}\mathscr{C}^{(s)}\\
+\frac{1}{2}\gamma_A\Bigg\{&\Bigg[-\beta_L\beta_A\frac{\partial}{\partial\beta_L}\frac{\partial}{\partial\beta_A} - \beta_L\left(\frac{1-s}{2} + \frac{\partial}{\partial\beta_A}\frac{\partial}{\partial\beta_A^*}\right)\frac{\partial}{\partial\beta_L}\\
&+\beta_A\left(-\frac{1+s}{2} + \frac{\partial}{\partial\beta_L}\frac{\partial}{\partial\beta_L^*}\right)\frac{\partial}{\partial\beta_A}\\
&+\frac{1-s^2}{4}\left(-|\beta_A|^2\beta_L\frac{\partial}{\partial\beta_L} + |\beta_L|^2\beta_A\frac{\partial}{\partial\beta_A}\right) + \text{c.c.}\Bigg]\\
&+|\beta_L|^2\left[\frac{1-s^2}{2} + (1+s)\frac{\partial}{\partial\beta_A}\frac{\partial}{\partial\beta_A^*}\right]\\
&+|\beta_A|^2\left[-\frac{1-s^2}{2} + (1-s)\frac{\partial}{\partial\beta_L}\frac{\partial}{\partial\beta_L^*}\right]\Bigg\}\mathscr{C}^{(s)}\\
+\Bigg\{&\frac{1}{2}\gamma_{SA}\exp(-2\mathrm{i}\Delta\Omega\,\Delta t)\Bigg[\left(\beta_A\frac{\partial}{\partial\beta_S^*} - \beta_S\beta_A - \beta_S\frac{\partial}{\partial\beta_A^*}\right)\frac{\partial^2}{\partial\beta_L^2}\\
&+\beta_L^*\left((1-s)\beta_A\frac{\partial}{\partial\beta_S^*} + (1+s)\beta_S\frac{\partial}{\partial\beta_A^*}\right)\frac{\partial}{\partial\beta_L}\\
&-\beta_L^{*2}\left(\frac{1-s^2}{4}\beta_A\frac{\partial}{\partial\beta_S^*} + \frac{\partial}{\partial\beta_S^*}\frac{\partial}{\partial\beta_A^*} - \frac{1-s^2}{4}\beta_S\frac{\partial}{\partial\beta_A^*}\right)\Bigg] + \text{c.c.}\Bigg\}\mathscr{C}^{(s)}
\end{aligned}
$$

$$+ \gamma_S \langle \hat{n}_V \rangle \left\{ -\left(\frac{1}{2} \beta_L \frac{\partial}{\partial \beta_L} + \beta_L \beta_S \frac{\partial}{\partial \beta_L} \frac{\partial}{\partial \beta_S} + \frac{1}{2} \beta_S \frac{\partial}{\partial \beta_S} + \text{c.c.} \right) \right.$$

$$\left. + |\beta_S|^2 \frac{\partial}{\partial \beta_L} \frac{\partial}{\partial \beta_L^*} + |\beta_L|^2 \frac{\partial}{\partial \beta_S} \frac{\partial}{\partial \beta_S^*} \right\} \mathscr{C}^{(s)}$$

$$+ \gamma_A \langle \hat{n}_V \rangle \left\{ -\left(\frac{1}{2} \beta_L \frac{\partial}{\partial \beta_L} + \beta_L \beta_A \frac{\partial}{\partial \beta_L} \frac{\partial}{\partial \beta_A} + \frac{1}{2} \beta_A \frac{\partial}{\partial \beta_A} + \text{c.c.} \right) \right.$$

$$\left. + |\beta_A|^2 \frac{\partial}{\partial \beta_L} \frac{\partial}{\partial \beta_L^*} + |\beta_L|^2 \frac{\partial}{\partial \beta_A} \frac{\partial}{\partial \beta_A^*} \right\} \mathscr{C}^{(s)}$$

$$- \left\{ \frac{1}{2} \gamma_{AS} \langle \hat{n}_V \rangle \exp(2i \Delta \Omega \, \Delta t) \left[\beta_S \beta_A \frac{\partial^2}{\partial \beta_L^2} + \beta_L^{*2} \frac{\partial}{\partial \beta_S^*} \frac{\partial^2}{\partial \beta_A^*} \right. \right.$$

$$\left. \left. - \beta_L^* \left(\beta_S \frac{\partial}{\partial \beta_A^*} + \beta_A \frac{\partial}{\partial \beta_S^*} \right) \frac{\partial}{\partial \beta_L} \right] + \text{c.c.} \right\} \mathscr{C}^{(s)}$$

$$(63)$$

Here, we come upon similar difficulties in the way of obtaining an analytical solution of (63) as in the FPE case (61), inherent in the nonlinearity of the coefficients of the terms with first- and second-order derivatives as well as the presence of terms with third-order derivatives. Nevertheless, in contradistinction to $\mathscr{W}^{(s)}$, the existence of a solution for $\mathscr{C}^{(s_1)}$ implies, in view of the property (27), the existence of a solution for any other parameter s_2.

In view of the particularly complicated structure of Eqs. (61) and (63) or equivalent equations of motion derived within the completely quantum model of scattering into both the Stokes and anti-Stokes modes, it would seem that a solution in exact closed form cannot be obtained.[23] It is necessary to apply further restrictions or approximations in the model to achieve an analytical solution of (61). In Section VI.A.2 we present a strict analytical solution of the two-mode ME (including pump depletion) in terms of Fock states by applying the Laplace transform. In Section V.B we present solutions of two-mode linearized FPEs for $\mathscr{W}^{(s)}(\alpha_S, \alpha_A, t)$ and solutions of equivalent equations of motion for $\mathscr{C}^{(s)}(\beta_S, \beta_A, t)$ in the Raman scattering model under parametric approximation. In Appendix A we give the solution of a linearized form of the three-mode FPE (61) for $\mathscr{W}^{(-1)}(\alpha_L, \alpha_S, \alpha_A, t)$ properly describing the evolution of the radiation fields valid only on the assumption of small fluctuations of the fields around their mean values. There, we restrict our considerations to the

Q-function ($s = -1$) to avoid problems of the existence of the quasidistribution $\mathscr{W}^{(s)}(\alpha_L, \alpha_S, \alpha_A, t)$ (particularly important in the case of s close or equal to 1) and to simplify the third-order FPE (61) to second-order, which takes place for $s = \pm 1$. Within a similar model of Raman scattering from a single-phonon mode, Szlachetka et al.[58-61] (see also Ref. 27) and Tänzler and Schütte[63] have solved the equations of motion in the short-time approximation up to the second power in time. Within the latter (single phonon mode) model Peřina and Křepelka[67, 68] have obtained approximate solutions using the approximation of small fluctuations around a stationary solution.

B. Raman Scattering Without Pump Depletion

Here, to find a solution of the ME (18) we apply the parametric approximation, so no allowance for pump depletion is included. The trilinear Hamiltonians $\hat{H}_A$, $\hat{H}_S$ (2) can be reduced to bilinear functions as a result of the replacement of the annihilation operator $\hat{a}_L$, representing the quantum pump field, by the classical complex amplitude of the pump field, e_L. This approximation effectively linearizes our model of Raman scattering. Then, the Fokker-Planck equation for the two-mode s-parametrized quasidistribution $\mathscr{W}^{(s)}(\alpha_S, \alpha_A, t)$ takes the form

$$
\begin{aligned}
\frac{\partial}{\partial t}\mathscr{W}^{(s)}(\alpha_S, \alpha_A, t) = \Bigg\{ &-\left[\left(\frac{\kappa_S}{2} + i\Delta\Omega\right)\frac{\partial}{\partial\alpha_S}\alpha_S + \text{c.c.}\right] \\
&+\left[\left(\frac{\kappa_A}{2} - i\Delta\Omega\right)\frac{\partial}{\partial\alpha_A}\alpha_A + \text{c.c.}\right] \\
&-\left[\frac{\kappa_{SA}}{2}\left(\alpha_A^*\frac{\partial}{\partial\alpha_S} - \alpha_S^*\frac{\partial}{\partial\alpha_A}\right) + \text{c.c.}\right] \\
&+\kappa_S\left(\langle\hat{n}_V\rangle + \frac{s+1}{2}\right)\frac{\partial^2}{\partial\alpha_S\partial\alpha_S^*} \\
&+\kappa_A\left(\langle\hat{n}_V\rangle + \frac{1-s}{2}\right)\frac{\partial^2}{\partial\alpha_A\partial\alpha_A^*} \\
&-\left[\kappa_{SA}\left(\langle\hat{n}_V\rangle + \frac{1}{2}\right)\frac{\partial^2}{\partial\alpha_S\partial\alpha_A} + \text{c.c.}\right]\Bigg\} \\
&\times\mathscr{W}^{(s)}(\alpha_S, \alpha_A, t)
\end{aligned}
\tag{64}
$$

on applying the rules (59) to the ME (18) with complex classical amplitude e_L instead of the annihilation operator $\hat{a}_L$ and transforming the variables $\alpha_k \rightarrow \exp(-i\Delta\Omega\Delta t)\alpha_k$ and $\beta_k \rightarrow \exp(-i\Delta\Omega\Delta t)\beta_k$ with the frequency mismatch $\Delta\Omega$ defined by (12). For brevity, we have incorporated the complex amplitude e_L into the coupling constants $\kappa_S = \gamma_S |e_L|^2$, $\kappa_A = \gamma_A |e_L|^2$, and $\kappa_{SA} = \kappa_{AS}^* = \gamma_{SA} e_L^2$. Equation (64) is a generalization for any parameter s ($s \in \langle -1, 1 \rangle$) of the FPE given by Walls for the P-function ($s = 1$)[122] and by Peřina for the P- and Q-functions ($s = \pm 1$).[53, 54] If we consider production of the Stokes radiation only, neglecting anti-Stokes scattering, then Eq. (64) reduces to the s-parametrized FPE obtained by Peřinova et al.[189]

We can interpret the FPE (64) in the same manner as the ME (18). The first term in (64) describes the amplification of the Stokes beam, whereas the second term describes the attenuation of the anti-Stokes beam; the third term shows the coupling between the Stokes and anti-Stokes fields; the remaining three terms account for the noise diffusion from the "noisy" (for nonzero temperature) reservoir into the system. Contrary to the former equations of motion (18), (61), and (60), we lose all information about the depletion of the laser field. It is seen that the FPE (64) for any quasidistribution $\mathscr{W}^{(s)}$, even if related to the field ordering $s \neq \pm 1$, does not contain third-order derivatives, contrary to the FPE (61) without parametric approximation. Let us note that (64) describes an Ornstein-Uhlenbeck process[191] since the components of the drift vector are linear and those of the diffusion matrix are constant. Various methods have been developed for solving the equations of motion for Ornstein-Uhlenbeck processes.[23, 188] For instance, expressing the quasidistribution $\mathscr{W}^{(s)}(\alpha_S, \alpha_A, t)$ by its Fourier transform (22) with respect to the variables α_S, α_A, we obtain the following first-order differential equation for the Fourier transform, i.e., for the characteristic function $\mathscr{C}^{(s)}(\beta_S, \beta_A, t)$:

$$
\begin{aligned}
\frac{\partial}{\partial t} \mathscr{C}^{(s)}(\beta_S, \beta_A, t) = \Bigg\{ &\left[\left(\frac{\kappa_S}{2} - i\Delta\Omega \right) \beta_S \frac{\partial}{\partial \beta_S} + \text{c.c.} \right] \\
&- \left[\left(\frac{\kappa_A}{2} + i\Delta\Omega \right) \beta_A \frac{\partial}{\partial \beta_A} + \text{c.c.} \right] \\
&+ \left[\frac{\kappa_{SA}}{2} \left(\beta_A^* \frac{\partial}{\partial \beta_S} - \beta_S^* \frac{\partial}{\partial \beta_A} \right) + \text{c.c.} \right] \\
&- \kappa_S \left(\langle \hat{n}_V \rangle + \frac{s+1}{2} \right) |\beta_S|^2 \\
&- \kappa_A \left(\langle \hat{n}_V \rangle + \frac{1-s}{2} \right) |\beta_A|^2 \\
&- \left[\kappa_{AS} \left(\langle \hat{n}_V \rangle + \frac{1}{2} \right) \beta_S \beta_A + \text{c.c.} \right] \Bigg\} \mathscr{C}^{(s)}(\beta_S, \beta_A, t)
\end{aligned}
$$
(65)

Again, to obtain (22), one might use the rules (60) applied to (18) with $\hat{a}_L \to e_L$. Our further results presented in this section are mainly based on very extensive studies carried out by Peřina,[53, 54] Peřinová and Peřina,[48] and Kárská and Peřina[56] (see also Ref. 23 and references therein). However, their solutions of the equations of motion for the Raman effect under parametric approximation hold only for quasidistributions $\mathscr{W}^{(1)}(\alpha_S, \alpha_A, t)$ or $\mathscr{W}^{(-1)}(\alpha_S, \alpha_A, t)$ and characteristic functions $\mathscr{C}^{(\pm 1)}(\beta_S, \beta_A, t)$ related to normal and/or antinormal ordering of the field operators. We generalize their results to functions related to s-ordering of the field operators, i.e., to an s-parametrized quasidistribution $\mathscr{W}^{(s)}(\alpha_S, \alpha_A, t)$ and s-parametrized characteristic function $\mathscr{C}^{(s)}(\beta_S, \beta_A, t)$. Let us use, after Ref. 23, the following simplified notation for functions characterizing the quantum noise, i.e., the Wigner covariances and variances as well as the mean values of the annihilation operators $\hat{a}_S$ and $\hat{a}_A$:

$$B_k^{(s)}(t) = \frac{1}{2}\langle\{\Delta\hat{a}_k^+(t), \Delta\hat{a}_k(t)\}\rangle - \frac{s}{2}$$

$$D_{kl}(t) = D_{lk}(t) = \frac{1}{2}\langle\{\Delta\hat{a}_k(t), \Delta\hat{a}_l(t)\}\rangle$$

$$\overline{D}_{kl}(t) = \overline{D}_{lk}^*(t) = -\frac{1}{2}\langle\{\Delta\hat{a}_k^+(t), \Delta\hat{a}_l(t)\}\rangle \qquad (66)$$

$$C_k(t) = \langle(\Delta\hat{a}_k(t))^2\rangle$$

$$\xi_k(t) = \langle\hat{a}_k(t)\rangle$$

where $k = S, A$ and $\{\ldots, \ldots\}$ is an anticommutator. Assuming the initial condition that the Stokes and anti-Stokes fields are stochastically independent, the solution of (65) for the s-parametrized characteristic function exists for any parameter s and is equal to

$$\mathscr{C}^{(s)}(\beta_S, \beta_A, t) = \exp\Big\{ \sum_{k=S, A} \Big[-B_k^{(s)}(t)|\beta_k|^2$$

$$+ \big(\tfrac{1}{2}C_k^*(t)\beta_k^2 + \text{c.c.}\big) + \big(\beta_k\xi_k^*(t) - \text{c.c.}\big)\Big]$$

$$+ \Big[D_{SA}(t)\beta_S^*\beta_A^* + \overline{D}_{SA}(t)\beta_S\beta_A^* + \text{c.c.}\Big]\Big\} \qquad (67)$$

where

$$B_S^{(s)}(t) = \left(B_S^{(s)} + \langle \hat{n}_V \rangle + \frac{1+s}{2}\right)|U_S(t)|^2$$

$$+ \left(B_A^{(s)} - \langle \hat{n}_V \rangle - \frac{1-s}{2}\right)|V_S(t)|^2 - \langle \hat{n}_V \rangle - \frac{1+s}{2}$$

$$B_A^{(s)}(t) = \left(B_A^{(s)} - \langle \hat{n}_V \rangle - \frac{1-s}{2}\right)|U_A(t)|^2$$

$$+ \left(B_S^{(s)} + \langle \hat{n}_V \rangle + \frac{1+s}{2}\right)|V_A(t)|^2 + \langle \hat{n}_V \rangle + \frac{1-s}{2}$$

$$D_{SA}(t) = \left(B_S^{(s)} + \langle \hat{n}_V \rangle + \frac{1+s}{2}\right)U_S(t)V_A(t) \qquad (68)$$

$$+ \left(B_A^{(s)} - \langle \hat{n}_V \rangle - \frac{1-s}{2}\right)V_S(t)U_A(t)$$

$$\overline{D}_{SA}(t) = C_S U_S(t)V_A^*(t) + C_A^* U_A^*(t)V_S(t)$$

$$C_S(t) = C_S U_S^2(t) + C_A^* V_S^2(t)$$

$$C_A(t) = C_A U_A^2(t) + C_S^* V_A^2(t)$$

$$\xi_S(t) = U_S(t)\xi_S + V_S(t)\xi_A^*$$

$$\xi_A(t) = U_A(t)\xi_A + V_A(t)\xi_S^*$$

The solution (67) for $\mathscr{C}^{(s)}(\beta_S, \beta_A, t)$ with any parameter s from $\langle -1, 1 \rangle$ is, in view of the property (27), a straightforward generalization of the solutions given by Kárská and Peřina[56] (see also Ref. 23) for $\mathscr{C}^{(\pm 1)}(\beta_S, \beta_A, t)$ related with normal or antinormal field operator ordering. Setting initial values $C_A = C_S = 0$, which implies that $\overline{D}_{SA}(t) = C_A(t) = C_S(t) = 0$ for any time t, the solution (67) reduces to that of Peřina.[53] The time-dependent functions $U_k(t)$, $V_k(t)$ ($k = S, A$) appearing in (68) can be expressed as

$$V_S(t) = \frac{\kappa_{SA}}{2}Q_1$$

$$V_A(t) = -\frac{\kappa_{SA}}{2}Q_1^*$$

$$U_S(t) = Q_2 + \left(\frac{\kappa_A}{2} + i\,\Delta\Omega\right)Q_1 \qquad (69)$$

$$U_A(t) = Q_2^* - \left(\frac{\kappa_S}{2} - i\,\Delta\Omega\right)Q_1^*$$

in terms of the auxiliary functions

$$Q_1 = \frac{\exp(P_1 \Delta t) - \exp(P_2 \Delta t)}{P_1 - P_2}$$

$$Q_2 = \frac{\partial Q_1}{\partial t} = \frac{P_1 \exp(P_1 \Delta t) - P_2 \exp(P_2 \Delta t)}{P_1 - P_2}$$

(70)

$$P_{1,2} = \frac{1}{2} \left\{ \frac{\kappa_S - \kappa_A}{2} \pm \left[\left(\frac{\kappa_S - \kappa_A}{2} \right)^2 - 4 \left((\Delta\Omega)^2 - i \frac{\kappa_S + \kappa_A}{2} \Delta\Omega \right) \right]^{1/2} \right\}$$

(71)

It is seen that for the initial moment of time t_0 the functions $V_k(t_0)$ vanish and the $U_k(t_0)$ are equal to unity, so that the initial Wigner covariances $D_{k,l}$, and $\bar{D}_{k,l}$ $(k, l = S, A)$ also vanish as a result of the initial condition of zero stochastical correlation between the scattered modes. Let us note that on the assumption of the frequency resonant condition $\Delta\Omega = 0$, the functions (69)–(71) simplify considerably[53] since $P_1 = \frac{1}{2}(\kappa_S - \kappa_A)$ and $P_2 = 0$. This leads, in particular, to the relations

$$V_S(t) + V_A(t) = 0,$$

$$U_S(t) + U_A(t) = 1 + \exp\left(\frac{\kappa_S - \kappa_A}{2} \Delta t \right)$$

(72)

To obtain the solution of the FPE (64) we perform the Fourier transform (21) of $\mathscr{C}^{(s)}(\beta_S, \beta_A, t)$, which leads to the s-parametrized quasidistribution $\mathscr{W}^{(s)}(\alpha_S, \alpha_A, t)$ in the form

$$\mathscr{W}^{(s)}(\alpha_S, \alpha_A, t) =$$

$$\frac{1}{L^{(s)}} \exp\left\{ (L^{(s)})^{-2} \left[-E_1 |\alpha_S - \xi_S(t)|^2 - E_2 |\alpha_A - \xi_A(t)|^2 \right.\right.$$

$$+ \frac{1}{2} E_3 (\alpha_S^* - \xi_S^*(t))^2 + \frac{1}{2} E_4 (\alpha_A^* - \xi_A^*(t))^2$$

$$+ E_5 (\alpha_S^* - \xi_S^*(t))(\alpha_A^* - \xi_A^*(t))$$

$$\left.\left. + E_6 (\alpha_S - \xi_S(t))(\alpha_A^* - \xi_A^*(t)) + \text{c.c.} \right] \right\}$$

(73)

which generalizes the Kárská and Peřina result for antinormal ordering[56] going over into s-ordering of the field. The time-dependent functions E_i ($i = 1, \ldots, 6$) and $L^{(1)}$ have been calculated by Peřinová[47] in her analysis of quadratic optical parametric processes. Here, we have the following generalized s-parametrized functions E_i and $L^{(s)}$ occurring in (73):

$$
\begin{aligned}
E_1 &= B_S^{(s)}(t) K_A^{(s)}(t) - B_A^{(s)}(t) K_+(t) \\
&\quad + \left(C_A^*(t) D_{SA}(t) \overline{D}_{SA}(t) + \text{c.c.} \right) \\
E_2 &= B_A^{(s)}(t) K_S^{(s)}(t) - B_S^{(s)}(t) K_+(t) \\
&\quad + \left(C_S(t) D_{SA}^*(t) \overline{D}_{SA}(t) + \text{c.c.} \right) \\
E_3 &= C_S(t) K_A^{(s)}(t) + 2 B_A^{(s)}(t) D_{SA}(t) \overline{D}_{SA}^*(t) \\
&\quad + C_A^*(t) D_{SA}^2(t) + C_A(t) \overline{D}_{SA}^{*2}(t) \\
E_4 &= C_A(t) Q(t) K_S^{(s)}(t) + 2 B_S^{(s)}(t) D_{SA}(t) \overline{D}_{SA}(t) \\
&\quad + C_S(t) \overline{D}_{SA}^2(t) + C_S^*(t) D_{SA}^2(t) \\
E_5 &= D_{SA}(t) \left[B_S^{(s)}(t) B_A^{(s)}(t) - K_-(t) \right] \\
&\quad + B_S^{(s)}(t) C_A(t) \overline{D}_{SA}^*(t) + B_A^{(s)}(t) C_S(t) \overline{D}_{SA}(t) \\
&\quad + C_S(t) C_A(t) D_{SA}^*(t) \\
E_6 &= -\overline{D}_{SA}(t) \left[B_S^{(s)}(t) B_A^{(s)}(t) + K_-(t) \right] \\
&\quad - B_S^{(s)}(t) C_A(t) D_{SA}^*(t) - B_A^{(s)}(t) Q(t) C_S^*(t) D_{SA}(t) \qquad (74) \\
&\quad - C_S^*(t) C_A(t) \overline{D}_{SA}^*(t) \\
\left(L^{(s)} \right)^2 &= K_S^{(s)}(t) K_A^{(s)}(t) - 2 B_S^{(s)}(t) B_A^{(s)}(t) K_+(t) \\
&\quad + \left[C_S(t) C_A(t) D_{SA}^{*2}(t) + C_S(t) C_A^*(t) \overline{D}_{SA}^2(t) \right. \\
&\quad + 2 B_S^{(s)}(t) C_A^*(t) D_{SA}(t) \overline{D}_{SA}(t) \\
&\quad \left. + 2 B_A^{(s)}(t) C_S^*(t) D_{SA}^*(t) \overline{D}_{SA}(t) + \text{c.c.} \right] + K_-^2(t)
\end{aligned}
$$

(75)

with

$$
\begin{aligned}
K_{S,A}^{(s)}(t) &= \left(B_{S,A}^{(s)}(t) \right)^2 - \left| C_{S,A}(t) \right|^2 \\
K_\pm(t) &= \left| D_{SA}(t) \right|^2 \pm \left| \overline{D}_{SA}(t) \right|^2
\end{aligned}
$$

(76)

The two-mode functions $\mathscr{W}^{(s)}(\alpha_S, \alpha_A, t)$ (73) and $\mathscr{C}^{(s)}(\beta_S, \beta_A, t)$ (67) reduce to the single-mode functions $\mathscr{W}^{(s)}(\alpha_k, t)$ and $\mathscr{C}^{(s)}(\beta_k, t)$ ($k = S, A$) simply by setting either $\alpha_S = \beta_S = 0$ or $\alpha_A = \beta_A = 0$, implying that the coefficients $V_S(t)$, $V_A(t)$, $D_{SA}(t)$, and $\overline{D}_{SA}(t)$ vanish and, for instance, $L^{(s)}$ reduces to $\sqrt{K_k^{(s)}(t)}$.

Contrary to the solution (67) for the characteristic function $\mathscr{C}^{(s)}(\beta_S, \beta_A, t)$, the solution (73) for the quasidistribution $\mathscr{W}^{(s)}(\alpha_S, \alpha_A, t)$ may be absent for some s-ordering of the field operators, depending on the choice of initial field. The condition for the existence of the QPD (73), i.e., the existence of the Fourier transform (21) of $\mathscr{C}^{(s)}(\beta_S, \beta_A, t)$ (67), is that the function $K_A^{(s)}(t)$, $L^{(s)}(t)$, Re $C_A(t) + B_A^{(s)}(t)$, and

$$\overline{L}^{(s)} \equiv \left(K_A^{(s)}(t) \right)^{1/2} \left[\text{Re } C_S(t) + B_S^{(s)}(t) \right] + \left(K_A^{(s)}(t) \right)^{-1/2}$$
$$\times \left[\text{Re } C_A^*(t) \left(\overline{D}_{SA}(t) - D_{SA}(t) \right)^2 - B_A^{(s)}(t) \left| \overline{D}_{SA}(t) - D_{SA}(t) \right|^2 \right] > 0$$

$$(77)$$

should be positive. If any of the four functions $K_A^{(s_1)}(t)$, $L^{(s_1)}(t)$, $\overline{L}^{(s_1)}(t)$, and Re$C_A(t) + B_A^{(s_1)}(t)$ (for a particular parameter s_1) is not positive definite everywhere, the equation of motion (64) for the s_1-parametrized quasidistribution cannot be interpreted as a FPE describing the Brownian motion, i.e., the equation is not a "true" FPE. The quasidistribution $\mathscr{W}^{(s_1)}(\alpha_S, \alpha_A, t)$ does not exist as a positive well-behaved function; still it does exist as a generalized function according to the Klauder–Sudarshan theorem.[198] This property is a signature of quantum effects.[98–100] Let us note that it is possible to use generalized P-representations (positive P-representations) by doubling the phase space, as has been proposed by Drummond and Gardiner.[124] The generalized P-representations have been applied successfully to solve master equations of various nonlinear problems (see, e.g., Ref. 124, 127, 128, and 200). This method, if applied to our model, requires us to handle eight real variables (not counting time), instead of four.

For initially coherent Stokes and anti-Stokes fields, i.e., satisfying $C_S = C_A = \overline{D}_{SA} = 0$, the rather complicated expressions for $\overline{L}^{(s)}$ (77) and $L^{(s)}$ (75) reduce to

$$\overline{L}^{(s)} = B_S^{(s)}(t) B_A^{(s)}(t) - \left| D_{SA}(t) \right|^2 \tag{78}$$

$$L^{(s)} = \left| B_S^{(s)}(t) B_A^{(s)}(t) - \left| D_{SA}(t) \right|^2 \right| \tag{79}$$

It is seen that, in the case of initially coherent fields, the sufficient condition for the existence of $\mathscr{W}^{(s)}(\alpha_S, \alpha_A, t)$ (73) is only that the function $\bar{L}^{(s)}$ (78) shall be positive. One obtains further simplifications of the problem under the assumption of negligible frequency mismatch ($\Delta\Omega = 0$). The functions $B_{S,A}^{(s)}(t)$ and $D_{SA}(t)$ (68) now reduce to

$$B_S^{(s)}(t) = \frac{\kappa_S}{\kappa_S - \kappa_A} f_- \left(\frac{\kappa_S}{\kappa_S - \kappa_A} f_+ - 2 \frac{\kappa_A}{\kappa_S - \kappa_A} + \langle \hat{n}_V \rangle f_+ \right)$$

$$+ \frac{1-s}{2} \geq 0$$

$$(80)$$

$$B_A^{(s)}(t) = \frac{\kappa_A}{\kappa_S - \kappa_A} f_- \left(\frac{\kappa_S}{\kappa_S - \kappa_A} f_- + \langle \hat{n}_V \rangle f_+ \right) + \frac{1-s}{2} \geq 0$$

$$|D_{SA}(t)| = \frac{\sqrt{\kappa_S \kappa_A}}{\kappa_S - \kappa_A} f_- \left(\frac{\kappa_S}{\kappa_S - \kappa_A} f_- + \langle \hat{n}_V \rangle f_+ + 1 \right)$$

with

$$f_\pm = \exp\left(\frac{\kappa_S - \kappa_A}{2} \Delta t \right) \pm 1 \qquad (81)$$

In particular, the Wigner function exists, since

$$\bar{L}^{(0)} = \frac{1}{4} + \frac{1}{2} \frac{\exp[(\kappa_S - \kappa_A)\Delta t] - 1}{\kappa_S - \kappa_A} [\langle \hat{n}_V \rangle (\kappa_S + \kappa_A) + \kappa_S] > 0 \quad (82)$$

contrary to the P-function, which does not exist for $t > t_0$, since[53]

$$\bar{L}^{(1)} = - \frac{\kappa_S \kappa_A}{(\kappa_S - \kappa_A)^2} \left[\exp\left(\frac{\kappa_S - \kappa_A}{2} \Delta t \right) - 1 \right]^2 < 0 \qquad \text{for } \Delta t > 0 \quad (83)$$

In general, the solution (73) at a given time t exists for parameters s less than

$$s < B_S^{(1)}(t) + B_A^{(1)}(t) + 1 - \sqrt{\left(B_S^{(1)}(t) + B_A^{(1)}(t) \right)^2 - 4\bar{L}^{(1)}} \quad (84)$$

Assuming that the damping constant γ_A is equal to the gain constant γ_S, or equivalently $\kappa_A = \kappa_S = \kappa$, we arrive at

$$\overline{L}^{(s)} = \tfrac{1}{4}\Big[(1-s)^2 + 2(1-s)\big(1 + 2\langle \hat{n}_V \rangle\big)\kappa\,\Delta t - s\kappa^2(\Delta t)^2\Big] \quad (85)$$

which is greater than zero for parameters s less than

$$s < \frac{1}{2} + \frac{(\kappa\,\Delta t + 1)^2}{2} + \left\{ 2\langle \hat{n}_V \rangle - \left[\left(1 + 2\langle \hat{n}_V \rangle + \frac{\kappa\,\Delta t}{2}\right)^2 + 1\right]^{1/2}\right\}\kappa\,\Delta t$$

$$(86)$$

In particular, $\overline{L}^{(1)}$ (83) for the P-function and $\overline{L}^{(0)}$ (82) for the Wigner function respectively reduce to

$$\overline{L}^{(1)} = -\left(\frac{\kappa\,\Delta t}{2}\right)^2 < 0 \qquad \text{for } \Delta t > 0$$

$$(87)$$

$$\overline{L}^{(0)} = \tfrac{1}{4} + \left(\langle \hat{n}_V \rangle + \tfrac{1}{2}\right)\kappa\,\Delta t > 0$$

The condition for s fulfilling $\overline{L}^{(s)} > 0$ cannot be expressed explicitly in a simple form in cases with frequency mismatch $\Delta\Omega \neq 0$. As another example, let us assume that the Stokes and anti-Stokes fields are initially chaotic, which mathematically differs from our former example of initially coherent state by the presence of nonzero initial coefficients $B_k^{(-1)} = \langle \hat{n}_{\mathrm{ch}\,k} \rangle$ ($k = S, A$). By virtue of the relations (68), the functions $B_k^{(s)}(t)$, $D_{SA}(t)$ for chaotic field are the same as for a coherent field with extra terms. Here, the function $\overline{L}^{(s)}$ (75) is found to be

$$\overline{L}^{(s)} = \left(B_S^{(s)}(t) + \langle \hat{n}_{\mathrm{ch}\,S} \rangle |U_S(t)|^2 + \langle \hat{n}_{\mathrm{ch}\,A} \rangle |V_S(t)|^2\right)$$

$$\times \left(B_A^{(s)}(t) + \langle \hat{n}_{\mathrm{ch}\,A} \rangle |U_A(t)|^2 + \langle \hat{n}_{\mathrm{ch}\,S} \rangle |V_A(t)|^2\right) \quad (88)$$

$$+ \big|\,|D_{SA}| + \langle \hat{n}_{\mathrm{ch}\,S} \rangle U_S(t)V_A(t) + \langle \hat{n}_{\mathrm{ch}\,A} \rangle U_A(t)V_S(t)\big|^2$$

which has a form similar to (79) with the same function $B_k^{(s)}(t)$ and $D_{SA}(t)$

given by (80). In the case of equal damping and gain constants we obtain

$$
\begin{aligned}
\bar{L}^{(s)} = & -\left(\frac{\gamma \Delta t}{2}\right)^2 s\big(\langle \hat{n}_{\mathrm{ch}\,A}\rangle + \langle \hat{n}_{\mathrm{ch}\,S}\rangle - 1\big) \\
& + \gamma \Delta t \Bigg[\langle \hat{n}_{\mathrm{ch}\,A}\rangle \left(\langle \hat{n}_V\rangle + \frac{1+s}{2}\right) \\
& + \langle \hat{n}_{\mathrm{ch}\,S}\rangle \left(\langle \hat{n}_V\rangle + \frac{1-s}{2}\right) + \left(\langle \hat{n}_V\rangle + \frac{1}{2}\right)(1-s) \Bigg] \\
& + \langle \hat{n}_{\mathrm{ch}\,A}\rangle \left(\langle \hat{n}_{\mathrm{ch}\,S}\rangle + \frac{1-s}{2}\right) + \langle \hat{n}_{\mathrm{ch}\,S}\rangle \frac{1-s}{2} + \left(\frac{1-s}{2}\right)^2
\end{aligned}
\tag{89}
$$

It is seen that the Wigner function always exists, since

$$
\begin{aligned}
\bar{L}^{(0)} = & \; \gamma \Delta t \big(\langle \hat{n}_{\mathrm{ch}\,S}\rangle + \langle \hat{n}_{\mathrm{ch}\,A}\rangle + 1\big)\big(\langle \hat{n}_V\rangle + \tfrac{1}{2}\big) \\
& + \langle \hat{n}_{\mathrm{ch}\,A}\rangle\big(\langle \hat{n}_{\mathrm{ch}\,S}\rangle + \tfrac{1}{2}\big) + \tfrac{1}{2}\langle \hat{n}_{\mathrm{ch}\,S}\rangle + \tfrac{1}{4} > 0
\end{aligned}
\tag{90}
$$

whereas the P-function exists only for times shorter than

$$
\begin{aligned}
\Delta t < & \; \frac{2}{\gamma}\big(\langle \hat{n}_{\mathrm{ch}\,A}\rangle + \langle \hat{n}_{\mathrm{ch}\,S}\rangle + 1\big)^{-1} \Big\{\langle \hat{n}_{\mathrm{ch}\,A}\rangle^2 \big[\langle \hat{n}_{\mathrm{ch}\,S}\rangle + (\langle \hat{n}_V\rangle + 1)^2\big] \\
& + \langle \hat{n}_{\mathrm{ch}\,A}\rangle\langle \hat{n}_{\mathrm{ch}\,S}\rangle\big[\langle \hat{n}_{\mathrm{ch}\,S}\rangle + \langle \hat{n}_V\rangle^2 + (\langle \hat{n}_V\rangle + 1)^2\big] \\
& + \langle \hat{n}_{\mathrm{ch}\,S}\rangle^2 \langle \hat{n}_V\rangle^2 \Big\}^{1/2} \\
& + \langle \hat{n}_{\mathrm{ch}\,A}\rangle\big(\langle \hat{n}_V\rangle + 1\big) + \langle \hat{n}_{\mathrm{ch}\,S}\rangle\langle \hat{n}_V\rangle
\end{aligned}
\tag{91}
$$

The relation (89) is quadratic in s and readily gives an analytic expression for the largest parameter s ($s \leq 1$) for which the quasidistribution $\mathscr{W}^{(s)}(\alpha_S, \alpha_A, t)$ exists at a given time of the evolution $\Delta t = t - t_0$.

In Fig. 1 we present the function $\bar{L}^{(s)}(t)$ for different values of the frequency mismatch $\Delta\Omega$, of the mean number of photons $\langle \hat{n}_V\rangle$, and of the damping (κ_A) and gain (κ_S) constants. We assume that the Stokes and anti-Stokes modes are initially coherent. Thus, for all discussed cases (Figs. 1a–d), the condition of a positive definite function $\bar{L}^{(s)}(t)$ is sufficient for the existence of the corresponding s-parametrized QPD. For clarity, the dashed lines in Fig. 1 are depicted for $\bar{L}^{(s)}(t) = 0$.

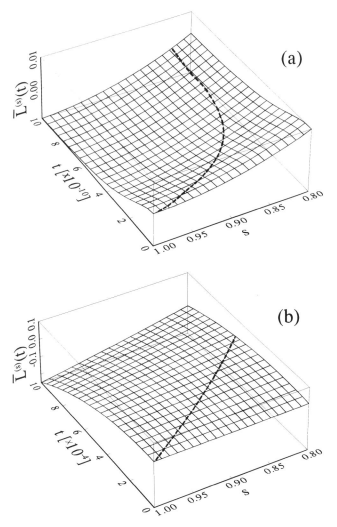

Figure 1a. The time and parameter s dependence of the function $\overline{L}^{(s)}(t)$, related to the existence of the QPD $W^{(s)}(\alpha_S, \alpha_A, t)$, for (a) $\kappa_S = 10^8$, $\kappa_A = 10^{10}$, $|\Delta\Omega| = 1 \div 10^6$ (the surfaces coincide in this range of $|\Delta\Omega|$), $\langle \hat{n}_V \rangle = 0$; (b) $\kappa_S = \kappa_A = 10^3$, $|\Delta\Omega| = 1$, $\langle \hat{n}_V \rangle = 0 \div 100$; (c) $\kappa_S = \kappa_A = 10^8$, $|\Delta\Omega| = 10^6$, $\langle \hat{n}_V \rangle = 10$; and (d) $\kappa_S = \kappa_A = 10^8$, $|\Delta\Omega| = 10^6$, $\langle \hat{n}_V \rangle = 0$. The Stokes and anti-Stokes fields are initially coherent. The dashed lines on the surfaces are depicted for $\overline{L}^{(s)}(t) = 0$.

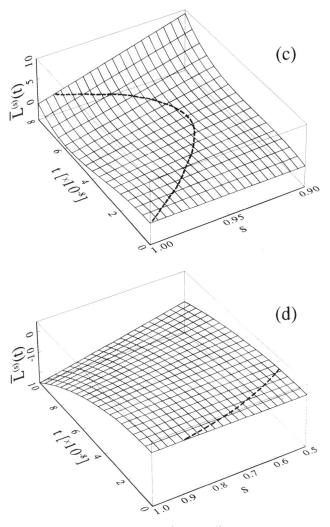

Figure 1b.   *(Continued)*

We shall briefly analyze a more general situation, which comprises the above cases and others. Let us assume after Refs. 56 and 66 (for a general analysis see Refs. 23 and 219) that the Stokes ($k = S$) as well anti-Stokes (A) modes are initially in squeezed states characterized by complex amplitudes ξ_k, parameters r_k, and phases ϕ_k, superposed with a chaotic field, characterized by the mean number of chaotic photons $\langle \hat{n}_{\text{ch}\,k} \rangle$. The initial s-parametrized quasidistribution $\mathscr{W}^{(s)}(\alpha_S, \alpha_A, t_0)$ is then given by

$$\mathscr{W}^{(s)}(\alpha_S, \alpha_A, t_0) = \prod_{k=S, A} \left(K_k^{(s)}\right)^{-1/2}$$

$$\times \exp\left\{-\frac{1}{K_k^{(s)}}\left[B_k^{(s)}|\alpha_k - \xi_k|^2 - \text{Re}\left(C_k^*(\alpha_k - \xi_k)^2\right)\right]\right\}$$

$$(92)$$

with

$$B_k^{(s)} = B_k^{(s)}(t_0) = (\cosh r)^2 + \langle n_{\text{ch }k}\rangle - \frac{s+1}{2}$$

$$C_k = C_k(t_0) = \tfrac{1}{2}\exp(i\phi_k)\sinh(2r_k)$$

$$(93)$$

which trivially reduces to the quasidistributions of a pure squeezed state ($\langle \hat{n}_{\text{ch }k}\rangle = 0$), a coherent state ($\langle \hat{n}_{\text{ch }k}\rangle = r_k = 0$), or a chaotic state ($r_k = \xi_k = 0$). In Section VI.B we analyze another special case of (93) with $r_k = 0$ describing a general superposition of coherent and chaotic fields.

The Raman effect model under parametric approximation is fully specified either by the s-parametrized characteristic function $\mathscr{C}^{(s)}(\beta_S, \beta_A, t)$ (67) or the s-parametrized quasidistribution $\mathscr{W}^{(s)}(\alpha_S, \alpha_A, t)$ (73). In particular, by virtue of the relations presented in Section IV, one can obtain complete information about the photon-counting statistics and squeezing properties of the scattered fields from (67) or (73).

One can calculate the photon-counting probability distribution $p(n)$ from the quasidistribution $\mathscr{W}^{(s)}(\alpha_S, \alpha_A, t)$ or integrated quasidistribution $\mathscr{W}^{(s)}(W, t)$ (30) by means of (31), or equivalently from the generating function $\langle\exp(-\lambda W(t))\rangle_{(s)}$ (32) by virtue of (33). We apply the latter method, which gives us, after insertion of $\mathscr{W}^{(s)}(\alpha_S, \alpha_A, t)$ (73) or $\mathscr{C}^{(s)}(\beta_S, \beta_A, t)$ (67) into (32), the following time-dependent s-parametrized generating function:

$$\langle\exp(-\lambda W)\rangle_{(s)} = \lambda^{-2}\left(\mathscr{L}_1^{(s)}\right)^{-1/2}\exp\left(\frac{\mathscr{L}_2^{(s)}}{\mathscr{L}_1^{(s)}}\right)$$

$$(94)$$

where the $\mathscr{L}_1^{(s)}$ ($\mathscr{L}_2^{(s)}$) are polynomials of the fourth (third) order in λ^{-1}:

$$\mathscr{L}_1^{(s)} = \sum_{j=0}^{4} \left(\lambda^{-1} + \frac{1-s}{2}\right)^j b_j$$

$$\mathscr{L}_2^{(s)} = \sum_{j=0}^{3} \left(\lambda^{-1} + \frac{1-s}{2}\right)^j a_j \tag{95}$$

Adapting the results of Peřinová and Peřina[48] for the coefficients a_j, b_j ($j = 0, 1, \ldots$) occurring in (95), one obtains

$$a_0 = \left[-B_S^{(1)}K_A^{(1)} + B_A^{(1)}K_+ + \left(C_A D_{SA}\overline{D}_{AS} + \text{c.c.}\right)\right]|\xi_S|^2$$
$$+ \left\{\left[B_A^{(1)}D_{SA}\overline{D}_{SA} + \tfrac{1}{2}\left(C_S^* K_A^{(1)} + C_A D_{SA}^2 + C_A^*\overline{D}_{SA}^2\right)\right]\xi_S^2\right.$$
$$+ \tfrac{1}{2}\left[B_S^{(1)}B_A^{(1)}D_{SA} + 2B_S^{(1)}C_A^*\overline{D}_{SA} + C_S^* C_A^* D_{SA}^* - D_{SA}K_-\right]\xi_S\xi_A$$
$$- \tfrac{1}{2}\left[B_S^{(1)}B_A^{(1)}\overline{D}_{SA} + 2B_S^{(1)}C_A D_{SA} + C_S^* C_A \overline{D}_{AS}\right.$$
$$\left.\left. + \overline{D}_{SA}K_-\right]\xi_S\xi_A^* + \text{c.c.}\right\} + [S \leftrightarrow A]$$

$$a_1 = \left[-2B_S^{(1)}B_A^{(1)} - K_A^{(1)} + K_+\right)|\xi_S|^2 \tag{96}$$
$$+ \left[\left(B_A^{(1)}C_S^* + D_{SA}\overline{D}_{SA}\right)\xi_S^2 + \left(B_S^{(1)}D_{SA} + C_S^*\overline{D}_{AS}\right)\xi_S\xi_A\right.$$
$$\left. - \left(B_S^{(1)}\overline{D}_{SA} + C_S^* D_{SA}^*\right)\xi_S\xi_A^* + \text{c.c.}\right]$$
$$+ [S \leftrightarrow A]$$

$$a_2 = -\left(B_S^{(1)} + 2B_A^{(1)}\right)|\xi_S|^2$$
$$+ \tfrac{1}{2}\left(C_S^*\xi_S^2 + D_{SA}\xi_S\xi_A - \overline{D}_{SA}\xi_S\xi_A^* + \text{c.c.}\right) + [S \leftrightarrow A]$$

$$a_3 = -|\xi_S|^2 - |\xi_A|^2$$

$$b_0 = \tfrac{1}{2}K_S^{(1)}K_A^{(1)} - B_S^{(1)}B_A^{(1)}K_+ - 2B_S^{(1)}\left(C_A^* D_{SA}^*\overline{D}_{AS} + \text{c.c.}\right) + \tfrac{1}{2}K_-^2$$
$$- \tfrac{1}{2}\left[C_A\left(C_S D_{SA}^2 + C_S^*\overline{D}_{AS}^2\right) + \text{c.c.}\right] + [S \leftrightarrow A]$$

$$b_1 = 2B_S^{(1)}\left(K_A^{(1)} - K_+\right) - 2\left(C_S D_{SA}\overline{D}_{SA} + \text{c.c.}\right) + [S \leftrightarrow A] \tag{97}$$

$$b_2 = 2B_S^{(1)}B_A^{(1)} + K_S^{(1)} - K_+ + [S \leftrightarrow A]$$

$$b_3 = 2\left(B_S^{(1)} + B_A^{(1)}\right)$$

$$b_4 = 1$$

where $[S \leftrightarrow A]$ stands for the preceding terms albeit with interchanged subscripts S and A. For brevity, we have omitted the time dependence of Eqs. (96) and (97). Our formulas (94) and (95) are generalizations of the results given in Refs. 23, 48, and 56 for $s = -1$ to any s. It is seen that the simplest form of (95) is for normal ordering of the field operators; hence, here, we use only this ordering. Peřinová and Peřina[48] have shown that if the polynomial $\mathscr{L}_1^{(1)}$ has four single roots $\lambda_k = -(1/\lambda)_k$, the generating function $\langle \exp(-\lambda W(t)) \rangle_{(1)}$ has the form of the fourfold generating function for Laguerre polynomials

$$\left\langle \exp(-\lambda W(t)) \right\rangle_{(1)} = \prod_{k=1}^{4} (1 + \lambda \lambda_k)^{-1/2} \exp\left(-\frac{\lambda A_k}{1 + \lambda \lambda_k}\right) \quad (98)$$

The field is described by a superposition of signal components

$$A_k = \prod_{\substack{l=1 \\ l \neq k}}^{4} \left(\lambda_k^{-1} - \lambda_l^{-1}\right)^{-1} \sum_{l=0}^{3} a_l \left(-\lambda_k^{-1}\right)^l \quad (99)$$

and the noise components λ_k. With Eq. (98) available, one obtains[48, 56] the following photocount distribution

$$
\begin{aligned}
p(n, t) = \sum_{\substack{k_1, k_2, k_3, k_4 \\ k_1 + k_2 + k_3 + k_4 = n}} \prod_{l=1}^{4} \exp\left(-\frac{A_l}{1 + \lambda_l}\right) \\
\times \frac{\lambda_l^{k_l}}{(1 + \lambda_l)^{k_l + 1/2} \Gamma\left(k_l + \frac{1}{2}\right)} L_{k_l}^{-1/2}\left(\frac{-A_l}{\lambda_l(1 + \lambda_l)}\right)
\end{aligned}
\quad (100)
$$

and its factorial moments

$$\left\langle W^k(t) \right\rangle_{(1)} = k! \sum_{k_1 + k_2 + k_3 + k_4 = k} \prod_{l=1}^{4} \frac{\lambda_l^{k_l}}{\Gamma\left(k_l + \frac{1}{2}\right)} L_{k_l}^{-1/2}\left(-\frac{A_l}{\lambda_l}\right) \quad (101)$$

by applying well-known properties of the generating function of the generalized Laguerre polynomials $L_k^\alpha(x)$ to the definition relations (33) for $p(n, t)$ with $s = 1$ and to the relations (34) for $\langle W^k(t) \rangle_{(1)}$. Much simpler expressions are found in the special case when the radiation fields are initially superpositions of coherent and chaotic fields. From relations (93) with $r_k = 0$ and (68) it is seen that $C_S(t) = C_A(t) = \overline{D}_{SA}(t) = 0$. The fourfold generating function (98) reduces to a twofold generating function in the form of (98), where the upper limit of the product should be replaced by 2 and the square root in $(1 + \lambda \lambda_k)^{-1/2}$ should be omitted. Then the photocount distribution $p(n, t)$ and its factorial moments $\langle W^k(t) \rangle_{(1)}$ become

$$
p(n, t) = (n!)^{-1} \exp\left(-\frac{A_1}{1 + \lambda_1} - \frac{A_2}{1 + \lambda_2} \right) \sum_{l=0}^{n} \binom{n}{l} \lambda_1^l \lambda_2^{n-l} (1 + \lambda_1)^{-(l+1)}
$$
$$
\times (1 + \lambda_2)^{-n+(l+1)} L_l\left(-\frac{A_1}{\lambda_1(1 + \lambda_1)} \right) L_{n-l}\left(-\frac{A_2}{\lambda_2(1 + \lambda_2)} \right)
$$

$$(102)$$

$$
\langle W^n(t) \rangle_{(1)} = \sum_{l=0}^{n} \binom{n}{l} \lambda_1^l \lambda_2^{n-l} L_l\left(-\frac{A_1}{\lambda_1} \right) L_{n-l}\left(-\frac{A_2}{\lambda_2} \right) \qquad (103)
$$

where $L_n(x) = L_n^0(x)$ and the roots λ_k and coefficients A_k are[48]

$$
\lambda_{1,2} = \tfrac{1}{2}\left\{ B_S^{(1)}(t) + B_A^{(1)}(t) \mp \left[\left(B_S^{(1)}(t) - B_A^{(1)}(t) \right)^2 + 4|D_{SA}(t)|^2 \right]^{1/2} \right\}
$$
$$
A_{1,2} = \pm \left[\left(B_S^{(1)}(t) - B_A^{(1)}(t) \right)^2 + 4|D_{SA}(t)|^2 \right]^{-1/2}
$$
$$(104)$$
$$
\times \left[\tfrac{1}{2}\left(B_S^{(1)}(t) - B_A^{(1)}(t) \right)\left(|\xi_A|^2 - |\xi_S|^2 \right) - \left(D_{SA}\xi_S\xi_A^* + \text{c.c.} \right) \right]
$$
$$
+ \tfrac{1}{2}\left(|\xi_S|^2 + |\xi_A|^2 \right)
$$

The photon-counting statistics of scattering either into the Stokes or anti-Stokes mode can be calculated from formulas (98)–(101). In the single-mode case the moments $D_{SA}(t)$ and $\bar{D}_{SA}(t)$ vanish, considerably simplifying the polynomial $\mathscr{L}_1^{(1)}$ (95), with coefficients b_j, to the form

$$
\mathscr{L}_1^{(1)} = \prod_{k=S, A} \left(\lambda^{-2} + 2\lambda^{-1}B_k^{(1)}(t) + K_k^{(1)}(t) \right) \qquad (105)
$$

with the roots $\lambda_{1, 2 S, A} = -(\lambda^{-1})_k$

$$
\lambda_{1, 2k} = B_k^{(1)}(t) \mp |C_k(t)| \qquad (k = S, A) \qquad (106)
$$

The notation $\lambda_{1S, A}$ and $\lambda_{2S, A}$, instead of $\lambda_{1, 2, 3, 4}$, emphasizes the dependence on the single-mode variables in accordance with the assumption of alternative scattering into the Stokes or anti-Stokes mode. Analogously, it is seen that

$$A_{1,2k} = \tfrac{1}{2}|\xi_k(t)|^2 \mp \tfrac{1}{4}|C_k(t)|^{-1}\left(C_k^*(t)\xi_k^2(t) + \text{c.c}\right) \qquad (107)$$

On insertion of (106) into twofold functions (98)–(101) one immediately obtains

$$\langle \exp(-\lambda W_k)\rangle_{(1)} = \left[(1+\lambda\lambda_{1k})(1+\lambda\lambda_{2k})\right]^{-1/2}$$
$$\times \exp\left[-\lambda A_{1k}(1+\lambda\lambda_{1k})^{-1} - \lambda A_{2k}(1+\lambda\lambda_{2k})^{-1}\right] \qquad (108)$$

$$p_k(n) = \left[(1+\lambda_{1k})(1+\lambda_{2k})\right]^{-1/2}\left(1+\lambda_{2k}^{-1}\right)^{-n}$$
$$\times \exp\left(-\frac{A_{1k}}{1+\lambda_{1k}} - \frac{A_{2k}}{1+\lambda_{2k}}\right)$$
$$\times \sum_{l=0}^{n} \frac{1}{\Gamma(l+\tfrac{1}{2})\Gamma(n-l+\tfrac{1}{2})}\left(\frac{1+\lambda_{2k}^{-1}}{1+\lambda_{1k}^{-1}}\right)^{l} \qquad (109)$$
$$\times L_l^{-1/2}\left(-\frac{A_{1k}}{\lambda_{1k}(1+\lambda_{1k})}\right) L_{n-l}^{-1/2}\left(-\frac{A_{2k}}{\lambda_{2k}(1+\lambda_{2k})}\right)$$

$$\langle W_k^n(t)\rangle_{(1)} = n!\lambda_{2k}^n \sum_{l=0}^{n} \frac{1}{\Gamma(l+\tfrac{1}{2})\Gamma(n-l+\tfrac{1}{2})}\left(\frac{\lambda_{1k}}{\lambda_{2k}}\right)^{l}$$
$$\times L_l^{-1/2}\left(-\frac{A_{1k}}{\lambda_{1k}}\right) L_{n-l}^{-1/2}\left(-\frac{A_{2k}}{\lambda_{2k}}\right) \qquad (110)$$

To obtain the results of Refs. 23, 56, and 189, one should replace λ_{1k} by $E_k - 1$ and λ_{2k} by $F_k - 1$. In particular, assuming that a scattered (Stokes or anti-Stokes) mode is initially in a coherent state (thus $C_k = 0$) the mean photon numbers $\langle \hat{n}_k \rangle$ ($k = S, A$) are

$$\langle \hat{n}_k(t)\rangle = \langle W_k(t)\rangle_{(1)} = |\xi_k(t)|^2 + B_k^{(1)}(t) \qquad (111)$$

or explicitly

$$\langle \hat{n}_S(t)\rangle = |\xi_S|^2 \exp(\kappa_S \,\Delta t) + \left(\langle \hat{n}_V\rangle + 1\right)\left[\exp(\kappa_S \,\Delta t) - 1\right] \qquad (112)$$
$$\langle \hat{n}_A(t)\rangle = |\xi_A|^2 \exp(-\kappa_A \,\Delta t) + \langle \hat{n}_V\rangle\left[1 - \exp(-\kappa_A \,\Delta t)\right] \qquad (113)$$

whereas the mean-square photon-numbers $\langle \hat{n}_k^2 \rangle$ are

$$
\begin{aligned}
\langle \hat{n}_k^2 \rangle &= \langle W_k^2 \rangle_{(1)} + \langle W_k \rangle_{(1)} \\
&= |\xi_k(t)|^4 + |\xi_k(t)|^2 (4B_k(t) + 1) + 2B_k^2(t) + B_k(t)
\end{aligned}
\tag{114}
$$

Then, the normalized second-order factorial moments (35) are equal to

$$
\begin{aligned}
\gamma_k^{(2)}(t) &= B_k^{(1)}(t) \Big[|\xi_k(t)|^2 + B_k^{(1)}(t) \Big]^{-1} \\
&\quad \times \Big\{ |\xi_k(t)|^2 \Big[|\xi_k(t)|^2 + B_k^{(1)}(t) \Big]^{-1} + 1 \Big\}
\end{aligned}
\tag{115}
$$

Let us proceed to analyze squeezing along the lines presented in Section IV. We focus our attention on single- and two-mode squeezed light according to the definition of "usual" squeezing and principal squeezing of Lukš et al.[132, 211, 213] Using the definitions (66) of the functions $B_k^{(s)}(t)$, $D_{kl}(t)$, $\overline{D}_{kl}(t)$, and $C_k(t)$ we readily obtain expressions for the moments of the quadratures $\hat{X}_k$ and $\hat{X}_{kl}$

$$
\left\langle \left(\Delta \hat{X}_{k1,k2} \right)^2 \right\rangle = \pm 2 \operatorname{Re} C_k(t) + 2B_k^{(s)}(t) + s
\tag{116}
$$

$$
\left\langle \left(\Delta \hat{X}_{k\pm} \right)^2 \right\rangle = \pm 2 |C_k(t)| + 2B_k^{(s)}(t) + s
\tag{117}
$$

$$
\left\langle \left\{ \Delta \hat{X}_{k1}, \Delta \hat{X}_{k2} \right\} \right\rangle = 4 \operatorname{Im} C_k(t)
$$

$$
\begin{aligned}
\langle \Delta \hat{X}_{k1} \Delta \hat{X}_{l1} \rangle &= 2 \operatorname{Re}\big[D_{kl}(t) - \overline{D}_{kl}(t) \big] \\
\langle \Delta \hat{X}_{k2} \Delta \hat{X}_{l2} \rangle &= -2 \operatorname{Re}\big[D_{kl}(t) + \overline{D}_{kl}(t) \big] \\
\langle \Delta \hat{X}_{k1} \Delta \hat{X}_{l2} \rangle &= 2 \operatorname{Im}\big[D_{kl}(t) - \overline{D}_{kl}(t) \big] \\
\langle \Delta \hat{X}_{k2} \Delta \hat{X}_{l1} \rangle &= 2 \operatorname{Im}\big[D_{kl}(t) + \overline{D}_{kl}(t) \big]
\end{aligned}
\tag{118}
$$

where, as usual, $k, l = S, A$ and $k \neq l$. Thus, the two-mode quadrature variances now have the form

$$
\left. \begin{aligned}
\left\langle \left(\Delta \hat{X}_{SA1} \right)^2 \right\rangle \\
\left\langle \left(\Delta \hat{X}_{SA2} \right)^2 \right\rangle
\end{aligned} \right\} = \begin{aligned}
&\pm 2 \operatorname{Re}\big[C_S(t) + C_A(t) + 2D_{SA}(t) \big] \\
&+ 2 \big[B_S^{(s)}(t) + B_A^{(s)}(t) - 2 \operatorname{Re} \overline{D}_{SA}(t) + s \big]
\end{aligned}
\tag{119}
$$

and the extremal variances are

$$
\left\langle \left(\Delta \hat{X}_{SA\pm}\right)^2 \right\rangle = \pm 2\left| C_S(t) + C_A(t) + 2D_{SA}(t) \right|
$$
$$
+ 2\left[B_S^{(s)}(t) + B_A^{(s)}(t) - 2\,\mathrm{Re}\,\overline{D}_{SA}(t) + s \right]
\tag{120}
$$

The single-mode squeezing, defined in standard manner, and the single-mode principal squeezing require, respectively, that

$$
\left.\begin{array}{c} \left|\mathrm{Re}\,C_k(t)\right| \\ \left|C_k(t)\right| \end{array}\right\} > B_k^{(s)}(t) + \frac{s}{2} \qquad (k = S, A)
\tag{121}
$$

whereas the conditions for the two-mode squeezing are, respectively,

$$
\left.\begin{array}{c} \left|\mathrm{Re}[C_S(t) + C_A(t) + 2D_{SA}(t)]\right| \\ \left|C_S(t) + C_A(t) + 2D_{SA}(t)\right| \end{array}\right\} > B_S^{(s)}(t) + B_A^{(s)}(t) - 2\,\mathrm{Re}\,\overline{D}_{SA}(t) + s
$$
$$
\tag{122}
$$

Examples of the time evolution of $\langle \hat{n}_S(\tau) \rangle$, $\langle \hat{n}_S^2(\tau) \rangle$, $\langle \hat{a}_S(\tau) \rangle$, and $\langle \hat{a}_S^2(\tau) \rangle$ are given by curves C in Figs. 2, 3, 7, and 8, respectively. We assume that the Stokes fields are initially in a coherent state (stimulated Raman scattering) or in a vacuum state (spontaneous Raman scattering). The rescaled time τ is defined by $t \to \tau = t\gamma_S$. Anti-Stokes scattering is neglected. The phonon bath is at very low temperature, so we put $\langle \hat{n}_V \rangle = 0$. In Fig. 9 we present the time evolution of the extremal variances $\langle (\Delta X_{S\pm}(\tau))^2 \rangle$ for fields initially coherent with amplitudes equal to $\alpha_L = \sqrt{2}$, $\alpha_S = \sqrt{0.2}$ and assuming that the heat bath is "quiet" (i.e., $\langle \hat{n}_V \rangle = 0$). In the model under discussion, the variance for the Stokes mode, $\langle (\Delta X_S(\theta, \tau))^2 \rangle$ (curve C in Fig. 9), is independent of θ, i.e., $\langle (\Delta X_{S+}(\tau))^2 \rangle = \langle (\Delta X_{S-}(\tau))^2 \rangle$. Hence, squeezing is not observed if the initial Stokes mode is in a coherent state. Even if the Stokes field is initially squeezed and $\gamma_S > \gamma_A$ (not necessarily $\gamma_A = 0$), squeezing will rapidly vanish due to strong amplification of this mode, which leads to a strong increase in quantum noise.[56] The results of this section (curves C) are compared with the exact solutions (without parametric approximation) derived in Section VI.A.2. (curves A) and the short-time solutions of SectionVI.A.1.

VI. MASTER EQUATION IN FOCK REPRESENTATION

The parametric approximation, applied in the previous section, introduces linearization into our Raman scattering model described by the Hamiltonians (1)–(3). Here, we shall search for a solution to the nonlinear problem, thus including pump depletion. The generalized Fokker-Planck equation (61) and the corresponding equation of motion (63) for the characteristic function reveal the difficulties to be overcome in the complete analysis of Raman scattering into simultaneously both the Stokes and anti-Stokes fields from phonons treated as a "noisy" ($\langle \hat{n}_V \rangle \neq 0$) reservoir. Let us assume that the temperature of the medium is low. Under this assumption it is quite reasonable to neglect the anti-Stokes scattering ($\gamma_A = \gamma_{SA} = \gamma_{AS} = 0$) and, with regard to Eq. (14), to assume that the reservoir is "quiet" ($\langle \hat{n}_V \rangle = 0$). Under these approximations the master equation (18) reduces to the simple form[73]:

$$\frac{\partial \hat{\rho}}{\partial \tau} = \frac{1}{2}\left(\left[\hat{a}_L \hat{a}_S^+, \hat{\rho}\hat{a}_L^+ \hat{a}_S\right] + \left[\hat{a}_L \hat{a}_S^+ \hat{\rho}, \hat{a}_L^+ \hat{a}_S\right]\right) \tag{123}$$

where we have introduced a rescaled time $t \to \tau = \gamma_S t$. Let us denote the matrix elements of the reduced density operator $\hat{\rho}$ in Fock representation by

$$\left\langle n_L, n_S | \hat{\rho}(\tau) | n'_L, n'_S \right\rangle \equiv \left\langle n, m | \hat{\rho}(\tau) | n + \nu, m + \mu \right\rangle \equiv \rho_{n,m}(\nu, \mu, \tau) \tag{124}$$

where for simplicity we identify $n_L = n$, and $n_S = m$; μ is the degree of off-diagonality for the elements of the matrix $\hat{\rho}$ for the Stokes mode, whereas ν is the degree of off-diagonality for the pump laser mode elements. The master equation for the matrix elements (124) readily follows from Eq. (123) and can be written as

$$\frac{\partial}{\partial \tau}\rho_{nm}(\nu\mu\tau) = -\frac{1}{2}\left[n(m+1) + (n+\nu)(m+\mu+1)\right]\rho_{nm}(\nu\mu\tau)$$

$$+ \left[(n+1)(n+\nu+1)m(m+\mu)\right]^{1/2}\rho_{n+1,\,m-1}(\nu\mu\tau) \tag{125}$$

The equation (125) for the diagonal matrix elements $\rho_{nm}(00\tau)$ reduces to the rate equations of Loudon,[30] and McNeil and Walls.[73] Simaan[75] (cf. Ref. 30) analyzed Raman scattering from a gas of two-level atoms. On the assumption that almost all the atoms are in their ground state, the Simaan rate equation of Ref. 75 takes the form of Eq. (125) for $\nu = \mu = 0$.

A. Raman Scattering Including Pump Depletion

1. Short-time Solutions

Before proceeding to derive an exact solution of (125) we shall present the short-time solutions calculated with the help of the relation $\langle \hat{A}(\tau) \rangle = \mathrm{Tr}\{\hat{A}[\hat{\rho}(\tau_0) + \hat{\rho}'(\tau_0)\,\Delta\tau + \hat{\rho}''(\tau_0)(\Delta\tau)^2/2]\}$, where $\hat{\rho}''(\tau_0)$ is found by differentiating Eq. (125) with respect to τ. The solutions for the mean $\langle \hat{n} \rangle$ and mean-square number of photons $\langle \hat{n}^2 \rangle$ in the laser mode up to $\Delta\tau$ squared are

$$\langle \hat{n}(\tau) \rangle = \langle \hat{n} \rangle - \langle \hat{n} \rangle (\langle \hat{m} \rangle + 1)\,\Delta\tau$$

$$- \Big[\langle \hat{n}^2 \rangle (\langle \hat{m} \rangle + 1) - \langle \hat{n} \rangle (\langle \hat{m}^2 \rangle + 3\langle \hat{m} \rangle + 2) \Big] \frac{(\Delta\tau)^2}{2} \quad (126)$$

$$\langle \hat{n}^3(\tau) \rangle = \langle \hat{n}^2 \rangle - (2\langle \hat{n}^2 \rangle - \langle \hat{n} \rangle)(\langle \hat{m} \rangle + 1)\,\Delta\tau$$

$$- \Big[2\langle \hat{n}^2 \rangle (\langle \hat{m} \rangle + 1) - \langle \hat{n}^2 \rangle (4\langle \hat{m}^2 \rangle + 13\langle \hat{m} \rangle + 9)$$

$$+ 3\langle \hat{n} \rangle (\langle \hat{m}^2 \rangle + 3\langle \hat{m} \rangle + 2) \Big] \frac{(\Delta\tau)^2}{2} \quad (127)$$

where for brevity we set $\langle \hat{n}^p(\tau_0) \rangle = \langle \hat{n}^p \rangle$ and $\langle \hat{m}^p(\tau_0) \rangle = \langle \hat{m}^p \rangle$ ($k = 1, 2, 3$) as well as $\Delta\tau = \tau - \tau_0$). Then the normalized second-order factorial moment, $\gamma_L^{(2)}(\tau)$, defined by (35), is equal to

$$\gamma_L^{(2)}(\tau) = \gamma_L^{(2)} + \Big[\big(\langle \hat{n}^2 \rangle^2 - \langle \hat{n}^3 \rangle \langle \hat{n} \rangle \big)(1 + \langle \hat{m} \rangle)$$

$$+ \langle \hat{n} \rangle (\langle \hat{n}^2 \rangle - \langle \hat{n} \rangle)\big(1 + \langle \hat{m} \rangle - \langle \hat{m} \rangle^2 + \langle \hat{m}^2 \rangle \big) \Big] \langle \hat{n} \rangle^{-3}(\Delta\tau)^2$$

$$(128)$$

which reduces to the Simaan result[75]:

$$\gamma_L^{(2)}(\tau) = \eta_L^{(2)} + \Bigg[\langle \hat{n}^2 \rangle \bigg(\frac{\langle \hat{n}^2 \rangle}{\langle \hat{n} \rangle} + 1 \bigg)$$

$$- \langle \hat{n}^3 \rangle - \langle \hat{n} \rangle \Bigg] \langle \hat{n} \rangle^{-2}(\Delta\tau)^2 \quad (129)$$

in the special case in which no scattered photons are excited initially, i.e., $\langle \hat{m} \rangle = \langle \hat{m}^2 \rangle = 0$. For the initially coherent Stokes and laser modes the factorial moment (128) reduces to the simple form $\gamma_L^{(2)} = |\alpha_S|^2 (\Delta \tau)^2$.

Our short-time solutions for the Stokes mode are

$$\langle \hat{m}(\tau) \rangle = \langle \hat{m} \rangle + \langle \hat{n} \rangle (\langle \hat{m} \rangle + 1)\, \Delta \tau$$

$$- \Big[\langle \hat{m}^2 \rangle \langle \hat{n} \rangle + \langle \hat{m} \rangle (3 \langle \hat{n} \rangle - \langle \hat{n}^2 \rangle) + 2 \langle \hat{n} \rangle - \langle \hat{n}^2 \rangle \Big] \frac{(\Delta \tau)^2}{2}$$

$$+ \Big[\langle \hat{m}^3 \rangle \langle \hat{n} \rangle + \langle \hat{m}^2 \rangle (7 \langle \hat{n} \rangle - 4 \langle \hat{n}^2 \rangle) + \langle \hat{m} \rangle (14 \langle \hat{n} \rangle$$

$$- 12 \langle \hat{n}^2 \rangle + \langle \hat{n}^3 \rangle) + 8 \langle \hat{n} \rangle - 8 \langle \hat{n}^2 \rangle + \langle \hat{n}^3 \rangle \Big] \frac{(\Delta \tau)^3}{6}$$

$$\langle \hat{m}^2(\tau) \rangle = \langle \hat{m}^2 \rangle + \langle \hat{n} \rangle (2 \langle \hat{m}^2 \rangle + 3 \langle \hat{m} \rangle + 1)\, \Delta \tau$$

$$- \Big[2 \langle \hat{m}^3 \rangle \langle \hat{n} \rangle + \langle \hat{m}^2 \rangle (9 \langle \hat{n} \rangle - 4 \langle \hat{n}^2 \rangle)$$

$$+ \langle \hat{m} \rangle (13 \langle \hat{n} \rangle - 9 \langle \hat{n}^2 \rangle) + 6 \langle \hat{n} \rangle - 5 \langle \hat{n}^2 \rangle \Big] \frac{(\Delta \tau)^2}{2}$$

$$+ \Big[2 \langle \hat{m}^4 \rangle \langle \hat{n} \rangle + \langle \hat{m}^3 \rangle (21 \langle \hat{n} \rangle - 14 \langle \hat{n}^2 \rangle)$$

$$+ \langle \hat{m}^2 \rangle (73 \langle \hat{n} \rangle - 72 \langle \hat{n}^2 \rangle + 8 \langle \hat{n}^3 \rangle)$$

$$+ \langle \hat{m} \rangle (102 \langle \hat{n} \rangle - 118 \langle \hat{n}^2 \rangle + 21 \langle \hat{n}^3 \rangle)$$

$$+ 48 \langle \hat{n} \rangle - 60 \langle \hat{n}^2 \rangle + 13 \langle \hat{n}^3 \rangle \Big] \frac{(\Delta \tau)^3}{6}$$

(130)

(131)

On adding Eqs. (126) and (130) we note that the sum of the mean number of photons in both the Stokes and laser modes is constant (at least up to $(\Delta \tau)^2$):

$$\langle \hat{n}(\tau) \rangle + \langle \hat{m}(\tau) \rangle = \langle \hat{n} \rangle + \langle \hat{m} \rangle \tag{132}$$

Taking a closer look at Eq. (125), which contains only terms with ρ_{nm} and $\rho_{n+1, m-1}$, one can draw the more fundamental conclusion that the property (132) holds for any times, in particular for the steady solutions for $\tau \rightarrow \infty$. Equation (132) is a special case of (62). Actually, we note in view of the master equation (123) that the operator $\hat{a}_L^+(\tau) \hat{a}_L(\tau) + \hat{a}_S^+(\tau) \hat{a}_S(\tau)$ is a constant of motion.

The time evolution of the mean values $\langle \hat{n}(\tau) \rangle$, $\langle \hat{m}(\tau) \rangle$, $\langle \hat{n}^2(\tau) \rangle$, and $\langle \hat{m}^2(\tau) \rangle$ is shown in Figs. 2 and 3 for initially coherent distributions. Curves B are obtained from Eqs. (126), (127), (130), and (131).

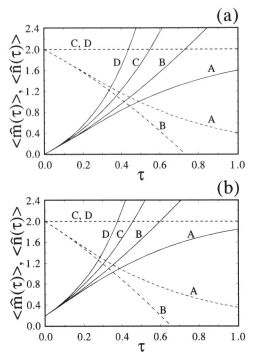

Figure 2. Time behavior of the mean number of the Stokes photons $\langle \hat{m}(t) \rangle$ (solid lines) and the laser photons $\langle \hat{n}(t) \rangle$ (dashed lines) for the initial fields: (a) $|\alpha_L = \sqrt{2}\,\rangle$, $|\alpha_S = 0\rangle$, and (b) $|\alpha_L = \sqrt{2}\,\rangle$, $|\alpha_S = \sqrt{0.2}\,\rangle$. Numerical results with exact solutions of Section VI.A.2 (curves A); short-time approximation of Section VI.A.1 (curves B); parametric approximation of Section V.B (curves C); approximate solutions of Section VI.B (curves D).

The factorial moment $\gamma_S^{(2)}(\tau)$, in the case of nonzero $\langle \hat{m} \rangle$, is equal to

$$
\begin{aligned}
\gamma_S^{(2)}(\tau) = {}& \gamma_S^{(2)} - 2\big(\langle \hat{m}^2 \rangle - 2\langle \hat{m} \rangle^2 - \langle \hat{m} \rangle\big)\langle \hat{n} \rangle\langle \hat{m} \rangle^{-3}\,\Delta\tau \\
& - \Big[\langle \hat{m}^3 \rangle\langle \hat{m} \rangle^2\langle \hat{n} \rangle - \langle \hat{m}^2 \rangle^2\langle \hat{m} \rangle\langle \hat{n} \rangle \\
& + \langle \hat{m}^2 \rangle\langle \hat{m} \rangle^2\big(\langle \hat{n} \rangle^2 + 2\langle \hat{n} \rangle - \langle \hat{n}^2 \rangle\big) \\
& - \langle \hat{m}^2 \rangle\langle \hat{m} \rangle\big(2\langle \hat{n} \rangle^2 + 2\langle \hat{n} \rangle - \langle \hat{n}^2 \rangle\big) \\
& - 3\langle \hat{m}^2 \rangle\langle \hat{n} \rangle^2 + \langle \hat{m} \rangle^3\big(7\langle \hat{n} \rangle^2 + 8\langle \hat{n} \rangle - 5\langle \hat{n}^2 \rangle\big) \\
& + \langle \hat{m} \rangle^2\big(10\langle \hat{n} \rangle^2 + 4\langle \hat{n} \rangle - 3\langle \hat{n}^2 \rangle\big) \\
& + 3\langle \hat{m} \rangle\langle \hat{n} \rangle^2\Big]\langle \hat{m} \rangle^{-4}(\Delta\tau)^2
\end{aligned}
\tag{133}
$$

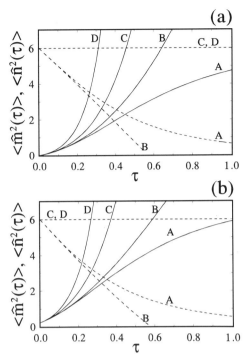

Figure 3. Time behavior of the mean-square number of the Stokes photons $\langle \hat{m}^2(t) \rangle$ (solid lines) and the laser photons $\langle \hat{n}^2(t) \rangle$ (dashed lines) for the same cases as in Fig. 2.

whereas in the case when all moments $\langle \hat{m}^k \rangle$ (for $k = 1, 2, \dots$) are zero, $\gamma_S^{(2)}(\tau)$ can be expressed as

$$\gamma_S^{(2)}(\tau) = 2\gamma_L^{(2)} + 1 + \left(6\langle \hat{n}^3 \rangle - \frac{6\langle \hat{n}^2 \rangle^2}{\langle \hat{n} \rangle} - 8\langle \hat{n}^2 \rangle + 8\langle \hat{n} \rangle \right) \langle \hat{n} \rangle^{-2} \frac{\Delta\tau}{3}$$

$$(134)$$

To obtain a correct time dependence of the factorial moment (134), it is clearly necessary to include in Eqs. (130) and (131) terms at least up to third order in τ. An equation similar to (134) has been obtained by Simaan.[75] In Fig. 4 we compare, in particular, our result for the factorial moments of photon number in the Stokes mode calculated with Eq. (133) and (134) (curves B) with that obtained from the exact solution (curves A) of the master equation (125) discussed in Section VI.A.2. Our Eq. (134) gives much better approximation to the exact results than Simaan's

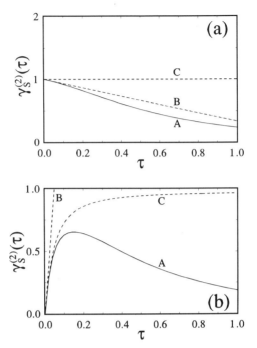

Figure 4. Time behavior of the normalized factorial moments $\gamma_S^{(2)}(\tau)$ for the Stokes mode for the same cases (except for curve D) as in Fig. 2.

formula (33) in Ref. 75. Analogously in Fig. 5, the factorial moments for the laser mode calculated with Eqs. (128) and (129) (curves B) are compared, in particular, with the exact solutions (curves A).

The Eqs. (133) and (134) reduce, respectively, to

$$\gamma_S^{(2)}(\tau) = 2|\alpha_L|^2|\alpha_S|^{-2}\Delta\tau$$
$$- \left(2|\alpha_S|^4 + 3|\alpha_S|^2 + 3|\alpha_L|^2 + |\alpha_S|^2|\alpha_L|^2\right)|\alpha_L|^2|\alpha_S|^{-4}(\Delta\tau)^2$$

$$(135)$$

$$\gamma_S^{(2)}(\tau) = 1 - \tfrac{2}{3}\Delta\tau \qquad (136)$$

for initially coherent radiation fields.

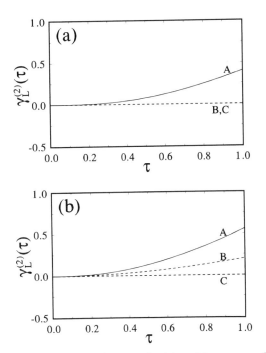

Figure 5. Same as Fig. 4, but for the normalized factorial moments $\gamma_L^{(2)}(\tau)$ of the laser mode.

The corresponding short-time dependence of the cross-correlation (interbeam) function is

$$
\begin{aligned}
\langle \hat{n}(\tau)\hat{m}(\tau)\rangle &= \langle \hat{n}\rangle\langle \hat{m}\rangle + \Big[\langle \hat{n}^2\rangle(\langle \hat{m}\rangle + 1) - \langle \hat{n}\rangle(\langle \hat{m}^2\rangle + 2\langle \hat{m}\rangle + 1)\Big]\Delta\tau \\
&\quad + \Big[\langle \hat{n}^3\rangle(\langle \hat{m}\rangle + 1) - \langle \hat{n}^2\rangle(4\langle \hat{m}^2\rangle + 11\langle \hat{m}\rangle + 7) \\
&\qquad + \langle \hat{n}\rangle(\langle \hat{m}^3\rangle + 6\langle \hat{m}^2\rangle + 11\langle \hat{m}\rangle + 6)\Big]\frac{(\Delta\tau)^2}{2} \\
&\quad - \Big[3\langle \hat{n}^4\rangle(2\langle \hat{m}\rangle + \langle \hat{m}^2\rangle + 1) - \langle \hat{n}^3\rangle(3\langle \hat{m}\rangle + 4\langle \hat{m}^2\rangle - 1) \\
&\qquad - \langle \hat{n}^2\rangle(68\langle \hat{m}\rangle + 43\langle \hat{m}^2\rangle + 11\langle \hat{m}^3\rangle + 36) \\
&\quad + \langle \hat{n}\rangle(66\langle \hat{m}\rangle + 47\langle \hat{m}^2\rangle + 14\langle \hat{m}^3\rangle + \langle \hat{m}^4\rangle + 32)\Big]\frac{(\Delta\tau)^3}{6}
\end{aligned}
$$

$$(137)$$

On inserting (126), (130), and (137) into the definition (38) of the inter-beam degree of second-order coherence, $g_{LS}^{(2)}(\tau)$, we obtain the following relation for the case when photons are initially present in the Stokes mode:

$$
\begin{aligned}
g_{LS}^{(2)}(\tau) = {}& \Big[\langle \hat{n}^2 \rangle (\langle \hat{m} \rangle + 1) + \langle \hat{n} \rangle \big(\langle \hat{m} \rangle^2 - \langle \hat{m} \rangle - \langle \hat{m}^2 \rangle - 1 \big) \\
& \qquad - \langle \hat{n} \rangle^2 (\langle \hat{m} \rangle + 1) \Big] \big(\langle \hat{n} \rangle \langle \hat{m} \rangle \big)^{-1} \Delta\tau \\
& + \Big\{ \langle \hat{n}^3 \rangle \langle \hat{m} \rangle (\langle \hat{m} \rangle + 1) + \langle \hat{n}^2 \rangle \langle \hat{m} \rangle \\
& \qquad \times \big(3\langle \hat{m} \rangle^2 - 6\langle \hat{m} \rangle - 4\langle \hat{m}^2 \rangle - 5 \big) \\
& \qquad - \langle \hat{n}^2 \rangle \langle \hat{n} \rangle \big(3\langle \hat{m} \rangle^2 + 5\langle \hat{m} \rangle + 2 \big) \\
& \qquad + \langle \hat{n} \rangle \langle \hat{m} \rangle \Big[2\langle \hat{m} \rangle^3 - 3\langle \hat{m} \rangle^2 + \langle \hat{m} \rangle (5 - 3\langle \hat{m}^2 \rangle) \\
& \qquad\qquad + 4\langle \hat{m}^2 \rangle + \langle \hat{m}^3 \rangle + 4 \Big] \\
& \qquad - \langle \hat{n} \rangle^2 \Big[2\langle \hat{m} \rangle^3 - 3\langle \hat{m} \rangle^2 - 3\langle \hat{m} \rangle (\langle \hat{m}^2 \rangle + 2) - 2(\langle \hat{m}^2 \rangle + 1) \Big] \\
& \qquad + 2\langle \hat{n} \rangle^3 \big(\langle \hat{m} \rangle^2 + 2\langle \hat{m} \rangle + 1 \big) \Big\} \langle \hat{m} \rangle^{-2} \langle \hat{n} \rangle \frac{(\Delta\tau)^2}{2}
\end{aligned}
\tag{138}
$$

Otherwise, for the case $\langle \hat{m} \rangle = \langle \hat{m}^2 \rangle = \langle \hat{m}^3 \rangle = \langle \hat{m}^4 \rangle = 0$, we get

$$
\begin{aligned}
g_{LS}^{(2)}(\tau) = {}& \gamma_L^{(2)} + \big(\langle \hat{n}^3 \rangle - \langle \hat{n}^2 \rangle^2 / \langle \hat{n} \rangle \\
& \qquad - 2\langle \hat{n}^2 \rangle + 2\langle \hat{n} \rangle \big) \langle \hat{n} \rangle^{-2} \frac{\Delta\tau}{2} \\
& - \big(6\langle \hat{n}^4 \rangle \langle \hat{n} \rangle^2 + 5\langle \hat{n}^3 \rangle \langle \hat{n}^2 \rangle \langle \hat{n} \rangle \\
& \qquad - 12\langle \hat{n}^3 \rangle \langle \hat{n} \rangle^2 - 3\langle \hat{n}^2 \rangle^3 - 22\langle \hat{n}^2 \rangle^2 \langle \hat{n} \rangle \\
& \qquad + 26\langle \hat{n}^2 \rangle \langle \hat{n} \rangle^2 \big) \langle \hat{n} \rangle^{-4} \frac{(\Delta\tau)^2}{12}
\end{aligned}
\tag{139}
$$

Assuming that the Stokes and laser modes are initially coherent, Eqs.

(138) and (139) reduce respectively to

$$\gamma_{LS}^{(2)}(\tau) = -\Delta\tau + \left(|\alpha_S|^2 - 2|\alpha_S|^2|\alpha_L|^2 + |\alpha_L|^2\right)|\alpha_S|^{-2}\frac{(\Delta\tau)^2}{2} \quad (140)$$

$$\gamma_{LS}^{(2)}(\tau) = -\frac{\Delta\tau}{2} + \left(1 - 13|\alpha_L|^2 - 8|\alpha_L|^4\right)\frac{(\Delta\tau)^2}{12} \quad (141)$$

Equation (137), calculated up to the third order in $\Delta\tau$, enables us to determine the relation (139) correct up to $\Delta\tau$ squared only. Simaan[75] has calculated an expression similar to Eq. (139). Examples of the time evolution of $g_{LS}^{(2)}(\tau)$ for initially coherent fields are presented in Fig. 6. Curves B in Figs. 6a and b are calculated with Eqs. (139) and (138) (including terms up to $\Delta\tau$ only). Curve S in Fig. 6a is calculated from the Simaan short-time approximate solution (32) of Ref. 75. One can compare these results (curves B and S) with $g_{LS}^{(2)}(\tau)$ obtained from our numerical calculations utilizing the exact solution of the master equation (123) (curves A). We note the supremacy of our short-time approximation (141).

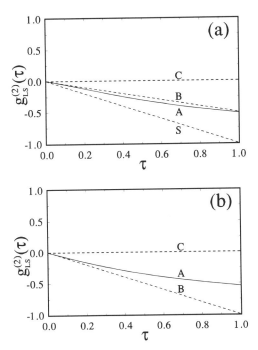

Figure 6. Same as Fig. 4, but for the interbeam degree of coherence $g_{LS}^{(2)}(\tau)$. Additional curve S is calculated with the Simaan short-time approximation (Eq. (32) of Ref. 75).

By analogy to the photon-number moments we calculate, in the short-time approximation, the mean and mean square of the annihilation operators, $\langle \hat{a}_k^+(\tau) \rangle$ and $\langle \hat{a}_k^{+2}(\tau) \rangle$ for both fields ($k = L, S$), as well as the cross-correlation functions $\langle \hat{a}_L^+(\tau)\hat{a}_S^+(\tau) \rangle$ and $\langle \hat{a}_L(\tau)\hat{a}_S^+(\tau) \rangle$. After some algebra, we arrive at

$$\langle \hat{a}_S^+(\tau) \rangle = \langle \hat{a}_S^+ \rangle + \langle \hat{a}_L^+\hat{a}_L \rangle \langle \hat{a}_S^+ \rangle \frac{\Delta\tau}{2}$$
$$+ \left(\langle \hat{a}_L^{+2}\hat{a}_L^2 \rangle \langle \hat{a}_S^+ \rangle - 2\langle \hat{a}_L^+\hat{a}_L \rangle \langle \hat{a}_S^{+2}\hat{a}_S \rangle - 3\langle \hat{a}_L^+\hat{a}_L \rangle \langle \hat{a}_S^+ \rangle \right) \frac{(\Delta\tau)^2}{8}$$
$$(142)$$

$$\langle \hat{a}_L^+(\tau) \rangle = \langle \hat{a}_L^+ \rangle - \langle \hat{a}_L^+ \rangle \left(\langle \hat{a}_S^+\hat{a}_S \rangle + 1 \right) \frac{\Delta\tau}{2} + \left[\langle \hat{a}_L^+ \rangle \langle \hat{a}_S^{+2}\hat{a}_S^2 \rangle \right.$$
$$\left. - 2\langle \hat{a}_L^{+2}\hat{a}_L \rangle \left(\langle \hat{a}_S^+\hat{a}_S \rangle + 1 \right) + \langle \hat{a}_L^+ \rangle \left(3\langle \hat{a}_S^+\hat{a}_S \rangle + 1 \right) \right] \frac{(\Delta\tau)^2}{8}$$
$$(143)$$

$$\langle \hat{a}_S^{+2}(\tau) \rangle = \langle \hat{a}_S^{+2} \rangle + \langle \hat{a}_L^+\hat{a}_L \rangle \langle \hat{a}_S^{+2} \rangle \Delta\tau$$
$$+ \left(\langle \hat{a}_L^{+2}\hat{a}_L^2 \rangle \langle \hat{a}_S^{+2} \rangle - \langle \hat{a}_L^+\hat{a}_L \rangle \langle \hat{a}_S^{+3}\hat{a}_S \rangle - 2\langle \hat{a}_L^+\hat{a}_L \rangle \langle \hat{a}_S^{+2} \rangle \right) \frac{(\Delta\tau)^2}{2}$$
$$(144)$$

$$\langle \hat{a}_L^{+2}(\tau) \rangle = \langle \hat{a}_L^{+2} \rangle - \langle \hat{a}_L^{+2} \rangle \left(\langle \hat{a}_S^+\hat{a}_S \rangle + 1 \right) \Delta\tau + \left[\langle \hat{a}_L^{+2} \rangle \langle \hat{a}_S^{+2}\hat{a}_S^2 \rangle \right.$$
$$\left. - \langle \hat{a}_L^{+3}\hat{a}_L \rangle \left(\langle \hat{a}_S^+\hat{a}_S \rangle + 1 \right) + \langle \hat{a}_L^{+2} \rangle \left(3\langle \hat{a}_S^+\hat{a}_S \rangle + 1 \right) \right] \frac{(\Delta\tau)^2}{2} \quad (145)$$

$$\langle \hat{a}_L^+(\tau)\hat{a}_S^+(\tau) \rangle$$
$$= \langle \hat{a}_L^+ \rangle \langle \hat{a}_S^+ \rangle + \left(\langle \hat{a}_L^{+2}\hat{a}_L \rangle \langle \hat{a}_S^+ \rangle - \langle \hat{a}_L^+ \rangle \langle \hat{a}_S^{+2}\hat{a}_S \rangle - 2\langle \hat{a}_L^+ \rangle \langle \hat{a}_S^+ \rangle \right) \frac{\Delta\tau}{2}$$
$$+ \left(\langle \hat{a}_L^{+3}\hat{a}_L^2 \rangle \langle \hat{a}_S^+ \rangle - 11\langle \hat{a}_L^{+2}\hat{a}_L \rangle \langle \hat{a}_S^+ \rangle - 6\langle \hat{a}_L^{+2}\hat{a}_L \rangle \langle \hat{a}_S^{+2}\hat{a}_S \rangle \right.$$
$$\left. + 5\langle \hat{a}_L^+ \rangle \langle \hat{a}_S^{+2}\hat{a}_S \rangle + \langle \hat{a}_L^+ \rangle \langle \hat{a}_S^{+3}\hat{a}_S^2 \rangle + 4\langle \hat{a}_L^+ \rangle \langle \hat{a}_S^+ \rangle \right) \frac{(\Delta\tau)^2}{8} \quad (146)$$

$$\langle \hat{a}_L(\tau)\hat{a}_S^+(\tau) \rangle$$
$$= \langle \hat{a}_L \rangle \langle \hat{a}_S^+ \rangle + \left(\langle \hat{a}_L^+\hat{a}_L^2 \rangle \langle \hat{a}_S^+ \rangle - \langle \hat{a}_L \rangle \langle \hat{a}_S^{+2}\hat{a}_S \rangle - 2\langle \hat{a}_L \rangle \langle \hat{a}_S^+ \rangle \right) \frac{\Delta\tau}{2}$$
$$+ \left(\langle \hat{a}_L^{+2}\hat{a}_L^3 \rangle \langle \hat{a}_S^+ \rangle - 11\langle \hat{a}_L^+\hat{a}_L^2 \rangle \langle \hat{a}_S^+ \rangle - 6\langle \hat{a}_L^+\hat{a}_L^2 \rangle \langle \hat{a}_S^{+2}\hat{a}_S \rangle \right.$$
$$\left. + 9\langle \hat{a}_L \rangle \langle \hat{a}_S^{+2}\hat{a}_S \rangle + \langle \hat{a}_L \rangle \langle \hat{a}_S^{+3}\hat{a}_S^2 \rangle + 12\langle \hat{a}_L \rangle \langle \hat{a}_S^+ \rangle \right) \frac{(\Delta\tau)^2}{8} \quad (147)$$

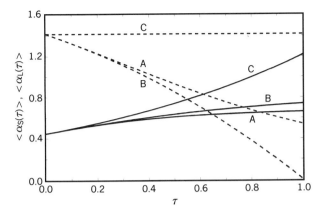

Figure 7. Time dependence of the expectation values of the field amplitudes $\langle \hat{a}_S(\tau) \rangle = \langle \alpha_S(\tau) \rangle$ (solid lines) and $\langle \hat{a}_L(\tau) \rangle = \langle \alpha_L(\tau) \rangle$ (dashed lines) for fields initially coherent $|\alpha_L = \sqrt{2} \rangle$ and $|\alpha_S = \sqrt{0.2} \rangle$. Curves A, B, C are calculated within the formalisms of Sections VI.A.2, VI.A.1, and V.B, respectively.

For brevity, here, we shall restrict our considerations to initially coherent states for the Stokes mode denoted as $\alpha_S = |\alpha_S| \exp(i\phi_S)$ and for the laser mode $\alpha_L = |\alpha_L| \exp(i\phi_L)$. In Figs. 7 and 8 we demonstrate the evolution of our short-time approximations for $\langle \hat{a}_S(\tau) \rangle$ (solid line B in Fig. 7), $\langle \hat{a}_L(\tau) \rangle$ (dashed line B in Fig. 7), $\langle \hat{a}_S^2(\tau) \rangle$ (solid line B in Fig. 8), and $\langle \hat{a}_L^2(\tau) \rangle$ (dashed line B in Fig. 8) for initially coherent radiation modes.

From the general relations (130) and (126), under the condition of initially coherent Stokes and laser fields, we get

$$\langle \hat{m}(\tau) \rangle = |\alpha_S|^2 + |\alpha_L|^2 \left(|\alpha_S|^2 + 1 \right) \Delta\tau + |\alpha_L|^2 \left[\left(|\alpha_L|^2 + 1 \right) \left(|\alpha_S|^2 + 1 \right) \right.$$
$$\left. - \left(|\alpha_S|^4 + 4|\alpha_S|^2 + 2 \right) \right] \frac{(\Delta\tau)^2}{2} \tag{148}$$

$$\langle \hat{n}(\tau) \rangle = |\alpha_L|^2 + |\alpha_S|^2 - \langle \hat{m}(\tau) \rangle \tag{149}$$

Inserting (148) as well as (142) and (144) with $\langle \hat{a}_k^{+p} \hat{a}_k^q \rangle = |\alpha_k|^{p+q} \exp[-i(p-q)\phi_k]$ $(k = S, A)$, into (43) we obtain the θ-dependent variance for the Stokes mode:

$$\left\langle \left(\Delta \hat{X}_S(\theta) \right)^2 \right\rangle = 1 + 2|\alpha_L|^2 \Delta\tau$$
$$+ |\alpha_L|^2 \left\{ |\alpha_L|^2 - \left[1 + \cos^2(\theta - \phi_S) \right] |\alpha_S|^2 - 1 \right\} (\Delta\tau)^2. \tag{150}$$

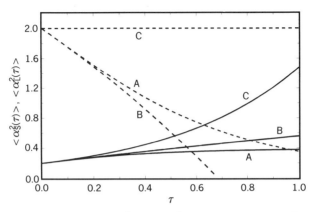

Figure 8. Time dependence of $\langle \hat{a}_S^2(\tau) \rangle = \langle \alpha_S^2(\tau) \rangle$ (solid lines) and $\langle \hat{a}_L^2(\tau) \rangle = \langle \alpha_L^2(\tau) \rangle$ (dashed lines) for the same cases as in Fig. 7.

The minimal variance $\langle (\Delta \hat{X}_{S-})^2 \rangle$, which follows from (46) or directly from (150), is equal to

$$\left\langle \left(\Delta \hat{X}_{S-} \right)^2 \right\rangle = 1 + 2|\alpha_L|^2 \Delta\tau + |\alpha_L|^2 \left(|\alpha_L|^2 - 2|\alpha_S|^2 - 1 \right) (\Delta\tau)^2 \quad (151)$$

Analogously, for the laser field we obtain the following θ-dependent variance:

$$\left\langle \left(\Delta \hat{X}_L(\theta) \right)^2 \right\rangle = 1 + \tfrac{1}{2} \left[\cos(2\theta - 2\phi_L) + 1 \right] |\alpha_L|^2 |\alpha_S|^2 (\Delta\tau)^2 \quad (152)$$

on insertion of Eqs. (149), (143), and (145) into (43). With regard to the relation (46), the minimal variance for the field, $\langle (\Delta \hat{X}_{L-})^2 \rangle$, is constant up to the second order in time:

$$\left\langle \left(\Delta \hat{X}_{L-} \right)^2 \right\rangle = 1 \quad (153)$$

One can readily deduce the maximal variances $\langle (\Delta \hat{X}_{L,S+})^2 \rangle$ from (150) and (152) or from (46). The time evolution of the single-mode extremal variances obtained from (150)–(153) is presented in Figs. 9 and 10 (for $\phi_L = 0$): $\langle (\Delta \hat{X}_{S-}(\tau))^2 \rangle = \langle (\Delta \hat{X}_{S2}(\tau))^2 \rangle$ (solid line B in Fig. 9), $\langle (\Delta \hat{X}_{S+}(\tau))^2 \rangle = \langle (\Delta \hat{X}_{S1}(\tau))^2 \rangle$ (dashed line B in Fig. 9), $\langle (\Delta \hat{X}_{L-}(\tau))^2 \rangle = \langle (\Delta \hat{X}_{L1}(\tau))^2 \rangle$ (solid line B in Fig. 10), $\langle (\Delta \hat{X}_{L+}(\tau))^2 \rangle = \langle (\Delta \hat{X}_{L2}(\tau))^2 \rangle$ (dashed line B in Fig. 10).

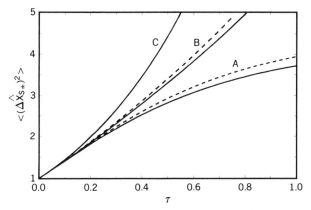

Figure 9. Time dependence of the extremal variances $\langle(\Delta \hat{X}_{S-})^2\rangle$ (solid lines) and $\langle(\Delta \hat{X}_{S+})^2\rangle$ (dashed lines) for the same cases as in Fig. 7.

The covariances for quadratures in the Stokes and laser mode, according to (52), are, respectively,

$$\left\langle\left\{\Delta \hat{X}_{S1}, \Delta \hat{X}_{S2}\right\}\right\rangle = |\alpha_L|^2|\alpha_S|^2 \sin(2\phi_S)(\Delta\tau)^2 \tag{154}$$

$$\left\langle\left\{\Delta \hat{X}_{L1}, \Delta \hat{X}_{L2}\right\}\right\rangle = -|\alpha_L|^2|\alpha_S|^2 \sin(2\phi_L)(\Delta\tau)^2 \tag{155}$$

The generalized Heisenberg uncertainty relation (51) with the covariances (154) and (155) inserted takes the following form for the Stokes mode in

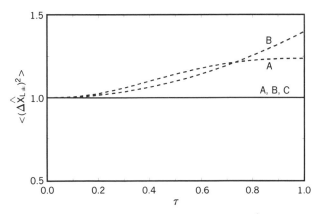

Figure 10. Time dependence of the extremal variances $\langle(\Delta \hat{X}_{L-})^2\rangle$ (solid lines) and $\langle(\Delta \hat{X}_{L+})^2\rangle$ (dashed lines) for the same cases as in Fig. 7.

our short-time approximation:

$$4|\alpha_L|^2 \Delta\tau + |\alpha_L|^2(6|\alpha_L|^2 - 3|\alpha_S|^2 - 2)(\Delta\tau)^2 \geq 0 \qquad (156)$$

and for the laser mode

$$2|\alpha_L|^2|\alpha_S|^2(\Delta\tau)^2 \geq 0 \qquad (157)$$

To obtain the two-mode variances and covariances of the quadratures one has to calculate, apart from the single-mode functions (150), (152), (154), and (155), the cross-correlations (57), which are obtained in the following form:

$$\left\langle \Delta\hat{X}_{L1}\,\Delta\hat{X}_{S1} \right\rangle$$

$$= -|\alpha_L|\,|\alpha_S|\left\{ 2\cos\phi_L \cos\phi_S\,\Delta\tau + \left[\cos(\phi_L - \phi_S)(4n_L - 6n_S - 11)\right.\right.$$

$$\left.\left. +\cos(\phi_L + \phi_S)(4|\alpha_L|^2 - 2|\alpha_S|^2 - 3)\right]\frac{(\Delta\tau)^2}{4}\right\}$$

$$(158)$$

$$\left\langle \Delta\hat{X}_{L2}\,\Delta\hat{X}_{S2} \right\rangle$$

$$= -|\alpha_L|\,|\alpha_S|\left\{ 2\sin\phi_L \sin\phi_S\,\Delta\tau + \left[\cos(\phi_L - \phi_S)(4n_L - 6n_S - 11)\right.\right.$$

$$\left.\left. -\cos(\phi_L + \phi_S)(4|\alpha_L|^2 - 2|\alpha_S|^2 - 3)\right]\frac{(\Delta\tau)^2}{4}\right\}$$

$$(159)$$

$$\left\langle \Delta\hat{X}_{L1}\,\Delta\hat{X}_{S2} \right\rangle$$

$$= |\alpha_L|\,|\alpha_S|\left\{ 2\cos\phi_L \sin\phi_S\,\Delta\tau + \left[\sin(\phi_S - \phi_L)(4n_L - 6n_S - 11)\right.\right.$$

$$\left.\left. +\sin(\phi_S + \phi_L)(4|\alpha_L|^2 - 2|\alpha_S|^2 - 3)\right]\frac{(\Delta\tau)^2}{4}\right\} \qquad (160)$$

$$\left\langle \Delta\hat{X}_{L2}\,\Delta\hat{X}_{S1} \right\rangle$$

$$= |\alpha_L|\,|\alpha_S|\left\{ 2\sin\phi_L \cos\phi_S\,\Delta\tau + \left[\sin(\phi_L - \phi_S)(4n_L - 6n_S - 11)\right.\right.$$

$$\left.\left. +\sin(\phi_L + \phi_S)(4|\alpha_L|^2 - 2|\alpha_S|^2 - 3)\right]\frac{(\Delta\tau)^2}{4}\right\} \qquad (161)$$

Thus, the two-mode Wigner covariance (56) of the quadratures $\hat{X}_{LS1}$ and $\hat{X}_{LS2}$ is

$$\left\langle\left\{\Delta\hat{X}_{LS1},\Delta\hat{X}_{LS2}\right\}\right\rangle = 4|\alpha_L|\,|\alpha_S|\sin(\phi_L+\phi_S)\,\Delta\tau$$
$$+\left[4|\alpha_L|^2|\alpha_S|^2\cos\tfrac{1}{2}(\phi_S+\phi_L)\sin\tfrac{1}{2}(\phi_S-\phi_L)\right.$$
$$+|\alpha_L|\,|\alpha_S|\sin(\phi_L+\phi_S)$$
$$\left.\times\left(4|\alpha_L|^2-2|\alpha_S|^2-3\right)\right](\Delta\tau)^2 \qquad (162)$$

The two-mode variances (55) of $\hat{X}_{LS1}$ and $\hat{X}_{LS2}$ are

$$\left\langle\left(\Delta\hat{X}_{LS1,2}\right)^2\right\rangle$$
$$= 2 + 2\left\{|\alpha_L|^2-|\alpha_L|\,\alpha_S|\left[\cos(\phi_L-\phi_S)\pm\cos(\phi_L+\phi_S)\right]\right\}\Delta\tau$$
$$+\left\{2|\alpha_L|^2\left(|\alpha_L|^2-|\alpha_S|^2-1\right)\pm|\alpha_L|^2|\alpha_S|^2(\cos 2\phi_L-\cos 2\phi_S)\right.$$
$$-|\alpha_S|\,|\alpha_L|\left[\cos(\phi_L-\phi_S)\left(4|\alpha_L|^2-6|\alpha_S|^2-11\right)\right. \qquad (163)$$
$$\left.\left.\pm\cos(\phi_L+\phi_S)\left(4|\alpha_L|^2-2|\alpha_S|^2-3\right)\right]\right\}\frac{(\Delta\tau)^2}{2}$$

whereas the extremal variances are

$$\left\langle\left(\Delta\hat{X}_{LS\pm}\right)^2\right\rangle = 2 + 2\left[|\alpha_L|^2-|\alpha_L|\,|\alpha_S|\cos(\phi_L-\phi_S)\right]\Delta\tau$$
$$+\left[|\alpha_L|^2\left(|\alpha_L|^2-|\alpha_S|^2-1\right)\right.$$
$$\left.-|\alpha_L|\,|\alpha_S|\cos(\phi_L-\phi_S)\left(2|\alpha_L|^2-3|\alpha_S|^2-\tfrac{11}{2}\right)\right](\Delta\tau)^2$$
$$\pm\left|2\alpha_L\alpha_S\,\Delta\tau+\left[|\alpha_L|^2\alpha_S^2-|\alpha_S|^2\alpha_L^2\right.\right. \qquad (164)$$
$$\left.\left.+\alpha_L\alpha_S\left(4|\alpha_L|^2-2|\alpha_S|^2-3\right)\right]\frac{(\Delta\tau)^2}{2}\right|$$

according to the general expression (58).

Equations (158)–(164) can be readily generalized to any initial distribution of the radiation fields.

2. Exact Solutions

Let us now proceed to the exact solution of the master equation (125). We apply the Laplace transform method. The method is readily applicable to nonlinear master equations for a variety of nonlinear optical phenomena (Refs. 30, 73, and 220 and references therein); in particular, it has been applied successfully to different multiphoton Raman processes in Refs. 73, 75, 78, 106, 107, 112, 115, and 116. The solution of (125) for diagonal terms of the density matrix $\rho_{nm}(00\tau)$ (i.e., for $\nu = \mu = 0$) was derived by McNeil and Walls[73] and then, in a more general form, by Simaan.[75] As the chief result of the present work we derive the time-dependence of the complete density matrix $\rho_{nm}(\nu\mu\tau)$, where the degrees of off-diagonality ν, μ are arbitrary. To the best of our knowledge, ours is the first derivation of a complete analytical solution to the Raman scattering model including depletion of the pump field.

As usual, we assume that the Stokes and laser beams are mutually independent at the initial time $\tau = \tau_0$. Thus, the initial joint distribution $\rho_{nm}(\nu\mu\tau_0)$ is a product of the distributions for the separate beams,

$$\rho_{nm}(\nu\mu\tau_0) = \rho_n^L(\nu\tau_0)\rho_m^S(\mu\tau_0) \tag{165}$$

Let us define the coefficient λ in terms of the integer-value function $[[x]]$ (the maximum integer $\leq x$):

$$\lambda = \left[\left[\frac{m - n + 1}{2} + \frac{\mu - \nu}{4}\right]\right] \tag{166}$$

The exact solution of (125) under the condition (165), derived in Appendix B, reads as follows for $\lambda < 0$:

$$\rho_{nm}(\nu\mu\tau) = \left[\frac{m!(m + \mu)!}{n!(n + \nu)!}\right]^{1/2} \sum_{l=0}^{m} \rho_{n+l}^L(\nu\tau_0)\rho_{m-l}^S(\mu\tau_0)$$

$$\times \left[\frac{(n + l)!(n + l + \nu)!}{(m - l)!(m - l + \mu)!}\right]^{1/2} \tag{167}$$

$$\times \sum_{q=0}^{l} \exp[-f(q)\,\Delta\tau] \prod_{\substack{p=0 \\ p \neq q}}^{l} [f(p) - f(q)]^{-1}$$

whereas for $\lambda \geq 0$ it is

$$\rho_{nm}(\nu\mu\tau) = \left[\frac{m!(m+\mu)!}{n!(n+\nu)!}\right]^{1/2}$$

$$\times \Bigg\{ \sum_{l=0}^{\lambda} \rho_{n+l}^{L}(\nu\tau_0)\rho_{m-l}^{S}(\mu\tau_0)$$

$$\times \left[\frac{(n+l)!(n+l+\nu)!}{(m-l)!(m-l+\mu)!}\right]^{1/2}$$

$$\times \sum_{q=0}^{l} \exp[-f(q)\,\Delta\tau] \prod_{\substack{p=0\\p\neq q}}^{l} [f(p)-f(q)]^{-1}$$

$$+(1-\delta_{m0}) \sum_{l=\lambda+1}^{m} \rho_{n+l}^{L}(\nu\tau_0)\rho_{m-l}^{S}(\mu\tau_0) \qquad (168)$$

$$\times \left[\frac{(n+l)!(n+l+\nu)!}{(m-l)!(m-l+\mu)!}\right]^{1/2}$$

$$\times \sum_{q=0}^{\lambda} \sum_{q'=\lambda+1}^{l} \prod_{\substack{p=0\\p\neq q}}^{\lambda} [f(p)-f(q)]^{-1} \prod_{\substack{p'=\lambda+1\\p'\neq q'}}^{l} [f(p')-f(q')]^{-1}$$

$$\times \Bigg(\delta_{f(q)f(q')}\,\Delta\tau\,\exp[-f(q)\,\Delta\tau]$$

$$+(\delta_{f(q)f(q')}-1)\frac{\exp[-f(q)\,\Delta\tau]-\exp[-f(q')\,\Delta\tau]}{f(q)-f(q')}\Bigg) \Bigg\}$$

where the function $f(x)$ is given by

$$f(x) = \tfrac{1}{2}[(n+x)(m-x+1)+(n+x+\nu)(m-x+\mu+1)] \quad (169)$$

The coefficient λ can be alternatively defined as

$$\lambda = \left[\left[\frac{m-n}{2}+\frac{\mu-\nu}{4}\right]\right] \qquad (170)$$

The solution (167) and (168) with consequent application of the coefficient λ of Eq. (170) reduces, as it should, to the Simaan solution (45) and (49) of Ref. 75 for the diagonal matrix elements ($\nu = \mu = 0$); McNeil and Walls have also obtained a solution of (125) for the diagonal matrix elements (Eqs. (6.5), (6.6), and (4.14) in Ref. 73); however, their solution is not in full agreement with Simaan's solution and is not a special case of ours for reasons given by Simaan.[75]

The solution (167) and (168) is very well adapted to numerical analysis; nonetheless, it is of a rather complicated form. Our solution of (125) can be rewritten more compactly. Following the method of Malakyan,[116] we find (for details, see Appendix B)

$$
\rho_{nm}(\nu\mu\tau)
$$

$$
= \left[\frac{m!(m+\mu)!}{n!(n+\nu)!}\right]^{1/2} \sum_{l=0}^{m} \rho_{n+l}^{L}(\nu\tau_0)\rho_{m-l}^{S}(\mu\tau_0)
$$

$$
\times \left[\frac{(n+l)!(n+l+\nu)!}{(m-l)!(m-l+\mu)!}\right]^{1/2} \tag{171}
$$

$$
\times \hat{\mathscr{D}} \sum_{\substack{q=0 \\ q \neq q_1', q_2', \ldots, q_d'}}^{l} \exp[-f(q)\,\Delta\tau] \prod_{\substack{p=0 \\ p \neq q, q_1', q_2', \ldots, q_d'}}^{l} [f(p)-f(q)]^{-1}
$$

The differential operator of the dth order, $\hat{\mathscr{D}}$, is defined as follows:

$$
\hat{\mathscr{D}} = (-1)^d \prod_{r=1}^{d} \frac{\partial}{\partial f(q_r)} \tag{172}
$$

The order d of the differential operator (172) is equal to the number of pairs of mutually equal factors occurring in the product of Eq. (171), $f(q_1)=f(q_1'), f(q_2)=f(q_2'),\ldots,f(q_d)=f(q_d')$. If there are no pairs of equal factors, then the operator $\hat{\mathscr{D}}$ is defined to be unity (see Appendix B). The solutions (167), (168), and (171) represent the chief result of our paper. In Section V.B in the Raman effect model under the parametric approximation we have analyzed, in particular, the single-mode solutions for either the Stokes mode or for the anti-Stokes mode. For completeness, we give in Appendix C the solution for anti-Stokes scattering without the parametric approximation. The degree of off-diagonality μ is assumed to be nonnegative (contrary to ν); nonetheless, the time dependence of the complete density matrix $\rho_{nm}(\nu\mu\tau)$ is determined by the simple relation for the inverse matrix elements:

$$
\rho_{nm}^*(\nu\mu\tau) = \rho_{n+\nu, m+\mu}(-\nu, -\mu, \tau) \tag{173}
$$

Thus, solutions (167), (168), and/or (171) provide an entire specification for all measurable properties of the light field under consideration.

The two-mode (joint) density matrix with elements $\rho_{nm}(\nu\mu\tau)$, (167) and (168), enables the calculation of the single-mode (separate) density matrix with elements $\rho_m^S(\mu\tau)$ and $\rho_m^L(\mu\tau)$. The Stokes mode matrix elements $\rho_m^S(\mu\tau)$ can be calculated from

$$\rho_m^S(\mu\tau) = \sum_{n=0}^{\infty} \sum_{\nu=-n}^{\infty} \rho_{nm}(\nu\mu\tau) \tag{174}$$

and the laser mode matrix elements $\rho_m^L(\mu\tau)$ can be found analogously, with the exception that for terms $\rho_{nm}(\nu\mu\tau)$ with $\mu < 0$ the property (173) must be used. The already mentioned solution of McNeil and Walls[73] corresponds to the separate diagonal density matrix (174).

There is yet another manner of expressing the two-mode solutions of the master equation (125), $\rho_{nm}(00\tau)$, for any initial distributions, via the density matrix elements for the initial number states in the Stokes and laser fields:

$$\rho_{nm}(00\tau) = \sum_{n_0=0}^{\infty} \sum_{m_0=0}^{\infty} \rho_{nm}^{(n_0, m_0)}(00\tau)\rho_{n_0}^L(0\tau_0)\rho_{m_0}^S(0\tau_0) \tag{175}$$

where $\rho_{nm}^{(n_0, m_0)}(00\tau)$ is the solution (167) and (168) for $\rho_{nm}(00\tau)$, under the initial conditions that the laser field is in the number state $|n_0\rangle$ and the Stokes mode is in the number state $|m_0\rangle$. The weighting functions in (175) are arbitrary initial distributions of the laser, $\rho_n^L(0\tau_0)$, and the Stokes field, $\rho_n^S(0\tau_0)$. Here, for brevity, we restrict our considerations to diagonal terms (with $\nu = \mu = 0$). Otherwise, instead of $\rho_{nm}^{(n_0, m_0)}(00\tau)$ we would have to use the solution $\rho_{nm}^{(n_0, m_0, \nu_0, \mu_0)}(\nu\mu\tau)$ and perform two extra summations in Eq. (175) over ν, μ. McNeil and Walls[73] have presented their solution of Eq. (125) in this manner. Analogously, we can express the single-mode distributions $\rho_n^L(0\tau)$ ($\rho_m^S(0\tau)$) for arbitrary initial states using the solutions $\rho_n^{(m_0)}(\tau)$ ($\rho_m^{(n_0)}(\tau)$) for the initial photon-number states $|m_0\rangle\langle|n_0\rangle)$. For instance, for the Stokes mode solution we apply the formula

$$\rho_m^S(0\tau) = \sum_{m_0=0}^{\infty} \rho_n^{(n_0)}(\tau)\rho_{n_0}^L(0\tau_0) \tag{176}$$

We shall make use of this procedure for the diagonal approximate solutions (194).

Having the solutions (167), (168), or (171) available we can, at least numerically, analyze, e.g., the single- and two-mode photocount statistics and quadrature squeezing. The expectation values in the relations describing squeezing and photocount statistics (see Section IV) are readily expressed in terms of the density matrix elements $\rho_{nm}(\nu\mu\tau)$ by way of

$$\langle \hat{n}^k(t) \rangle = \sum_{n,m=0}^{\infty} n^k \rho_{n,m}(0,0,t) \tag{177}$$

$$\langle \hat{m}^k(t) \rangle = \sum_{n,m} m^k \rho_{n,m}(0,0,t) \tag{178}$$

$$\langle \hat{a}_L^{+k}(t) \rangle = \sum_{n,m} \left[\frac{(n+k)!}{n!} \right]^{1/2} \rho_{n,m}(k,0,t) \tag{179}$$

$$\langle \hat{a}_S^{+k}(t) \rangle = \sum_{n,m} \left[\frac{(m+k)!}{m!} \right]^{1/2} \rho_{n,m}(0,k,t) \tag{180}$$

$$\langle \hat{a}_L^+(t)\hat{a}_S^+(t) \rangle = \sum_{n,m} [(n+1)(m+1)]^{1/2} \rho_{n,m}(1,1,t) \tag{181}$$

$$\langle \hat{a}_L^+(t)\hat{a}_S(t) \rangle = \sum_{n,m} [(n+1)m]^{1/2} \rho_{n,m}(1,-1,t) \tag{182}$$

In Section III we defined the s-parametrized quasiprobability distribution $\mathscr{W}^{(s)}(\{\alpha_k\})$ and the s-parametrized characteristic function $\mathscr{C}^{(s)}(\{\beta_k\})$ and adduced relations between them for any parameter s. Here, we deal with matrix elements in Fock basis of the density operator, $\rho_{nm}(\nu\mu\tau)$. To achieve consistency between our analysis of the Raman effect presented in this section with the analysis of section V, we shall present some relations between the functions $\mathscr{W}^{(s)}(\{\alpha_k\})$ or $\mathscr{C}^{(s)}(\{\beta_k\})$ and the density operator $\hat{\rho}(\{\hat{a}_k\})$. We restrict the general formulas for the M-mode fields to our two-mode situations, so that $(\{\alpha_k\}) = (\alpha_L, \alpha_{S,A})$. These formulas are in complete analogy with the results of Cahill and Glauber[197] for the single-mode case. By virtue of the operator $\hat{T}(\alpha_L, \alpha_{S,A})$, which is the Fourier transform of the s-parametrized displacement operator $\hat{D}^{(s)}(\beta_L, \beta_{S,A})$ (see Eq. (19)),

$$\hat{T}^{(s)}(\alpha_L, \alpha_{S,A}) = \frac{1}{\pi^2} \int \hat{D}^{(s)}(\beta_L, \beta_{S,A})$$
$$\times \exp(\alpha_L \beta_L^* + \alpha_{S,A} \beta_{S,A}^* - \text{c.c.}) d^2\beta_L \, d^2\beta_{S,A} \tag{183}$$

the density matrix $\hat{\rho}(\hat{a}_L, \hat{a}_{S,A})$ can be obtained from the s-parametrized quasidistribution $\mathscr{W}^{(s)}(\alpha_L, \alpha_{S,A})$, (21),

$$\hat{\rho}(\hat{a}_L, \hat{a}_{S,A}) = \frac{1}{\pi^2} \int \mathscr{W}^{(s)}(\alpha_L, \alpha_{S,A}) \tag{184}$$
$$\times \hat{T}^{(-s)}(\alpha_L, \alpha_{S,A}) d^2\alpha_L \, d^2\alpha_{S,A}$$

The inverse relation,

$$\mathscr{W}^{(s)}(\alpha_L, \alpha_{S,A}) = \text{Tr}\left[\hat{\rho}(\hat{a}_L, \hat{a}_{S,A})\hat{T}^{(s)}(\alpha_L, \alpha_{S,A})\right] \tag{185}$$

resembles expression (20) for the characteristic function $\mathscr{C}^{(s)}(\beta_L, \beta_{S,A})$, which is the average value of the displacement operator $\hat{D}^{(s)}(\alpha_L, \alpha_{S,A})$. We are interested in the relations for the matrix elements of $\hat{\rho}(\hat{a}_L, \hat{a}_{S,A})$. They immediately follow from (184) and (185):

$$\rho_{n,m}(\nu, \mu) = \frac{1}{\pi^2} \int \mathscr{W}^{(s)}(\alpha_L, \alpha_{S,A}) \tag{186}$$
$$\times \langle n, m | \hat{T}^{(-s)}(\alpha_L, \alpha_{S,A}) | n + \nu, m + \mu \rangle d^2\alpha_L \, d^2\alpha_{S,A}$$

$$\mathscr{W}^{(s)}(\alpha_L, \alpha_{S,A}) = \sum_{n=0}^{\infty} \sum_{m=0}^{\infty} \sum_{\nu=-n}^{\infty} \sum_{\mu=-m}^{\infty} \rho_{n,m}^*(\nu, \mu) \tag{187}$$
$$\times \langle n, m | \hat{T}^{(s)}(\alpha_L, \alpha_{S,A}) | n + \nu, m + \mu \rangle$$

The Fock matrix elements for the two-mode field,

$$\langle n, m | \hat{T}^{(s)}(\alpha_L, \alpha_{S,A}) | n + \nu, m + \mu \rangle = \langle n | \hat{T}^{(s)}(\alpha_L) | n + \nu \rangle \tag{188}$$
$$\langle m | \hat{T}^{(s)}(\alpha_{S,A}) | m + \mu \rangle$$

are simply products of the two single-mode Fock matrix elements given by Cahill and Glauber[197]:

$$\langle n | \hat{T}^{(s)}(\alpha_L) | n + \nu \rangle = \sqrt{\frac{n!}{(n+\nu)!}} \left(\frac{2}{1-s}\right)^{\nu+1} \left(\frac{s+1}{s-1}\right)^n$$
$$\times \exp\left(-\frac{2}{1-s}|\alpha_L|^2\right) L_n^{(\nu)}\left(\frac{4|\alpha_L|^2}{1-s^2}\right)(\alpha_L^*)^\nu \tag{189}$$

where $L_n^{(\nu)}(x)$ is the generalized Laguerre polynomial. The above equations show equivalency of the two apparently different formalisms we have been dealing with: On the one side, the s-parametrized quasiprobability distribution functions $\mathscr{W}^{(s)}$ obtained within the Fokker-Planck equation formalism presented in Section V, and, on the other side, the density matrix operator formalism under discussion in this section.

Solutions (167), (168), or (171) reduce to rather simple expressions in special cases, for instance, on the one side, for long periods of time when the laser beam is totally depleted, and on the other, for an intense laser beam almost unaffected (undepleted) during the process of scattering. We now discuss these two cases.

3. Long-time Solutions

After a sufficiently long time, the system settles down to a steady state as a result of the total depletion of the laser pump. The steady-state solutions can be readily deduced from (167) and (168). Indeed, in the time limit $(\tau \to \infty)$, the nonzero matrix elements $\rho_{nm}(\nu\mu\infty)$ must satisfy the condition for the function $f(x)$ (169) that $f(q) = 0$, which implies that $q = 0$. Hence, we have

$$\rho_{nm}(\nu\mu, \tau = \infty) = 0 \qquad \text{for } n, \nu \neq 0 \tag{190}$$

for arbitrary $m, m + \mu$ ranging from zero to infinity. All photon-number and annihilation operator moments for the laser beam vanish in the time limit

$$\begin{aligned}
\langle \hat{n}^p(\infty) \rangle &= 0 \\
\langle \hat{a}_L^p(\infty) \rangle &= 0 \qquad \text{for } p > 0
\end{aligned} \tag{191}$$

which reflects the fact that the laser beam is totally depleted. The normalization condition takes the form

$$\sum_{m=0}^{\infty} \rho_{0m}(00\infty) = 1 \tag{192}$$

In the model of hyper-Raman scattering, as was shown by Malakyan,[116] there intervene in the limit $\tau \to \infty$ the nonzero density matrix elements $\rho_{0m}(0\mu\infty)$, $\rho_{1m}(0\mu\infty)$, $\rho_{0m}(1\mu\infty)$, and $\rho_{1m}(-1, \mu\infty)$. Hence, the photon-number moments and annihilation operator moments of the laser mode do not vanish, contrary to the model of Raman scattering under consideration in view of (191). If we assume that initially there are no photons in the Stokes mode, then $\rho_n^L(0\tau_0) = \rho_n^S(0\infty)$ (Ref. 75), which implies that an arbitrary photon-number moment $\langle \hat{m}^p(\infty) \rangle$ (with any p) for the Stokes mode in the time limit is identical with the corresponding moment for the laser mode, $\langle \hat{n}^p(\tau_0) \rangle$, at the time $\tau = \tau_0$.

B.   Raman Scattering Without Pump Depletion

Compact approximate solutions can be obtained from Eqs. (167) under the condition that the initial laser beam is much more intense than the Stokes beam, i.e., $\langle \hat{n} \rangle \gg \langle \hat{m} \rangle$. The depletion of the laser beam and amplification of the Stokes beam restrict the validity of this approximation to short evolution times τ ($\tau \ll 1$). This approximation implies that the density matrix elements $\rho_{nm}(\nu\mu\tau)$ for $\lambda \geq 0$ given by Eq. (168) are negligible. Moreover, we can simplify the remaining solution (167) by setting $n \approx n \pm m$. In the analysis of the phenomena described by the density $\rho_{nm}(\nu\mu\tau)$ with small degree of off-diagonality ν (such as quadrature squeezing), we can set $n \approx n + \nu$. Alternatively, in order to, for instance, investigate phase properties[129, 130, 131] (which require summation over ν ranging from zero to infinity) one might assume that the fluctuations in the laser beam are small in comparison to their mean value, i.e., $\langle \hat{n} \rangle \gg \sqrt{\langle (\Delta \hat{n})^2 \rangle}$. Under these approximations the solution of (125) takes the form

$$
\begin{aligned}
\rho_{nm}(\nu\mu\tau) &\approx \left[m!(m + \mu)! \right]^{1/2} \\
&\times \sum_{l=0}^{m} \rho_n^L(\nu\tau_0) \rho_{m-l}^S(\mu\tau_0) \left[(m - l)!(m - l + \mu)! \right]^{-1/2} \\
&\times \sum_{q=0}^{l} \exp\left[-n(m - q + 1 + \mu/2)\, \Delta\tau \right] \prod_{\substack{p=0 \\ p \neq q}}^{l} (q - p)^{-1}
\end{aligned}
$$

(193)

Applying the binomial theorem we rewrite (193) as

$$
\begin{aligned}
\rho_{nm}(\nu\mu\tau) &\approx \sum_{l=0}^{m} \left[\binom{m}{l}\binom{m + \mu}{l} \right]^{1/2} (e^{n\,\Delta\tau} - 1)^l \\
&\times \exp\left[-n(m + 1 + \mu/2)\, \Delta\tau \right] \rho_n^L(\nu\tau_0) \rho_{m-l}^S(\mu\tau_0)
\end{aligned}
$$

(194)

which, for $\mu = 0$ and $\nu = 0$, goes over into Simaan's equation of Ref. 75. The density matrix (193), applied to relations (177)–(182), enables the calculation of the expectation values and variances for the Stokes mode and the laser mode; however, in the latter case, as a result of the approximations assumed, we find no time dependence of the laser field photon-number moments for a Stokes beam initially in a number state containing

$$
\langle f[\hat{n}(\tau)]\rangle
$$

$$
= \sum_{n=0}^{\infty} f(n)\rho_n^L(0\tau_0)\exp[-n(m_0+1)] \sum_{m=0}^{\infty} \binom{m+m_0}{m_0}(1-e^{-n\tau})^m
$$

$$
= \sum_{n=0}^{\infty} f(n)\rho_n^L(0\tau_0) = \langle f(\hat{n})\rangle \tag{195}
$$

The result (195a) is valid for any initial number-state Stokes beam, so we conclude that the pump beam is time-independent for arbitrary initial Stokes beam. The photon-number moments for the Stokes mode calculated from (194) are of particularly simple form. For instance, we have

$$
\langle \hat{m}(\tau)\rangle = \langle \hat{m}\rangle \sum_{n=0}^{\infty} \exp(n\,\Delta\tau)\rho_n^L(0\tau_0)
$$

$$
+ \left(\sum_{n=0}^{\infty} \exp(n\,\Delta\tau)\rho_n^L(0\tau_0) - 1 \right)
$$

$$
\langle \hat{m}^2(\tau)\rangle = (\langle \hat{m}^2\rangle + 3\langle \hat{m}\rangle + 2) \sum_{n=0}^{\infty} \exp(2n\,\Delta\tau)\rho_n^L(0\tau_0)
$$

$$
- 3(\langle \hat{m}\rangle + 1) \sum_{n=0}^{\infty} \exp(n\,\Delta\tau)\rho_n^L(0\tau_0) + 1 \tag{197}
$$

$$
\langle \hat{m}(\tau)\hat{n}(\tau)\rangle = (\langle \hat{m}\rangle + 1) \sum_{n=0}^{\infty} n \exp(n\,\Delta\tau)\rho_n^L(0\tau_0) - \langle \hat{n}\rangle \tag{198}
$$

Equations (196) and (197) were obtained by Simaan.[75] Equation (196) is in agreement with the Shen relation in Ref. 26. Equations (196)–(198) reduce to Loudon's results of Ref. 30 for the simpler special case in which no scattered photons are excited initially. The sum of the mean photon numbers for the laser and Stokes mode (196) is not a constant of motion, contrary to our former considerations (132). Nonetheless, in view of the intense laser field approximation, the conservation of the total number of photons is at least approximately fulfilled. It is easy to find a physical interpretation of Eq. (196). The first term of (196) describes the amplification of the initial Stokes beam with $\langle \hat{m}\rangle$ photons at the time τ_0 and can be identified as *sensu stricto* stimulated Raman scattering. The second term of (196) corresponds to an amplification of the vacuum fluctuations and can be interpreted as spontaneous Raman scattering, which occurs even in the case when the Stokes field contains initially no photons ($\langle \hat{m} \rangle = 0$). Note that even in the model of scattering from phonons at zero temperature ("quiet" reservoir), spontaneous scattering does take place. The coefficients $\gamma_S^{(2)}(\tau)$ and $g_{LS}^{(2)}(\tau)$, readily obtained from (35) and (38) by insertion of (196)–(198), can be explicitly compared to the coefficients calculated from other, corresponding relations. Assuming that initially there are no photons in the Stokes beam, $\langle \hat{m} \rangle = \langle \hat{m}^2 \rangle = 0$, we obtain from small-time expansions of the exponential functions in Eqs. (196) and (197) the following simple expressions for the normalized factorial moment $\gamma_S^{(2)}(\tau)$:

$$\gamma_S^{(2)}(\tau) = 2\gamma_L^{(2)} + 1 + 2\big(\langle \hat{n}^3 \rangle - \langle \hat{n}^2 \rangle^2/\langle \hat{n} \rangle\big)\langle \hat{n} \rangle^{-2} \Delta\tau \qquad (199)$$

as well as the normalized cross-correlation function $g_{LS}^{(2)}(\tau)$:

$$g_{LS}^{(2)}(\tau) = \gamma_L^{(2)} + \big(\langle \hat{n}^3 \rangle - \langle \hat{n}^2 \rangle^2/\langle \hat{n} \rangle\big)\langle \hat{n} \rangle^{-2} \Delta\tau/2 \qquad (200)$$

Equations (199) and (200) can be equivalently obtained from the short-time expansions (134) and (139), respectively, on omitting the expressions $1/\langle \hat{n} \rangle$ and $\langle \hat{n}^2 \rangle/\langle \hat{n} \rangle^2$ in the terms proportional to $\Delta\tau$, which are negligible in comparison with the terms $\langle \hat{n}^3 \rangle/\langle \hat{n} \rangle^2$ and $\langle \hat{n}^2 \rangle^2/\langle \hat{n} \rangle^3$. The Simaan approximate relations for $g_{LS}^{(2)}(\tau)$ (32) and $\gamma_S^{(2)}(\tau)$ (33) in Ref. 75, rewritten in our notation (with extra -1 in view of (38) and (35)), do not reduce exactly to our Eqs. (199) and (200), respectively.

By substituting Eq. (194) with $\nu = \mu = 0$ into (174) one can obtain solution (176), for any initial distribution of the laser mode, with the following distribution $\rho_m^{(n_0)}(\tau)$:

$$\rho_m^{(n_0)}(\tau) = \exp\big[-(m + 1)n_0 \Delta\tau\big]$$
$$\times \sum_{l=0}^{m} \binom{m}{l}(e^{n_0 \Delta\tau} - 1)^l \rho_{m-l}^S(0\tau_0) \qquad (201)$$

calculated for the laser field initially in a number state containing n_0 photons. In this case the mean ($\langle \hat{m}(\tau) \rangle$) and mean-square number of Stokes photons ($\langle \hat{m}^2(\tau) \rangle$),

$$\langle \hat{m}(\tau) \rangle = (\langle \hat{m} \rangle + 1)\exp(n_0 \Delta\tau) - 1 \qquad (202)$$

$$\langle \hat{m}^2(\tau) \rangle = (\langle \hat{m}^2 \rangle + 3\langle \hat{m} \rangle + 2)\exp(2n_0 \Delta\tau) \qquad (203)$$
$$- 3(\langle \hat{m} \rangle + 1)\exp(n_0 \Delta\tau) + 1$$

can be immediately obtained either from (196) and (197) or from (201). Assuming that the Stokes beam is initially in a coherent state $|\alpha\rangle$, we can perform summation in (201) which leads to

$$
\rho_m^{(n_0)}(\tau) = \exp\left[-|\alpha|^2 - n_0\,\Delta\tau\right]\left(1 - e^{-n_0\Delta\tau}\right)^m
$$
$$
\times {}_1F_1\left[-m; 1; -|\alpha|^2(e^{n_0\Delta\tau} - 1)^{-1}\right]
\tag{204}
$$

where ${}_1F_1$ is a confluent hypergeometric function. The density matrix elements $\rho_m^{(n_0)}(\tau_0)$ (204) describe a superposition of coherent and chaotic fields.[23, 222] This will be more transparent if we rewrite Eq. (204) in terms of the average number of Stokes photons in the chaotic part,

$$
\langle \hat{m}_{\text{ch}}(\tau)\rangle = \exp(n_0\,\Delta\tau) - 1
\tag{205}
$$

and the mean number of photons in the coherent part alone,

$$
\langle \hat{m}_c(\tau)\rangle = |\alpha|^2 \exp(n_0\,\Delta\tau)
\tag{206}
$$

Then, one obtains, using the Laguerre polynomial L_m, the standard form of the distribution (204)[23, 42, 75, 221]:

$$
\rho_m^{(n_0)}(\tau) = \frac{\langle \hat{m}_{\text{ch}}(\tau)\rangle^m}{\left(1 + \langle \hat{m}_{\text{ch}}(\tau)\rangle\right)^{1+m}} \exp\left(-\frac{\langle \hat{m}_c(\tau)\rangle}{1 + \langle \hat{m}_{\text{ch}}(\tau)\rangle}\right)
$$
$$
\times L_m\left(-\frac{\langle \hat{m}_c(\tau)\rangle}{\langle \hat{m}_{\text{ch}}(\tau)\rangle\left(1 + \langle \hat{m}_{\text{ch}}(\tau)\rangle\right)}\right)
\tag{207}
$$

Similarly, by expressing the relation (202) in terms of the mean values (205) and (206) it is seen that

$$
\langle \hat{m}(\tau)\rangle = \langle \hat{m}_c(\tau)\rangle + \langle \hat{m}_{\text{ch}}(\tau)\rangle
\tag{208}
$$

The general moment of the pth order $\langle \hat{m}^p\rangle$ can be found by repeated use of the recursion relation[221]:

$$
\langle m^{r+1}(\tau)\rangle = \langle m_{\text{ch}}(\tau)\rangle\left(\langle m_{\text{ch}}(\tau)\rangle + 1\right)\frac{\partial\langle m^r(\tau)\rangle}{\partial\langle m_{\text{ch}}(\tau)\rangle}
$$
$$
+ \langle m_c(\tau)\rangle\left(2\langle m_{\text{ch}}(\tau)\rangle + 1\right)\frac{\partial\langle m^r(\tau)\rangle}{\partial\langle m_c(\tau)\rangle} + \langle m^r(\tau)\rangle\langle m(\tau)\rangle
\tag{209}
$$

with the help of (208) or using the following explicit expression[23, 42]:

$$\langle m^r(\tau)\rangle = r!\langle m_{ch}^r(\tau)\rangle L_r\left(-\frac{\langle m_c(\tau)\rangle}{\langle m_{ch}(\tau)\rangle}\right) \qquad (210)$$

The factorial moments of pth order can readily be calculated from (209) or (210). In particular, the second-order factorial moment reads as follows:

$$\gamma_S^{(2)}(\tau) = 1 - \left(\frac{\langle m_c(\tau)\rangle}{\langle m(\tau)\rangle}\right)^2 \qquad (211)$$

which takes the minimal value, equal to zero, for the initial time τ_0, since only then $\langle \hat{m}_{ch}\rangle = 0$.

For the Stokes beam initially in a vacuum state $|0\rangle$ the distributions (204) and (207) reduce to the Bose-Einstein distribution

$$\rho_m^{(n_0)}(\tau) = \frac{\langle \hat{m}_{ch}(\tau)\rangle^m}{\left(1 + \langle \hat{m}_{ch}(\tau)\rangle\right)^{1+m}} \qquad (212)$$

describing a chaotic field (cf. (14)). In this case, in the absence of stimulated scattering ($\langle \hat{m}\rangle = 0$), the chaotic field is generated in spontaneous Raman scattering as an amplification of the vacuum fluctuations.

To compare the results for the expectation values of the Stokes mode obtained in Section V.B with the present results, we assume that the laser and Stokes beams are initially in a coherent state $|\alpha_L\rangle$ and $|\alpha_S\rangle$, respectively. Performing summation in Eqs. (196)–(198) with the coherent weight function $\rho_n^L(0\tau_0)$ one readily arrives at

$$\langle \hat{m}(\tau)\rangle = \left(|\alpha_S|^2 + 1\right)\exp\left[|\alpha_L|^2(e^{\Delta\tau} - 1)\right] - 1 \qquad (213)$$

$$\langle \hat{m}^2(\tau)\rangle = \left(|\alpha_S|^4 + 4|\alpha_S|^2 + 2\right)\exp\left[|\alpha_L|^2(e^{2\Delta\tau} - 1)\right]$$
$$- 3\left(|\alpha_S|^2 + 1\right)\exp\left[|\alpha_L|^2(e^{\Delta\tau} - 1)\right] + 1 \qquad (214)$$

$$\langle \hat{m}(\tau)\hat{n}(\tau)\rangle = |\alpha_L|^2\left(|\alpha_S|^2 + 1\right)\exp\left[|\alpha_L|^2(e^{\Delta\tau} - 1) + \Delta\tau\right] - |\alpha_L|^2 \qquad (215)$$

Within the Fokker-Planck equation approach under parametric approximation (Section V.B) we have obtained Eqs. (111) and (114), which can be rewritten, using the notation of this section, i.e., $\kappa_S t = |e_L|^2\gamma_S t = |\alpha_L|^2\tau$, and assuming that the mean number of phonons is zero, in the following form:

$$\langle \hat{m}(\tau) \rangle = \left(|\alpha_S|^2 + 1 \right) \exp\left(|\alpha_L|^2 \Delta\tau \right) - 1 \tag{216}$$

$$\langle \hat{m}^2(\tau) \rangle = \left(|\alpha_S|^4 + 4|\alpha_S|^2 + 2 \right) \exp\left(2|\alpha_L|^2 \Delta\tau \right) \\ - 3\left(|\alpha_S|^2 + 1 \right) \exp\left(|\alpha_L|^2 \Delta\tau \right) + 1 \tag{217}$$

We also note that

$$\langle \hat{m}(\tau)\hat{n}(\tau) \rangle = \langle \hat{m}(\tau) \rangle \langle \hat{n} \rangle \tag{218}$$

For short times of evolution, $\Delta\tau \ll 1$, and intense pump beams, $|\alpha_L|^2 \gg 1$, Eqs. (213), (214), and (215) go over into Eqs. (216), 217), and (218), respectively. Indeed, the short-time expansions of (213) and (214) are

$$\langle \hat{m}(\tau) \rangle = |\alpha_S|^2 + |\alpha_L|^2\left(1 + |\alpha_S|^2 \right) \Delta\tau \\ + |\alpha_L|^2\left(|\alpha_L|^2 + 1 \right)\left(1 + |\alpha_S|^2 \right)\frac{(\Delta\tau)^2}{2} \tag{219}$$

$$\langle \hat{m}^2(\tau) \rangle = |\alpha_S|^2\left(1 + |\alpha_S|^2 \right) + |\alpha_L|^2\left(1 + 5|\alpha_S|^2 + 2|\alpha_S|^4 \right) \Delta\tau \\ + |\alpha_L|^2\left(|\alpha_L|^2 + 1 \right)\left(5 + 13|\alpha_S|^2 + 4|\alpha_S|^4 \right)\frac{(\Delta\tau)^2}{2} \tag{220}$$

whereas Eqs. (216) and (217) obtained within the formalism of Section V.B reduce to

$$\langle \hat{m}(\tau) \rangle = |\alpha_S|^2 + |\alpha_L|^2\left(1 + |\alpha_S|^2 \right) \Delta\tau \\ + |\alpha_L|^4\left(1 + |\alpha_S|^2 \right)\frac{(\Delta\tau)^2}{2} \tag{221}$$

$$\langle \hat{m}^2(\tau) \rangle = |\alpha_S|^2\left(1 + |\alpha_S|^2 \right) + |\alpha_L|^2\left(1 + 5|\alpha_S|^2 + 2|\alpha_S|^4 \right) \Delta\tau \\ + |\alpha_L|^4\left(5 + 13|\alpha_S|^2 + 4|\alpha_S|^4 \right)\frac{(\Delta\tau)^2}{2} \tag{222}$$

respectively. It is seen that for high intensity of the pump field, (219) goes over into (221), and (220) into (222) by setting $|\alpha_L|^2(|\alpha_L|^2 + 1) \approx |\alpha_L|^4$.

Hence, the factorial moment $\gamma_S^{(2)}(\tau)$,

$$\gamma_S^{(2)}(\tau) = 2|\alpha_L|^2|\alpha_S|^{-2}\Delta\tau - \big[|\alpha_L|^2\big(3 + |\alpha_S|^2\big) \\ - |\alpha_S|^4 - 5|\alpha_S|^2 - 2\big]|\alpha_L|^2|\alpha_S|^{-4}(\Delta\tau)^2 \tag{223}$$

calculated from (219) and (220) in the case of nonzero α_S and an intense pump beam, goes over into

$$\gamma_S^{(2)}(\tau) = 2|\alpha_L|^2|\alpha_S|^{-2}\Delta\tau - \big(3 + |\alpha_S|^2\big)|\alpha_L|^4|\alpha_S|^{-4}(\Delta\tau)^2 \tag{224}$$

obtained from (221) and (222). If the initial field contains no Stokes photons, we obtain the following factorial moments $\gamma_S^{(2)}(\tau)$:

$$\gamma_S^{(2)}(\tau) = 1 + 2|\alpha_L|^{-2} + 2\,\Delta\tau + \big(2 + \tfrac{5}{6}|\alpha_L|^2\big)(\Delta\tau)^2 \tag{225}$$

$$\gamma_S^{(2)}(\tau) = 1 \tag{226}$$

within the formalisms of this section and Section V.B, respectively. The differences between the factorial moments $\gamma_S^{(2)}(\tau)$ are more pronounced in the case $\alpha_S = 0$ since the expansion of $\langle \hat{m}(\tau)\rangle$ and $\langle \hat{m}^2(\tau)\rangle$ correct to the third order in τ is required in the derivation of (225). The interbeam degree of coherence $g_{LS}^{(2)}(\tau)$ (38), as expected, vanishes for the model of Section V.B. The short-time expansion of $g_{LS}^{(2)}(\tau)$ obtained from (213)–(215), for $\alpha_S \neq 0$, is

$$g_{LS}^{(2)}(\tau) = \big(1 + |\alpha_S|^{-2}\big)\Delta\tau \\ - \big(1 + |\alpha_S|^2\big)\big(2|\alpha_L|^2 - |\alpha_S|^2\big)|\alpha_S|^{-4}\frac{(\Delta\tau)^2}{2} \tag{227}$$

Otherwise, $\alpha_S = 0$, we get

$$g_{LS}^{(2)}(\tau) = |\alpha_L|^{-2} + \frac{1}{2}\,\Delta t + \frac{1}{12}\big(|\alpha_L|^2 + 3\big)\frac{(\Delta\tau)^2}{2} \tag{228}$$

It is seen that the approaches of Sections V.B and VI.B give similar predictions for the Stokes beam.

The evolution of the photon-number moments is demonstrated in Figs. 2 and 3: $\langle \hat{m}(\tau)\rangle$ calculated with Eq. (221) is depicted by solid line C or with (219) by solid line D in Fig. 2; $\langle \hat{m}^2(\tau)\rangle$ obtained from Eq. (222) is given by solid line C or from (220) by solid line D in Fig. 3. No time dependence of $\langle \hat{n}(\tau) \rangle$ and $\langle \hat{n}^2(\tau) \rangle$ is observed for the results of this section and Section V.B; i.e., we obtain straight lines C and D in Figs. 2 and 3. Similar notation is used in Figs. 4–6 for the normalized factorial moments $\gamma_S^{(2)}(\tau)$ (Fig. 4), $\gamma_L^{(2)}(\tau)$ (Fig. 5), and the degree of interbeam coherence $g_{LS}^{(2)}(\tau)$ (Fig. 6). We have chosen rather small initial numbers of laser photons ($|\alpha_L|^2 = 2$) for numerical reasons. In this case, the factorial moments, calculated from (223), (225), (227), and (228), differ significantly from the exact numerical results. So we omit them (curves D) in Figs. 4 and 6.

VII. CONCLUSIONS

Raman scattering from a great number of phonon modes is described from a quantum-statistical point of view within the standing-wave model. The master equation for the completely quantum case, including laser pump depletion and stochastic coupling of Stokes and anti-Stokes modes, is derived and converted to classical equations: either into a generalized Fokker-Planck equation and an equation of motion for the characteristic function or into the master equation in Fock representation. These two approaches are developed both in linear and nonlinear régime. A detailed analysis of scattering into Stokes and anti-Stokes modes in linear régime, i.e., under parametric approximation, is presented. The existence of s-parametrized quasiprobability distributions, in particular the Glauber-Sudarshan P-function, is investigated. An analysis of Raman scattering into separate Stokes and anti-Stokes modes in nonlinear régime, thus including pump depletion, is given. The master equation in Fock representation is solved exactly for the complete density matrix using the Laplace transform method. Short-time solutions, steady-state solutions and approximate compact form solutions are obtained. Relations between the quasidistribution approach based on the Fokker-Planck equation and the density matrix approach based on the master equation in Fock representation are presented. The photocount distribution and its factorial moments as well as variances and extremal variances of quadratures are calculated in both approaches giving the basis for the analysis of the quantum properties of radiation such as sub-Poissonian photon-counting statistics and squeezing. A comparison of various statistical moments obtained from numerical calculations utilizing our exact solution of the master equation and from the approximate relations for short times, as well as obtained under parametric approximation, is presented graphically.

In Figs. 2–9 we compared various statistical moments obtained (1) from numerical calculations utilizing the exact solutions without the parametric approximation (Eqs. (167), (168), and/or (171) derived in Appendix B), (2) from the short-time solutions of Section VI.A.1, (3) from the solutions obtained in Section V.B within the framework of the FPE approach under the parametric approximation, and (4) from the approximate solutions derived in Section VI.B within the density-matrix formalism.

In Figs. 2, 3, 7, and 8 we demonstrated that the initial and short-time behavior of the approximate functions (curves B, C, and D) is consistent with the exact evolution (curves A). We have shown analytically that our expressions for the Stokes scattering formalisms presented in Sections V.B, VI.A.1, VI.A.2, and VI.B are equivalent for short times and high initial intensities of the pump field. Nevertheless, it is seen that the equations derived in Section VI.A.1 give the best, whereas those derived in Section VI.B give the worst approximation to the exact solution of Section VI.A.2 for small initial intensities of the laser field.

We showed in Fig. 4 (curve A) that the Stokes mode photon-number fluctuations vary from initially chaotic to Poissonian in asymptotics (for $\langle \hat{m} \rangle = 0$ and $\langle \hat{n} \rangle = |\alpha_L|^2$) or from Poissonian, through super-Poissonian, to Poissonian for large times (if $\langle \hat{m} \rangle = |\alpha_S|^2 \neq 0$ and $\langle \hat{n} \rangle = |\alpha_L|^2$). The asymptotic behavior of $\gamma_S^{(2)}(\tau)$ is consistent with our predictions in Section VI.A.3. The short-time behavior of $\gamma_S^{(2)}(\tau)$ for hyper-Raman scattering[116] is similar to that presented in Fig. 4 for Raman scattering. However, for long times the Stokes hyper-Raman photon-number fluctuations become sub-Poissonian for reasons given in Section VI.A.3.

In Fig. 5 we demonstrated that the normalized factorial moment for the laser mode, $\gamma_L^{(2)}(\tau)$, changes from initially Poissonian to super-Poissonian. The differences in $\gamma_L^{(2)}(\tau)$ between Fig. 5a (spontaneous Stokes scattering) and Fig. 5b (stimulated Stokes scattering) are only quantitative. We note that for hyper-Raman scattering the photon-number fluctuations in the initially coherent laser mode become sub-Poissonian.[116] The time behavior of the interbeam degree of coherence was presented in Figs. 6a and b. Curve A in Fig. 6a coincides with the Simaan exact solution.[75] Sub-Poissonian statistics in the compound laser-Stokes mode is observed.

It is thought (see, for instance, Ref. 22, p. 192) that the Simaan approach[75] is the most rigorous application of master equations to the Raman problem. However, Simaan's solution is restricted to the diagonal matrix elements in number representation. Only these terms are needed to obtain the mean photon numbers and their higher moments, whereby the photon correlation effects can be investigated. To investigate squeezing properties and phase correlations it is necessary to obtain the off-diagonal elements of the density matrix. We generalized the solution of the master equation obtained by McNeil and Walls[73] and Simaan[75] to comprise all the off-diagonal matrix elements as well. Our derivation of the complete density matrix represents the main result of this paper.

Our intention was to cite an extensive literature related to our Raman scattering approaches. Nevertheless, we realize that the cited literature is not complete. We include only those references that are most relevant for the purposes of our article.

Acknowledgments

In the preparation of this paper we have benefitted greatly from contacts with Ryszard Tanaś, Przemysław Szlachetka, and Krzysztof Grygiel. Our warm appreciation must also go to Jan Peřina.

APPENDIX A

Here, we give a simple formal solution of the generalized FPE (61) for the quadidistribution $\mathscr{W}^{(-1)}(\alpha_L, \alpha_S, \alpha_A, t)$ (Q-functions) as well the corresponding characteristic function $\mathscr{C}^{(-1)}(\beta_L, \beta_S, \beta_A, t)$—the solution of the simplified equation of motion (63). We choose antinormal order ($s = -1$) to avoid the problems of existence of the quasiprobability distributions and to reduce the FPE (61) containing the third-order derivatives (for $s \neq \pm 1$) to a second-order FPE. It is by no means easy to find an exact solution of (61) or (63) even for particular orders, because the drift coefficients are not linear and the diffusion coefficients are not constant. An often employed method to solve problems of this kind is to assume that the fluctuations of the radiation fields are small compared to their mean values; i.e., the quasidistribution describing the fields is sharply peaked.[19, 23, 188, 223, 224] This will be the case for suitably chosen input state and the initial output states. Under these restrictions we can rewrite our FPE (61) related to antinormal order in the linearized form:

$$
\frac{\partial}{\partial t}\mathscr{W}^{(-1)}(\alpha_L, \alpha_S, \alpha_A, t)
$$

$$
= \frac{1}{2}\gamma_S \Bigg\{ \Bigg[-(D_{LS} + \xi_L\xi_S)\frac{\partial}{\partial \alpha_L}\frac{\partial}{\partial \alpha_S}
$$

$$
+ \langle \hat{n}_S \rangle \frac{\partial}{\partial \alpha_L}\alpha_L - (\langle \hat{n}_L \rangle - 1)\frac{\partial}{\partial \alpha_S}\alpha_S + \text{c.c.} \Bigg]
$$

$$
+ 2\langle \hat{n}_S \rangle \frac{\partial}{\partial \alpha_L}\frac{\partial}{\partial \alpha_L^*} \Bigg\}\mathscr{W}^{(-1)}
$$

$$+ \frac{1}{2}\gamma_A \Bigg\{ \Bigg[-(D_{LA} + \xi_L\xi_A)\frac{\partial}{\partial\alpha_L}\frac{\partial}{\partial\alpha_A}$$

$$- (\langle\hat{n}_A\rangle - 1)\frac{\partial}{\partial\alpha_L}\alpha_L + \langle\hat{n}_L\rangle\frac{\partial}{\partial\alpha_A}\alpha_A + \text{c.c.}\Bigg]$$

$$+ 2\langle\hat{n}_L\rangle\frac{\partial}{\partial\alpha_A}\frac{\partial}{\partial\alpha_A^*}\Bigg\}\mathscr{W}^{(-1)}$$

$$- \Bigg\{ \frac{1}{2}\gamma_{SA}e^{-2i\Delta\Omega\Delta t}\Bigg[(C_L + \xi_L^2)\Bigg(\alpha_A^*\frac{\partial}{\partial\alpha_S} + \frac{\partial}{\partial\alpha_S}\frac{\partial}{\partial\alpha_A} - \alpha_S^*\frac{\partial}{\partial\alpha_A}\Bigg)$$

$$- 2(\overline{D}_{SL} - \xi_L\xi_S^*)\frac{\partial}{\partial\alpha_L^*}\frac{\partial}{\partial\alpha_A} - (D_{SA}^* + \xi_S^*\xi_A^*)\frac{\partial^2}{\partial\alpha_L^{*2}}\Bigg] + \text{c.c.}\Bigg\}\mathscr{W}^{(-1)}$$

$$+ \gamma_S\langle\hat{n}_V\rangle\Bigg\{\Bigg(\frac{1}{2}\frac{\partial}{\partial\alpha_L}\alpha_L - (D_{LS} + \xi_L\xi_S)\frac{\partial}{\partial\alpha_L}\frac{\partial}{\partial\alpha_S} + \frac{1}{2}\frac{\partial}{\partial\alpha_S}\alpha_S + \text{c.c.}\Bigg)$$

$$+ \langle\hat{n}_S\rangle\frac{\partial}{\partial\alpha_L}\frac{\partial}{\partial\alpha_L^*} + \langle\hat{n}_L\rangle\frac{\partial}{\partial\alpha_S}\frac{\partial}{\partial\alpha_S^*}\Bigg\}\mathscr{W}^{(-1)}$$

$$+ \gamma_A\langle\hat{n}_V\rangle\Bigg\{\Bigg(\frac{1}{2}\frac{\partial}{\partial\alpha_L}\alpha_L - (D_{LA} + \xi_L\xi_A)\frac{\partial}{\partial\alpha_L}\frac{\partial}{\partial\alpha_A} + \frac{1}{2}\frac{\partial}{\partial\alpha_A}\alpha_A + \text{c.c.}\Bigg)$$

$$+ \langle\hat{n}_A\rangle\frac{\partial}{\partial\alpha_L}\frac{\partial}{\partial\alpha_L^*} + \langle\hat{n}_L\rangle\frac{\partial}{\partial\alpha_A}\frac{\partial}{\partial\alpha_A^*}\Bigg\}\mathscr{W}^{(-1)}$$

$$- \Bigg\{ \gamma_{AS}\langle\hat{n}_V\rangle e^{2i\Delta\Omega\Delta t}\Bigg((D_{SA}^* + \xi_S^*\xi_A^*)\frac{\partial^2}{\partial\alpha_L^{*2}} + (C_L + \xi_L^2)\frac{\partial}{\partial\alpha_S}\frac{\partial}{\partial\alpha_A}$$

$$+ (\overline{D}_{AL} - \xi_L\xi_A^*)\frac{\partial}{\partial\alpha_L^*}\frac{\partial}{\partial\alpha_S} + (\overline{D}_{SL} - \xi_L\xi_S^*)\frac{\partial}{\partial\alpha_L^*}\frac{\partial}{\partial\alpha_A}\Bigg) + \text{c.c.}\Bigg\}\mathscr{W}^{(-1)}$$

$$\tag{A.1}$$

where the coefficients D_{kl}, $\overline{D}_{kl}$, C_k, ξ_k for $k, l = L, S, A$ are defined by (66) at the initial moment t_0. Equation (A.1) is the generalization of the FPE given in Ref. 218 for the case of nonzero γ_S and γ_{AS}. It is seen that the Raman effect under the approximations applied can be treated as an Ornstein-Uhlenbeck process.[191] The FPE (A.1) can be solved exactly by various techniques; see, e.g., Ref. 188. For instance, using the inverse Fourier transform (22), one can transform the FPE (A.1) into the corresponding equation of motion for the characteristic function

$\mathscr{C}^{(-1)}(\beta_L, \beta_S, \beta_A, t)$, which is a first-order differential equation. The method of characteristics applied to the latter equation leads to the solution

$$
\begin{aligned}
\mathscr{C}^{(-1)}(\beta_L, \beta_S, \beta_A, t) = \Big\langle \exp\Big\{ &- \sum_{k=L,S,A} \Big[B_k^{(-1)}(t) |\beta_k|^2 \\
&+ \big(\tfrac{1}{2} C_k^*(t)\beta_k^2 + \text{c.c.}\big) + \big(\beta_k \xi_k^*(t) - \text{c.c.}\big) \Big] \\
&+ \Big[D_{LS}(t)\beta_L^*\beta_S^* + \overline{D}_{LS}\beta_L\beta_S^* \\
&\quad + D_{LA}(t)\beta_L^*\beta_A^* + \overline{D}_{LA}\beta_L\beta_A^* \\
&\quad + D_{SA}(t)\beta_S^*\beta_A^* + \overline{D}_{SA}\beta_S\beta_A^* + \text{c.c.} \Big] \Big\} \Big\rangle
\end{aligned}
\tag{A.2}
$$

The angle brackets mean averaging over the complex amplitudes ξ_k ($k = L, S, A$) with the initial distribution $\mathscr{W}^{(-1)}(\alpha_L, \alpha_S, \alpha_A, t_0)$. They represent the influence of the initial photon statistics of the pump and scattered fields on the evolution of the system. The solution of Eq. (A.1) can be readily obtained by applying the Fourier transform (21) to solution (A.2), and has the form of a shifted seven-dimensional (including time) Gaussian distribution involving correlation between the radiation fields,

$$
\begin{aligned}
\mathscr{W}^{(-1)}&(\alpha_1, \alpha_2, \alpha_3, t) \\
&= \Big\langle \frac{1}{L^{(-1)}} \exp\Big\{ - (L^{(-1)})^{-2} \sum_{j=1}^{3} \Big[|\alpha_j - \xi_j(t)|^2 E_j^{(-1)} \\
&\quad + \frac{1}{2}\big((\alpha_j^* - \xi_j^*(t))^2 E_{j+3}^{(-1)} + \text{c.c.}\big) \Big] \\
&\quad + (L^{(-1)})^{-2} \sum_{j=1}^{2} \sum_{k=j+1}^{3} \Big[\big(\alpha_j^* - \xi_j^*(t)\big)\big(\alpha_k^* - \xi_k^*(t)\big) E_{j+k+4}^{(-1)} \\
&\quad + \big(\alpha_j - \xi_j(t)\big)\big(\alpha_k^* - \xi_k^*(t)\big) E_{j+k+7}^{(-1)} \Big] + \text{c.c.} \Big\} \Big\rangle
\end{aligned}
\tag{A.3}
$$

where, for simplicity, we have identified the subscripts in $\alpha_1 = \alpha_L$, $\alpha_2 = \alpha_S$, $\alpha_3 = \alpha_A$, $\xi_1 = \xi_L$, $\xi_2 = \xi_S$, $\xi_3 = \xi_A$. The functions $E_1^{(-1)}, \ldots, E_{12}^{(-1)}$, and $L^{(-1)}$, which are time-dependent, are connected with the functions

$B_k^{(-1)}, C_k, D_{kl}, \overline{D}_{kl}$ appearing in (A.2) in a manner similar to (74) and (75), respectively. We do not adduce explicit formulas for the coefficients listed, since solutions (A.2) and (A.3) serve only as an example of how one can deal with Eqs. (61) and (63). The validity of solutions (A.2) and (A.3) is restricted by strong approximations, which are actually equivalent to the parametric approximation and the short-time approximation.

APPENDIX B

Here, we solve the equation of motion (125). To eliminate the square root appearing in Eq. (125) for off-diagonal terms, it is convenient to introduce the transformation

$$\psi_{nm}(\nu\mu\tau) = \left[\frac{n!(n+\nu)!}{m!(m+\mu)!} \right]^{1/2} \rho_{nm}(\nu\mu\tau) \qquad (B.1)$$

where the degree of off-diagonality μ is restricted to nonnegative integers, whereas the degree ν is $\geq -n$. On insertion of (B.1) into (125), the equation of motion for the transformed matrix elements $\psi_{nm}(\nu\mu\tau)$ takes the form

$$\dot{\psi}_{nm}(\nu\mu\tau) = -\tfrac{1}{2}[n(m+1) + (n+\nu)(m+\mu+1)]\psi_{nm}(\nu\mu\tau)$$
$$+ \psi_{n+1,m-1}(\nu\mu\tau) \qquad (B.2)$$

We apply the Laplace transform method to (B.2), which readily leads to the solution

$$\overline{\psi}_{nm}(\nu\mu s) = \sum_{l=0}^{m} \psi_{n+l,m-l}(\nu\mu\tau_0) \prod_{p=0}^{l} [s + f(p)]^{-1} \qquad (B.3)$$

for $\overline{\psi}_{nm}(\nu\mu\tau)$, the Laplace transform of $\psi_{nm}(\nu\mu\tau)$. The function $f(p)$ occurring in (B.3) is given by (169). If there are no equal terms among the elements of the set $f(0), f(1), \ldots, f(l)$ the inverse transform, after retaining the $\rho_{nm}(\nu\mu\tau)$ notation, yields (167). If there are repeated elements in the denominator of (B.3), the inverse transforms will involve convolutions. We apply two general procedures essentially equivalent to that of Simaan[75] and Malakyan.[116] It is convenient to split (B.3) into two terms, as follows:

$$\bar{\psi}_{nm}(\nu\mu s) = \sum_{l=0}^{\lambda} \psi_{n+l,m-l}(\nu\mu\tau_0) \prod_{p=0}^{l} [s + f(p)]^{-1}$$

$$+ (1 - \delta_{m0}) \sum_{l=\lambda+1}^{m} \psi_{n+l,m-l}(\nu\mu\tau_0) \qquad (B.4)$$

$$\times \prod_{p=0}^{\lambda} [s + f(p)]^{-1} \prod_{p'=\lambda+1}^{l} [s + f(p')]^{-1}$$

with λ defined by (166) (or equivalently by (170)). Let us note that a parabola $f(q)$ = const takes its maximum value for $q_0 = (2m - 2n + \mu - \nu + 2)/4$. This value, q_0 or better λ, the maximum integer $\leq q_0$, can serve as a criterion to split (B.3) in such a manner that a convolution theorem can be easily applied. The first term in (B.4) has no mutually equal factors in the denominator, so the inverse Laplace transform has the form of (167) with the proper upper limit of summation. The denominator of the second term of (B.4) contains repeated factors, which are separated, so that we can readily apply the convolution theorem finally obtaining the solution (168). Equations (167) and (168) have a rather complicated structure. We can rewrite (167) and (168) in a more compact form. If we assume that there is only one pair of equal factors among the elements of the set $f(0), f(1), \ldots, f(l)$, i.e., if

$$\bigvee_{\substack{q_1 \neq q_1' \\ q_1, q_1' \in \{0, \ldots, l\}}} f(q_1) = f(q_1') \bigwedge_{\substack{q=0,\ldots,l \\ q \neq q_1, q_1'}} f(q_1) \neq f(q)$$

then we can express the solution (B.3) for $\bar{\psi}_{nm}(\nu\mu\tau)$ as

$$\bar{\psi}_{nm}(\nu\mu s) = \sum_{l=0}^{m} \psi_{n+l,m-l}(\nu\mu\tau_0)[s + f(q_1)]^{-2} \prod_{\substack{p=0 \\ p \neq q_1, q_1'}}^{l} [s + f(p)]^{-1}$$

$$(B.5)$$

The inverse transform of (B.5) is

$$\psi_{nm}(\nu\mu\tau) = \sum_{l=0}^{m} \psi_{n+l,\,m-l}(\nu\mu\tau_0)$$

$$\times \left\{ \sum_{\substack{q=0 \\ q \neq q_1, q_1'}}^{l} \exp[-f(q)\,\Delta\tau] \prod_{\substack{p=0 \\ p \neq q}}^{l} [f(p) - f(q)]^{-1} \right.$$

$$+ \left(\Delta\tau - \sum_{\substack{k=0 \\ k \neq q_1, q_1'}}^{l} [f(k) - f(q_1)]^{-1} \right) \exp[-f(q_1)\,\Delta\tau] \qquad \text{(B.6)}$$

$$\left. \times \prod_{\substack{p=0 \\ p \neq q_1, q_1'}}^{l} [f(p) - f(q_1)]^{-1} \right\}$$

which is a derivative of the solution (167) over $f(q_1)$ with extra minus,

$$\psi_{nm}(\nu\mu\tau) = \sum_{l=0}^{m} \psi_{n+l,\,m-l}(\nu\mu\tau_0)$$

$$\times \left(-\frac{\partial}{\partial f(q_1)} \right) \sum_{\substack{q=0 \\ q \neq q_1'}}^{l} \exp[-f(q)\,\Delta\tau] \prod_{\substack{p=0 \\ p \neq q, q_1'}}^{l} [f(p) - f(q)]^{-1} \qquad \text{(B.7)}$$

In the case of d equal pairs, i.e., $f(q_1) = f(q_1'), \ldots, f(q_d) = f(q_d')$, the Laplace transform solution (B.3) can be rewritten as

$$\bar{\psi}_{nm}(\nu\mu s) = \sum_{l=0}^{m} \psi_{n+l,\,m-1}(\nu\mu\tau_0)$$

$$\times \prod_{r=1}^{d} [s + f(q_r)]^{-2} \prod_{\substack{p=0 \\ p \neq q_1, q_1', \ldots, q_d, q_d'}}^{l} [s + f(p)]^{-1} \qquad \text{(B.8)}$$

finally leading to

$$
\psi_{nm}(\nu\mu\tau) = \sum_{l=0}^{m} \psi_{n+l,\,m-l}(\nu\mu\tau_0)\left((-1)^d \prod_{r=1}^{d} \frac{\partial}{\partial f(q_r)}\right)
$$

$$
\times \sum_{\substack{q=0 \\ q \neq q_1', \ldots, q_d'}}^{l} \exp[-f(q)\,\Delta\tau] \prod_{\substack{p=0 \\ p \neq q, q_1', \ldots, q_d'}}^{l} [f(p) - f(q)]^{-1}
$$

$$(\text{B.9})$$

or equivalently to the solution (171) with the dth order differential operator $\hat{\mathscr{D}}$ (172).

APPENDIX C

In Section VI we have given an analysis of Stokes scattering. For completeness, in this appendix, we present the solution describing the anti-Stokes effect including laser depletion, but neglecting the Stokes generation and assuming that the reservoir is "quiet," i.e.,

$$
\gamma_S = \gamma_{AS} = \gamma_{SA} = 0
$$
$$
\langle \hat{n}_V \rangle = 0
$$

$$(\text{C.1})$$

Under these conditions the master equation (18) in Fock representation is

$$
\frac{\partial}{\partial\tau}\rho_{nm}(\nu\mu\tau) = -\frac{1}{2}[(n+1)m + (n+\nu+1)(m+\mu)]\rho_{nm}(\nu\mu\tau)
$$

$$
+ [n(n+\nu)(m+1)(m+\mu+1)]^{1/2}\rho_{n-1,\,m+1}(\nu\mu\tau)
$$

$$(\text{C.2})$$

where, for brevity, we have set $n_L = n$, $n_L' = n + \nu$ (as in Section VI) and $n_A = m$, $n_A' = m + \mu$. If we define λ as follows:

$$
\lambda = \left[\left[\frac{n-m+1}{2} + \frac{\nu-\mu}{4}\right]\right]
$$

$$(\text{C.3})$$

then the solution of (C.2) for $\lambda < 0$ becomes

$$
\begin{aligned}
\rho_{nm}(\nu\mu\tau) = {} & \left[\frac{n!(n+\nu)!}{m!(m+\mu)!}\right]^{1/2} \sum_{l=0}^{n} \rho_{n-l}^{L}(\nu\tau_0)\rho_{m+l}^{A}(\mu\tau_0) \\
& \times \left[\frac{(m+l)!(m+l+\mu)!}{(n-l)!(n-l+\nu)!}\right]^{1/2} \quad\quad\quad\quad (C.4) \\
& \times \sum_{q=0}^{l} \exp[-g(q)\,\Delta\tau] \prod_{\substack{p=0 \\ p\neq q}}^{l} [g(p)-g(q)]^{-1}
\end{aligned}
$$

whereas for $\lambda \geq 0$ it becomes

$$
\begin{aligned}
\rho_{nm}(\nu\mu\tau) = {} & \left[\frac{n!(n+\nu)!}{m!(m+\mu)!}\right]^{1/2} \Bigg\{ \sum_{l=0}^{\lambda} \rho_{n-l}^{L}(\nu\tau_0)\rho_{m+l}^{A}(\mu\tau_0) \\
& \times \left[\frac{(m+l)!(m+l+\mu)!}{(n-l)!(n-l+\nu)!}\right]^{1/2} \\
& \times \sum_{q=0}^{l} \exp[-g(q)\,\Delta\tau] \prod_{\substack{p=0 \\ p\neq q}}^{l} [g(p)-g(q)]^{-1} \\
& + (1-\delta_{n0}) \sum_{l=\lambda+1}^{n} \rho_{n-l}^{L}(\nu\tau_0)\rho_{m+l}^{A}(\mu\tau_0) \\
& \times \left[\frac{(m+l)!(m+l+\mu)!}{(n-l)!(n-l+\nu)!}\right]^{1/2} \\
& \times \sum_{q=0}^{\lambda} \sum_{q'=\lambda+1}^{l} \prod_{\substack{p=0 \\ p\neq q}}^{\lambda} [g(p)-g(q)]^{-1} \prod_{\substack{p'=\lambda+1 \\ p'\neq q'}}^{l} [g(p')-g(q')]^{-1} \\
& \times \Bigg(\delta_{g(q)g(q')}\,\Delta\tau \exp[-g(q)\,\Delta\tau] \\
& + (\delta_{g(q)g(q')}-1)\frac{\exp[-g(q)\,\Delta\tau]-\exp[-g(q')\,\Delta\tau]}{g(q)-g(q')} \Bigg) \Bigg\} \quad (C.5)
\end{aligned}
$$

with

$$g(x) = \tfrac{1}{2}[(m + x)(n - x + 1) + (m + x + \mu)(n - x + \nu + 1]] \quad (C.6)$$

Alternatively, we can express solution (C.4) and (C.5) as

$$\rho_{nm}(\nu\mu\tau) = \left[\frac{n!(n + \nu)!}{m!(m + \mu)!}\right]^{1/2} \sum_{l=0}^{n} \rho_{n-l}^{L}(\nu\tau_0)\rho_{m+l}^{A}(\mu\tau_0)$$

$$\times \left[\frac{(m + l)!(m + l + \mu)!}{(n - l)!(n - l + \nu)!}\right]^{1/2} \quad (C.7)$$

$$\times \hat{\mathscr{D}} \sum_{\substack{q=0 \\ q \neq q_1', q_2', \ldots, q_d'}}^{l} \exp[-g(q)\,\Delta\tau] \prod_{\substack{p=0 \\ p \neq q, q_1', q_2', \ldots, q_d'}}^{l} [g(p) - g(q)]^{-1}$$

using the differentiation operator $\hat{\mathscr{D}}$ given by (172).

References

1. C. V. Raman and K. S. Krishnan, *Nature* **121**, 501 (1928).
2. C. V. Raman, *Indian J. Phys.* **2**, 387 (1928).
3. G. Landsberg and L. I. Mandel'stamm, *Naturwissenschaften* **16**, 557 (1928).
4. E. J. Woodbury and W. K. Ng, *Proc. IRE* **50**, 2347 (1962).
5. G. Eckhardt, R. W. Hellwarth, F. J. McClung, S. E. Schwarz, D. Weiner, and E. J. Woodbury, *Phys. Rev. Lett.* **9**, 455 (1962).
6. N. Bloembergen, *Am. J. Phys.* **35**, 989 (1967).
7. W. Kaiser and M. Maier, in F. T. Arecchi and E. O. Schultz-Dubois (Eds.), *Laser Handbook*, North-Holland, Amsterdam, 1972, p. 1077.
8. J. A. Koningstein, *Introduction to the Theory of the Raman Effect*, Reidel, Dordrecht, 1972.
9. A. Z. Grasyuk, *Kvant. Elektron.* **1**, 485 (1974) [*Sov. J. Quantum Electron.* **4**, 269 (1974)].
10. A. Z. Grasyuk, in N. G. Basov (Ed.), Lasers and Their Applications, Proc. of the Lebedev Physics Institute, Vol. 76, Nauka, Moscow, 1976.
11. C. S. Wang, in M. Rabin and C. L. Tang (Eds.), *Quantum Electronics: A Treatise*, Vol. 1, Part A, Academic, New York, 1975, p. 447.
12. C. S. Wang, *Phys. Rev.* **182**, 482 (1969).
13. M. Cardona, (Ed.), *Light Scattering in Solids*, Topics Appl. Phys., Vol. 8, Springer, Berlin, 1975.
14. D. A. Long, *Raman Spectroscopy*, McGraw-Hill, New York, 1977.
15. W. Hayes and R. Loudon, *Scattering of Light by Crystals*, Wiley, New York, 1978.
16. A. Penzkofer, A. Laubereau, and W. Kaiser, *Prog. Quantum Electron.* **6**, 55 (1979).
17. S. Kielich, *Prog. Opt.* **20**, 156 (1983).
18. S. Kielich, *Proc. Indian Acad. Sci.* (*Chem. Sci.*) **94**, 403 (1985).

19. S. Kielich, *Nonlinear Molecular Optics*, Nauka, Moscow, 1981 (in Russian).

20. Y. R. Shen, *Nonlinear Optics*, Wiley, New York, 1984, Chapter 10.

21. Yu. E. D'yakov and S.Yu. Nikitin, *Kvant. Elektron.* **14**, 1925 (1987) [*Sov. J. Quantum Electron.* **17**, 1227 (1987)].

22. M. G. Raymer and I. A. Walmsley, *Prog. Opt.* **28**, 182 (1990).

23. J. Peřina, *Quantum Statistics of Linear and Nonlinear Optical Phenomena*, Reidel, Dortrecht, 1984 (2d enlarged ed., 1991).

24. J. Mostowski and M. G. Raymer, in G. P. Agrawal and W. Boyd (Eds.), in *Contemporary Nonlinear Optics*, Academic, Boston, 1992, p. 187.

25. Special issue of *J. Opt. Am. Soc. B* **3** (10) (1986).

26. Y. R. Shen, *Phys. Rev.* **155**, 921 (1967).

27. P. Szlachetka, *Quantum Fluctuations in Multi-Photon Raman Scattering Processes*, Thesis, Poznań University, 1980.

28. I. Abram, *Phys. Rev. A* **35**, 4661 (1987).

29. I. Abram and E. Cohen, *Phys. Rev. A* **44**, 500 (1987).

30. R. Loudon, *The Quantum Theory of Light*, Clarendon, Oxford, UK, 1973 (2nd enlarged ed., 1983).

31. A. Schenzle and H. Brand, *Phys. Rev. A* **20**, 1628 (1979).

32. C. Mavroyannis, *Mol. Phys.* **48**, 4847 (1983).

33. C. Mavroyannis, *J. Phys. Chem.* **88**, 4868 (1984).

34. C. Mavroyannis, *J. Chem. Phys.* **47**, 2554 (1986).

35. H. S. Freedhoff, *J. Chem. Phys.* **47**, 2554 (1967).

36. N. Bloembergen and Y. R. Shen, *Phys. Rev. Lett.* **13**, 720 (1964).

37. N. Bloembergen and Y. R. Shen, *Phys. Rev.* **133**, A37 (1964).

38. Y. R. Shen and N. Bloembergen, *Phys. Rev.* **137**, A1787 (1965).

39. J. A. Armstrong, N. Bloembergen, J. Ducuing, and P. S. Pershan *Phys. Rev.* **127**, 1918 (1962).

40. E. A. Mishkin and D. F. Walls, *Phys. Rev.* **185**, 1618 (1969).

41. W. H. Louisell, A. Yariv, and A. E. Siegman, *Phys. Rev.* **124**, 1646 (1961).

42. B. R. Mollow and R. J. Glauber, *Phys. Rev. A* **160**, 1076, 1097 (1967).

43. R. Graham and H. Haken, *Z. Phys.* **210**, 276 (1968).

44. J. Tucker and D. F. Walls, *Phys. Rev.* **178**, 2036 (1969).

45. M. E. Smithers and E. Y. C. Lu, *Phys. Rev. A* **10**, 1874 (1974).

46. S. K. Srinivasan and S. Udayabaskaran, *Opt. Acta* **26**, 1535 (1979).

47. V. Peřinová, *Opt. Acta* **28**, 747 (1981).

48. V. Peřinová and J. Peřina, *Opt. Acta* **28**, 769 (1981).

49. C. M. Caves and B. L. Schumaker, *Phys. Rev. A* **31**, 3068 (1985).

50. B. L. Schumaker and C. M. Caves, *Phys. Rev. A* **31**, 3093 (1985).

51. D. F. Walls, *Z. Phys.* **234**, 231 (1970).

52. D. F. Walls, *J. Phys. A* **4**, 813 (1971).

53. J. Peřina, *Opt. Acta* **28**, 325 (1981).

54. J. Peřina, *Opt. Acta* **28**, 1529 (1981).

55. J. Peřina, *Acta Phys. Pol. A* **78**, 173 (1990).

56. M. Kárská and J. Peřina, *J. Mod. Opt.* **37**, 195 (1990).

57. D. F. Walls, *Z. Phys.* **237**, 224 (1970).

58. P. Szlachetka, S. Kielich, J. Peřina, and V. Peřinová, *J. Phys. A* **12**, 1921 (1979).

59. P. Szlachetka, S. Kielich, J. Peřina, and V. Peřinová, *J. Mol. Spectrosc.* **61**, 281 (1980).

60. P. Szlachetka, S. Kielich, J. Peřina, and V. Peřinová, *Opt. Acta* **27**, 1609 (1980).

61. P. Szlachetka and S. Kielich, in W. F. Murphy (Ed.), *Photon Co- and Anti-correlations in Two-photon Raman Scattering*, VII International Conference on Raman Spectroscopy, Ottawa, 1980, North Holland, Amsterdam, 1980.

62. T. V. Trung and F.-J. Schütte, *Ann. Phys.* (*Leipzig*) **7**, 216 (1978).

63. W. Tänzler and F.-J. Schütte, *Ann. Phys.* (*Leipzig*) **7**, 73 (1981).

64. F. G. Reis and M. P. Sharma, *Opt. Commun.* **41**, 341 (1982).

65. J. Peřina, V. Peřinová, and J. Kodousek, *Opt. Commun.* **49**, 210 (1984).

66. J. Peřinová, M. Kárská, and J. Křepelka, *Acta Phys. Pol. A* **79**, 817 (1991).

67. J. Peřina and J. Křepelka, *J. Mod. Opt.* **38**, 2137 (1991).

68. J. Peřina and J. Křepelka, *J. Mod. Opt.* **39**, 1029 (1992).

69. M. D. Levenson, M. J. Holland, D. F. Walls, P. J. Manson, P. T. H. Fisk, and H. A. Bachor, *Phys. Rev. A* **44**, 2023 (1991).

70. G. P. Agrawal and C. L. Mehta, *J. Phys. A* **7**, 607 (1974).

71. M. Schubert and B. Wilhelmi, *Nonlinear Optics and Quantum Electronics*, Wiley, New York, 1986, Chapter 1.

72. Y. R. Shen, in R. J. Glauber (Ed.), *Quantum Theory of Nonlinear Optics*, Proc. Int. School Phys. "Enrico Fermi," Vol. 42, Academic, New York, 1969, p. 473.

73. K. J. McNeil and D. F. Walls, *J. Phys. A* **7**, 617 (1974).

74. D. F. Walls, *Z. Phys.* **244**, 117 (1971).

75. H. D. Simaan, *J. Phys. A* **8**, 1620 (1975).

76. H. Iwasawa, *Z. Phys.* **23**, 399 (1976).

77. D. Eimerl, *Phys. Rev. A* **20**, 369 (1979).

78. P. S. Gupta and B. K. Mohanty, *Czech. J. Phys. B* **30**, 1127 (1980).

79. B. K. Mohanty and P. S. Gupta, *Czech. J. Phys. B* **31**, 275 (1981).

80. B. K. Mohanty, N. Nayak, and P. S. Gupta, *Opt. Acta* **29**, 1017 (1982).

81. N. N. Bogolubov, Jr., F. Le Kien, and A. S. Shumovsky, *Physica* **130A**, 273 (1985).

82. K. Grygiel and S. Kielich, *J. Mod. Opt.* **34**, 61 (1987).

83. K. Grygiel, *Quantum Theory of Stimulated Raman Scattering Processes* Thesis, Poznań University, 1987 (in Polish).

84. V. A. Andreev and P. B. Lerner, *Opt. Commun.* **84**, 323 (1991).

85. Th. Hansch and P. Toschek, *Z. Phys.* **236**, 213 (1970).

86. P. G. Brewer and E. L. Hahn, *Phys. Rev. A* **11**, 1641 (1975).

87. G. S. Agarwal and S. Jha, *Z. Phys. B* **35**, 391 (1979).

88. N. N. Bogolubov, Jr., A. S. Shumovsky, and T. Quang, *Physica* **144A**, 503 (1987).

89. A. S. Shumovsky, T. Quang, and R. Tanaś, *Physica* **149A**, 283 (1988).

90. P. S. Gupta and J. Dash, *Czech. J. Phys.* **40**, 432 (1990).

91. A. A. Villaeys, A. Boeglin, and S. H. Lin, *Phys. Rev. A* **44**, 4660, 4671 (1991).

92. H. Ritsch, M. A. M. Marte, and P. Zoller, *Europhys. Lett.* **19**, 7 (1992).

93. S. Kielich and P. Piątek, The squeezed states of fields in propagation and harmonic generation of laser beam, paper of this compilation, Part I, 1993.

94. S. Kielich, R. Tanaś, and R. Zawodny, Self-squeezing of elliptically polarized light propagating in a Kerr-like optically active medium, paper of this compilation, Part I, 1993.

95. J. Peřina, J. Bajer, V. Peřinová, and Z. Hradil, Photon statistics of nonclassical fields, paper of this compilation, Part I, 1993.

96. S. Kielich, M. Kozierowski, and R. Tanaś, *Opt. Acta* **32**, 1023 (1985).

97. G. Leuchs, in G. T. Moore and M. O. Scully (Eds.), *Frontiers of Nonequilibrium Statistical Physics*, Plenum, New York, 1986.

98. R. Loudon and P. L. Knight, *J. Mod. Opt.* **34**, 709 (1987).

99. M. C. Teich and B. E. A. Saleh, *Prog. Opt.* **26**, 1 (1988).

100. M. C. Teich and B. E. A. Saleh, *Quantum Opt.* **1**, 153 (1989).

101. K. Zaheer and M. S. Zubairy, *Adv. At. Mol. Opt. Phys.* **28**, 143 (1990).

102. Special issue of *J. Opt. Soc. Am. B* **4** (10) (1987).

103. Special issue of *J. Mod. Opt.* **34** (6/7) (1987).

104. K. Germey, F.-J. Schütte, and R. Tiebel, *Ann. Phys.* (*Leipzig*) **38**, 80 (1981).

105. P. Král, *Czech. J. Phys.* **40**, 1226 (1990).

106. P. S. Gupta and J. Dash, *Indian J. Pure and Appl. Phys.* **29**, 606 (1991).

107. H. D. Simaan, *J. Phys. A* **11**, 799 (1978).

108. P. Szlachetka and S. Kielich, *VIII Conf. on Quantum Electronics and Nonlinear Optics*, Poznań, 24–27 April 1978, Section B, p. 33.

109. G. S. Agarwal, *Opt. Commun.* **31**, 325 (1979).

110. V. Peřinová, J. Peřina, P. Szlachetka, and S. Kielich, *Acta Phys. Pol. A* **56**, 267, 275 (1979).

111. W. Tänzler and F.-J. Schütte, *Opt. Commun.* **37**, 447 (1981).

112. B. K. Mohanty and P. S. Gupta, *Czech. J. Phys. B* **31**, 857 (1981).

113. V. Peřinová and R. Tiebel, *Opt. Commun.* **50** 401 (1984).

114. P. S. Gupta and J. Dash, *Opt. Commun.* **88**, 273 (1992).

115. Yu.P. Malakyan, *Opt. Commun.* **78**, 67 (1990).

116. Yu.P. Malakyan, *J. Mod. Opt.* **39**, 509 (1992).

117. S. Kielich, *J. Phys. Lett.* **43**, L-389 (1982).

118. S. Singh, *Opt. Commun.* **44**, 254 (1983).

119. X. T. Zhou and L. Mandel, *Phys. Rev. A* **41**, 475 (1990).

120. M. Le Berre-Rousseau, E. Ressayre, and A. Tallet, *Phys. Rev. Lett.* **43**, 1314 (1979).

121. Z. Bialynicka-Birula, I. Bialynicki-Birula, and G. M. Salomone, *Phys. Rev. A* **43**, 3696 (1991).

122. D. F. Walls, *J. Phys. A* **6**, 496 (1973).

123. M. Smoluchowski, *Ann. Phys.* **25**, 332 (1908).

124. P. D. Drummond and C. W. Gardiner, *J. Phys. A* **13**, 2353 (1980).

125. S. J. Carter and P. D. Drummond, *Phys. Rev. Lett.* **67**, 3757 (1991).

126. N. R. Davis and C. D. Cantrell, *J. Opt. Am. Soc. B* **6**, 74 (1989).

127. S. L. Braunstein, C. M. Caves, and G. J. Milburn, *Phys. Rev. A* **43**, 1153 (1991).

128. R. Schack and A. Schenzle, *Phys. Rev. A* **44**, 682 (1991).

129. D. T. Pegg and S. M. Barnett, *Phys. Rev. A* **39**, 1665 (1989).

130. R. Tanaś, Ts. Gantsog, A. Miranowicz, and S. Kielich, *J. Opt. Soc. Am. B* **8**, 1576 (1991).

131. Ts. Gantsog, A. Miranowicz, and R. Tanaś, *Rev A* **46**, 2870 (1992).

132. R. Tanaś, A. Miranowicz, and S. Kielich, *Phys. Rev. A* **43**, 4014 (1991).

133. P. L. Knight, *Phys. Scr. T* **12**, 51 (1986).

134. S. J. D. Phoenix and P. L. Knight, *J. Opt. Soc. Am. B* **7**, 116 (1990).

135. L. Schoendorff and H. Risken, *Phys. Rev. A* **41**, 5147 (1990).

136. G. S. Agarwal and R. R. Puri, *Phys. Rev. A* **43**, 3949 (1991).

137. R. R. Puri and G. S. Agarwal, *Phys. Rev. A* **45**, 5073 (1992).

138. C. C. Gerry and J. H. Eberly, *Phys. Rev. A* **42**, 6805 (1990).

139. C. C. Gerry, *Phys. Lett. A* **161**, 9 (1991).

140. C. C. Gerry and H. Huang, *Phys. Rev. A* **45**, 8037 (1992).

141. G. Gangopadhyay and D. S. Ray, *Phys. Rev. A* **45**, 1843 (1992).

142. J. Mostowski and M. G. Raymer, *Opt. Commun.* **36**, 237 (1981).

143. M. G. Raymer and J. Mostowski, *Phys. Rev. A* **24**, 1980 (1981).

144. S. Ya. Kilin, *Europhys. Lett.* **5**, 419 (1988).

145. Z. Z. Li, C. Radzewicz, and M. G. Raymer, *J. Opt. Soc. Am. B* **5**, 2340 (1988).

146. S. A. Akhmanov, A. P. Sukhorukov, and A. S. Chirkin, *Zh. Eksp. Teor. Fiz.* **55**, 1430 (1968) [*Sov. Phys.-JETP* **28**, 748 (1969)].

147. T. von Foerster and R. J. Glauber, *Phys. Rev. A* **3**, 1484 (1971).

148. S. A. Akhmanov, K. N. Drabovich, A. P. Sukhorukov and A. S. Chirkin, *Zh. Eksp. Teor. Fiz.* **59**, 485 (1970) [*Sov. Phys.-JETP* **32**, 266 (1971)].

149. V. I. Emel'yanov and V. N. Seminogov, *Zh. Eksp. Teor. Fiz.* **76**, 34 (1979) [*Sov. Phys.-JETP* **49**, 17 (1979)].

150. E. Haake, H. King, G. Schröder, J. W. Haus, and R. Glauber, *Phys. Rev. A* **20**, 2047 (1979); **23**, 1322 (1981).

151. M. G. Raymer, K. Rzążewski, and J. Mostowski, *Opt. Lett.* **7**, 71 (1982).

152. F. Haake, *Phys. Lett. A* **90**, 127 (1982).

153. J. Mostowski and B. Sobolewska, *Phys. Rev. A* **30**, 610 (1984).

154. I. A. Walmsley and M. G. Raymer, *Phys. Rev. Lett.* **50**, 962 (1983).

155. M. G. Raymer and I. A. Walmsley, in L. Mandel and E. Wolf (Eds.), *Proc. 5th Rochester Conf. on Coherence and Quantum Optics*, Plenum, New York, 1984.

156. N. Fabricius, K. Nattermann, and D. van der Linde, *Phys. Rev. Lett.* **52**, 113 (1984).

157. D. C. MacPherson, R. C. Swanson, and J. L. Carlsten, *Phys. Rev. A* **40**, 6745 (1989).

158. R. C. Swanson, D. C. MacPherson, P. R. Battle, and J. L. Carlsten, *Phys. Rev. A* **45**, 450 (1992).

159. J. C. Englund and C. M. Bowden, *Phys. Rev. Lett.* **57**, 2661 (1986).

160. J. C. Englund and C. M. Bowden, *Phys. Rev. A* **42**, 2870 (1990).

161. S. G. Rautian and B. M. Chernobrod, *Zh. Eksp. Teor. Fiz.* **72**, 1342 (1977) [*Sov. Phys.-JETP* **54**, 642 (1977)].

162. Yu. M. Kirin, Yu. N. Popov. S. G. Rautian, V. P. Safonov, and B. M. Chernobrod, *Kvant. Elektron.* **1**, 430 (1974) [*Sov. J. Quantum Electron.* **4**, 244 (1974)].

163. V. S. Pivtsov, S. G. Rautian, V. P. Safonov, K. G. Folin, and B. M. Chernobrod, *JETP Lett.* **30**, 317 (1979); Zh. Eksp. Teor. Fiz. **81**, 468 (1981) [Sov. Phys.-JETP **54**, 250 (1981)].

164. R. W. Terhune, P. D. Maker, and C. M. Savage, *Phys. Rev. Lett.* **14**, 681 (1965).

165. P. D. Maker, *Physics and Quantum Electronics*, McGraw-Hill, New York, 1966, p. 60.

166. T. Neugebauer, *Acta. Phys. Hungar.* **16**, 217, 227 (1963).

167. S. Kielich, *Bull. Acad. Pol. Sci.* (*Sér. Math. Astron. Phys.*) **11**, 201 (1963); **12**, 53 (1964).

168. S. Kielich, *Acta Phys. Pol.* **26**, 135 (1964).

169. Y. Li, *Acta Phys. Sinica* **20**, 164 (1964).

170. S. Kielich, *Proc. Phys. Soc. London* **86**, 709 (1965).

171. S. Kielich, *Acta Phys. Pol.* **30**, 393 (1966).

172. D. C. Hanna, M. A. Yuratich, and D. Cotter, *Nonlinear Optics of Free Atoms and Molecules*, Springer, Heidelberg, 1979.

173. Ch. Kittel, *Quantum Theory of Solids*, Wiley, New York, 1963.

174. R. H. Pantell and H. E. Puthoff, *Fundamentals of Quantum Electronics*, Wiley, New York, 1969.

175. H. H. Ritze and A. Bandilla, *Opt. Commun.* **29**, 126 (1979).

176. R. Tanaś and S. Kielich, *Opt. Commun.* **30**, 443 (1979).

177. R. Tanaś and S. Kielich, *Opt. Commun.* **45**, 351 (1983).

178. R. Tanaś and S. Kielich, *Opt. Acta* **31**, 81 (1984).

179. Ts. Gantsog and R. Tanaś, *J. Mod. Opt.* **38**, 1537 (1991).

180. R. J. Glauber, *Phys. Rev.* **130**, 2529 (1963).

181. G. Rivoire, *C. R. Acad. Sci. Paris* **260**, 5743 (1965).

182. J. Peřina, J. Peřinová, P. Szlachetka, and S. Kielich, *Approximate Photon and Phonon Statistics and Anticorrelation in Raman and Brillouin Scattering*, Commun. Palacký University, Olomouc, 1979.

183. W. H. Louisell, *Quantum Statistical Properties of Radiation*, Wiley, New York, 1973.

184. C. W. Gardiner, in H. Haken (Ed.), *Springer Series in Synergetics*, Vol. 13, *Handbook of Stochastic Methods*, Springer, Berlin, 1984.

185. G. S. Agarwal, *Prog. Opt.* **11**, 1 (1973).

186. H. Haken, *Handbuch der Physik*, Vol. 25/2c, Springer, Berlin, 1970.

187. M. W. Evans, G. J. Evans, W. T. Coffey, and P. Grigolini, *Molecular Dynamics and Theory of Broad Band Spectroscopy*, Wiley, New York, 1982.

188. H. Risken, *The Fokker-Planck Equation*, Springer, Berlin, 1984.

189. V. Peřinová, J. Křepelka, and J. Peřina, *Opt. Acta* **33**, 1263 (1986).

190. M. Sugawara, M. Hayashi, and Y. Fujimura, *Chem. Phys. Lett.* **189**, 346 (1992).

191. G. E. Uhlenbeck, L. S. Ornstein, *Phys. Rev.* **36**, 823 (1930).

192. R. J. Glauber, *Phys. Rev. A* **131**, 2766 (1963).

193. R. J. Glauber, *Phys. Rev. Lett.* **10**, 84 (1963).

194. K. E. Cahill, *Phys. Rev.* **138**, B1566 (1965).

195. J. R. Klauder, J. McKenna, and D. G. Currie, *J. Math. Phys.* **6**, 743 (1965).

196. G. S. Agarwal and E. Wolf, *Phys. Lett.* **26A**, 485 (1968).

197. K. E. Cahill and R. J. Glauber, *Phys. Rev.* **177**, 1857, 1882 (1969).

198. J. R. Klauder and E. C. G. Sudarshan, *Fundamentals of Quantum Optics*, Benjamin, New York, 1968.

199. V. I. Tatarsky, *Usp. Fiz. Nauk* **139**, 587 (1983).

200. M. Hillery, R. F. O'Connell, M. O. Scully, and E. P. Wigner, *Phys. Rep.* **106**, 121 (1984).

201. A. Miranowicz, R. Tanaś and S. Kielich, *Quantum Opt.* **2**, 253 (1990).

202. U. M. Titulaer and R. J. Glauber, *Phys. Rev.* **140**, B676 (1965).

203. M. Hillery, *Phys. Rev. A* **35**, 725 (1987).

204. S. Stenholm, *Eur. J. Phys.* **1**, 244 (1980).

205. L. Mandel, *Proc. Phys. Soc.* **72**, (1958) 1037; **74**, 233 (1959).

206. L. Mandel, *Prog. Opt.* **2**, 181 (1963).

207. J. Peřina and R. Horak, *Opt. Commun.* **1**, 91 (1969).

208. R. J. Glauber, in C. DeWitt, A. Blandin, and C. Cohen-Tannoudji (Eds.), *Quantum Optics and Electronics*, Gordon & Breach, New York, 1965, p. 144.

209. B. Saleh, *Springer Ser. Opt. Sci.*, *Vol.* 6 *Photoelectron Statistics*, Springer, Berlin, 1978.

210. H. Paul, *Rev. Mod. Phys.* **54**, 1061 (1982).

211. L. Lukš, V. Peřinová, and J. Peřina, *Opt. Commun.* **67**, 149 (1988).

212. R. Loudon, *Opt. Commun.* **70**, 109 (1989).

213. A. Lukš, V. Peřinová, and Z. Hradil, *Acta Phys. Pol. A* **74**, 713 (1988).

214. E. Schrödinger, *Sitzungsber. Preuss. Akad. Wiss. Phys. Math. K* **1**, 296 (1930).

215. M. Kozierowski and V. I. Man'ko, *Opt. Commun.* **69**, 71 (1988).

216. K. Vogel and H. Risken, *Phys. Rev. A* **39**, 4675 (1989).

217. Ning-Lu, Shi-Yao Zhu, and G. S. Agarwal, *Phys. Rev. A* **40**, 258 (1989).

218. A. Miranowicz, *Master Equation Approach to Raman Scattering*, Thesis, Poznań University, 1988 (in Polish).

219. H. Fearn and M. J. Collet, *J. Mod. Opt.* **35**, 553 (1988).

220. M. O. Scully and W. E. Lamb, *Phys. Rev.* **159**, 208 (1967).

221. G. Lachs, *Phys. Rev.* **138**, B1012 (1965).

222. V. Peřinová, A. Lukš, and P. Szlachetka, *J. Mod. Opt.* **36**, 1435 (1989).

223. D. A. Long and L. Stanton, *Proc. R. Soc. London Ser. A* **318**, 441 (1970).

224. S. M. Barnett, S. Stenholm, and D. T. Pegg, *Opt. Commun.* **73**, 314 (1989).

FIELD-PERTURBED WATER

SHENG-BAI ZHU

IBM Research Division, Almaden Research Center, San Jose, California

SURJIT SINGH

and

G. WILSE ROBINSON

SubPicosecond and Quantum Radiation Laboratory, Texas Tech University, Lubbock, Texas

CONTENTS

Modern Nonlinear Optics, Part 3, Edited by Myron Evans and Stanisław Kielich. Advances in Chemical Physics Series, Vol. LXXXV.
ISBN 0-471-30499-9 © 1994 John Wiley & Sons, Inc.

I. INTRODUCTION

The connotation "field-perturbed water" is a many-faceted concept. Important from environmental,[1] technological,[2] electrochemical,[3–5] and biological[6–9] viewpoints are perturbations by surfaces and dissolved solutes. The surfaces or the solutes may be molecular or ionic; they may be a biological membrane or a macromolecule; they may be metallic; and they may be electrically charged. At a molecular level, of course, there are always charges, and very often the extent of such charge distributions is comparable to the size of a water molecule, giving rise to intricate close-range quantum interactions between the liquid and its perturber. Liquid water may also be measurably perturbed by an external field, such as that originating from these surfaces or from a strong laser pulse. A question immediately arises as to the effect of such perturbations on properties of the more distant liquid, considering the fact that water is such a highly structured material compared with most other liquids. For example, in an aqueous phase, how is chemistry, such as an acid–base or an oxidation–reduction reaction, affected at various distances from a perturbing surface or a solute? What about physical properties such as self-diffusion, the density, the viscosity, and the electrical conductivity— how do they change? An essential requirement for many biological processes is the presence of a layer of ordered water adjacent to a membrane or a protein surface.[8, 9] Understanding surface-perturbed water is thus essential if biological function at a molecular level of knowledge is to be gained.

The interaction of water with clays, surfactants, and metal surfaces is important in oil recovery, mining, catalysis, corrosion inhibition, and other technologically important concerns.[2] Water is also a commonly employed liquid for generating a supercontinuum for the production of intense ultrafast broadband "white light" pulses.[10] The addition of ions to the water may considerably enhance such effects.[11] Can a better picture of liquid water at the molecular level help in the understanding and development of these diverse areas of technological importance. If so, probing into the essence of field-perturbed water would be of fundamental utility, not only in many branches of pure science, but also in engineering and industry.

A separate facet of this topic concerns the properties of liquid water itself in the absence of perturbations. No other liquid has such abnormal properties. Though numerous experimental measurements of the properties of water at normal[12, 13] and supercooled temperatures[13–18] have been carried out, little is understood about the molecular-level origins of many of these properties, let alone what happens to them in the presence of perturbations. Unlike other substances that have a comparable molecular weight, such as nitrogen, carbon dioxide, methane, and ammonia, water exhibits anomalously high melting, boiling, and critical temperatures. Besides these unusual thermal properties, water displays a large dielectric constant and surface tension. The density maximum near 277 K for H_2O,[13] the much larger than expected heat capacity,[14–17] and the minimum in the isothermal compressibility[13] are celebrated properties of water that do not have a well-accepted explanation. How then will it be possible to learn details about field-perturbed water when properties of the unperturbed substance are so poorly understood?

Attempts at purely theoretical descriptions of liquids can be made.[19] However, when applied to water, such descriptions, while useful for suggesting a qualitative picture, lack sufficient molecular detail and many-body properties to constitute an accurate theory of this substance. More promising are computer simulations.[20-22] Such methods have been applied to a variety of liquids, both in the pure and perturbed states. Typically, it is possible to treat about a thousand or so atomic particles, such as argon or xenon, using the relatively simple Lennard-Jones interatomic interaction potential.[23] Presently, however, fewer than half this number of molecular particles can be handled by computational methods, depending on the number of constraints imposed on the intramolecular degrees of freedom and the simplicity of the molecule–molecule intermolecular potential functions employed.

The computational approach to unperturbed liquid water ("bulk water") or perturbed liquid water, though there are but three atoms per molecule, presents a special challenge because of the need for complicated angle- and space-dependent intermolecular interactions. Furthermore, when time-varying electromagnetic fields are present, accurate water modeling requires the presence of rapidly responsive electronic polarization. Even without an applied field, electrical disturbances by surface charges, ions, and the surrounding polar water molecules themselves, as the positions and orientations of these disturbances constantly fluctuate in time, require the presence of polarization if the ultimate water model is ever to be achieved. Adding intramolecular vibrational degrees of freedom may also be essential in the ultimate water model.

During the past twenty odd years a large volume of effort has been expended on computer simulations of various liquid water models.[24-76] However, in many ways this work must still be classed as a failure, since it has provided at most only a small crack in the fortress that separatesthe understanding of this complex problem from realization. Experiments[12, 14, 17] on liquid water have ranged over temperatures from deep supercooled regions near 238 K to the critical temperature near 650 K; and from atmospheric to very high pressures. Computational efforts have not come close to matching these conditions. Temperatures well below the freezing point create difficulties in flexibly bonded computational models because of inevitable time scale separation problems[77] between hydrogen motions and the motions of the entire molecule. Furthermore, realistic pressures of water ensembles have often been difficult to attain if other properties are preserved. The RWK-type potentials,[27, 71] the Evans potential,[43] SPC,[36] and TIP4P,[40] together with some of the more recent flexible/polarizable models[73-75] do produce acceptable pressures, at least at specific temperatures. Without realistic pressure and temperature effects, the most important keys to the understanding of water are missing.

Many attempts to develop a liquid model that is based on the properties of the water monomer, the dimer, or other small polymers have been made.[24, 27, 34, 57, 66, 69, 71, 73, 78–87] More sparingly[27, 55, 69, 88, 89] have the interaction models been checked to see whether and how well they reproduce the structures of the thermodynamically most stable forms of ice—ice Ih, II, III, and V.[12] Because of the close relationships between intermolecular bonding in the liquid and solid phases of most molecular substances, such connections would seem to be paramount before embarking on the extensive use of any model in the investigation of bulk or perturbed water.

Finally, quantum effects[90] must be included for any system, such as water, where spectral frequencies above kT are present, or when electronic phenomena occur, such as in the case of water perturbed by a very intense laser field.

Of course, the underlying reason behind all the failings of water computations is expense, computer expense, so there is the expectation of much better simulations of water in the future as the evolution of new computer generations continues. Rather than sitting and waiting for this evolution to occur, water scientists have continued to improve their methods, not only as a preparation for things to come, but also with the hope that "the crack in the fortress" can be widened to reveal glimmers of a better understanding necessary for a more rapid future progress in all fundamental studies of liquid water. It is with these considerations in mind that the following review material was contemplated and composed.

II. ELECTROMAGNETICS

A. Units and Conversion Factors

Of all the fields of physics, electromagnetism has suffered the greatest confusion from reformulations by various international bodies of notation and systems of units. Since the world is always fragmented in its acceptance of each subsequent set of recommendations, a plethora of symbols and units now exists. For instance, the long-standing dipole moment symbol μ can no longer be used because of possible confusion with the symbol for the magnetic permeability. The confusion is often mentioned in prefaces to texts on electromagnetic theory[91] and was anticipated by Jackson[92] in his book *Classical Electrodynamics*. In this book an entire appendix is dedicated to the subject, which includes specific connectivity relationships between the Gaussian (cgs, esu), the rationalized MKSA (SI), and other systems.

More recently, we find other recommendations,[93] such as that intensity, the traditional term for photon flux, should now be used only for qualitative descriptions; irradiance, with symbol E, should henceforth be used for this quantity. In this review, we will ignore the latter directive and continue to use the symbols $\mathbf{E}(\mathbf{r}, t) = \mathbf{i}_x E_0 \exp[i\mathbf{k} \cdot \mathbf{r} - i\omega t]$ with dimensions V/m and $\mathbf{B}(\mathbf{r}, t) = \mathbf{i}_y B_0 \exp[i\mathbf{k} \cdot \mathbf{r} - i\omega t]$ with dimensions Wb/m^2 to denote the plane wave electric field and magnetic induction vectors at the position $\mathbf{r}$ from a suitably chosen origin. The symbol I, which is proportional to E_0^2, with dimensions W/m^2, will be used for intensity, since it is by far the most commonly used and familiar term in the laser physics community. In the above discussion, E_0 and B_0 are complex amplitudes, constant in time and space, ω is the angular frequency, and $\mathbf{i}_x$ and $\mathbf{i}_y$ are unit vectors perpendicular to the propagation vector $\mathbf{k}$ of the light. The present review utilizes the MKSA (SI) system of units, but may additionally use or refer to the Gaussian cgs system, not only because fragments of this system are still in common usage, but also because some quantities have their best remembered values in cgs units. Mixed descriptions in the literature are common, so it is not wise for a reader to assume Gaussian only because cm is used as the unit of length. Even notation may not be consistent. Though ε_0 and μ_0 are universally used for the vacuum permittivity and permeability, some authors use the symbols ε and μ in a medium, while others use these same symbols for the dimensionless quantities relative to ε_0 and μ_0. We will use the latter notation in this review. Since magnetic fields of between 10^8 and 10^9 G are required for energy changes of the order of kT in a rotating water molecule, no treatment of macroscopic magnetic effects will be necessary,[94] simplifying the discussion of units.

B. Laser Fields and Intensity

In a homogeneous nonconducting medium, with general values, rather than vacuum values, of the electric susceptibility and the magnetic permeability, and remembering that the relative value ε is the dielectric constant, the Poynting vector,

$$\mathbf{S} = [\mathbf{E} \times \mathbf{H}^*] = \left(\frac{\varepsilon \varepsilon_0}{\mu \mu_0} \right)^{1/2} E_0^2 \mathbf{i}_z \tag{1}$$

gives the energy flow, which has dimensions W/m^2. Here, $\mathbf{H}(\mathbf{r}, t)$ is the magnetic field vector and $\mathbf{i}_z$ is a unit vector in the propagation direction. The above results[91, 92] follow from the plane wave relationship $B_0 = (\mu\mu_0 \varepsilon\varepsilon_0)^{1/2} E_0$ and $\mathbf{B} = \mu\mu_0 \mathbf{H}$ for nonferromagnetic media. There will be no need in the present review to go beyond this linearly polarized description.

The time-averaged intensity, termed merely *intensity* or I in this review, is taken to be one-half the absolute value $|S|$ of the Poynting vector. Since μ is typically unity for optical materials, the factor $(\varepsilon/\mu)^{1/2}$ can often be replaced[91] by the refractive index n, which, in liquid water at ambient pressures and temperatures, is about 1.33 for frequencies in the optical region.[12] The square-root prefactor in Eq. (1) then becomes $nc\varepsilon_0$, since the velocity of light $c = (\varepsilon_0\mu_0)^{-1/2}$. Thus, with all units MKSA, the relationship between E_0 (V/m) and I (W/m^2) is $E_0 = (2I/nc\varepsilon_0)^{1/2}$, which very closely equals $23.8 \times I^{1/2}$.

The high-frequency dielectric constant $\varepsilon(\infty)$ of liquid water used in the above equations was 1.77. In reality, however, because of relatively strong infrared absorption bands, the actual index of refraction varies[12] between 1.2 and 2.2 in the wavelength range 2×10^3 to 2×10^5 nanometers (nm), and directly measured values of $\varepsilon(\infty)$ are actually larger than 1.77, ranging up to around 6.0. Thus, variations in the I vs. E_0 relationship of perhaps 30% exist for liquid water, depending on the radiation frequency.

It has been stated[95] that 5×10^9 V/cm is approximately equal to the "atomic field" and that therefore any applied field very much less than this would have a negligible effect. This magnitude of field is that caused by a unit of electronic charge q at a distance equal to the Bohr radius (0.52918 Å). In the Gaussian system of units, the charge q is 4.80321×10^{-10} esu, and thus $E = 4.80321 \times 10^{-10}/(0.52918 \times 10^{-8})^2 = 1.71524 \times 10^7$ statvolt/cm. Multiplying[92] this by 2.99792×10^4, obtained from the velocity of light in vacuum, gives 5.1421×10^{11} V/m, or the commonly used 5.1421×10^9 V/cm, which is the result of Ref. 95.

Another simple exercise, albeit an important one, is to determine the magnitude of the field under laser illumination of a given intensity. Modern lasers are almost routinely able to produce pulses as brief as 20 fs, and new advances in titanium–sapphire[96] and other solid-state laser materials have made the availability of these and even shorter pulses still more routine. A 20-fs pulse contains only around 10 optical cycles of light in the visible region of the spectrum and is typically in the 100-nJ range of energy per pulse. If such pulses can be appropriately amplified[97] so that the energy per optical cycle reaches 1×10^{-4} J, peak powers of 50 GW would be available. If, furthermore, the pulse can be focused down to its diffraction limited spot size of about 10 μm diameter,[98] then the intensity would be approximately 6×10^{20} W/m^2 over the brief pulse duration. According to the earlier discussion, the electromagnetic field strength E_0 in such a pulse is 6×10^{11} V/m, which would be comparable to the magnitude of the atomic field mentioned above. Under these circumstances, as emphasized by Delone and Fedorov,[95] "treating the atom as a single system requires separate justification." In other words, in such fields, atoms or molecules should not be looked upon as isolated quantum-mechanical objects to which a field perturbation has been applied. Rather, the atom and the field must be treated on equal terms.[99, 100] The very concepts of atomic or molecular energy levels, multiphoton ionization, etc., become indistinct. To our knowledge, however, laser fields thus far studied with regard to liquid water fall 2–3 orders of magnitude short of these.

In dielectrics, the polarization **P** is defined as the macroscopic dipole moment **M** per unit volume. Ignoring all higher pole contributions to **P** beyond the dipole density,[92] the electric displacement vector **D** is related to **E** through the fundamental relationship,

$$\mathbf{D} = \varepsilon_0 \mathbf{E} + \mathbf{P} \tag{2}$$

where, in general, **D**, **E**, and **P** are functions of space and time. To study the frequency-dependent properties, it is sufficient to apply a spatially uniform but time-dependent field. Furthermore, if the field is uniaxial, it can be expressed as $E(t) = E_0 \sin \omega t$. For such a periodic field, **D** must also be periodic in time. However, **D** is not necessarily in phase with **E** because of the inertial response of the dielectric to the fast alternation of the electric field. For an isotropic system, the polarization has the same direction as the applied field and can be written[92]

$$P(t) = \frac{1}{V} \sum_{j=1}^{N} \sum_{k=1}^{n} q_{jk}(t) X_{jk}(t) \tag{3}$$

where $q_{jk}(t)$ is the instantaneous charge at site k on molecule j, and $X_{jk}(t)$ is the component of the position vector parallel to the applied field. By writing the real part of the polarization,[101]

$$P(t) = P_0 + P_1 \sin \omega t + P_2 \cos \omega t \tag{4}$$

where the constant term is from the permanent dipoles, one can obtain the real and imaginary parts of the dielectric constant. From these, the optical constants, namely the real n_r and imaginary n_i parts of the refractive index,[102] can be found. As outlined in Section VII.C, in a computer molecular dynamics (MD) simulation of liquid water in an intense laser field,[103, 104] one can find the time-dependent polarization and thus can evaluate these optical constants.

III.   EXPERIMENTAL OVERVIEWS

To appreciate effects that can occur when liquid water is subjected to a perturbing field, one must be aware of some of the properties of an individual, unperturbed water molecule. Moreover, to understand the intermolecular interactions between water molecules in both the unperturbed and perturbed liquid states, it is worthwhile to have as much knowledge as possible about well-defined ensembles of water molecules, from the dimer to the various crystalline phases.

A.   Monomer and Dimer

The equilibrium structure and intramolecular vibrational frequencies of monomeric water are known to remarkable accuracy from the rotation–vibration spectra of both normal and isotopic water vapor.[12, 105] Equilibrium values for the O–H bond lengths and H–O–H bond angles are 0.9572 Å and 104.52°, respectively. In the standard normal mode description, the symmetric O–H stretching fundamental frequency in H_2O lies at 3657.05 cm^{-1}, the antisymmetric stretching fundamental at 3755.79 cm^{-1}, while the H–O–H bending vibration has a frequency of 1594.78 cm^{-1}. Using Fourier transform spectroscopy, Camy-Peyret and coworkers[106] have measured the rotational–vibrational spectrum of overtone and combination bands, together with many intensities, of vapor-phase H_2O to energies of 3.13 eV (25 250 cm^{-1}). In addition, the analysis of much of this spectrum in terms of the normal modes has been successfully carried out by these authors. An interesting *local mode* description for some of the vibrations has also been presented.[107] The properties of these highly excited vibrational levels are useful in single or multiphoton photochemistry using powerful laser excitation of the water molecule either in the gaseous or the condensed phase.[108, 109] No experimental evidence for the predicted hyperspherical[110, 111] modes in highly excited water molecules has yet been reported.

Water is a polar molecule with a permanent dipole moment $\mathbf{m}_0$ of -1.8546 ± 0.0006 debye (D = 1×10^{-18} esu).[112] Using beam-maser Zeeman spectra, Verhoeven and Dymanus[113] have measured the permanent quadrupole moments of water. With the center of mass as origin, their values are

$$Q_{xx} = -(2.50 \pm 0.02) \times 10^{-26} \text{ esu}$$

$$Q_{yy} = +(2.63 \pm 0.02) \times 10^{-26} \text{ esu}$$

$$Q_{zz} = -(0.13 \pm 0.03) \times 10^{-26} \text{ esu}$$

with the z axis along the symmetry axis pointing in a direction from the H–H axis toward the oxygen atom and the x axis perpendicular to the molecular plane. Quantum-mechanical calculations,[113, 114] tend to confirm these values. The octopole moments R_{xxz}, etc., have not yet been measured, though theoretical values are available.[115] Note, however, that not only the choices of origin but also the actual definitions of the quadrupole and octopole moments in some important reference books[12] are not uniform.

When the water molecule is placed in an electric field, its dipole moment changes,

$$\mathbf{m} = \mathbf{m}_0 + \alpha \mathbf{E} \tag{5}$$

where $\alpha \mathbf{E} = \mathbf{m}_{ind}$. Here, α denotes the polarizability, which in general is a tensor having a component along each of three principal axes. These components are still unknown for water vapor. Since the dipole polarizability of water mainly derives from the oxygen atom, the anisotropy of the polarization is not large. Therefore, it is reasonable to treat α as a scalar. A reasonably recent experimental value of α, 1.4288×10^{-24} cm^3, was derived through a detailed analysis that combined refractivity, dispersion, and dipole oscillator strength data in the visible and ultraviolet spectral regions.[116]

Thermal conductivity measurements on the vapor[117] show that the dimer has an association energy of -22.8 ± 2.9 kJ/mol. This is at an $O \cdots O$ separation of 2.98 Å,[118] about 0.2 Å greater than the closest $O \cdots O$ separation in ice Ih. The internal energy for ice Ih is -58.9 kJ/mol,[119] much deeper than double the binding energy of the water dimer, indicating sizable contributions from nonpairwise additive interactions in the crystal. This is partly caused by the strongly increased dipole moment of the water molecule in condensed phases, as discussed in the next section. This fact, together with the known complexity and fragility of the potential surfaces for small water polymers,[86] ruins straightforward extensions of dimer properties to those of the condensed phrase. More extensive details about the dimer spectroscopy and its potential energy surfaces are becoming available through tunable far infrared laser spectroscopy by Saykally and his coworkers[120] at Berkeley.

B. Ice

In condensed phase water, ice is much more easily understandable than the liquid. In fact, ice is especially important since its crystal structure can often provide hints for interpreting the unusual characteristics of the liquid.[121]

It has been well established that the basic structure of normal ice, ice Ih, is hexagonal,[12] with disordered hydrogen bonds, which contribute 3.408 J K^{-1} mol^{-1} of stabilizing Pauling entropy to this form.[122] In the ice Ih lattice,[12] every oxygen atom resides at the center of a tetrahedron formed by four nearest-neighbor oxygens whose positions are separated by 2.74 Å at 98 K (2.76 Å near the melting point). Each water molecule forms two nearly linear O–H $\cdots$ O hydrogen bonds and two nearly linear O $\cdots$ H–O hydrogen bonds with its four nearest neighbors, averaging to two hydrogen bonds per molecule over the complete ice lattice.

The locations of the hydrogen nuclei are relatively less certain than those of the oxygens because they scatter x-rays and electrons in a less effective way. Neutron diffraction (D_2O)[123, 124] and NMR (H_2O) data[125] have provided evidence that the hydrogen nuclei lie near the line joining the oxygen atoms, but have not given unequivocal information about the distortion of the intramolecular H–O–H bond angle compared with its monomer value. This lies between $-2°$ and the full tetrahedral distortion of $+5°$. We note here that in the liquid the distortion is probably within $\pm 2°$, both from experiment[126, 127] and recent MD simulations using flexibly bonded water models.[54, 57, 73-75]

According to the neutron diffraction data,[12, 123, 124] each oxygen atom in ice Ih is intramolecularly bonded to two hydrogen atoms at a distance of roughly 1.01 Å. The longer intramolecular bond as compared with the monomer is expected in the presence of the strong hydrogen bonding in ice Ih. However, this reported bond length seems perhaps 0.01 to 0.02 Å too long compared with a recently reported value for the liquid (~ 0.989 Å).[127]

A metastable modification of ice I, ice Ic (cubic), exists[12] and has a local structure similar to that of ice Ih. The structures, II, III, and V[12] are extremely interesting, since differential x-ray scattering data[128, 129] seem to suggest that a similar bonding arrangement may exist in the liquid. The arrangement[12, 130, 131] of hydrogen bonds in ice II resembles that of ice Ih, except that the hexagonal tunnels are packed in a more economical manner, and the hydrogen atom positions are ordered, so the Pauling entropy[122] is absent. It is interesting that, without this ordering, ice II,[130] with a density of 1.17 g cm^{-3}, would have roughly equal stability as ice I, which would, of course, have had an immeasurable impact on life forms on earth. Ice III consists of helical chains of hydrogen-bonded oxygens, whereas ice V shows signs of further distortions. In both these structures the hydrogen bonds are disordered. It is most important to note that, while the nearest-neighbor hydrogen-bonded O $\cdots$ O distances in all these forms of ice—Ih, Ic, II, III, V—are nearly the same, ranging from about 2.74 to 2.87 Å, non-hydrogen-bonded O $\cdots$ O neighbors have very different spacings. They are shortened from about 4.5 Å in ice Ih and ice Ic to 3.24–3.47 Å in ice III, and V. This is the main reason for the huge density difference in these structures, around 0.94 g cm^{-3} in ice Ih and Ic and 1.14–1.23 g cm^{-3} in the three other forms (all at $T = 98$ K).[12] These bonding variations in the high- and low-density forms of ice could also account for the density anomalies in the liquid (see Section III.C). In addition to these thermodynamically more stable forms of ice, there also exist the higher pressure polymorphs VI, VII, and VIII and other forms, such as ice IX,[131] which is an ordered hydrogen-bonded form of ice III. The crystallographic properties of the most commonly known ice polymorphs are well summarized in Tables 3.3 and 3.4 of Eisenberg and Kauzmann.[12]

Infrared and Raman spectra[12, 119] of the various polymorphs of ice, and also the liquid, are complicated because of the couplings among the intramolecular and intermolecular vibrations. See Ref. 132 for a good discussion of this extensive research as it applies to the high-frequency region of the ice spectrum. In H_2O, broad bands at frequencies less than 1600 cm^{-1} are fundamentals, overtones, or combinations of intermolecular translational and intermolecular librational modes, while bands having a higher frequency than about 1600 cm^{-1} are associated with intramolecular H–O–H bending and O–H stretching modes, in addition to overtone and combination bands involving the intermolecular vibrations. In particular, there is a relatively intense association band in liquid water[133, 134] and in the ice polymorphs[12] near 2200 cm^{-1} (1600 cm^{-1}, D_2O) that is attributed to a combination band involving the H–O–H intramolecular bending mode and librational modes. A variety of weak bands appearing in the frequency region above 4000 cm^{-1} represent further overtones and combinations.

A comprehensive discussion of the electrical properties of ice can be found in the book by Eisenberg and Kauzmann.[12] Values of the low-frequency dielectric constant $\varepsilon(0)$ are useful for the determination of the dipole moment in the condensed phase. Experimental measurements of $\varepsilon(0)$ in polycrystalline ice were carried out by Auty and Cole.[135] Their data showed that $\varepsilon(0)$ decreases with increasing temperature,[136]

$$\varepsilon(0) = \frac{2.50 \times 10^4}{T} \tag{6}$$

An approximate theoretical justification for the inverse T dependence derives from the Kirkwood-Fröhlich theory[12, 136, 137] of dielectrics for a highly polar substance,

$$\varepsilon(0) = \frac{2\pi N^* m_s^2 g_K}{kT} \tag{7}$$

where N^* is the number of molecules per unit volume, m_s is the magnitude of the effective dipole moment per water molecule in the crystal, and g_K is the Kirkwood correlation parameter, which is a measure of the correlation between the dipole direction of one molecule with those of all other molecules in the crystal. See Section VII.C for a more detailed discussion of this important quantity. The value of g_K is unity for no correlation and may range[136, 137] up to nearly 3 for strongly correlated dipoles that exist in some of the ice forms. Equation (6) combined with (7) indicates that $m_s^2 g_K$ has a value of 17.9 D^2 in ice Ih. This would be consistent with the value of $m_s \sim 2.6$ D calculated by Coulson and Eisenberg[138] with $g_K = 2.65$. Using this value of m_s in Eq. (5) indicates that the field[138] experienced by a water molecule in the ice Ih structure is about 5.2×10^5 esu/cm^2 or 1.56×10^{10} V/m, only about 3% of the so-called atomic field discussed in Section II.

C. Liquid

In keeping with the main topic of this review, this subsection is divided into two parts. The first part briefly summarizes some of the salient experimental knowledge about water in its bulk liquid state, while the second part is devoted to experimental results on the perturbed liquid.

1. Bulk Liquid

The intramolecular structure of water in the liquid state is changed from that in the gas phase and is slightly different from that in ice Ih. According to Thiessen and Narten,[126] who carried out a detailed neutron diffraction study of a series of H_2O, HDO, and D_2O mixtures, the H–O–H bond angle in the liquid is about 102.8° and the O–H bond length is 0.97 Å. These values are an average over the isotopic species present in the mixtures. Interestingly, they agree very well with the most reliable MD values so far obtained, 102.3° and 0.96 Å,[75] indicating a measurable decrease of the bond angle in the liquid, thus probably in ice Ih as well. This would cast doubt on the widely presumed bond perturbation toward the tetrahedral angle in condensed water (see also Ref. 124). However, a very recent reexamination of neutron diffraction data (D_2O) using a novel correction method for recoil and inelasticity effects[127] gives 106.3° and 0.989 Å for these structural parameters, leaving the situation with respect to the intramolecular structure in the condensed phases of water a bit up in the air.

The intramolecular vibrational spectrum of bulk liquid water[12] is very closely related to the infrared and Raman spectra of the ice polymorphs.[12, 119] The stretching frequencies ν_1 and ν_3 are known to decrease from the gas phase to these condensed phases, while the bending frequency ν_2 increases. More interesting, perhaps, is the fact that the liquid phase spectra are sensitive to the temperature.[12, 134, 139-152] In keeping with expectations of decreased hydrogen bonding with increasing temperature, the intramolecular bending frequency decreases while the stretching frequencies increase toward their gas-phase values as the temperature is raised. For example, the stretching vibration peak moves[144] from ~ 3220 cm^{-1} at 249 K in the supercooled liquid to ~ 3440 cm^{-1} at 368 K, a 7% change. Roughly consistent with these observations are inelastic neutron scattering data extended to high-energy transfer.[145]

The extremely careful Raman work of Walrafen and his colleagues[146-148] has also left little doubt about the existence of a nearly[149-151] exact isosbestic point in the stretching region 3000–3700 cm^{-1}, at least for temperatures not too deep into the supercooled regime.[144, 149-151] Figure 5 of Ref. 147 provides a particularly clear picture of the temperature effects and the isosbestic point in this spectral region. Isosbestic points also occur in other frequency regions in the liquid water spectrum.[146] Such temperature-invariant intensity points, lying between two temperature-dependent intensity maxima, usually signify the presence of two spectral components, which vary in intensity, but not in shape, position or half-width, as the temperature changes. The MC calculations of Reimers and Watts[55] provide insights into the possible origins of the two types of structures and the approximate isosbestic points (see Fig. 5 of Ref. 55).

Some of the intermolecular vibrations exhibit relatively even larger temperature effects. Inelastic neutron scattering data[143] have indicated that a librational frequency, which is probably the A_2 mode around the molecular C_2 axis,[134] whose value is about 550 cm^{-1} near the freezing point, declines to about 400 cm^{-1} at the boiling point. Some fairly recent neutron diffraction work[145] shows this frequency to be about 597 cm^{-1} at 258 K.

In Raman work,[147] two relatively sharp intermolecular translational frequencies (termed acoustic and optic modes in Ref. 148) were found to decrease smoothly from about 56 cm^{-1} and 180 cm^{-1} at the freezing point to 40 and 143 cm^{-1}, respectively, at the boiling point. The huge frequency changes with temperature, ranging up to 30%, may indicate, as first suggested by Segré,[139] a massive restructuring (breakdown) of the liquid water intermolecular potentials (structure) as the temperature changes.

This affects not only the effective potential surfaces on which the inter-molecular vibrations take place, but also, through hydrogen bonding and various mode-coupling mechanisms, the intramolecular vibrations and other dynamical processes in water.

Most liquids exhibit an Arrhenius temperature dependence of viscosity, diffusion, and other dynamic relaxation processes. The rates of such processes are proportional to an inverse exponential function of a temperature-independent activation energy $\Delta E^\dagger$ divided by kT. Water is an exception.[12] It shows anomalous temperature dependence, particularly in the supercooled regions,[17] of the transport phenomena,[153–156] NMR spin–lattice relaxation times,[157–161] and dielectric relaxation.[162, 163] These anomalies can be interpreted as being caused by a temperature-dependent $\Delta E^\dagger$. A review of some of this work as it applies to hydrogen bonding in water has been given by Bertolini et al.[164] The non-Arrhenius behavior in liquid water, as well as the heat capacity anomaly[15–17] and the minimum in the isothermal compressibility,[13] have all been ascribed[121] to the same effect as the large temperature dependence of the $\nu_L(A_2)$ mode: a "softening" and flattening of the intermolecular potential surfaces, thus a lowering of the ν_L frequency, and an attendant lowering of the $\Delta E^\dagger$ values with increasing temperature. In fact, a nearly linear correlation between librational frequencies and activation barriers is already known to exist for solid hydrates,[165] so this general concept may have utility in the interpretation of other water systems, both in the bulk and perturbed states.

As is the case for the high-frequency intramolecular vibrations, the electrical properties of water in the liquid state are expected to be similar to those in the ice Ih crystal. The anomalies associated with these properties seem mild. In fact, according to the data summarized by Eisenberg and Kaufmann,[12] the static dielectric constant of the liquid for temperatures up to 373 K is a rough extrapolation of Eq. (6) for ice Ih,

$$\varepsilon(0) = \frac{2.5 \times 10^4}{T(1 - 8.90 \times 10^{-4}\, T + 3.84 \times 10^{-6}\, T^2)} \tag{8}$$

For various reasons,[57] the dipole moment in the liquid is taken to be 2.4 D. This is only slightly smaller than the 2.6-D value[138] most often used for ice Ih. Since the field caused by neighboring molecules at the typical intermolecular water–water distance and orientations in the liquid is expected to be rather large (Section III.B), this is an important quantity with respect to computer modeling of the liquid.

A variety of scattering techniques are currently used to characterize the structure and dynamics of liquid water at the molecular level. X-ray diffraction extracts information mainly for the O $\cdots$ O radial distribution function, whereas neutron diffraction, in principle, is capable of exposing all the pair correlations of oxygen and hydrogen atoms, as well as the angular correlations between molecules. Furthermore, inelastic neutron scattering enables one to probe dynamical properties associated with both single-particle and collective processes, including, as we have seen earlier,[143, 145] vibrational characteristics. Light scattering detects density fluctuations in the hydrodynamic limit, while depolarized light scattering can provide information about rotational relaxation times.

A portrayal of many important characteristics of the short-range molecular order in the liquid is provided by nuclear pair correlation functions $g_{ab}(r)$, which give the probability of occurrence at the separation r of pairs of nuclei of the subscripted species. Accordingly, for water, there are three nuclear pair correlation functions, g_{OO}, g_{OH}, and g_{HH}. In principle, information about these correlation functions, often referred to as radial distribution functions (RDFs), may be extracted by combining results from x-ray and neutron diffraction experiments. In practice, an accurate determination of the RDFs for liquid water is not easy.[127, 166-168] Different experimental arrangements and data processing may result in quite different functions because of the unique properties of hydrogen as a scatterer.[166] For example, the earlier g_{OH} and g_{HH} reported by Narten[169] were much less structured than the later ones of Narten, Thiessen, and Blum.[170] Discordancies exist even for the presumably[167] more easily obtainable g_{OO} pair correlation function: The first peak of g_{OO} observed by Soper and Phillips[166] using neutron scattering is significantly higher than that found from x-ray results.[169, 171, 172]

While relatively good agreement for the structure of liquid water now seems to have been achieved,[166, 167] in the coming few years, when computer simulations improve still further, it will be relatively easy to obtain good RDFs over wide ranges of temperature and pressure by computational methods.

In spite of the problems in obtaining good experimental RDFs, g_{OO} has been extremely informative in providing views about the structure of liquid water. The most striking feature of the g_{OO} curves is the persistence of tetrahedral icelike order into the liquid phase. Though[166] the first peak around 2.85 Å and the second one around 4.50 Å are broader than in ice Ih, the peak positions are nearly identical. Furthermore, though not entirely a unanimous opinion,[172] both x-ray[171] and neutron diffraction[166] methods seem to show that the average number of nearest neighbors of an oxygen atom in the liquid is about 4.5, consistent with the picture of a general tetrahedral structure mixed with other configurations, such as those that would arise from bifurcated structures[173] or random networks of tetrahedral structures.[174] Fairly recent review articles on the diffraction work have been published by Egelstaff[175] and by Chen and Teixeira.[129]

Bosio, Teixeira, and coworkers[127-129] have measured the diffraction patterns of D_2O at supercooled temperatures. A fairly distinct peak near 3.4 Å could be brought out very clearly using *isochoric* differential x-ray diffraction data,[128, 129] obtained at different temperatures for pairs of liquid systems having the same density. This peak was also observed by Hajdu et al.[172] The 3.4-Å peak grows in as the temperature increases (see in particular Fig. 5, Fig, 8, or Fig. 4, respectively, of Ref. 128, 129, or 176). This structural change occurs, not in the local hydrogen-bonding with $O \cdots O$ distances near 2.75 Å as in all the lower pressure ice forms, but rather in the $O \cdots O$ *next*-nearest-neighbor region. Structure in this region can be considered to be a "fingerprint" that distinguishes next-nearest-neighbor $O \cdots O$ structure of the type in ice Ih and ice Ic (~ 4.5 Å) from that in the higher density ice forms (~ 3.25–3.45 Å).[12] Possibly, in the liquid, with the curtailment of the Pauling entropy,[122] the main factor stabilizing ice Ih with respect to ice II, a number of these energetically similar structures play a role. In fact, the grow-in with increasing temperature of the peak at 3.4 Å, counteracting ordinary thermal expansion, is almost certainly related to the density maximum in the liquid.[121] There may also be a relationship between the temperature dependence of this high-density feature and the temperature dependence of the isothermal compressibility of liquid water,[121, 177] a property that has not yet been explained well by any of the current water models.[41, 178] Could it be that the breakdown of the liquid water structure, suggested years ago by Segré[139] on the basis of some of the earliest known low-frequency Raman work, is actually a transformation in the liquid from ice I-type bonding at lower temperatures to the more collapsed next-nearest-neighbor bonding of the higher density forms of ice as the temperature is raised? If so, then computational models of water should surely have this feature built into them.

A part of the research related to the structure of the highly supercooled liquid concerns the similarities with amorphous solid water (ASW), which is thought to be a continuous random network (CRN) of basic 5-water tetrahedral structures.[179, 180] A very brief review from the neutron diffraction perspective of work as it is related to hydrogen bonding in supercooled water and the amorphous solid has recently been presented by Dore.[176] Other neutron diffraction work, performed on emulsified water droplets "a few μm" in size,[181] have indicated that the correlations of the hydrogen-bond assemblies are less developed in supercooled water than in

ASW at 241.6 K, the limit of the experimental range. These small droplets are probably large enough to avoid structural changes in the liquid that are known to occur when water is near a surface (see next subsection), but one must certainly be aware of the possibility of problems in such experiments.

2. Perturbed Liquid

Liquids can be perturbed by applied fields; by surfaces and interfaces, ranging from metallic or semi-conductor surfaces to the vapor–liquid interface; or by dissolved solutes such as atoms, ions, or complex molecules. Perturbed water is particularly interesting, since there are seemingly so many local and far-ranging structural effects that can occur under a perturbation. These structural effects may influence the energy, the entropy, and the dynamics of the system in surprising ways.

Solutes that are relatively soluble in water are termed hydrophilic, while those with less solubility are referred to as being hydrophobic.[6, 182] The former includes atoms or molecules with electrically charged groups, or molecules with neutral groups having a substantial electric dipole moment. Hydrophobic solutes are usually neutral and nonpolar, and they experience energetically repulsive interactions with water molecules. Hydrophobic solutes and solutes such as *amphiphiles*,[183] which are molecules having a hydrophobic chainlike molecular grouping ("tail") attached to a more compact hydrophilic one ("head group"), when placed in water, tend to cluster together to "hide" their least polar regions from the surrounding strong dielectric medium. This type of binding is important for biomolecular membrane formation.

Entropy can be a major factor controlling the thermodynamic properties of solutions,[184] as the solvent molecules gain or lose the structural and orientational freedom they possessed in the bulk state, and as the solute molecules change conformation. The presence of these entropy changes often ruins intuition about the way molecules dissolve and react in water, since entropy works in an opposite direction from the energy. Structuring in a solution causes the energy to be lower (more stable) but the entropy to be lower also (less stable), and these effects can be particularly prominent in water. Much of the hydrogen–deuterium isotope effect accompanying a hydration reaction, such as acid dissociation, may arise from the entropy,[185] since it is entropically more costly to "freeze out" the lower frequency intermolecular vibrations of D_2O than those of H_2O. In fact, this isotope effect could act as a "fingerprint" for the presence of solvent entropy changes accompanying an aqueous reaction.

Entropy is also particularly important as an influence in the immiscibility of nonpolar substances with water. In fact, the concept of a *hydropho-*

bic effect, which includes the entropy as an important contributor, has been extensively employed in the biological literature to describe a variety of processes, such as protein folding and enzyme specificity,[6-9, 186] and the formation of biological functional units in a living system. It has also been claimed that hydrophobic interactions are important in the formation of clathrate hydrates,[187] in micelle formation,[188] and in many other phenomena. Recently, however, the wide role of hydrophobic interactions has been questioned.[189-191] It was found that the solvent-induced hydrophilic interaction between two functional groups that can form hydrogen bonds is much stronger than the corresponding hydrophobic binding. In addition, solubility experiments[192] as a function of pressure on the arch hydrophobic systems, rare gas atoms dissolved in water, indicate no strong hydrophobic tendencies for clustering, at least for argon and krypton at the concentrations employed in these experiments. Instead, the data can be accurately described simply in terms of the increased PV work to place the atomic solute in the solvent.

The introduction of ions into water produces a large perturbation on the hydrogen-bonded network and thus has a marked effect on both the structure and the dynamical properties of the solvent. Understanding these modifications is a classical problem in electrochemistry.[3-5] Inorganic ions, such as Li^+ or Mg^{2+}, with a high charge density or a small physical size are strongly hydrated. The waters of hydration surrounding such ions are electrically ordered in a tightly bound primary shell, which is able to move with the ion through the solution, solvent residence times being often much greater than 1 ns. Beyond the primary shell, water molecules become gradually disordered as a consequence of the competition between the orienting forces from the primary shell and the forces from the surrounding bulk water. As the ionic charge increases or the size of the ion decreases, the primary hydration shell becomes more strongly ordered, and the effects of this ordering may extend to large distances. For example, experiments by Lee[193, 194] have revealed large effects on the solvation rates of protons caused by ionic hydration perturbations of the aqueous media. His measurements indicate a retardation by two to three orders of magnitude of the proton hydration rates in moderately concentrated solutions of cations. The anions seemed to have much less of an effect, probably because in these studies they were larger and more polarizable. Particularly effective perturbers were found to be Li^+, Mg^{2+}, or Ca^{2+} ions, with K^+ being rather ineffective. Equilibrium constants and pH need not be so severely affected because the reverse recombination reaction rate also seems to be strongly perturbed,[195] probably for the same reason: Needed for recombination is an untangling of the tight water network structure around the proton and a reconstruction of a different solvent configuration around the neutral molecule. In this regard, however, Bhattacharyya et al.[196] have reported more than an order of magnitude effect on the equilibrium constant (in favor of the neutral) of proton dissociation from p-nitrophenol at another type of interface, the water liquid–vapor interface. The proton is simply not comfortable being hydrated by the strongly orientationally perturbed water molecules near probably any surface. For large organic polyatomic ions having small or delocalized charges, hydration energies become much weaker and such ions may even behave hydrophobically. This is why it is usual for proton hydration dynamical effects to dominate those of the large organic anions in weak acid dissociation processes: Hydration properties of the neutral and the anion are not significantly different.

Surprisingly long-range attractive forces between hydrophobic surfaces separated by thin films of water have been detected by the direct surface force measurement technique.[197] Ranges up to 0.1 μm have been cited. This effect is not understood, though in studies of the effect of various surfaces[198] on laser-induced proton dissociation reactions in water, perturbations persist a few hundred angstroms from the surface.[199] Since the presence of ethylene glycol[200] in the water substantially reduced the range in the surface force experiments, these perturbations may very well have to do with long-range distortions of the water structure near a surface. On the other hand, they may be caused by adventitious electrostatic effects,[201] or they may not exist at all.[202] However, a recent paper[203] on a non-aqueous liquid between mica plates, using either a hard sphere or a Lennard-Jones theoretical model, did indicate a correlation between the molecular diameter of the liquid molecules and the observed oscillatory forces, which extended about eight molecular diameters from the surface. In the case of water, assuming the same mechanism, these forces might be expected to extend nearly 25 Å from a surface, and, with water's special molecular interaction properties, could actually extend farther.

If there is an understanding of the long-range surface perturbations in terms of fundamental phenomena for the case of aqueous solutions, it might also help to explain experimentally based suggestions[204] by some chemists and biophysicists that the properties of "vicinal water" fairly far from a surface can affect the physics, chemistry, and biological function in aqueous systems. Largely because no computational study of water has produced such long-range effects, these ideas have been discounted by most chemical physicists. However, without a better way of assessing the effect of the boundary conditions (a cubic sample containing 512 water molecules is only about 25 Å across), together with the present lack of really good water interaction models containing systematically developed orientational features, it may be premature to discount these long-range effects on the basis of computational studies alone.

Dynamic properties of water molecules near a surface can be studied by measurement of free-radical-induced NMR proton relaxation. Some of the first experiments of this kind[205, 206] concerning a protein surface indicated that the translational diffusion coefficient of the water molecules may be slowed perhaps by up to an order of magnitude near this type of surface. Fluorescence quenching experiments[207] may also be used to measure diffusion at an interface. See Ref. 9 for other results of this type. Questions about the dynamical properties of water under the influence of a surface perturbation are just the ones most likely to be eventually resolved through computational methods.

Experimentally determined properties of water very near the vapor–liquid interface have been briefly reviewed in the MD paper of Matsumoto and Kataoka[208] and in the paper by Yang et al.[209] Surface structure can be investigated by x-ray reflection techniques, which can measure surface roughness and density profiles[210, 211]; second-harmonic generation (SHG),[209, 212] which in principle can measure the most probable distribution of water dipole orientations[209] relative to the surface plane; infrared–visible sum-frequency generation (SFG),[213] which measures intramolecular vibrational frequencies, thus revealing surface hydrogen bond strengths; and the surface potential, which also can give information about the preferred surface orientations of water molecules. In the x-ray experiments, the density at the air–water interface was found to decline gradually within the very short range of about 3.3 Å,[211] a bit more than one molecular diameter. This is small compared with the results for nonpolar liquids and would certainly imply that any long-range surface effects could not affect the density. However, such a result may not say very much about the all-important orientational perturbations, which could be responsible for large effects on the physics and chemistry of water near surfaces. Orientational perturbations are better studied by SHG experiments, but, for various reasons,[209] the interpretation of these experiments is not straightforward. Therefore, it is still not entirely clear what the orientational properties of surface water molecules are. At the vapor–liquid interface, however, many workers tend to believe that, as the water molecules approach the surface from the liquid side, their dipoles tend to align parallel to the surface plane with one hydrogen preferentially directed outward toward the vapor.

Another way that water can be perturbed experimentally is by the application of an external electric field. Strong laser fields[214–217] have been applied, mainly in an effort to learn more about the discrete and continuous energy levels in the liquid state, about the nature of the ensuing multiphoton processes, and about the production of hydrated ions and free radicals, including the hydrated electron. Of interest in this regard are nonlinear optical effects such as the nonlinear index of refraction and supercontinuum production.[10] Some of this work[214] also holds great interest in ophthalmological surgery.[217]

In an effort to understand dielectric breakdown in water near electrodes in terms of energy level diagrams, an interesting study of the effect of pulsed electrical fields has been published.[218] One of the conclusions concerning the breakdown potential was that it is the potential at which the Fermi level of the cathode equals the energy corresponding to the bottom of the conduction band of liquid water. While this type of work and the laser work would seem to have many questions in common, they are currently fairly well separated in their goals.

IV.   THEORETICAL METHODS

Theoretical studies of bulk and perturbed liquid water have four motivations: reproducing experimental knowledge under normal thermodynamic conditions; predicting or understanding the behavior of water under extreme conditions of temperature, pressure, and external fields; probing at the molecular level to reveal microscopic properties that cannot be observed in real laboratory experiments; and providing a guide for suggesting new experiments and analyzing data.

A.   Analytical Approaches

Analytical approaches for a system as complex as liquid water always involve approximations. Angell[17] and Speedy[219] have suggested a number of theoretical avenues of approach[179, 220] to the liquid water problem through their experimental studies of amorphous solid water (ASW), the supercooled liquid, and the now renowned "singular temperature" near 227.4 K. Some promising theoretical work of a completely different type is being carried out by Kusalik[221] on dipolar soft spheres, using reference hypernetted chain theories[19] (RHNC and RLHNC). Comparisons with MD are made in this paper. Perhaps, in the future, analytical theories of this latter type can be used to help solve some of the problems concerned with liquid water.

Application of random network models to liquid water and ASW was first introduced by Rice and his coworkers.[220, 222–224] With the help of MD, a fairly recent critical overview of these types of theories has been presented by Belch and Rice.[224] It is probably fair to say that, at least at the present time, the future does not look very bright for these approaches. First of all, water may indeed be a hydrogen-bonded network, but it is far from random. Second, such a variety of ring transformations are required, which are supposed to represent collective excitations and eventually relate to thermodynamic properties, that mathematical difficulties get in the way of making realistic extensions to the theories. Belch and Rice do conclude in their paper that a "two-state" interpretation of water, with the 5-ring concentration as a monitor for one of the states,[174] is inconsistent with their findings.

A semiempirical theory, attempting quantitatively to correlate thermodynamic and dynamic properties of bulk liquid water is being developed by one of the authors of this review.[121, 177] This theory is a modernization of so-called "mixture" or "two-state" models of the liquid. However, instead of "broken" and "intact" hydrogen bonds[225, 226] or specific bonding arrangements[174] comprising the two states, one "state" has an unperturbed ice Ih local structure and the other is considered to be a mixture of higher density forms: most prominently, ice III, ice V, and ice II (but lacking the ordered proton arrangement in the crystal). These higher density structures grow in with increasing temperature, in accord with experimental findings.[128, 129] Thus, the investigation of (random) hydrogen bonding by analytical methods is bypassed through the direct use of known physical properties of the various ice polymorphs. Remarkably good quantitative correlations are being found that explain the density maximum, the minimum of the isothermal compressibility, the non-Arrhenius behavior of the viscosity, the heat capacity, the dielectric and NMR spin–lattice relaxation times, along with the temperature dependence of the intermolecular vibrations from the deep supercooled regime to the critical temperature and for elevated pressures. This and some of the other theories mentioned above would be consistent with the large local density fluctuations in water that have been observed experimentally[227] and in computer simulations,[228] and in fact the theories may be related.

In the case of perturbed water, except for solute perturbations, which will be discussed below, very little analytical work has been carried out. Suggestions for future work on waterlike liquids perturbed by a surface have been published recently by Quintana et al.[229] These authors used Ornstein-Zernike theory with the Percus-Yevick approximation to study molecular orientations near a hard wall. At about the same time, Bérard and Patey,[230, 231] using Ornstein-Zernike theory and the RHNC approximation, studied liquid dipole orientations near conducting and nonconducting surfaces. The liquid model consisted of rigid dipolar spherical particles. The latter paper seems particularly relevant to the behavior of multipolar liquids near electrode surfaces. By employing a mean field theory, which reduces the many-body wall–solvent interaction to an effective pair potential, these authors were able to obtain self-consistent solutions for the solvent structure near polarizable walls having dielectric constants ranging from unity (insulator) to infinity (conductor). Found were oscillations of the dipole directions with distance from the wall. In the highest approximation, where image charges of all the particles were included, i.e., other (OI) as well as self (SI) images, the particles in contact with the conducting wall showed a distribution that strongly maximized for dipoles parallel to the wall, similar to the case for the insulating wall. This similarity between conducting and insulating walls was said to be caused by cancellation effects from the OI. The second layer of particles showed a tendency to have their dipoles perpendicular to the insulating surface, but this preference was largely washed out for the conducting surface.

Early theories of electrolyte solutions, such as the Debye-Hückel model,[232] treated the dissolved ions as point charges in a dielectric continuum. Adding a hard sphere radius to the ions produced the so-called "primitive model."[3] Because of their presumed intuitive value, these types of models persist even today for perturbed water ensembles. For instance, water is still often viewed as a dielectric continuum characterized by its bulk, frequency-dependent dielectric function, while the solute is viewed as a spherical Onsager cavity[233] carrying a point charge or point multipole. Nonspherical cavities have also been considered.[234] Dynamical extensions of such models have employed an equation of the Smoluchowski-Vlasov type.[25, 236]

Other theories have treated inhomogeneous continuum models in which the molecular nature of the solvent is considered under the mean spherical approximation (MSA).[237, 238] More recently, a series of important papers by Wei and Patey[239-241] used the van Hove angular space–time correlation function, expanded in rotational invariants, to obtain relaxation times for rigid nonpolarizable aspherical molecules in a dielectric medium. By generalizing[239] the Kerr approximation[242] to nonspherical particles, interesting connections with the Debye model[243] and with Neumann's extensive work[244-249] on finite-system dielectric effects and boundary conditions were obtained. Another important conclusion reached in this paper, through the investigation of the nonspherical solute particles, was that coupling between rotational and translational degrees of freedom could have a large effect on the relaxation phenomena. This conclusion would also imply that general coupling among other degrees of freedom, such as those inherent in polarizable, flexibly bonded molecules, might play a strong role in dynamical processes in water. In later work,[240] using related methods, possible improvements to the MSA results for dielectric relaxation and ion solvation times were discussed.

Assuming no correlation between their positions, the concept of a solvent-mediated solute–solute interaction can be introduced by bringing two infinitely separated solutes dissolved in the solvent to a distance r at constant temperature and pressure.[189] An alternative route starts with the two infinitely separated solutes in a vacuum, then brings them to the required distance and transfers the dimer from the gas phase to the solvent. The free energy must experience the same change $\Delta G(r)$ along these two routes. According to classical statistical mechanical theory,[250] there is an important relationship between the gradient of $\Delta G(r)$, with respect to r, and the mean force $f_{ss}(r)$ operating between the two solutes,

$$f_{ss}(r) = -\frac{\partial \Delta G(r)}{\partial r} \qquad (9)$$

Therefore, the function $\Delta G(r)$ is referred to as the potential of mean force (PMF), which consists of a direct part caused by the solute–solute interaction and an indirect part originating from the presence of the solvent. The PMF plays a central role in theoretical studies of chemical reaction rate constants[251] and in traditional approaches to solvation effects at the McMillan-Mayer level,[252, 253] where the ionic but not the solvent particles are explicitly considered. A very good discussion of some of these theories as applied to dilute and concentrated solutions of Na^+Cl^-, K^+Cl^-, and generic M^+Br^-, with comparisons and relevant references, can be found in the recent paper by Kusalik and Patey.[254] An important conclusion was that the McMillan-Mayer level of theory is unlikely to be reliable when relatively "small" ions such as Na^+ are present. The molecular nature of the solvent seems to be of utmost importance for such ions. The association constant of ion pairs in an ideal solution can be expressed in terms of the solvent-mediated PMF.[255–257] When water is the solvent, nonpolar molecules tend to attract each other through the indirect force, which defines the hydrophobic interaction.

One semiempirical approach solves the Ornstein-Zernike-type equations based on a (reference) interaction site model, popularly called RISM.[258, 259] This method has been used to analyze the short-ranged structure of several polar molecular liquids perturbed by dissolved ions,[260–264] in the context of a correction to the continuum picture.[262] Kusalik and Patey[265–267] have investigated the static properties of electrolyte solutions with a polarizable solvent using an r-dependent mean field theory, which considers the average local electric field experienced by a solvent particle as a function of its separation from an ion.

These analytical theories are expected to approach validity most closely at low solute concentrations, where the radius of the ion, including its perturbed outer solvation shells, is small compared with the average interionic distance. As the concentration of the electrolyte is increased, an ever greater fraction of water molecules becomes associated with the perturbed solvent shells of the ions. At such concentrations, PMFs begin to become concentration dependent, because of solvent-mediated many-body effects among the ions. Particularly in water, perturbations on the solvent by ions can extend to large distances and can involve a significant fraction of the solvent molecules.[193, 194] This fact, combined with the usual neglect of intramolecular bond flexibility and polarization, severely limits the validity of any of the analytical theories. Related to this problem, is the fact that such theories are presently incapable of accurately incorporating important effects caused by the plethora of orientationally dependent bonding configurations that exist in real water. A molecule consisting of a few interacting points may give an incomplete picture of the angular dependent forces. This is, of course, also a problem in computer simulations of water.

Even though these analytical models cannot contain the molecular details required to describe accurately bulk or perturbed liquid water, they are still useful for creating semiquantitative theoretical models, particularly for the description of ions or neutral molecules dissolved in a generalized polar solvent (which water is not). For example, by treating the solute in terms of an effective interparticle interaction or PMF, thermodynamic relationships can be derived and nonequilibrium dynamics can be investigated for such model systems. Such results may help to improve intuition about the real water system.

B.   Computer Methods

Recent advances in modern computer technology have made it possible to investigate complex systems in much more detail. Two of the most commonly used methods[20] are Monte Carlo (MC) and molecular dynamics (MD). Given an interaction model, these methods in principle can provide an exact thermodynamic description of a classical many-body system and can provide a direct route for converting information at a microscopic level to macroscopic properties of interest. A wide range of physical, chemical, and biological phenomena may be studied by carrying out these types of computer simulations.

Typically, as described by Lie and Clementi,[57] in both MC and MD, the number of molecules N is held constant. However, MC[46] and a recent MD, called grand molecular dynamics (GMD),[47, 48] have been used for calculating thermodynamic properties, in particular the Gibb's free energy, of ensembles containing simple rigid water models in open systems. In fact, besides the above paper, there now exist a number of other recent papers on free energy calculations of water using various models and methods,[49-51] indicating that computational methodology is becoming much more versatile than in the past, when mainly NVE or NVT ensembles were considered. In MD, the temperature T, i.e., the average kinetic energy, always fluctuates, while the total energy E remains constant. In all cases, either the pressure P or the volume V is allowed to vary by keeping the other quantity fixed. Strictly speaking, then, standard MD simulations deal with microcanonical NVE ensembles, though it is always possible, for example, through "thermostating" by continuously rescaling velocities, to keep the temperature fixed[20] as in NVT and NPT ensembles. Also, for a subensemble of N_b solute particles dissolved in a solvent, energy is no longer conserved because of energy exchange between the solute and the solvent bath. For sufficiently large systems of this type, the subensemble is N_bVT or N_bPT. In MC, the temperature and/or pressure is a fixed input parameter, while the potential energy fluctuates. The first isothermal–isobaric (NPT) computation for liquid water was an MC carried out by Owicki and Scheraga,[35] and the first MD of this type was performed by Ruff and Diestler.[63]

While MC sampling is computationally efficient for analyzing static properties, it relies on statistics based on an equilibrium ensemble. Therefore, it cannot be applied to nonequilibrium molecular systems or provide dynamical information. In addition to the thermodynamics, MD generates *trajectories*, configurations of the system as a function of time. Moreover, for perturbed water, in addition to the thermodynamics, MD is able to reproduce a wide variety of experimental conditions, is adaptable to various types of statistical ensembles, and most importantly provides detailed information about both structural and dynamic properties of the system as a function of the perturbation strength. One of the most important aspects of field-perturbed water is the behavior of various time correlation functions, which depend sensitively on the strength and the type of perturbation. Thus, MD methods have been by far the most popular for studying perturbed water ensembles.

Intrinsic problems of MD for the study of water lie in the difficulties inherent in treating large systems of molecules with elaborate intramolecular structure. In particular, the choice of realistic interaction potentials for water molecules containing flexible bonds and polarization, and the currently prohibitive computer expense of accurately including the ultimately necessary quantum effects, all contribute to these difficulties. These problems do not seem insurmountable for work in the future.

MD procedures that are most commonly employed for water are various modifications of an algorithm due originally to Verlet.[268] Newton's equations of motion are numerically integrated over a sufficiently long series of finite time steps Δt. To avoid errors and to prevent computational overruns so that stable solutions are maintained, the time steps for a flexibly bonded water model must be short compared with the hydrogen atom intramolecular vibrational period of about 10 fs. Most modern water MDs[60, 73-75] of this type use a Δt in the vicinity of 0.25 fs, but even shorter Δt would be desirable when computational efficiencies improve. The position of an oxygen or hydrogen atom, thus the forces to which it is subjected, is computed at a future time step from the position, velocity, and/or acceleration of the particle at a previous time step. For the study of an extended liquid, periodic boundary conditions are commonly employed, where particles near one boundary of an imaginary container are caused, through a "wraparound" algorithm, to interact with and diffusively exchange with particles near the opposite boundary, as if they were neighbors. An excellent discussion of these general computational methods can be found in the book by Allen and Tildesley,[20] and more recent discourses from somewhat different perspectives have been given by van Gunsteren and Berendsen[21] and by Straatsma and McCammon.[22]

1. Historical Overview

Over forty years have passed since the first MC computer simulations of liquids were performed on the Mechanical and Numerical Integrator and Calculator (MANIAC) at Los Alamos National Laboratories.[269] The earliest representations of molecular liquids were highly idealized within the framework of hard disks or hard spheres, for which the particles move between collisions without acceleration. In fact, the first MD simulation was accomplished by Alder and Wainwright[270] for a hard sphere system. In 1957 the Lennard-Jones potential was introduced into MC simulations by Wood and Parker.[271] This success was later extended to MD systems containing Lennard-Jones particles,[272] and to diatomic molecules.[273] Computer simulations on liquid water were initiated in 1969 by Barker and Watts[24] using the MC approach with the Rowlinson water–water interaction potential.[79] This work was closely followed by the celebrated MD work of Rahman and Stillinger.[28] Both of these pioneering efforts used rigid 5-site nonpolarizable models for the water molecule. The development and testing of other rigid models followed.[25-27, 29-42, 46] The most recent advances in bulk water simulations have included the presence of intramolecular bond flexibility,[52-63] specific non-pair-additive contributions to the intermolecular potentials or rapidly responsive electronic polarization,[64-72] or both flexibility and polarization.[73-76]

Past MD simulations on perturbed liquids other than water have been very numerous. A few of these are mentioned here: binary mixtures,[274, 275] the liquid–vapor interface,[276-279] fluids near surfaces,[280-291] a nonpolar liquid perturbed by a polar solute,[292] and molecular liquids in intense laser fields.[293-295]

The systems studied for water are even more varied. Liquid water near the ice interface has recently been considered by Karim and Haymet.[296]

See also the recent review on the general problem of the crystal–liquid interface.[297] Much effort has been devoted to the important problem of the vapor–liquid interface of water.[208, 298–305] This problem seems particularly difficult since subtle angular-dependent phenomena play a role, and these undoubtably depend sensitively on the water interaction model, that is, where the charges are placed and the presence or absence of electronic and vibrational flexibility. However, a recent paper by Motakabbir and Berkowitz,[305] using a polarizable extension of the standard 4-site model of water,[40] concludes that "polarization effects are of secondary importance in predicting the orientational structure" in the interface. It is probably fair to say that no real concensus has yet been reached concerning the structure and dynamics, or the importance of polarization and bond flexibility, near this type of surface, either from an experimental or a theoretical point of view. Ions near the vapor–liquid interface have also been studied.[306–308] In the work by Benjamin,[308] using a flexible SPC model of water, it was found that ions tended to retain the same first solvation shell as in the bulk. This is expected for the strongly hydrated "small" ions studied in this work, and is in essential agreement with studies performed at other types of surfaces.[309–311]

Beginning with the early work of Watts, Clementi, and coworkers,[312–314] many MD and MC computations of aqueous electrolyte solutions have now been carried out. The long series of papers by Heinzinger's group is particularly noteworthy. These references are too numerous to discuss here individually, but descriptions of the subject with many references[315–318] cover all work up to about 1986. A more modern review has been written by Heinzinger,[319] and many recent references on this subject can also be found in the Zhu and Robinson paper.[320] The dynamics of aqueous solvation has been addressed,[321] and water near various types of solid, liquid, or biomolecular surfaces has been given a great deal of attention.[322–331] Of electrochemical interest is the growing number of papers on water or electrolyte solutions near neutral or electrically charged metallike surfaces.[309–311, 332–343] The effect of laser fields on liquid water has also recently been studied.[103, 104, 344]

These studies of perturbed water will form the main emphasis of the remainder of this review. However, before this central issue is addressed, it is necessary to discuss some of the unperturbed water models in more detail.

2. Role of Flexibility

Because of computational efficiency, a vast majority of water models initially developed, and still in common usage, were rigid models,[24–50] with respect to both electronic polarization and the intramolecular degrees of freedom. Any such model, when used in a computational ensemble, gives rise to only pairwise-additive interactions and thus cannot give a complete description of the condensed phase. In fact, quantum chemistry calculations[82, 345-347] have indicated that non-pairwise-additive interactions play a significant role in water, and it has already been remarked that about 23% of the internal energy of ice Ih cannot be attributed to dimer-based pairwise additivity.

More recently, however, a state-of-the-art quantum-mechanical description[348] of three different trimer configurations has somewhat confused the issue with respect to classical simulations. In certain regions of high orbital overlap, it was found that the Heitler-London exchange contribution to overall nonadditivity exceeds the polarization effect. It was therefore concluded that classical polarization is too poor a model. However, in defense of the classical models, one must still account for the nonnegligible polarizability of water molecules in the bulk liquid, which is what modern empirical methods have been attempting to do. Perhaps too much detailed quantum-mechanical input at the present very restrictive level of cluster size and structural constraints in the quantum chemistry calculations is not that helpful in the liquid water problem, particularly since all these rather fragile structures are expected to be modified in the bulk liquid anyway.

To attempt a correction for the failing of pairwise additive models, both the molecular structural parameters and the dipole moment in a rigid model can be distorted from the gas phase and fixed at their averaged condensed phase values. Such approximations cannot of course predict any instantaneous geometrical changes of water molecules from the vapor to the perturbed or unperturbed liquid phases, nor, without electronic polarizability, can they follow state dependent variations of the electrical properties under the influence of time varying fields. These deformations, in conjunction with the induced coupling between the intra- and intermolecular degrees of freedom, may considerably alter structural and transport properties in the liquid state.

3. Electronic Polarization

In view of the 30–40% increase[12] of the average dipole moment in the condensed phases of water, the electrical properties of water at interfaces, near solutes, or under strong external fields are expected to vary strongly. Rigid water models take the polarization effect into account in an average way by introducing fixed electric charges discretely distributed at certain points so as to reproduce the condensed phase dipole moment of water. In reality, however, when the environment is not spatially isotropic, such as near an interface, or a protein molecule, the dipole moment of each individual water molecule fluctuates around its average value, and the instantaneous direction of its dipole vector seldom coincides with the bisector of the H–O–H bond angle. To embody these effects, rapidly responsive electronic polarization must be explicitly accounted for.

In a computer simulation, electronic contributions to non-pairwise-additive many-body interactions may arise either from the inclusion of explicit many-body terms added onto the pair interactions,[66, 67] or automatically from the inclusion of electrical flexibilities in the water model.[64, 68–76] To model electronic polarizations, it may be safely assumed that water behaves as an isotropic charge distribution, since the oxygen atom is electronically dominant. One then writes the induced dipole moment as in Eq. (5), where the field $\mathbf{E}$, which includes contributions from fluctuating neighboring charges and dipoles, denotes the local instantaneous electric field at the polarization center of the molecule. For a system of N water molecules, the induced dipole moments are given as the solution of $3N$ simultaneous linear equations. Theoretically, this can be carried out through a $3N \times 3N$ matrix inversion. For condensed phases, where N is of the order of several hundreds, it is too time-consuming and not really worthwhile to undertake such computations exactly. Conventionally, these equations are solved by a more economical yet still time-consuming iteration procedure, starting from an initial guess of $\mathbf{m}_{ind}$, then terminating the iteration when self-consistency is achieved.

A number of schemes have been proposed to reduce the computational effort for determining the induced dipole moment in MD simulations. van Belle et al.[349] have proposed carrying out only one interation in each time step, starting from the induced dipole moment in the previous step. Ahlström et al.[68] have devised a predictor–corrector algorithm based on slowly time-varying electric fields. This approximation can usually be justified, even for the case of hydrogen motions in water, by keeping the time steps sufficiently short. In the above work, the electric field autocorrelation function (EFAF) was found to decay only 1% after roughly 10 fs, corresponding to 10 time steps in this calculation.

An alternative approach[64] retains the polarization as an explicit degree of freedom and treats the fluctuating polarizability as a Drude oscillator.[23] By introducing an artificial inertial mass associated with the fluctuating polarization, Sprik and Klein[64] were then able to evaluate the induced dipole moment by integrating the equation of motion using standard MD techniques. This result averaged over a sufficiently long time should reproduce the equilibrium statistical mechanics of the Drude model.

A much simpler and less time-consuming scheme for incorporating polarization has recently been developed at Texas Tech University.[74, 75] In this method the polarization effect is taken into account by allowing the values of the charges to vary according to the instantaneous local electric field strength. The intramolecular positions of the charges vary only in accord with the vibrations of the hydrogen or oxygen atom to which they are associated. The charges thus consist of unperturbed contributions q_0, which reproduce the permanent dipole moment in the gas phase, plus a fluctuating part, which gives rise to the induced moment according to Eq. (5). In principle, the resulting set of equations of motion can be solved by iterations. However, it was found that, without introducing error, the fluctuating charge can be calculated from the electric field in the previous time step. This is equivalent to one iteration in the predictor–corrector algorithm of Ahlström and coworkers.[68] It can also be viewed as a special case of the Drude model, where the artificial mass is chosen to be constant in the corresponding equations of motion, so that the response of the dipole moment to the electric perturbation occurs in a single time step. The method is stable as long as the time step is appropriately short (~ 0.25 fs), as it must already be for an MD simulation of liquid water with intramolecular hydrogen vibrational motions. The approximation of keeping the positions of the charge sites fixed to the atoms, instead of allowing them to distort independently, as they actually must do, is justified because of the large intermolecular distances compared with the size of a water molecule. This simple scheme has allowed simultaneous inclusion of intramolecular bond flexibility and electronic polarization in a statistically meaningful MD algorithm, and led to the first such computations on both bulk[73] and perturbed water.[342]

4. Bond Flexibility

Causing the same types of effects, but perhaps of lesser importance than the flexibility caused by electronic polarization, is the flexibility of the intramolecular bonds in a water molecule. It has been found,[58, 74, 75, 350] perhaps not unexpectedly, that added intramolecular bond flexibility reduces the effect of intermolecular repulsive forces. This leads to a greater potential energy stabilization; it also reduces the liquid structure and probably speeds up the transport properties as well.[351] The molecules become "softer" or "more slippery,"[74, 75] giving rise to a less rapid decay of the velocity autocorrelation function (VACF) and thus to larger diffusion coefficients. However, this general notion, at least as far as flexibility alone is concerned, has been questioned recently in a paper by Wallqvist and Teleman,[59] who state that earlier results[58] from their laboratory were flawed because equilibrium had not been reached. This is certainly an important point, which is easy to overlook in these already extremely long computations. However, the lack of an effect of intramolecular flexibility on the transport properties does seem a bit strange, since the rapidly moving intramolecular hydrogen motions should easily be able to respond to outside disturbances and avoid the hard collisions that tend to reduce the diffusion coefficients.

Various angular velocity autocorrelation functions (AVACF) have been found to decay more rapidly when intramolecular flexibility in the water model is employed (see Fig. 4 of Ref. 74). Therefore, if the parameters in a water MD model employing rigid molecules are optimized for transport properties or radial distribution functions, the addition of bond (or electronic) flexibility clearly requires a "back correction" of the originally optimized parameters toward a stiffer interaction model.

The combined presence of molecular flexibility and electronic polarization also gives rise to an additional coupling mechanism between the intramolecular and the intermolecular vibrational modes. This effect must be present in real liquid water. Additionally, bond flexibility is an extremely useful component of any computational model because the frequencies of the intramolecular vibrations may be calculated. These vibrations are sensitive to all types of perturbations, including those from the neighboring molecules. The frequency shifts so calculated then act as a probe that can be compared directly with condensed phase experimental infrared and Raman frequency shifts.[213, 352—355] This is the great asset of flexibly bonded models in liquid water computations.

5. Dielectric Effects

A stringent test of the reliability of a given potential function for water is the comparison of the calculated with the experimental dielectric properties: the mean dipole moment per molecule, the dielectric constant, and various relaxation phenomena. These properties depend on hydrogen bonding and orientational correlations between molecules or within groups of molecules, and are expected to be a sensitive function of the detailed intermolecular interactions. One of the most fascinating characteristics of water is its high dielectric constant, which, in turn, affects long-range electrostatic interactions in the water model being used. Any reasonable water model should correctly reproduce this quantity.

Long-range interactions between a dipole and its surroundings are usually taken into account through a cavity method introduced for gases by Lorentz and extended by Debye (see pp. 9–11 of Ref. 243). Strides for extending these ideas to polar liquids were made by Onsager,[233] who complemented Debye's use of the Clausius-Mosotti internal field with the idea of a reaction field (RF), which tends to enhance the electrical asymmetry. However, the method still had to be suitably generalized to take into account the shorter range forces produced by the permanent dipoles of the water molecules, as was first discussed by Kirkwood.[137, 356]

In the end, neither the Onsager nor the Kirkwood theories are really valid. First of all, they are still continuum models and many molecular details are omitted. Related to this feature is the fact that they are both in essence mean field theories. This failing may be important even in assessing the average properties of dipolar fluids, since it is known to be for magnetization in an Ising ferromagnet.[357] An extremely interesting discussion of the history of this entire problem, including a good survey of approximations and possible flaws in the Onsager-Kirkwood theory, has been presented by R. H. Cole.[358]

For finite samples, as encountered in computational work, even the Onsager-Kirkwood theories cannot be used without modification. The reason is that these theories assume the dipole to be embedded in a medium whose dielectric constant is equal to that of the bulk sample. On the other hand, in a computational model, one places the sample in a medium having a different dielectric constant. This dielectric constant depends on the sample size and geometry used in the simulation (see, for example, Eqs. (1) and (2) of Ref. 244). In Neumann's work,[244–246] general equations were derived for treating this problem rigorously, at least within the Onsager-Kirkwood theoretical framework. Neumann's results reduce to all the previous cases under special conditions. Prior to his work, this important correction was neglected. Consequently, errors up to factors of two in the calculated dielectric constant were present.[247]

For describing long-range electrostatic interactions in computational models of water, one has to deal properly with the missing molecules at the surface and beyond. The various treatments described by Neumann and coworkers,[244–249] have been summarized starting on p. 155 in the book by Allen and Tildesley.[20] The usual method treats the interactions exactly in a cavity of a certain shape, then chooses one of a number of available techniques to include the remainder of the system. The *spherical cutoff* and *minimum image* methods neglect the interactions beyond the cavity. In the spherical cutoff method, the interactions are taken into account only within a sphere and are assumed to be zero outside of it. In the minimum image method, only the first periodic image from the periodic boundary condition algorithm is taken into account.

More accurate methods for evaluating the interactions attempt to approximate the contributions from the more distant molecules. In the *spherical reaction field* (RF) method, the interactions of the rest of the system with the cavity are taken into account by going back to the Onsager approach[223]: A molecule in the solvent is assumed to be enclosed in a cavity immersed in a dielectric medium; and the cavity produces a reaction field in its vicinity, which can be calculated by using macroscopic theory,

$$\frac{1}{4\pi\varepsilon_0 r_c^3} \frac{2(\varepsilon_{RF} - 1)}{2\varepsilon_{RF} + 1} \mathbf{m} \tag{10}$$

where ε_{RF} is the assumed dielectric constant of the surrounding continuum, r_c is the cutoff radius, and $\mathbf{m}$ is the total dipole moment in the cavity.

In the *Ewald lattice summation* method (LS)[20], originally devised for the ionic crystalline state, one tries to sum the interactions exactly. The charges in a central region are surrounded by sets of image regions having identical charge distributions, except that the polarities of the charges are reversed in each successive region (see Fig. 5.7 of Ref. 20). This screens the interactions which are now short-ranged and can be summed. To compensate one also introduces distributions having the same signs as those in the central region. This part is summed in reciprocal space. The trick usually is to find the appropriate cutoff for the distribution so as to do the least amount of computation in either real or reciprocal space.

Earlier inconsistencies in obtaining the dielectric constant by RF and LS methods were addressed by Neumann and Steinhauser[247] for the Stockmayer dipolar hard sphere system. Neumann's group has also compared[245, 246] the dielectric properties of some of the rigid water models and found notable deviations from the experimental expectations. Work coming out of Rutgers University by Levy and coworkers[45, 359] further clarifies these sorts of problems, particularly as they apply to water. In Ref. 45, following an idea of Watts,[360] these authors were able to compute the dielectric constant of water from the direct effect of a homogeneous external field from the slope of the linear portion of the field response. Using the rigid SPC model, Alper and Levy[45] found that this method is computationally more efficient and gives a dielectric constant in good agreement with the one obtained from the average squared total dipole moment.[244]

In two important papers, Kusalik,[361, 362] using the Ewald LS method and the dipolar soft sphere model,[221] has investigated the dielectric properties of highly polar fluids through computer methods. The goal of this work was to examine the effects of both the ensemble size and ε_{RF} on the evaluation of the dielectric constant and other properties. In Ref. 361, ensemble sizes from 4 to 1372 particles and ε_{RF} from 15 to ∞ were considered. Such a systematic study has been long overdue. Kusalik finds that the static dielectric constant, and to a lesser degree, thermodynamic properties showed significant variation, particularly in the smaller systems ($N < 256$), both with respect to sample size and ε_{RF}. Generally, ε was found to increase with increasing N and decreasing ε_{RF}, an effect most noticeable for the smaller sample sizes. In the second paper,[362] Kusalik introduces a new method for computing the dielectric constant, namely the determination of the long-range asymptotic behavior of dipole–dipole correlations. This method, as expected, requires rather large ensemble sizes, 4000 particles being used in this study. Concluded is that the description that was becoming comfortable to workers in the field, namely the one espoused in the paper by Neumann and Steinhauser,[247] "may fail both qualitatively and quantitatively." Another recent advance, particularly important for computational studies of perturbed water ensembles, has been made by Hautman and Klein,[363] who have presented an LS method for systems having periodicity in only two dimensions and a finite third dimension.

V. INTERACTION POTENTIALS

Developing good models for the water monomer and the dimer would seem to be a central problem for computer simulations of water in any of its condensed phases. However, even if the characteristics of the monomer and the dimer were known to high precision, these small entities would provide a false security in the formulation of bulk phase water models, since the main problem in transferring information from them to condensed phases remains one of assessing the effects of the many-body forces. For this reason, interaction models for water ensembles have mainly been developed empirically, though some notable attempts to devise potentials based purely on ab initio quantum calculations have been made by Clementi and his coworkers.[33, 34, 67, 82] A background and summary of the interaction potentials used in MD simulations are presented below.

Since so many water models have been used or are now being used, nomenclature is becoming confusing, particularly to those not completely immersed in this field. First of all, the models can be divided into groups, depending on the total number of distinct sites employed. It would be useful (perhaps) for future workers if a nomenclature scheme based on this aspect could be devised. However, names of the now well-recognized models—ST2, MCY, TIP4P, CF, SPC, PE, RWK, etc.—would all have to be changed into less recognizable names, and, like so many other directives about units and conversion factors (Section II.A), would be ignored by a great fraction of workers. Thus, we will retain the above names in this review. However, following Stillinger and Rahman,[30] we will attach the numeral 2 to the latest improved version of the above models. This is consistent with most, but not all,[54] current notation. A suffix F will be attached for all flexibly bonded models, but it is important to distinguish whether the intramolecular bonding anharmonically couples the bending and stretching modes. If rapidly responsive electronic polarization is included in the model, the suffix P will be attached. The central force (CF) models always contain anharmonic coupling between the atoms, so our notation here is a bit redundant, but hopefully not confusing. For example, the final "best" CF model produced by Stillinger and coworkers[53] will be called CF-F, while the further improved CF model of Bopp et al.[54] will henceforth be called CF2-F.

A. Ab Initio Potentials

These potentials surfaces are usually obtained by making detailed quantum-mechanical calculations of the ground-state energies of water monomers, dimers, trimers, or tetramers, then fitting the resulting data for different configurations to a convenient analytic form. A survey of quantum-mechanical calculations on small water polymers within the self-consistent field (SCF) approximation was given some time ago by Del Bene and Pople.[364] There is still some controversy[365-367] about the subtler details of the dimer interaction potential energy surface, such as whether or not the metastable ($\sim +5$–8 kJ mol^{-1}) bifurcated structure has a "global minimum" with all six intermolecular vibrations nonimaginary.

An analytically fitted Hartree-Fock potential surface for the water–water interaction was initially developed in 1973 by Popkie et al.[33] from 190 configurations of the water dimer. This dimer potential was refined by Kistenmacher et al.[368] Later, Matsuoka et al.,[82] using a configuration-interaction (CI) method, fitted the computed energies of 66 different configurations to an analytical expression derived for a 4-site model on each molecule. Included were nine pairs of Coulomb interactions and thirteen exponential terms. This model is referred to as MCY. The flexible bond extension[57] of MCY is usually called MCYL, but, in keeping with the notation used in this review, it will be referred to as MCY-F. Generally speaking, the MCY and MCY-F models reproduce many properties of water reasonably well. However, the resulting potentials are complicated, which limits their application. In addition, the MCY model yielded too little cohesive energy, giving excessively high pressures,[40] and gave disappointing dielectric properties.[245] Probably the worst feature of these models with respect to liquid water modeling is the fact that the dimer geometry is totally inappropriate for the liquid state. Since many-body interactions play a very important role in condensed phases of water, we note that a pair potential representing the correct dimer geometry would yield strong repulsions in the liquid state as the molecular pairs with a 2.98-Å O $\cdots$ O dimer separation are forced by the density constraint to move to a closer distance ($\sim 2.76-2.85$ Å) in the liquid. This is the reason why the ab initio MCY potential produces such high pressures. Scaling the MCY parameters to match the effective liquid-phase potential and using a simpler analytical form would have helped, but this is what the empirical potentials do. The inclusion of intramolecular bond flexibility and polarization are really required so that the dimer potential surface can continuously adjust itself in the liquid state to a more realistic many-body form.

Since the idea of a nonempirical pair potential function is so appealing, there have been many efforts to improve the MCY model. The earliest attempts extended the sampling for the parameter determination.[369, 370] Considered also were explicit three-[83] and four-body[84] corrections, but these lead to a huge computational effort. More recently, the MCY model was further modified[66] by adding new exponential repulsion terms at the two-body level, five between the negative charge points and the hydrogen atoms and two between the negative charge points and the oxygen atoms. In addition, a polarization term was explicitly included. Altogether, 350 dimer configurations and 250 trimer configurations were used in the construction of the analytic potential. Adding the repulsions helped to correct some of the deficiencies of the original two-body model at short distances. The latter potential was used in an MD simulation with reasonable success for the study of a limited number of properties.[67]

As already alluded to from time to time in this review, the construction of a usable potential function for bulk liquid water, starting from even the most accurate experimental or quantum-mechanical descriptions of the dimer and small polymers, presents formidable problems. While extremely useful in a broad sense for suggesting pathways to be followed in the construction of empirical potentials, and even leading to reasonably good many-body potential functions[67] for the liquid state if computational costs were not limited, there are just too many undesirable details in the ab initio potentials. These details lead to impractical computational difficulties in MD or MC calculations, particularly for the perturbed water problem. Worst than this is the high probability that minute details in the small polymers will be modified in the liquid state anyway. This is part of the reason why purely empirical interaction models have been so popular.

B. Empirical Potentials

Most empirical models for water place hydrogen and oxygen masses at sites roughly equivalent to their positions in a water molecule. To model electrical effects, charge centers are placed either coincident with the mass sites, or at other locations within the molecular frame, in such a way as to retain overall electrical neutrality but to reproduce either the dipole moment of the free monomer or some effective molecular dipole moment in the condensed phase.

1. Brief Comments About Dimer

Dimer interactions are functions of the intermolecular separation and the mutual molecular orientation. Even in the bare water dimer, as discovered through modern quantum theory,[365-367] the picture is not a simple one. Empirically speaking, intermediate range forces are derived from induction and dispersion mechanisms. The induction energy is related to the molecular polarization. Dispersion forces, i.e., van der Waals forces or London forces,[371] have their origin in the correlated motion of electrons in neighboring molecules. These forces may be represented[23] by a series with a leading term proportional to r^{-6}, which provides the attractive term in the Lennard-Jones 6-12 potential used in water MD models. Electrostatic forces arise from interactions between the electric moments of the water molecules. Hydrogen bond interactions in the linear dimer, again empirically speaking, are caused by bonding electrons mainly associated with an oxygen atom in one molecule, delocalizing into the intermolecular space near one hydrogen atom of its neighbor. Conventionally, the latter molecule, that is, the one whose hydrogen atom is shared in (donated to) the intermolecular bond, is called the donor, while the molecule with both hydrogen atoms free is called the acceptor (see Fig. 1 of Ref. 86 or Fig. 6 of Ref. 120).

2. One-Site Models

The water molecule represented as a Stockmayer particle, a hard sphere with an embedded point dipole, or more realistically as a dipolar[221] or multipolar soft sphere provides the simplest empirical model for water. Clearly, many desirable details are missing from such a model. However, to achieve the highest computational efficiency for either a preliminary evaluation or for use in biological systems, where already the computations stretch modern technology to the limit, this would be the model of choice. The best example of this sort of potential is the polarizable electropole model (PE) of Barnes and coworkers,[372, 373] in which the water–water interactions are composed of an electrostatic term to account for the permanent dipole and quadrupole moments, a polarizable term accounting for nonadditivity, and a 6-12 Lennard-Jones potential. A detailed and interesting comparison of this model with rigid MCY and ST2 models has been made with respect to the dimer energy surfaces.[374]

3. Three-Site Models

The simple point charge (SPC) model[36] is very popular because of its simplicity and its resulting economy in MD computations. It is often used in perturbed water computations, where the other complicating factors in the problem prohibit the use of a more detailed water model. A partially shielded negative charge site is located at the oxygen atom position, while the two hydrogen atom positions carry partially shielded positive charges. The O–H bond length of SPC water is taken to be exactly 1 Å, and the H–O–H bond angle is assigned the precise tetrahedral angle of 109°28′. The intermolecular interactions are equivalently simplified, consisting only of a Lennard-Jones potential between oxygen atoms and intermolecular Coulomb potentials between the charge points. A revised SPC model,[37] termed SPC/E by its authors, but renamed SPC2 here, includes a self-energy correction. This model was found to provide a much improved effective pair potential, and gave better density, RDFs and diffusion constants than the original SPC model. TIP3P[39] (transferable intermolecular potential with three points) is modeled similarly to SPC. However, the molecular geometry is chosen to be exactly the same as that of an isolated water molecule. In these basic 3-site models, the fixed dipole moments from the charge points are 2.274 D (SPC), 2.351 D (SPC2), and 2.347 D (TIP3P).

Further extensions of the SPC model, including bond flexibilities or electric polarization or both, have been made. For example, with an anharmonic flexible version SPC-F, Toukan and Rahman[56] were able to investigate perturbations in the liquid state on the high-frequency internal vibrational modes of water. Ahlström et al.[68] have also investigated the effect of adding point polarization to the rigid SPC model, SPC-P. Another flexible version of SPC has been created by Anderson et al.[61] using anharmonic intramolecular coupling derived from the Morse potential.[375]

An SPC model with both harmonic bond flexibility and instantaneously responsive polarizability (SPC-FP) has been created by Zhu et al.[74] This potential has been used in perturbed water MD studies.[376, 377] The SPC-FP model, as mentioned earlier, adds variable charges onto the oxygen and hydrogen atoms to introduce the dipole polarization. The parameters in this model were then reoptimized for best agreement with the properties of liquid water.

Another 3-site model that is well worth mentioning here utilizes a central force (CF-F) potential function,[52, 53, 81] which takes into account the flexible bonding automatically. This model treats the constituent oxygen and hydrogen atoms as dynamically distinct mass points carrying suitable electric charges. Interactions within a molecule are regarded as indistinguishable from those between molecules. However, the intramolecular pair potentials are chosen to form stable water molecules with the proper geometry, about which vibrational motions can take place. These pair potentials are intended to produce hydrogen bonding of the approximately correct energy and geometry between neighboring water molecules. This central force model has been modified[54] by incorporating a more appropriate potential for the internal vibrations in order to give better vibrational frequency shifts from the gas to the liquid phase. We will see later in this review that an appropriate form of the anharmonic intramolecular potential function is absolutely essential for obtaining even qualitatively correct frequency shifts in the condensed phase.

A rather complicated 3-site model[73] in historically important, since it was the first to contain both flexible intramolecular harmonic bonding and polarization. A Morse potential was used to describe the interaction between oxygen and hydrogen atoms on different molecules participating in a hydrogen bond. To model the polarization, variable charge magnitudes as a function of the hydrogen-bond strength were used. This latter idea is based on the proposition that at any instant the molecule's dipole moment, which varies from 1.85 D in gaseous water to 2.6 D in ice Ih, is related to its energy, which varies from zero for gaseous water to -58.9 kJ/mol in ice Ih. Whether this idea, which gave good overall results but which ignores the electrical properties of the molecule per se, is worth pursuing farther remains to be seen. The latter 3-site model has not yet been used in the study of perturbed water.

4. Four-Site Models

TIP4P[40] is a planar four-site model, modified from TIP3P by moving the negative charge 0.5 Å off the oxygen atom toward the hydrogens to a point on the bisector of the H–O–H bond angle. TIP4P is nearly as simple as SPC or TIP3P, and is quite a good rigid model, having been widely used in various types of perturbed water studies. The Drude model algorithm of Sprik and Klein[64] for treating many-body polarization was tested using a TIP4P-P model with three additional charge sites.

A more complicated four-site model, RWK2, based on the earlier 3-site potential of Watts,[85] but with the negative charge associated with the oxygen atom shifted 0.26 Å along the symmetry axis toward the hydrogen atoms, was developed by Reimers et al.[27] The parameters in this model were determined from the second virial coefficient of the vapor, the static lattice energy, the bulk moduli of ices Ih, VII, and VIII, dipole and quadrupole moments of the isolated water molecule, as well as accurate calculations of the long-range dispersion coefficients. The molecular geometry is similar to that of MCY and TIP4P. The approach used here for determining the potential parameters is a very sensible one, but the resulting potential turns out to be fairly complicated, limiting its use. A very good summary of the properties of some of the previous rigid models is contained in Ref. 27.

Reimers and Watts[86] have extended the RWK2 model, RWK2-F, by adding an intramolecular potential surface consisting of Morse-type functions. A somewhat modified RWK2 potential, including a nonadditive term, was created by Lybrand and Kollman,[87] and a polarizable version, RWK2-P, has been proposed by Cieplak et al.[69] on the basis of isotropic atomic polarizabilities.[378] Vectorization strategies for improving computational efficiencies for the Lybrand-Kollman model[87] have been published by Sauniere et al.[379]

Recently, a new model, which is much simpler, yet retains the essence of RWK2, has been proposed by Sprik.[71] This model converts the more complicated exponential interactions of RWK into a Lennard-Jones interaction between oxygen atoms. The model is similar to TIP4P, except for charge sizes and placements. The negative charge is moved farther from the oxygen atom in the RWK models, and, to keep the same dipole moment, the charge must be correspondingly larger. As a matter of interest, as the negative charge nears the H–H axis, the charge magnitudes have to become huge, approaching infinity as this axis is approached. In this limit, the dielectric constant accordingly verges to unity for a nonpolarizable model, since an external field would have no effect on such strongly interacting charges in this system. As seen in Fig. 5 of Sprik's paper, this is exactly what happens when the TIP4P charge placements are changed to RWK2 charge placements: The resulting dielectric constant is "far too small." Sprik has also described the effects of adding polarization to this and other models. This will, of course, increase the dielectric constant. The discussion in this paper[71] seems to reflect the fact that none of these planar models can really give a correct picture of all the properties of water. It seems that when you push in one place, something undesirable pops out another. Things can only get worse when temperature and pressure effects come into play. This is why future simulations need to look at these effects in a more systematic manner.

5. Five-Site Models

The historically important Rowlinson potential[79] is based on a point charge tetrad, plus a neutral oxygen atom. As usual, the two positive charges are placed on the hydrogen atoms. The molecular geometry of this early model is only slightly different than that used in other rigid models, $r(O–H) = 0.96$ Å and H–O–H = 105°. However, a major difference lies in the placement of the two negative charge points, only 0.2539 Å directly above and below the oxygen atom. Of course, to achieve the correct dipole moment (Rowlinson used the gas phase value) with this placement of charges, the charges had to be more robust than in other 5-site models. In fact, the Rowlinson model resembles 3-site models because of the nearness of the negative charges to the oxygen atom. Therefore, many of the disadvantages of these models are present, without the advantage of the simpler 3-site computational algorithm. Nevertheless, the essence of Rowlinson's model has carried over to the more modern 5-site models of water.

The first widely useful water model for MD computations was ST2.[30] After twenty years, this model or modifications of it are still being used for calculations on bulk and perturbed water. The prototype[28] of ST2 was based on a tetrahedral potential model (BNS) suggested by Ben-Naim and Stillinger.[80] It is a 5-site model, which places the four charges in a tetrad geometry with the neutral oxygen atom symmetrically positioned at a central point. In the final version of ST2, there are two sites carrying the positive charges, which are identified as partially shielded protons. These are located exactly 1 Å from the oxygen atom, forming an H–O–H angle of 109.47°, as compared with the now presumed experimental value in the liquid of $< 106.3°$ (see Section III.C.1). The other two sites carry negative charges that must, of course, have an identical magnitude as the positive charges in order to keep the molecule neutral. The two negative sites each are separated from the oxygen atom by 0.8 Å to emulate the oxygen $2p$ lone pairs. Included with the Coulomb terms is a single intermolecular Lennard-Jones potential with its center on the oxygen atom. The inter-molecular Coulomb interactions, 16 in all, between any two molecules have to be multiplied by a modulation function, which smoothly varies from 0 to 1 between about 2 and 3 Å, in order to avoid Coulombic energy divergencies at close interparticle separations. A modified version of ST2 has been proposed by Evans.[43] Since this model includes atom–atom Lennard-Jones interactions between both the oxygen and hydrogen atoms on different molecules, there is no need to use a modulation function to keep the charges apart.

Recently, Zhu et al.[75] have developed a new 5-site water model, called MST-FP in this review, which is based on ST2 with various modifications: The neutral oxygen atom and the two partially shielded protons constitute the normal geometry of a free water molecule; molecular polarization, according to the earlier variable charge prescription of Zhu and Robinson,[74] is explicitly incorporated; a modified[380] Urey-Bradley poten-tial[381] was used to determine the intramolecular vibrational motions; a Morse type function to represent the intermolecular O $\cdots$ H interaction was introduced to obtain better orientational properties; an exp-6 was substituted for the Lennard-Jones 12-6 O $\cdots$ O interaction; and, finally, instead of the modulation function used in ST2, the point charges were replaced by ones distributed according to a squared Lorentzian. The Urey-Bradley intramolecular potential function, or some other well-designed anharmonic model,[54] seems essential if good gas to liquid phase vibrational spectral shifts are to be achieved in MD calculations. From earlier studies[73, 74] it was also felt that a sharper angular dependence of the hydrogen-bond interactions, as provided by the Morse potential, was desirable. The squared Lorentzian charge distribution introduces a $(2/\pi)$ arctan(ar) factor into the Coulomb potential, where $a = 5$ Å^{-1} and r is the distance from the charge center. With this modification, the resulting electrostatic potential becomes a constant $3.183q$ at distances less than about 0.1 Å, then smoothly becomes a normal Coulomb potential beyond about 1.5 Å. This idea is similar to one in the earlier algorithm of Sprik and Klein,[64] who used Gaussian charge distributions. The arctan dependence in the modified Coulomb interaction is somewhat more economical in computer time than the error function resulting from Gaussian charge distributions. In spite of the seeming complexity of the MST-FP model, modern supercomputers were found to be capable of handling the problem in a statistically meaningful manner for both pure and perturbed water ensembles.

VI.   COMPUTER SIMULATIONS: BULK WATER

In this section, we lay emphasis on the description of water models that have played the greatest role in the study of field perturbed water. These include the perfectly rigid models,[24-51] together with variations of these models containing flexible bonding (F),[51-63] rapidly responsive electronic polarization (P),[64-72] or both (FP).[73-76] For quick reference, a summary of some of these models and their properties are displayed, along with the experimental data, in Tables I–IV.

A.   Structure

It will be assumed that the best experimental RDFs are those obtained from the neutron diffraction work of Soper and Phillips.[166] Generally speaking,[69] the 3-site models are unable to provide sufficient liquid structure beyond the first peak in the O $\cdots$ O RDF, specifically in the neighborhood of 4.5 Å, which marks the location of the closest non-hydrogen-bonded oxygen neighbor in tetrahedrally packed water molecules. One exception is the 3-site Zhu-Robinson model,[73] which stiffens the angular dependence with Morse-type hydrogen-bond interactions. A similar behavior can be found for MCY,[34] where the second and third RDF maxima are also prominent. It does not seem therefore that the origin of inadequate $g_{OO}(r)$ structure lies solely in the 3-site nature of these models. Adding the softening effects of bond flexibility or polarization to a 3-site model usually makes matters worse.[74] However, the flexible version of MCY[57] seems to have slightly more pronounced O $\cdots$ O RDF structure, probably because of the introduction of stiffer interaction terms in that model.

TABLE I
Rigid Models

	SPC	SPC2	TIP4P	RWK2	ST2
Main reference[a]	37	37	40	71	30
$R(OH)$ (Å)	1.0	1.0	0.9572	0.9572	1.0
$\theta(HOH)$	109.47°	109.47°	104.52°	104.52°	109.47°
$q(H)$ ($|e|$)	0.41	0.4238	0.52	0.60	0.2357
$R(q)$ (Å)	0	0	0.15	0.26	0.8
$\theta(q)$	—	—	360°	360°	109.47°
$\varepsilon(LJ)$ (kJ mol^{-1})	0.6502	0.6502	0.6487	(0.6487)	0.3169
$\sigma(LJ)$ (Å)	3.166	3.166	3.154	(3.154)	3.10
$|m_0|$ (D)	2.274	2.351	2.177	1.878	2.353
Q_{xx} (B)[b]	-1.830	-1.892	-2.091	-2.323	-1.703
Q_{yy} (B)[b]	2.110	2.180	2.203	2.630	2.010
Q_{zz} (B)[b]	-0.279	-0.288	-0.112	-0.307	-0.306
Method[c]	MD-NVT	MD-NVT	MC-NPT	MD-NVT	MD-NVT
T (K)	308	306	298	300	283
ρ (g cm^{-3})	0.970	0.998	1.002[e]	1.000	0.997
U (kJ mol^{-1})	-37.7	-41.4	-42.1	-39.3	-37.4
C_V (kJ mol^{-1} K^{-1})	0.091	—	0.067	0.069[g]	0.126
P (atm)	-0.99	5.92	1	-1283^g	118
ε	72[d]	70[d]	61[d, f]	23	—
D (10^{-5} cm^2 s^{-1})	4.3	2.5	3.3[c]	—	1.9
$r_1(OO)$ (Å)	2.77	2.75	2.76	2.75	2.84
$g_1(OO)^{-1}$	1.78	1.96	2.00	1.95	2.16
$\bar{r}_1(OO)$ (Å)	3.55	3.31	3.37	3.35	3.53
$1 - \bar{g}_1(OO)$	0.10	0.16	0.18	0.15	0.32
$r_2(OO)$ (Å)	4.58	4.64	4.42	4.41	4.65
$g_2(OO) - 1$	0.09	0.14	0.13	0.09	0.18

[a]Other references are in the footnotes.
[b]Quadrupole elements recalculated in all cases using center-of-mass coordinate system and axis designations in Ref. 113. B = Buckingham = 10^{-26} esu.
[c]MD-NVT = MD-NVE with "thermostat."
[d]Ref. 65.
[e]Ref. 41.
[f]Neumann,[246] calculates $\varepsilon = 53$ for TIP4P at $T = 293$ K.
[g]Ref. 27. Other entries are from the parametrized version of RWK2 in Ref. 71.

While rigid TIP4P produces $g_{OO}(r)$ in reasonable agreement with the neutron diffraction data,[166] the accord with experimental $g_{OH}(r)$ and $g_{HH}(r)$ is not very good, as is the case for most other water models. Before drawing too many conclusions about the agreement between the experimental and calculated RDFs, one must be cognizant of the difficulties involved in an accurate experimental determination of these functions. It is interesting that the recently reported polarizable water model (TIP4P-P) of Sprik and Klein[64] gives RDFs as strongly structured as those obtained

TABLE II
Flexible Models

	SPC-F	SPC-F	SPC-F	SPC-F	RWK2-F
Main reference	56	61	58, 59	58, 59	55
Input	a	b	e	f	g
Method	MD-NVT	MD-NVT	MD-NVT	MD-NVT	MC-NPT
T (K)	325	300	~300	~300	298
$R(OH)$ (Å)	—	—	1.017	1.022	0.979
$\theta(HOH)$	—	—	104.9°	105.6°	104.6°
$\lvert m \rvert$ (D)	—	2.42	2.44	2.43	—
U (kJ mol^{-1})	—	-40.2^c	-40.7	-40.9	-31.0
C_V (kJ mol^{-1} K^{-1})	—	—	—	—	0.084
P (atm)	—	—	—	—	-987
ε	—	82.5	—	—	—
D (10^{-5} cm^2 s^{-1})	—	2.54	3.6	3.1	—
$r_1(OO)$ (Å)	—	2.8	2.75	2.73	2.78
$g_1(OO) - 1$	—	1.8	1.86	1.84	1.82
$\bar{r}_1(OO)$ (Å)	—	—	3.31	3.34	3.43
$1 - g_1(OO)$	—	—	0.14	0.17	0.14
$r_2(OO)$ (Å)	—	—	4.46	4.46	4.43
$g_2(OO) - 1$	—	—	0.05	0.05	0.16
$\nu_1(g)$ (cm^{-1})	3828^a	—	—	—	3657^h
$\nu_2(g)$ (cm^{-1})	1655^a	—	—	—	1595^h
$\nu_3(g)$ (cm^{-1})	3958^a	—	—	—	3756^h
$\nu_1(l)$ (cm^{-1})	3842^a	3501^d	3676	3462	—
$\nu_2(l)$ (cm^{-1})	1794^a	1831^d	1591	1858	~1665^i
$\nu_3(l)$ (cm^{-1})	3968^a	3645^d	3729	3626	3330^j
$\nu_L(C_2)$ (cm^{-1})	540^a	446^d	—	—	~800

[a] Input parameters identical to Ref. 36 except for addition of purely harmonic potential for O–H and H $\cdots$ H, plus harmonic cross terms. For some reason, the force constants for the harmonic model were chosen about 5% too large, giving comparably large monomer frequencies $\nu_i(g)$ and liquid state frequencies $\nu_i(L)$ as well. $\nu_L(C_2)$ is A_2 librational mode around the C_2 molecular axis.

[b] Same input as Morse potential calculation in Ref. 56, but $T = 300$ K.

[c] Cited in Table 2 of Ref. 68.

[d] These frequencies are from Ref. 56.

[e] Input parameters identical to those in Ref. 36, except for addition of purely harmonic potential for O–H and H–O–H bend; no cross terms.

[f] Same input as Ref. 36, except for addition of Morse potentials for O–H, harmonic potential for H $\cdots$ H, plus harmonic cross terms.

[g] Same input as Ref. 27, plus sum of three Morse oscillators for intramolecular potential surface.

[h] The potential surface was fitted to the experimental data. See Ref. 107.

[i] From statement in Ref. 55 that the monomer bending mode is blue shifted about 70 cm^{-1} in the liquid. However, a crude estimate from Fig. 4 of Ref. 55 shows this frequency to be closer to 1620 cm^{-1}.

[j] Estimated from Fig. 5 of Ref. 55 at 298 K. This figure correctly shows the temperature dependence of the O–H stretching vibrations, a low-temperature (100 K) peak at 3310 cm^{-1} from hydrogen-bonded oscillators, evolving into a high-temperature (1100 K) peak at 3660 cm^{-1}, arising from O–H oscillators "not so directed" toward an oxygen atom on a neighboring molecule. It is not clear from Ref. 55 whether this "oscillator" derives primarily from the symmetric or the antisymmetric stretching frequency. Since the method used is a dipole method, which gives the infrared spectrum, it can be presumed that the oscillator is primarily the antisymmetric stretch, but see footnote n of Table IV.

TABLE III
Polarizable Models

	SPC-P	SPC2-P	TIP4P-P	RWK2-P	RWK2-P
Main reference	68	70	64	69	71
$q(H)$ ($|e|$)	0.3345	0.3650	0.4428	0.6	0.6
ε(LJ) (kJ mol^{-1})	0.5418	0.6502	0.6487	g	0.8394
σ(LJ) (Å)	3.263	3.160^c	3.154	g	3.193
$|m_0|$ (D)	1.85	2.025	1.85	1.878	1.878
$|m|$ (D)	2.9	2.516	2.85	2.20	2.63
Q_{xx} (B)	-1.493	-1.629	-1.781	-2.323	-2.323
Q_{yy} (B)	1.721	1.877	1.876	2.630	2.630
Q_{zz} (B)	-0.228	-0.248	-0.095	-0.307	-0.307
Other input	a	c	e	h	j
Method	MD-NVT	MD-NVT	MD-NVT	MC-NPT	MD-NVT
T (K)	302	303	300	298	300
ρ (g cm^{-3})	1.004	0.991	1.0	0.964	0.997
U (kJ mol^{-1})	-38.0	-41.1	-42.2	-41.7	-46.6
C_V (kJ mol^{-1} K^{-1})	—	—	—	0.0721	—
P (atm)	—	—	—	1	592
ε	—	—	—	—	86
D (10^{-5} cm^2 s^{-1})	2.0	3.1	—	—	2.4
r_1(OO) (Å)	$\sim 2.82^b$	$\sim 2.79^d$	2.75	2.98	2.82
g_1(OO) $- 1$	$\sim 1.8^b$	$\sim 1.95^d$	2.55	1.78	2.33
$\bar{r}_1$(OO) (Å)	$\sim 3.6^b$	$\sim 3.26^d$	3.26^f	i	3.35
$1 - \bar{g}_1$(OO)	b	$\sim 0.13^d$	0.34^f	i	0.16
r_2(OO) (Å)	b	$\sim 4.6^d$	4.5	i	4.52
g_2(OO) $- 1$	b	$\sim 0.15^d$	0.08	i	0.11

^{a}Structure and charge placements same as Ref. 36.

^{b}Difficult to estimate from published RDFs. Seems similar to rigid SPC model, except for slightly larger r_1(OO).

^{c}Same input as Ref. 37 except that cited σ for SPC2 of 3.160 Å is 0.006 Å smaller than σ of Ref. 37.

^{d}Very similar to O $\cdots$ O RDF for rigid SPC2 plotted in this paper and also in Ref. 37.

^{e}Structure and charge placements same as Ref. 40, except that negative charge site is divided into close tetrahedral array, giving a 7-site model.

^{f}Estimate. The full RDF shown only for case where attractive interaction was reduced by 30%.

^{g}Not Lennard-Jones.

^{h}Approximately the same form of the potential as in Ref. 27, but many parameters changed. Charge placements and molecular structure same as in Ref. 27.

^{i}Virtually no structure beyond first peak.

^{j}This is the parameterized version of RWK2. It is interesting that the RDF structure has returned.

from rigid models. Noted in this paper is the fact that, in nearly all computational models, the first maximum in the O $\cdots$ O RDF is at too small a distance compared with the experimental value. In ST2,[30] too much structure was said to be encountered,[40] but remember that in the older computational work comparisons were made with the x-ray results of Narten and Levy,[171] which show a somewhat flatter first peak.

Characteristic values of the O $\cdots$ O pair correlation functions, namely the positions of the first and second maxima, r_1 and r_2, the first minimum, $\bar{r}_1$, and the amplitudes, $g_1 - 1$, $g_2 - 1$, and $1 - \bar{g}_1$, from a variety of water models, together with data observed from neutron[166] scattering experiments, are summarized in the tables.

Temperature effects on the pair correlation functions have been investigated by Stillinger and Rahman for the BNS potential[29] and by Jorgensen and Madura[41] for TIP4P water. As the temperature increases, the height of the first $g_{OO}(r)$ peak decreases. This is accompanied by a slight shift of this peak to a larger distance, while the structure beyond this region tends to be eliminated. These effects indicate the expected loss of liquid structure with increasing temperature.

B.   Hydrogen Bonding

Conventionally, we refer to a pair of molecules as being hydrogen bonded if their mutual interaction energy is below a preassigned cutoff value.

TABLE IV
Flexible/Polarizable Models

	SPC-FP	MST-FP	Experiment
Main reference	74	75	
$q(H)\,(|e|)$	0.325	0.164	—
$R(q)\,(\text{Å})$	0	1	—
$\theta(q)$	—	109.47°	—
$\varepsilon(LJ)\,(kJ\,mol^{-1})$	0.5418	b	—
$\sigma(LJ)\,(\text{Å})$	3.241	b	—
A. *Monomer*			
$R(OH)\,(\text{Å})$	0.9572	0.9572	0.9572[c]
$\theta(HOH)$	104.52°	104.52°	104.52°[c]
$|m_0|\,(D)$	1.829	1.833	1.855[d]
$\nu_1\,(cm^{-1})$	~3828[a]	3656	3657[c]
$\nu_2\,(cm^{-1})$	~1655[a]	1594	1595[c]
$\nu_3\,(cm^{-1})$	~3958[a]	3755	3756[c]
$Q_{xx}\,(B)$	−1.325	−1.390	−2.50[e]
$Q_{yy}\,(B)$	1.536	1.539	2.63[e]
$Q_{zz}\,(B)$	−0.212	−0.149	−0.13[e]
$\alpha\,(10^{-24}\,cm^3)$	1.271	1.444	1.429[f]

TABLE IV (Continued)

	SPC-FP	MST-FP	Experiment
B. *Liquid*			
Method	MD-NVT	MD-NVT	
T (K)	298	298	298
ρ (g cm^{-3})	0.997	0.997	0.997[i]
U (kJ mol^{-1})	-41.0	-40.3	-41.5[j]
C_V (kJ mol^{-1} K^{-1})	—	—	0.0745[i]
P (atm)	~ 1	~ 0	1
ε	54.6	58.7	78.4[i]
D (10^{-5} cm^2 s^{-1})	7.1	1.6	2.4[k]
R(OH) (Å)	0.966	0.963	0.966[l]
θ(HOH)	101.0°	102.3°	102.8°[o][l]
$|m|$ (D)	2.44	2.45	$\lesssim 2.6$[m]
ν_1 (cm^{-1})	3875	3251	~ 3280[n]
ν_2 (cm^{-1})	1802	1713	~ 1645[n]
ν_3 (cm^{-1})	4080	3418	~ 3490[n]
$\nu_L(C_2)$ (cm^{-1})	—	474	425–450°
τ_c (ps)	6.0	6.8	7.3[p]
τ_s (ps)	1.3	2.0	2.5[p]
r_1(OO) (Å)	2.82	2.85	2.87[q]
g_1(OO) -1	1.34	2.31	2.10[q]
$\bar{r}_1$(OO) (Å)	g	3.22	3.31[q]
$1 - \bar{g}$(OO)	g	0.28	0.27[q]
r_2(OO) (Å)	g	h	4.50[q]
g_2(OO) -1	g	h	0.14[q]

[a] Except for the equilibrium HOH angle change from tetrahedral to 104.52°, the force constants are the same as those in Ref. 56. They thus give monomer frequencies that are all about 5% too high compared with experiment.

[b] Not Lennard-Jones.

[c] Ref. 105.

[d] Ref. 112.

[e] Ref. 113.

[f] Ref. 116.

[g] No structure beyond first peak.

[h] Broad maximum near 4.0 Å with little amplitude.

[i] Ref. 483, p. 6–8. C_V has been obtained from C_P in the standard way.

[j] See discussion, Ref. 40.

[k] Ref. 386.

[l] Ref. 126. See, however, Ref. 127 which gives R(OH) = 0.989 Å and θ(HOH) = 106.3°.

[m] Ref. 138.

[n] These frequencies are from Table 4.10 of Ref. 12. However, the actual identification of ν_1 and ν_3 is open to interpretation. Mode coupling and ordinary liquid state broadening give rise to a very complex structure in the infrared and Raman spectra in the 3200- to 3600-cm^{-1} range. In addition, there are two components whose intensities and shapes are temperature dependent. These effects to some extent should also be present in the MD simulations.

[o] Refs, 134, 148.

[p] See Refs. 391–394 and discussion in this review.

[q] From Table 5, Ref. 166.

Stillinger and Rahman[382] have shown that if this energy lies between -1.7 and -4.5 kcal/mol, then the ST2 potential reproduces the conventional hydrogen bonding pattern in ice. At high temperatures, any residual ice structure progressively disappears. Consequently, the mean number of hydrogen bonds per water molecule decreases and the average strength of the hydrogen bonds becomes weaker. This effect has been studied in detail by Jorgensen et al.[40, 41] By setting the cutoff potential at -2.25 kcal/mol, these authors first analyzed the hydrogen bonds in water at 298 K and 1 atm pressure for various models.[40] According to their definition, the average number of hydrogen bonds ranges from 3.50 to 3.73, much less than the value of 4.4 obtained by integrating the first peak of the experimental O–H radial distribution functions.[166] They then carried out the same kind of study for TIP4P water at a series of temperatures.[41] Perhaps the most fascinating feature of this work is that there is no significant occurrence of free non-hydrogen-bonded water molecules at any temperature (see Table IV of Ref. 40 and Table 4 of Ref. 41). There is also a sharp increase of the fraction with four hydrogen bonds in the supercooled region at the expense of coordination numbers of all other species. This indicates the formation of an icelike structure in supercooled water. However, there may be some ambiguity in these values because of the choice of hydrogen-bond energy. It was also found in this paper that 80.5% in TIP4P and 81.5% in ST2 of monomers participate in three or four hydrogen bonds. Clearly, this type of study needs to be carried out with water models that contain bond flexibility and polarization.

C. Vibrations

In MD simulations, intramolecular vibrational frequencies can be determined by taking the Fourier transform of the VACFs for the oxygen and hydrogen atoms. It is therefore more convenient to use the change in the H–H distance, instead of the change in the bond angle, to describe the intramolecular potential.

Though it has been claimed[383] that an accurate picture of condensed phase perturbations on the intramolecular vibrational frequencies requires quantum-mechanical considerations, the problem seems to lie more in the choice of a good intramolecular potential function. The modified Urey-Bradley harmonic restoring potential, with anharmonic cross terms coupling the bending and stretching vibrations,[380] used in the formulation of MST-FP water, was found to have good success in providing relatively accurate liquid phase frequency shifts, though because of the temperature shifts, mode coupling and the suggestion of more than one vibrating species, the experimental values are not certain. What is certain is that the bending vibration is shifted to a higher frequency, while the stretching vibrations are shifted to lower frequencies in the liquid compared with the gas phase. Most other computer simulations on flexibly bonded water have predicted stretching frequencies that are too high (see Table II). The potential function in the MST-FP calculations was chosen so as to give the unperturbed molecular geometry and vibrational frequencies equivalent to the experimental values for the gas phase water molecule. Undoubtably, with a few more refinements, it should be possible to attain essentially exact agreement with the experimental bulk phase frequency shifts without resorting to quantum-mechanical input. After all, it is the potential surfaces that are perturbed by the liquid state environment, and the same type of effects apply whether classical or quantum mechanics is eventually used to calculate the frequencies.

Molecular librational frequencies were also determined using the MST-FP model. One can resolve the velocity vector of the hydrogen atoms into molecular translational and rotational components. By taking the Fourier transform of the translational VACF, intermolecular peaks are found around 60 and 200 cm^{-1}. These two modes are associated with oscillations of the center of mass. At 293 K, the experimental values[147] for these modes were found to lie at about 50 and 170 cm^{-1}, so the agreement is fairly good. From OACFs for MST-FP water, two relatively sharp librational bands were found centered at 474 and 788 cm^{-1}. The third librational mode is broader and was found to lie near 563 cm^{-1}. These frequencies are in pretty good agreement with the experimental[134] observations: the libration around the twofold C_{2v} molecular axis, ~ 425–450 cm^{-1}, the in-plane (rocking) libration, ~ 550 cm^{-1}, and the out-of-plane libration, ~ 720–740 cm^{-1}.

D. Thermodynamics

To obtain the experimental intermolecular energy U, Jorgensen and Madura[41] used the following relationship:

$$\Delta H_{\text{vap}} = -U + P[V(g) - V(l)] + \Delta Q - (H^0 - H) \qquad (11)$$

where the second term is the work of expanding the gas, which is approximately kT. ΔQ denotes the quantum correction between the gas and liquid for quantization of the intra- and intermolecular degrees of freedom, and the last term gives the enthalpy difference between the ideal and real gases, which can be evaluated by using the virial equation of state. Often, the last two terms are neglected. It was found that the calculated heats of vaporization agree with the observed values fairly well over a broad range of temperatures from 248 to 373 K. However, agreement for the liquid densities is not good, and for the isothermal compressibilities and heat capacity it is even poorer.

One of the most serious shortcomings for many water models is the poor prediction of the pressure in NVT simulations or the poor prediction of density in NPT simulations. The use of rigid models with potentials having a minimum at the isolated dimer separation is often the cause. In fact, of all the rigid water models, the Evans pair potential model,[43] is the only one that reproduces the pressure satisfactorily at high temperatures (to 1273 K).[44]

Good pressures are obtained in all three of the flexible-polarizable models studied at Texas Tech.[73-75] Apparently, the added softening effects of polarization and intramolecular bond flexibility automatically keep the molecules away from the strongly repulsive parts of the intermolecular potentials.

E. Dielectric Properties

One of the most fascinating characteristics of water is its high dielectric constant. Any reasonable water model should correctly reproduce this property. However, most water models have produced static dielectric constants that are too small relative to the experimental value.

Using the RF geometry, benchmark calculations of the static dielectric constants of MCY and TIP4P water have been presented by Neumann.[245, 246] At 292 K, ε for MCY was only 34 and the derivative of g_K with temperature was positive rather than negative, as it is known to be for water. For the computationally much simpler TIP4P model, the results were somewhat better, $\varepsilon = 53$ (293 K), but the derivative of g_K still had the wrong sign. A reasonable result for ε was found by Anderson et al.[61] who employed a flexible SPC model with intramolecular Morse potential coupling. These authors also obtained a fairly satisfactory temperature dependence for ε, within 10% of the experimental value at 300 and 350 K, and within 15% in the supercooled state. Considering their simplicity, the rigid SPC and SPC2 models provide good values of ε, $\sim 72^{36, 65}$ and $70.7^{65, 384}$ at 298 K. Not only this, but the temperature dependence of g_K for SPC2 behaves in the proper manner.[384]

The low values of ε for the SPC-FP and MST-FP models (54.6 and 58.7, respectively, ~ 295 K) were a bit of a disappointment considering the excellent average liquid state dipole moments obtained (~ 2.45 D, perturbed from the input monomer value of 1.83 D). Some of the deviation in these simulations, however, may have been caused by insufficient run times, which were only about 60 ps, or other causes.[362] On the other hand, the smaller than desirable vapor phase quadrupole moment in MST-FP signals that further modification of the charge placements in that model is desirable, and this could improve ε. It is worth pointing out here that the effective molecular quadrupole tensor is unknown in liquid water. Just like the dipole moment, it may have much larger components than in the vapor. In fact, if the scaling is similar, MST-FP would have a quadrupole tensor in the liquid state that isn't much different than that used in most successful rigid models.

With the SPC and TIP4P potentials, Alper and Levy[45] have found that the reaction field treatment does not significantly affect the radial distribution functions or the binding energy. They also obtained consistent results from calculating the dielectric constant from fluctuations in the mean square dipole moment of the system and from the effects of an applied external field, as mentioned previously. More recently, Belhadj et al.[359] have investigated the orientational structure and energetic properties of an ensemble of 126 SPC molecules using the Ewald summation. They claimed that the anticorrelation between orientations of next-nearest neighbors deduced from simulations using the Ewald summation, with either a spherical cutoff or a minimum image cutoff, was smaller than that from reaction field boundary conditions. However, consistent values were obtained for the dielectric constant. They also found that the dielectric constant is insensitive to both the number of water molecules and the cutoff radius for system sizes from $N = 126$, a $(15.6 \text{ Å})^3$ cube, to $N = 345$, a $(21.8 \text{ Å})^3$ cube.

Thus, there seems to be no way, if run times are sufficiently long and if everything else is done correctly, to bring calculated dielectric constants of water into agreement with the experimental value by merely changing the boundary condition or increasing the size of the ensemble or the cutoff radius. Rather, the water structure, the placement of the charge sites, its geometric and electronic flexibility, the nature of the intermolecular potentials, and the ability of these input parameters to match the long-range structural integrity found in real water would seem to be the answer to the achievement of good dielectric properties.

Even though the use of mean-field continuum theories to make corrections for finite size effects appears tainted with some danger, continued analytical work by Patey and coworkers seems to be putting this approach on a more understandable footing. For example, Wei and Patey[239] remind us that "truncating the dipolar potential at some cutoff radius and calculating a correction for the missing long-range forces by assuming the truncation sphere is surrounded by a dielectric continuum characterized by a frequency independent dielectric constant amounts to replacing the true dipole–dipole interaction with an effective short-range potential." In any case, in the future, it may be useful to compare the results of these methods with polarization response results[45, 360] computed as a function of ensemble size.

F. Dynamical Features

One advantage of using the MD method is the capability of revealing the particle-averaged short time details of all types of molecular relaxation phenomena. The translational self-diffusion coefficient in MD simulations may be evaluated either by integrating the translational VACF or simply from the mean-square displacements of the molecules. Generally speaking, 3-site water models are relatively more mobile than models with a larger effective molecular size, such as 5-site models. The extra girth of the 5-site models tends to retard the translational motion. Thus, the achievement of good diffusional behavior in a 3-site model requires increased interactions with neighbors compared with 5-site models. A direct comparison, except for scale differences, between translational VACFs in flexible SPC models and in the flexible–polarizable 5-site MST-FP model can be made by looking at Fig. 2a of Ref. 59 and Fig. 2 of Ref. 75. Both of these functions decay in a similar way from unity to a shallow minimum, which develops at around 70 or 80 fs. A weak reflection of the VACF to about −0.08 then seems to occur—near 300 fs in Ref. 59 and closer to 175 fs in Ref. 75. The diffusion coefficients from the two sets of calculations, accordingly, are different. It is about 30% too large from the SPC data and about 30% too small for MST-FP. One would then have to surmise that the reflection, which can only retard the diffusion, occurs too early in the MST-FP calculation and too late in the SPC calculations, compared with what probably occurs in real water. This provides yet another example of subtleties in the evaluation of properties from water MD calculations that arise from the use of different models.

Any coupling between the internal vibrational motion and the polarization with external degrees of freedom is of course ignored in MD computations using rigid or exclusively flexible or polarizable water models. However, this hidden approximation might be more severe than many others, as pointed out by Dinur.[385] Described earlier was the work of Wallqvist and Teleman,[59] who performed MD using both rigid and flexible SPC models. They found that transport properties are not significantly changed by added intramolecular flexibility. On the other hand, Anderson et al.[61] using an SPC-F model, found a self-diffusion coefficient that is 30% *smaller* than the result from the rigid SPC model,[36] which puts it in good agreement with the experimental value.[386] Could it be that there is competition between two effects here, with the "softening" effect described in the Zhu et al. papers[74, 75] speeding up transport properties, and the mode coupling effect discussed by Dinur,[385] which could slow down diffusion. Clearly there are still unresolved issues here.

The rotational mobility is described by the appropriate AVACFs. According to the MST-FP data,[75] these functions display pronounced oscillations as a result of the strong angular dependences of the force field created by the hydrogen bonding. It should be remembered that the directionality of this bonding was enhanced in MST-FP through the addition of Morse-type intermolecular potentials. In models that have a less severe angular dependence, oscillations would be expected to be less pronounced.

Another description of the rotational mobility resides in the three orientational autocorrelation functions (OACFs): of the unit vectors along the intramolecular H–H axis, along the H–O–H bisector, and of the vector normal to the molecular plane. For the MST-FP model, these three functions were found to have essentially identical slow and nonoscillatory decay properties, decaying about a factor of two more slowly than for the SPC-FP model. For these motions, intramolecular flexibility from the polarization and anharmonic bonding seems to allow molecular orientations to survive longer. However, no systematic study has yet been performed to evaluate any of these effects.

The temperature dependence of the VACFs and AVACFs in water was first investigated by Stillinger and Rahman.[29] At low temperatures, a water molecule is held in position for long periods of time by its neighbors, which form a rather rigid network of hydrogen bonds. The molecule executes pronounced oscillatory motions, as evidenced by strong oscillations in the center-of-mass VACF. As the temperature increases, the hydrogen-bond network becomes flexible and is no longer able to constrain the molecule, allowing diffusion to proceed more freely.

Using the Evans[43] potential, Evans et al.[44] have studied a number of cross-correlation functions over a wide range of thermodynamic conditions. The MD simulation of translational motions in a modified CF-F water model over a temperature range from 353 K down to 243 K was reported by Ullo.[66] The modifications to the original CF-F model[53] were of the intramolecular potential surface in order to match the experimentally realistic potential of Carney et al.[387] Ullo's study was aimed at investigating the dramatic temperature dependence of the dynamic behavior of liquid water as probed by neutron scattering. The calculation qualitatively, and in some cases quantitatively, reproduced the considerable narrowing of the translational linewidth in these experiments with decreasing temperature.[388] This is an example of yet another experimental quantity that can be considered when testing water models. In this paper,[66] Ullo also calculated the temperature dependence of diffusion, which he found to be in "fair" agreement with experiment.

Also capable of evaluation is the relaxation time τ_s for the single dipole moment $\mathbf{m}$ and the relaxation time τ_c for the collective dipole moment $\mathbf{M}$.

In computer simulations, these quantities are defined as being the integral of the respective autocorrelation functions. The collective dipole moment autocorrelation function (CDACF) determines the frequency dependent dielectric constant at zero wave vector.[239] By analyzing the respective frequency-dependent dielectric constants, Neumann[246] concluded that both the MCY and TIP4P models are capable of explaining many processes observed in the experimental microwave and far-infrared spectrum of water,[102, 147, 389, 390] though TIP4P[40] makes a significant improvement over MCY.[82] However, the experimental 200 cm^{-1} translational libration could not be reproduced by either MCY or TIP4P. Neumann[246] attributed its absence to the lack of nonadditive interactions. In agreement with this idea, it has already been pointed out in Section VI.C that MST-FP, which contains rather realistic nonadditivity effects, gives a reasonable accounting of all the intermolecular and intramolecular vibrational modes in the liquid, including the 200 cm^{-1} mode.

For comparison of the computed τ_s and τ_c with experimental relaxation times, some clarification is necessary. Wei and Patey[239, 241] have described the origins of the various relationships and have presented a good summary of earlier work. Though an approximation, the experimental reorientation time[391] for bulk water seems a good choice for the experimental τ_s; and the value of $\tau_c \sim \tau_D$ (the Debye relaxation time) is obtained from τ_s through the relationship $\tau_c \sim g_k\tau_s$, where g_k is the Kirkwood infinite system g factor with a value of about 2.9 for water.[246] Also, with respect to these matters, see the discussion concerning Fig. 4 in Ref. 74 and other views expressed by experimentalists.[392, 393] Interpolation of the experimental data[391] at 1 bar pressure and at temperatures between 283 and 303 K indicates that τ_s is 2.5 ps at 298 K, giving a τ_c of about 7.3 ps. This is close to but slightly smaller than the dielectric relaxation time τ_d, in accordance with the discussion of Hasted.[394] A recent MD study of collective motions and interparticle correlations in liquid water has been presented by Bertolini and coworkers.[395]

VII. COMPUTER SIMULATIONS: PERTURBED LIQUID

A. Aqueous Solutions

1. Atomic Solutes

Computer simulation techniques have been extensively applied to the study of hydrophobic interactions. Neutral atoms, including inert gases, which can be modeled as Lennard-Jones particles, are prototype hydrophobic solutes because they are nonpolar and their interactions with water are very weak.[192] To date, a number of computer studies of two or more such atoms or spherical methanelike molecules dissolved in water have been reported.[396-406] Good recent summaries of the state of this problem, from somewhat different points of view, can be found in the articles of Smith and Haymet[405] and Wallqvist.[406] The various concepts surrounding the hydrophobic interaction have indeed had a long and bumpy road, as evidenced by many exchanges, some heated, in the literature.[407, 408]

Geiger et al.[396] carried out a molecular dynamics simulation for a system comprising two Lennard-Jones particles and 214 ST2 water molecules. Their results, as do the results of some of the later authors, indicate the existence of two relatively stable configurations for atomic solutes in water: a near-neighbor contact pair and a pair separated by a single solvent molecule. The first peak in the solute–solute correlation function is therefore expected to be located near the minimum of the bare atom–atom potential.[182] However, the structure and relative stability of the solvent-separated pair has been controversial. Ben-Naim[189] has claimed that the outer minimum occurs when the two solutes are bridged by a water molecule whose preferential orientation is one that allows other water molecules to preserve a hydrogen-bonded network, leaving empty, clathratelike spaces to accommodate the solute. In fact, because of this type of structuring, it has been widely believed that the solvent separated pair is more stable than the contact pair (see the discussion by Wallqvist[406]).

From a glance at any thermodynamics table, there is no doubt that entropy must be considered in the solvation of hydrophobic solutes—see, for example, Table IV of Abraham's paper.[409] One argument for clustering of a hydrophobic solute in water derives from entropy considerations[405]: The entropy penalty for solvating two well-separated solutes should be greater than that for solvating a contact solute pair, for which the total nonpolar surface area is reduced substantially. This prediction is, in fact, verified in the molecular dynamics simulations of Haymet and coworkers.[404, 405] In a study of methanelike solutes, the latter authors decomposed the calculated PMF into an energetic component and an entropic component. They then found that the entropy promotes an attraction between methane solutes in SPC water at separations less than about 5.5 Å. By substracting the contribution caused by the entropy of association from the free energy, one obtains the enthalpy change, which is repulsive at distances less than about 5 Å (see Fig. 1 of Ref. 404).

All else being equal, the overall increase in entropy on aggregation would also mean that the tendency for aggregation increases with increasing temperature.[182] However, evidence supporting this effect is lacking from both experimental and theoretical points of view. For example, in the early paper[410] entitled "Theory of the Hydrophobic Effect," Pratt and

Chandler state that "there appears to be no microscopic interpretation" that is compatible with this claim. The situation becomes confused because the PMF changes with temperature and with the theoretical model employed.

The required small ensemble sizes and the resulting unrealistically high concentrations in any of the hydrophobic computations could distort the picture from laboratory reality. We recall that Kennan and Pollack[192] found no experimental evidence for aggregation of rare gas atoms when they are dissolved in water, and the earlier paper of Abraham[409] reported "no unusual effect.". Perhaps, a very large rare gas atom, such as xenon, or an almost equally large methane molecule would behave differently experimentally than the smaller rare gases that Kennan and Pollack were able to study well. Of course, we also note here that most of the computer calculations have used the simplest rigid, nonpolarizable water models, so only recently[406] have we started to hear that the omission of such features may have undesirable effects on these rather fragile results. It would also seem that polarization of the solute should be considered.

There have in fact been many computations that show no aggregational tendencies of hydrophobic systems in water. Rapaport and Scheraga[399] found no aggregation in a computer simulation on a system composed of four Lennard-Jones particles and 339 MCY water molecules. This study was extended by Watanabe and Andersen[400] by simulating a system of five krypton atoms and 195 slightly modified CF–F water molecules.[53] After a long (1.2-ns) run, these authors found that the probability for close encounters for the Kr atoms is even less than expected for a random distribution. The probability was a bit larger for the solvent-separated arrangement, but still within expected statistical limits. The latter two results are consistent with the theory of Pratt and Chandler,[410, 411] who found that hydrophobic correlations are sensitive to relatively long-range, slowly varying interactions, and are thus sensitive to the specific form of the attractive interactions employed in the model (compare Fig. 3 and Fig. 6 of Ref. 411). According to this theory, the question of aggregation is not so straightforward. Along with the recent papers by Ben-Naim[189] and by Mezei and Ben-Naim,[191] and the recent Kennan-Pollack experiments[192] discussed in Section III.C.2, the whole question of the importance of the hydrophobic interaction seems to have been reopened. At this stage, the computer simulations have not really helped to resolve this problem.

2. Fe^{2+}, Fe^{3+} System

A number of simulations on the Fe^{2+}, Fe^{3+} system have been reported because of the interest in the electron transfer reaction in water involving this ion pair. These studies also hold a general interest for the modeling of other aqueous electrolyte solutions. For any ionic solution, aside from the water–water interaction selected from one of the water models, the other required interactions are combinations of non-Coulombic and Coulombic terms. The Coulombic terms are determined by the ionic charges and the partial charges assigned to the water molecule. The non-Coulombic terms may be Lennard-Jones interactions, they may be derived by fitting data obtained from quantum-mechanical ab initio or semiempirical calculations, or they may be inferred from experimental quantities such as crystal radii and electric multipole moments. A variety of ion–oxygen and ion–hydrogen pair potentials have been suggested by Heinzinger.[317]

The modified central force model of water, CF2–F,[54] was used by Kneifel et al.[412] in a study mainly directed toward the solvent isotope effect on the hydration thermodynamics and the structure of the water molecules around Fe^{2+} and Fe^{3+}. They also reported some vibrational frequency shifts: red shifts for the intramolecular stretching vibrations and blue shifts for the librational modes. This is in agreement with earlier conclusions from quantum-chemical studies.[413] However, it was seen in Section III.C.1 that, as the hydrogen-bond structure of bulk water breaks down, vibrational shifts in just the opposite direction would be expected. Does the direction of these shifts in the Fe^{2+}, Fe^{3+} system imply that more hydrogen bonding in the neighborhood of the ions occurs? An important observation in Ref. 412 was that the CF2-F model gives too low a dipole moment for water in the bulk phase, 2.0 D vs. ~ 2.4 D, a fact that could ruin conclusions, such as those concerning vibrational spectral shifts, about the structure and the dynamics in simulated ionic solutions. Also, and part of the same problem, rapidly responsive polarization is missing in the CF models.

Using the flexible model SPC-F of Toukan and Rahman,[56] Guàrdia and Padró have carried out molecular dynamics simulations of a single Fe^{2+} or Fe^{3+} ion in the aqueous phase.[414] A large variety of results was obtained. Indicated, as expected, was that the influence of a dissolved cation on the orientation of the neighboring water molecules increases with the ionic charge. Moreover, for polyvalent ions, the influence remains effective beyond the first hydration shell. Both the translational and reorientational motions of water molecules are retarded by the presence of such ions. All these results agree with the conclusions from the experimental work of Lee.[193, 194] Guàrdia and Padró did not observe any significant changes in the bond lengths and bond angles of water, even when the solvent molecules were nearest neighbors of an ion. This is surprising, but may have to do with the nearly harmonic intramolecular potential used in their SPC–F model. However, using the more realistic anharmonic MST-FP potential, Zhu and Robinson[320] also found a very slight change in the solvent geometry, but these results are uncertain because they had to be averaged over all the water molecules in a 1.791-molal Na^+Cl^- solution. Guàrdia and Padró concluded that, because of the loss of mode coupling effects, omission of the internal vibrational motions in MD computations may have a nonnegligible influence on the PMF between ions.

Bader and Chandler[415] investigated the important question of cutoff effects for long-range forces in SPC water containing two iron ions. Found was the fact that the Ewald lattice summation (LS) method gave reasonable results, while complete truncation of the forces beyond a cutoff of 11.73 Å gave an unphysical picture, namely a deep effective attractive potential between the two positive ions. Bader and Chandler concluded that such a cutoff procedure, as in spherical cutoff or minimum image boundary conditions (see Section IV.B.5), should not be used. These results are pertinent to other studies as discussed in the next subsection.

A more recent paper,[416] again using SPC water, focused attention mainly on the hydration structure of water molecules near the Fe^{2+}, Fe^{3+} ion-pair, whose separation in the solution was fixed at 5 Å. This paper also provides a good summary and list of references of past work.

3. Other Ions

Using the ion–water potential derived from Hartree-Fock computations,[417] Clementi and Barsotti[314] carried out Monte Carlo simulations for single ions of Li^+, Na^+, K^+, F^-, or Cl^- surrounded by a cluster of 200 MCY molecules at $T = 298$ K. Impey et al.[315] performed a similar study for Li^+ and Cl^- at various temperatures using the same ab initio potential function. However, the charge distribution of the MCY molecules was modified by six fractional charges to enhance the dipole moment caused by the polarization effect from the field of the ion. An important feature of this study was the observation that the breakup of the first coordination shell is initiated by molecular orientations rather than translations.

Using the TIP4P model, Chandrasekhar et al.[316] have carried out Monte Carlo simulations on dilute aqueous solutions of Li^+, Na^+, F^-, and Cl^- in TIP4P water, with the primary goal of determining heats of solution. The ion–water potential function was chosen so as to reproduce the experimental interaction energies and the Hartree-Fock geometries of ion–water complexes. The heat of solution is defined as the energy for transferring the ion from the state of an ideal gas into the solvent. This quantity is composed of a solute–solvent interaction energy E_{sx} and a solvent reorganization energy ΔE_{ss}, which is the difference between the water–water interaction energies in the dilute solution and in the pure liquid. Mezei and Beveridge[418] had pointed out earlier that the hydration energetics for ions could vary by $\sim 15\%$ in going from 125 to 215 water molecules. The heat of hydration of F^- was found by Chandrasekhar et al.[316] to display an even greater change, being lowered from -464 to -611 kJ/mol. Mainly responsible for this difference was ΔE_{ss}. It decreased from 423 to 289 kJ/mol, owing to diminished solvent disruption in the larger system. The component E_{sx} was lowered by only 13 kJ/mol. These edge effects were found to be far less significant for the larger anion Cl^- and for the cations. Another interesting conclusion from this work is that the binding of water to an ion in the first solvation shell is so strong that it changes the water–water interaction to a net mutual repulsion.

Computational work on ion–water clusters has been inspired by the experiments of Kebarle and his coworkers[419–421] and by Keesee and Castleman.[422, 423] Jordan and his colleagues[424, 425] used a polarizable version of the electropole model of Barnes et al.[372, 373] in simulations of aqueous microclusters of alkali metal cations Li^+, Na^+, K^+, Rb^+, and Cs^+ and the halogen anions F^-, Cl^-, Br^-, and I^- at 150, 300, and 400 K. The ions were treated as polarizable charged spheres, using the polarizabilities of Gowda and Benson.[426]

More recently, somewhat larger clusters, ranging up to 20 water molecules and containing one sodium or chloride ion, have been studied by Perera and Berkowitz.[427, 428] Both TIP4P and SPC-P were used to model the water–water interactions. The polarizability of the ions was neglected. Large differences in the results obtained from the two water models were found. In particular, for the polarizable water model, but not for the rigid model, the Cl^- ion had a propensity to reside on the surface of these moderate-sized clusters in an incompletely hydrated state. Ion–water interactions are reduced because of the polarization response, which should also be a concern when calculating heats of hydration. Another interesting finding[428] was that merely reversing the charge on the chloride ion caused it to undergo a more complete hydration away from the surface. Clearly, this is not only an ion size effect, but it must have to do with the charge placements and the polarization modeling of the water model.

A detailed molecular dynamics study by Kollman and his colleagues[70, 429, 430] has been made of Na^+ and Cl^- ion–$(H_2O)_n$ complexes and bulk solutions containing a single ion and 215 SPC2-P water molecules. Explicit nonadditive polarization and three-body exchange repulsion energies were added to this three-site water–water potential. The polarizabilities of the ions were also included. The local structures of water molecules around the ions seemed to be well represented by the computational data, which indicated a significant difference in local hydration structure between the cluster and the solution phase. The fact that, with the polarizable model, a much higher density was found between the first and second maxima of the Na $\cdots$ O RDF than in nonpolarizable models implied that water molecules were able to exchange rapidly between the first and second hydration shells in the clusters. Again, because of the polarization response, interactions near the ions are very likely reduced over what they would be in a rigid model, where Coulomb interactions have a harder time readjusting. As long as the polarization term was augmented by the three-body exchange term, the Kollman et al. calculations have been the only ones where a near quantitative agreement between the simulation data and the experimental hydration enthalpies has been obtained.

As mentioned in Section IV.B.1, an extensive series of papers describing molecular dynamics simulations of aqueous electrolyte solutions has been published by Heinzinger and coworkers,[317-319, 431-445] using either the ST2 model[30] or the improved CF2–F model.[54] The solvation structures and some dynamic properties for a large number of alkali halides, including LiF, LiCl, LiI, NaF, NaCl, NaI, KF, KCl, CsF, CsF, CsCl, CsI, $CaCl_2$, MgF_2, and $MgCl_2$ at various concentrations have been studied.

A concentrated solution of LiF, where the hydration shells of the ions considerably overlap, has been studied by Zhu and Robinson[376] using the harmonically bound flexible-polarizable SPC-FP model of water. The orientational distribution of water molecules around the ions was found to be a sensitive function of the particular water model being used. This is an expected result, since the method by which the various charge centers are modeled as well as the nature of other angular dependent intermolecular bonding could cause large effects. Zhu and Robinson[320] pointed out that properties, such as hydration energies, hydrogen bonding, and perturbations caused by surfaces should also be sensitive to these angular dependent effects.

In a study using a much better water model,[75] Zhu and Robinson[320] employed a system containing eight sodium ions, eight chloride ions, and 248 MST-FP water molecules, corresponding to a 1.791-molal aqueous NaCl solution. The interionic pair potential was described through contributions from Coulombic terms, core repulsions and induced-dipole–induced-dipole attractions. Polarization of the ions was unfortunately not included, which, according to the recent paper by Karim,[446] would cause the calculated hydration numbers to be too large, particularly for the more polarizable Cl^- ion. Correcting for this omission would probably also change the hydration energies, which were not far from the experimental values in the original paper. From an estimated running integration number in the Zhu-Robinson paper, at the first minimum of the Na^+–Cl^- RDF positioned around 3.5 Å, it was found that on the average an ion has an 80% chance to be separated by at least one solvent molecule from one of its counterions. This tendency is not much different than would be expected statistically considering roughly six available nearest-neighbor sites. It can also be compared with the overall tendency for aggregation (of an unspecified type) implied by the experimental mean ionic activity coefficient of 0.67 in a 2.0-molal NaCl solution.[5] The observed first peak of the Na^+–Cl^- RDF lies near 2.6 Å, and the bottom of the Na^+Cl^- effective potential well was found to lie ~ 6.2 kJ mol^{-1}, or about 2.5 kT, below the first potential maximum located near 3.5 Å. An equally deep, but broader, second minimum is evident near 4.9 Å in the 1.791-molal solution. This is about 0.5 Å nearer than the sum of most probable Na^+–O and a Cl^-–O separations. The first Na^+Cl^- RDF peak occurs at a greater distance than the equilibrium bond length of 2.29 Å for the isolated ion pair, showing, as expected, that the solvent weakens the interaction and tends to pull the Na^+ and Cl^- ions apart.

Since the first maxima of the Na^+–Na^+ and Cl^-–Cl^- RDFs are located near the second maximum of the O $\cdots$ O RDF in bulk water, it would appear as if pairs of identical ions tend to be coordinated to the same water molecule. In fact, in the 1.791-molal solution,[320] the first hydration shells of two ions, whether they have the same or opposite charges, very likely overlap one another. This effect emphasizes one of the flaws in continuum models for solutions of moderate concentration.

Considerable effort, not without its pitfalls, has been expended on attempts to determine the PMF at infinite dilution for pairs of ions having the same charge. Using the TIPS three-site water model and the extended RISM integral equation approximation, Pettitt and Rossky[262] found a shallow minimum near 4.0 Å for the effective potential of the Na^+–Na^+ pair, but a rather deep well near 3.5 Å for the potential between two Cl^- ions. This result would imply the existence of a fairly stable Cl^-–Cl^- contact ion pair. These results were confirmed in the MD calculations of Dang and Pettitt[447–449] using the same interaction potential. Adopting the four-site TIP4P model,[40] together with ion–water and ion–ion interaction potentials slightly different from those used by Pettitt and Rossky,[262] Buckner and Jorgensen[256] performed MC simulations on an analogous system. Although their Cl^-–Cl^- PMF also displayed a deep well, its location at 4.8 Å almost coincides with the second maximum of the Pettitt-Rossky PMF curve. On the other hand, Zhu and Robinson,[320] using the more realistic MST-FP water–water potential, obtained ion–ion PMFs that agreed with the ones obtained earlier by Guàrdia et al.[450, 451] At about the same time, Karim,[452] using the polarizable TIP4P model of Sprik and Klein,[64] also found a very shallow well for the Cl^-–Cl^- PMF. Clearly, there is some sort of a problem here.

A shallow potential well for the Cl^-–Cl^- ion pair, as found in the last three cited papers is a reasonable result. In fact, Friedman et al.,[453] as well as Xu and Friedman,[454] have compared the mixing coefficients from model calculations with experimental data and concluded that the Cl^--Cl^- potential well of Pettitt and Rossky[262] is unrealistically deep. A shallow Cl^--Cl^- minimum, by the way, is also consistent with neutron diffraction experiments on $NiCl_2^{455}$ and $LiCl.^{456}$ Rather than the form of the interaction potentials causing these disagreements, as was originally believed,[320, 450-452] it is much more likely that the previously described considerations of Bader and Chandler[415] are at the heart of the matter. Corrections for the long-range forces must be made. In Ref. 320 the RF with conducting boundary conditions was used, and, in the other two papers, the Ewald LS method was used. However, it would be premature to conclude that all structural or dynamic effects can be reproduced by either of these methods, which are after all still approximations.

To understand the way that ions affect the solvent structure, one can examine the $O \cdots O$, $O \cdots H$, and $H \cdots H$ RDFs. By comparing[320] these RDF curves with those obtained in pure water, it is clear that the water molecules in solution are somewhat less structured. This is evident from the lower amplitudes of the first maxima in all three of these curves, the clearly lower second maxima in the $O \cdots H$ and $H \cdots H$ RDFs, and the concomitant movement inward of all the maxima, giving rise to the higher density expected for broken-down hydrogen-bonding structure. These features are qualitatively consistent with the effects observed by Vogel and Heinzinger[432] using rigid ST2 water.

MD calculations can also provide information about orientational ordering of water molecules around the ions. This ordering has been found[317, 320] to depend sensitively on the manner in which the negative charge sites are modeled: extended "lone pairs" as in 5-site models or at a single point as in the common 3- or 4-site models. If $\cos \theta$ is defined as the cosine of the angle between the oxygen–ion vector and the bisector vector of the H–O–H bond angle, then one may plot the average $\langle \cdots \rangle$ of this function for a series of distances from the ion. Near an ion, one would expect this function to be structured, while at large distances there are no orientational preferences and $\langle \cos \theta \rangle$ should trend to zero. Using MST-FP, over the whole range of the first peak of the Na^+–O RDF, $\langle \cos \theta \rangle$ is nearly a constant, -0.65. For the Cl^- ion, the function decreases to unity more rapidly than for the cation over the range of the first peak, showing that only water molecules very near the anion form linear hydrogen bonds with that ion, an expected result for the larger anion. The $\langle \cos \theta \rangle$ distribution of the water molecules in the first hydration shell of the chloride ion does exhibit a maximum near one-half the average H–O–H bond angle. This implies the predominance of configurations where an

O–H intramolecular bond is pointing directly toward the anion. In MST-FP water, the cationic $\langle \cos \theta \rangle$ configuration is essentially a mirror image of the anionic one. If the lone-pair vector exactly pointed toward a sodium ion, $\cos \theta$ would have to be around -0.58, slightly less than the observed maximum near -0.65. These distributions also indicate that there is a greater probability for both hydrogen atoms to interact simultaneously with a Cl^- ion than for the two lone pairs to interact simultaneously with a sodium ion. The influence of both ions on the orientation of water molecules beyond the first hydration shell was found to be of minor importance for the Na^+Cl^- system in MST-FP water. Very likely, with smaller ions, such as Li^+ and F^-, a longer range effect would be found. It is doubtful that the addition of ion polarizability to these models would change this qualitative picture, though certainly there will be quantitative effects.

B. Water at Interfaces

Studying the structure and properties of liquid water having finite dimensions along one direction, such as in a bubble or a film, is attractive from either a theoretical or an experimental point of view. The outcome of such studies could have an important impact on our intuition about chemical and biological processes involving water near a surface. Generally speaking, water molecules near interfaces differ from those in the bulk phase. There exists a transition zone which makes a small but sometimes perceptible contribution to the mechanical, thermodynamic, and dielectric behavior of the system.[457]

In the case of water, there is an important correspondence between structure and density: Broken hydrogen-bond structure gives rise to a higher density than intact ice Ih-type hydrogen bonding. Density profiles are therefore convenient quantities for monitoring the liquid structure near a surface. However, there are still many unanswered questions in this seemingly straightforward problem. Near any interface, intact hydrogen bonding would be expected to give way to a different water–water binding, and this would generally lead to a higher density. Are there experiments or computer simulations that actually bear this out? Are there density oscillations? How far from the interface do these surface effects persist? How do the chemical, physical, and electrical properties near the surface depend on the answers to these questions? Needless to say, there is a lot of controversy about these matters. For instance, a recent paper by Wallqvist[406] concluded by stating that "properties of interfacial water presented here belie the notion of any specific character of bound water associated with hydrophobic solvation." It is " 'liquid' water everywhere."

These statements are based on calculations of a mixture of 18 Lennard-Jones methane particles and 107 rigid SPC water molecules.[36] Many experimental results would say otherwise.

Another of the important and unanswered questions about interfacial water concerns the orientation of the molecules as a function of distance from the interface. Orientational structure and dynamics have been found[185] to have a great influence on important chemical processes, such as proton dissociation from an acid. Thus, many earlier studies[458-461] have laid emphasis on the question of molecular orientations at an interface. One of the pioneers in this area was Frenkel,[458] who suggested that the quadrupole moment of the water molecule is responsible for a preferred orientation. His analysis was based on the concept that molecules in the surface layer tend to be oriented so as to immerse their electric field as much as possible in the region of highest dielectric constant, thereby minimizing the free energy. This discussion was extended by Stillinger and Ben-Naim,[460] and more recently by Croxton.[461] However, a variety of conclusions were reported. Croxton[461] concluded that the orientation at the interface sensitivity depends on the sign and values of the components of the molecular quadrupole moment. Tables I–IV show that these electrical properties vary with the water model used.

1. Liquid–Vapor Interface

If there is molecular orientational ordering of a polar molecule near a surface, an electric double layer can be formed, across which a discontinuity in the potential appears. Such is the situation for liquid water. The resulting effect, usually called the *surface potential*, is the electrostatic potential difference between coexisting liquid and vapor phases. As might be expected, the surface potential is a characteristic of the molecular structure at the interface, and in MD computations of liquid water it depends sensitively on the model employed. Again, because of the dependence on angular dependent terms, planar 3- and 4-site models would be expected to behave qualitatively differently than 5-site models in this respect. A detailed description of the surface potential would provide an important perspective for chemical and electrochemical processes at aqueous surfaces. It would also constitute an additional testing ground for water modeling. However, to date, few aspects of the spatial variation of the electric field at an interface have been directly or unambiguously measured in laboratory experiments. Therefore, future computer simulation techniques will need to be combined with improved laboratory experiments to straighten out this issue.

A free-standing thin water film can be modeled by MD through modification of the ordinary periodic boundary condition in one direction.

If this boundary condition is merely removed, there is a severe possibility of evaporation. To prevent this, the film can be subjected to an external potential. This procedure adds a weak inward force to the interfacial molecules and is designed to have a negligible effect on all molecules throughout the film. However, an inappropriate choice of the artificial force may bring in unwanted distortions of the entire film.

An alternative boundary condition for MD studies of the water–vapor interface uses hard walls placed somewhere beyond the film. These walls elastically reflect molecules escaping the film. This procedure is actually a special case of the first boundary condition.

A third method applies a different type of periodic boundary condition. A molecule that evaporates from one side of the film reenters the other side with the same velocity vector. Since a molecule on the opposite surface is not in close interaction with the escaped molecule from the other surface, there is the possibility that, when the escaping molecule suddenly appears, the two molecules collide in an unprepared manner, creating huge repulsive forces and computer overruns.

A number of computer studies of self-supporting water films,[208, 300–305, 462] free clusters,[298, 299, 463] ions,[306–308] or polar molecules[464] in a water film have been published. Structural features related to the directionality and strength of the hydrogen bonding in the interfacial region are still controversial issues. For example, using ST2 and CF-F, Rice and his coworkers[298, 299] observed an argonlike[208] density profile at the surface of clusters up to 1000 molecules in size: The profile monotonically decreased from about 90 to 10% of the bulk density in a transition region width of 3.45 Å. This width matched well the experimentally observed width.[209, 210] On the other hand, Zhu et al.,[304] based on studies, not of a cluster but of a thin film, using the Toukan and Rahman SPC-F water model,[56] found a somewhat broader transition region, and furthermore claimed that there are weak density oscillations near the surface. These were said to "have to do with the size of the oxygen atom and with packing considerations of directionally bound water molecules near a structural discontinuity." The density oscillations in Ref. 304, though not distinct, did roughly match those found for SPC-FP[74] water near a Lennard-Jones wall.[342] Statistics were not very good in any of the above studies, since run times were not very long. For example, the longest Townsend and Rice[299] run time, and this for their smallest cluster, was only 20 ps. The run time in the Zhu et al. paper[304] was about 100 ps.

A very detailed comparison, though without improvement in run times, of a Lennard-Jones film and a water film, using a rather different pair potential,[370] was carried out by Matsumoto and Kataoka.[208] Their results for the water film generally agreed with those of Rice et al.[298, 299] While

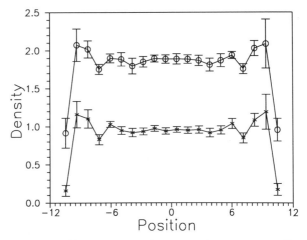

Figure 1. Density profiles of H atoms (upper curve) and O atoms (lower curve) of double-sided liquid–vapor film of MST-FP water. Note the reasonably good agreement in the profiles on each side of the film, the density maxima about 2 Å from each edge, and the strongly damped density oscillations.

there seems to be broad agreement that density oscillations do exist near solid surfaces (see next subsection), their presence at the vapor–liquid interface is still controversial.

In an effort to try to clarify the question of oscillations, Zhu and Robinson have recently completed some studies on MST-FP water, which are reported here for the first time. The time step used was 0.25 fs and run times were 350 ps. MST-FP should be fairly similar to the ST2 water used by Rice et al.[298, 299] However, the conclusions reached differ from all the above. The results showed (see Fig. 1) a transition zone (bulk to near zero density) of about 4–5 Å, which matches the result of Townsend and Rice.[299] Also found was (1) a pronounced local density maximum near the surface for both the oxygen and hydrogen densities, (2) density oscillations that die out 5 to 6 Å from the surface, and (3) a propensity for hydrogen atoms to congregate in the very low density part of the surface. Compare the oxygen and hydrogen densities in the outermost regions of Fig. 1, noting that the ratio is about two everywhere else. The density maximum could be the expected result because of hydrogen-bond breaking and molecular reorientation, which allows the water molecules to pack in a more dense manner, a well-known scenario for water.

Because of the good inter- and intramolecular vibrational properties obtained with the MST-FP model and the presence also in this model of dynamic polarization that leads to the correct liquid state dipole moment, plus the far better statistics, as evidenced by the good agreement between the independently calculated density profiles on both sides of the film, it is difficult to discount these results. Note, however, from Tables I and IV that the quadrupole moment components are somewhat smaller in MST-FP water than in ST2 water, at least for the gas phase molecule, though the signs are the same. As mentioned earlier, this is because smaller charges were used at the ST2 positions in order to reproduce the gas phase dipole moment. The resulting surface potential also seems too large in the MST-FP calculation and with the "wrong" sign. On the other hand, the problem of relating the various ways of calculating surface potential with the experiments, together with the difficulty itself in the measurement and interpretation of the experimental data, leaves this question a bit cloudy. The sign ambiguity was addressed earlier by Wilson et al.[303]

Whatever the outcome of all this, one lesson learned is that from a computational point of view the results for the vapor–liquid interface must strongly depend on the water model used. The fact that the Morse hydrogen-bonding term in MST-FP may be a bit too directional and the liquid state quadrupole moment components may be too small could have a sizable effect on Fig. 1. We do believe that Fig. 1 is correct for the model used. Thus, the question of whether density oscillations and a maximum higher than that of the bulk are possible at a vapor–liquid interface should be shifted to whether they occur in water. The MST-FP results certainly show that there is a possibility that they do. Experiments are most difficult, since unavoidable supramicroscopic surface motions would tend to average out the oscillations.

Another example of the disagreement between structural results using different potential models concerns the molecular orientation at the surface. Townsend and Rice[299] found that the interfacial molecules tend to align with their H–O–H bisector generally lying in a plane parallel to the surface, with one O–H bond directed toward the vapor and the other pointing inward. On the other hand, the results of Zhu and Robinson for MST-FP water show a tendency for this bisector to point 20–40° out of the surface plane, a result which may be more in line with the recent SFG experiments.[213] Clearly, the problem of the vapor–liquid interface remains unsolved and controversial. Thus, all these questions need to be examined more systematically in future work using realistic anharmonic FP water models.

One of the most important thermodynamic properties associated with the liquid–vapor interface is the surface tension. There are two principal routes for calculating this quantity from the molecular distributions[465]: the virial theorem and the density gradient. Since the virial of the pair potential is readily calculated at each step of a computer simulation, and the molecular distributions are generated naturally, the first route is more efficient and more convenient for such studies. According to this method, the surface tension is determined from an integration over the interfacial region of the difference between the normal and the tangential components of the pressure tensor. It should be realized that the precision of such calculations is low, since the result is deduced from a small difference between two large quantities. As a collective quantity, the calculated surface tension sensitively depends on the boundary conditions of the system, the number of particles, and the method by which the potentials are truncated. In fact, Nijmeijer et al.[279] have compared the reduced surface tension of a Lennard-Jones fluid calculated by different authors and found large variations in the results.

The Nijmeijer et al. group in other work has addressed through a combination of analytical and computer methods a number of other interesting surface problems. These concern nonaqueous systems, but will certainly act as models for future computational efforts on water. Covered were the surface and line tensions of a fluid near a hard wall and near a structured soft wall[466]; wetting, drying, and prewetting phenomena[467, 468]; and a most interesting description[469] of size effects on the surface tension of a Lennard-Jones droplet with an estimate of the elusive Tolman length,[470-472] which is the lowest-order correction coefficient for the surface tension in terms of inverse droplet size. Also in Ref. 469, plots of density profiles, with much better statistics than in any of the earlier work on the vapor–liquid interface, including error bars for the smallest stable droplet considered, were presented for droplets containing 5165, 9295, and 12 138 Lennard-Jones particles. Smaller droplets were unstable in the MD procedure used. The half-width of all the above surfaces is about three particle diameters. There are certainly no density oscillations at the surface of a Lennard-Jones droplet!

In another type of study, Wilson and coworkers,[306, 307] adopting the ion–water potential model of Chandrasekhar et al.[316] have performed MD simulations on a series of ions near the liquid–vapor interface. This system consisted of 342 TIP4P molecules and one ion, either Na^+, F^-, or Cl^-. It was found that the free energy required to move ions to the interface depends on the sign, but not the size, of the ionic charge. Specifically, the anions are able to approach to within two molecular layers of the interface without a significant change in the free energy. On the other hand, the cost to bring the Na^+ ion to this distance is about 10.5 kJ/mol. The difference arises because of firmer orientational ordering near the cation, which requires a greater degree of solvent rearrangement. A more detailed but concurring picture of the energetics has been presented by

Benjamin[308] who used an SPC-F model for water. Benjamin also found that τ_s near an ion was faster at the interface.

2. Nonmetallic Solid Surfaces

From a computational point of view, the investigation of liquid water between two simple hydrophobic or hydrophilic solid surfaces is a good starting point for learning about water near more complex surfaces, such as metals or biologically important membranes or macromolecules. Applying an electric charge to the surfaces provides further hints about effects to be expected when these membranes or macromolecules are ionic. The surface may be treated as an assembly of individual atoms or by an averaged gas-crystal potential. For example, the particle–wall interaction caused by a surface constructed from Lennard-Jones atoms can be obtained by integration.[473–475]

To the best of our knowledge, the computer simulation of water between flat hydrophobic surfaces was initiated by Jönsson (MCY water)[476] using the Monte Carlo method and by Marchesi (ST2 water)[322] with the MD method. This work was followed by a more detailed picture presented by Lee et al. (ST2 water),[323] where the walls were composed of idealized hydrocarbon molecules. Observed was an oscillatory density profile that propagated at least 10 Å into the liquid, with slight maxima for both the hydrogen and oxygen densities near the surface, caused by the competition between atomic packing and the intermolecular Coulomb forces from the charge sites on ST2. In addition, significant molecular orientational preferences out to about 7–8 Å were found. The hydrogen atoms generally pointed toward the surface and were characterized as "dangling hydrogen bonds."[477] Later, a continuation of this work was carried out by Rossky and Lee (TIP4P water),[324] where the nature of the surfaces was varied from purely hydrophobic to a fully hydroxylated silica surface. Dynamical properties were also investigated. An interesting observation was that water molecules were able to retain about 75% of their hydrogen bonding by completely sacrificing one such bond so that the other three can point into the liquid. A similar effect was noticed by Wallqvist[406] for water at a fluid Lennard-Jones surface. Discovered by Rossky and Lee[324] was a relatively mild perturbation on the dynamics by all the surfaces they studied. More recent advances include simulations using flexible and polarizable water models. Wallqvist[478] compared the results from a nonempirical water–water potential[479] with and without polarization. He claimed that the structural orderings are similar for these two cases, except for the layers nearest to the surfaces. On the other hand, the increased strength of intermolecular hydrogen bonds induced by the polarization retards

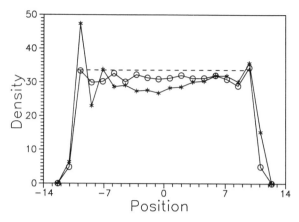

Figure 2. H-atom density profiles for SPC-FP water film between "plates." ○, Plates uncharged; *, 10^9 V m^{-1} dc field applied. The positively charged plate is to the left and the negatively charged plate is to the right in the figure. The plate surfaces lie at about ± 13 Å. Note the high density of H atoms about 3 Å from the positively charged plate caused by the attraction of O atoms to this plate. Very close to the negatively charged plate on the right, a large H-atom buildup can be seen.

diffusion along the surface. It also reduces the reorientational correlation time compared with the nonpolarizable model. It was concluded that the many-body polarization effects resulted in quantitative but not qualitative differences for water in contact with a hydrophobic wall.

Zhu and Robinson[342] carried out an MD study of SPC-FP water between infinitely thick neutral and charged Lennard-Jones walls. When the walls were not charged, modest density oscillations extended 2 or 3 molecular diameters into the liquid layer. When a dc charge was applied across the two walls, both the oxygen and hydrogen atom densities near the oppositely charged wall increased as expected. The density oscillations increased in amplitude and their period decreased in the presence of the field. The effect on the hydrogen atom density, shown in Fig. 2, was much more pronounced because of its small atomic size. Contrary, perhaps, to intuition, the most noticeable increase in hydrogen atom density occurred at a position just removed from the positive wall, on the left of Fig. 2. This is because hydrogen atoms have no choice but to accompany the oxygens as they are attracted to this wall. Interestingly, in this study, both transverse and longitudinal translational diffusion was found to be faster near the surface of the wall. This was ascribed to the breakup of the hydrogen-bond structure and consequent weaker water–water interactions. The decreased hydrogen bonding was evident in the intramolecular O–H bond length, which near the wall was closer to the vapor phase value. However, all the various orientational correlations decayed faster near the wall, indicating slower orientational dynamics. The normal tendency for the formation of bifurcated intermolecular bonds in an electric field was opposed by the presence of the wall.

Following the appearance of the first report on a molecular level simulation of aqueous LiI solution between two 12-6 Lennard-Jones walls,[480] a considerable extension of the work has emerged. Very recently, for example, simulations of electrolyte solutions between Lennard-Jones walls were carried out at almost the same time by Matsui and Jorgensen[343] and by Glosli and Philpott.[311] In the former case, image interactions were added as in a classical metal, and, in the latter, the electrodes were charged, as in the Zhu-Robinson[342] study discussed above. More will be said about these results in Section VII.B.5, which treats the topic of metallic surfaces.

3. Water–Ice Interface

Computer simulations by Broughton and Gilmore,[481] who modeled melting at the interface of a simple Lennard-Jones crystal, have indicated the existence of a quasi-liquid layer of 2 to 5 molecular diameters in extent. The surface melting could not be classed as a phase transition and the melting temperature in this region is lower than that in the bulk. This type of study was applied to the ice–liquid interface by Karim and Haymet[296] in order to examine the structure and dynamics of the stable interface between the basal plane of ice Ih and liquid water. The temperature chosen for these studies was 240 ± 1.5 K and the pressure was -105 ± 118 atm. The ordered ice Ih sample was constructed using 36 translational copies of a unit cell each with eight TIP4P water molecules positioned in an antiferromagnetic fashion. The liquid sample, separately constructed, also consisted of 288 TIP4P molecules. Two such liquid samples and three such ice samples were placed end-to-end along a specific direction to construct the interface. Altogether, the system consisted of 1440 water molecules, 60% of which were originally in the ice structure. From the density profiles, the molecular orientation, and the self-diffusion constant, the authors concluded that the transition layer extended about 10 Å from the surface. Experimental data,[199] though not directly on the water ice interface, has been interpreted to imply that interfacial perturbations on the collective orientational properties of water may extend two to three times farther from the surface than this. Perhaps the necessary small size of the computer ensembles prevent a realistic assessment of these collective phenomena. Similar problems with computer simulations of water near biologically relevant interfaces have arisen in the apparent failure to predict the correct diffusion coefficients (see Section VII.B.6). Again, we sense loose ends in the surface perturbed water problem that need to be tied to a firmer base through the adoption of more advanced modeling.

4. The Liquid Interface

Computer studies of water near rigid hydrophobic surfaces have been extended to systems with flexible surfaces, such as that which one would encounter with an immiscible liquid phase. Many important industrial processes,[2] such as phase-transfer catalysis and surfactant technology have to do with reactions occurring at liquid–liquid interfaces.

Benzene may be viewed as a prototype hydrophobic *molecular* liquid. Linse[482] has investigated the interfacial structure in the water–benzene system at a temperature of 308 K. He used ab initio potentials based on the Hartree-Fock SCF method for the benzene–benzene and benzene–water interactions, and the MCY potential[82] for water–water interactions. According to this study, the interfacial water dipoles showed a preference for aligning themselves parallel to the surface. The estimated surface tension of the benzene–water interface was 55 ± 45 dyne cm^{-1} (millinewton m^{-1}). The experimental surface tensions, with respect to air, of water and benzene at this temperature are 70.4 and 26.9 mN m^{-1}, respectfully.[483] Roughly speaking, according to the old Antonoff rule,[484] the "experimental" surface tension between water and benzene would be the difference, 43.5 mN m^{-1}.

Another study of this type was carried out by Carpenter and Hehre[485] who used the MD approach on a system comprising 187 hexane and 1200 SPC molecules. These authors found an interface with a width of approximately 10 Å, significantly greater than the dimension of a single molecule, but similar to widths found in most studies of other surfaces. This finding, however, was different from that of Linse[482] for the benzene–water interface, who attributed the intersurface infringement, within an otherwise molecularly sharp boundary, to capillary waves. Also found in the Carpenter and Hehre work was that a small number of hexane molecules became detached from their bulk phase and were completely surrounded by water molecules. This solubility greatly exceeded expectations from the experimental solubility data and was therefore considered an artifact of the simulation. The conformational change (*gauche–trans*) of hexane in the water was also poorly represented by the computational data, because of poor statistics among other things. Probably, as these authors imply, more sophisticated intermolecular modeling needs to be used in this type of calculation. It would seem that the water should have a measure of rapidly responsive polarization, which would tend to increase the water–water interactions at the expense of water–hydrocarbon interactions. Also, both types of molecules should probably be modeled in a full three-dimensional context, instead of SPC, to help resist the unrealistically rapid diffusional mixing of the two molecular species.

5. Metallic Surfaces

A detailed insight into microscopic phenomena occurring in liquid water in contact with metal surfaces is of fundamental significance for electro-chemistry, catalysis, and corrosion studies.[2] Molecular level information about water adsorption and orientational structure, and the dynamic properties of hindered translations, rotations, and vibrations is available from various experiments.[486–492]

The first reports on the MD of water–metal interfaces were published a few years ago by Parsonage and Nicholson,[332, 333] using MC, and by Spohr and Heinzinger,[325] using MD. To perform a computer simulation, one has to establish the water–metal interaction appropriately. An image charge method based on continuum electrostatics[332] was chosen for this purpose. However, these computer simulations, as well as those of later workers using similar methods,[493, 494] failed to predict the correct molecular orien-tations at the metallic surface. The computational studies gave a strong preference for the dipole moment of the water molecule to lie parallel to the surface. On the other hand,[490] the decrease in the work function of the metal when water is adsorbed on a Pt(111) surface, as well as the results of infrared absorption and scattering measurements, rules out this orienta-tion, as well as the one where hydrogen atoms are preferentially bonded to the surface. Experimental results on other metallic surfaces[495–497] provide the same conclusion: It is the oxygen atom lone pairs that are bound to the metal surface. This is in agreement with an early quantum-chemistry calculation.[498] More recent quantum-mechanical calculations[499, 500] have confirmed this result and also that the water molecules are bound "on top" of the metal atom rather than at "hollow sites." This latter result was also in disagreement with the early MC and MD calculations. A brief review of these points has been given recently by Heinzinger.[501]

To correct for the failings of the early simulation studies, Spohr[334, 335] devised a cluster-model[502, 503] ansatz for O–Pt and H–Pt pair interactions based on extended Hückel molecular orbital calculations of a water molecule at the top of a five-atom platinum cluster. With the aid of the CF2-F water model,[54] Spohr was able to investigate dynamical features,[335] including hindered translational motions, librations, self-diffusion, reorien-tation, and intramolecular vibrations of the water lamina bound by Pt(100) surfaces of the face-centered-cubic crystal. He found that adsorbed water formed a well-ordered immobile overlayer even at the high temperature of 349 K. Beyond the first layer (~ 4 Å from the surface), perturbations on the dynamics rapidly diminished until they completely disappeared at about the third layer. There was a slight reduction of hydrogen bonding relative to the bulk in this region. In contrast to water near Lennard-Jones surfaces (see Section VII.B.2), where pronounced density oscillations of the water are found, surprisingly, perhaps, there was no such tendency near the platinum surface. Spohr attributed the different behavior to the strength of the intermolecular hydrogen bonds compared with the weaker water–platinum interactions, but this is also the case for other surfaces that show the oscillations. It could be that the missing oscillations could be caused by the method of treating the image charges,[231] or by the nature of the CF2-F water model. Spohr also concluded that the severely distorted orientational structure that was observed was derived from the water–platinum potential. According to what is now known about the exceedingly large effects of orientational confinement on acid dissociation reactions,[199] this effect, if present in the real system, could play an important role in surface chemistry.

An analytical form of the interaction potential between rigid water molecules and a platinum metal surface, which takes into account the surface symmetry, has recently been suggested.[341] In this treatment, the water–surface mean potential was represented by the sum of the interactions between individual atoms in water and the hexagonal Pt(111) surface, the parameters being adjusted by fitting to the Spohr potential.[334, 504] The water lamina were restricted by two Pt surfaces. Observed were patches of icelike structure embedded in the water layer adjacent to the metallic surfaces. Because of hydrogen bonding, this structure extended through the next two layers, beyond which a transition to bulk behavior was noticeable. In still more recent work, Heinzinger[501] summarized the changes in the structural and dynamic behavior of water between Pt(100) plates with and without a homogeneous electric field of 10^{10} V/m. Particularly surprising was the conclusion that the effects of the surface perturbation did not extend appreciably beyond the adsorbed layer.

Important systems concern electrolyte solutions near metal surfaces.[309, 505] Although the interfacial water structure may differ greatly from that of the bulk liquid, the change of structure around an ion is relatively small, except that one or more water molecules may be replaced by the metal atoms. This effect is caused by the extremely strong ion–water binding, particularly for small ions having a high charge. In fact, Rose and Benjamin[309] have found that the adsorption free energy profiles do not vary smoothly with distance from the metal wall; and, because of the differences in binding strengths and liquid structural perturbations, the profiles have qualitatively different behaviors for the cation and the anion.

These authors have compared their results with similar work on ion solvation at the liquid–vapor interface.[307, 308]

Free energy profiles for Na$^+$ adsorption on a metal electrode was determined through MC methods by Matsui and Jorgensen,[343] using TIP4P water that was modified by the addition of hard sphere interactions on the hydrogen atoms. In addition to the standard Lennard-Jones terms, ion–wall and water–wall interactions were supplemented by image interactions, self as well as other.[231] As expected, hydrogen and oxygen density profiles of the water were found to be oscillatory, with strong peaks near the surface. In agreement with other simulations,[427, 428] the Na$^+$ ion was found to be fully hydrated and thus always separated from the wall by water molecules in its first hydration shell.

Another very interesting study of the adsorption of hydrated ions on charged Lennard-Jones "electrodes" has been conducted by Glosli and Philpott,[311] employing rigid ST2 water. Image charges were not considered in this work, so the charged surface is similar to the one considered by Zhu and Robinson.[342] Observed was a decreasing trend for solvation with increasing size of the halide ion. Specifically, the fluoride ion was found to remain fully solvated and was at all times separated from the surface by its primary solvation shell. The chloride ion spent part of its time in physical contact with the electrode, while the bromide ion favored physisorption over hydration most of the time. There was almost no possibility for water molecules to be interposed between the iodide ion and the electrode. This ion size dependence is consistent with a variety of other surface results, and primarily has to do with the energetic competition between the ion–water and ion–surface interactions. Related to the tendency for adsorption in the Glosli and Philpott computations[311] was a variation in residence time. Whereas the time spent by a water molecule in the hydration shell of the lithium ion was much longer than any other ion, the first solvation shell of the larger anions was very dynamic. These are expected results for the weaker binding of water molecules around the larger ions and certainly bear on the problem of electrochemical ion–electrode structure involving more complex ions. For example, if the ionic charge is strongly localized, a completely different solvation structure would be expected than for a delocalized ionic charge.

6. Biomolecular Surfaces

Recently, it has been recognized that the behavior of proteins may be significantly influenced by the present of water.[328, 329, 506–518] Protein structures simulated in an aqueous environment more closely resemble the known experimental forms. Moreover, in the presence or absence of surrounding and interpenetrating water molecules,[510] dielectric properties and reorientational correlation times for backbone and side-chain vectors in both the surface and internal residues behave differently. An ever-increasing number of computer simulations on biologically relevant systems are therefore explicitly considering the water molecules. It should be remembered, however, that with the present state of computer technology and the intrinsic complexity that already exists in biosystems, even in the absence of water, these types of studies are still in their infancy, with only the most simple water models, such as SPC or TIP4P, currently having practical utility.

An interesting study of dielectric effects at different sites in a hydrated protein has recently been carried out by King et al.[515] The effects of boundary conditions were carefully discussed. It was concluded that the dielectric constant of a site could not be accurately evaluated unless the effects of surrounding water molecules were included. Computational studies of DNA-type systems, in the presence of water[516, 517] and of ions,[518] have also recently begun to emerge. Strong hydration forces between the macromolecular surfaces were found to be caused by the fact that the dipoles of the surface water molecules align perpendicular to the axes of the DNA helix.[516] It was found[518] too that metal ions had a large effect on the stability of base-pairing because of long-range polarization effects on the hydrogen-bond stability.

Aside from the exceedingly important problem of protein structure and dynamics, and the behavior of DNA in the presence of water and ionic solutions, biomembrane–water interfaces are known to play a significant role in biological activity. From a computational point of view, these structures, which are quasi-two-dimensional and relate closely to other computationally practical interfacial studies, are much easier to deal with than proteins and have therefore received a high level of attention. Prototypical membranes may be constructed either experimentally or computationally from a long-chain hydrophobic molecule, such as a alkyl hydrocarbon, to which a smaller hydrophilic group, the head group, has been attached. As mentioned in Section III.C.2, such molecules are termed *amphiphilic*, from the Greek "affinity for both sides." At an oil–water interface, simplistically speaking, the hydrophilic part of the molecule dissolves in the water, while the other part of the molecule dissolves in the oil. A monolayer is thereby produced. Two such quasiplanar monolayers with a hydrophobic medium sandwiched between them form a bilayer structure, which is an important constituent of biological membranes. Other types of microsystems, which are spherical or nearly so, can also be synthesized. In a surrounding water environment, small pools of oil can be encapsulated in a micelle composed of amphiphiles, a familiar requirement in surfactant technology. Of perhaps more interest in water research, is the fact that small pools of water (up to about 150 Å diameter[519]) can be enclosed in a reverse micelle when the outside environment is hydrophobic. It is the chemical and physical nature of the water at such interfaces that is of particular interest to biophysicists.[520] There is no doubt that water near such interfaces is a far different substance than it is in the bulk.

Commonly believed from crystallographic studies is that, to minimize the number of unfulfilled hydrogen bonds, there exists a layer of ordered solvent structure adjacent to the surface of a macromolecule or a membrane. Such a water layer would be expected to exhibit exceedingly slow diffusion compared with bulk phase water. However, simulations[328-330, 342, 508, 509, 521, 522] have given only modest perturbations on diffusion, possibly indicating some sort of a flaw in the computational modeling thus far used for these studies.

According to the elegant theory of Marčelja and Radić,[523] the hydration force[524] is measured by an order parameter. In an effort to understand the origin of this force, Kjellander and Marčelja,[525, 526] and, more recently, Berkowitz and Raghavan[527] have performed molecular dynamics simulations of water molecules interacting with a bilayer system. All these results revealed that, in contrast to the situation from more conventional calculations,[528, 529] the orientational polarization of water molecules displayed an oscillatory decay, even in the presence of thermal motions, and consequently was not appropriate to serve as the order parameter for water structure near a membrane. A better choice is the hydrogen-bond deficiency, which is also amenable to study by computational methods. The simulation results[527] indicated that strong repulsive forces acting on the water between electrically neutral amphiphilic bilayers[530] were caused by the work required to break the hydrogen bond network present in water.

To gain an insight into the structure and dynamics of water near a membrane, Raghavan et al.[531] have performed two molecular dynamics simulations on a system consisting of 437 water molecules between bilayers having 16 amphiphilic molecules on each side. One simulation treated the head groups as flexible and the other as rigid. These head groups were arranged to be parallel to the interfaces. There appeared a clear distinction between the bulk and the head group solvent regions, as indicated by the density profile of water. Although the thermal motion of the polar head groups had no significant influence on the orientational polarization of water, it considerably accelerated the dynamics of water molecules between the bilayer surfaces.

Water was also explicitly included in the studies of Watanabe et al.[532] and of Watanabe and Klein.[533] Using the SPC model for the solvent, the potential parameters of Jorgensen[38, 39] to describe the carbon atom, the methyl, methylene, and polar carboxylate head groups, the potential of Chandrasekhar et al.[316] for the sodium ion, together with the harmonic function of van der Ploeg and Berendsen[534] to handle the C–C–C–C dihedral angle torsion and the C–C–C angle bending, Watanabe et al.[532, 533] performed full molecular dynamics simulations on the sodium octanoate micelle in aqueous solution. Particular attention was paid to the local structure, the micelle shape, and chain conformations. According to these calculations, the micelle was stable and could be described by a prolate spheroid that underwent shape fluctuations on a characteristic time scale of about 30 ps. In accord with the results obtained from the more simplified simulations of O'Connell et al.,[535–537] which used boundary forces instead of explicitly including the solvent, an increased proportion of the *trans* conformation in the alkyl chains was produced by micellization. Moreover, many of the other results obtained by Watanabe et al.[532, 533] were found to be in remarkable agreement with these simplified simulations and also with the corresponding Monte Carlo calculations[538] based on a simple lattice model.

Even though small spherically shaped reverse micelles containing from hundreds to thousands of water molecules have been studied exclusively experimentally[519, 520] and would therefore seem to lend themselves well to computational investigations, Brown and Clarke[539] have been the only workers to publish an MD on this type of system. Their highly simplified model contained 72 SPC water molecules within a reverse micelle composed of 36 amphiphilic-type molecules. Each head group was composed of two oppositely charged spheres connected by a harmonic potential. Outside the reverse micelle, there were 1079 hydrophobic solvent molecules represented by single Lennard-Jones particles. All these distinct particles were then confined within a spherical cavity of solvent continuum. Through an analysis of the calculated radial density profiles of the equilibrated assembly, it was evident that the head groups formed a coating with substantial roughness at the aqueous interface. Furthermore, a considerable number of water molecules penetrated the hydrophobic region, with the surface water molecules adopting a specific orientation. These outermost water molecules were not hydrogen bonded but were strongly coordinated to the anions in the ionic layer. In future computational work on reverse micelles, it would be useful if at least 130 water molecules were considered, since this is the minimum size usually reported in experimental studies.[519]

C. External Fields

Reported here for the first time are some results of new MD simulations on MST-FP water subjected to a sinusoidally oscillatory field.[104] Two field strengths were considered, $E_0 = 10^{10}$ V m^{-1} and $E_0 = 10^9$ V m^{-1}, both considerably lower than the atomic field discussed in Section II.B, and thus amenable to experimental investigations with modern laser light sources. A wavelength of $\lambda = 2.92$ μm was used in these calculations, corresponding to an oscillation frequency of about 2×10^{14} s^{-1}. This laser wavelength is within the first prominent absorption band of liquid water in the infrared region.[102]

Some time ago, the first results of this type were reported for the nonpolar, but polarizable, carbon disulfide molecule in a strong laser field.[295] A paper concerning laser field effects on SPC-FP water has also been published.[103]

1. Structure and Dynamics

Before discussing the MD results, a brief derivation is given of the Kirkwood relation[137, 356] between dielectric constant and dipole moment fluctuations. This is used in simulations to calculate the dielectric constant. For low-frequency dielectric phenomena in a finite sample, the first step is to put the sample in a constant homogeneous external electric field $\mathbf{E}_0$. This field attempts to align all the dipoles, but there will be randomization because of thermal effects. The average value of the total dipole moment in the direction of the field can be calculated easily using the Langevin theory.[540] For weak fields and an isotropic medium one obtains,

$$\langle \mathbf{M} \rangle = \frac{\langle \mathbf{M}^2 \rangle}{3kT} \mathbf{E}_0 + \cdots \tag{12}$$

where $\mathbf{M}$ is the total dipole moment of the system in the presence of $\mathbf{E}_0$. By assuming the fields are weak, all the nonlinear effects such as saturation are ignored, which is the required state of affairs for dielectric constant calculations.

The second step is to relate the total dipole moment at any point to the external field $\mathbf{E}_0$. If the medium is assumed to have a dielectric constant ε_{RF}, one finds

$$\langle \mathbf{M} \rangle = 3\varepsilon_0 V \frac{\varepsilon - 1}{\varepsilon + 2} \left[1 - \frac{\varepsilon - 1}{\varepsilon + 2} \frac{2(\varepsilon_{RF} - 1)}{2\varepsilon_{RF} + 1} \right]^{-1} \mathbf{E}_0 \tag{13}$$

where ε is the actual dielectric constant of the system. Eliminating $\mathbf{E}_0$, the final relationship is

$$G_K = \frac{\langle \mathbf{M}^2 \rangle}{Nm^2} = \frac{9kT\varepsilon_0 V}{Nm^2} \frac{\varepsilon - 1}{\varepsilon + 2} \left[1 - \frac{\varepsilon - 1}{\varepsilon + 2} \frac{2(\varepsilon_{RF} - 1)}{2\varepsilon_{RF} + 1} \right]^{-1} \tag{14}$$

where the finite-system Kirkwood factor $G_K = \langle \mathbf{M}^2 \rangle / Nm^2$, for a cubic sample with periodic boundary conditions, has been introduced. Sometimes the other Kirkwood factor g_K is used. It is related to G_K by the relation

$$g_K = \frac{(2\varepsilon + 1)(2\varepsilon_{RF} + \varepsilon)}{3\varepsilon(2\varepsilon_{RF} + 1)} G_K \qquad (15)$$

The exact forms of the above equations depend on the geometry and boundary conditions used.[244] It also turns out that $\langle \mathbf{M}^2 \rangle$ is a very slowly developing quantity in a computer simulation,[71, 249] so that extremely long run times are required to obtain an accurate dielectric constant. Ruff and Diestler[63] have pointed out an additional assumption related to the absence of an explicit polarization term in many pair potential models. Neumann and his colleagues[248, 249] have also addressed the problem of calculating the frequency-dependent dielectric constant through computational methods.

The calculated $g_{OH}(r)$ in SPC-FP water subjected to a strong electrical field[294] shows a very prominent first maximum near 1.75 Å, possibly indicating a preponderance of bifurcated H_2O:::H_2O water–water bonds, where both hydrogen atoms of one molecule are equivalently attached to the oxygen atom of a neighbor. There appears to be some evidence[148, 173] for this type of bonding in bulk water, and, because of the large dipole moment of such a dimer arrangement, such structures may be particularly prevalent when water is subjected to high electrical fields.

As a consequence of the molecular alignment in the field, the normal liquid structure is destroyed. As mentioned earlier, the second peak of the oxygen–oxygen RDF is too flat in the SPC-FP model,[74] so this model is not very accurate for the non-nearest-neighbor structure. However, for first neighbors, the results indicate very strong field effects on the latent nearest-neighbor bonding in the normal liquid. This effect is evidenced by the fact that the first peaks of the O $\cdots$ O, O $\cdots$ H, and H $\cdots$ H RDFs become more prominent and shift toward shorter distances in the strong field. The molecular alignment apparently tends to destroy the existing structures in ordinary tetrahedral water groupings and forms new groupings with different configurations, possibly including various bifurcated structures. A 70% enhancement of the integrated first peak of the O $\cdots$ O RDF is observed. To be noted is the fact that the intensity would be about 100% greater if all the local tetrahedral structures in the normal liquid were replaced by bifurcated structures in the perturbed liquid. This is indeed a very large effect, indicating once again the fragility of structures in liquid water, a feature that makes any type of computer analysis so difficult.

The enhancements and shifts of the radial distribution functions in the presence of an externally applied field can also be found in nonpolar, but polarizable, liquids such as carbon disulfide.[295] This compression phenomenon, which tends to squeeze out molecular atoms from regions normally having low radial density, seems therefore to be a general tendency.

Using flexible models[74, 75] also allows the distortion of the intramolecular geometry upon application of the field to be studied. On average, the O–H bond length is stretched while the H–O–H bond angle contracts. Again, the net effect increases the molecular dipole moment in the field.

Evans et al. (see Section II of Ref. 344) found that the mean internal energy of their rigid 5-site water model at equilibrium was raised by 40% from field-off to field-on conditions. However, the data from the MST-FP computations[75] provide an opposite conclusion. This difference is caused directly by the inclusion of electric polarization and bond flexibility. While partial molecular alignment in the field has a tendency to bring like charges closer together, and thus to raise the internal energy, the simultaneously increased molecular dipole moment and the greater ability to form bifurcated bonds in the flexible model result in a net energy stabilization. If the lowering of the entropy of water in the field is not too great, the rather strong stabilization of the internal energy would mean that field-on water has a lower vapor pressure than normal liquid water. This may be an experimentally observable property.

2. Nonlinear Refractive Indices

In Section II.B, Eq. (3), it was seen that for a spatially uniform, uniaxial field $E(t) = E_0 \sin \omega t$ the polarization $P(t)$ can be written in terms of the time-dependent values and coordinates of the charges. This expression leads directly into a computational algorithm for the determination of the polarization. By a least-squares fit of this computationally derived $P(t)$ to Eq. (4), one can obtain the in-phase and out-of-phase components of $P(t)$. It is known that in Fourier space the electrical displacement $D(\omega)$ is related to the field through the permittivity of the medium,

$$D(\omega) = \varepsilon(\omega)\varepsilon_0 E(\omega) \qquad (16)$$

Because of the phase difference, the dielectric constant appearing in Eq. (16) is in general complex:

$$\varepsilon(\omega) = \varepsilon_1(\omega) + i\varepsilon_2(\omega) \qquad (17)$$

the real part of which represents the dielectric dispersion, while the imaginary part gives the dielectric loss. Eliminating D from these equations, then comparing the real parts, gives

$$\varepsilon_1(\omega) = 1 + \frac{P_1}{\varepsilon_0 E_0}$$

$$\varepsilon_2(\omega) = \frac{P_2}{\varepsilon_0 E_0}$$

$$(18)$$

Since $\varepsilon(\omega) = n(\omega)^2$, the real n_r and imaginary n_i parts of the refractive index can then be obtained:

$$\varepsilon_1(\omega) = n_r^2 - n_i^2$$

$$\varepsilon_2(\omega) = 2n_r n_i$$

$$(19)$$

The experimental value of n_r measured at $\lambda = 2.92$ μm and $T = 293$ K is 1.327. This is also close to the value found for visible light. Also known is that throughout the visible and infrared regions of the spectrum, $n_i \ll n_r$.[102]

From the MST-FP computations, it was found from the above procedure that, for the weaker field, $n_r = 1.587$, and for the more intense one, $n_r = 1.604$. The difference is expected because of nonlinear effects. Neglecting higher-order nonlinear contributions, one has

$$n_r = n_r(0) + \tfrac{1}{2}n_r(1)E_0^2 + \cdots \qquad (20)$$

from which it is seen that $n_r(0) = 1.587$ and $n_r(1) = 3.43 \times 10^{22}$ (V m^{-1})$^{-2}$. Converting to commonly used esu units, the latter value is 3.09×10^{-13} esu. The imaginary part of the refractive index is also found from Eqs. (19); it is ~ 0.019. The comparison of these values with the experimental quantities,[102] $n_r(0) = 1.274$ and $n_i(0) = 0.281$, shows that the calculated real part is too large by about 25%, while the imaginary part is more than an order of magnitude too small, corresponding to what is found experimentally at longer wavelengths. This very likely means that the response time of the liquid MST-FP molecules to the field is not exactly right, and this undoubtedly has to do with the electronic polarization modeling that was used. In any case, the results are not bad and provide a challenge for future work in this field.

3.   Supercontinuum Production

Related to the optical nonlinear effects are the phenomena of self-focusing and supercontinuum generation in condensed matter.[10] These give rise to the generation of a broad-band white light continuum when intense ultrafest laser pulses propagate through media. Water is a common liquid used to generate such a continuum.[11] When a laser beam propagates through a medium with large high-order susceptibilities, the index of refraction of the medium at the central part of the beam profile becomes larger than that near the periphery. This disturbance acts like a convex lens that refracts the laser beam toward the center, producing a self-focusing effect. Self-focusing significantly lowers the thresholds for Raman scattering and supercontinuum generation.

Even though electronic effects are mostly missing, it is still possible to obtain useful information from a classical MD computation about the response of liquid water to an intense laser field. As seen in the last subsection, the dielectric properties of a substance can be significantly changed in such a field. We saw that a difference in the index of refraction in weak and intense light arose because of the medium's nonlinear response, which is field-intensity dependent. When there is no external field, the induced dipole moment arises only from fields derived from neighboring molecules. Under the influence of the applied field, the average dipole moment per molecule is increased still farther. For SPC-FP water, this was about 6% for the 10^{10} V m^{-1} sinusoidal field applied in that calculation. This increment, mainly caused by the partially aligned neighboring molecules, gives rise to the positive nonlinear index of refraction, and thus to the self-focusing phenomenon.

For substances with spatially inhomogeneous polarizability, such as carbon disulfide, the molecular alignment contributes additionally to the enhancement of the index of refraction because of more efficient polarization along the molecular axis than normal to this axis. Compounded with the large primary polarizability, this is presumably the reason why self-focusing occurs in liquid carbon disulfide even at moderate laser power levels.

In the laser field, not only the longitudinal autocorrelation function, but also, somewhat less distinctly and because of mode coupling, the transverse ones oscillate with the period of the laser light oscillations. A small time lag occurs for the transverse oscillations, since mode coupling involves nuclear motions that are intrinsically slower than the electronic response. These kinds of effects were also found in earlier simulations of FP carbon disulfide in laser fields.[295]

The value $n_r(1)$ of the first nonlinear contribution to the real part n_r of the refractive index of MST-FP water was calculated to be 3.09×10^{-13} esu. The experimental value for this quantity is not known, but has been estimated to be about 220 times smaller than that for CS_2,[10] or roughly 5×10^{-14} esu. This is only about 15% of the above calculated value. A value in better agreement with this experimental value was obtained for the SPC-FP model.[377] Of course, there is a large statistical uncertainty in these calculations as the difference between two large numbers.

VIII.   CONCLUDING REMARKS

In gases, where kinetic energy is dominant, one can solve the ideal gas model exactly and then add the potential energy via a virial expansion.[541] Acceptable results are obtained without too difficult an analysis. In solids, where the potential energy is dominant, one can start with a perfectly crystalline solid at absolute zero, then study vibrations and electronic motions using quantum-statistical mechanics.[542] Even a one-electron approximation, which neglects correlations, does a pretty good job. For cooperative effects, such as occur in superconductivity and other types of phase transitions, the solution of a genuine many-body problem is of course required.

In liquids there is no natural starting point where something can be neglected. Everything is important at the same time. One type of idea, represented by lattice theories[543] and random network models,[220] considers the liquid to be like a solid. Another type, such as those that use the BBGKY hierarchy of equations,[544] argues that a liquid is like a frozen gas. Actually, a liquid, particularly a molecular liquid such as water, is neither a gas nor a solid, so none of these ideas really works. As soon as the exact BBGKY equations are decoupled or truncated, the critical properties are ruined.[545] How then can such theories work well away from the critical point, say at room temperature for water?

For the above reasons, an MD or MC simulation approach is traditionally used for liquid water. The requirements are equations of motion and forces. However, this is not quite as simple as it sounds. Should quantum mechanics be used for the equations of motion? Except for quantum liquids such as helium at very low temperatures, probably the explicit incorporation of quantum mechanics is not all that necessary at the present state of liquid water simulations. Even high-frequency molecular vibrations can be reproduced quite well[75] by classical MD, as emphasized for water in this review.

The potential energy of interaction between molecules in the liquid state is also required. Of course, this property originates from a quantum-mechanical calculation, but, as discussed in this review, there are many complications. Through the study of small clusters of water, quantum-mechanical calculations are indeed capable of telling us something about the true many-body forces that exist in the real liquid. They allow detailed tables of interaction potentials to be presented as a function of molecular geometry and the distance between nuclear centers. However, such tabular material is too inconvenient and expensive to use in a computer simulation, so attempts to parameterize the data are made, but the very act of parameterizing these potentials, using combinations of exponentials and power laws, robs them of any superiority over the purely empirical potentials using the same type of formulation. Besides this, the act of extrapolating to the liquid state from the small clusters that are currently amenable to ab initio quantum computations is fraught with dangers. An ice Ih crystal composed of pure dimer interactions, with rigidly fixed $O \cdots O$ nearest-neighbor distances of 2.98 Å,[118] would have a density of 0.73 g cm^{-3}, compared with the actual density,[12] 0.92 g cm^{-3} (273 K), of ice Ih, where $O \cdots O$ distances are 2.76 Å.

Since the calculation of nonelectrical properties converges rapidly as a function of system size N, many properties of water can be readily determined in an MD or an MC computation using the wide variety of standard relationships between the macroscopic thermodynamic quantities and statistical averages, fluctuations, etc. of the microscopic variables. For instance, it would not be necessary to carry out computations at a series of pressures to obtain the isothermal compressibility. Rather, a more economical and probably more accurate way is to use the connection[546] with the fluctuation in the number of molecules in a given volume, which in turn is related to the intensity of scattered light at zero angle.

In contrast with analytical theories, MD and MC calculations of liquid water can be carried out only on a finite system. For short-range forces, which are dominant for the aforementioned thermal properties, this causes a problem only near a surface. It is only there that the interacting particles find out that they are in a finite system. On the other hand, for long-range Coulomb forces, the situation worsens considerably. Not only particles near a surface, but even those in the middle of the sample, discover that they are in a finite system. They interact with everything. The strength of the interaction does decrease as the reciprocal of the distance, and, because of screening, the sum of such interactions eventually converges with increasing distance. However, the dielectric constant is an almost nonconvergent function of system size, far too long range[20] to hope to obtain a stable result by present day computational methods.

Thus, for electrical properties, one needs a different type of statistical-mechanical strategy, which is available in the inexact mean-field theories of Lorentz,[243] Onsager,[233] and Kirkwood.[356] Noted, however, is that an exact formula relating the dielectric constant to microscopic quantities does exist. It was worked out in a very interesting paper by Lax[547] for the spherical model, who found that the Onsager result does agree with the exact one, but only up to terms of order T^{-3} in a weak-field expansion.

Even though the Lax model allows larger than realistic fluctuations of the individual dipole vectors, it may contain useful guidelines for handling finite samples of a dipolar material.

In summary then, there are really no exact ways for dealing with the problem of the long-range forces. There are mean-field-type corrections, the RF technique, and there are exact-sounding methods such as the Ewald lattice summation. The latter is indeed exact in principle, if it were really possible to sum accurately over the infinite set of periodic images, but cursorily this seems to offer few improvements over the problem faced in the first place. However, the LS method owes its superiority to the fact that the resulting sums do eventually converge sufficiently well for this method to be used in computational work. Though it is certainly much more efficient than trying to sum all the screened interactions directly, hidden dangers may still exist. Various workers[244, 359] have indeed found that most, but not all, of the answers from the RF and LS schemes do agree pretty well.

Of course, when using RF, the dependence of the results on the dielectric constant chosen for the missing part of the system should be closely examined, and not necessarily chosen to be of the conducting type, as is most commonly done. Also, if the LS method is used, the dependence of the results on the cutoff distance, usually arbitrarily chosen, should always be considered. See Refs. 361 and 362 for a fuller discussion of these concerns.

What are the major findings from MD and MC computations for water? Presently, there seems to be no best model for everything. One can optimize any of the popular models to get decent values for many quantities, such as total energy, dielectric constant, and pressure. Of course, the more complicated the model, with the more adjustable parameters, the better the fits that can eventually be obtained. Is this the only reason that a 5-site FP model[75] has been found to be superior to 3-site rigid models? Or, is this model really intrinsically better, as the authors claim?

An astute outsider to the field may ask further questions: Have the MD/MC computations on liquid water done much more than to verify that, using Newton's laws and reasonable potentials with adjustable parameters, some of the properties of water can be reproduced? Has anything not previously known to experimentalists actually been predicted? Have new insights about the water problem been gained from these methods? Have new techniques that can be used on other systems been discovered? Have better approximations to any analytic model been found? Has a systematic search for the simplest MD/MC interaction model that fits the most data ever been made? Has anything surprising been discovered about water? Have any old questions been settled, that could not have been settled otherwise ... things, such as the so-called anomalous properties, that have been argued about for a long time?

Unfortunately, the answers to these questions are in the negative or, at most, very weakly in the affirmative, which takes us back to comments made in the Introduction. The reason behind failings of MD/MC simulations on liquid water is computer expense, which severely limits systematic investigations. Even though few of the above critical questions have more than a weak affirmative answer, the methods are being improved, and glimmers of understanding of the problem of liquid water are indeed being created. In fact, the MD/MC calculations have already guided experimentalists in the quest for improved RDFs of liquid water, and they are close to helping experimentalists sort out the structure at surfaces, such as the liquid–vapor interface.

Continued use of the most simple models, SPC2[37] and TIP4P,[40] is still justified for otherwise complicated problems, such as biosystems, where a certain amount of intuition about the effect of the water presence on these systems has already been gained. However, it should be remembered that many of the results using simple water models are not entirely trustworthy, so perhaps it is time to begin systematic searches for better water models that give a good accounting of the temperature and pressure dependence of the thermodynamic and electric properties, yet, by modern standards, remain computationally efficient.

It would certainly seem that the search for a better computational model of water should start with 5-site models that contain both bond flexibility and rapidly responsive electronic polarization. Such models may even have to contain Lennard-Jones centers on the hydrogen atoms, as in the Evans potential.[43] It would also seem essential that not only the monomer structure, its vibrational frequencies, and its dipole moment, as in the MST-FP model,[75] but also its quadrupole components, be faithfully incorporated into this model. It can readily be established that, for the standard ST2-type model, given bond angle and lengths equal to those in the monomer, and with positive charges located at the hydrogen atom positions, it is mathematically possible to find just two structures with positions of the negative charges and charge magnitudes such that the experimentally known dipole moment and electric quadrupole components are reproduced. Structure I, perhaps the more realistic one, uses charge magnitudes of $0.1139|e|$ at each of the four charge sites, with the negative charges lying 1.9455 Å from the oxygen atom, forming an angle of 110.44°, close to the usually chosen tetrahedral angle of 109.47° for ST2 structures. For structure II, these parameters are $0.5715|e|$, 0.3339 Å, and 275.92°, meaning that the negative charges are bent back toward the hydrogens, very similar to the RWK2 model. However, since planar models are a special case of the geometry considered, clearly there can be no planar model that satisfies all the required constraints.

Acknowledgments

The authors dedicate this review to Walter Kauzmann, a very good friend, who a quarter of a century ago with David Eisenberg composed the best book[12] on water ever written. If only one thing is accomplished by this review, we hope it is the encouragement for these authors to write another edition of this wonderful book. GWR also wants to thank his wife, Ellen J. Robinson, and a friend, De Alva Holmes, for enduring the many hours of "study hall" required to finish this much larger than originally anticipated task. He also appreciates the hospitality afforded him by the Chemistry School, University of Melbourne, Australia, during the initial writing stages of the review. Acknowledged for financial assistance, without which this work could not have been completed, is the National Science Foundation and the Robert A. Welch Foundation.

References

1. F. van der Leeden, F. L. Troise, and D. K. Todd, *The Water Encyclopedia*, Lewis, Chelsea, MI, 1990.

2. F. Franks (Ed.), *Water: A Comprehensive Treatise*, Vol. 4, Plenum, New York, 1973.

3. R. A. Robinson and R. H. Stokes, *Electrolyte Solutions*, 2d ed., Butterworths, London, 1959.

4. J. O'M. Bockris and A. K. N. Reddy, *Modern Electrochemistry*, Vol. 1, Plenum, New York, 1970.

5. H. S. Harned and B. B. Owen, *The Physical Chemistry of Electrolytic Solutions*, Reinhold, New York, 1958.

6. W. Kauzmann, *Adv. Protein Chem.*, **14**, 1 (1959).

7. Workshop on Water, *J. Phys.* (Colloque C7, supplément au no. 9), **45** (1984). This collection of papers also contains work on the physics of water.

8. D. Vasilescu, J. Jaz, L. Packer and B. Pullman (Eds.), *Water and Ions in Biomolecular Systems*, Birkhauser, Basel, 1990.

9. J. A. McCammon and S. C. Harvey, in *Dynamics of Proteins and Nucleic Acids*, Cambridge University Press, Cambridge, UK, 1987.

10. R. R. Alfano (Ed.), *The Supercontinuum Laser Source*, Springer, Berlin, 1989.

11. T. Jimbo, V. L. Caplan, Q. X. Li, Q. Z. Wang, P. P. Ho, and R. R. Alfano, *Opt. Lett.*, **12**, 477 (1987).

12. D. Eisenberg and W. Kauzmann, *The Structure and Properties of Water*, Oxford University Press, London, 1969.

13. G. S. Kell, *J. Chem. Eng. Data* **20**, 97 (1975).

14. R. J. Speedy and C. A. Angell, *J. Chem. Phys.* **65**, 851 (1976).

15. C. A. Angell, M. Oguni, and W. J. Sichina, *J. Phys. Chem.* **86**, 998 (1982).

16. M. Oguni and C. A. Angell, *J. Chem. Phys.* **78**, 7334 (1983).

17. C. A. Angell, *Annu. Rev. Phys. Chem.* **34**, 593 (1983).

18. D. E. Hare and C. M. Sorensen, *J. Chem. Phys.* **84**, 5085 (1986).

19. S. A. Rice and P. Gray, Supplement in I. Z. Fisher, *Statistical Theory of Liquids*, University of Chicago Press, Chicago, 1964. Also, see the more recent paper, L. J. Root, F. H. Stillinger, and G. E. Washington, *J. Chem. Phys.* **88**, 7791 (1988).

20. M. P. Allen and D. J. Tildesley, *Computer Simulation of Liquids*, Clarendon, Oxford, UK, 1987.

21. W. F. van Gunsteren and H. J. C. Berendsen, *Angew. Chem. Int. Ed. Engl.* **29**, 992 (1990).

22. T. P. Straatsma and J. A. McCammon, *Annu. Rev. Phys. Chem.* **43**, 407 (1992).

23. J. O. Hirschfelder, C. F. Curtiss, and R. B. Bird, *Theory of Gases and Liquids*, Wiley, New York, Chapman & Hall, London, 1954.

24. J. A. Barker and R. O. Watts, *Chem. Phys. Lett.* **3**, 144 (1969).

25. J. A. Barker and R. O. Watts, *Mol. Phys.* **26**, 789 (1973).

26. R. O. Watts, *Mol. Phys.* **28**, 1069 (1974).

27. J. R. Reimers, R. O. Watts, and M. L. Klein, *Chem. Phys.* **64**, 95 (1982).

28. A. Rahman and F. H. Stillinger, *J. Chem. Phys.* **55**, 3336 (1971).

29. F. H. Stillinger and A. Rahman, *J. Chem. Phys.* **57**, 1281 (1972).

30. F. H. Stillinger and A. Rahman, *J. Chem. Phys.* **60**, 1545 (1974).

31. F. H. Stillinger and A. Rahman, *J. Chem. Phys.* **61**, 4973 (1975).

32. G. N. Sarkisov and V. G. Dashevskii, *Z. Strukturnoi Khim.* **13**, 199 (1972).

33. H. Popkie, H. Kistenmacher, and E. Clementi, *J. Chem. Phys.* **59**, 1325 (1973).

34. G. C. Lie, E. Clementi, and M. Yoshimine, *J. Chem. Phys.* **64**, 2314 (1976).

35. J. C. Owicki and H. A. Scheraga, *J. Am. Chem. Soc.* **99**, 7403 (1977).

36. H. J. C. Berendsen, J. P. M. Postma, W. F. von Gunsteren, and J. Hermans, in B. Pullman (Ed.), *Intermolecular Forces*, Reidel, Dordrecht, 1981.

37. H. J. C. Berendsen, J. R. Grigera, and T. P. Straatsma, *J. Phys. Chem.* **91**, 6269 (1987).

38. W. L. Jorgensen, *J. Am. Chem. Soc.* **103**, 335 (1981).

39. W. L. Jorgensen, *J. Chem. Phys.* **77**, 4156 (1982).

40. W. L. Jorgensen, J. Chandrasekhar, J. D. Madura, R. W. Impey, and M. L. Klein, *J. Chem. Phys.* **79**, 926 (1983).

41. W. L. Jorgensen and J. D. Madura, *Mol. Phys.* **56**, 1381 (1985).

42. R. W. Impey, M. L. Klein, and I. R. McDonald, *J. Chem. Phys.* **74**, 647 (1981).

43. M. W. Evans, *J. Mol. Liquids* **32**, 173 (1986).

44. M. W. Evans, G. C. Lie, and E. Clementi, *J. Chem. Phys.* **88**, 5157 (1988).

45. H. E. Alper and R. M. Levy, *J. Chem. Phys.* **91**, 1242 (1989).

46. M. Mezei, *Mol. Phys.* **47**, 1307 (1982); erratum: **67**, 1205 (1989).

47. T. Çagin and B. M. Pettitt, *Mol. Phys.* **72**, 169 (1991).

48. J. Ji, T. Çagin, and B. M. Pettitt, *J. Chem. Phys.* **96**, 1333 (1992).

49. Z. Li and H. A. Scheraga, *Chem. Phys. Lett.* **154**, 516 (1989).

50. W. L. Jorgensen, J. F. Blake, and J. K. Buckner, *Chem. Phys.* **129**, 193 (1989).

51. J. Quintana and A. D. J. Haymet, *Chem. Phys. Lett.* **189**, 273 (1992).

52. A. Rahman, F. H. Stillinger, and H. L. Lemberg, *J. Chem. Phys.* **63**, 5223 (1975).

53. F. H. Stillinger and A. Rahman, *J. Chem. Phys.* **68**, 666 (1978).

54. P. Bopp, G. Janscó, and K. Heinzinger, *Chem. Phys. Lett.* **98**, 129 (1983).

55. J. R. Reimers and R. O. Watts, *Chem. Phys.* **91**, 201 (1984).

56. K. Toukan and A. Rahman, *Phys. Rev. B* **31**, 2643 (1985).

57. G. C. Lie and E. Clementi, *Phys. Rev. A* **33**, 2679 (1986).

58. O. Teleman, B. Jönsson, and S. Engström, *Mol. Phys.* **60**, 193 (1987).

59. A. Wallqvist and O. Teleman, *Mol. Phys.* **74**, 515 (1991). Problems with equilibration in Ref. 58 are discussed, and corrected results are presented in this reference.

60. J. J. Ullo, *Phys. Rev. A* **36**, 816 (1987).

61. J. Anderson, J. J. Ullo, and S. Yip. *J. Chem. Phys.* **87**, 1726 (1987).

62. D. E. Smith and A. D. J. Haymet, *J. Chem. Phys.* **96**, 8450 (1992).

63. I. Ruff and D. J. Diestler, *J. Chem. Phys.* **93**, 2032 (1990).

64. M. Sprik and M. L. Klein, *J. Chem. Phys.* **89**, 7556 (1988).

65. K. Watanabe and M. L. Klein, *Chem. Phys.* **131**, 157 (1989).

66. U. Niesar, G. Corongiu, M.-J. Huang, M. Dupuis, and E. Clementi, *Int. J. Quant. Chem. Symposium* **23**, 421 (1989).

67. U. Niesar, G. Corongiu, E. Clementi, G. R. Kneller, and D. K. Bhattacharya, *J. Phys. Chem.* **94**, 7949 (1990).

68. P. Ahlström, A. Wallqvist, S. Engström, and B. Jönsson, *Mol. Phys.* **68**, 563 (1989).

69. P. Cieplak, P. Kollman, and T. Lybrand, *J. Chem. Phys.* **92**, 6755 (1990).

70. J. Caldwell, L. X. Dang, and P. A. Kollman, *J. Am. Chem. Soc.* **112**, 9144 (1990).

71. M. Sprik, *J. Chem. Phys.* **95**, 6762 (1991).

72. R. E. Kozack and P. C. Jordan, *J. Chem. Phys.* **96**, 3120 (1992).

73. S.-B. Zhu and G. W. Robinson, in L. P. Kartashev and S. I. Kartashev (Eds.), *Proc. 4th Internat. Conf. Supercomputing* Vol. 2, 1989, pp. 189–197.

74. S.-B. Zhu, S. Yao, J.-B. Zhu, S. Singh, and G. W. Robinson, *J. Phys. Chem.* **95**, 6211 (1991).

75. S.-B. Zhu, S. Singh, and G. W. Robinson, *J. Chem. Phys.* **95**, 2791 (1991).

76. A. Wallqvist, *Chem. Phys.* **148**, 439 (1990).

77. M. Tuckerman, G. Martyna, and B. J. Berne, *J. Chem. Phys.* **93**, 1287 (1990); M. E. Tuckerman, B. J. Berne, and G. J. Martyna, *J. Chem. Phys.* **94**, 6811 (1991).

78. J. D. Bernal and R. H. Fowler, *J. Chem. Phys.* **1**, 515 (1933).

79. J. S. Rowlinson, *Trans. Faraday Soc.* **47**, 120 (1951).

80. A. Ben-Naim and F. H. Stillinger, in R. A. Horne (Ed.), *Structure and Transport Processes in Water and Aqueous Solutions*, Wiley-Interscience, New York, 1972.

81. H. L. Lemberg and F. H. Stillinger, *J. Chem. Phys.* **62**, 1677 (1975).

82. O. Matsuoka, E. Clementi, and M. Yoshimine, *J. Chem. Phys.* **64**, 1351 (1976).

83. E. Clementi and G. Corongiu, *Int. J. Quant. Chem., Symposium* **10**, 31 (1983).

84. J. Detrich, G. Corongiu, and E. Clementi, *Chem. Phys. Lett.* **11**(2), 426 (1984).

85. R. O. Watts, *Chem. Phys.* **26**, 367 (1977).

86. J. R. Reimers and R. O. Watts, *Chem. Phys.* **85**, 83 (1984).

87. T. P. Lybrand and P. A. Kollman, *J. Chem. Phys.* **83**, 2923 (1985).

88. M. D. Morse and S. A. Rice, *J. Chem. Phys.* **76**, 650 (1982).

89. S. Kuwajima and A. Warshel, *J. Chem. Phys.* **94**, 460 (1990).

90. G. S. Del Buono, P. J. Rossky, and J. Schnitker, *J. Chem. Phys.* **95**, 3728 (1991).

91. F. N. H. Robinson, *Macroscopic Electromagnetism*, Pergamon, New York, 1973, Preface. See also the Introduction of this book for an interesting and pertinent discussion of the generally unappreciated difficulties encountered when attempts are made to associate modern microscopic and traditional macroscopic concepts in this field.

92. J. D. Jackson, *Classical Electrodynamics*, Wiley, New York, 1962.

93. Newsletter 25, European Photo-chemistry Association, November 1985, p. 28.

94. G. Herzberg, *Electronic Spectra of Polyatomic Molecules*, Van Nostrand, Princeton, NJ, 1966. See the discussions on pp. 122–124 and 272.

95. N. B. Delone and M. V. Fedorov, *Usp. Fiz. Nauk* **158**, 215 (1989); *Sov. Phys. Usp.* **32**, 500 (1989).

96. F. Krausz, M. E. Fermann, T. Brabec, P. F. Curley, M. Hofer, M. H. Ober, C. Spielmann, E. Wintner, and A. J. Schmidt, *IEEE J. Quantum Electron.* **28**, 2097 (1992).

97. J. Squier, F. Salin, G. Mourou, and D. Harter, *Opt. Lett.* **16**, 324 (1991).

98. H. E. Lessing and A. Von Jena, in M. L. Stitch (Ed.), *Laser Handbook*, Vol. 3, North-Holland, Amsterdam, 1979, pp. 764–765.

99. A. Szöke, in A. Bandrauk (Ed.), *Atomic and Molecular Processes with Short Intense Laser Pulses*, NATO Advanced Study Institute, Series B, Vol. 171, Plenum, New York, 1988, p. 207; S. L. Chin, C. Rolland, P. B. Corkum, and P. Kelly, *Phys. Rev. Lett.* **61**, 153 (1988).

100. L. V. Keldysh, *Zh. Eksp. Teor. Fiz.* **47**, 1945 (1964); *Sov. Phys. JETP* **20**, 1307 (1965).

101. R. O. Watts, *Chem. Phys. Lett.* **80**, 211 (1981).

102. A. N. Rusk, D. Williams, and M. R. Querry, *J. Opt. Soc. Am.* **61**, 895 (1971).

103. S.-B. Zhu, J.-B. Zhu, and G. W. Robinson, *Phys. Rev. A* **44**, 2602 (1991).

104. S.-B. Zhu and G. W. Robinson, MST-FP Water in Strong Laser Fields, not published.

105. W. S. Benedict, N. Gailar, and E. K. Plyler, *J. Chem. Phys.* **24**, 1139 (1956).

106. C. Camy-Peyret, J.-M. Flaud, J.-Y. Mandin, J.-P. Chevillard, J. Brault, D. A. Ramsay, M. Vervloet, and J. Chauville, *J. Mol. Spectrosc.* **113**, 208 (1985).

107. J. R. Reimers and R. O. Watts, *Mol. Phys.* **52**, 357 (1984).

108. W. C. Natzle, C. B. Moore, D. M. Goodall, W. Frisch, and J. F. Holzwarth, *J. Phys. Chem.* **85**, 2882 (1981).

109. D. N. Nikogosyan, A. A. Oraevsky, and V. I. Rupasov, *Chem. Phys.* **77**, 131 (1983).

110. T. Joseph, T.-M. Kruel, J. Manz, and I. Rexrodt, *Chem. Phys.* **113**, 223 (1987).

111. W. Jakubetz, J. Manz, and V. Mohan, *J. Chem. Phys.* **90**, 3686 (1989).

112. S. A. Clough, Y. Beers, G. P. Klein, and L. S. Rothman, *J. Chem. Phys.* **59**, 2254 (1973).

113. J. Verhoeven and A. Dymanus, *J. Chem. Phys.* **52**, 3222 (1970).

114. D. W. Schwenke, S. P. Walch, and P. R. Taylor, *J. Chem. Phys.* **94**, 2986 (1991).

115. D. Hankins, J. W. Moskowitz, and F. H. Stillinger, *J. Chem. Phys.* **53**, 4544 (1970).

116. G. D. Zeiss and W. J. Meath, *Mol. Phys.* **33**, 1155 (1977).

117. L. A. Curtis, D. J. Frurip, and M. Blander, *J. Chem. Phys.* **71**, 2703 (1979).

118. T. R. Dyke, K. M. Mack, and J. S. Muenter, *J. Chem. Phys.* **66**, 498 (1977).

119. E. Whalley, in E. Whalley, S. J. Jones and L. W. Gold (Eds.), *Physics and Chemistry of Ice*, Ottawa, Royal Society of Canada, 1973.

120. N. Pugliano and R. J. Saykally, *J. Chem. Phys.* **96**, 1832 (1992).

121. M.-P. Bassez, J. Lee, and G. W. Robinson, *J. Phys. Chem.* **91**, 5818 (1987).

122. J. F. Nagle, *J. Math. Phys.* **7**, 1484 (1966).

123. S. W. Peterson and H. A. Levy, *Acta Crystallogr.* **10**, 70 (1957).

124. R. Chidambaram, *Acta Crystallogr.* **14**, 467 (1961).

125. K. Kume, *J. Phys. Soc. Jpn.* **15**, 1493 (1960).

126. W. E. Thiessen and A. M. Narten, *J. Chem. Phys.* **77**, 2656 (1982).

127. M. C. Bellissent-Funel, L. Bosio, and J. Teixeira, *J. Phys.: Condens. Matter* **3**, 4065 (1991).

128. L. Bosio, S.-H. Chen, and J. Teixeira, *Phys. Rev. A* **27**, 1468 (1983).

129. S.-H. Chen and J. Teixeira, *Adv. Chem. Phys.* **64**, 1 (1986).

130. B. Kamb, *Acta Crystallogr.* **17**, 1437 (1964).

131. B. Kamb, W. C. Hamilton, S. J. LaPlaca, and A. Prakash, *J. Chem. Phys.* **55**, 1934 (1971).

132. J. P. Devlin, P. J. Wooldridge, and G. Ritzhaupt, *J. Chem. Phys.* **84**, 6095 (1986).

133. J. G. Bayly, V. B. Kartha, and W. H. Stevens, *Infrared Phys.* **3**, 211 (1963).

134. G. E. Walrafen, M. S. Hokmabadi, and W.-H. Yang, *J. Phys. Chem.* **92**, 2433 (1988).

135. R. P. Auty and R. H. Cole, *J. Chem. Phys.* **20**, 1309 (1952).

136. G. T. Hollins, *Proc. Phys. Soc. (London)* **84**, 1001 (1964).

137. H. Fröhlich, *Theory of Dielectrics*, Oxford University Press, Oxford, UK, 1958.

138. C. A. Coulson and D. Eisenberg, *Proc. R. Soc. London Ser. A* **291**, 445 (1966). For the electric field calculation, Coulson ánd Eisenberg used different values of the polarizability and dipole moment than we would use. However, the overall result is unaffected because of this.

139. E. Segré, *Rend. Lincei* **13**, 929 (1931).

140. G. Bolla, *Nuovo Cimento* **9**, 290 (1932).

141. G. Bolla, *Nuovo Cimento* **10**, 101 (1933).

142. G. Bolla, *Nuovo Cimento* **12**, 243 (1934).

143. K. E. Larsson and U. Dahlborg, *J. Nucl. Energy* **B16**, 81 (1962).

144. G. d'Arrigo, G. Maisano, F. Mallamace, P. Migliardo, and F. Wanderlingh, *J. Chem. Phys.* **75**, 4264 (1981).

145. S.-H. Chen, K. Toukan, C.-K. Loong, D. L. Price, and J. Teixeira, *Phys. Rev. Lett.* **53**, 1360 (1984).

146. G. E. Walrafen, M. S. Hokmabadi, and W.-H. Yang, *J. Chem. Phys.* **85**, 6964 (1986).

147. G. E. Walrafen, M. R. Fisher, M. S. Hokmabadi, and W.-H. Yang, *J. Chem. Phys.* **85**, 6970 (1986).

148. G. E. Walrafen, *J. Phys. Chem.* **94**, 2237 (1990).

149. D. E. Hare and C. M. Sorensen, *J. Chem. Phys.* **93**, 25 (1990).

150. D. E. Hare and C. M. Sorensen, *J. Chem. Phys.* **93**, 6954 (1990).

151. D. E. Hare and C. M. Sorensen, *J. Chem. Phys.* **96**, 13 (1992).

152. D. E. Hare and C. M. Sorensen, *Chem. Phys. Lett.* **190**, 605 (1992).

153. B. V. Zheleznyi, *Russ. J. Phys. Chem. (Engl. Transl.)* **43**, 1311 (1969).

154. Yu. A. Osipov, B. V. Zheleznyi, and N. F. Fondarenko, *Russ. J. Phys. Chem. (Engl. Transl.)* **51**, 748 (1977).

155. L. D. Eicher and B. J. Zwolinski, *J. Phys. Chem.* **75**, 2016 (1971).

156. H. R. Pruppacher, *J. Chem. Phys.* **56**, 101 (1972).

157. E. Lang and H. D. Lüdemann, *J. Chem. Phys.* **67**, 718 (1977).

158. E. Lang and H.-D. Lüdemann, *Ber. Bunsen-Ges. Phys. Chem.* **84**, 462 (1980).

159. E. W. Lang and H.-D. Lüdemann, *Ber. Bunsen-Ges. Phys. Chem.* **85**, 603 (1981).

160. E. W. Lang and H.-D. Lüdemann, *NMR Basic Principles and Progress*, Vol. 24, Springer, Berlin, 1990, pp. 129–187.

161. E. W. Lang, D. Girlich, H.-D. Lüdemann, L. Piculell, and D. Müller, *J. Chem. Phys.* **93**, 4796 (1990).

162. D. Bertolini, M. Cassettari, and G. Salvetti, *J. Chem. Phys.* **76**, 3285 (1982).

163. M. A. Floriana and C. A. Angell, *J. Chem. Phys.* **91**, 2537 (1989).

164. D. Bertolini, M. Cassettari, M. Ferrario, P. Grigolini, and G. Salvetti, *Adv. Chem. Phys.* **62**, 277 (1985).

165. K. Larsson, J. Tegenfeldt, and K. Hermansson, *J. Chem. Soc. Faraday Trans.* **87**, 1193 (1991).

166. A. K. Soper and M. G. Phillips, *Chem. Phys.* **107**, 47 (1986).

167. J. C. Dore, in F. Franks (Ed.), *Water Science Reviews*, Vol. 1, Cambridge University Press, Cambridge, UK, 1985.

168. J. C. Dore, *J. Mol. Struct.* **250**, 193 (1991).

169. A. H. Narten, *J. Chem. Phys.* **56**, 5681 (1972).

170. A. H. Narten, W. E. Thiessen, and L. Blum, *Science* **217**, 1033 (1982).

171. A. H. Narten and H. A. Levy, *J. Chem. Phys.* **55**, 2263 (1971).

172. F. Hajdu, S. Lengyel, and G. Pálinkás, *J. Appl. Crystallogr.* **9**, 134 (1976).

173. P. A. Giguére, *J. Chem. Phys.* **87**, 4835 (1987).

174. F. Sciortino, A. Geiger, and H. E. Stanley, *Phys. Rev. Lett.* **65**, 3452 (1990).

175. P. A. Egelstaff, *Adv. Chem. Phys.* **53**, 1 (1983).

176. J. C. Dore, *J. Mol. Struct.* **237**, 221 (1990).

177. G. W. Robinson, to be published.

178. K. A. Motakabbir and M. Berkowitz, *J. Phys. Chem.* **94**, 8359 (1990).

179. H. E. Stanley and J. Teixeira, *J. Chem. Phys.* **73**, 3404 (1980).

180. M. R. Chowdhury, J. C. Dore, and J. T. Wenzel, *J. Non-Cryst. Solids* **53**, 247 (1982).

181. M. C. Bellissent-Funel, J. Teixeira, and L. Bosio, *J. Phys.: Condens. Matter* **1**, 7123 (1989).

182. A. Ben-Naim, *Hydrophobic Interactions*, Plenum, New York, 1980.

183. P. L. Luisi, and B. E. Straub (Eds.), *Reverse Micelles*, Plenum, New York, 1984.

184. G. L. Pollack, *Science* **251**, 1323 (1991).

185. G. W. Robinson, P. J. Thistlethwaite, and J. Lee, *J. Phys. Chem.* **90**, 4224 (1986).

186. G. Némethy, W. J. Peer, and H. A. Scheraga, *Annu. Rev. Biophys. Bioeng.* **10**, 459 (1981).

187. E. D. Sloan, *Clathrate Hydrates of Natural Gases*, Dekker, New York, 1990.

188. C. Tanford, *The Hydrophobic Effects: Formation of Micelles and Biological Membranes*, Wiley, New York, 1990.

189. A. Ben-Naim, *J. Chem. Phys.* **90**, 7412 (1989).

190. M. Mezei and A. Ben-Naim, *J. Chem. Phys.* **92**, 1359 (1990).

191. A. Ben-Naim, *J. Chem. Phys.* **93**, 8196 (1990).

192. R. P. Kennan and G. L. Pollack, *J. Chem. Phys.* **93**, 2724 (1990).

193. J. Lee, *J. Am. Chem. Soc.* **111**, 427 (1989).

194. J. Lee, *J. Phys. Chem.* **94**, 258 (1990).

195. S.-H. Yao, J. Lee, and G. W. Robinson, *J. Am. Chem. Soc.* **112**, 5698 (1990).

196. K. Bhattacharyya, E. V. Sitzmann, and K. B. Eisenthal, *J. Chem. Phys.* **87**, 1442 (1987).

197. J. N. Israelachvili, *Surf. Sci. Rep.* **14**, 109 (1992).

198. F. Grieser, P. Thistlethwaite, and P. Triandos, *J. Am. Chem. Soc.* **108**, 3844 (1986).

199. T. G. Fillingim, S.-B. Zhu, S. Yao, J. Lee, and G. W. Robinson, *Chem. Phys. Lett.* **161**, 444 (1989).

200. J. L. Parker and P. M. Claesson, *Langmuir* **8**, 757 (1992).

201. H. K. Christenson, P. M. Claesson, and J. L. Parker, *J. Phys. Chem.* **96**, 6725 (1992).

202. J. D. Porter and A. S. Zinn, *J. Phys. Chem.* **97**, 1190 (1993).

203. P. Attard and J. L. Parker, *J. Phys. Chem.* **96**, 5086 (1992).

204. For a recent overview of these interesting but controversial problems, see W. Drost-Hansen and J. L. Singleton, *Fundamentals of Medical Cell Biology* (JAI Press, Inc.), **3A**, 157 (1992).

205. C. F. Polnaszek and R. G. Bryant, *J. Am. Chem. Soc.* **106**, 428 (1984).

206. C. F. Polnaszek and R. G. Bryant, *J. Chem. Phys.* **81**, 4038 (1984).

207. F. Caruso, F. Grieser, A. Murphy, P. Thistlethwaite, R. Urquhart, M. Almgren, and E. Wistus, *J. Am. Chem. Soc.* **113**, 4838 (1991).

208. M. Matsumoto and Y. Kataoka, *J. Chem. Phys.* **88**, 3233 (1988).

209. B. Yang, D. E. Sullivan, B. Tjipto-Margo, and C. G. Gray, *J. Phys. Condens. Matter* **3**, F109 (1991).

210. A. Braslau, M. Deutsch, P. S. Pershan, and A. H. Weiss, *Phys. Rev. Lett.* **54**, 114 (1985).

211. A. Braslau, P. S. Pershan, G. Swislow, B. M. Ocko, and J. Als-Nielsen, *Phys. Rev. A* **38**, 2457 (1988).

212. M. C. Goh, J. M. Hicks, K. Kemnitz, G. R. Pinto, K. Bhattacharyya, K. B. Eisenthal, and T. F. Heinz, *J. Phys. Chem.* **92**, 5074 (1988).

213. Q. Du, R. Superfine, E. Freysz, and Y. R. Shen, *Phys. Rev. Lett.* **70**, 2313 (1993).

214. C. A. Sacchi, *J. Opt. Soc. Am. B* **8**, 337 (1991); F. Docchio, A. Avigo, and R. Palumbo, *Europhys. Lett.* **15**, 69 (1991).

215. Y. Gauduel, S. Pommeret, A. Migus, N. Yamada, and A. Antonetti, *J. Opt. Soc. Am. B* **7**, 1528 (1990).

216. C. Pépin, D. Houde, H. Remita, T. Goulet, and J.-P. Jay-Gerin, *Phys. Rev. Lett.* **69**, 3389 (1992).

217. F. Docchio, *Nuovo Cimento D* **13**, 87 (1991).

218. M. Szklarczyk, R. C. Kainthla, and J. O'M. Bockris, *J. Electrochem. Soc.* **136**, 2512 (1989).

219. R. J. Speedy, *J. Phys. Chem.* **96**, 2322 (1992), and references cited.

220. M. G. Sceats and S. A. Rice, in F. Franks (Ed.), *Water: A Comprehensive Treatise* Vol. 7, Plenum, New York, 1982. This 387-page review contains 706 references up to the early 1980s. It remains a very useful review.

221. P. G. Kusalik, *Mol. Phys.*, **67**, 67 (1989).

222. S. A. Rice, W. G. Madden, R. McGraw, M. G. Sceats, and M. S. Bergen, *J. Glaciol.* **21**, 509 (1978).

223. M. G. Sceats, M. Stavola, and S. A. Rice, *J. Chem. Phys.* **70**, 3927 (1979).

224. A. C. Belch and S. A. Rice, *J. Chem. Phys.* **86**, 5676 (1987).

225. G. Némethy and H. A. Scheraga, *J. Chem. Phys.* **36**, 3382 (1962).

226. G. Némethy and H. A. Scheraga, *J. Chem. Phys.* **41**, 680 (1964).

227. L. Bosio, J. Teixeira and M.-C. Bellissent-Funel, *Phys. Rev. A* **39**, 6612 (1989).

228. H. Tanaka and I. Ohmine, *J. Chem. Phys.* **87**, 6128 (1987).

229. J. Quintana, D. Henderson, and A. D. J. Haymet, *J. Chem. Phys.* **98**, 1486 (1993).

230. D. R. Bérard and G. N. Patey, *J. Chem. Phys.* **95**, 5281 (1991).

231. D. R. Bérard and G. N. Patey, *J. Chem. Phys.* **97**, 4372 (1992).

232. P. Debye and E. Hückel, *Z. Phys.* **24**, 195 (1923).

233. L. Onsager, *J. Am. Chem. Soc.* **58**, 1486 (1936).

234. B. Honig, K. Sharp, and A.-S. Yang, *J. Phys. Chem.* **97**, 1101 (1993).

235. D. F. Calef and P. G. Wolynes, *J. Phys. Chem.* **87**, 3387 (1983).

236. B. Bagchi, *Annu. Rev. Phys. Chem.* **40**, 115 (1989).

237. E. W. Castner, Jr., G. R. Fleming, B. Bagchi, and M. Maroncelli, *J. Chem. Phys.* **89**, 3519 (1988).

238. B. Bagchi and A. Chandra, *J. Chem. Phys.* **90**, 7338 (1989).

239. D. Wei and G. N. Patey, *J. Chem. Phys.* **91**, 7113 (1989).

240. D. Wei and G. N. Patey, *J. Chem. Phys.* **93**, 1399 (1990).

241. D. Wei and G. N. Patey, *J. Chem. Phys.* **94**, 6795 (1991).

242. W. C. Kerr, *Phys. Rev.* **174**, 316 (1968).

243. P. Debye, *Polar Molecules*, Dover, New York, 1929.

244. M. Neumann, *Mol. Phys.* **50**, 841 (1983).

245. M. Neumann, *J. Chem. Phys.* **82**, 5663 (1985).

246. M. Neumann, *J. Chem. Phys.* **85**, 1567 (1986).

247. M. Neumann and O. Steinhauser, *Chem. Phys. Lett.* **95**, 417, (1983).

248. M. Neumann and O. Steinhauser, *Chem. Phys. Lett.* **102**, 508 (1983).

249. M. Neumann, O. Steinhauser, and G. S. Pawley, *Mol. Phys.* **52**, 97 (1984).

250. D. A. McQuarrie, *Statistical Mechanics*, Harper & Row, New York, 1976.

251. J. T. Hynes, in M. Baer (Ed.), *The Theory of Chemical Reaction Dynamics*, CRC, Boca Raton, FL, 1985.

252. W. G. McMillan and J. E. Mayer, *J. Chem. Phys.* **13**, 276 (1945).

253. H. Friedman, *A Course in Statistical Mechanics*, Prentice-Hall, New York, 1985.

254. P. G. Kusalik and G. N. Patey, *J. Chem. Phys.* **89**, 7478 (1988).

255. B. E. Conway, *Ionic Hydration in Chemistry and Biophysics*, Elsevier, Amsterdam, 1981.

256. J. K. Buckner and W. L. Jorgensen, *J. Am. Chem. Soc.* **111**, 2507 (1989).

257. J. Trullàs, A. Giró, and J. A. Padró, *J. Chem. Phys.* **91**, 539 (1989).

258. D. Chandler and H. C. Andersen, *J. Chem. Phys.* **57**, 1930 (1972).

259. L. J. Lowden and D. Chandler, *J. Chem. Phys.* **59**, 6587 (1973).

260. B. M. Pettitt and P. J. Rossky, *J. Chem. Phys.* **77**, 1451 (1982).

261. B. M. Pettitt and P. J. Rossky, *J. Chem. Phys.* **78**, 7296 (1983).

262. B. M. Pettitt and P. J. Rossky, *J. Chem. Phys.* **84**, 5836 (1986).

263. R. A. Chiles and P. J. Rossky, *J. Am. Chem. Soc.* **106**, 6867 (1984).

264. B. M. Pettitt and M. Karplus, *J. Chem. Phys.* **83**, 781 (1985).

265. P. G. Kusalik and G. N. Patey, *J. Chem. Phys.* **88**, 7715 (1988).

266. P. G. Kusalik and G. N. Patey, *J. Chem. Phys.* **92**, 1345 (1990).

267. P. G. Kusalik, *Mol. Phys.* **76**, 337 (1992).

268. L. Verlet, *Phys. Rev.* **159**, 98 (1967).

269. N. Metropolis, A. W. Rosenbluth, M. N. Rosenbluth, A. H. Teller, and E. Teller, *J. Chem. Phys.* **21**, 1087 (1953).

270. B. J. Alder and T. E. Wainwright, *J. Chem. Phys.* **27**, 1208 (1957).

271. W. W. Wood and F. R. Parker, *J. Chem. Phys.* **27**, 720 (1957).

272. A. Rahman, *Phys. Rev.* **136**, A405 (1964).

273. G. D. Harp and B. J. Berne, *J. Chem. Phys.* **48**, 1249 (1968).

274. B. J. Alder, *J. Chem. Phys.* **40**, 2724 (1964).

275. K. C. Mo, K. E. Gubbins, G. Jacucci, and I. R. McDonald, *Mol. Phys.* **27**, 1173 (1974).

276. J. K. Lee, J. A. Barker, and G. M. Pound, *J. Chem. Phys.* **60**, 1976 (1974).

277. K. S. Liu, *J. Chem. Phys.* **60**, 4226 (1974).

278. A. C. L. Opitz, *Phys. Lett. A* **47**, 439 (1974).

279. M. J. P. Nijmeijer, A. F. Bakker, C. Bruin, and J. H. Sikkenk, *J. Chem. Phys.* **89**, 3789 (1988).

280. W. E. Alley and B. J. Alder, *J. Chem. Phys.* **66**, 2631 (1977).

281. I. K. Snook and D. Henderson, *J. Chem. Phys.* **68**, 2134 (1978).

282. I. K. Snook and W. van Megan, *J. Chem. Phys.* **72**, 2907 (1980).

283. V. Ya. Antonchenko, V. V. Ilyin, N. N. Makovsky, A. N. Pavlov, and V. P. Sokhan, *Mol. Phys.* **52**, 345 (1984).

284. J. J. Magda, M. Tirrell, and H. T. Davis, *J. Chem. Phys.* **83**, 1888 (1985).

285. A. K. MacPherson, Y. P. Carignan, and T. Vladimiroff, *J. Chem. Phys.* **86**, 4228 (1987).

286. S. H. Lee, J. C. Rasaiah, and J. B. Hubbard, *J. Chem. Phys.* **85**, 5232 (1986).

287. M. Schoen, D. J. Diestler, and J. H. Cushman, *J. Chem. Phys.* **87**, 5464 (1987).

288. M. Schoen, J. H. Cushman, D. J. Diestler, and C. L. Rhykerd, Jr., *J. Chem. Phys.* **88**, 1394 (1987).

289. S.-B. Zhu, J. Lee, and G. W. Robinson, *Mol. Phys.* **67**, 321 (1989).

290. S.-B. Zhu and G. W. Robinson, *Chem. Phys.* **134**, 1 (1989).

291. S.-B. Zhu, J.-B. Zhu, and G. W. Robinson, *Mol. Phys.* **68**, 1321 (1989).

292. S.-B. Zhu and G. W. Robinson, *J. Chem. Phys.* **90**, 7127 (1989).

293. M. W. Evans, *Adv. Chem. Phys.* **62**, 183 (1985).

294. S.-B. Zhu, J. Lee, and G. W. Robinson, *Phys. Rev. A* **38**, 5810 (1988).

295. S.-B. Zhu, J. Lee, and G. W. Robinson, *J. Opt. Soc. Am. B* **6**, 250 (1989).

296. O. A. Karim and A. D. J. Haymet, *J. Chem. Phys.* **89**, 6889 (1988).

297. B. B. Laird and A. D. J. Haymet, *Chem. Rev.* **92**, 1819 (1992).

298. R. M. Townsend, J. Gryko, and S. A. Rice, *J. Chem. Phys.* **82**, 4391 (1985).

299. R. M. Townsend and S. A. Rice, *J. Chem. Phys.* **94**, 2207 (1991).

300. N. I. Christou, J. S. Whitehouse, D. Nicholson, and N. G. Parsonage, *Mol. Phys.* **55**, 397 (1985).

301. M. A. Wilson, A. Pohorille, and L. R. Pratt, *J. Chem. Phys.* **91**, 4873 (1987).

302. M. A. Wilson, A. Pohorille, and L. R. Pratt, *J. Chem. Phys.* **88**, 3281 (1988).

303. M. A. Wilson, A. Pohorille, and L. R. Pratt, *J. Chem. Phys.* **90**, 5211 (1989).

304. S.-B. Zhu, T. G. Fillingim, and G. W. Robinson, *J. Phys. Chem.* **95**, 1002 (1991).

305. K. A. Motakabbir and M. L. Berkowitz, *Chem. Phys. Lett.* **176**, 61 (1991).

306. M. A. Wilson, A. Pohorille, and L. R. Pratt, *Chem. Phys.* **129**, 209 (1989).

307. M. A. Wilson and A. Pohorille, *J. Chem. Phys.* **95**, 6005 (1991).

308. I. Benjamin, *J. Chem. Phys.* **95**, 3698 (1991).

309. D. A. Rose and I. Benjamin, *J. Chem. Phys.* **95**, 6856 (1991).

310. J. Seitz-Beywl, M. Poxleitner, and K. Heinzinger, *Z. Naturforsch. Teil A* **46**, 876 (1991).

311. J. N. Glosli and M. R. Philpott, *J. Chem. Phys.* **96**, 6962 (1992).

312. R. O. Watts, E. Clementi, and J. Fromm, *J. Chem. Phys.* **61**, 2550 (1974).

313. J. Fromm, E. Clementi and R. O. Watts, *J. Chem. Phys.* **62**, 1388 (1975).

314. E. Clementi and R. Barsotti, *Chem. Phys. Lett.* **59**, 21 (1978).

315. R. W. Impey, P. A. Madden, and I. R. McDonald, *J. Phys. Chem.* **87**, 5071 (1983).

316. J. Chandrasekhar, D. C. Spellmeyer, and W. L. Jorgensen, *J. Am. Chem. Soc.* **106**, 903 (1984).

317. K. Heinzinger, *Pure Appl. Chem.* **57**, 1031 (1985).

318. P. Bopp, *Pure Appl. Chem.* **59**, 1071 (1987).

319. K. Heinzinger, in C. R. A. Catlow, S. C. Parker and M. P. Allen (Eds.), *Computer Modelling of Fluids, Polymers and Solids*, Kluwer Academic, Dordrecht, 1990.

320. S.-B. Zhu and G. W. Robinson, *J. Chem. Phys.* **97**, 4336 (1992).

321. M. Maroncelli and G. R. Fleming, *J. Chem. Phys.* **89**, 5044 (1988).

322. M. Marchesi, *Chem. Phys. Lett.* **97**, 224 (1983).

323. C. Y. Lee, J. A. McCammon, and P. J. Rossky, *J. Chem. Phys.* **80**, 4448 (1984).

324. P. J. Rossky and S. H. Lee, *Chem. Scr. A* **29**, 93 (1989).

325. E. Spohr and K. Heinzinger, *Chem. Phys. Lett.* **123**, 218 (1986).

326. E. Spohr and K. Heinzinger, *Electrochim. Acta* **33**, 1211 (1988).

327. N. N. Makovsky, *Mol. Phys.* **72**, 235 (1991).

328. C. F. Wong and J. A. McCammon, *Isr. J. Chem.* **27**, 211 (1986).

329. C. F. Wong and J. A. McCammon, *J. Am. Chem. Soc.* **108**, 3830 (1986).

330. P. Ahlström, O. Teleman, B. Jönsson, and S. Forsen, *J. Am. Chem. Soc.* **109**, 1541 (1987).

331. P. Ahlström, O. Teleman, and B. Jönsson, *Chem. Scr. A* **29**, 97 (1989).

332. N. G. Parsonage and D. Nicholson, *J. Chem. Soc., Faraday Trans. 2* **82**, 1521 (1986).

333. N. G. Parsonage and D. Nicholson, *J. Chem. Soc., Faraday Trans. 2* **83**, 663 (1987).

334. E. Spohr, *J. Phys. Chem.* **93**, 6171 (1989).

335. E. Spohr, *Chem. Phys.* **141**, 87 (1990).

336. K. Heinzinger, *Electrochem. Acta* **34**, 1849 (1989).

337. G. Nagy and K. Heinzinger, *J. Electroanal. Chem.* **296**, 549 (1990).

338. G. Nagy and K. Heinzinger, *J. Electroanal. Chem.* **327**, 25 (1992).

339. K. Foster, K. Raghavan, and M. Berkowitz, *Chem. Phys. Lett.* **162**, 32 (1989).

340. K. Raghavan, K. Foster, and M. Berkowitz, *Chem. Phys. Lett.* **177**, 426 (1991).

341. K. Raghavan, K. Foster, K. Motakabbir, and M. Berkowitz, *J. Chem. Phys.* **94**, 2110 (1991).

342. S.-B. Zhu and G. W. Robinson, *J. Chem. Phys.* **94**, 1403 (1991).

343. T. Matsui and W. L. Jorgensen, *J. Am. Chem. Soc.* **114**, 3220 (1992).

344. M. W. Evans, G. C. Lie, and E. Clementi, *J. Chem. Phys.* **87**, 6040 (1987).

345. E. Clementi, W. Kolos, C. G. Lie, and G. Ranghino, *Int. J. Quant. Chem.* **17**, 377 (1980).

346. J. E. H. Koehler, W. Saenger, and B. Lesyng, *J. Comp. Chem.* **8**, 1090 (1987).

347. K. Hermansson, *J. Chem. Phys.* **89**, 2149 (1988).

348. G. Chalasinski, M. M. Szczesniak, P. Cieplak, and S. Scheiner, *J. Chem. Phys.* **94**, 2873 (1991).

349. D. van Belle, I. Couplet, M. Prevost, and S. J. Wodak, *J. Mol. Biol.* **198**, 721 (1987).

350. E. Guàrdia and J. A. Padró, *J. Phys. Chem.* **94**, 6049 (1990).

351. S.-B. Zhu and C. F. Wong, *J. Chem. Phys.* **98**, 8892 (1993).

352. M. M. Probst and K. Hermansson, *J. Chem. Phys.* **96**, 8995 (1992).

353. K. Hermansson and L. Ojamäe, *Int. J. Quant. Chem.* **42**, 1251 (1992).

354. L. Ojamäe and K. Hermansson, *J. Chem. Phys.* **96**, 9035 (1992).

355. L. Ojamäe and K. Hermansson, *Chem. Phys.* **161**, 87 (1992).

356. J. G. Kirkwood, *J. Chem. Phys.* **7**, 911 (1939).

357. R. K. Pathria, *Statistical Mechanics*, Pergamon, Oxford, UK, 1972, Chapter 12.

358. R. H. Cole, *Annu. Rev. Phys. Chem.* **40**, 1 (1989).

359. M. Belhadj, H. E. Alper, and R. M. Levy, *Chem. Phys. Lett.* **179**, 13 (1991).

360. R. O. Watts, *Chem. Phys.* **57**, 185 (1981).

361. P. G. Kusalik, *J. Chem. Phys.* **93**, 3520 (1990).

362. P. G. Kusalik, *Mol. Phys.* **73**, 1349 (1991).

363. J. Hautman and M. L. Klein, *Mol. Phys.* **75**, 379 (1992).

364. J. E. Del Bene and J. A. Pople, *J. Chem. Phys.* **58**, 3605 (1973).

365. B. J. Smith, D. J. Swanton, J. A. Pople, and H. F. Schaefer III, *J. Chem. Phys.* **92**, 1240 (1990).

366. F. F. Muguet, M.-P. Bassez-Muguet, and G. W. Robinson, *Int. J. Quantum. Chem.* **39**, 449 (1991).

367. F. F. Muguet, Thesis, Texas Tech University, Lubbock, 1992.

368. H. Kistenmacher, G. C. Lie, H. Popkie, and E. Clementi, *J. Chem. Phys.* **62**, 546 (1974).

369. E. Clementi and B. Habitz, *J. Chem. Phys.* **87**, 2815 (1983).

370. V. Carravetta and E. Clementi, *J. Chem. Phys.* **81**, 2646 (1984).

371. F. London, *Trans. Faraday Soc.* **33**, 8 (1937).

372. P. Barnes, in C. A. Croxton (Ed.), *Progress in Liquid Physics*, Wiley-Interscience, New York, 1978, Chapter 9.

373. P. Barnes, J. L. Finney, J. D. Nicholas, and J. E. Quinn, *Nature* **282**, 459 (1979).

374. B. J. Gellatly, J. E. Quinn, P. Barnes, and J. L. Finney, *Mol. Phys.* **50**, 949 (1983).

375. P. M. Morse, *Phys. Rev.* **34**, 57 (1929).

376. S.-B. Zhu and G. W. Robinson, *Z. Naturforsch. Teil A* **46**, 221 (1990).

377. P. K. L. Drude, *The Theory of Optics*, Longmans, Green, London, 1933.

378. J. Applequist, J. R. Carl and K.-K. Fung, *J. Am. Chem. Soc.* **94**, 2952 (1972).

379. J. C. Sauniere, T. P. Lybrand, J. A. McCammon, and L. D. Pyle, *Comput. Chem.* **13**, 313 (1989).

380. L. X. Dang and B. M. Pettitt, *J. Phys. Chem.* **91**, 3349 (1987).

381. H. C. Urey and C. A. Bradley, *Phys. Rev.* **38**, 1969 (1931).

382. A. Rahman and F. H. Stillinger, *J. Am. Chem. Soc.* **95**, 7943 (1973).

383. L. Ojamäe, K. Hermansson, and M. Probst, *Chem. Phys. Lett.* **191**, 500 (1992).

384. M. R. Reddy and M. Berkowitz, *Chem. Phys. Lett.* **155**, 173 (1989).

385. U. Dinur, *J. Phys. Chem.* **94**, 5669 (1990).

386. K. Krynicki, C. D. Green, and D. W. Sawyer, *Faraday Discuss. Chem. Soc.* **66**, 199 (1978).

387. G. D. Carney, L. A. Curtis, and S. R. Langhoff, *J. Mol. Spectrosc.* **61**, 371 (1976).

388. J. Teixeira, M.-C. Bellissent-Funel, S.-H. Chen, and A. J. Dianoux, *Phys. Rev. A* **31**, 1913 (1985).

389. U. Kaatze and V. Uhlendorf, *Z. Phys. Chem. N.F.* **126**, 151 (1981).

390. M. N. Afsar and J. B. Hasted, *J. Opt. Soc. Am.* **67**, 902 (1977).

391. J. Jonas, T. DeFries, and D. J. Wilber, *J. Chem. Phys.* **65**, 582 (1976).

392. H. G. Hertz, in F. Franks (Ed.), *Water: A Comprehensive Treatise*, Vol. 3, Plenum, New York, 1973, Chapter 7. It is the correlation time of the first-order spherical harmonics on p. 358 of this reference that we associate with the Debye relaxation time τ_D.

393. R. Pottel, in F. Franks (Ed.), *Water: A Comprehensive Treatise*, Vol. 3, Plenum, New York, 1973, Chapter 8.

394. J. B. Hasted, *Aqueous Dielectrics*, Chapman & Hall, London, 1973.

395. D. Bertolini, A. Tani, and R. Vallauri, *Mol. Phys.* **73**, 69 (1991).

396. A. Geiger, A. Rahman, and F. H. Stillinger, *J. Chem. Phys.* **70**, 263 (1979).

397. C. Pangali, M. Rao, and B. J. Berne, *J. Chem. Phys.* **71**, 2982 (1979).

398. G. Ravishanker, M. Mezei, and D. L. Beveridge, *Faraday Symp. Chem. Soc.* **17**, 79 (1982).

399. D. C. Rapaport and H. A. Scheraga, *J. Phys. Chem.* **86**, 873 (1982).

400. K. Watanabe and H. C. Andersen, *J. Phys. Chem.* **90**, 795 (1986).

401. A. Wallqvist and B. J. Berne, *Chem. Phys. Lett.* **145**, 26 (1988).

402. W. L. Jorgensen, J. K. Buckner, S. Boudon, and J. Tirado-Rives, *J. Chem. Phys.* **89**, 3742 (1988).

403. B. Guillot, Y. Guissani, and S. Bratos, *J. Chem. Phys.* **95**, 3643 (1991).

404. D. E. Smith, L. Zhang, and A. D. J. Haymet, *J. Am. Chem. Soc.* **114**, 5875 (1992).

405. D. E. Smith and A. D. J. Haymet, *J. Chem. Phys.*, **98**, 6445 (1993).

406. A. Wallqvist, *J. Phys. Chem.* **95**, 8921 (1991).

407. D. H. Wertz, *J. Am. Chem. Soc.* **102**, 5316 (1980).

408. A. Ben-Naim and Y. Marcus, *J. Chem. Phys.* **81**, 2016 (1984).

409. M. H. Abraham, *J. Am. Chem. Soc.* **104**, 2085 (1982).

410. L. R. Pratt and D. Chandler, *J. Chem. Phys.* **67**, 3683 (1977).

411. L. R. Pratt and D. Chandler, *J. Chem. Phys.* **73**, 3434 (1980).

412. C. L. Kneifel, H. L. Friedman, and M. D. Newton, *Z. Naturforsch. Teil A* **44**, 385 (1989).

413. M. D. Newton and H. L. Friedman, *J. Chem. Phys.* **83**, 5210 (1985).

414. E. Guàrdia and J. A. Padró, *Chem. Phys.* **144**, 353 (1990).

415. J. S. Bader and D. Chandler, *J. Phys. Chem.* **96**, 6423 (1992).

416. P. V. Kumar and B. L. Tembe, *J. Chem. Phys.* **97**, 4356 (1992).

417. H. Kistenmacher, H. Popkie, and E. Clementi, *J. Chem. Phys.* **59**, 5842 (1973).

418. M. Mezei and D. L. Beveridge, *J. Chem. Phys.* **74**, 6902 (1981).

419. P. Kebarle, M. Arshadi, and J. Scarborough, *J. Chem. Phys.* **49**, 817 (1968).

420. S. K. Searles and P. Kebarle, *Can. J. Chem.* **47**, 2619 (1969).

421. M. Arshadi, R. Yamaguchi, and P. Kebarle, *J. Phys. Chem.* **74**, 1475 (1970).

422. R. G. Keesee and A. W. Castleman, Jr., *Chem. Phys. Lett.* **74**, 139 (1980).

423. R. G. Keesee and A. W. Castleman, Jr., *J. Phys. Chem. Ref. Data* **15**, 1011 (1986).

424. S.-S. Sung and P. C. Jordan, *J. Chem. Phys.* **85**, 4045 (1986).

425. S. Lin and P. C. Jordan, *J. Chem. Phys.* **89**, 7492 (1988).

426. B. T. Gowda and S. W. Benson, *J. Phys. Chem.* **86**, 1544 (1982).

427. L. Perera and M. L. Berkowitz, *J. Chem. Phys.* **95**, 1954 (1991).

428. L. Perera and M. L. Berkowitz, *J. Chem. Phys.* **96**, 8288 (1992).

429. P. Cieplak, T. P. Lybrand, and P. A. Kollman, *J. Chem. Phys.* **86**, 6393 (1987).

430. L. X. Dang, J. E. Rice, J. Caldwell, and P. A. Kollman, *J. Am. Chem. Soc.* **113**, 2481 (1991).

431. K. Heinzinger and P. C. Vogel, *Z. Naturforsch. Teil A* **29**, 1164 (1974).

432. P. C. Vogel and K. Heinzinger, *Z. Naturforsch. Teil A* **30**, 789 (1975).

433. K. Heinzinger and P. C. Vogel, *Z. Naturforsch. Teil A* **31**, 463 (1976).

434. P. C. Vogel and K. Heinzinger, *Z. Naturforsch. Teil A* **31**, 476 (1976).

435. G. Pálinkás, W. O. Riede and K. Heinzinger, *Z. Naturforsch. Teil A* **32**, 1137 (1977).

436. Gy. I. Szász and K. Heinzinger, *Z. Naturforsch. Teil A* **34**, 840 (1979).

437. T. Radnai, G. Pálinkás, Gy. I. Szász, and K. Heinzinger, *Z. Naturforsch. Teil A* **36**, 1076 (1981).

438. W. Dietz, W. O. Riede, and K. Heinzinger, *Z. Naturforsch. Teil A* **37**, 1038 (1982).

439. Gy. I. Szász and K. Heinzinger, *Z. Naturforsch. Teil A* **38**, 214 (1983).

440. M. M. Probst, T. Radnai, K. Heinzinger, P. Bopp, and B. M. Rode, *J. Phys. Chem.* **89**, 753 (1985).

441. K. Heinzinger, *Physica B* **131**, 196 (1985).

442. M. Migliore, S. L. Fornili, E. Spohr, G. Pálinkás, and K. Heinzinger, *Z. Naturforsch. Teil A* **41**, 826 (1986).

443. K. Tanaka, N. Ogita, Y. Tamura, I. Okada, H. Ohtaki, G. Pálinkás, E. Spohr, and K. Heinzinger, *Z. Naturforsch. Teil A* **42**, 29 (1987).

444. E. Spohr, G. Pálinkás, K. Heinzinger, P. Bopp, and M. M. Probst, *J. Phys. Chem.* **92**, 6754 (1988).

445. M. M. Probst, E. Spohr, and K. Heinzinger, *Chem. Phys. Lett.* **161**, 405 (1989).

446. O. A. Karim, *Chem. Phys. Lett.* **184**, 560 (1991).

447. L. X. Dang and B. M. Pettitt, *J. Chem. Phys.* **86**, 6560 (1987).

448. L. X. Dang and B. M. Pettitt, *J. Am. Chem. Soc.* **109**, 5531 (1987).

449. L. X. Dang and B. M. Pettitt, *J. Phys. Chem.* **94**, 4303 (1990).

450. E. Guàrdia, R. Rey, and J. A. Padró, *J. Chem. Phys.* **95**, 2823 (1991).

451. E. Guàrdia, R. Rey, and J. A. Padró, *Chem. Phys.* **155**, 187 (1991).

452. O. A. Karim, Private communication.

453. H. L. Friedman, F. O. Rainieri, and H. Xu, *Pure Appl. Chem.* **63**, 1347 (1991).

454. H. Xu and H. L. Friedman, *J. Solution Chem.* **19**, 1155 (1990).

455. A. P. Copestake, G. W. Neilson, and J. E. Enderby, *J. Phys. C* **18**, 4211 (1985).

456. G. W. Neilson and J. E. Enderby, *Proc. R. Soc. London Ser. A* **390**, 353 (1983).

457. S. Ono and S. Kondo, in S. Flugge (Ed.), *Encyclopedia of Physics*, Vol. 10, Springer, Berlin, 1960.

458. J. Frenkel, *Kinetic Theory of Liquids*, Dover, New York, 1955.

459. N. H. Fletcher, *Philos. Mag.* **7**, 255 (1962).

460. F. H. Stillinger and A. Ben-Naim, *J. Chem. Phys.* **47**, 4431 (1967).

461. C. A. Croxton, *Physica A* **106**, 239 (1981).

462. G. Aloisi, R. Guidelli, R. A. Jackson, and P. Barnes, *Chem. Phys. Lett.* **133**, 343 (1987).

463. E. N. Brodskaya and A. I. Rusanov, *Mol. Phys.* **62**, 251 (1987).

464. A. Pohorille and I. Benjamin, *J. Chem. Phys.* **94**, 5599 (1991).

465. J. S. Rowlinson and B. Widom, *Molecular Theory of Capillarity*, Clarendon, Oxford, UK, 1982.

466. M. J. P. Nijmeijer and J. M. J. van Leeuwen, *J. Phys. A: Math. Gen.* **23**, 4211 (1990).

467. M. J. P. Nijmeijer, C. Bruin, A. F. Bakker, and J. M. J. van Leeuwen, *Phys. Rev. A* **42**, 6052 (1990).

468. M. J. P. Nijmeijer, C. Bruin, A. F. Bakker, and J. M. J. van Leeuwen, *Mol. Phys.* **72**, 927 (1991).

469. M. J. P. Nijmeijer, C. Bruin, A. B. van Woerkom, A. F. Bakker, and J. M. J. van Leeuwen, *J. Chem. Phys.* **96**, 565 (1992).

470. R. C. Tolman *J. Chem. Phys.* **17**, 118 (1949).

471. R. C. Tolman, *J. Chem. Phys.* **17**, 333 (1949).

472. F. P. Buff, *J. Chem. Phys.* **23**, 419 (1955).

473. W. A. Steele, *The Interaction of Gases with Solid Surfaces*, Pergamon, Oxford, UK, 1974.

474. F. F. Abraham, *J. Chem. Phys.*, **68**, 3713 (1978).

475. V. Vlachy and A. D. J. Haymet, *Chem. Phys. Lett.* **146**, 32 (1988).

476. B. Jönsson, *Chem. Phys. Lett.* **82**, 520 (1981).

477. B. Rowland, M. Fisher, and J. P. Devlin, *J. Chem. Phys.* **95**, 1378 (1991).

478. A. Wallqvist, *Chem. Phys. Lett.* **165**, 437 (1990).

479. A. Wallqvist, P. Ahlström, and G. Kalström, *J. Phys. Chem.* **94**, 1649 (1990).

480. E. Spohr and K. Heinzinger, *J. Chem. Phys.* **84**, 2304 (1986).

481. J. Q. Broughton and G. H. Gilmer, *J. Phys. Chem.* **91**, 6347 (1987).

482. P. Linse, *J. Chem. Phys.* **86**, 4177 (1987).

483. *Handbook of Chemistry and Physics*, 71st ed., D. R. Lide (Ed.), CRC Press, Boca Raton, FL, 1990, pp. 6–8 and 6–104.

484. G. Antonoff, *J. Chim. Phys.* **5**, 372 (1907).

485. I. L. Carpenter and W. J. Hehre, *J. Phys. Chem.* **94**, 531 (1990).

486. P. A. Thiel and T. E. Madey, *Surf. Sci. Rept.* **7**, 211 (1987).

487. H. Ibach and S. Lehwald, *Surf. Sci.* **91**, 187 (1980).

488. B. A. Sexton, *Surf. Sci.* **94**, 435 (1980).

489. G. B. Fisher and J. L. Gland, *Surf. Sci.* **94**, 446 (1980).

490. E. Langenbach, A. Spitzer, and H. Lüth, *Surf. Sci.* **147**, 179 (1984).

491. J. Fusy and R. Ducros, *Surf. Sci.* **176**, 157 (1986).

492. F. T. Wagner and T. E. Moylan, *Surf. Sci.* **191**, 121 (1987).

493. J. P. Valleau and A. A. Gardner, *J. Chem. Phys.* **86**, 4162 (1987).

494. A. A. Gardner and J. P. Valleau, *J. Chem. Phys.* **86**, 4171 (1987).

495. D. E. Peebles and J. M. White, *Surf. Sci.* **144**, 512 (1984).

496. C. Nöbl and C. Benndorf, *Surf. Sci.* **182**, 499 (1987).

497. K. Bange, T. E. Madey, J. K. Sass, and E. M. Stuve, *Surf. Sci.* **183**, 334 (1987).

498. S. Holloway and K. H. Bennemann, *Surf. Sci.* **101**, 327 (1980).

499. M. W. Ribarsky, W. D. Luedtke, and U. Landman, *Phys. Rev. B* **32**, 1430 (1985).

500. C. W. Bauschlicher, Jr., *J. Chem. Phys.* **83**, 3129 (1985).

501. K. Heinzinger, *Pure Appl. Chem.* **63**, 1733 (1991).

502. R. P. Messmer, in T. N. Rhoden and G. Ertl (Eds.), *The Nature of the Surface Chemical Bond*, North-Holland, Amsterdam, 1979, p. 51.

503. G. Pacchioni, P. S. Bagus, M. R. Philpott, and C. J. Nelin, *Int. J. Quantum Chem.* **38**, 675 (1990).

504. E. Spohr and K. Heinzinger, *Ber. Bun. Phys. Chem.* **92**, 1358 (1988).

505. J. Seitz-Beywl, M. Poxleitner, M. M. Probst, and K. Heinzinger, *Int. J. Quant. Chem.* **42**, 1141 (1992).

506. W. F. van Gunsteren and H. J. C. Berendsen, *J. Mol. Biol.* **176**, 559 (1984); W. F. van Gunsteren and M. Karplus, *Biochemistry* **21**, 2259 (1982).

507. I. Ghosh and J. A. McCammon, *J. Phys. Chem.* **91**, 4878 (1987).

508. M. Levitt and R. Sharon, *Proc. Natl. Acad. Sci. USA* **85**, 7557 (1988).

509. P. Ahlström, O. Teleman, and B. Jönsson, *J. Am. Chem. Soc.* **110**, 4198 (1988).

510. R. Pethig, *Annu. Rev. Phys. Chem.* **43**, 177 (1992).

511. S. Swaminathan, T. Ichiye, W. van Gunsteren, and M. Karplus, *Biochemistry* **21**, 5230 (1982).

512. W. F. van Gunsteren, H. J. C. Berendsen, J. Hermans, W. G. J. Hol, and J. P. M. Postma, *Proc. Natl. Acad. Sci. USA* **80**, 4315 (1983).

513. G. Otting, E. Liepinsh, and K. Wüthrich, *Science* **254**, 974 (1991).

514. P. A. Bash, U. C. Singh, F. K. Brown, R. Langridge, and P. A. Kollman, *Science* **235**, 574 (1987).

515. G. King, F. S. Lee, and A. Warshel, *J. Chem. Phys.* **95**, 4366 (1991).

516. M. R. Reddy and M. Berkowitz, *Proc. Natl. Acad. Sci. USA* **86**, 3165 (1989).

517. M. R. Reddy, K. Foster, and M. Berkowitz, *J. Mol. Liq.* **41**, 181 (1989).

518. E. H. S. Anwander, M. M. Probst, and B. M. Rode, *Biopolymers* **29**, 757 (1990).

519. A. Maitra, *J. Phys. Chem.* **88**, 5122 (1984).

520. S.-H. Chen and R. Rajagopalan (Eds.), *Micellar Solutions and Microemulsions*, Springer, New York, 1990.

521. O. Teleman and P. Ahlström, *J. Am. Chem. Soc.* **108**, 4333 (1986).

522. J. Tirado-Rives and W. L. Jorgensen, *J. Am. Chem. Soc.* **112**, 2773 (1990).

523. S. Marčelja and N. Radić, *Chem. Phys. Lett.* **42**, 129 (1976).

524. J. N. Israelachvili and H. Wennerström. *J. Phys. Chem.* **96**, 520 (1992).

525. R. Kjellander and S. Marčelja, *Chem. Scripta* **25**, 73 (1985).

526. R. Kjellander and S. Marčelja, *Chem. Phys. Lett.* **120**, 393 (1985).

527. M. L. Berkowitz and K. Raghavan, *Langmuir* **7**, 1042 (1991).

528. D. W. R. Gruen and S. Marčelja, *J. Chem. Soc., Faraday Trans. 2* **79**, 211 (1983).

529. D. W. R. Gruen and S. Marčelja, *J. Chem. Soc., Faraday Trans. 2* **79**, 225 (1983).

530. D. M. LeNeveu, R. P. Rand, and V. A. Parsegian, *Nature* **259**, 601 (1976).

531. K. Raghavan, M. R. Reddy, and M. L. Berkowitz, *Langmuir* **8**, 233 (1992).

532. K. Watanabe, M. Ferrario, and M. L. Klein, *J. Phys. Chem.* **92**, 819 (1988).

533. K. Watanabe and M. L. Klein, *J. Phys. Chem.* **93**, 6897 (1989).

534. P. van der Ploeg and H. J. C. Berendsen, *J. Chem. Phys.* **76**, 3271 (1982).

535. J. M. Haile and J. P. O'Connell, *J. Phys. Chem.* **88**, 6363 (1984).

536. M. C. Woods, J. M. Haile, and J. P. O'Connell, *J. Phys. Chem.* **90**, 1875 (1986).

537. S. Karaborni and J. P. O'Connell, *J. Phys. Chem.* **94**, 2624 (1990).

538. S. W. Haan and L. R. Pratt, *Chem. Phys. Lett.* **79**, 436 (1981); erratum, *Chem. Phys. Lett.* **81**, 386 (1981).

539. D. Brown and J. H. R. Clarke, *J. Phys. Chem.* **92**, 2881 (1988).

540. R. H. Fowler, *Statistical Mechanics*, 2d ed., Cambridge University Press, Cambridge, UK, 1966; see Section 12.2.

541. See Ref. 357, Chapter 9.

542. C. Kittel, *Quantum Theory of Solids*, Wiley, New York, 1963.

543. *The International Encyclopedia of Physical Chemistry and Chemical Physics*, E. A. Guggenheim, J. E. Mayer, and F. C. Tompkins (Eds.), Vol. 1, *Lattice Theories of the Liquid State*, Macmillan, New York, 1963.

544. A. Isihara, *Statistical Physics*, Academic, New York, 1971, Chapter 6.

545. H. E. Stanley, *Introduction to Phase Transitions and Critical Phenomena*, Oxford University Press, Oxford, UK, 1971.

546. See Ref. 544, Section 6.7.

547. M. Lax, *J. Chem. Phys.* **20**, 1351 (1952).

OPTICAL NONLINEARITY AND ATOMIC COHERENCES

WOJCIECH GAWLIK

Instytut Fizyki, Uniwersytet Jagielloński, Kraków, Poland

CONTENTS

I. INTRODUCTION

Light-induced polarization is the quantity that reflects the reaction of atomic/molecular media to electromagnetic fields. The bulk polarization **P** can easily be evaluated as an average of individual atomic polarizations **p** over the whole medium. For an atom possessing some structure of the lower state (sublevels $|\mu\rangle$) and some of the upper state (sublevels $|m\rangle$) the

Modern Nonlinear Optics, Part 3, Edited by Myron Evans and Stanisław Kielich. Advances in Chemical Physics Series, Vol. LXXXV.
ISBN 0-471-30499-9 © 1994 John Wiley & Sons, Inc.

light-induced polarization **p** is given as an expectation value of an electric dipole operator **d** in the state described by density matrix ρ:

$$\mathbf{p} = \mathrm{Tr}(\mathbf{d}\rho) = \sum_{\mu,m} d_{\mu m}\rho_{m\mu} + \mathrm{c.c.} \tag{1}$$

On the other hand, the induced polarization is related to the electric field **E** of the light wave by the polarizability α:

$$\mathbf{p} = \alpha\mathbf{E} \tag{2}$$

In general α is a tensor. Its specific, analytical form can be derived by comparing expressions (1) and (2) and solving the equations describing the evolution of the atomic density matrix under given conditions. The optical properties of a material are usually described by its refractive index n and absorption coefficient κ, which are, respectively, the real and imaginary parts of a complex refractive index. The complex refractive index is related to the susceptibility χ of the medium by the well-known relation

$$\eta = n + i\kappa \propto 1 + 2\pi\chi = 1 + 2\pi N\alpha \tag{3}$$

N being the atomic density.

Knowing the susceptibility of the medium, one can readily describe such fundamental effects of the atomic interaction with light as absorption and dispersion. In particular, modifications of n and κ due to nonlinearities in the interaction with intense light beams can be related to the density matrix of an interacting system on the basis of Eqs. (1)–(3).

II. ATOMIC COHERENCE

The density matrix used for our description of the atom–field interaction has a very simple physical interpretation: In a given reference frame, determined by the eigenstates of a given observable, its diagonal elements represent the populations of these states while the nondiagonal elements are the measure of their phase correlations, i.e., their coherence. For the simplest, two-level system the density matrix has two diagonal, real elements—populations ρ_{gg} and ρ_{ee} of the lower (g) and upper (e) levels and two complex conjugate, nondiagonal elements ($\rho_{eg} = \rho_{ge}^*$) representing the so-called *optical coherences*, which, according to Eq. (1), are related to an optical dipole or electrical polarization induced by the interaction.

If the atomic system under consideration has a more complex energy-level structure then its density matrix contains correspondingly more diagonal and off-diagonal elements. In addition to the optical coherences which represent phase correlations of levels linked by an optical transition, i.e., levels of opposite parity, other types of coherences are also possible which might occur between levels of the same parity, e.g., Zeeman or hyperfine sublevels. Such coherences are termed *Zeeman* or *hyperfine coherences*, respectively. Since the range of eigenfrequencies of these coherences is typically much lower than the optical frequency range, they are sometimes called *Hertzian coherences*. Coherences among the lower-state sublevels due to Raman scattering are often described in the literature as *Raman coherences*. An excellent monograph on the density matrix formalism by Omont[1] describes in a very detailed way the role of various coherences in the linear and nonlinear interaction processes.

III. ORIGIN OF NONLINEARITY IN SIMPLE SYSTEMS

If the light perturbation is weak the matrix elements $\rho_{\mu m}$ are linearly dependent on **E** and, as will be shown below, χ, n, and κ are material constants which do not depend on E but only on the atomic energy structure and transition probabilities in addition to the atomic density. This is the situation of linear optics, in which the light wave transmitted through the medium has the same frequency and polarization as the incident wave and its intensity is directly proportional to the incident intensity. The situation changes, however, when the intensity of the perturbing wave is so strong that $\rho_{\mu m}$ exhibits a nonlinear dependence on E. In this case χ, n, and κ are no longer constant in E and give rise to a variety of nonlinear phenomena. To determine the optical properties of the system under consideration one has thus to know its density matrix for which the solution of the following equation is required:

$$\dot{\rho} = -\frac{i}{\hbar}[H,\rho] + L(\rho) \tag{4}$$

where H is the Hamiltonian of the system interacting with the field and L describes the system losses. Below we shall see how nonlinearity occurs in the simplest cases of two- and three-level structures.

A. Two-Level Systems

For a two-level atomic system perturbed by a nearly resonant light wave, the mechanism responsible for optical nonlinearity is simply saturation. Eq. (4) can be solved accurately for the two-level case, yielding the steady-state optical coherence and population difference which describe the atomic response close to a single-photon resonance:

$$\rho_{eg} = \beta N \frac{\delta - i\gamma_k}{\delta^2 + \Gamma_B^2} \tag{5}$$

$$\rho_{gg} - \rho_{ee} = N \frac{\delta^2 + \gamma_k^2}{\delta^2 + \Gamma_B^2} \tag{6}$$

where $\beta = \mathbf{E} \cdot \mathbf{D}/2\hbar$ is the measure of the atom–field perturbation known as the Rabi frequency, N is the atomic density, $\delta = \omega_{eg} - \omega_L$, γ_k and Γ_e are the relaxation rates of the coherence ρ_{eg} and upper-state population ρ_{ee}, respectively, while the power-broadened linewidth Γ_B is defined as

$$\Gamma_B^2 = \gamma_k^2 \left(1 + \frac{4\beta^2}{\Gamma_e \gamma_k} \right) \tag{7}$$

For $\beta \ll \Gamma_e \gamma_k$ the intensity dependence of the denominators in Eqs. (5) and (6) is negligible and one obtains the linear-optics case of constant χ, n, κ, whereas the nonlinear regime is obtained if β^2 is comparable to or bigger than $\Gamma_e \gamma_k$. In the latter case, interaction with intense light causes a decrease in the optical response of the medium, i.e., saturation. As seen from Eqs. (5) and (6), saturation depends resonantly on the light frequency. This is the essence of some very important practical applications of nonlinear laser spectroscopy.[2, 3]

The degree of saturation is measured by the dimensionless saturation parameter G. For a genuine (closed) two-level system G is defined as the ratio of the square of the Rabi frequency to the product of the relaxation rates Γ_e and γ_k:

$$G = \frac{2\beta^2}{\Gamma_e \gamma_k} \tag{8}$$

But if the two-level system under consideration is not closed, i.e., if only a fraction b of N returns via spontaneous decays back to the lower level while the rest, $(1 - b)N$, decays out of the system (to other, unperturbed states), G is defined as

$$G = \left(\frac{2\beta^2}{\Gamma_e \gamma_k} \right) \left[\frac{\Gamma_e(1 - b) + \Gamma_g}{\Gamma_g} \right] \tag{9}$$

where the relaxation rate of the ground state has been denoted as Γ_g. For $G \ll 1$ there is no significant saturation and the atomic response to the optical field is linear, while for $G \geq 1$ the response is nonlinear and saturation sets in.

Though Eqs. (5) and (6) are exact it is instructive to see their perturbative expansions. For small values of G, Eq. (5) can be written as

$$\rho_{eg} = \frac{\beta N}{\delta + i\gamma_k}(1 - GL) \tag{10}$$

and

$$\rho_{gg} - \rho_{ee} = N(1 - GL) \tag{11}$$

where $L = \gamma_k^2/(\delta^2 + \gamma_k^2)$. The results show that the saturation characteristics depend strongly on atomic properties (dipole matrix elements, relaxation constants, and spontaneous decay branching ratios) and not only on the light intensity and frequency.

B. Three-Level Systems with Magnetic Degeneracy

The two-level atom is the simplest possible system in which resonant processes in light–matter interactions can be analyzed. Although it does provide a sufficient basis for discussion of many phenomena, there are numerous situations when more than two levels need to be considered. One of the most obvious of such cases is when the polarization properties of the light fields are important. This is always the case when there are nonzero dc magnetic or electric fields.

In this paper we concentrate on such situations, where one or both levels of a two-level system have magnetic (Zeeman) degeneracy, i.e., when they are characterized by nonzero angular momenta. The presence of degenerate substates permits many interference effects to occur, which could severely affect nonlinear-optical phenomena. The quantities responsible for such interferences are nondiagonal (more precisely near-diagonal) elements of the density matrix $\rho_{\alpha\beta}$ describing the interacting system. These nonzero elements represent a given phase correlation, i.e., coherence, in the evolution of the particular states α, β.

Below, we focus our attention on the most simple examples of such situations. We consider a two-level system in which one of the levels is single ($J_g = 0$) while the second has $J_e = 1$, hence it has Zeeman sublevels $m = 0, \pm 1$ (Fig. 1). If this system interacts with linearly polarized light and the quantization axis z is taken along the direction of the light beam, the light polarization is a superposition of two orthogonal circular polar-

Figure 1. The simplest atomic system in which interference effects associated with atomic coherence can be discussed. The lower level is a singlet ($J_g = 0$) while the upper $J_e = 1$ level has three Zeeman sublevels, two of which can be coherently coupled by the light beam with $\sigma^{\pm}$ polarization components. Such an excitation creates coherence of the $m = \pm 1$ sublevels.

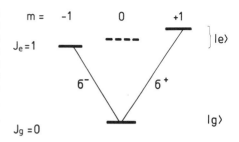

izations σ^+ and σ^-. As can be seen from Fig. 1 such an interaction might seem to be a simple mixture of two, two-level systems of the type discussed above, perturbed by the σ^+ and σ^- light fields. However, this situation is much more interesting since light components σ^+, σ^- are not independent but form a coherent superposition. It is this superposition that is responsible for establishing the coherent coupling of the $m = \pm 1$ sublevels and for creating the $\rho_{+-} \neq 0$ coherence. We shall see how this coherence affects the nonlinear optical response of the system. We present the equations of motion for the density matrix of the system in Fig. 1. Their solutions ρ_{eg} are necessary for calculation of the polarization associated with each of the polarization components of the field.

Assuming steady-state conditions and making the rotating-wave approximation, we have

$$
\begin{aligned}
(\delta_+ - i\gamma_k)\rho_{+0} &= -\beta(\rho_{00} - \rho_{++} - \rho_{+-}) \\
(\delta_- - i\gamma_k)\rho_{-0} &= -\beta(\rho_{00} - \rho_{--} - \rho_{-+}) \\
(\omega_{+-} - i\gamma_k)\rho_{+-} &= \beta(\rho_{+0} - \rho_{0-}) \\
\Gamma_e\rho_{++} &= -i\beta(\rho_{0+} - \rho_{+0}) \\
\Gamma_e\rho_{--} &= -i\beta(\rho_{0-} - \rho_{-0}) \\
\rho_{00} &= N - \rho_{++} - \rho_{--}
\end{aligned}
\tag{12}
$$

where $\delta_{\pm} = \omega_{\pm 0} - \omega_L$, $\omega_{ik} = (E_i - E_k)/\hbar$, and $\gamma_k, \gamma, \Gamma_e$ are relaxation rates of the optical coherences, Zeeman coherence, and excited state populations, respectively. We have also assumed that the Rabi frequencies associated with the $\sigma^{\pm}$ transitions are real and equal.

Though Eqs. (12) could, in principle, be solved analytically, their solutions have rather complicated forms. Nevertheless, one can use simple perturbative solutions to demonstrate that the saturation of the optical coherences $\rho_{\pm 0}$ depends not only on decay rates of the levels involved and on the field strength, as in the two-level case, but also on a mutual coherence between levels $| + \rangle$ and $| - \rangle$. Including terms up to the third order in the laser electric field, the optical coherences are

$$\rho_{+0}^{(3)} = -N\frac{\beta}{\Omega_+}\left[1 - G(2L_+ + L_-) + \frac{\beta^2}{\Omega}\left(\frac{1}{\Omega_+} - \frac{1}{\Omega_-^*}\right)\right] \quad (13)$$

$$\rho_{-0}^{(3)} = -N\frac{\beta}{\Omega_-}\left[1 - G(L_+ + 2L_-) + \frac{\beta^2}{\Omega^*}\left(\frac{1}{\Omega_+^*} - \frac{1}{\Omega_-}\right)\right] \quad (14)$$

where

$$\Omega_\pm = \omega_{\pm 0} - \omega_L - i\gamma_k, \ \Omega = \omega_{+-} - i\gamma, \ L_\pm = \gamma_k^2/\left[(\omega_{\pm 0} - \omega_L)^2 + \gamma_k^2\right],$$

and G is the saturation parameter defined by Eqs. (8) and (9).

Comparing expressions (13) and (14) with an analogous expansion for a two-level system given by Eq. (10), one can easily see that the two first terms in Eqs. (13) and (14) correspond to the populations-saturation contribution in the two-level case, whereas the last terms in (13) and (14) do not have any counterpart in Eq. (10). These are the contributions that are due to the Zeeman coherence ρ_{+-}. For the case where there are no collisions and the relaxation of the system is purely radiative, $\gamma = \Gamma_e = 2\gamma_k$ and the coherence contribution in Eq. (13) is proportional to the cross product $1/(\Omega_+\Omega_-^*)$ (and to its complex conjugate in Eq. (14)). This illustrates the relation between coherence ρ_{+-} and interference of the two excitation channels: $|0\rangle - |+\rangle$ and $|0\rangle - |-\rangle$. As seen from Eqs. (13) and (14) the degree of saturation (i.e., nonlinearity) in the three-level system of Fig. 1 very strongly depends on correlation of the sublevels $|+\rangle$ and $|-\rangle$, i.e., on the degree of their coherence.

Below we concentrate on the specific situation where the nonlinear response of the atomic medium arises from the atomic coherences in this way. We show that the results of many nonlinear processes are strongly modified by these coherences and that some phenomena are specifically due to such coherences.

IV. SPECIFIC SITUATIONS

A. Nonlinear Magneto-optical Effects

Since the early measurements of Faraday and Voigt it has been known that the polarization of a light wave propagating through an atomic medium changes in the presence of an external magnetic field. The rotation of the polarization plane in a longitudinal field is known as the Faraday effect, whereas rotation in the perpendicular field is the essence of the Voigt effect. The magneto-rotation in the longitudinal field is sometimes called the Macaluso–Corbino effect, and the name Cotton–Mouton effect is used for the rotation in the transverse field.

1. Typical Experimental Arrangement

A convenient arrangement for studying the two effects is to place the atomic medium between two crossed polarizers (Fig. 2). In this way the direct, unscattered light from the light source, which does not change its polarization, is eliminated and only the light that has changed its polarization when interacting with the atoms can be detected (provided the polarizers are ideal).[4] An important modification of this method uses slightly uncrossed polarizers which allow a small, controlled fraction of the incident light field heterodyne the atomic contributions. This greatly increases detection sensitivity.[5] Often the uncrossing angle is periodically changed by an appropriate modulator and phase-sensitive detection is used. This enables the dispersive and absorptive contributions to be separated.[6, 7] With the method of either exactly, or nearly crossed polarizers very small rotations (of the order of 10^{-7} rad),[8] as well as giant ones (of the order of 10π),[9] have been precisely measured.

The intensity of the light detected after the crossed analyzer can be obtained as the result of projecting the intensity of the field $\mathbf{E}_L$ transmitted through the medium onto the direction $\hat{\mathbf{e}}_a$ defined by the analyzer:

$$I_\perp \propto |\mathbf{E}_L \cdot \hat{\mathbf{e}}_a|^2 \tag{15}$$

It is convenient to expand $\mathbf{E}_L$ into eigenwave components. Depending on the direction of the magnetic field, the eigenwaves are either circularly or linearly polarized. For the geometry of the Faraday effect with the magnetic field directed along the light beam the eigenwaves are circularly σ^+ and σ^- polarized, whereas for the geometry of the Voigt effect they are linearly polarized along (π polarization), or perpendicular to the magnetic

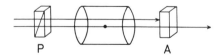

P   A

Figure 2. Isolation of the forward-scattering radiation from unscattered primary light using crossed polarizers. The primary beam is rejected but the scattered beam is transmitted if the polarization has been changed in the process of scattering.

field (σ polarization) This yields explicit expressions for the signals of the Faraday effect

$$I^F_\perp \propto \left(e^{-2l\kappa_+} - e^{-2l\kappa_-}\right)^2 + 4e^{-l(\kappa_+ + \kappa_-)}\sin^2 l(n_+ - n_-) \qquad (16)$$

and of the Voigt effect

$$I^v_\perp \propto \left(e^{-2l\kappa_\pi} - e^{-2l\kappa_\sigma}\right)^2 + 4e^{-l(\kappa_\pi + \kappa_\sigma)}\sin^2 l(n_\pi - n_\sigma) \qquad (17)$$

where $l = (\omega/c)L$, L being the sample length, and κ_α, n_α denote absorption coefficients and refractive indices, respectively, for the particular (α) polarization component.

The two parts in Eqs. (16) and (17) reflect contributions due to atomic absorption and dispersion and describe the effects of dichroism and birefringence, respectively. In the Faraday effect dispersion anisotropy $(n_+ - n_-)$ is related to the rotation of the polarization plane of the transmitted light beam and dichroism (absorption anisotropy) is related to the light ellipticity. In the Voigt effect the roles played by the real and imaginary parts are interchanged.

The constants κ_α and n_α in Eqs. (16) and (17), which characterize propagation of the α-polarization component of the light, could be modified by the light itself, giving rise to a variety of nonlinear magneto-optical effects.

2.   Nonlinear Susceptibility of an Atom in a Magnetic Field

As pointed out above, atomic susceptibility is widely used to describe magneto-optical effects since it is directly related to the atomic density matrix which describes atom–field interactions in the most general way. Historically, the relation between atomic polarizability and coherence was first pointed out by Series[10] and later by Happer and Mathur[11] and Laloë.[12] In these early papers, however, the light sources were treated as classical ones, too weak to establish atomic coherences, which had to be induced by a second perturbation, e.g., a radio-frequency field. Thus, the susceptibility was independent of the light intensity and could be related to linear magneto-optical effects alone. The susceptibility formalism has been generalized by Gawlik for the specific purpose of describing the influence of light-induced atomic coherences.[13]

In principle, any mechanism leading to optical nonlinearity, e.g., saturation, may also be responsible for nonlinear magneto-optical effects. In this paper, however, we concentrate specifically on nonlinearities that are exclusively due to atomic coherences and, consequently, do not discuss here much other work on nonlinear dispersion, the inverse Faraday effect, etc.

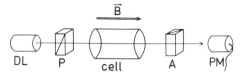

Figure 3. Experimental setup used by Gawlik et al.[14, 15, 31] for observation of the nonlinear Faraday effect. The light from the dye laser (DL) is linearly polarized by the polarizer (P), interacts with atoms in the cell placed within the magnetic field **B**, and is detected by the photomultiplier (PM) after passing through the analyzer (A) crossed with the polarizer (P).

3. Observations of the Nonlinear Faraday and Voigt Effects

The first observation of the nonlinear, resonant Faraday effect (NLFE) was in the experiments of Gawlik et al. with sodium vapor illuminated by tunable dye lasers, both pulsed[14] and cw.[15] The outline of the experimental arrangement used by Gawlik and coworkers is depicted in Fig. 3 and some typical experimental results are shown in Fig. 4a, b.

If the laser frequency is tuned to the atomic resonance the dichroic contribution to the forward-scattered intensity (the first term in Eq. (16)) vanishes and the signal observed with crossed polarizers is (for small rotation angles) a measure of the birefringence contribution alone, i.e., of the square of the Faraday angle. Thus, this arrangement with crossed polarizers is very convenient since no modulation techniques are required to separate the pure dispersive part. In general, however, with broadband or nonresonant laser light the signal consists of both dichroic and birefringent contributions.

With a low light intensity such as that of conventional spectral lamps, the Faraday effect is linear in E and the signal intensity as a function of magnetic field is represented by a broad curve because of the Doppler effect.[4] If laser light is used, coherences between atomic magnetic sublevels (Zeeman coherences) affect the Faraday effect and are responsible for the appearance of narrow resonances around $B = 0$. This has been clearly demonstrated by Gawlik et al.[15] (Figs. 4). In the magnetic field dependence of the FS intensity the contributions due to the linear and nonlinear FE are represented by Doppler-broadened and narrow curves, respectively, all centered at $B = 0$. While the nonlinear lineshapes are of a rather simple form for the Na D_1 line, they are much more complex for the Na D_2 transitions. In both cases there is strong power-broadening and fast, nonlinear increase of the signal amplitude with the incident light power. The signals occurring at the Na D_2 transition depend more strongly on I_0 than those associated with the D_1 line. Also the signal-shapes

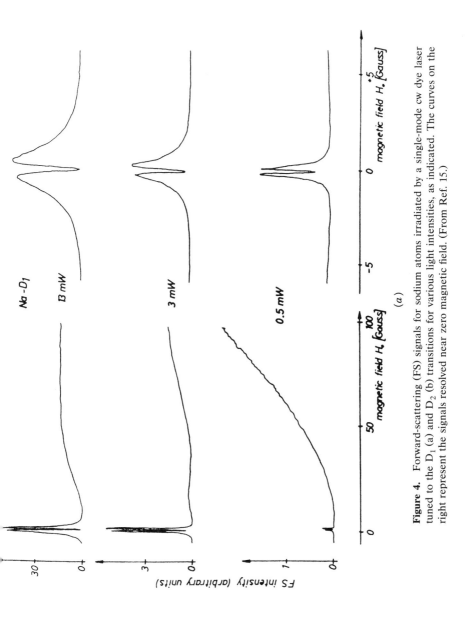

Figure 4. Forward-scattering (FS) signals for sodium atoms irradiated by a single-mode cw dye laser tuned to the D_1 (a) and D_2 (b) transitions for various light intensities, as indicated. The curves on the right represent the signals resolved near zero magnetic field. (From Ref. 15.)

743

Figure 4. (Continued)

for the Na D_2 signals change rapidly with I_0. In the first interpretation of these signals[15] it was assumed that higher-order Zeeman coherences including the hexadecapole moment (the coherence of the sublevels with $|\Delta m| = 4$) should be responsible for the observed nonlinearities. This interpretation was based on (1) the specific form of the signals, which suggested that different components with different width and intensity dependences contribute (2) a nonlinear (faster than I_0^3) dependence of the signal amplitudes for the D2 line, (3) rapid destruction of the signals by a small amount of buffer gas, indicating that they are coupled to fragile optical coherences by many-photon interactions, and (4) an apparent similarity of the signal characteristics with those observed in fluorescence and attributed to the $\Delta m = 4$ coherences by Ducloy et al.[16]

NLFE has also been observed in the microwave domain.[17] These observations differ substantially from those in the optical region in the principle of detection. In the optical experiments the detectors are sensitive to intensity, i.e., to $|E|^2$, whereas in the microwave experiments the E field is detected directly. This results in a different symmetry for the observed signals.

A full theoretical analysis of the nonlinear Faraday effect for transitions with hfs and arbitrary angular momenta is very complicated because many states are involved and there are mutual, nonlinear couplings of all the coherences and populations. For this reason the observations of Gawlik et al.[14, 15] remained without detailed, quantitative interpretation for a long time. The first attempt to provide such an interpretation was by Giraud-Cotton et al.[18, 19] who performed perturbative, third-order calculations of the NLFE for the sodium lines with the aim of explaining the observations of Gawlik et al. Indeed, the calculated lineshapes were qualitatively similar to some of the experimental ones. From this agreement, Giraud-Cotton et al. claimed that their perturbative analysis with $\Delta m = 2$ coherences, without invoking higher-order ($\Delta m = 4$) ones, provides a satisfactory explanation. This started a long-lasting controversy, since, as pointed out by Gawlik,[20] the agreement was not full and the theory of Giraud et al. was far from complete. In particular, the restoration of coherence and population by spontaneous emission was neglected in this treatment. Moreover, under the conditions of the experiment of Gawlik et al.[14, 15] the saturation parameter was well above unity, making the perturbative approach inapplicable.

Zakrzewski and Dohnalik[21] have made a profound contribution to the theory of the NLFE. They made a nonperturbative analysis of forward-scattering (and polarization spectroscopy) experiments for a $J_g = 0 - J_e = 1$ transition driven by partially coherent fields. They pointed out that the standard formalism of susceptibility may not be applied when the light is not perfectly coherent. They worked out two particular models of partly coherent laser fields and showed that the NLFE is sensitive to the statistical characteristics of the light field.

The lack of a satisfactory theoretical explanation for the NLFE stimulated further studies of this effect, but they were mainly limited to simple energy level structures, 0-1, 1-0, 1-1, for which exact solutions could be obtained. There has been also much interest in this field because the NLFE must be taken into account in measurements of the very small rotations caused by parity nonconservation.[22] Performing a nonperturbative analysis, Schuller et al. discussed in much detail how saturation and coherence affects the NLFE spectra, with particular emphasis on collisional effects for the $J_g = 0 - J_e = 1$ (Ref. 23) and $J_g = 1 - J_e = 0$ (Ref. 24) structures. They found that the two cases have significantly different characteristics, despite the fact that the expressions given by the theory are superficially similar. This is because the collisional destruction of orientation and alignment of the state $J_e = 1$ affects the Faraday rotation only when saturation is significant, while in the case of $J_g = 1$ it has a profound effect at all intensities. Later, Schuller et al. extended their formalism, developed in Ref. 23, to analyze also the nonlinear Voigt effect.[25]

A significant difference between the NLFE in Λ and V systems has also been found by other authors. Baird and coworkers[26, 27] also obtained analytical solutions by a nonperturbative analysis of such effects in the $J_g = 0 - J_e = 1$ structure and also performed experiments on Sm that agreed well with the theory.

Samarium, with its 0-1, 1-0, 1-1 transitions, has been a playground for magneto-optical studies performed by several groups. Drake et al.[28] analyzed the NLF and Voigt effects for the 651-nm line very thoroughly and found satisfactory agreement with their perturbative calculations. The Novosibirsk group also studied this effect in samarium and observed a dramatic increase of the rotation (by about four orders of magnitude) due to a nonlinear contribution.[29, 30] By an appropriate choice of experimental conditions they distinguished two contributions to the NLFE: one due to the Zeeman coherence, and another related to the optical coherence or "Bennet hole" in the velocity distribution of the population. (While papers 23–27, 29, and 30 deal with rotation spectra, i.e., the wavelength dependences of the Faraday angles, papers 14, 15, 17–21, and 28 are devoted to the forward-scattering signals, i.e., to the magnetic dependence of the Faraday rotation, mainly for near-resonant excitation).

Another work on the NLFE, by Gawlik and Zachorowski,[31] describes experiments on sodium with a powerful, pulsed laser and gives a perturbative (up to third order) analysis of the effect that takes into account all optical pumping effects (depopulation as well as repopulation, which was overlooked in Refs. 18 and 19), and additionally the effect of a finite laser bandwidth. The results for broadband laser excitation are interpreted in terms of "hfs uncoupling"[32, 33] and the theoretical curves obtained for nearly monochromatic excitation are in apparent agreement with the experimental signals of Ref. 15. A perturbative analysis was also performed by Jungner, Ståhlberg, and coworkers,[34–36] who measured the NLFE/FS signals for many lines in neon and found qualitative agreement of the theoretical and experimental signals. In fact, the agreement is quite convincing and it seems that the hexadecapole coherence could by no means produce dramatic changes in the NLFE/FS signals. Its influence might be of a rather indirect nature and could affect such characteristics as the dependences of amplitude and width on the incident intensity.

An important study of the NLFE has been performed by Weis et al.[37] who made elegant, precision experiments on cesium together with some related analysis. In their first paper on the NLFE, like Gawlik in Ref. 20, they realized the missing terms in Giraud-Cotton's treatment[18, 19] and made calculations with these terms included. Surprisingly, they did not obtain better agreement until some of these terms were arbitrarily ignored. Most likely this was due to omitting some of the contributions describing the hfs pumping in the spontaneous emission terms.

Later papers by Weis et al.[38, 39] present a very interesting approach to the NLFE in which the quantization axis is taken along the E-field vector rather than along the magnetic field. In this representation no Zeeman coherences can be created by the light and the terms describing the light perturbation are substantially simplified. On the other hand, however, the terms describing the interaction with the magnetic field are more complicated in this representation and the approach is practical only for weak light intensities, but may be used for arbitrary angular momenta.

B. Four-Wave Mixing and Phase Conjugation

One of the most important examples of nonlinear wave mixing is four-wave mixing (FWM). In a perturbative approach it is a fourth-order process described by $\chi^{(3)}$ and the lowest-order nonlinear process that is observable for media with an inversion symmetry, e.g., atomic gases. Depending on the particular frequencies and directions of the mixed waves, one can distinguish many kinds of FWM, having very different properties and applications. They are extensively described in the literature.[40, 41] Here we shall limit our interest to backward degenerate FWM (DFWM), i.e., the case when all four interacting waves have the same frequency, because of its importance for optical phase conjugation (wave-front reversal). A typical setup for such mixing is shown in Fig. 5.

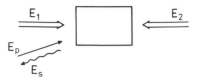

Figure 5. Experimental setup for studying backward degenerate FWM. The nonlinear medium is perturbed by two counterpropagating pump beams E_1 and E_2 and the probe beam E_p. When all beams have the same frequency and $\mathbf{k}_1 = -\mathbf{k}_2$, FWM produces the conjugate, fourth signal wave $E_s = E_p^*$ with $\mathbf{k}_s = -\mathbf{k}_p$.

It follows from the phase-matching condition the fourth wave E_s in backward DFWM is conjugate to the probe wave $E_s = E_p^*$ and has $\mathbf{k}_s = -\mathbf{k}_p$. Since the effect is well known and well described we concentrate here, as elsewhere in this chapter, solely on the role of atomic coherences in this phenomenon. We refer the reader to other papers[42] for more general information. We shall thus assume that the medium has an energy-level structure that permits Zeeman and/or hyperfine or Raman coherences. It should be realized that the interaction of an atom with more than two light beams of different polarizations must necessarily involve more than two levels, according to Zeeman degeneracy and polarization selection rules. A tensor character of the susceptibility must also be considered, since a scalar formalism of DFWM is not adequate in such cases. The first tensorial analyses of DFWM were performed by Zeldovich et al.[43, 44] and Steel and Lam.[45] However, although a complete spatio-polarizational (i.e., vectorial) phase conjugation was discussed, no effects associated with Zeeman coherences were explicitly analyzed in these papers. One of the consequences of the vectorial phase conjugation is the possibility of observing the conjugate wave in the case when the probe is polarized orthogonally to the pump beams. The first experiment of this kind was performed by Liao et al.,[46] but the authors did not discuss it in the context of FWM. The role of Zeeman coherences in phase conjugation with DFWM was pointed out for the first time by Lam et al.[47, 48] and Jabr et al.[49] They noticed that the nonlinearity responsible for phase conjugation depends not only on the population grating known from earlier studies, but also on the contribution of the coherence grating. The nonlinear polarization $\mathbf{P}^{(3)}$ induced in the media by beams E_1, E_2, and E_p polarized along $\hat{\mathbf{e}}_1, \hat{\mathbf{e}}_2$, and $\hat{\mathbf{e}}_p$, respectively, can be represented as

$$\mathbf{P}^{(3)} \propto \left(A\hat{\mathbf{e}}_2 \hat{\mathbf{e}}_3^* \cdot \hat{\mathbf{e}}_1 + B\hat{\mathbf{e}}_1 \hat{\mathbf{e}}_3^* \cdot \hat{\mathbf{e}}_2 + C\hat{\mathbf{e}}_3^* \hat{\mathbf{e}}_1 \cdot \hat{\mathbf{e}}_2 \right) E_1 E_2 E_p^* \qquad (18)$$

The coefficients A, B, and C have resonances at the atomic transition. The first two contributions to $\mathbf{P}^{(3)}$ describe two spatial population gratings induced by waves E_1 and E_p, as well as E_2 and E_p, whereas the third is due to the Zeeman coherence grating induced by E_1 and E_2. In this interpretation, the phase-conjugation signal results from the diffraction of the probe beam by one or more of the above gratings. It is clear that the coherence grating may be induced only if appropriate polarizations of the pump beams are used. Even when $\hat{e}_p \perp \hat{e}_1, \hat{e}_2$, so that no population gratings are possible, the coherence grating may contribute with an efficiency comparable to that of the population gratings. Since collisions affect populations and coherences differently, the population and coherence gratings also relax in different ways, and that is why the phase-conjugation signals in DFWM, with various appropriate beam polarizations, may yield different, complementary information on collisions. In particular, the FWM signals are well suited for studying depolarizing collisions because coherences also depend sensitively on such processes.

As an example, three different signals are shown in Fig. 6 (from Ref. 49). These were obtained with sodium atoms and various combinations of beam polarizations while scanning the laser frequency around the D_1 and D_2 lines. As can be seen, a given wave-mixing contribution can be selected in this way by an appropriate polarization combination. The phase-conjugation signals in Fig. 6 are background-free, which gives good detection sensitivity.

Interpretation of DFWM in cases where the relevant states are degenerate was first given in terms of a perturbative third-order theory.[49] Lam and Abrams[48] discussed three mechanisms related to the three terms in Eq. (18) and called them normal population, cross population, and Zeeman coherence contributions, respectively. The specific interactions represented by these contributions are shown in Fig. 7. A perturbative formalism was also used by Saikan who calculated elements of the susceptibility tensor for transitions between states with arbitrary angular momenta.[50] He noticed that, in contrast to polarization spectroscopy in which differently polarized light beams are also used, the backward DFWM signals allow detection of individual components of $\chi^{(3)}$. In his subsequent work Saikan[51] obtained experimental results that agreed well with his calculations.

The first nonperturbative analysis of the influence of Zeeman coherences on phase-conjugation resonances in DFWM was performed by Agrawal.[52, 53] He considered the situation in which the counterpropagating pump beams are circularly polarized in opposite senses and the probe is also circularly polarized, in the same sense as the backward pump (Fig. 8). Due to conservation of angular momentum, the conjugate wave must also be circularly polarized but orthogonally to the backward beam. This could be very useful for a complete discrimination of the background associated with the pump when the pump and probe beams are perfectly overlapped.

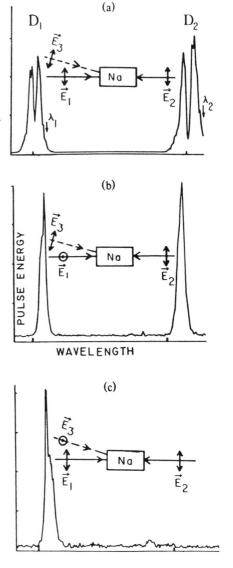

Figure 6. Degenerate FWM signals as a function of wavelength (increasing from right to left) obtained by Jabr et al. with sodium atoms for three cases of beam polarization as shown. (From Ref. 49.)

In this way strong signals can be obtained from the largest possible interaction volume. In other situations with the same polarization of the probe and backward pump, a spatial discrimination between these two beams is necessary. On the one hand, this decreases the interaction volume and, on the other, it reduces the mixing efficiency due to motional washout of the grating.[54] In Agrawal's theory,[52, 53] the model of atomic

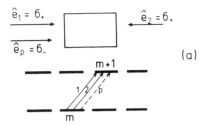

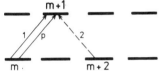

(a)

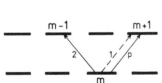

(b)

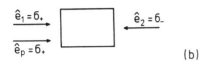

Figure 7. Quantum-mechanical paths giving rise to various mechanisms contributing to degenerate FWM depending on beam polarizations (after Ref. 48). For the case of equal σ^+ polarizations of E_1, E_2, and E_p the normal population mechanism (a) is responsible for the appearance of E_s, while the cross-population and Zeeman-coherence mechanisms (b) occur in the case when E_1 and E_2 are orthogonally polarized (see Fig. 8).

structure considered consists of a Zeeman-degenerate lower level with $J_g = 1$ and a single upper level $J_e = 0$ (Fig. 9), and the beam polarizations are as shown in Fig. 8. Since the pump beams have different polarizations (σ^+ and σ^-), no standing wave is formed and no related spatial-hole burning occurs. Also, the contribution of self-saturation (normal population) disappears. As to the two remaining contributions in Eq. (18), the cross-population is the only nonvanishing one when relaxation damps the

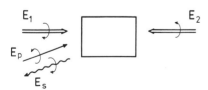

Figure 8. The case of opposite circular polarizations of the pump beams permits Zeeman coherence to contribute to the FWM signals. Due to conservation of angular momentum and appropriate selection rules, the conjugate wave E_s is orthogonally polarized to E_p.

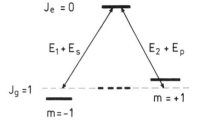

Figure 9. The atomic structure considered by Agrawal[52, 53] in his nonperturbative analysis of the influence of Zeeman coherence on the degenerate FWM, with polarizations of the light beams as shown in Fig. 8.

Zeeman coherence. If, however, this is not the case, and if the upper state relaxes much faster than the lower one, the Zeeman coherence contribution dominates the whole signal.

The nonperturbative theory of Agrawal enables saturation effects to be discussed. In particular, it predicts that for structures with very slow relaxation of the lower states, as in the case of stable ground states, saturation could be achieved with much lower light intensities than for two-level systems. Similar behavior is observed in velocity-selective optical pumping and polarization spectroscopy.[55–58] This enables large values of the phase-conjugate reflectivity $R = |E_s^*/E_p|^2$ to be obtained with light intensities far smaller than those required for two-level systems.

Very interesting effects occur when the degeneracy of the $m = \pm 1$ sublevels is shifted by an external magnetic field, thereby changing the degree of coherence of these levels. In the case under consideration, for equal, resonant frequencies of the beams the coherence reaches its maximum value when the sublevels are fully degenerate, i.e., for $B = 0$. Under such conditions population trapping takes place in the lower state (see Section IV. C), which manifests itself by a dip in the $R(B)$ dependence at $B = 0$ (Fig. 10). As in the case of the Hanle effect,[59] the width of the dip yields information on the coherence decay rate. Taking atomic movement and the Doppler effect into account results in a change in the relations between the population and coherence contributions to the DFWM by washing out the population grating and leaving only the coherence grating. In this way the coherence dominates phase conjugation in systems with Doppler broadening.

Relation (18), in which the phase-conjugate signal is proportional to the scalar product of two field amplitudes is not always true. In the case when the probe and one of the pump beams are orthogonally polarized, Eq. (18) erroneously links the third contribution with a two-photon susceptibility oscillating as 2ω (Ref. 60) and ignores the process in which phase-conjugate emission could be induced by the Zeeman coherence grating. For Eq. (18) to remain generally correct, it should be understood as a tensorial

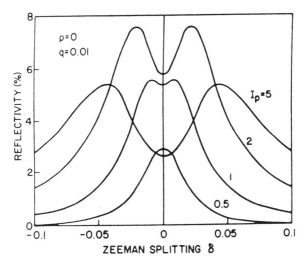

Figure 10. Phase-conjugate reflectivity as a function of the magnetic field intensity, i.e., of the splitting of the Zeeman sublevels $m = \pm 1$ for various values of the pump intensities I_p (from Ref. 53). The values of I_p are given in units of the saturation intensity, while those of the Zeeman splitting are in units of the homogeneous linewidth. In the saturation regime ($I_p \cong 1$) a narrow central resonance related to the Zeeman coherence occurs and its width exhibit power broadening with a further increase in the pump intensity.

one, i.e., the constants A, B, and C should be taken as tensors, not scalars.[61] After such a generalization, the coherence grating can be properly accounted for even when $\mathbf{E}_p \perp \mathbf{E}_1$ or $\mathbf{E}_2$ and for arbitrary values of J (Ref. 56). One can also analyze in detail the influence of relaxation on the various gratings, as well as their dependence on the angle ϑ between the probe and pump beams. When atomic movement is taken into account, this dependence affects not only the interaction volume but also the degree of motional washout of the population grating. For instance, for very small angles $\vartheta \ll 1$ only the grating formed by waves with $\mathbf{k}_+, \mathbf{k}_p$ becomes important because the one with $\mathbf{k}_-, \mathbf{k}_p$ is motionally quenched, i.e., washed out due to the Doppler effect. This directional anisotropy was studied by Ducloy and Bloch with a three-level, lambda-like atomic model in which one of the pumping beams was assumed to be strong.[54] The interaction with the strong pump was described to all orders, while other, weak beams were treated perturbatively.

A convenient description of atomic interactions with a strong field is offered by the dressed atom model.[63] Using this approach, Pinard et al.[64] described results of their experiment with DFWM with three-level, lambda-like neon atoms (607.4-nm transition in Ne) placed in a transverse magnetic field. One of the pumps (dressing field) was strong and the other was weak. The observed resonances were interpreted as the effects of crossings of energy levels of the dressed atoms. The same authors performed a similar analysis of a phase-conjugation experiment with orthogonally polarized pump beams for transitions in neon atoms with Λ- and V-like structures.[65] They observed a dip in the center of the signal ($\delta = 0$) that was due to Zeeman coherence. The interpretation of the experiment[65] was given in terms of the dressed-atom model where the Zeeman coherence does not appear directly as in, e.g., Agrawal's theory.[52, 53] In spite of this, the dressed-atom model makes prediction of the frequencies and shapes of the FWM resonances easier than the tedious standard approach.

Performing nonperturbative calculations for the case where both pumps are strongly interacting with a V-like, $J = 0 - J = 1$ system, Saxena and Agarwal[66] demonstrated the appearance of doublet structures, even for low powers of the light beams. Although they resemble the ac-Stark splitting, they are associated with coherence contributions in an analogous way to the Hanle effect. An interesting feature of these signals is their subnatural width; similar lineshapes, also related to Zeeman coherences, were analyzed for Λ-like structures by Agrawal.[52, 53]

The notion of Zeeman coherence is strongly linked with nonequilibrium population distributions in Zeeman sublevels or, in other words, with multipole moments in these levels. An appropriate choice of quantization axis allows interpretation either in terms of individual sublevel populations or in terms of nondiagonal elements of the density matrix, i.e., coherences (see also Ref. 1).

Some very interesting effects associated with nonconservation of various observables were studied theoretically and experimentally by Berman et al.,[67] who showed that when an atomic system is not closed, new

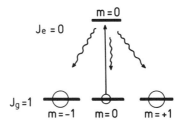

Figure 11. The $J_e = 0 - J_g = 1$ system interacting with linearly, π-polarized light. Even if the total population of the upper and lower levels is conserved, the population of the two-level subsystem consisting of $|J_e = 0, m = 0\rangle$ and $|J_g = 1, m = 0\rangle$ is not, because of redistribution of population to the $m = \pm 1$ sublevels by optical pumping which creates alignment of the lower magnetic states. This illustrates that the total magnetic state alignment may not be conserved despite conservation of the total population.

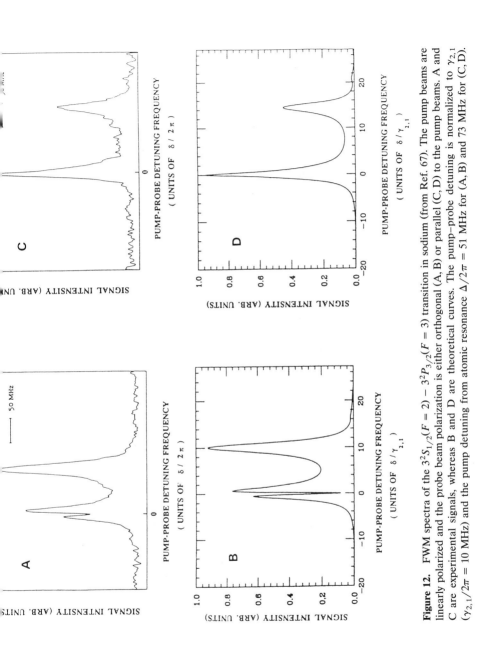

Figure 12. FWM spectra of the $3^2S_{1/2}(F = 2) - 3^2P_{3/2}(F = 3)$ transition in sodium (from Ref. 67). The pump beams are linearly polarized and the probe beam polarization is either orthogonal (A, B) or parallel (C, D) to the pump beams. A and C are experimental signals, whereas B and D are theoretical curves. The pump–probe detuning is normalized to $\gamma_{2,1}$ ($\gamma_{2,1}/2\pi = 10$ MHz) and the pump detuning from atomic resonance $\Delta/2\pi = 51$ MHz for (A, B) and 73 MHz for (C, D).

755

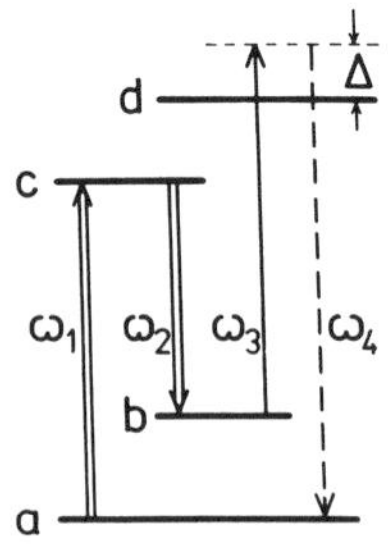

Figure 13. The four-level system used by de Oliveira et al.[68] for analysis of FWM when two strong fields of frequencies ω_1 and ω_2 create the Raman coherence ρ_{ab}. The field of frequency ω_4 is generated by FWM with the pump beams of ω_1, ω_2 and the probe of ω_3 and is proportional to ρ_{ab}.

resonances appear in the FWM signals. These resonances are affected in a very specific way by optical pumping processes, which, depending on the branching ratio of the excited state decay into lower states, could either conserve or not population, alignment, or orientation. Let us focus, for example, on the $J_g = 1$, $J_e = 0$ system interacting with linearly π-polarized light. Optical pumping (absorption followed by spontaneous emission) creates alignment and changes populations of the $m_g = 0$, $m_e = 0$ subsystem (Fig. 11), although the total populations of the ground and excited states remain constant. The signals depicted in Figs. 12a–d are examples of such resonances. They have the form of a narrow dip at $\delta = 0$ for the configuration that depends on alignment and orientation (Figs. 12a, b) and a peak when the population term dominates (Figs. 12c, d).

When studying coherence effects in FWM the situation is often encountered in which Raman coherences are created between levels that are not necessarily Zeeman sublevels, e.g., in an energy-level system such as that depicted in Fig. 13 where transitions driven by two strong fields are labeled by double arrows and a weak wave probes transition $b \to d$, marked with a single line. De Oliveira et al.[68] showed that the field generated by FWM is proportional to a two-photon Raman coherence ρ_{ab}. They also discussed an analogy between population trapping, where absorption cancellation is observed when the coherence is maximal, and the case considered in Fig. 13 when zero emission corresponds to the minimum of the coherence resulting from destructive interference, i.e., competition of two optical coherences ρ_{bc} and ρ_{ca}.

C. Dark Resonances, Coherent Population Trapping, Electromagnetically Induced Transparency, and Highly Dispersive, Nonabsorbing Media

One of the most spectacular effects associated with atomic coherence is the phenomenon of coherent population trapping. It was observed for the first time by Alzetta et al.[69, 70] in an experiment with a multimode dye laser tuned to the sodium D lines (Fig. 14). The sodium atoms were in a resonance cell placed in a longitudinal magnetic field. When the splitting

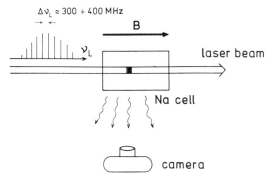

Figure 14. Experimental setup used by Alzetta et al.[69, 70] for observation of the dark resonances due to coherent population trapping. Sodium atoms submitted to the nonuniform magnetic field B with a longitudinal gradient are irradiated by a multimode cw dye-laser beam, the separation $\Delta\nu_L$ between its longitudinal modes being variable between 300 and 400 MHz. The laser-induced fluorescence is observed laterally. The patterns (bright and dark places along the field gradient) can be either recorded by a camera or imaged from a specific place within the cell onto a photomultiplier and scanned as a function of the magnetic field.

of some of the Zeeman sublevels of the ground-state hyperfine structure just corresponded to the frequency difference between some of the laser modes (Fig. 15a), a distinct decrease of fluorescence light was visible in the cell, hence the name "dark" resonances. The magnetic field in this experiment was purposely made inhomogeneous, with a longitudinal gradient. This permitted excellent spatial resolution, since the resonance condition and the resulting dark resonance occurred only in a given place within the cell.

To explain the observed behavior of the atomic fluorescence it is sufficient to consider a simple structure of three levels coupled by two laser modes[71, 72] (Fig. 15b). The fluorescence intensity is proportional to the population ρ_{ee} of the state excited by the two modes. With the help of a simple calculation it can be shown that under stationary conditions the population ρ_{ee} consists of three contributions:

$$\rho_{ee} \propto I_1\rho_{11} + I_2\rho_{22} - \sqrt{I_1 I_2}\,\mathrm{Re}\,\rho_{12} \tag{19}$$

The first and second terms in Eq. (19) arise from independent excitations of the state e by each of the modes. The third contribution is associated with the coherence created by both modes ω_1 and ω_2 between states 1 and 2. When the modes have equal intensities, formula (19) takes a simpler form

$$\rho_{ee} \propto \rho_{11} + \rho_{22} - 2\,\mathrm{Re}\,\rho_{12} \tag{20}$$

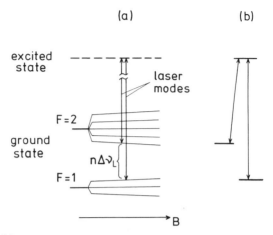

Figure 15. (a) Energy levels of sodium indicating the resonance condition when the frequency difference $n\Delta\nu_L$ of some longitudinal modes matches the separation of given Zeeman sublevels of the ground state. For the sake of simplicity the structure of the excited state has been ignored. (b) A simplified three-level system used for explanation of population trapping by a superposition of two sublevels of the ground state coupled by different laser modes.

Assuming that $\rho_{11} + \rho_{22} = 1$, that the two-photon resonance condition is fulfilled, $\omega_1 + \omega_2 = \omega_{12}$, and the induced coherence ρ_{12} is of its maximum possible amplitude, it can be shown that $\rho_{ee} = 0$, which leads to zero fluorescence intensity. This vanishing of fluorescence results from the cancellation of the absorption probability in the two states 1 and 2. It might seem surprising, since each of the individual modes is strongly absorbed. It is only their joint action that creates coherence ρ_{12} and causes the destructive interference of transition amplitudes $1 \rightarrow e$ and $2 \rightarrow e$ and a corresponding inhibition of absorption. Even when state e can relax spontaneously to states other than 1 and 2, i.e., the system is open, the loss of population is eliminated because it is trapped in a coherent superposition of states 1 and 2. Although the phenomenon of population trapping is not one of the most typical examples of nonlinear optics it is highly nonlinear because the intensity of the fluorescence (population ρ_{ee}) is a strongly nonlinear function of the intensity of both modes when $\rho_{12} \neq 0$.

A profound physical interpretation of population trapping was given by Gray et al.,[73] who transformed the unperturbed states $|1\rangle$, $|2\rangle$ to states $\psi_{C,NC} = (|1\rangle \pm |2\rangle)/\sqrt{2}$ and showed that population trapping in this new basis is equivalent to a redistribution of the population (optical pumping) to the trap state ψ_{NC} that is not perturbed by the light and cannot therefore absorb light (Fig. 16). The coupled state ψ_C that can absorb will

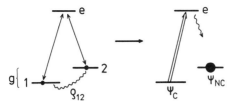

Figure 16. Two laser fields interacting with a three-level system (excited state e and two sublevels 1 and 2 of the ground state g) create coherence ρ_{12}. The coherence is responsible for trapping the population in a coherent superposition of states 1 and 2. After transformation of states 1, 2 to $\psi_{C,NC}$ the phenomenon of population trapping can be interpreted as transferring the population to the uncoupled state ψ_{NC}, which cannot absorb light. The coupled state ψ_C, which interacts with the light, is depleted, hence there is no absorption and no fluorescence.

be depleted by pumping and hence, excitation of state e, and subsequent fluorescence, will be zero.

Coherent population trapping has found many applications, among the most attractive of which we may include laser cooling below the recoil limit[74] and proposals for lasers without population inversion (described in Section IV.D) or the preparation of highly dispersive, nonabsorbing media.

Closely related to population trapping is the phenomenon of electromagnetically induced transparency (EIT).[75] The mechanism for this can be discussed on the basis of the three-level model in Fig. 17 (Ref. 76). State 1 is taken to have greater population than both states 2 and 3 ($\rho_{11} > \rho_{22} + \rho_{33}$). The question is, what is the absorption on the $1 \to 3$ transition, under the conditions of driving the coupled transition 2-3 with a strong, coherent field? Simplistic arguments could suggest that there will be strong absorption on the 1-3 transition because of the high population ρ_{11}. It turns out, however, that the action of a coherent field on the 2-3 transition drastically reduces the absorption, i.e., causes the EIT. This phenomenon can be explained in terms of a quantum interference of two absorption channels: the direct transition $1 \to 3$ and the transition associated with an exchange (absorption and stimulated emission) of photons of the strong-field acting

Figure 17. A three-level system in which the electromagnetically induced transparency can occur on the 1-3 transition due to coherent coupling of states 2-3 by a strong field.

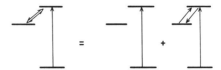

Figure 18. An illustration of the quantum interference process responsible for cancellation of absorption on the 1-3 transition: The two destructively interfering quantum paths between states 1 and 3 are the direct transition 1-3 and the transition with an exchange of photons from the coherent field that drives the coupled transition 3-2.

on the 2-3 transition (Fig. 18). When the frequencies of the two fields fulfill the two-photon resonance condition $\omega_1 - \omega_2 = \omega_{12}$ (as in population trapping), and there is no efficient relaxation 1-2, the interference becomes fully destructive, and the medium is rendered transparent on the 1-3 transition.

Cancellation of absorption due to destructive quantum interference in the population trapping or EIT schemes could be applied as a means of obtaining exotic and very attractive lossless media with very high refractive indices. It is known that the index of refraction rises resonantly (as $\Delta\nu/[\Gamma^2 + (\Delta\nu)^2]$) in proximity to an absorption line. Such a resonant increase, however, is not very practical, since the related very strong increase of absorption makes the medium opaque at resonance frequencies. Still, as noticed by Scully,[77] it should be realistic to chose the relevant parameters of a three-level system in such a way that the refractive index would reach its maximum value while absorption would be canceled. Since n scales proportionally to the density of the medium one could think of preparing dense, yet lossless media with gigantic values of n. Before this objective can be reached, it is be necessary to consider whether the possible influence of local fields in such dense media would not impose some limitations on the expected increase of n.[78]

D. Lasers Without Inversion

It is generally believed that population inversion is a necessary condition for laser action. Indeed, the intensity of a light beam transmitted through a medium consisting of two-level atoms changes according to Beer's law, $I = I_0 e^{-\kappa L}$, where L denotes the length of the medium and κ its absorption coefficient. For a two-level atom $\kappa \propto \text{Im}\,\rho_{eg}$, where ρ_{eg} is an atomic coherence related to populations

$$\rho_{eg} = \beta \frac{\rho_{ee} - \rho_{gg}}{\delta - i\Gamma/2} \tag{21}$$

As we have seen, when there is no inversion ($\rho_{ee} < \rho_{gg}$), $\kappa > 0$, and the light beam is attenuated after traversing the medium, whereas it is amplified when there is inversion ($\rho_{ee} > \rho_{gg}$).

In spite of these dependences, there are special cases when even in a two-level system amplification without inversion is possible. These cases are described in detail in a recent review.[79] Here we shall concentrate on some very complex but more practical situations where amplification and laser action are based on the phenomenon of quantum interference.

As is known from the experiment of Alzetta et al.[69, 70] for a Λ-like system, described above, an interference-based cancellation of absorption and population trapping in a superposition state is possible. We now consider such a system with a trapped population and an excited state that is weakly, possibly incoherently, populated. There are allowed transitions from this state to the trapping superposition. An example of such a situation is depicted in Fig. 19, where double lines indicate transitions induced by the coherent laser field which establish the superposition of states 1 and 2. This system $\{1, 2, e\}$ is the same Λ-system as the one considered above in the discussion of population trapping. Single lines in Fig. 19 indicate transitions from state e′ to sublevels 1 and 2 of the lower state g. As before, given that the matrix elements of transitions e′-1 and e′-2 are equal, the absorption coefficient for the e′-g transition can be easily calculated:

$$\rho_{e'g} \propto n_{e'} - n_1 - n_2 - 2\,\mathrm{Re}\,\rho_{12} \tag{22}$$

As can be seen from Eq. (22), even if there is no inversion in the medium, $n_{e'} < n_1 + n_2$, amplification of the light intensity can be achieved on transition e′-g when $2\,\mathrm{Re}\,\rho_{12} > 0$. Since a large number, $n_1 + n_2$, of the potential absorbers are "frozen," even without inversion emission of n_e, emitters could dominate a zero absorption. This is the essence of some of the proposals[80–82] for lasing without inversion (LWI). Let us note that level e′ is quite independent of e, in particular, it can be of much higher energy than e. This means that a laser working according to this principle could generate radiation with a much shorter wavelength than the pump-

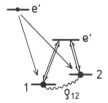

Figure 19. A possible scheme for lasing without inversion based on coherent population trapping. The Λ-system of states 1, 2, and e perturbed by two strong fields is used to create coherence ρ_{12} and to trap the the ground-state population. Any population in another, possibly incoherently, excited state e′ would then give rise to emission to the superposition of 1 and 2 without losses due to reabsorption even though there is no inversion: $n_{e'} < n_1 + n_2$.

ing radiation, which has to create the coherence ρ_{12} on the e-g transitions. This offers the possibility of generating radiation in very interesting, new spectral regions of the VUV, or even x-rays, with pumping by available, strong lasers working in the VIS or IR domains. Moreover, it is not essential for the coherence ρ_{12} to be induced via optical Raman-like transitions. It could equally well be created via microwave fields resonant with the 1-2 transition. (Due to selection rules and same parity of states 1 and 2 this has to be a magnetic-dipole transition.) Such possibilities have been considered in papers by Scully and coworkers.[83]

Most papers on LWI deal with an amplification of a weak probe wave. The first model of a real laser, i.e., light generator, without inversion in which the role of the cavity was fully taken into account was discussed by Karawajczyk et al.[82]

The inhibition of absorption in EIT could be used for LWI in much the same way as its elimination in coherent population trapping. For this purpose, appropriate, incoherent pumping is necessary in order to have nonzero net emission.

In Fig. 20, which depicts the closed system considered by Imamoglu et al.,[76] such pumping with rates R_1, R_2 is indicated with thick arrows, the strong, coherent field with a double arrow, and the transition with amplification (lasing) with a single arrow. The scheme of Imamoglu et al.[76] might seem to be just another version of the Raman laser where gain is due to stimulated Raman scattering. However, in conventional Raman lasers, gain occurs only when a Raman inversion exists, i.e., when the final state of the laser transition has lower population than the state from which the pumping originates (Fig. 21). Thus, the condition of no inversion $(n_1 > n_2 + n_e)$ in this scheme does not permit us to treat it as a standard Raman laser and makes a more thorough analysis necessary. On close inspection it turns out that stimulated Raman scattering is indeed responsible for the gain (driving field scattered on the 2-e transition yields the

Figure 20. The closed, three-level scheme for lasing without inversion used by Imamoglu et al.[76] Transition e-1 is the lasing (amplifying) transition while the e-2 one is coherently driven by a strong field $\Omega_{1,d}$ are the Rabi frequencies of the lasing (probe) field and the strong, coherent field, respectively. Γ_1 and Γ_2 are the decay rates of e into 1 and 2, and R_1 and R_2 are the incoherent pump rates from the states 1 and 2, respectively. There is no inversion, i.e., $n_1 > n_2 + n_e$.

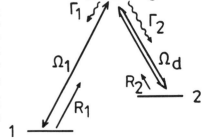

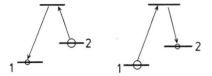

Figure 21. Two possible situations where gain can occur in the stimulated Raman scattering due to an initial Raman inversion of population: $n_1 > n_2$ and $n_2 > n_1$.

laser photon on the 1-e one), whereas an inverse process is the main loss factor. Gain due to stimulated Raman scattering turns out to be possible even if $n_1 > n_2 + n_e$, i.e., when there is no Raman inversion, $n_1 < n_2$. In this case, however, the joint action of the incoherent pumping and spontaneous relaxation must provide recycling of the population between the final and initial states that is sufficiently efficient for the Raman process to occur, which yields the condition $R_1\Gamma_2 > R_2\Gamma_1$. On the other hand, the first step of the inverse Raman scattering, which is the reabsorption of a laser photon by the atom in state 1, could be canceled by destructive quantum interference of the two transition paths, shown in Fig. 22. The two amplitudes are related to direct absorption of the laser photon and to absorption of the laser photon followed by interaction with two driving field photons (stimulated emission and absorption). Such interference occurs when the driving field is sufficiently intense ($\Omega_d \gg \Omega_1$) and is a manifestation of the Raman coherence ρ_{12} or of coherent population trapping.[69-73] Thus, although the principle of the laser system of Imamoglu et al.[76] cannot be interpreted simply as population trapping, both of these schemes are linked by destructive quantum interference. Therefore, quantum interference is the underlying mechanism for lasing without population inversion.

The above ideas have been verified experimentally by Harris and coworkers.[84, 85] In Fig. 23 we present the energy level scheme of a

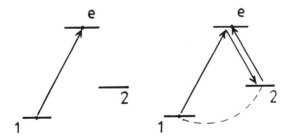

Figure 22. Two quantum paths interfering destructively in the scheme of Imamoglu et al.[76] The second path associated with the exchange of photons of the coherent field during the transition from 1 to e is important when there is sufficient coherence ρ_{12}.

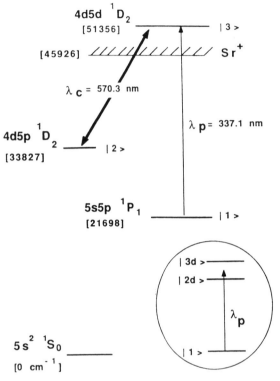

Figure 23. Energy-level diagram of strontium with the transitions used in the EIT experiment of Boller et al.[84] The inset shows the related dressed-state picture. (From Ref. 84.)

strontium atom with relevant transitions driven by strong and probe lasers in their experiment.[84] Figure 24a shows the absorption spectrum obtained by scanning the probe-laser frequency without the coherent field acting on the 2-3 transition, and Fig. 24b depicts the absorption spectrum when the coherent field was on. A very strong decrease of absorption (EIT) is clearly visible in the latter case and is a manifestation of the interference mechanism discussed above.

The first experiment where not only EIT, i.e., reduction of absorption, but also gain was observed was performed by Gao et al.[86] They worked with sodium vapors in a cell equipped with electrodes, allowing incoherent excitation of sodium energy levels by electric discharge. Absorption on the D_2 ($3^2S_{1/2} - 3^2P_{3/2}$) transition was probed by a low-power, single-mode cw dye laser, and the D_1 ($3^2S_{1/2} - 3^2P_{1/2}$) transition sharing the same

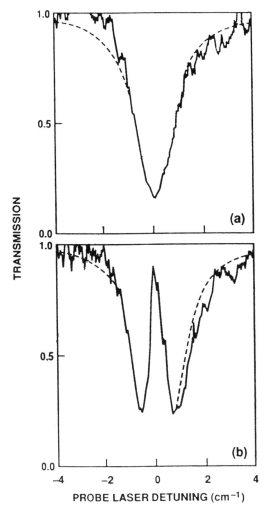

Figure 24. Transmission versus probe beam detuning for (a) no coherent field driving the coupled 3-2 transition; and (b) for coherent field driving the transition 3-2 with $\Omega_2 = 1.3$ cm^{-1}. (From Ref. 84.)

ground state, was coherently driven by another powerful, pulsed dye laser (Fig. 25). When no coherent field was applied to the D_1 line, the ordinary absorption spectrum was measured when the frequency of the probe laser was scanned across the D_2 resonance line (Fig. 26a). This line is considerably broadened by collisions with buffer gas and electrons. A spectacular change of sign of the absorption occurred when the powerful laser drove

Figure 25. Energy-level diagram of sodium with the transitions used in the experiment of Gao et al.[86] A strong, pulsed laser drives the transition D_2 while a weak, single-mode cw dye laser probes the absorption on the D_1 transition. An electric discharge incoherently populates the states $3^2P_{1/2}$ and $3^2P_{3/2}$, yet there is no inversion in the system.

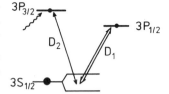

the D_1 transition coherently. Within a time interval corresponding to the powerful laser pulse, selected by appropriate time gating by a boxcar averager, significant gain occurred around the center of the D_2 line (Fig. 26b). The gain spectrum was in qualitative agreement with theoretical predictions,[81] i.e., it was peaked at two components originating from the ac-Stark splitting of dressed states.

The experiment of Gao et al.[86] provides spectacular illustration of the feasibility of amplification without population inversion. Still, a detailed comparison of this experiment with theory is difficult. In principle, the energy level structure applied could allow both schemes for LWI, i.e., the EIT and the one based on the coherence between the hfs components of the ground state, to be active in the experiment discussed. The experimental conditions, however, were not sufficiently well defined to enable these two mechanisms to be distinguished. The main complications are the pulsed nature of the powerful laser, making a steady-state theory impractical and necessitating direct integration of density-matrix equations[87]; the lack of control over the discharge current, as a result of which the

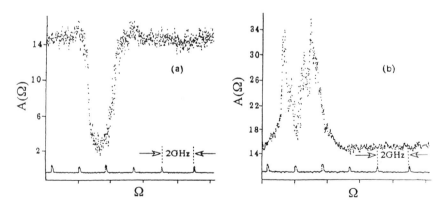

Figure 26. Absorption/gain spectrum $A(\Omega)$ of a weak probe in the vicinity of the D_2 (589.0-nm) line of sodium. The lower traces show calibration frequency marks with 2-GHz separation. (a) Absorption spectrum without coherent driving pulse; (b) gain profile in the presence of a coherent driving pulse with an intensity of 7 MW/cm^2. (From Ref. 86.)

incoherent population of the excited state could not be varied to test the theoretical models; and the wealth of hfs components in the sodium levels and their poor spectral resolution, which considerably complicates the interpretation.

E. Lasers in a Magnetic Field

The simplest, most widely used laser theories generally neglect magnetic degeneracy of the laser transitions. As long as no magnetic fields are used, this simplification is justified. Another, more complicated, situation arises, however, when the laser cavity is polarization anisotropic (e.g., when it contains Brewster windows) and an active medium or some part of it is placed in a dc magnetic field. In such cases the effects related to possible coherence between atomic Zeeman sublevels might become relevant. Lasers operating under such conditions are sometimes termed Zeeman lasers. They became an object of active investigation shortly after the development of the first lasers.[88-93] This work led to the development of a particularly useful method for studying coherence effects, mainly the Hanle and level-crossing effects in an intracavity cell. In general, the cell contained the same gas as used for generation, but, because the cell is placed independently from the main laser-discharge tube, more flexibility is gained for changing such parameters as the content and pressure of a gas mixture or the discharge current. Such studies developed fruitfully. They go beyond the scope of the present paper but have been widely discussed in an excellent review by Dumont et al.[94] Here we limit ourselves to a presentation of the main effects occurring due to Zeeman coherences in laser media.

1. The Zeeman Laser

A laser placed in a magnetic field emits light in which intensities I_i and frequencies ν_i of each of the polarization components ($i = +, -$, corresponding to polarizations σ^+ and σ^-) are mutually coupled. This can be illustrated by simple relations which take the first and third order of the laser perturbation into account[93]:

$$\dot{I}_+ = 2I_+ \left(a_+ - \beta_+ I_+^2 - \vartheta_{+-} I_-^2 \right) \qquad (23a)$$

$$\nu_+ + \dot{\phi}_+ = \dot{\Omega} + \sigma_+ - \rho_+ I_+^2 - \tau_{+-} I_-^2 \qquad (23b)$$

where a_i and β_i ($i = +, -$) are the linear gain and self-saturation coefficients, respectively, and ϑ_{+-} is the cross-saturation coefficient. In Eq. (23b) ϕ_i denotes phase of the ith component, Ω is the passive cavity frequency, σ_i and ρ_i are the linear mode pulling and self-pushing coefficients, respectively, and τ_{+-} represents the cross-pushing. The relations for I_- and ν_- are identical but with signs $+$ and $-$ replaced.

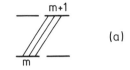

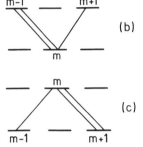

Figure 27. Diagrams depicting possible third-order combinations of couplings of various Zeeman sublevels by laser modes with $\sigma^{\pm}$ circular polarization components. Interaction with three laser photons of the same σ^{+} polarization contributes to self-saturation (a) represented by the second term (β_{+}) in Eq. (23a), whereas the interaction with photons of opposite polarizations gives rise to the ordinary cross-saturation term (b) and the quadrupole cross-saturation term (c) represented by the third term (ϑ_{+-}) in Eq. (23b).

Equations (23) have the same form as the equations for the case of coupling of two counterpropagating waves in a ring laser.[95] The difference is that in the present case both modes differ in their polarizations and possible Zeeman detunings, whereas in a ring laser the two coupled modes propagate in different directions and their detunings are due to the Doppler effect, which shifts the frequencies in opposite senses. This coupling between the modes is accounted for by the last terms of Eqs. (23). The nonlinear parts of Eq. (23a) can be illustrated by simple diagrams representing the interaction with individual polarization components (Fig. 27). The superposition of Zeeman sublevels, characterized by Zeeman coherences, leads to a coupling between laser modes of circular polarization σ^{+}, σ^{-}, which can be decreased by an external magnetic field. Such coupling depends also, in an obvious way, not only on values of the magnetic field and the angular momenta of the states involved, but also on any possible anisotropy of the cavity. Depending on the strength of the coupling, either an independent oscillation of the differently polarized modes is possible, or strong competition between them occurs, leading to a bistable behavior. Such effects, related to couplings of polarization modes and the influence of cavity anisotropy have been the subject of much theoretical and experimental study,[90, 92, 96-99] and were reviewed by Voitovich, whose monograph[100] contains many relevant references.

2. Spontaneous Emission from Laser Media

Another class of phenomena associated with lasers in a magnetic field can be studied in spontaneous emission from the laser medium. Here two kinds of effects that are due to coherences of atomic Zeeman states can be specified: the mode-crossing effect and the level-crossing effect. The mode-crossing effect is the phenomenon that manifests itself by a resonant change of intensity of spontaneous emission from a level perturbed by a multimode laser when the magnetic splitting of the Zeeman sublevels equals the frequency difference between the laser modes. The first observation of such an effect was reported, and interpreted as a fourth-order process, by Fork et al.[90] A detailed experimental and theoretical analysis of mode crossing in an inhomogeneously broadened medium can be found in papers by Dumont and coworkers[101] (see also Ref. 94 and references therein).

Dumont specified two mechanisms yielding resonances in spontaneous emission from an atom within laser cavity when the magnetic splitting was equal to the mode-frequency difference. The first is a population effect (PE) which is essentially related to saturation of populations rather than to coherence of magnetic sublevels. It occurs when the σ^+ component of one given laser mode and the σ^- component of another interact with atoms of same velocity. Two holes are then independently burned in the population distribution by the σ^+, σ^- components overlap. Due to such hole overlap (hole crossing) the total population of a given level exhibits a resonance resulting in a corresponding resonant change in the spontaneous emission.

The second mechanism, the Zeeman coherence effect (ZCE), is due to coherent superposition of Zeeman sublevels by the joint action of different circular components of a pair of modes. This distinction between PE and ZCE is quite analogous to the distinction between the self-saturation and cross-saturation contributions occurring in a Zeeman laser, as discussed above, and cross-population and Zeeman-coherence contributions to FWM signals (described in Section IV.B).

Another class of effects that appear in spontaneous emission from laser media in a magnetic field are resonant effects originating from the crossing of energy levels in a given magnetic field. The Hanle effect, i.e., zero-field level crossing,[59] is a particular example of such a situation. As is well known, resonant coherence effects occur at level crossings even for weak light intensities. A strong laser field, however, causes several characteristic modifications sometimes termed the *stimulated* or *nonlinear Hanle effect*. There is an extensive literature on these effects[102-106] (see also Refs. 107 and 108 and references therein) where an interested reader will find detailed descriptions of such phenomena. Some of them have already been presented in preceding paragraphs. It suffices here to point out that the main differences between the linear and nonlinear level crossings and Hanle effects rely on the fact that many more sublevels could be coherently coupled, due to the multiphoton character of interactions with strong fields and that the corresponding coherences are strongly mutually coupled. Due to strong mixing of atomic states the level-crossing signals are no longer good signatures of individual atomic levels, but, in general, contain contributions from all mixed states.

5. CONCLUSIONS

We have presented a review of nonlinear optical effects in which the nonlinear response of the medium is due to a specific interference mechanism arising from a coherent superposition of atomic levels. Generally, nonlinear optics is considered to require powerful light beams. This is true in many cases when the nonlinearity is due to such phenomena as optical saturation in the gas phase, or thermal or structural changes in liquids or solids. There is, on the other hand, a broad class of specific nonlinear phenomena which do not require very high powers. There are the effects reported in this paper, where the nonlinearity is based on atomic coherences. They occur at powers that are often orders of magnitude lower than in other, more typical cases. Apart from this feature, which has considerable practical importance, these effects are very interesting manifestations of quantum interference which may occur in various situations involving complex (i.e., more than two-level) atomic energy level structures.

Acknowledgments

It is a pleasure to thank S. Lea for careful reading of the manuscript. The author is also grateful to the American Institute of Physics and to the Elsevier Scientific Publishers B.V. for permission to reproduce figures (Figs. 4, 6, 10, 12, 23, 24, 26). This work was supported in part by KBN Grant 20139101.

References

1. A. Omont, *Prog. Quant. Electron.* **5**, 69 (1977).

2. V. S. Letokhov and V. P. Chebotayev, *Nonlinear Laser Spectroscopy*, Springer, Berlin, 1977.

3. M. D. Levenson, *Introduction to Nonlinear Laser Spectroscopy*, Academic, New York, 1982.

4. A. Corney, B. P. Kibble, and G. W. Series, *Proc. Roy. Soc. London Ser. A* **293**, 70 (1966).

5. C. Wieman and T. W. Hänsch, *Phys. Rev. Lett.* **36**, 1170 (1976).

6. S. J. Williamson, J. M. Weingart, and R. D. Andrews, *J. Opt. Soc. Am.* **54**, 337 (1964).

7. C. Delsart and J. C. Keller, *J. Appl. Phys.* **49**, 3662 (1978).

8. e.g., E. N. Fortson and L. L. Lewis, *Phys. Rep.* **113**, 289 (1984).

9. W. Gawlik, J. Kowalski, R. Neumann, H. B. Wiegemann, and K. Winkler, *J. Phys. B* **12**, 3873 (1979).

10. G. W. Series, *Proc. Phys. Soc.* **88**, 995 (1966).

11. W. Happer and B. S. Mathur, *Phys. Rev.* **163**, 12 (1967).

12. F. Laloë, *Ann. Phys.* (*Paris*) **6**, 5 (1971).

13. W. Gawlik, *J. Phys. B* **10**, 2561 (1977).

14. W. Gawlik, J. Kowalski, R. Neumann, and F. Träger, *Phys. Lett.* **48A**, 283 (1974).

15. W. Gawlik, J. Kowalski, R. Neumann, and F. Träger, *Opt. Commun.* **12**, 400 (1974).

16. M. Ducloy, M. P. Gorza, and B. Decomps, *Opt. Commun.* **8**, 21 (1973).

17. B. Segard and B. M. Carpentier, *J. Phys. E* **14**, 442 (1981).

18. S. Giraud-Cotton, V. P. Kaftandjian, and L. Klein, *Phys. Lett.* **88A**, 453 (1982).

19. S. Giraud-Cotton, V. P. Kaftandjian, and L. Klein, *Phys. Rev. A* **32**, 2223 (1985).

20. W. Gawlik, *Phys. Lett.* **89A**, 278 (1982).

21. J. Zakrzewski and T. Dohnalik, *J. Phys. B* **16**, 2119 (1983).

22. P. E. G. Baird, in A. C. P. Alvez, J. M. Brown, and J. M. Hollas (Eds.), *Frontiers of Laser Spectroscopy of Gases* (NATO ACI Series), Kluver Academic, Dordrecht, 1987, p. 187.

23. F. Schuller, M. J. D. Macpherson, and D. N. Stacey, *Physica* **147C**, 321 (1987).

24. F. Schuller, M. J. D. Macpherson, and D. N. Stacey, *Opt. Commun.* **71**, 61 (1989).

25. F. Schuller, M. J. D. Macpherson, D. N. Stacey, R. B. Warrington, and K. P. Zetie, *Opt. Commun.* **86**, 123 (1991).

26. I. O. Davies, P. E. G. Baird, and J. L. Nicol, *J. Phys. B* **20**, 5371 (1987).

27. P. E. G. Baird, M. Irie, and T. D. Wolfenden, *J. Phys. B* **22**, 1733 (1989).

28. K. H. Drake, W. Lange, and J. Mlynek, *Opt. Commun.* **66**, 315 (1988); K. H. Drake, Thesis, University of Hannover, 1986.

29. L. M. Barkov, M. S. Zolotorev, and D. A. Melik-Pashayev, *Pisma v Zh. Eksp. Teor. Phys.* [*Sov. Phys. JETP Pis'ma*] **48**, 134 (1988).

30. L. M. Barkov, D. A. Melik-Pashayev, and M. S. Zolotorev, *Opt. Commun.* **70**, 467 (1989).

31. W. Gawlik and J. Zachorowski, *J. Phys. B* **20**, 5939 (1987).

32. A. Ekert and W. Gawlik, *Phys. Lett.* **121A**, 175 (1987).

33. W. Gawlik, *Am. J. Phys.* **59**, 706 (1991).

34. P. Jungner, B. Ståhlberg, and M. Lindberg, *Phys. Scr.* **38**, 550 (1988).

35. P. Jungner, T. Fellman, B. Ståhlberg, and M. Lindberg, *Opt. Commun.* **73**, 38 (1989).

36. B. Ståhlberg, P. Jungner, T. Fellman, and M. Lindberg, *Appl. Phys. B* **50**, 547 (1990).

37. X. Chen, V. OL. Telegdi, and A. Weis, *Opt. Commun.* **74**, 301 (1990).

38. A. Weis, J. Wurster, and S. I. Kanorsky, *J. Opt. Soc. Am. B.* **10**, 716 (1993).

39. S. I. Kanorsky, A. Weis, J. Wurster, and T. W. Hänsch, *Phys. Rev. A* **47**, 1220 (1993).

40. N. Bloembergen, in H. Walther and K. W. Rother (Eds.), *Laser Spectroscopy IV*, Springer, Berlin, 1979, p. 340.

41. Y. R. Shen, *The Principles of Nonlinear Optics*, Wiley, New York, 1984.

42. e.g., R. A. Fisher (Ed.), *Optical Phase Conjugation*, Academic, New York, 1983.

43. B. Ya. Zel'dovich and V. V. Shkunov, *Kvant. Elektron.* [*Sov. J. Quant. Electron.*] **6**, 629 (1979).

44. V. N. Blashchuk, B. Ya. Zel'dovich, A. V. Mamaev, N. F. Pilipetskii, and V. V. Shkunov, *Kvant. Elektron.* [*Sov. J. Quant. Electron.*] **7**, 627 (1980).

45. D. G. Steel and J. F. Lam, *Phys. Rev. Lett.* **43**, 1588 (1979).

46. P. F. Liao, N. P. Economou, and R. R. Freeman, *Phys. Rev. Lett.* **39**, 1473 (1977).

47. J. F. Lam, D. G. Steel, R. A. McFarlane, and R. C. Lind, *Appl. Phys. Lett.* **38**, 977 (1981).

48. J. F. Lam and R. L. Abrams, *Phys. Rev. A* **26**, 1539 (1982).

49. S. N. Jabr, L. K. Lam, and R. W. Hellwarth, *Phys. Rev. A* **24**, 3264 (1981).

50. S. Saikan, *J. Opt. Soc. Am.* **72**, 514 (1982).

51. S. Saikan and M. Kiguchi, *Opt. Lett.* **7**, 555 (1982).

52. G. P. Agrawal, *Opt. Lett.* **8**, 359 (1983).

53. G. P. Agrawal, *Phys. Rev. A* **28**, 2286 (1983).

54. D. Bloch and M. Ducloy, *J. Opt. Soc. Am.* **73**, 635 (1983).

55. W. Gawlik and G. W. Series, in H. Walther and K. W. Rothe (Eds.), *Laser Spectroscopy IV*, Springer, Berlin, 1979, p. 210.

56. W. Gawlik, *Acta Phys. Pol. A* **66**, 401 (1984).

57. D. E. Murnick, M. S. Feld, M. M. Burns, T. U. Kühl, and P. G. Pappas, in H. Walther and K. W. Rothe (Eds.), *Laser Spectroscopy IV*, Springer, Berlin, 1979, p. 195.

58. P. G. Pappas, M. M. Burns, D. D. Hinshelwood, M. S. Feld, and D. E. Murnick, *Phys. Rev. A* **21**, 1955 (1980).

59. G. Moruzzi and F. Strumia (Eds.), *The Hanle Effect and Level-Crossing Spectroscopy*, Plenum, New York, 1991, and references therein.

60. D. G. Steel, R. C. Lind, and J. F. Lam, *Phys. Rev. A* **23**, 2513 (1981).

61. M. Ducloy and D. Bloch, *Phys. Rev. A* **30**, 3107 (1984).

62. D. Bloch and M. Ducloy, *J. Phys. B* **14**, L471 (1981).

63. e.g., C. Cohen-Tannoudji, J. Dupont-Roc, and G. Grynberg, *Atom-Photon Interactions*, Wiley, New York, 1992, Chapter VI.

64. M. Pinard, P. Verkerk, and G. Grynberg, *Opt. Lett.* **9**, 399 (1984).

65. M. Pinard, P. Verkerk, and G. Grynberg, *Phys. Rev. A* **35**, 4679 (1987).

66. R. Saxena and G. S. Agarwal, *Phys. Rev. A* **31**, 877 (1985).

67. P. R. Berman, D. G. Steel, G. Khitrova, and J. Liu, *Phys. Rev. A* **38**, 252 (1988).

68. F. A. M. de Oliveira, C. B. de Araújo, and J. R. Rios Leite, *Phys. Rev. A* **38**, 5688 (1988).

69. G. Alzetta, A. Gozzini, L. Moi, and G. Orriols, *Nouvo Cimento* **36B**, 5 (1976).

70. G. Alzetta, L. Moi, and G. Orriols, *Nouvo Cimento* **52B**, 209 (1979).

71. E. Arimondo and G. Orriols, *Nouvo Cimento Lett.* **17**, 333 (1976).

72. G. Orriols, *Nuovo Cimento* **53B**, 1 (1979).

73. H. R. Gray, R. M. Whitley, and C. R. Stroud, Jr., *Opt. Lett.* **3**, 218 (1978).

74. A. Aspect, E. Arimondo, R. Kaiser, N. Vansteenkiste, and C. Cohen-Tannoudji, *Phys. Rev. Lett.* **61**, 826 (1988); *J. Opt. Soc. Am. B* **6**, 2112 (1989).

75. S. E. Harris, *Phys. Rev. Lett.* **62**, 1033 (1989).

76. A. Imamoglu, J. E. Field, and S. E. Harris, *Phys. Rev. Lett.* **66**, 1154 (1991).

77. M. O. Scully, *Phys. Rev. Lett.* **67**, 1855 (1991); *Phys. Rep.* **219**, 191 (1992).

78. J. P. Dowling and C. M. Bowden, *Phys. Rev. Lett.* **70**, 1421 (1993); see also the discussion session in *Phys. Rep.* **219**, 203 (1992).

79. W. Gawlik, *Proc. SPIE* **1711**, 11 (1993); *Comments on At. Mol. Phys.*, in press.

80. O. Kocharovskaya and Ya. I. Khanin, *Pis'ma Zh. Eksp. Theor. Fiz.* **48**, 581 (1988) [*JEPT Lett.* **48**, 630 (1988)]; O. Kocharovskaya and P. Mandel, *Phys. Rev. A* **42**, 523 (1990).

81. L. M. Narducci, H. M. Doss, P. Ru, M. O. Scully, S. Y. Zhu, and C. Keitel, *Opt. Commun.* **81**, 379 (1991); L. M. Narducci, M. O. Scully, C. Keitel, S. Y. Zhu, and H. M. Doss, *Opt. Commun.* **86**, 324 (1991).

82. A. Karawajczyk, J. Zakrzewski, and W. Gawlik, *Phys. Rev. A* **45**, 420 (1992).

83. M. O. Scully, S. Y. Zhu, and A. Gavrielides, *Phys. Rev. Lett.* **62**, 2813 (1989).

84. K. J. Boller, A. Imamoglu, and S. E. Harris, *Phys. Rev. Lett.* **66**, 2533 (1991).

85. J. E. Field, K. H. Hahn, and S. E. Harris, *Phys. Rev. Lett.* **67**, 3062 (1991).

86. J. Y. Gao, C. Guo, X. Z. Guo, G. X. Jin, Q. W. Wang, J. Zhao, H. Z. Zhang, Y. Jiang, D. Z. Wang, and D. M. Jiang, *Opt. Commun.* **93**, 323 (1992).

87. H. M. Doss, L. M. Narducci, M. O. Scully, and J. Y. Gao, *Opt. Commun.* **95**, 57 (1993).

88. W. Culshaw and J. Kannelaud, *Phys. Rev.* **126**, 1747 (1962); **133**, A691 (1964); **136**, A1209 (1964).

89. W. Culshaw, *Phys. Rev.* **135**, A316 (1964).

90. R. L. Fork, L. E. Hargrove, and M. A. Pollack, *Phys. Rev. Lett.* **12**, 705 (1964).

91. M. I. Dyakonov and V. I. Perel, *Opt. Spektrosk.* **20**, 472 (1966) [*Opt. Spectrosc.* (Engl. Transl.) **20**, 257 (1966)].

92. W. van Haeringen, *Phys. Rev.* **158**, 256 (1967).

93. M. Sargent III, W. E. Lamb, Jr., and R. L. Fork, *Phys. Rev.* **164**, 436; 450 (1967).

94. B. Decomps, M. Dumont, and M. Ducloy, in H. Walther (Ed.), *Topics in Applied Physics*, Vol. 2, *Laser Spectroscopy of Atoms and Molecules*, Springer, Berlin, 1976, p. 284.

95. e.g., M. Sargent III, M. O. Scully, and W. E. Lamb, Jr., *Laser Physics*, Addison-Wesley, Reading, MA, 1974.

96. W. J. Tomlinson and R. L. Fork, *Phys. Rev.* **164**, 466 (1967).

97. H. Greenstein, *Phys. Rev.* **178**, 585 (1969).

98. A. le Floch and R. le Naour, *Phys. Rev. A* **4**, 290 (1971).

99. A. le Floch and G. Stephan, *Phys. Rev. A* **6**, 845 (1972).

100. A. P. Voitovich, *Magnitoptika gasovykh laserov* (Magnetooptics of gas lasers), Nauka i Tekhnika, Minsk, 1984 (in Russian); *J. Sov. Laser Res.* (Engl. Transl.) **8**, 551 (1987).

101. M. Dumont, *J. Phys.* (*Paris*) **33**, 971 (1972).

102. T. Krupennikova and M. Tchaika, *Opt. Spektrosk.* **20**, 1088 (1966) [*Opt. Spectrosc.* (Engl. Transl.) **20** (1966)].

103. B. Decomps and M. Dumont, *J. Phys.* (*Paris*) **29**, 443 (1968); *IEEE J. Quant. Electron.* **QE-4**, 916 (1968).

104. M. Tsukakoshi and K. Shimoda, *J. Phys. Soc. Jpn.* **26**, 758 (1969).

105. M. S. Feld, A. Sanchez, A. Javan, and B. J. Feldman, in *Methodes de Spectroscopie sans Largeur Doppler de Niveaux Excites de Systems Moleculaires Simples*, Colloques Internationaux du CNRS, No. 217 (1974), p. 87.

106. C. Cohen-Tannoudji, in G. zu Putlitz, E. W. Weber, and A. Winnacker (Eds.), *Atomic Physics IV*, Plenum, New York, 1975, p. 589; and in R. Balian, S. Haroche, and S. Liberman (Eds.), *Frontiers in Laser Spectroscopy*, Vol. 1 (Proc. Les Houches Summer School on Theoretical Physics, Session XXVII, 1975), North-Holland, Amsterdam, 1977, p. 3.

107. W. Gawlik, D. Gawlik, and H. Walther, in Ref. 59, p. 47.

108. G. Moruzzi, F. Strumia, and N. Beverini, in Ref. 59, p. 123.

AUTHOR INDEX

Numbers in parentheses are reference numbers and indicate that the author's work is referred to although his name is not mentioned in the text. Numbers in *italic* show the pages on which the complete references are listed.

SUBJECT INDEX